TABLE A-6
Student's *t* distribution

Example For 15 degrees of freedom, the t value that corresponds to an area of 0.05 in both tails combined is 2.131.

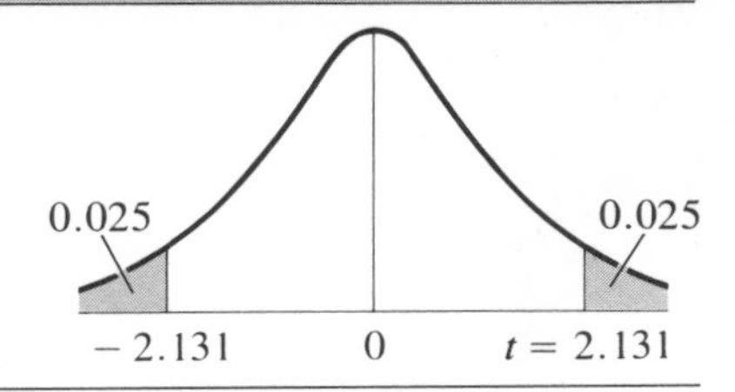

Degrees of Freedom	Area in Both Tails Combined			
	0.10	0.05	0.02	0.01
1	6.314	12.706	31.821	63.657
2	2.920	4.303	6.965	9.925
3	2.353	3.182	4.541	5.841
4	2.132	2.776	3.747	4.604
5	2.015	2.571	3.365	4.032
6	1.943	2.447	3.143	3.707
7	1.895	2.365	2.998	3.499
8	1.860	2.306	2.896	3.355
9	1.833	2.262	2.821	3.250
10	1.812	2.228	2.764	3.169
11	1.796	2.201	2.718	3.106
12	1.782	2.179	2.681	3.055
13	1.771	2.160	2.650	3.012
14	1.761	2.145	2.624	2.977
15	1.753	2.131	2.602	2.947
16	1.746	2.120	2.583	2.921
17	1.740	2.110	2.567	2.898
18	1.734	2.101	2.552	2.878
19	1.729	2.093	2.539	2.861
20	1.725	2.086	2.528	2.845
21	1.721	2.080	2.518	2.831
22	1.717	2.074	2.508	2.819
23	1.714	2.069	2.500	2.807
24	1.711	2.064	2.492	2.797
25	1.708	2.060	2.485	2.787
26	1.706	2.056	2.479	2.779
27	1.703	2.052	2.473	2.771
28	1.701	2.048	2.467	2.763
29	1.699	2.045	2.462	2.756
30	1.697	2.042	2.457	2.750
40	1.684	2.021	2.423	2.704
60	1.671	2.000	2.390	2.660
120	1.658	1.980	2.358	2.617
Normal Distribution	1.645	1.960	2.326	2.576

Source: From Table III of Fisher and Yates: *Statistical Tables for Biological, Agricultural and Medical Research*, published by Longman Group, Ltd., London (1974) 6th edition (previously published by Oliver and Boyd, Ltd., Edinburgh), and by permission of the authors and publishers.

STATISTICAL ANALYSIS FOR DECISION MAKING

4th EDITION

STATISTICAL ANALYSIS FOR DECISION MAKING

4th EDITION

Morris Hamburg

The Wharton School
University of Pennsylvania

HARCOURT BRACE JOVANOVICH, PUBLISHERS
and its subsidiary, Academic Press
San Diego New York Chicago Austin
London Sydney Tokyo Toronto

ISBN: 0-15-583453-3

Library of Congress Catalog Card Number: 86-80751

Printed in the United States of America

Preface

The objective of this Fourth Edition of *Statistical Analysis for Decision Making* is to present the fundamental concepts and methods of statistics in a clear and straightforward way—thus enabling students to develop critical judgment and decision making ability using quantitative tools. This textbook is designed primarily for a first course in statistics for students of business and economics, but the development of topics is also suitable for students in public administration, social science, and liberal arts. The reasoning and logic, the underlying concepts and techniques, and the interpretation and use of statistical results are stressed. Statistics is presented as an exciting field that deals with a scientific method for acquiring, analyzing, and using numerical data for decision making and inference.

Many of the applications, examples, and exercises pertain to the analysis and solution of problems in managerial decision making, but there are also problems in the fields of market research, quality control, accounting, polling, education, and psychology. The only mathematical background required is high school algebra. Emphasis is on showing the *power* of modern statistical reasoning and the *scope* and *versatility* of the methods, without bogging down the discussion in mathematical formalities. Focus is on the meaning, interpretation, and limitations of the methods discussed, rather than on the mechanics of calculation. Mathematical derivations have not been included in the body of the text, but some are given in footnotes and in Appendix C. Points for which an explanation in the language of calculus is particularly illuminating also appear in footnotes.

The structure of the book gives the instructor considerable flexibility in designing a course.

- Chapters 1–10 cover the fundamentals of classical statistics—descriptive statistics, probability and random variables, sampling, statistical inference, and regression and correlation analysis.

- Chapters 11, 12, and 13 deal with time series, index numbers, and nonparametric statistics.
- Chapters 14, 15, and 16 deal with statistical decision analysis.
- Chapter 17 compares classical statistics with Bayesian decision analysis.

Many changes and revisions have been made in this edition. Most of the exercises are new, and many of the examples have been revised throughout each chapter. Review exercises at the ends of Chapters 5, 8, and 10 cover material that appears in more than one section or chapter. Because these review exercises require the student to decide which method or technique to use (as well as to carry out the necessary calculations), the review exercises provide more challenging questions than the exercises that appear at the ends of sections within the chapters.

Computer exercises have been included at appropriate places to provide an opportunity for students who have access to computers to work with larger and more realistic data sets. A data base for these computer exercises, consisting of demographic and economic data for 75 families, is provided in Appendix E.

Chapter 10, on multiple regression and correlation analysis, reflects the book's emphasis on modern statistical methodology by focusing on the interpretation of computer output. More exercises have been included at the end of Chapter 10, containing data sets and questions that can be assigned to students who have access to computers. In Chapters 7, 8, and 10, computer output is displayed for examples of hypothesis testing, chi-square tests, analysis of variance, and multiple regression analysis.

A new optional section on Tchebycheff's inequality has been added to Chapter 3; this strengthens the discussion of expected values and variances. In Chapter 7, the material on hypothesis testing has been rewritten for greater clarity. New exercises have been introduced in Chapter 7 on tests to control both Type I and Type II errors. These exercises can be assigned by instructors who wish to teach the optional section (7.5) covering this topic.

Solutions to all even-numbered text exercises are given at the back of the book. The endpaper tables are four of the most useful tables from the text and Appendix A, repeated here for easy reference. A glossary of symbols follows Appendix E, with each symbol keyed to the section in which that symbol is introduced.

For the first time with this Fourth Edition, several supplementary pedagogical aids are available in separate publications.

- A computer program called Easystat is available for students who have access to an IBM-PC. The program works within the framework of a spreadsheet and includes a data set for 1,000 urban families. Students can select random samples for carrying out exercises. The accompanying manual provides problems with data sets that are larger and more realistic than the exercises in the text.

- The Study Guide provides supplementary exercises, as well as worked-out problems and explanations.
- Solutions to all of the text exercises appear in the Solutions Manual. (Only the even-numbered solutions appear in the text.)
- Additional exercises (with solutions) suitable for homework and for examinations are available in a Test Book, which also has a set of multiple-choice and true-false questions for examinations.

I would like to thank the many individuals and organizations that have made contributions to this book. Joseph Fuhr of Widener University deserves special mention for his constructive comments and recommendations in reviewing the manuscript for this Fourth Edition. I express grateful appreciation to Richard W. Gideon, president of Dick Gideon Enterprises, for the television and other data and estimates that he so kindly made available for use in the chapters on time series analysis.

My deep appreciation is again expressed to those who aided in the development of the earlier editions. I am especially grateful to the reviewers of the previous edition, Richard W. Andrews (University of Michigan), Roger L. Wright (University of Michigan), and Patrick McKeown (University of Georgia) for their constructive comments, suggestions, and criticisms.

I have continued to benefit greatly from the comments of teachers and students who have used the previous editions. The following instructors have been particularly helpful: David Ashley (University of Missouri), Eileen C. Boardman (Colorado State University), William D. Coffey (St. Edward's University), F. Damanpour (La Salle College), Maynard M. Dolecheck (Northeast Louisiana University), Fred H. Dorner (Trinity University), Linda W. Dudycha (University of Wisconsin), David Eichelsdorfer (Gannon University), David Frew (Gannon University), James C. Goodwin (University of Richmond), Charles R. Gorman (La Salle College), I. Greenberg (George Mason University), John B. Guerard, Jr. (Lehigh University), Jack A. Holt (University of Virginia), Yutaka Horiba (Tulane University), Gary Kern (University of Virginia), Burton J. Kleinman (Widener University), Ming-Te Lu (St. Cloud State University), William F. Matlack (University of Pittsburg), Elias Alphonse Parent (George Mason University), Chander T. Rajaratnam (Rutgers University), S.R. Ruth (George Mason University), James R. Schaefer (University of Wisconsin), Stanley R. Schultz (Cleveland State University), Marion G. Sobol (Southern Methodist University), William R. Stewart, Jr. (College of William and Mary), Chris A. Theodore (Boston University), Charles F. Warnock (Colorado State University), Thomas A. Yancey (University of Illinois).

I thank Suchat Boonbanjerdsri for his excellent and effective assistance in the preparation of exercises and solutions for the Fourth Edition. An especially warm acknowledgment of gratitude goes to Delores Johnson, Kristine Massenburg, and Tanya Winder for their cheerful, helpful, and exceptionally competent performance of the secretarial and typing work. My thanks go, again, to my colleagues in the Statistics Department of The Wharton School of the University of Pennsylvania, who were responsible for many of the ideas involved, particularly in the review exercises. I am grateful to the staff of Harcourt Brace Jovanovich. Johannah McHugh, Richard Bonacci, and Mickey Cox all provided helpful counsel and sound overall management; Marji James did an outstanding job as editor of this edition, bringing skillful improvements to every chapter; then the production staff put it all together—Cathy Reynolds (designer), Rebecca Lytle (art editor), and Lynn Edwards (production manager).

I am grateful to the Literary Executor of the Late Sir Ronald A. Fisher, F.R.S., to Dr. Frank Yates, F.R.S., and to Longman Group, Ltd., London, for permission to reprint Tables III and IV from their book *Statistical Tables for Biological, Agricultural, and Medical Research* (6th edition, 1974). My gratitude also goes to the other authors and publishers whose generous permission to reprint tables or excerpts from tables has been acknowledged at appropriate places.

As in previous editions, I dedicate this book to my wife, June, and my children, Barbara and Neil.

Morris Hamburg

Contents

5 SAMPLING DISTRIBUTIONS 183

6 ESTIMATION 221

7 HYPOTHESIS TESTING 253

8 CHI-SQUARE TESTS AND ANALYSIS OF VARIANCE 307

9 REGRESSION ANALYSIS AND CORRELATION ANALYSIS 367

10 MULTIPLE REGRESSION AND CORRELATION ANALYSIS 417

Introduction

The Nature of Statistics

This is a book about statistics; but what *is* "statistics"? We are all familiar with statistics that are collections of numerical data pertaining to sports, population, the economy, and the stock market. In this book, however, **statistics** is both a body of theory and methods of analysis. The subject matter of statistics covers a wide range—extending from the *planning* of experiments and other studies that generate data to the *collection*, *analysis*, *presentation*, and *interpretation* of the data. Numerical data constitute the raw material of the subject matter of statistics.

The most widely known statistical methods are those that summarize numerical data in terms of **averages** and other descriptive measures.

> If we are interested in the incomes of a group of 1,000 families chosen at random in a particular city, important characteristics of these incomes may be described by calculating an average income and a measure of the spread, or dispersion, of these incomes around the average.

The essence of modern statistics, however, is the theory and the methodology of **drawing inferences** that extend beyond the particular set of data examined and of **making decisions** based on appropriate analyses of such inferential data.

> We are probably not so much interested in the incomes of the particular 1,000 families included in the sample as we are interested in an inference about the incomes of all families in the city from which the sample was drawn.

- Such an inference might be in the form of a **test of a hypothesis** that the average income of all families in the city is $26,000 or less.
- The inference could also be in the form of a single figure, an **estimate** of the average income of all families in the city based on the average income observed in the sample of 1,000 families.
- Or the marketing department of a company may want the information in order to decide among different types of advertising programs based on the identification of the city as a low-, medium-, or high-income area.

The mathematical theory of probability provides the logical framework for the mental leap from the sample of data studied to the inference about all families in the city and for decisions such as the type of advertising program to be used.

The preceding example presents three points:

- We may have wanted an inference about the incomes of *all* families in the city. The totality of families in the city (or, more generally, the totality of the elements about which the inference is desired) is referred to in statistics as the **universe** or **population**.
- Because it would have been too expensive and too time-consuming to obtain the income data for every family in the city, only the sample of 1,000 families was observed. The 1,000 families, which represent a collection of only some elements of the universe, are referred to as a **sample**.
- In statistics, sample data are collected in order to make **inferences** or **decisions** concerning the populations from which samples are drawn.

Note that the sample was drawn "at random" from the population. A **random sample** is one drawn in such a way that the probability, or likelihood, of inclusion of every element in the population is known. However, even though these probabilities of inclusion may be known, the average income that would be observed for a random sample of 1,000 families would vary from sample to sample. These sample-to-sample variations are known as **chance sampling fluctuations**.

Although we cannot predict with certainty what the average income will be for any particular sample, the **theory of probability** enables us to compute how often these different sample results occur in the long run. It is an intriguing and remarkable fact that even though there is *uncertainty* concerning which particular sample may have been drawn, probability theory provides a rational basis for inference and decision making about the population from which the sample was taken. This textbook deals with the theory and methods by which inferences and decisions are made.

The Role of Statistics

Statistical concepts and methods are applied in many areas of human activity. They are used in the physical, natural, and social sciences, in business and public administration, and in many other fields.

In the sciences, the applications extend from the design and analysis of experiments to the testing of new and competing hypotheses. In industry, statistics makes its contributions in both short- and long-range planning and decision making. Many firms use statistical methods to analyze patterns of change and to forecast economic trends for the firm, the industry, and the economy as a whole. Such forecasts provide the foundation for corporate planning and control; purchasing, production, and inventory control depend on short-range forecasts, while capital investment and long-term development decisions depend on long-range forecasts. Statistical methods are employed in production control, inventory control, and quality control. To control the quality of manufactured products, for example, statistical methods are used to differentiate between variation attributable to chance causes and variation too great to be considered a result of chance; the latter type of variation can be analyzed and remedied. Application of statistical quality control methods results in substantial improvement in the quality of products and in lower costs because of reduction in rework and spoilage. Such statistical quality control methods have been a major factor in the improvement of the quality of Japanese-manufactured products since World War II.*

Over the past four decades, a body of **quantitative techniques and procedures** has been developed in the fields of business and government to aid and improve managerial decision making. The field of statistics has provided many fruitful ideas and techniques in this development. Applications of statistics are evident in most activities of business firms, including production, financial analysis, distribution analysis, market research, research and development, and accounting. Statistical methods are used often and with increasing sophistication in every field in which they have been introduced, and they are an integral part of the development of rational and quantitative approaches to the solution of business problems. One characteristic of this development has been the increased adoption of scientific decision-making approaches using **mathematical models**. These models are mathematical formulas or equations that state the relationship among the important factors or variables in a problem or system. (For example, an equation may be developed to represent the relationship between a company's sales and the economic and other variables that influence sales.) Chapters 9 and 10 of this text discuss the methods for deriving one such mathematical model. The statistical methods discussed in those and other chapters bring a logical, objective, and systematic approach to decision making in business and other fields.

* The American statistician Dr. W. Edwards Deming was a major contributor to the introduction of these methods in Japan. An annual award in Deming's name is made to a Japanese firm singled out for distinguished achievements in quality control.

Extensive statistical activities are conducted by federal, state, and local governments. There are many applications of statistical ideas and methods in governmental administration; governments collect and disseminate a great deal of statistical data. The most highly organized and extensive **statistical information systems** are those of the federal government. Such information systems—which include national income and product accounts, input–output accounts, flow of funds accounts, balance of payments accounts, and national balance sheets—depend on massive statistical collection and distribution systems. Statistical methods are applied to the resulting data to assess past trends and current status and to project future economic activity. These methods provide measures of human and physical resources and of economic growth, well-being, and potential. They are essential tools for appraising the performance and for analyzing the structure and behavior of an economy.

Data collection and dissemination activities are also carried out by governmental and private agencies in fields such as population, vital statistics, education, labor force, employment and earnings, business and trade, prices, housing, medical care, public health, agriculture, natural resources, welfare services, law enforcement, area and industrial development, construction, manufacturing, transportation, and communications.

Statistical analysis constitutes a body of theory and methods that plays an important role in this wide variety of human activities. It is extremely useful for communicating information, for drawing conclusions from data, and for the guidance of planning and decision making.

1

Frequency Distributions and Summary Measures

Variation is a basic fact of life. As individuals, we differ in age, sex, height, weight, and intelligence as well as in the quantities of the world's goods we possess, the amount of our good or bad luck, and countless other characteristics. In the business world, variations are observed in the articles produced by manufacturing processes, in the yields of the economic factors of production, in production costs, financial costs, marketing costs, and so forth.

The methods discussed in this chapter are useful for describing patterns of variation in data. Such variations occur in data observed at a particular time and in data occurring over a period of time.

Summarizing Variations in Data

Group by class and record frequency of occurrence

We begin our discussion by asking how you might go about summarizing the variation in a large body of numerical data. Suppose that data had been collected on the ages of all individuals in the United States and that you wanted to describe these approximately 240 million figures in some generally useful manner. How might you go about it, assuming that adequate resources for processing the data were available? Since it would be difficult to see important characteristics of the data by merely listing the data, you probably would group the figures into classes. For example, you might set up age classes of under five years, at least five but under 10 years, and so forth. You could list the number of persons in each class. If you divide these numbers by the total population, you have the proportion of the population in each class. You would then find it relatively easy to summarize the general characteristics of the age distribution of the population. If you compared similar distributions, say, for the years

1915 and 1985, a number of important features would be observable without any further statistical analysis. For example, the range of ages in both distributions would be clear at a glance. The higher proportions of persons under the age of 20 and of persons over the age of 65 in 1985 than in 1915 would stand out. Also, smaller percentages of persons in the age categories from 45 to 55 years would be observed for 1985, reflecting the decline in births during the 1930–1940 decade. Thus, the simple device of grouping the age figures into classes and recording the frequencies of occurrence in these classes shows us some of the underlying characteristics of the nation's age composition for each year. Generalizations about age patterns thus become easier to make.

Calculate an average and a measure of spread

You might also want to continue your description of the age distributions by calculating one or more types of average. For example, you might be interested in computing an average age for the population in 1915 and in 1985 to determine whether this average had increased or decreased. Furthermore, if you wished to give a more exact description of the fact that in the later period there were heavier concentrations of persons in the younger and older age groups, you might attempt to construct a measure of how the ages were spread around an average age. Because of these concentrations, the measure of spread or dispersion around the average would tend to be larger in 1985 than in the earlier period.

Statistical terms

The types of statistical techniques that might be used to summarize and describe the characteristics of the age data constitute the subject of this chapter.

- The table into which the data are grouped is referred to as a **frequency distribution**.
- The average or averages that can be computed are measures of **central tendency** or **central location** of the data.
- The measure of spread around the average is a measure of **dispersion**.

These and other techniques that group, summarize, and describe data are referred to as **descriptive statistics**. If the data treated by descriptive statistics represent a sample from a larger group or population, as noted in the Introduction, inferences may be desired about this larger group. Ways of making such statistical inferences are discussed in subsequent chapters.

Significance of order of occurrence of data

The order of occurrence of data is sometimes significant in data analysis. **Cross-sectional data** refers to data observed at a point in time, whereas **time series data** are sets of figures that vary over a period of time.

Frequency distribution analysis is concerned with cross-sectional data; in particular, such analysis deals with data in which the order of recording observations is of no importance (for example, the ages of the present members of the labor force in the United States, the present wage distribution of employees in the automobile industry, or the distribution of U.S. corporations by net worth on a given date).

Time series analysis would be used to record quality control data for a manufactured product, as we would be very much concerned with the order in which the articles were produced. If a run of defective articles was produced, we would want to know when this run occurred and what the general time pattern of production of defective and good articles was. Similarly, in the study of economic growth, we might be interested in the variation over time of such data as real income per person or real gross national product per person. General methods of time series analysis are treated in Chapter 11.

1.1 FREQUENCY DISTRIBUTIONS

When we are confronted with masses of ungrouped data (that is, listings of individual figures), it is difficult to generalize about the information the masses contain. However, if a frequency distribution of the figures is formed, many features become readily discernible.

*A **frequency distribution** or **frequency table** records the number of cases that fall in each class of the data.*

The numbers in each class are referred to as **frequencies**; hence the term "frequency distribution." When the numbers of items are expressed by their proportion in each class, the table is usually referred to as a **relative frequency distribution** or a **percentage distribution**.

Describing the classes of a frequency distribution

How the classes of a frequency distribution are described depends on the nature of the data. In all cases, data are obtained either by counting or by measuring. For example, individuals have characteristics such as race, nationality, sex, and religion, and counts can be made of the number of persons who fall in each of the relevant categories. If a classification of infants by feeding method is used, the frequency distribution may be shown as in Table 1-1.

TABLE 1-1
Distribution of infants by feeding method in an urban hospital

Method	Number of Cases
Bottle feeding	692
Breast feeding	353
Total	1,045

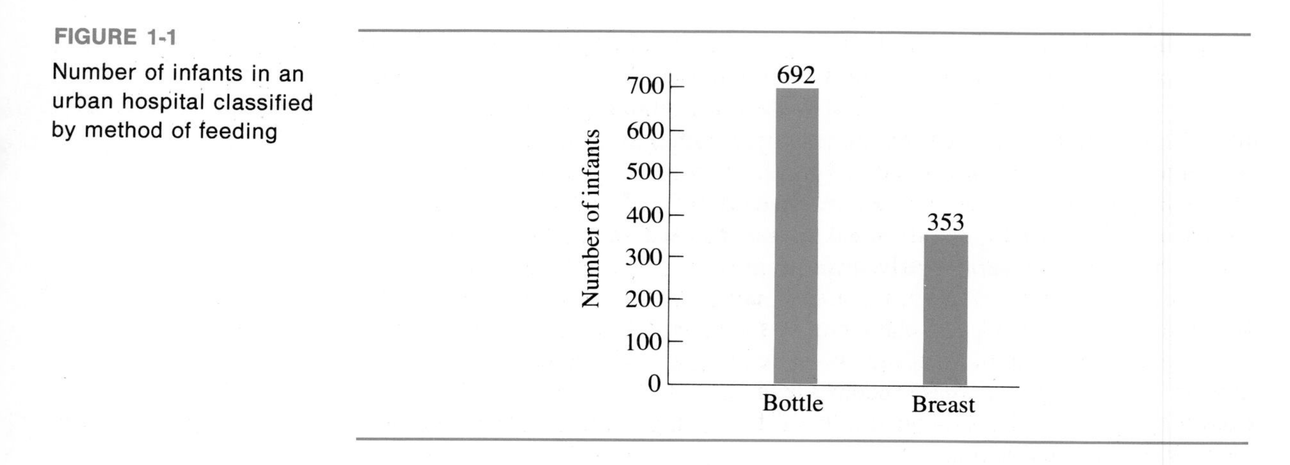

FIGURE 1-1
Number of infants in an urban hospital classified by method of feeding

Counting discrete variables

Characteristics such as nativity, color, sex, and religion that can be expressed in qualitative classifications or categories are often referred to as **attributes** or **discrete variables**. It is always possible to encode the attribute classifications to make them numerical. Thus, in our illustration, "bottle feeding" could have been denoted 0 and "breast feeding" 1. In certain cases the data seem to fall naturally into simple numerical classifications. For example, families may be grouped according to number of children; the classes could be labeled 0, 1, 2, and so forth.

Data for qualitative characteristics or discrete variables can be presented graphically in terms of simple bar charts. Figure 1-1 gives a bar chart representation of the data given in Table 1-1.

Measuring continuous variables

To obtain **continuous variables**, or **continuous data** (that is, data that can assume any value in a given range), we perform numerical measurements rather than counts. When large numbers of measurements are made, it is convenient to use intervals or groupings of values and to list the number of cases in each class. With this procedure, a few problems have to be resolved concerning the number of class intervals, the size of these intervals, and the manner in which class limits should be stated.

1.2 CONSTRUCTION OF A FREQUENCY DISTRIBUTION

The decisions about the number and size of the classes in a frequency distribution are essentially arbitrary. However, these two choices are clearly interrelated.

Selecting size and number of class intervals

The smaller the intervals chosen, the more intervals will be needed to cover the range of the scores.

Frequency distributions generally are constructed with from 5 to 20 classes. When class intervals are of equal size, comparisons of classes are easier and subsequent calculations from the distribution are simplified. However, this is not always a practical procedure. For example, with data on the annual incomes of families, in order to show the detail for the portion of the frequency distribution where the majority of incomes lie, class intervals of $4,000 may be used up to about $20,000; then intervals of $5,000 may be used up to $35,000 and a final class of $35,000 and over. Clearly, maintaining equal-sized classes of, say, $1,000 throughout the entire range of income would result in too many classes. On the other hand, if much larger class intervals were used, too many families would be lumped together in the first one or two classes, and we would lose the information concerning how these incomes were distributed. The use of unequal class intervals and an open-ended interval for the highest class provides a simple way out of the dilemma.

Open-ended class interval

Unequal class sizes

An **open-ended class interval** is one that contains only one specific limit and an "open" or unspecified value at the other end, as for example *$35,000 and over* or *110 pounds and under*. The use of **unequal class sizes** and open-ended intervals generally becomes necessary when most of the data are concentrated within a certain range, when gaps appear in which relatively few items are observed, and when there are a very few extremely large or extremely small values. Open-ended intervals are sometimes also used to retain confidentiality of information. For example, the identity of the small number of individuals or companies in the highest class may be general knowledge, and stating an upper limit for the class might be considered excessively revealing.

We illustrate the construction of a frequency distribution by considering the figures in Table 1-2, which represents the weights in ounces of 100 bottles filled at Basic Bottling Company in one test run of a bottling machine. Although the data have been arrayed from lowest weight to highest, it is difficult to discern patterns in the ungrouped figures. However, when a frequency distribution is constructed, the nature of the data clearly emerges. There is no single

TABLE 1-2
Weights (in ounces) of 100 bottles filled at Basic Bottling Company in one test run of a bottling machine

14.02	14.57	15.02	15.38	15.65	15.95	16.16	16.41	16.68	17.23
14.14	14.59	15.04	15.40	15.66	16.01	16.17	16.43	16.71	17.28
14.18	14.64	15.06	15.43	15.75	16.02	16.21	16.47	16.73	17.38
14.35	14.74	15.11	15.51	15.76	16.04	16.23	16.53	16.76	17.44
14.42	14.77	15.15	15.52	15.78	16.05	16.25	16.54	16.91	17.49
14.46	14.79	15.24	15.54	15.88	16.07	16.26	16.56	16.95	17.59
14.47	14.81	15.27	15.54	15.89	16.11	16.28	16.59	16.98	17.65
14.48	14.89	15.29	15.60	15.90	16.13	16.28	16.60	17.00	17.73
14.50	15.00	15.35	15.62	15.92	16.14	16.33	16.63	17.10	17.83
14.51	15.01	15.36	15.64	15.93	16.16	16.36	16.64	17.16	17.96

perfect frequency distribution for a given set of data. Several alternative distributions with different class interval sizes and different highest and lowest values may be equally appropriate.

Let us assume that we would like to set up a frequency distribution with 8 classes for the list of figures shown in Table 1-2 and that we want the classes to be of equal size. A simple formula to obtain an estimate of the appropriate interval size is

Formula for class interval size

$$i = \frac{H - L}{k}$$

where i = the size of the class intervals
H = the value of the highest item
L = the value of the lowest item
k = the number of classes

This formula for class interval size simply divides the total range of the data (that is, the difference between the values of the highest and lowest observations) by the number of classes. The resulting figure indicates how large the class intervals would have to be in order to cover the entire range of the data in the desired number of classes. Other considerations involved in determining an appropriate number of classes are discussed in section 1.4.

In the case of the bottle weights data,

$$i = \frac{(17.96 - 14.02)}{8} = 0.49$$

Rounding off for convenient class intervals

Since it is desirable to have convenient sizes for class intervals, the 0.49 figure may be rounded to 0.5, and the distribution may be tentatively set up on that basis. The frequency distribution shown in Table 1-3 results from a tally of the number of items that fall in each 0.5 ounce class interval.

Some important features of these data are immediately seen from the frequency distribution. The approximate value of the range, or the difference between the values of the highest and lowest items, is revealed. (Of course, since the identity of the individual items is lost in the grouping process, we cannot tell from the frequency table alone what the exact values of the highest and lowest items are.) Also, the frequency distribution gives at a glance some notion of how the elements are clustered. For example, more of the bottle weights fall in the interval from 16.0 to 16.5 ounces than in any other single class. When the frequencies in the classes immediately preceding and following the 16.0 to 16.5 grouping are added to the 22 in that interval, a total of 54 of the weights are accounted for. Furthermore, the distribution shows how the data are spread or dispersed throughout the range from the lowest to the highest value. We can quickly determine whether the items are bunched near the center of the distribution or spread rather evenly throughout. Also, we can see whether the

TABLE 1-3
Frequency distribution of weights of 100 bottles filled at Basic Bottling Company in one test run of a bottling machine

Weights	Number of Bottles
14.0 and under 14.5 ounces	8
14.5 and under 15.0 ounces	10
15.0 and under 15.5 ounces	15
15.5 and under 16.0 ounces	18
16.0 and under 16.5 ounces	22
16.5 and under 17.0 ounces	14
17.0 and under 17.5 ounces	8
17.5 and under 18.0 ounces	5
Total	100

frequencies fall away symmetrically on both sides of the center of the distribution or whether they tend to fall mostly to one side of the center. We now consider various statistical measures for describing these characteristics more precisely, but much information can be gained by simply studying the distribution itself.

1.3 CLASS LIMITS

The way in which class limits of a frequency distribution are described depends on the nature of the data. Figures on ages provide a good illustration of this point. Suppose that ages were recorded as of the *last* birthday. A clear and unambiguous way of stating the class limits is as follows: 15 and under 20, 20 and under 25 and so on. (Of course, there are other ways of wording the limits, such as *at least 15 but under 20* or *from 15 up to but not including 20*.)

Describing class limits

Consider the first class interval, "15 and under 20." Since ages have been recorded as of the last birthday, this class encompasses individuals who have reached at least their fifteenth birthday but not their twentieth birthday. If you are 19.999 years of age, that is, a fraction of a day away from your twentieth birthday, you fall in the first class. However, upon attaining your twentieth birthday, you fall in the second class, "20 and under 25." Thus, these class intervals are five years in size. The **midpoints** of the classes—that is, the values located halfway between the class limits—are 17.5, 22.5, 27.5, 32.5, 37.5, and so on. These values are used in computations of statistical measures for the distribution. Note that with class limits established and stated this way, the **stated limits** are in fact the true boundaries, or **real limits**, of the classes.

Stated limits and real limits

Suppose, on the other hand, that age data were rounded to the *nearest* birthday. We could follow a widely used convention and state the class limits as follows: 15–19, 20–24, and so on.

> Even though the stated limits in each class are only 4 years apart, the size of these class intervals is still 5 years.

For example, since the ages are given as of the nearest birthday, everyone between 14.5 and 19.5 years of age falls in the class 15–19. Thus, when data recorded to the nearest unit are grouped into frequency distribution classes, the lower real limit or lower boundary of any given class lies one-half unit below the lower stated limit and the upper real limit or upper boundary lies one-half unit above the upper stated limit. The midpoints of the class intervals may be obtained by averaging the lower and upper real limits or the lower and upper stated limits. For example, the midpoint of the class 15–19 is 17, which is the same figure obtained by averaging 14.5 and 19.5.

Summary

- When raw data are rounded to the *last* unit, the stated class limits and real class limits are identical.
- When raw data are rounded to the *nearest* unit, the real limits are one-half unit removed from the stated limits.
- With both types of data, the midpoints of classes are halfway between the stated limits or, equivalently, halfway between the real limits.

Class intervals should always be mutually exclusive, and the class each item falls into should always be clear. If class limits are stated as 30–40, 40–50, and so on, for example, it is not clear whether 40 belongs to the first class or the second.

Unfortunately, conventions are not universally observed. Often, one must use a frequency distribution constructed by others, and the nature of the raw data may not be clearly indicated. The producer of a frequency distribution should always indicate the nature of the underlying data.

1.4 OTHER CONSIDERATIONS IN CONSTRUCTING FREQUENCY DISTRIBUTIONS

Location of midpoint

A number of other points should be taken into account in the construction of a frequency distribution. If there are concentrations of particular values, it is desirable that these values be the midpoints of the class intervals. For example, assume that data are collected on the amounts of the lunch checks in a student cafeteria. Suppose these checks predominantly occur in multiples of five cents, although not exclusively so. If class intervals are set up as $1.70–$1.74, $1.75–$1.79, and so on, a preponderance of items would be concentrated at the lower limits.

> In calculating certain statistical measures from the frequency distribution, the assumption is made that the midpoints of classes are average (arithmetic mean) values of the items in these classes.

If, in fact, most of the items lie at the lower limits of the respective classes, a systematic error will be introduced by this assumption, because the actual averages within classes will typically fall below the midpoints.

Smooth progression of frequencies

Another factor to be considered in constructing a frequency distribution is the desirability of having a relatively smooth progression of frequencies. In many frequency distributions of business and economic data, one class contains more items than any other single class and the frequencies drop off more or less gradually on either side of this class. Table 1-3 is an example of such a distribution. (As indicated in section 1.2, the distribution may not be at all symmetrical.) However, erratic increases and decreases of frequencies from class to class tend to obscure the overall pattern, and such erratic variations often arise from the use of class intervals that are too small. Increasing the size of class intervals usually results in a smoother progression of frequencies, but wider classes reveal less detail than narrower classes. A compromise must be made in the construction of every frequency distribution.

- If we use class interval sizes of one unit each, every item of raw data is assigned to a separate class.
- If we use only one class interval as wide as the range of the data, all items fall in the single class.

Within the limits of these considerations, some freedom exists for the choice of an appropriate class interval size.

1.5 GRAPHIC PRESENTATION OF FREQUENCY DISTRIBUTIONS

Histogram

The use of graphs for displaying frequency distributions will be illustrated for the data on bottle weights shown in Table 1-3. One method is to represent the frequency of each class by a rectangle or bar. Such a chart is generally referred to as a **histogram**. A histogram for the frequency table given in Table 1-3 is

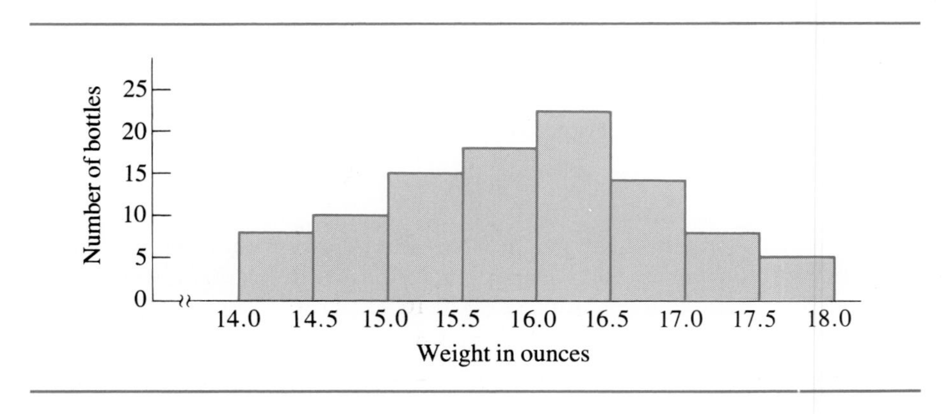

FIGURE 1-2
Histogram of frequency distribution of weights of 100 bottles filled at Basic Bottling Company during one test run of a bottling machine

shown in Figure 1-2. In agreement with the usual convention, values of the variable are depicted on the horizontal axis and frequencies of occurrence are shown on the vertical axis.

Frequency polygon

An alternative method for the graphic presentation of a frequency distribution is the **frequency polygon**. In this type of graph, the frequency of each class is represented by a dot above the midpoint of each class at a height corresponding to the frequency of the class. The dots are joined by line segments to form a many-sided figure, or polygon. A frequency polygon can also be thought of as the line graph obtained by joining the midpoints of the tops of the bars in a histogram. By convention, the polygon is connected to the horizontal axis by line segments drawn from the dot representing the frequency in the lowest class to a point on the horizontal axis one half a class interval below the lower limit of the first class, and from the dot representing the frequency in the highest class to a point one half a class interval above the upper limit of the last class. A frequency polygon for the distribution given in Table 1-3 is shown in Figure 1-3. It is important to realize that the line segments are drawn only for convenience in reading the graph and that the only significant points are the plotted frequencies for the given midpoints. Interpolation for intermediate values between such points would be meaningless. Often, midpoints of classes are shown on the horizontal axis rather than class limits as were shown in Figure 1-3.

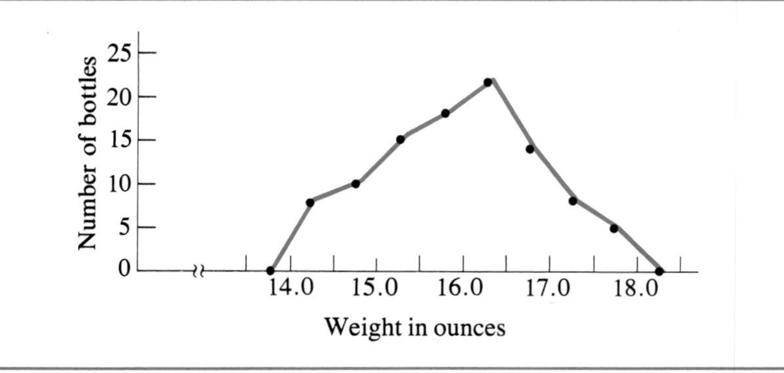

FIGURE 1-3
Frequency polygon of distribution of weights of 100 bottles filled at Basic Bottling Company during one test run of a bottling machine

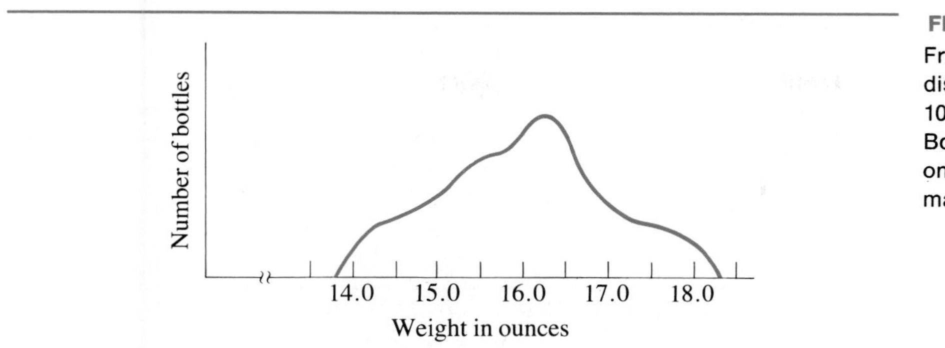

FIGURE 1-4
Frequency curve for distribution of weights of 100 bottles filled at Basic Bottling Company during one test run of a bottling machine

If the class sizes in a frequency distribution were gradually reduced and the number of items increased, the frequency polygon would approach a smooth curve more and more closely. Thus, as a limiting case, the variable of interest may be viewed as continuous rather than discrete, and the polygon would assume the shape of a smooth curve. The frequency curve approached by the polygon for the bottle weights would appear as shown in Figure 1-4.

1.6 CUMULATIVE FREQUENCY DISTRIBUTIONS

When interest centers on the number of cases that lie below or above specified values rather than within intervals, it is convenient to use a **cumulative frequency distribution** instead of the usual frequency distribution. Table 1-4 shows a so-called "less than" cumulative distribution for the bottle weights data shown in

TABLE 1-4
Cumulative frequency distribution of weights of 100 bottles filled at Basic Bottling Company during one test run of a bottling machine

Weight (in ounces)	Number of Bottles
Less than 14.0	0
Less than 14.5	8
Less than 15.0	18
Less than 15.5	33
Less than 16.0	51
Less than 16.5	73
Less than 17.0	87
Less than 17.5	95
Less than 18.0	100

FIGURE 1-5
Ogive for the distribution of weights of 100 bottles filled at Basic Bottling Company during one test run of a bottling machine

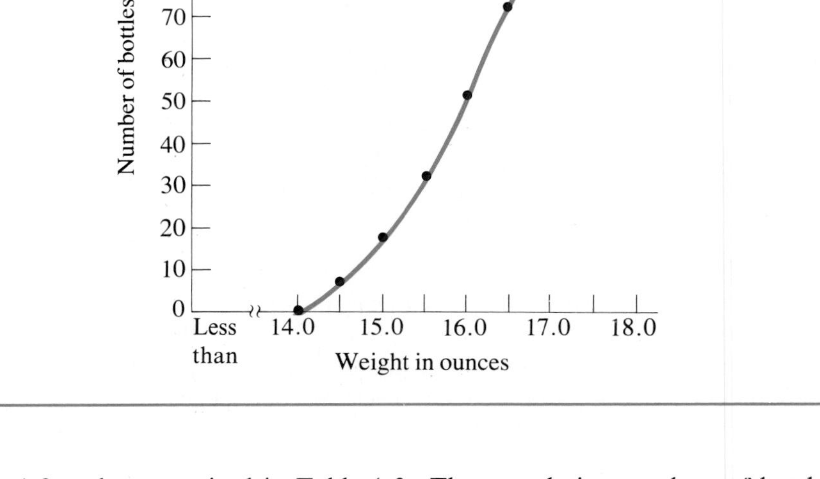

Table 1-2 and summarized in Table 1-3. The cumulative numbers of bottles with weights less than the lower class limits of 14.0, 14.5, and so on are given. Thus, there were no bottles with weights less than 14.0 ounces, eight bottles with weights less than 14.5 ounces, $8 + 10 = 18$ bottles with weights less than 15.0 ounces, and so on.

Ogive

The graph of a cumulative frequency distribution is referred to as an **ogive** (pronounced "ōjive"). The ogive for the cumulative distribution shown in Table 1-4 is given in Figure 1-5. The plotted points represent the number of bottles having weights less than the figure shown on the horizontal axis. The vertical coordinate of the last point represents the sum of the frequencies (100 in this case). The S-shaped configuration depicted in Figure 1-5 is quite typical of the appearance of a "less than" ogive. A "more than" ogive for the bottle weights distribution would have class limits reading "more than 14.0 ounces" and so on. In this case, a reverse S-shaped figure would have been obtained, sloping downward from the upper left to the lower right.

Exercises 1.6

1. The number of employees at Endicott Associates in 1985 and 1986 with a salary greater than \$30,000 but less than \$45,000 has been tabulated.
   a. Convert the numbers to percentages for each year.
   b. From part (a), construct "more than" cumulative frequency distributions.
   c. From part (b), plot "more than" ogives of both years on the same graph.

Salaries (in dollars)	Number of Employees 1985	1986
$30,000 and under $35,000	46	55
35,000 and under 40,000	11	18
40,000 and under 45,000	1	8

2. The dollar value of common stock owned on a given day by the stockholders of Textronics, Inc., are listed in a table.
   a. Convert the percentage distributions into "less than" cumulative frequency distributions.
   b. If "less than" ogive curves were plotted for the percentage distributions that you obtained in part (a), why would these curves not reach the 100% point?

Size of Holdings (in dollars)	Number of Stockholders	Percentage	Holdings (in dollars)	Percentage
1– 199	19,837	45.2	2,014,075	2.2
200– 399	8,266	18.8	2,457,692	2.7
400– 599	4,302	9.8	2,142,268	2.3
600– 999	5,472	12.5	4,344,090	4.8
1,000– 4,999	5,335	12.1	10,173,344	11.1
5,000–19,999	406	0.9	3,951,454	4.3
20,000–49,999	93	0.2	3,092,873	3.4
50,000–99,999	83	0.2	6,273,075	6.8
100,000 & over	141	0.3	57,101,205	62.4
	43,935	100.0	91,550,076	100.0

3. The lifetimes of a sample of 50 experimental electronic components are listed in days. You may assume that the lifetimes were recorded to the last full day.
   a. Construct a frequency distribution using 10 days as the size of class interval.
   b. Construct a frequency distribution with 10 classes.
   c. Construct a frequency distribution given that the midpoints of intervals are 50, 60, and so forth.

59	73	87	65	89	85	77	94	69	97
56	80	68	68	95	96	50	63	88	91
90	96	92	93	79	74	65	74	89	83
51	74	79	91	92	93	70	87	86	54
92	67	94	80	87	72	62	76	86	73

TABLE 1-5

State	Employment (in thousands)	State	Employment (in thousands)
Virgin Islands	35.6	South Carolina	1,164.6
Vermont	204.7	Oklahoma	1,195.7
Wyoming	214.6	Colorado	1,330.1
Alaska	230.3	Alabama	1,331.6
South Dakota	234.4	Connecticut	1,407.0
North Dakota	252.2	Washington	1,570.2
Delaware	264.5	Louisiana	1,573.2
Montana	266.2	Maryland	1,673.7
Idaho	315.9	Tennessee	1,681.1
Rhode Island	394.5	Michigan	1,712.4
Hawaii	400.7	Wisconsin	1,851.3
New Hampshire	402.2	Missouri	1,909.2
Nevada	419.1	Indiana	1,988.8
Maine	428.0	Virginia	2,147.8
New Mexico	482.3	Georgia	2,236.7
Utah	559.3	North Carolina	2,329.3
West Virginia	591.4	Massachusetts	2,585.6
Nebraska	592.6	New Jersey	3,123.2
District of Columbia	600.3	Michigan	3,165.9
Arkansas	724.8	Florida	3,786.8
Mississippi	776.8	Ohio	4,085.6
Kansas	897.2	Pennsylvania	4,456.0
Oregon	948.3	Illinois	4,511.8
Iowa	993.3	Texas	6,112.9
Arizona	1,011.7	New York	7,161.6
Kentucky	1,155.9	California	9,800.3

Source: *Monthly Labor Review*; U.S. Department of Labor, Bureau of Labor Statistics, December 1983.

4. Nonagricultural employment data as of August 1983 are listed in Table 1-5.
   a. Construct a frequency distribution, using class intervals beginning with "under 250," "250 and under 500," "500 and under 1,000," and so forth, despite the way the raw data are given.
   b. Justify the way that the class intervals are stated and your choice of number of classes.

5. Construct the following table and graphs from the list of the commissions paid in dollars to 19 Integrity Assurance sales representatives for a one-week period.

a. a frequency table (assuming the data were recorded to the last cent)
b. a histogram
c. a frequency polygon

425.00	438.50	516.74	382.75	440.40
469.95	512.55	458.00	462.63	490.15
475.00	401.82	505.01	460.00	528.00
482.05	538.09	471.13	498.50	

6. Pusch Beer Company had 2 production lines working each day. The number of cases produced per 8-hour shift for a week in 1986 are listed for each line.
   a. Using 6 classes, beginning with 5500 as the lower limit of the first class, construct and graph the frequency distribution for the number of cases produced per 8-hour shift for the 2 production lines combined.
   b. Then, using the same 6 classes and lower limit as in part (a), construct and graph the frequency distribution for each production line separately.
   c. Which answer gives a better picture of the distribution of production of cases of beer at the Pusch Beer Company—part (a) or part (b)? Explain your answer.

Line 1					Line 2				
6921	8205	6658	6835	7830	5655	6138	7891	7951	6666
6935	7563	6503	6777	6250	7891	7589	7002	8131	7662
6000	6935	7000	7012	5530	7555	8495	7832	7477	6886

1.7 EXPLORATORY DATA ANALYSIS°

A useful tool for data description—developed primarily by Professor John W. Tukey[1] of Princeton—is **exploratory data analysis**, which uses so-called **stem and leaf displays** to describe sets of data. The stem and leaf displays combine some of the characteristics of histograms and portray the general trend of a batch of values as a whole. The nature of some aspects of exploratory data analysis, including the stem and leaf display, is illustrated in the following examples.

° Optional material is indicated with this symbol; such material may be omitted without interfering with the continuity of the discussion.

[1] See John W. Tukey, *Exploratory Data Analysis* (Philippines copyright: Addison-Wesley Publishing Company, Inc. 1977).

EXAMPLE 1-1

An automobile manufacturer, about to introduce a new car model measured the mileage obtained by 24 test versions of the model and rounded the results to the nearest mile. Present the data as a simple stem and leaf display.

30	33	18	27	32	40	26	28	21	28	35	20
27	19	32	29	36	29	30	22	25	16	17	30

Solution

The first line on the left of the display represents a tabulation of the four mileages from 10 to 19, namely 16, 17, 18, and 19. The asterisk (*) is a place holder, or place filler, indicating that we have a two-digit number, in this case, a 1 followed by 6, 7, 8, and 9, respectively. Each line is a **stem**, and each item of information on the stem is a **leaf**. Here, the label for the stem is the first part of a number, which is followed in turn by each leaf. The label 1* is called the **starting part**.

	Gas Mileage per Gallon	(#)
1*	6789	(4)
2	01256778899	(11)
3	00022356	(8)
4*	0	(1)
		(24,✓)

The efficiency of this type of notion is apparent. For example, 1*|6789 uses seven characters to represent 16, 17, 18, 19. The actual form requires eight characters, or 12 if the commas and period are counted. A tally notation for these four numbers, such as 10–19|xxxx, requires 10 characters, while the identity of the individual items is lost. Note that the leaves in this example have been arranged in ascending order, as in 1*|6789. This is not a necessary procedure, and for many purposes the observations might very well be recorded in the order in which they occur.

The figures on the right side of the display represent tallies or counts of the leaves. Taken together with the stems, they constitute a frequency distribution of the original data. By convention we use the symbol (#) to denote "count" or "frequency of occurrence." At the bottom of the "count" column is the "check count" (24,✓), signifying that 24 mileages were recorded. The check mark indicates that we have counted the number of leaves and that this figure agrees with the number of original observations. The need for checking is especially evident when we deal with large bodies of data.

EXAMPLE 1-2

Disturbed because of increasing complaints from retailers who have not been receiving promised shipments of radios, a manufacturer of mobile radios, decides to run a check on the current radio distribution network. Each of the 40 warehouses owned by the manufacturer throughout the United States is instructed to maintain at least 300 mobile radios in stock at all times. The inventory levels of the 25 warehouses that were checked are listed.

a. Present the data as a two-digit stem and leaf display.
b. Are the warehouses keeping the required inventories?

150	280	60	10	85	160	305	70
253	180	0	150	90	0	50	110
300	25	400	610	100	320	200	330
210							

Solution

a. Inventory of mobile radios:

	Unit = 1 Radio	(#)
0**	00, 00, 10, 25, 50, 60, 85, 90	(8)
1	00, 10, 50, 50, 60, 70, 80	(7)
2	00, 10, 53, 83	(4)
3**	00, 05, 20, 30	(4)
4	00	(1)
5		
6**	10	(1)
		(25, ✓)

b. No, because only 6 out of 25 warehouses have inventories of at least 300 units.

A few points may be noted in this solution. First, all displays of data are in terms of some unit. In this example, the unit is a radio. In another situation, perhaps in stating prices, $10 may be an appropriate unit. Then $360 becomes 36 and is stated as 3*|6.

In the present example, a two-digit leaf has been used. Hence, 3**|20 represents 320, 6**|10 is 610, and so forth. Note that 0**| is used to record values between zero and 99. Had negative numbers been possible, −0**| would have been used for tabulating values from zero to −100.

1.8 DESCRIPTIVE MEASURES FOR FREQUENCY DISTRIBUTIONS

As indicated in section 1.2, once a frequency distribution is constructed from a set of figures, certain features of the data become readily apparent. For most purposes, however, it is necessary to have a more quantitative description of these characteristics than can be ascertained by a casual glance at the distribution. Analytical measures are usually computed to describe such characteristics as the central tendency, dispersion, and skewness of the data.

Averages

Averages are the measures used to describe the characteristic of **central tendency** or **central location** of data. Averages convey with a single number the notion of "central location" or the "middle property" of a set of data. The most familiar average is the **arithmetic mean**; in fact, it is often referred to as "the average." In ordinary conversation or in print, we encounter such terms as "average income," "average growth rate," "average profit rate," and "average person." Actually, several different types of averages, or measures of central tendency, are implied in these terms. The type of average to be employed depends on the purpose of the application and the nature of the data being summarized. In this section, we consider only the most commonly employed and most generally useful averages.

Dispersion

Dispersion refers to the spread, or variability, in a set of data. One method of measuring this variability is in terms of the difference between the values of selected items in a distribution, such as the difference between the values of the highest and lowest items. Another more comprehensive method is in terms of some average of the deviations of all the items from an average. Dispersion is an important characteristic of data because we are frequently interested as much in the variability of a set of data as in its central frequency.

Skewness

Skewness refers to the symmetry or lack of symmetry in the shape of a frequency distribution. This characteristic is useful in judging the typicality of certain measures of central tendency.

We begin the discussion of averages or measures of location by considering the most familiar one, the arithmetic mean. We assume throughout this chapter that the term **set of data** means a set of observations on a single numerical variable.

1.9 THE ARITHMETIC MEAN

Probably the most widely used and most generally understood way of describing the central tendency, or central location, of a set of data is the average known as the arithmetic mean.

> The **arithmetic mean**, or simply the **mean**, is the total of the values of a set of observations divided by the number of observations.

For example, if $X_1, X_2, \ldots, X_n$ represent the values of n items or observations, the arithmetic mean of these items, denoted $\bar{X}$, is defined as

$$\bar{X} = \frac{X_1 + X_2 + \cdots + X_n}{n} = \frac{\sum_{i=1}^{n} X_i}{n}$$

For simplicity, subscript notation such as that given above will usually not be used in this book. (However, before continuing, you should turn to Appendix B and work out the examples given there.) Thus, when the subscripts are dropped, the formula becomes

$$\bar{X} = \frac{\Sigma X}{n} \tag{1.1}$$

Σ = the sum of

where the capital Greek letter Σ (sigma) means "the sum of."

For example, suppose that an accounting department established accounts receivable in the following amounts during a one-hour period: \$600, \$350, \$275,

\$430, and \$520. The arithmetic mean of the amounts of the accounts receivable is

$$\bar{X} = \frac{\$600 + \$350 + \$275 + 430 + 520}{5} = \frac{\$2{,}175}{5} = \$435$$

The mean, \$435, may be thought of as the size of each account receivable that would have been set up if the total of the five accounts was the same (\$2,175) but all the accounts were the same size. That is, the mean is the value each item would have if they were all identical and the total value and number of items remained unchanged.

Symbolism

Roman letters for sample statistics

In keeping with the standard statistical practice of denoting sample statistics by Roman letters, we use the symbol $\bar{X}$ to denote the mean of a sample of observations. The number of observations in the sample is represented by the lower-case letter n. A value such as $\bar{X}$ computed from sample data is referred to as a **statistic**. A statistic may be used as an estimate of an analogous population measure, known as a **parameter**. Thus, the statistic $\bar{X}$ (the sample mean) may be thought of as an estimate of a parameter, the mean of the population from which the sample was drawn. The number of observations in the population is represented by the capital letter N.

Greek letters for population parameters

Just as Roman letters represent sample statistics, Greek letters represent population parameters. Accordingly, we will denote a population mean by the lower-case Greek letter μ (mu).

If the population mean were calculated directly from the data collected from the entire population of N members, then

$$\mu = \frac{X_1 + X_2 + \cdots + X_N}{N} = \frac{\Sigma X}{N} \tag{1.2}$$

In practice, the population mean μ is often not calculated, because it may be infeasible or inadvisable to accomplish a complete enumeration of the population.

SP - 2

Grouped Data

Computing the arithmetic mean

When data have been grouped into a frequency distribution, the airthmetic mean can be computed by a generalization of the definition for the mean of ungrouped data. As given in equation 1.1, the formula for the mean of a set of ungrouped data is $\bar{X} = (\Sigma X)/n$. However, with grouped data, since the identity of the individual items has been lost, an estimate must be made of the total of the values of the observations, ΣX. This estimate is obtained by multiplying the midpoint of each class in the distribution by the frequency of that class and

TABLE 1-6
Calculation of the arithmetic mean for grouped data by the direct method: bottle weights data

Weight (in ounces)	Number of Bottles f	Midpoints m	fm
14.0 and under 14.5	8	14.25	114.0
14.5 and under 15.0	10	14.75	147.5
15.0 and under 15.5	15	15.25	228.75
15.5 and under 16.0	18	15.75	283.5
16.0 and under 16.5	22	16.25	357.5
16.5 and under 17.0	14	16.75	234.5
17.0 and under 17.5	8	17.25	138.0
17.5 and under 18.0	5	17.75	88.75
	$n = \Sigma f =$ 100		1,592.50

$$\bar{X} = \frac{\Sigma fm}{n} = \frac{1{,}592.5}{100} = 15.93 \text{ ounces}$$

Direct method

summing over all classes. In symbols, if m denotes the midpoint of a class and f denotes the frequency, the arithmetic mean of a frequency distribution may be estimated from the following formula, known as the **direct method**:

$$\bar{X} = \frac{\Sigma fm}{n} \tag{1.3}$$

The computation of $\bar{X}$ for the frequency distribution of the bottle weights data shown in Table 1-3 is given in Table 1-6. The mean weight figure of 15.93 ounces calculated from the frequency distribution is very close to the corresponding mean of 15.90 ounces for the ungrouped data given in Table 1-2. The small difference in these two figures illustrates the slight loss of accuracy involved in calculating statistical measures from frequency distributions rather than from ungrouped data. When there is a large number of observations, this loss of accuracy is offset by the fact that the calculations are far less tedious than when done from the original data.

Step-deviation method

Shortcut formulas are often useful for calculating the arithmetic mean and other measures for frequency distributions. One such shortcut formula, known as the **step-deviation method**, is explained in Appendix D.

In many instances, for a variety of reasons, frequency distributions are shown with open intervals, usually in the first or last class. For example, such classes may appear as "Losses of $5,000 or more" or "Sales of $1,000,000 and over." In these cases, assumptions must be made concerning the midpoints of the classes in order to calculate statistical measures such as the arithmetic mean.

1.10 THE WEIGHTED ARITHMETIC MEAN

Weighted average

In averaging a set of observations, it is often necessary to compute a **weighted average** in order to arrive at the desired measure of central location. For example, suppose a company consists of three divisions, all selling different lines of products. The ratios of net profit to sales (expressed as percentages) for these divisions for the year 1986 were 5% for Division A, 6% for Division B, and 7% for Division C. Assume that we want to find the net profit to sales percentage *for the three divisions combined*, or equivalently, *for the company as a whole.* This ratio is the figure that results from dividing total net profits by total sales for the three divisions combined. Clearly, if we have only the profit *percentages* for the three divisions, we do not have enough information to compute the required figure. However, if we are given the dollar sales for each of the three divisions (that is, the denominators of the three ratios of net profit to sales from which the percentages were computed), then these figures can be used as "weights" in calculating the desired figure for the entire company. Specifically, the **weighted arithmetic mean** would be calculated as shown in Table 1-7 by carrying out the following steps:

Weighted arithmetic mean

1. "Weight" (that is, multiply) the net profit to sales percentage for each division by the sales of that division. As indicated in Table 1-7, the resulting figures are the net profits of the divisions.
2. Sum the net profits obtained in step 1 to obtain total net profit for the three divisions combined.

TABLE 1-7

Calculation of a weighted arithmetic mean: net profit to sales percentage for the three divisions of a company combined, 1986.

Division	Net Profit to Sales Percentage X	Sales w	Net Profit wX
A	5	$10,000,000	$ 500,000
B	6	10,000,000	600,000
C	7	30,000,000	2,100,000
		$50,000,000	$3,200,000

$$\text{Weighted mean} = \bar{X}_w = \frac{\Sigma wX}{\Sigma w} = \frac{\$3{,}200{,}000}{\$50{,}000{,}000} = 6.4\%$$

3. Sum the sales figures (the weights) to obtain total sales for the three divisions combined.
4. Calculate the desired average by dividing the total net profits figure (obtained in step 2) by the total sales figure (found in step 3).

Symbolically, the weighted arithmetic mean is given by the formula

$$\bar{X}_w = \frac{\Sigma wX}{\Sigma w} \tag{1.4}$$

where $\bar{X}_w$ = the weighted arithmetic mean
X = the values of the observations to be averaged (net profits to sales percentages in this case)
w = the weights applied to the X values (sales, in this case)

Note that the weighted average of 6.4% computed in Table 1-7 is interpreted as the net profit to sales percentage for the three divisions combined (that is, for the company as a whole). Since dollar sales are in the denominator of the $\Sigma wX/\Sigma w$ ratio, the answer of 6.4% may be interpreted in terms of dollars of profit per dollars of sales, that is, an average of $0.064 profit per dollar of sales.

Unweighted arithmetic mean

On the other hand, suppose we had asked instead, "What is the arithmetic mean net profit to sales ratio *per division*, without regard to the sales size of these divisions?" The answer is given by

$$\bar{X} = \frac{\Sigma X}{n} = \frac{5\% + 6\% + 7\%}{3} = \frac{18\%}{3} = 6\% \text{ per division}$$

where X = the percentages for the three divisions
n = the number of the divisions

The result, $\bar{X}$, may be referred to as an **unweighted arithmetic mean** of the three ratios. In this computation, the net profit to sales percentages were totaled and the result divided by the number of the company's divisions. The result is therefore stated as 6% *per division*, because of the appearance of the number of divisions (3) in the denominator. Clearly, this calculation disregards differences that may exist in the amount of sales of the three divisions; however, this does not make the average meaningless. If we are interested in obtaining a "representative" or "typical" profit ratio and there are no extreme values to distort the representativeness of the unweighted mean, then this type of computation is a valid one. Of course, if one were seeking a typical or representative figure, it would be desirable to have more than the three observations present in this illustration, and other averages, such as the median or mode, may be preferable to the arithmetic mean in the determination of a typical or representative value. However, the unweighted arithmetic mean is not a meaningless figure; indeed, it is the correct answer to the question just posed.

Returning to weighted mean calculation, the weights that were applied to the three profit to sales percentages in this problem were the actual dollar amounts of sales for the three divisions. That is, the weights used were the values of the denominators of the original ratios. An alternative procedure would be to weight the ratios by a percentage breakdown of the denominators (in this case, a percentage breakdown of total sales). For example, in the computation shown in Table 1-7, weights of 20%, 20%, and 60% could have been applied instead of weights of $10 million, $10 million, and $30 million, and the same answer of 6.4% would have resulted. Indeed, any figures in the same proportions as $10 million, $10 million, and $30 million would have led to the same numerical answer. Note that the reason the weighted arithmetic mean of 6.4% exceeded the unweighted mean of 6.0% was that in the weighted mean calculation, greater weight was applied to the 7% profit figure for Division C than to the corresponding 5% figure for Division A. This had the effect of pulling the weighted average up toward the 7% figure.

Exercises 1.10

1. A small clothing manufacturing company does both retail and wholesale trade. For the first quarter, ending March 31, its profit per item was $2.10 from the retail trade and was $1.30 from the wholesale trade. If wholesale business accounted for 80% of the company's business, what is the arithmetic mean profit per item for the quarter?

2. The West End Bank reports bad debt ratios (dollar losses to total dollar credit extended) of 0.05 for personal loans and 0.03 for industrial loans for a one-year period. For the same year, the East End Bank reports bad debt ratios of 0.06 for personal loans and 0.04 for industrial loans. Can one conclude from this that West End Bank's overall bad debt ratio is less than East End's? Justify your response.

3. Use the daily sales (in dollars) of 21 retail outlets to solve parts (a) and (b).
   a. Calculate the arithmetic mean from the data.
   b. Calculate the arithmetic mean from a frequency distribution for these data. Use a class interval size of $25.
   c. Which is the true arithmetic mean value? Explain your answer.

153.17	244.56	222.22	236.00	273.66	207.79	265.01
183.04	113.42	200.00	146.89	199.23	276.18	178.00
257.93	194.33	282.49	188.88	172.71	249.65	214.80

4. Mortimer Hutton, a wealthy investor who spends a large amount of time watching the quote machine in his broker's office, is weighing the relative merits of two periodic investment strategies. The first method is to buy the same number of shares of stock each investment period. The second method requires an investment of a constant dollar amount each period regardless of the stock price.

TABLE 1-8
Comparison of two investment strategies

	First Strategy						Second Strategy					
	United Aerodynamics			Mitton Industries			United Aerodynamics			Mitton Industries		
Year	Price per Share	Number of Shares	Total Cost	Price per Share	Number of Shares	Total Cost	Price per Share	Number of Shares	Total Cost	Price per Share	Number of Shares	Total Cost
1	$54	25	$1,350	$25	25	$625	$54	19	$1,026	$25	40	$1,000
2	50	25	1,250	30	25	750	50	20	1,000	30	33	990
3	42	25	1,050	43	25	1,075	42	24	1,008	43	23	989
4	35	25	875	40	25	1,000	35	29	1,015	40	25	1,000
5	30	25	750	49	25	1,225	30	33	990	49	20	980

TABLE 1-9
Percentage of sales in selected industries

Industry	Percentage	Industry	Percentage
Tobacco	0.5	Machinery (machine tools, industrial mining)	2.6
Fuel	0.5	Electrical	2.8
Textiles, apparel	0.6	Conglomerates	2.8
Steel	0.7	Chemicals	2.9
Food & beverage	0.7	Machinery (farm, construction)	3.3
Containers	0.7	Electronics	3.8
Paper	1.0	Automotive (cars, trucks)	4.0
Metals & mining	1.2	Leisure time industries	4.8
Telecommunications	1.3	Information processing (office equipment)	5.1
Building materials	1.3	Aerospace	5.1
Appliances	2.0	Instruments (measuring devices, controls)	5.2
Oil service & supply	2.1	Drugs	6.0
Personal & home care products	2.3	Information processing (computers)	6.8
Tires & rubber	2.3	Information processing (peripherals, services)	7.2
Automative (parts & equipment)	2.3	Semiconductors	7.8
Miscellaneous manufacturing	2.4		

His broker demonstrates the result of using each strategy on two stocks, United Aerodynamics and Mitton Industries, for a 5-year period, as shown in Table 1-8. The purchases in each case are made at midyear at the prevailing stock price. The constant dollar amount for the second strategy is $1,000 (that is, the amount invested in the stock is as close to $1,000 as possible).

a. Calculate the average cost per share for each stock in each strategy.
b. Which strategy achieved the lower average cost for United Aerodynamics? For Mitton Industries?
c. Explain these differences in terms of the weights used in calculating the average cost per share of each stock and strategy.

5. Research and development expenditures as a percentage of sales in selected industries in a recent year are listed in Table 1-9.
   a. Calculate the arithmetic mean of the percentages from the ungrouped data.
   b. Construct a frequency distribution using classes stated as follows: 0–0.9, 1–1.9, and so forth.
   c. Calculate the arithmetic mean of the frequency distribution that you obtained in part (b).
   d. Compare your result in part (b) with the arithmetic mean determined in part (a).

1.11 THE MEDIAN

The median is a well known and widely used average. It has the connotation of the "middlemost" or "most central" value of a set of numbers.

Ungrouped data

For ungrouped data, the **median** is defined simply as the value of the central item when the data are arrayed by size.

If there is an odd number of observations, the median is directly ascertainable. If there is an even number of items, there are two central values, and by convention, the value halfway between these two central observations is designated as the median.

For example, suppose that a test of a brand of gasoline in five new small economy cars yielded the following numbers of miles per gallon: 27, 29, 30, 32, and 33. Then the median number of miles per gallon would be 30. If another car were tested and the number of miles per gallon obtained was 34, the array would now read 27, 29, 30, 32, 33, 34. The median would be designated as 31, the number halfway between 30 and 32.

> Another way of viewing the median is as a value below and above which lie an equal number of items.

Thus, in the preceding illustration involving five observations, two lie above the median and two below. In the example involving six observations, three fall above and three fall below the median. Of course, in the case of an array with an even number of items, any value lying between the two central items may, strictly speaking, be referred to as a median. However, as indicated earlier, the convention is to use the midpoint between the two central items. In the case of tied values at the center of a set of observations, there may be no value such that equal numbers of items lie above and below it. Nevertheless, the central value, as defined in the preceding paragraph, is still designated as the median. For example, in the array 52, 60, 60, 60, 60, 61, 62, the number 60 is the median, although unequal numbers of items lie above and below this value.

Frequency distribution

In a frequency distribution, the median is necessarily an estimated value, since the identity of the original observations is not retained. Because in a frequency distribution the data are arranged in order of magnitude, frequencies can be cumulated to determine the class in which the median observation falls. It is then necessary to make some assumption about how observations are distributed in that class. Conventionally, the assumption is made that observations are equally spaced, or evenly distributed, throughout the class containing the median. The value of the median is then established by a linear interpolation. The procedure is illustrated for the distribution of personal loans given in Table 1-10. First, the calculation of the median is explained without the use of symbols. Then the procedure is generalized by stating it as a formula.

There are 100 loans represented in the distribution shown in Table 1-10, so the median lies between the fiftieth and fifty-first loans. Since 49 loans occur

TABLE 1-10
Calculation of the median for a frequency distribution of personal loan data

Class Interval	Number of Loans f	
\$ 0 and under \$ 200	6	
200 and under 400	18	
400 and under 600	25	
		$\Sigma f_p = 49$
600 and under 800	20	
800 and under 1,000	17	
1,000 and under 1,200	14	
	100	

prior to the class "$600 and under $800," the median must be in that class. Assuming that the 20 loans are evenly distributed between $600 and $800, we can determine the median observation by interpolating $\frac{1}{20}$ of the distance through this $200 class. The median is calculated by adding $\frac{1}{20}$ of $200 to the $600 lower limit of the class containing the median. That is,

$$Md = \$600 + \left(\frac{50 - 49}{20}\right)\$200 = \$600 + \left(\frac{1}{20}\right)\$200 = \$610$$

Thus, the formula for calculating the median of a frequency distribution is

$$Md = L_{Md} + \left(\frac{n/2 - \Sigma f_p}{f_{Md}}\right)i \tag{1.5}$$

where Md = the median
L_{Md} = the (real) lower limit of the class containing the median
n = the total number of observations in the distribution
Σf_p = the sum of the frequencies of the classes preceding the one containing the median
f_{Md} = the frequency of the class containing the median
i = the size of the class interval

It may seem that we have located the value of the fiftieth observation rather than one falling midway between the fiftieth and fifty-first. However, the value determined is indeed one lying halfway between the fiftieth and fifty-first observations, if we use the assumption that items are evenly distributed within the class in which the median falls.

If there are 20 observations in the $200 class from $600 to $800, we may divide the class into 20 equal subintervals of $10 each, as depicted in Figure 1-6. Since the items are assumed to be evenly distributed in the class from $600 to $800, they must be located at the midpoints of these subintervals. An interpolation of $\frac{1}{20}$ through the class interval brings us to the end of the first subinterval, $610, which is the value halfway between the first and second items. Since 49 frequencies preceded this class, the median of $610 is a value lying midway between the fiftieth and fifty-first observations.

Item number
1 2 3 4 5 6 7 8 9 10 11 12 13 14 15 16 17 18 19 20
$600 $620 $640 $660 $680 $700 $720 $740 $760 $780 $800
Subintervals

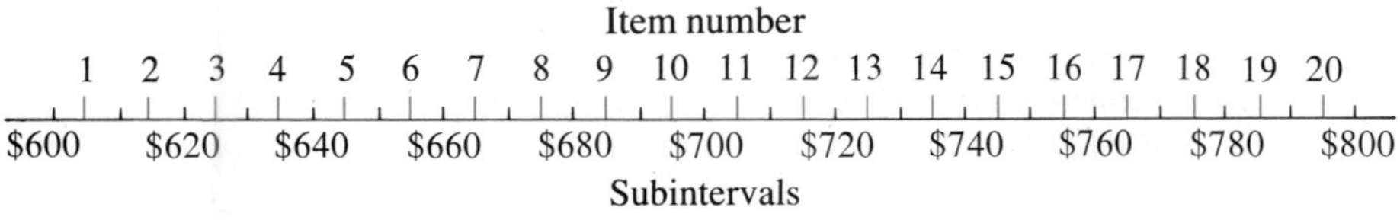

FIGURE 1-6 Diagram depicting the meaning of the assumption concerning an even distribution of observations

1.12

CHARACTERISTICS AND USES OF THE ARITHMETIC MEAN AND MEDIAN

The preceding sections have concentrated on the mechanics of calculating means and medians for ungrouped and grouped data. We now turn to a few of the characteristics and uses of these averages.

Arithmetic mean

The **arithmetic mean**—the most familiar measure of central tendency—is defined as the total of the values of a set of observations divided by the number of these observations.

It thus has the advantage of being a rigidly defined mathematical value that can be manipulated algebraically. For example, the means of two related distributions can be combined by suitable weighting. If the arithmetic mean income in 1986 of 10 marketing executives of a company was \$65,000 and the corresponding mean income of 19 marketing executives of another company was \$75,000, then the arithmetic mean income for the 29 executives combined was, by equation 1.4,

$$\bar{X}_w = \frac{10(\$65{,}000) + 19(\$75{,}000)}{10 + 19} = \$71{,}552$$

On the other hand, if we knew the median income of the 10 executives from the first company and the median income of the 19 executives from the second company, there would be no way of averaging those two numbers to obtain the median income for the 29 executives combined. Because of such mathematical properties, the arithmetic mean is used more often than any other average in advanced statistical techniques.

A disadvantage of the mean is its tendency to be distorted by extreme values at either end of a distribution. In general, it is pulled in the direction of these extremes. Thus, the arithmetic mean of the five figures \$110, \$126, \$132, \$157, and \$1,000 is

$$\frac{\$110 + \$126 + \$132 + \$157 + \$1{,}000}{5} = \$305$$

a value that is greater than four of the five items averaged. In such situations, the arithmetic mean may not be a typical or representative figure.

Median

The **median** is also a useful measure of central tendency. Its relative freedom from distortion by skewness in a distribution makes it particularly useful for conveying the idea of a typical observation. It is primarily affected by the number of observations rather than their size. Consider an array in which the median has been determined; if the largest item is multiplied by 100 (or any large number), the median remains unchanged. The arithmetic mean would, of course, be pulled toward the large extreme item.

The major disadvantage of the median is that it is an average of position and hence is more difficult to deal with than is the arithmetic mean. Thus, as indicated earlier, if one knows the medians of each of two distributions, there is no algebraic way of averaging the two figures to obtain the median of the combined distribution.

1.13 THE MODE

Another measure of central tendency that is sometimes useful but often not explicitly calculated is the mode. In French, to be "a la mode" is to be in fashion. The **mode** is the observation that occurs with the greatest frequency and thus is the most "fashionable" value. The mode is usually not determined for ungrouped data. The reason is that even when most of the data items are clustered toward the center of the array of observations, the item that occurs more often than any other may lie at the lower or upper end of the array and thus be an unrepresentative figure. Therefore, determination of the mode is generally attempted only for grouped data.

Ungrouped data

Grouped data

When data are grouped into a frequency distribution, it is not possible to specify the single observation that occurs most frequently, since the identity of the individual items is lost. However, we can determine the **modal class**, the class that contains more observations than any other. Of course, class intervals should be of the same size when this determination is made. When the location of the modal class is considered along with the arithmetic mean and median, much useful information is generally conveyed not only about central tendency but also about the skewness of a frequency distribution.

Location of mode

Several formulas have been developed for determining the location of the mode within the modal class. These usually involve the use of frequencies in the classes preceding and following the modal class as weighting factors that tend to pull the mode up or down from the midpoint of the modal class. We shall not present any of these formulas here. For our purposes, the midpoint of the modal class may be taken as an estimate of the mode. To understand the meaning of the mode clearly, let us visualize the frequency polygon of a distribution and the frequency curve approached as a limiting case when class size is gradually reduced. In the limiting situation, the variable under study may be thought of as continuous rather than as discrete. The mode may then be thought of as the value of the horizontal axis lying below the maximum point on the frequency curve (see Figure 1-7).

The mode of a frequency distribution has the connotation of a typical or representative value, a location in the distribution at which there is maximum

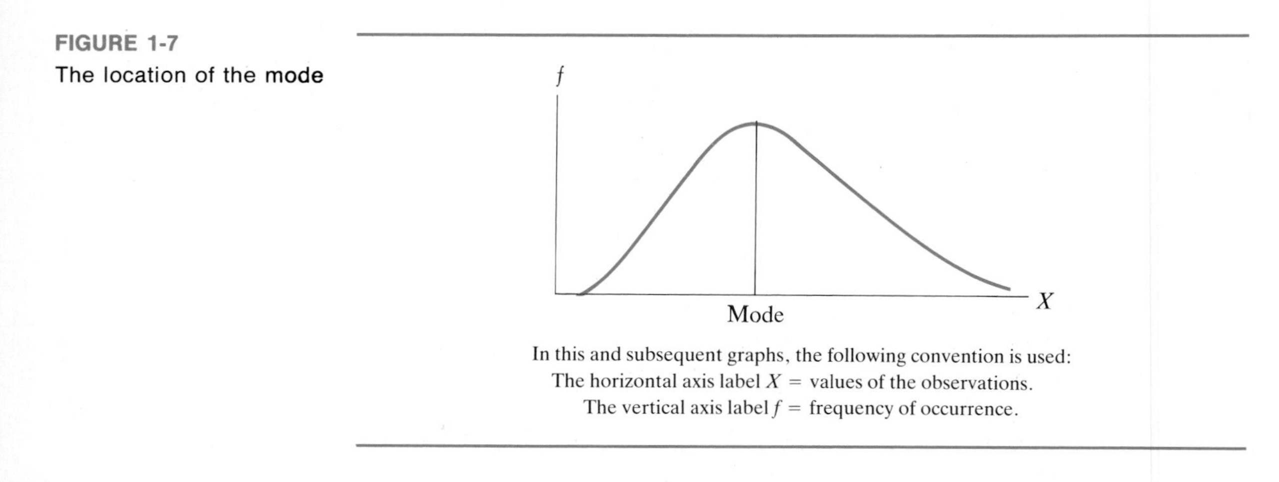

FIGURE 1-7
The location of the mode

In this and subsequent graphs, the following convention is used:
The horizontal axis label X = values of the observations.
The vertical axis label f = frequency of occurrence.

clustering. In this sense, it serves as a standard against which to judge the representativeness or typicality of other averages.

> If a frequency distribution is symmetrical, the mode, median, and mean coincide.

As noted earlier, extreme values in a distribution pull the arithmetic mean in the direction of these extremes. Stated somewhat differently, in a skewed distribution, the mean is pulled away from the mode toward the extreme values. The median also tends to be pulled away from the mode in the direction of skewness but is not affected as much as the mean. If the mean exceeds the median,[2] a distribution is said to have **positive skewness** or to be **skewed to the right**; if the mean is less than the median, the terms **negative skewness** and **skewed to the left** are used. The order in which averages tend to fall in skewed distributions is shown in Figure 1-8.

Positive skewness

Many distributions of economic data in the United States are skewed to the right. Examples include the distributions of incomes of individuals, savings of individuals, corporate assets, sizes of farms, and company sales within many industries. In many of these instances, the arithmetic mean is pulled so far from the median and mode as to be a very unrepresentative figure.

Multimodal Distributions

If more than one mode appears, the frequency distribution is referred to as **multimodal**; if there are two modes, it is referred to as **bimodal**. Extreme care must be exercised in analyzing such distributions. For example, consider a situation in which you want to compare the mean wages of workers in two different companies. Assume that the mean calculated for Company A exceeds that of Company B. If you conclude from this finding that workers in Com-

[2] Sometimes the mode, rather than the median, is used for this comparison.

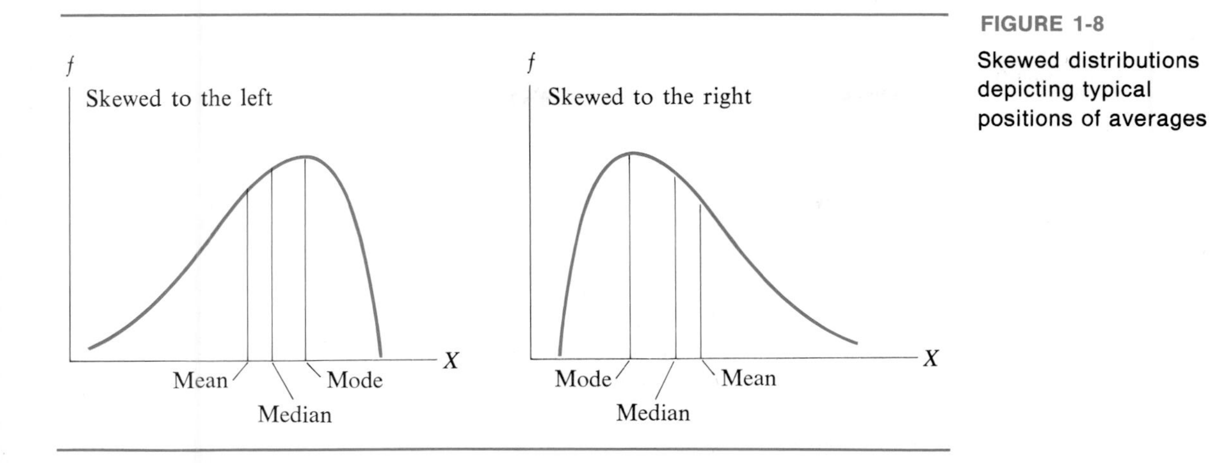

FIGURE 1-8
Skewed distributions depicting typical positions of averages

pany A earn higher wages, on the average, than those in Company B, without taking into account the fact that the wage distribution for each of these companies is bimodal, you may make serious errors of inference. To illustrate the principle involved, let us assume that the mean annual wage is \$10,000 for unskilled workers and \$20,000 for skilled workers at each of these companies. Let us also assume that the individual distributions of wages of unskilled and skilled workers are symmetrical and that there are the same total number of workers in each company. Further, let us assume that these companies have workers only in the aforementioned two skill classifications. However, suppose 75% of the Company A workers are skilled, whereas only 50% of the Company B workers are skilled. Figure 1-9 shows the frequency curves of the distributions of annual wages at the two companies. Clearly, the mean annual wage of workers in Company A exceeds that in Company B, simply because there is a higher percentage of skilled workers at Company A. However, if you were ignorant of this fact, you might be tempted to infer that workers at Company A earn more than those at Company B. The fact of the matter is that unskilled

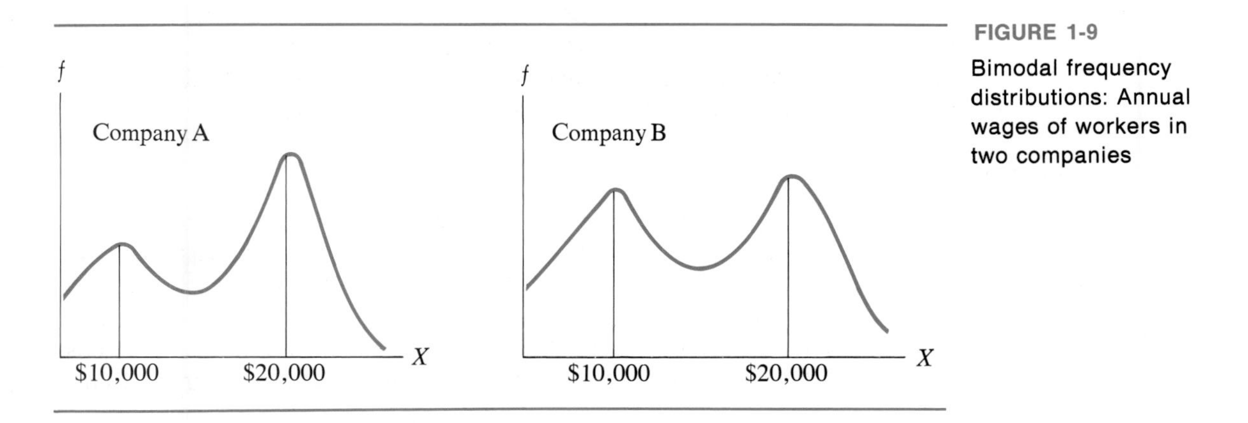

FIGURE 1-9
Bimodal frequency distributions: Annual wages of workers in two companies

Separate into two distributions

workers at both companies earn the same wages, on the average, and the same holds true for the skilled workers. What is required here is to separate two wage distributions at each company, one for skilled workers and one for unskilled. A comparison of the mean wages of unskilled workers (and of skilled workers) at the two companies would reveal their equality.

Nonhomogeneous data

The principle involved here is one of **homogeneity** of the basic data. The fact that a wage distribution is bimodal (that is, has two values around which frequencies are clustered) suggests that two different "cause systems" are present and that two distinct distributions should be recognized. The data on wages may be said to be **nonhomogeneous** with respect to skill level. Bimodal distributions would also result, for example, from the merging of height data for men and for women or weight data for men and for women.

> If a basis for separating a bimodal distribution into two distributions cannot be found, then extreme care must be used in describing the data.

In cases such as that shown for Company B in Figure 1-9, where the heights of the two modes are about equal, the arithmetic mean and median will probably fall between the modes and will not be representative of the large concentrations of values lying at the modes below and above these averages.

EXAMPLE 1-3

The following distribution gives the dollar cost per unit of output for 200 plants in the same industry:

Dollar Cost per Unit of Output	Number of Plants
\$1.00 and under \$1.02	6
1.02 and under 1.04	26
1.04 and under 1.06	52
1.06 and under 1.08	58
1.08 and under 1.10	39
1.10 and under 1.12	15
1.12 and under 1.14	3
1.14 and under 1.16	1
Total	200

a. Calculate the arithmetic mean of the distribution.
b. Calculate the median of the distribution.
c. Is the answer in part (a) the same as the one you would have obtained by computing the following ratio?

$$\frac{\text{Total dollar cost for the 200 plants}}{\text{Total number of units of output of the 200 plants}}$$

d. Can you say that 50% of the units produced cost less than the median calculated in part (b)? Explain.
e. Suppose that the last class had been "$1.14 and over." What effect, if any, would this have had on your calculation of the arithmetic mean and of the median? Briefly justify your answers.

Solutions

a. $\bar{X} = \dfrac{\$213.20}{200} = \$1.066$

b. Median class = $1.06 and under $1.08

$$\text{Median} = \$1.06 + \left(\frac{100 - 84}{58}\right)(\$0.02) = \$1.066$$

c. No. Conceptually, the ratio in part (c) is a weighted mean of 200 cost per unit output ratios. The mean calculated in part (a) is unweighted. It is an estimate of the figure that we would have obtained by adding up the original 200 cost per unit output ratios and dividing by 200 plants.
d. No. We have no way of telling how many units each plant produces. For example, the six plants with costs under $1.02 per unit might produce 60% of the total number of units in the industry. However, we can say that the grouped data procedure indicates that 50% of the plants had a cost per unit less than the median figure of $1.066.
e. The open-ended interval would have no effect on the median, which is concerned only with the order of the frequencies. However, the mean, which is concerned with the actual values, would have to be recalculated. A mean value for the items in the open interval would have to be assumed and multiplied by the frequency to obtain an estimate of the *fm* value for that class.

EXAMPLE 1-4

What is the proper average(s), if any, to use in the following situations? Briefly justify your answers.
a. Determining which members are in the upper half of a class with respect to their overall grade average.
b. Determining the average death rate for six cities combined, given the death rate of each.
c. Finding the average amount each worker is to receive in a profit-sharing plan in which a firm wishes to ensure equal distribution.
d. Determining how high to make a bridge (not a drawbridge). The distribution of the heights of boats expected to pass under the bridge is known and is skewed to the left.
e. Determining a typical wage figure for use in later arbitration for a company that employs 100 workers, several of whom are highly paid specialists.
f. Determining the average annual percentage rate of net profit to sales of a company over a 10-year period.

Solutions

a. The median, since (generally) one-half of the grade averages will fall above and below this figure.

b. The weighted arithmetic mean. The death rates of the six cities would be weighted by the population figures of these cities.
c. The arithmetic mean. Divide the total profits to be shared (ΣX) by the number of workers (n).
d. None of the averages would be appropriate, since they would result in a large number of boats being unable to pass under the bridge. The bridge should be high enough to allow all expected traffic to pass under it.
e. The median. Because of the tendency of the arithmetic mean to be distorted by a few extreme items, it would tend to be a less typical figure.
f. The weighted arithmetic mean. The profit rates would be weighted by the sales figures for each year.

Exercises 1.13

1. Five persons in a family have the following incomes: $20,000, $22,000, $50,000, $17,000, and $18,000.
   a. Find the arithmetic mean and median of these income figures.
   b. Which figure, the arithmetic mean or median, is a more typical average income for this family?
2. Find the mean, median, and mode for each of the following sets of dollar values:
   a. $8, $0, $6, $8, $10, $3, $2, $2, $8, $0
   b. $5, $3, $3, $1, $5, $7, $5, $9, $7
   c. $4, $5, $4, $120, $1, $2, $0, $4
3. A survey of the home addresses of 30 students at a college has been made, and the data are expressed by geographic region.
   a. Determine the number and percentage of students from each of the six geographic regions.
   b. What average would be most appropriate in this case to indicate typicality of the geographic regions of students' homes?
   c. Can you construct a cumulative frequency distribution for these data? Why or why not?

NE	MA	MA	PC	MA	MA
MA	NE	SE	MA	MW	SW
MA	NE	MA	MW	MA	MA
SE	PC	NE	PC	MA	PC
MA	MA	SW	PC	NE	NE

NE = New England
SE = Southeast
SW = Southwest
MA = Mid-Atlantic
MW = Midwest
PC = Pacific Coast

4. Consider Table 1-11, showing the frequency distribution of the passbook bank deposits of the Second National Bank and Trust Company on a certain day.
   a. Why do you think open-ended intervals were used in this distribution?
   b. Which is the modal class?
   c. Which is the median class?

TABLE 1-11
Passbook bank deposits on a certain day

Deposit Size	Number of Customers
$0– 499	8,178
500– 999	9,296
1,000–1,499	9,342
1,500–1,999	10,347
2,000–2,499	10,546
2,500–2,999	7,147
3,000–3,499	5,239
3,500–3,999	3,103
4,000–4,499	2,445
4,500 and over	1,676

5. Refer to Table 1-12 for data on the earnings of unskilled employees of the Surety Equipment Corporation for a certain month.
   a. Compute the arithmetic mean of the distribution.
   b. Compute the median of the distribution.
   c. In which direction are the data skewed? Why?
   d. The comptroller of the company stated that the total payroll for the month for these workers was $166,500. Do you have any reason to doubt this statement? Briefly support your position.
   e. Would you say that the arithmetic mean that you calculated in part (a) provides a satisfactory description of the typical earnings of these 150 unskilled employees for the given month? Why or why not?

TABLE 1-12
Monthly earnings for unskilled employees

Earnings for Month	Number of Employees
$900 and under $950	9
950 and under 1,000	14
1,000 and under 1,050	22
1,050 and under 1,100	26
1,100 and under 1,150	28
1,150 and under 1,200	20
1,200 and under 1,250	15
1,250 and under 1,300	10
1,300 and under 1,350	6
	150

1.14

DISPERSION: DISTANCE MEASURES

Central tendency, as measured by the various averages already discussed, is an important descriptive characteristic of statistical data. However, although two sets of data may have similar averages, they may differ considerably with respect to the spread, or dispersion, of the individual observations. Measures of dispersion describe this variation in numerical observations.

There are two types of measures of dispersion: distance measures and average deviation measures.

Distance measures describe the spread of data in terms of the distance between the values of selected observations.

Range

The simplest distance measure is the **range**, or the difference between the values of the highest and lowest items. For example, if the loans extended by a bank to five corporate customers are $82,000, $125,000, $140,000, $212,000, and $245,000, the range of these loans is $245,000 − $82,000 = $163,000. Such a measure of dispersion may be useful for obtaining a rough notion of the spread in a set of data, but it is certainly inadequate for most analytical purposes.

A disadvantage of the range is that it describes dispersion in terms of only two selected values in a set of observations. Thus, it ignores the nature of the variation among all other observations, which may be either tightly clustered in one small interval or spread out rather evenly between the extreme values. Furthermore, the two numbers used, the highest and the lowest, are extreme rather than typical values.

Quartiles

Other distance measures of dispersion, like the interquartile range, employ more typical values. **Quartiles** are special cases of general measures known as **fractiles**, which refer to values that exceed specified fractions of the data—the ninth decile exceeds $\frac{9}{10}$ of the items, the ninety-ninth percentile exceeds $\frac{99}{100}$ of the items in a distribution, and so on. Clearly, many arbitrary distance measures of dispersion could be developed, but they are infrequently used in practical applications.

Interquartile range

The **interquartile range** is the difference between the third-quartile and the first-quartile values. The **third quartile** is a figure such that three quarters of the observations lie below it; the **first quartile** is a figure such that one quarter of the observations lie below it. Thus, the distance between these two numbers measures the spread between the values that bound the middle 50% of the values in a distribution. A main disadvantage of such a measure is that it does not describe the variation *among* the middle 50% of the items (nor among the lower and upper fourths of the values).

The method of calculating quartile values will not be explicitly discussed here, but for frequency distribution data, its calculation proceeds in a manner completely analogous to that of the median, which is itself the second-quartile value (that is, two quarters of the observations in a distribution lie below the median and two quarters lie above it).

Calculating quartile values

1.15 DISPERSION: AVERAGE DEVIATION METHODS

The most comprehensive descriptions of dispersion are stated in terms of the **average deviation** from some measure of central tendency. The most important average deviation measures are the variance and the standard deviation.

The **variance** of the observations in a population—denoted by the Greek lower-case letter σ^2 (sigma, squared)—is the arithmetic mean of the squared deviations from the population mean. In symbols, if $X_1, X_2, \ldots, X_N$ represent the values of the N observations in a population and if μ is the arithmetic mean of these values, the population variance is defined by

Variance

(1.6)
$$\sigma^2 = \frac{(X_1 - \mu)^2 + (X_2 - \mu)^2 + \cdots + (X_N - \mu)^2}{N} = \frac{\Sigma(X - \mu)^2}{N}$$

As usual, subscripts have been dropped in the simplified form on the right side of equation 1.6.

Although the variance measures the extent of variation in the values of a set of observations, it is in units of squared deviations or squares of the original numbers. To obtain a measure of dispersion in terms of the units of the original data, the square root of the variance is taken. The resulting measure is known as the standard deviation. Thus, the standard deviation of a population is given by

(1.7)
$$\sigma = \sqrt{\frac{\Sigma(X - \mu)^2}{N}}$$

By convention, the positive square root is used.

The **standard deviation** is a measure of the spread in a set of observations.

Standard deviation

If all the values in a population were identical, each deviation from the mean would be 0, and the standard deviation would thus be equal to 0, its minimum value. On the other hand, as items are dispersed more and more widely from the mean, the standard deviation becomes larger and larger.

If we now consider the corresponding measures for a sample of n observations, it would seem logical to substitute the sample mean $\bar{X}$ for the population mean μ and the sample number of observations n for the population number N in equations 1.6 and 1.7. However, it can be shown that when the sample variance and standard deviation are defined with $n - 1$ in the divisors, better estimates are obtained of the corresponding population parameters.[3] Hence, in keeping with modern usage, we define the sample variance and sample standard deviation, respectively, as

Sample variance

(1.8)
$$s^2 = \frac{\Sigma(X - \bar{X})^2}{n - 1}$$

and

Sample standard deviation

(1.9)
$$s = \sqrt{\frac{\Sigma(X - \bar{X})^2}{n - 1}}$$

The term s^2 is usually referred to simply as the "sample variance" and s as the "sample standard deviation."

Computing the standard deviation

We will now concentrate on methods of computing the standard deviation, both for ungrouped and for grouped data. Then we discuss how this measure of dispersion is used. However, the major uses of the standard deviation are in connection with sampling theory and statistical inference, which are discussed in subsequent chapters.

Assume that the following numbers of miles were traveled each week in a five-week period by a traveling sales representative: 1,568, 1,330, 1,096, 1,200, and 1,056. The arithmetic mean is 1,250 miles. Regarding these data as a sample, the calculation of the standard deviation using equation 1.9 is illustrated in Table 1-13.

The resulting standard deviation of 207 miles is an absolute measure of dispersion, which means that it is stated in the units of the original data. Whether this is a great deal or only a small amount of dispersion cannot be immediately determined. This sort of judgment is based on the particular type of data analyzed (in this case, traveling sales representative data). Furthermore, as we shall see in section 1.16, relative measures of dispersion are preferable to absolute measures for comparative purposes.

[3] A brief justification for the division by $n - 1$: It can be shown mathematically that for an infinite population, when the sample variance is defined with divisor $n - 1$ as in equation 1.8, it is a so-called "unbiased estimator" of the population parameter σ^2. This means that if all possible samples of size n were drawn from a given population and the variances of these samples were averaged (using the arithmetic mean), this average would be equal to the population variance σ^2. Thus, when the sample variance is defined with divisor $n - 1$, on the average, it correctly estimates the population variance. Of course, for large samples, the difference in results obtained by using n rather than $n - 1$ as the divisor would tend to be very slight, but for small samples the difference can be rather substantial.

TABLE 1-13
Calculation of the standard deviation for ungrouped data by the direct method: traveling sales representative data

Miles Traveled X	Deviation from Mean $X - \bar{X}$	Squared Deviations $(X - \bar{X})^2$
1,568	1,568 − 1,250 = 318	101,124
1,330	1,330 − 1,250 = 80	6,400
1,096	1,096 − 1,250 = −154	23,716
1,200	1,200 − 1,250 = −50	2,500
1,056	1,056 − 1,250 = −194	37,636
$\bar{X}$ = 1,250	0	171,376

$$s = \sqrt{\frac{\Sigma(X - \bar{X})^2}{n - 1}} = \sqrt{\frac{171,376}{4}} = 207 \text{ miles}$$

In calculating the standard deviation for data grouped into frequency distributions, it is merely necessary to adjust the foregoing formulas to take account of this grouping. The defining equation 1.9 generalizes to

$$s = \sqrt{\frac{\Sigma f(m - \bar{X})^2}{n - 1}} \tag{1.10}$$

where, as usual for grouped data, m represents the midpoint of a class, f is the frequency in a class, $\bar{X}$ is the arithmetic mean, and n is the total number of observations. This calculation is illustrated in Table 1-14 for the frequency distribution of bottle weights data previously given in Tables 1-3 and 1-6. As shown in Table 1-14, the standard deviation is equal to 0.933 ounce.

Calculation of the sample standard deviation by the defining formula can be tedious, particularly if the class midpoints and frequencies contain several digits and the arithmetic mean is not a round number. A shortcut method of calculation often useful in such cases is given in Appendix D.

Uses of the Standard Deviation

The standard deviation of a frequency distribution is useful in describing the general characteristics of the data. For example, in the so-called normal distribution (a bell-shaped curve), which is discussed extensively in Chapter 5, 6, and 7, the standard deviation is used in conjunction with the mean to indicate the percentage of items that fall within specified ranges. Hence, if a population is in the form of a normal distribution, the following relationships apply:

$\mu \pm \sigma$ includes 68.3% of all of the items
$\mu \pm 2\sigma$ includes 95.5% of all of the items
$\mu \pm 3\sigma$ includes 99.7% of all of the items

TABLE 1-14
Calculation of the standard deviation for grouped data by the direct method: bottle weights data

Weight (in ounces)	Number of Bottles f	Midpoints m	$(m - \bar{X})$	Deviation $(m - \bar{X})^2$	$f(m - \bar{X})^2$
14.0 and under 14.5	8	14.25	−1.68	2.822	22.579
14.5 and under 15.0	10	14.75	−1.18	1.392	13.924
15.0 and under 15.5	15	15.25	−0.68	0.462	6.936
15.5 and under 16.0	18	15.75	−0.18	0.032	0.583
16.0 and under 16.5	22	16.25	0.32	0.102	2.253
16.5 and under 17.0	14	16.75	0.82	0.672	9.414
17.0 and under 17.5	8	17.25	1.32	1.742	13.939
17.5 and under 18.0	5	17.75	1.82	3.312	16.562
	100				86.190

$\bar{X} = 15.93$ ounces

$$s = \sqrt{\frac{\Sigma f(m - \bar{X})^2}{n-1}} = \sqrt{\frac{86.19}{99}} = \sqrt{0.8706}$$

$s = 0.933$ ounce

For example, if a production process is known to produce items that are normally distributed with a mean length of $\mu = 10$ inches and a standard deviation of 1 inch, then we can infer that 68.3% of the items have lengths between $10 - 1 = 9$ inches and $10 + 1 = 11$ inches. About 95.5% have lengths between $10 - 2 = 8$ and $10 + 2 = 12$ inches, and 99.7% have lengths between $10 - 3 = 7$ and $10 + 3 = 13$ inches. Thus, a range of $\mu \pm 3\sigma$ includes virtually all the items in a normal distribution. We shall see in Chapter 5 that the normal distribution is perfectly symmetrical. If the departure from a normal distribution is not too great, the rough generalization that virtually all the items are included within a range from 3σ below the mean to 3σ above the mean still holds.

The standard deviation is also useful in describing how far individual items in a distribution depart from the mean of the distribution. Suppose the population of students who took a certain aptitude test displayed a mean score of $\mu = 100$ with a standard deviation of $\sigma = 20$. Then a score of 80 on the examination can be described as lying one standard deviation below the mean. The terminology usually employed is that the standard score is -1; that is, if the examination score is denoted X, then

$$\frac{X - \mu}{\sigma} = \frac{80 - 100}{20} = -1$$

The **standard score** of an observation is the number of standard deviations the observation lies below or above the mean of the distribution.

Hence, the score of 80 deviates from the mean by -20 units, which is equal to -1 in terms of units of standard deviations away from the mean. If standard scores are computed from sample rather than universe data, the formula $(X - \bar{X})/s$ would be used instead.

Comparisons can thus be made for items in distributions that differ in order of magnitude or in the units employed. For example, if a student scored 120 on an examination in which the mean was $\mu = 150$ and $\sigma = 30$, the standard score would be $(120 - 150)/30 = -1$. Thus, the score of 120 is the same number of standard deviations below the mean as the 80 in the preceding example. We could also compare standard scores in a distribution of wages with comparable figures in a distribution of length of employment service, and so on.

The standard deviation is doubtless the most widely used measure of dispersion, and considerable use is made of it in later chapters of this text.

1.16 RELATIVE DISPERSION: COEFFICIENT OF VARIATION

Although the standard score discussed earlier is useful for determining how far an *item* lies from the mean of a set of data, we often are interested in comparing the dispersion of *an entire set of data* with the dispersion of another set. As observed earlier, the standard deviation is an absolute measure of dispersion, whereas a relative measure is required for purposes of comparison. A relative measure is essential when the sets of data to be compared are expressed in different units or when the data are in the same units but are of different orders of magnitude. Such a relative measure is obtained by expressing the standard deviation as a percentage of the arithmetic mean. The resulting figure is referred to as the **coefficient of variation** (CV) and is defined symbolically as

$$CV = \frac{s}{\bar{X}} \tag{1.11}$$

Thus, for the frequency distribution of weights of filled bottles, the standard deviation is 0.933 ounce with a mean of 15.93 ounces. The coefficient of variation is

$$CV = \frac{0.933}{15.93} = 5.86\%$$

Let us assume that the corresponding figures for a second bottle-filling operation revealed a standard deviation of 1.1 ounces with an arithmetic mean of 25.0 ounces. The coefficient of variation for this set of data is $CV = \frac{1.1}{25.0} = 4.4\%$. Therefore, the weights in the second bottle-filling operation were relatively more uniform—that is, they displayed relatively less variation than did the weights of the first operation. Note that the weights of the second operation had the larger standard deviation, but because of the higher average weight, relative dispersion was less.

Both absolute and relative measures of dispersion are widely used in practical sampling problems. For example, a question may arise about the sample size required to yield an estimate of a universe parameter with a specified degree of precision. Specifically, a finance company may want to know how large a random sample of its loans it must study in order to estimate the average dollar size of its delinquent loans. If the company wants this estimate accurate within a specified number of *dollars*, an absolute measure of dispersion is appropriate. On the other hand, if the company wants the estimate to be within a specified *percentage* of the true average figure, a relative measure of dispersion would be used.

1.17 ERRORS IN PREDICTION

In this chapter, we have discussed descriptive measures for empirical frequency distributions, with emphasis on measures of central tendency and dispersion. Some interesting relationships between these two types of measure are observable when certain problems of prediction are considered. Suppose we want to guess or "predict" the value of an observation picked at random from a frequency distribution. Let us refer to the penalty of an incorrect prediction as the "cost of error." If there were a **fixed cost error** on each prediction, no matter what the size of the error, we should guess the **mode** as the value of the random observation. This would give us the highest probability of guessing the *exact value* of the unknown observation. Assuming repeated trials of this prediction experiment, we would thus minimize the average (arithmetic mean) cost of error.

Fixed cost error

Suppose, on the other hand, that the **cost of error varies directly with the size of error regardless of its sign**, that is, regardless of whether the actual observation is above or below the predicted value. In this case, we would want a prediction that minimizes the average **absolute error**. The **median** would be the "best guess," since it minimizes average absolute deviations. The mean deviation about the median would be a measure of this minimum cost of error.

Absolute error

Finally, suppose **the cost of error varies according to the square of the error** (for example, an error of two units costs four times as much as an error of one unit). In this situation, the **mean** should be the predicted value, since it can be demonstrated mathematically that the average of the squared deviations about it is less than around any other figure. Here the variance, which may be interpreted as the average cost of error per observation, would represent a measure of this minimum error. Another point previously observed for the mean is that the average amount of error, taking account of sign, would be 0.

Minimum error

A practical business application of these ideas is in the determination of the optimum size of inventory to be maintained. Let us assume a situation in which the cost of overstocking a unit (cost of overage) is equal to the cost of being short one unit (cost of underage). Further, it may be assumed that the cost of error varies directly with the absolute amount of error. For example, having two units in excess of demand costs twice as much as one unit. In this situation, the optimum stocking level is the median of the frequency distribution of numbers of units demanded.

1.18 PROBLEMS OF INTERPRETATION

Many of the most common misinterpretations and misuses of statistics involve measures and concepts such as those discussed in this chapter: averages, dispersion, and skewness.

> Sometimes misleading interpretations are drawn from the use of averages that are not "typical" or "representative."

Reference was made in section 1.12 to the possible distortion of the typicality of the arithmetic mean because of the presence of extreme items at one end of a distribution. An interesting example of this distortion occurred in the case of a survey conducted by a popular periodical. One of the purposes of the survey was to determine the current status of persons who had graduated from college during the early Depression years. Among those included were the graduates of Princeton University for three successive years during the early 1930s. The results of the Princeton survey indicated that the arithmetic mean income of the respondents in the class that graduated in the second year was far higher than the corresponding mean income for the first- and third-year classes. The analysts attempted to rationalize this result in various ways. However, a reexamination of the data yielded a simple explanation, which precluded potential misinterpretations. It turned out that one of the graduates of the second-year class was a member of one of the wealthiest families in the United States and was an heir to an immense fortune. His very large income exerted an obvious upward pull on the mean income of his class, making it an unrepresentative average.

> Misinterpretations of averages often arise because dispersion is not taken into account.

Prospective college students are sometimes discouraged when they observe the mean scholastic aptitude test scores of classes admitted to colleges or universities in which they are interested. Admissions officers have commented that students sometimes erroneously assume they will not be admitted to a school because their test scores are somewhat below the published mean scores for that school. Of course, such students fail to take into account dispersion around this average. Assuming a roughly symmetrical distribution, about one half of the admitted students on whom the published means were based had test scores that fell below that average.

> Because of the shape of the underlying frequency distribution, sometimes no average will be typical.

In section 1.13, reference was made to bimodal distributions of wages of workers. Arithmetic means or medians for such distributions tend to fall somewhere between the two modes; they are not typical of the groups characterized by either of the modes. As indicated in section 1.13, the solution when nonhomogeneous data are present is to separate the distinct distributions. However, sometimes U-shaped frequency distributions are encountered in which the separation into different distributions is not warranted. In such distributions, frequencies are concentrated at both low and high values of the variable under consideration. For example, suppose the test scores of a mathematics class yield grades that are either very high (in the 90s) or very low (in the 60s). Means or medians, which might be about 75, would clearly be unrepresentative of the concentrations at either end of the distribution. When averages are presented for such distributions, without some indication of the nature of the underlying data, misinterpretations can occur quite easily.

Exercises 1.18

Note: In calculating the variance and standard deviation in the following exercises, use formulas with an $n - 1$ divisor.

1. There are five members in a family. The father's age is 45 and the mother's age is 42. The children are 21, 17, and 10 years old. At present, the mean age of the five people is 27 years and the standard deviation is 15.6 years.
   a. Calculate the mean and standard deviation of the ages of this family 6 years later.
   b. What generalizations concerning the mean and standard deviation are suggested by your results to part (a)?
2. Consider the closing prices of three common stocks traded on the Philadelphia Exchange for a two-week period.
   a. Compare the three stocks simply on the basis of measures of central tendency.
   b. Compute the standard deviation for each of the three stocks. What information do the standard deviations give concerning the price movements of the three stocks?

TABLE 1-15
Closing prices of three stocks for a two-week period

	Conservocorp	Mesocorp	Specucorp
Monday	$44	$44	$44
Tuesday	45	50	50
Wednesday	43	53	54
Thursday	45	48	56
Friday	46	46	53
Monday	48	42	50
Tuesday	49	41	46
Wednesday	49	45	42
Thursday	51	48	38
Friday	50	53	37

3. A trade association reported data on the distribution of sales volume in the preceding year for a sample of member firms.
   a. Compute the standard deviation and the arithmetic mean of the distribution.
   b. The trade association also constructed a frequency distribution of the number of employees in the 100 firms for the same year. The arithmetic mean was 30 persons and the standard deviation was 10 persons. Which distribution showed greater relative variability—the sales volume or the number of employees?
   c. Suppose that the last class in the sales volume distribution had read "300 and over." How, if at all, would this have affected your calculation of the standard deviation, the mean, and the median?

Sales Volume (in thousands of dollars)	Number of Firms
100 and under 150	4
150 and under 200	16
200 and under 250	40
250 and under 300	28
300 and under 350	12
	100

4. Based on a sample of 200 business days, a department store obtained a frequency distribution of the daily demand for a certain home furnishing item (see table on next page for the data).
   a. Compute the mean of this sample.
   b. Compute the variance and standard deviation of this sample.

Number of Units Demanded	Number of Days
0	73
1	74
2	37
3	12
4	4
	200

5. An applicant for a position received a score of 84 on aptitude test A and a score of 70 on aptitude test B. On aptitude test A, the mean and the standard deviation were 80 and 4, respectively. On aptitude test B, the mean and the standard deviation were 60 and 7, respectively. On which aptitude test was the applicant's performance relatively better?
6. The vice president of a Central Motors plant is interested in investigating whether production is lower on Mondays than other days. The production figures on six successive Mondays were 10, 15, 17, 21, 15, and 12.
   a. What is the mean of the six production figures? What is the median?
   b. Production for the same time period on the other days of the week was summarized in 5-unit groupings. Is the mean of the other days' production greater than the mean production on Mondays?
   c. Do the Monday production figures have greater relative dispersion than the other days?

Production	Number of Days
10 through 14	9
15 through 19	9
20 through 24	3
25 through 29	3
	24

Chapter 1 Computer Exercises

1. Find the arithmetic mean and standard deviation of the 75 annual food expenditures figures in Appendix E.
2. Find the arithmetic mean and standard deviation of the 75 annual income figures in Appendix E.
3. The annual food expenditures figure for the first family listed in Appendix E is 4.7: this family's annual income is 24 (both figures are in thousands of dollars). Comment on the typicality of this family's annual food expenditures and annual income by comparing the standard scores of the two figures.
4. Calculate the coefficients of variations for annual food expenditures and annual income and comment on the relative variability of these two series.

2

Introduction to Probability

In Chapter 1, we discussed the frequency distribution as a device for summarizing the variation in sets of numerical observations in convenient tabular form. It is often very useful to describe and draw generalizations about patterns of variation using the concept of probability. The discussions of probability in this chapter and in Chapter 3 lay the foundation for our treatment of statistical analysis for decision making.

2.1 THE MEANING OF PROBABILITY

Theory of probability

The development of a mathematical theory of probability began during the seventeenth century when the French nobleman Antoine Gombauld, known as the Chevalier de Méré, raised certain questions about games of chance. Specifically, he was puzzled about the probability of obtaining two 6s at least once in 24 rolls of a pair of dice. (This is a problem you should have little difficulty solving after reading this chapter.) De Méré posed the question to Blaise Pascal, a young French mathematician, who solved the problem. Subsequently, Pascal discussed this and other puzzlers raised by de Méré with the famous mathematician Pierre de Fermat. In the course of their correspondence, the mathematical theory of probability was born.

Methods of measuring probabilities

The several different methods of measuring probabilities represent different conceptual approaches and reveal some of the current controversy concerning the foundations of probability theory. In this chapter we discuss three conceptual approaches: **classical probability**, **relative frequency of occurrence**, and

subjective probability. Regardless of the definition of probability used, the same mathematical rules apply in performing the calculations (that is, measures of probability are always added or multiplied under the same general circumstances).

Classical Probability

Since probability theory had its origin in games of chance, it is not surprising that the first method developed for measuring probabilities was particularly appropriate for gambling situations. According to the so-called **classical** concept of probability, the probability of an event A is defined as follows: If there are a possible outcomes favorable to the occurrence of the event A and b possible outcomes unfavorable to the occurrence of A, and if all outcomes are equally likely and mutually exclusive, then the probability that A will occur, denoted $P(A)$, is

$$P(A) = \frac{a}{a+b} = \frac{\text{Number of outcomes favorable to occurrence of } A}{\text{Total number of possible outcomes}}$$

Thus, if a fair coin with two faces, denoted head and tail, is tossed into the air, the probability that it will fall with the head uppermost is $P(\text{Head}) = 1/(1+1) = \frac{1}{2}$. In this case, there is one outcome favorable to the occurrence of the event "head" and one unfavorable outcome. (The extremely unlikely situation that the coin will stand on end is defined out of the problem; that is, it is not classified as an outcome for the purpose of the probability calculation.)

The preceding equation can also be used to determine the probability that a certain face will show when a true die is rolled. (A die is a small cube with 1, 2, 3, 4, 5, or 6 dots on each of its faces.) A **true die** is one that is equally likely to show any of the six numbers on its uppermost face when rolled. The probability of obtaining a 1 if such a die is rolled is $P(1) = \frac{1}{1+5} = \frac{1}{6}$. Here, there is one outcome favorable to the event "1" and five unfavorable outcomes.

Terminology

Some of the terms used in classical probability require further explanation. The **event** whose probability is sought consists of one or more possible outcomes of the given activity (tossing a coin, rolling a die, or drawing a card). These activities are known in modern terminology as **experiments**, a term referring to processes that result in different possible outcomes or observations. The term **equally likely** in referring to possible outcomes is intuitively clear. Two or more outcomes are said to be **mutually exclusive** if when one of the outcomes occurs, the others cannot. Thus, the appearances of a 1 and a 2 on a die are mutually exclusive events, since if a 1 results, a 2 cannot. All possible results of an experiment are conceived of as a complete, or **exhaustive**, set of mutually exclusive outcomes.

Characteristics

These classical probability measures have two very interesting characteristics.

- The objects referred to as *fair* coins, *true* dice, or *fair* decks of cards are abstractions in the sense that no real-world object possesses exactly the features postulated. For example, in order to be a **fair coin** (equally likely to fall "head" or "tail") the object would have to be a perfectly flat, homogeneous disk—an unlikely object.
- In order to determine the probabilities in the above examples, no coins had to be tossed, no dice rolled, nor cards shuffled. That is, no experimental data had to be collected; the probability calculations were based entirely on logic.

In the context of this definition of probability, if it is *impossible* for an event A to occur, the probability of that event is said to be zero. For example, if the event A is the appearance of a 7 when a single die is rolled, then $P(A) = 0$. A probability of 1 is assigned to an event that is *certain* to occur. Thus, if the event A is the appearance of any one of the numbers 1, 2, 3, 4, 5, or 6 on a single roll of a die, then $P(A) = 1$. According to the classical definition, as well as all others,

> the probability of an event is a number between 0 and 1, and the sum of the probabilities that the event will occur and that it will not occur is 1.

Relative Frequency of Occurrence

Although the classical concept of probability is useful for solving problems involving games of chance, serious difficulties occur with a wide range of other types of problems. For example, it is inadequate for determining the probabilities that (a) a black male American, age 30, will die within the next year, (b) a consumer in a certain metropolitan area will purchase a company's product during the next month, or (c) a production process used by a particular firm will produce a defective item. In none of these situations is it feasible to establish a complete set of mutually exclusive outcomes, each equally likely to occur. For example, in (a), only two occurrences are possible: The individual will either live or die during the ensuing year. The likelihood that he will die is, of course, much smaller than the likelihood that he will live, but how much smaller? The probability that a 30-year-old black male American will live through the next year is greater than the corresponding probability for a 30-year-old black male inhabitant of India. However, how much greater is it and precisely what do these probabilities mean? Questions of this type require reference to data.

We know that the life insurance industry establishes mortality rates by observing how many of a sample of, say, 100,000 black American males, age 30, die within a one-year period. In this instance, the number of deaths divided by 100,000 is the **relative frequency of occurrence** of death for the 100,000 individuals studied. It may also be viewed as an estimate of the **probability** of death for Americans in the given color-sex-age group. This relative frequency of occurrence concept can also be illustrated by a simple coin-tossing example.

Suppose you are asked to toss a coin known to be biased (that is, not a fair coin). You are not told whether the coin is more likely to produce a head or a tail, but you are asked to determine the probability of the appearance of a head by means of many tosses of the coin. Assume that 10,000 tosses of the coin result in 7,000 heads and 3,000 tails. Another way of stating this is that the relative frequency of occurrence of heads is $\frac{7,000}{10,000}$, or 0.70. It certainly seems reasonable to assign a probability of 0.70 to the appearance of a head with this particular coin. On the other hand, if the coin had been tossed only three times and one head resulted, you would have little confidence in assigning a probability of $\frac{1}{3}$ to the occurrence of a head.

The **relative frequency concept of probability** may be interpreted as the proportion of times an event occurs in the long run under uniform or stable conditions.

In practice, past relative frequencies of occurrence are often used as probabilities. Hence, if 800 of the 100,000 30-year-old black male Americans died during the year, the relative frequency of death or probability of death is said to be $\frac{800}{100,000}$ for individuals in this group.

Subjective Probability

The **subjective** or **personalistic** concept of probability is a relatively recent development.[1] Its application to statistical problems has occurred almost entirely in the post–World War II period.

According to the **subjective** concept, the probability of an event is the degree of belief, or degree of confidence, placed in the occurrence of an event by a particular individual based on the evidence available.

This evidence may consist of data on relative frequency of occurrence and any other quantitative or nonquantitative information. The individual who considers it unlikely that an event will occur assigns a probability close to 0 to it; if an individual believes it is very likely an event will occur, he or she assigns it a probability close to 1. Thus, for example, in a consumer survey, an individual may assign a probability of $\frac{1}{2}$ to the event of purchasing an automobile during the next year. An industrial purchaser may assert a probability of $\frac{4}{5}$ that a future incoming shipment will have 2% or fewer defective items.

[1] The concept was first introduced in 1926 by Frank Ramsey, who presented a formal theory of personal probability in F.P. Ramsey, *The Foundation of Mathematics and Other Logical Essays* (London: Kegan Paul; New York: Harcourt Brace Jovanovich, 1931). The theory was developed primarily by B. de Finetti, B.O. Koopman, I.J. Good, and L.J. Savage.

Subjective probabilities should be assigned on the basis of all objective and subjective evidence currently available and should reflect the decision maker's current degree of belief. Reasonable persons might arrive at different probability assessments because of differences in experience, attitudes, values, and so on. Furthermore, these probability assignments may be made for events that will occur only once, in situations where neither classical probabilities nor relative frequencies appear to be appropriate.

This approach is a broad and flexible one, permitting probability assignments to events for which there are no objective data or for which there is a combination of objective and subjective data. However, the assignments of the probabilities must be consistent. For example, if the purchaser assigns a probability of $\frac{4}{5}$ to the event that a shipment will have no more than 2% defective items, then a probability of $\frac{1}{5}$ must be assigned to the event that a shipment will have more than 2% defective items. The concept of subjective, or personal, probability is a useful one, particularly in the context of business decision making.

Sample Spaces and Experiments

The concept of an experiment is a central one in probability and statistics. In this connection,

> an **experiment** is any process of measurement or observation of different **outcomes** (results).

Terms

The experiment may be real or conceptual. The collection or totality of the possible outcomes of an experiment is referred to as its **sample space**. Thus, the collection of outcomes of the experiment of tossing a coin once (or twice, or any number of times) is a sample space. The objects that constitute the sample space are referred to as its **elements**. The elements are usually enclosed within braces, and the symbol S is conventionally used to denote a sample space.

On a single toss of a coin, there are two possible outcomes, tail (T) and head (H). Therefore,

$$S = \{T, H\}$$

If the coin is tossed twice, there are four possibilities:

$$S = \{(T, T), (T, H), (H, T), (H, H)\}$$

In the first case, the experiment consists of **one trial**, a single toss of the coin; in the second case, the experiment contains **two trials**, the two tosses of the coin. In these examples, a physical experiment may actually be performed, or we may easily conceive of the possible outcomes of such an experiment.

In other situations, although no sequence of repetitive trials is involved, we may conceive of a set of outcomes as an experiment. These outcomes may simply be the result of an observational process and need not bear any resemblance to a laboratory experiment, as long as they are well defined. Thus, we may think of each of the following two-way classifications as constituting sample spaces:

0	1
Customer was granted credit.	Customer was not granted credit.
Employee elected a stock purchase plan.	Employee did not elect a stock purchase plan.
The merger will take place.	The merger will not take place.
The company uses direct mail advertising.	The company does not use direct mail advertising.

The elements in these two-element sample spaces may be designated 0 and 1, as indicated by the column headings. Therefore, each of the four illustrative sample spaces may be conveniently symbolized as

$$S = \{0, 1\}$$

Two methods of graphically depicting sample spaces are: (1) graphs using the conventional rectangular coordinate system and (2) tree diagrams. These methods are most useful for sample spaces with relatively small numbers of sample points (or elements). Tree diagrams are more manageable because of the obvious graphic difficulties encountered by the coordinate system method beyond three dimensions. These methods are illustrated in Example 2-1.

EXAMPLE 2-1

Aircraft can either land or depart on a single runway. If each operation takes one minute, depict graphically the sample space for possible operations during a two-minute period, assuming the runway is currently unoccupied.

Solution

The sample space is

$$S = \{(L, L), (L, D), (D, L), (D, D)\}$$

where L = land and D = depart, and the first and second elements in each ordered pair denote outcomes in the first and second minutes, respectively. Let $L = 0$ and $D = 1$. The sample space may now be written

$$S = \{(0, 0), (0, 1), (1, 0), (1, 1)\}$$

and graphed in two dimensions as shown in Figure 2-1(a). A tree diagram for this sample space is shown in Figure 2-1(b).

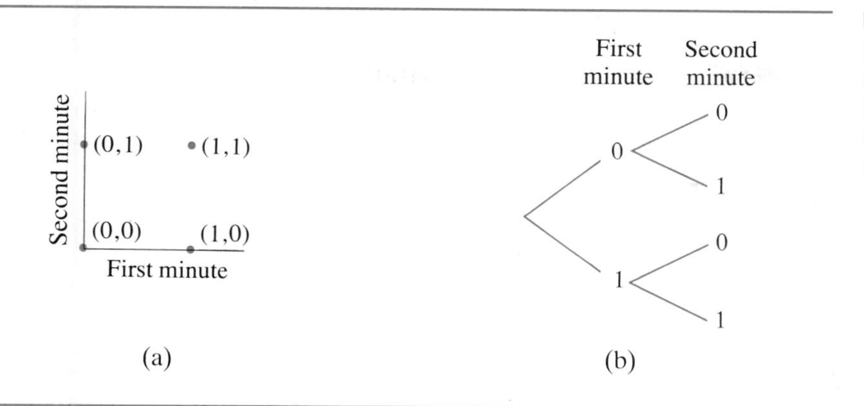

FIGURE 2-1
(a) Graph for aircraft operations example
(b) Tree diagram for aircraft operations example

Events

The meaning of the term "event" as used in ordinary conversation is usually clear. However, because the concept of an event is fundamental to probability theory, it requires an explicit definition.

Once a sample space S has been specified, an **event** is defined as a collection of elements each of which is also an element of S.

Elementary event

An **elementary event** is a single possible outcome of an experiment. It is an event that cannot be further divided into other events.

For example, a single roll of a die constitutes an experiment. The sample space generated is

$$S = \{1, 2, 3, 4, 5, 6\}$$

We may define an event A, say, as the "appearance of a 2 or a 3," which may be expressed as

$$A = \{2, 3\}$$

If either a 2 or a 3 appears on the upper face of the die when it is rolled, the event A is said to have occurred. Note that A is not an elementary event, since it can be divided into the two elementary events, $\{2\}$ and $\{3\}$.

The **complement** of an event A in the sample space S is the collection of elements that are not in A. We will use the symbol $\bar{A}$ (pronounced "A bar") for the complement of A. For example, in the experiment of rolling a die once, the complement of the event that a 1 appears on the uppermost face is the event that a 2, 3, 4, 5, or 6 appears.

Certain (sure) event

When a sample space S has been defined, S itself is referred to as the **certain event**, or the **sure event**, since in any single trial of the experiment that generates S, one or another of its elements must occur. Hence, in the experiment of rolling a die once, the certain event is that either a 1, 2, 3, 4, 5, or 6 appears.

Mutually exclusive event

Two events A_1 and A_2 are said to be **mutually exclusive events** if when one of these events occurs, the other cannot. On a single roll of a die, the event that a 1 appears and the event that a 2 appears are mutually exclusive. On the other hand, the appearance of a 1 and the appearance of an odd number (1, 3, or 5) are not mutually exclusive events, since 1 is an odd number.

Probability and Sample Spaces

Probabilities may be thought of as numbers assigned to points in a sample space.

Probabilities must have the following characteristics:

1. They are numbers greater than or equal to 0.
2. They must add up to 1.

EXAMPLE 2-2

In a study of the loans that were made to corporate borrowers and that matured during the past year, an officer of a commercial bank classified the loans into the following categories with respect to collection experience: excellent, good, fair, poor, and bad. The table gives the number of loans in each category. The relative frequencies in the table have been derived by dividing the number of loans in each category by the total number of loans.

Loan Collection Experience	Number of Loans	Relative Frequencies
Excellent	500	0.500
Good	400	0.400
Fair	50	0.050
Poor	25	0.025
Bad	25	0.025
	1,000	1.000

The loan officer could use the relative frequencies as probabilities for the collection experience for next year's loans to corporate borrowers. In the classification of next year's loans, there would thus be five elementary events (excellent, good, fair, poor, and bad) with respective probabilities of 0.50, 0.40, 0.05, 0.025, and 0.025. The two characteristics of probabilities are satisfied: Each probability is greater than or equal to 0, and the probabilities add up to 1. However, the events in a sample space need not be elementary events. If a "profitable loan" is defined as an excellent, good, or fair loan and an "unprofitable loan" as a poor or bad one, then the events "profitable loan" and "unprofitable loan" make up a sample space with $P(\text{profitable loan}) = 0.50 + 0.40 + 0.05 = 0.95$ and $P(\text{unprofitable loan}) = 0.025 + 0.025 = 0.05$. This classification of loans might be of more interest to the loan officer than the original one. Of course, the events defining the sample space must be a complete set of mutually exclusive events, and the sum of probabilities of the occurrence of an event A and its complement must be equal to 1, that is, $P(A) + P(\bar{A}) = 1$.

EXAMPLE 2-3

An investor is considering the likely trend in interest rates before deciding whether to invest his money in the stock market. He establishes the subjective probabilities shown in the table. The investor has assigned the value of 0.95 to the probability that the interest rates will not drop.

Events E_i	Probability of Events $P(E_i)$
E_1: Interest rates will rise	0.80
E_2: Interest rates will stay constant	0.15
E_3: Interest rates will drop	0.05
	1.00

Odds Ratios

Regardless of the definition used, we sometimes prefer to express probabilities in terms of **odds**. Thus, if the probability that a 6 will appear on the roll of a die is $\frac{1}{6}$, then the odds that it will appear are one to five, written 1:5. If an economist assesses the probability that the Federal Reserve System will adopt a tightened monetary policy as $\frac{1}{2}$, then this probability expressed in terms of odds is 1:1.

If the probability that an event A will occur is

$$P(A) = \frac{a}{n}$$

then the odds in favor of the occurrence of A are

$$\text{Odds in favor of } A = \frac{a}{n - a} = a:(n - a)$$

The odds against A are

$$\text{Odds against } A = \frac{n - a}{a} = (n - a):a$$

> Another way of viewing the odds ratio is as the ratio of the probability of the occurrence of an event A to the probability of its complement, $\bar{A}$.

Thus,

$$\text{Odds in favor of } A = \frac{P(A)}{P(\bar{A})} = \frac{a/n}{(n-a)/n} = \frac{a}{n-a} = a\!:\!(n-a)$$

$$\text{Odds against } A = \frac{P(\bar{A})}{P(A)} = \frac{(n-a)/n}{a/n} = \frac{n-a}{a} = (n-a)\!:\!a$$

Exercises 2.1

1. A student wants to see a movie on Sunday. There are 3 cinemas (A, B, and C), and on Sunday each cinema has shows at 1:00 P.M., 4:00 P.M., and 7:00 P.M. Draw a tree diagram to specify the possible choices.
2. A marketing manager told his marketing group that there was a 50-50 chance that a new advertising campaign would increase sales by 30%.
   a. What type of probability is this?
   b. Is it useful?
3. If "0" represents a nonresponse to a mailed questionnaire and if "1" represents a response, depict the set of outcomes representing responses to 3 out of 4 questionnaires.
4. A certain manufacturing process produces insect repellent: A worker tests a sample from each batch as it is produced and continues testing until he finds a batch that is ineffective. Specify the sample space of possible outcomes of the test process.
5. Write the sample space for tossing a fair coin 3 times. Draw the decision tree for this experiment.
6. An econometric model predicts whether national income will remain the same, increase, or decrease in the following year. Let X represent "the model's prediction, coded" and Y the "actual movement of national income, coded." Graph the possible outcomes of X and Y.
7. Which of the following pairs of events are mutually exclusive?
   a. On two rolls of a die, (1) a 3 occurs, (2) the sum of the faces is 4.
   b. On two rolls of a die, (1) a 4 occurs, (2) the sum of the two faces is 3.
   c. On a roll of three dice, (1) one of the dice displays a 6, (2) the sum of two faces is a 5.
   d. Five people—Barbara, Alex, Donna, Sam, and Sue—apply for one job opening: (1) Barbara is hired, (2) Barbara is not hired.
   e. In a group of three computer chips, (1) exactly two are defective, (2) two or fewer are defective.
   f. (1) Benjamin Dauntless is unsuccessful in business. (2) Benjamin Dauntless has ulcers.

2.2 ELEMENTARY PROBABILITY RULES

In most applications of probability theory, we are interested in combining probabilities of events that are related in some important way. In this section, we discuss two fundamental ways of combining probabilities: *addition* and *multiplication*.

Before considering the combining of probabilities by addition, we will define two new terms.

- The symbol $P(A_1 \text{ or } A_2)$ refers to the probability that *either* event A_1 *or* event A_2 occurs—*or* that they both occur. For example, if A_1 refers to the event that an individual is a male and A_2 refers to the event that the individual is a college graduate, then the symbol $P(A_1 \text{ or } A_2)$ denotes the probability that the individual is *either* a male *or* a college graduate. The term "or" is used inclusively; that is, it includes the case of a person who is both a male and a college graduate. If events A_1 and A_2 cannot both occur, $P(A_1 \text{ or } A_2)$ refers to the probability that *either* A_1 *or* A_2 occurs. For example, if A_1 and A_2 refer respectively to the obtaining of a head and a tail on a toss of a coin, then $P(A_1 \text{ or } A_2)$ denotes the probability of obtaining *either* a head *or* a tail.
- The symbol $P(A_1 \text{ and } A_2)$ is used to denote the probability that both events A_1 *and* A_2 will occur. $P(A_1 \text{ and } A_2)$ is called the **joint probability** of the events A_1 and A_2. Continuing the example in the preceding paragraph, $P(A_1 \text{ and } A_2)$ is the probability that the individual is both a male *and* a college graduate.

We now state the general addition rule for any two events A_1 and A_2 in a sample space S.

Addition rule for any two events A_1 and A_2

$$P(A_1 \text{ or } A_2) = P(A_1) + P(A_2) - P(A_1 \text{ and } A_2) \tag{2.1}$$

If A_1 and A_2 are **mutually exclusive events**, that is, if they cannot both occur, then

$$P(A_1 \text{ and } A_2) = 0$$

Addition rule for two mutually exclusive events A_1 and A_2

This leads to a special case of the general addition rule.

$$P(A_1 \text{ or } A_2) = P(A_1) + P(A_2) \tag{2.2}$$

Applying equation 2.2 to the rolling of a die, we note that the probability that either a 1 or a 2 will occur on a single roll is $P(A_1 \text{ or } A_2) = P(A_1) + P(A_2) = \frac{1}{6} + \frac{1}{6} = \frac{1}{3}$. In Example 2-4 we consider an application of equation 2.1 for two events that are not mutually exclusive.

EXAMPLE 2-4

What is the probability of obtaining a 6 on the first or second roll of a die or on both? Another way of wording this question is, "What is the probability of obtaining a 6 at least once in two rolls of a die?"

Solution

Let A_1 denote the appearance of a 6 on the first roll and A_2 the appearance of a 6 on the second roll. We want to find the value of $P(A_1 \text{ or } A_2)$. (As explained earlier, because of the inclusive meaning of "or," the symbol $P(A_1 \text{ or } A_2)$ means the probability that a 6 appears *either* on the first *or* on the second roll *or* on both rolls.) Consider the sample space of 36 equally likely elements listed below. These are all possible outcomes of two rolls of the die; the numbers in each element represent the outcomes on the first and second rolls, respectively.

1, 1	2, 1	3, 1	4, 1	5, 1	6, 1
1, 2	2, 2	3, 2	4, 2	5, 2	6, 2
1, 3	2, 3	3, 3	4, 3	5, 3	6, 3
1, 4	2, 4	3, 4	4, 4	5, 4	6, 4
1, 5	2, 5	3, 5	4, 5	5, 5	6, 5
1, 6	2, 6	3, 6	4, 6	5, 6	6, 6

The probability that a 6 will appear on both the first and second rolls is $\frac{1}{36}$, that is, $P(A_1 \text{ and } A_2) = \frac{1}{36}$. The probability that a 6 will appear on the first roll is $P(A_1) = \frac{1}{6}$, and on the second roll, $P(A_2) = \frac{1}{6}$. Hence, applying the addition rule, we have

$$\begin{aligned} P(A_1 \text{ or } A_2) &= P(A_1) + P(A_2) - P(A_1 \text{ and } A_2) \\ &= \frac{1}{6} + \frac{1}{6} - \frac{1}{36} \\ &= \frac{11}{36} \end{aligned}$$

The term $P(A_1 \text{ and } A_2)$ must be subtracted in this calculation in order to avoid double counting. That is, if we incorrectly solved the problem by using the addition theorem for mutually exclusive events, computing $P(A_1 \text{ or } A_2) = P(A_1) + P(A_2) = \frac{1}{6} + \frac{1}{6} = \frac{12}{36}$, we would have counted the event (6, 6) twice, because (6, 6) is an elementary event both of A_1 (6 on the first roll) and of A_2 (6 on the second roll). Note that 11 elements in the sample space listed above represent the event "6 at least once in two trials." Therefore, the same result could have been obtained by using the classical definition of probability. The ratio of outcomes favorable to the event "6 at least once" to the total number of outcomes is $\frac{11}{36}$.

The ideas involved in the use of the addition rule are portrayed in Figure 2-2. The interiors of the rectangles represent the sample space, and the two events A_1 and A_2 are displayed as circles. The event "A_1 or A_2," whose probability is to be found, is shown as the tinted region in Figure 2-2(a). If we added

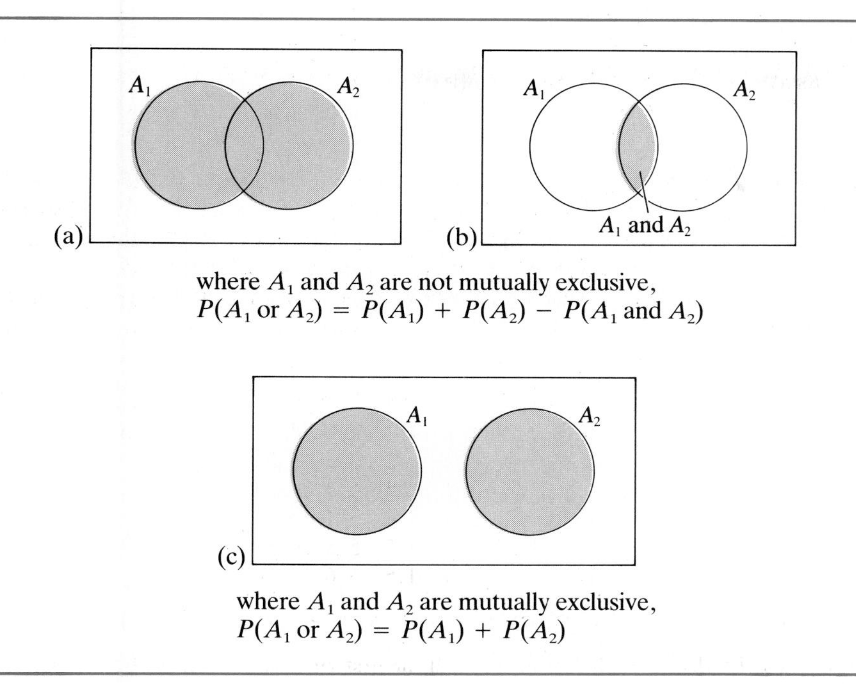

FIGURE 2-2
Symbolic portrayal of the addition rules

all the outcomes in A_1 to those in A_2, which would be implied when we add $P(A_1)$ and $P(A_2)$, we would count the outcomes associated with the event "A_1 and A_2" twice; the latter event is shown in Figure 2-2(b) as the tinted region where the circles overlap. That is why we must subtract the term $P(A_1 \text{ and } A_2)$ from the sum of $P(A_1)$ and $P(A_2)$ to obtain the desired probability, $P(A_1 \text{ or } A_2)$.

On the other hand, if A_1 and A_2 are mutually exclusive events, they cannot occur jointly. This is indicated in Figure 2-2(c), which depicts these events as circles that do not overlap. The event "A_1 or A_2" is represented by the tinted area of the sample space. Since the circles do not intersect, there is no double counting when $P(A_1)$ is added to $P(A_2)$ to obtain $P(A_1 \text{ or } A_2)$. (Note again that since A_1 and A_2 cannot occur together, $P(A_1 \text{ and } A_2) = 0$.)

The addition rule can, of course, be extended to more than two events. The generalization for n mutually nonexclusive events will not be given here because of its complexity. If the n events $A_1, A_2, \ldots, A_n$ are mutually exclusive, then we have the following rule:

Addition rule for n mutually exclusive events

$$P(A_1 \text{ or } A_2 \text{ or} \ldots \text{or } A_n) = P(A_1) + P(A_2) + \cdots + P(A_n) \tag{2.3}$$

This addition rule is applicable whenever we are interested in the probability that any one of several mutually exclusive events will occur. For example, returning to Example 2-2, the computation of the probability that the collection experience of a profitable loan will be excellent, good, or fair illustrates this rule.

In probability notation, we have

$$P(\text{excellent or good or fair}) = P(\text{excellent}) + P(\text{good}) + P(\text{fair})$$
$$= 0.50 + 0.40 + 0.05 = 0.95$$

Joint Probability Tables

In many applications, we are interested in the probability of the **joint occurrence** of two or more events. To illustrate joint probabilities, consider the data in Table 2-1. These figures represent the results of a market research survey in which 1,000 persons were asked which of two competitive products they preferred, product ABC or product XYZ. To simplify the discussion, as shown in Table 2-1, we designate A_1, A_2, B_1, and B_2, respectively, as "male," "female," "prefers product ABC," and "prefers product XYZ." Hence, the joint outcome that an individual is male and prefers product ABC is denoted as "A_1 and B_1," and the joint probability that a randomly selected individual is male and prefers product ABC is $P(A_1 \text{ and } B_1)$. Analogous notation is used for the other possible joint outcomes and joint probabilities.

Joint probabilities may be illustrated in the following manner. If a person is selected at random from this group of 1,000, the joint probability that the individual is a male and prefers product ABC is

$$P(A_1 \text{ and } B_1) = \frac{200}{1{,}000} = 0.20$$

Similarly, the probability that a randomly selected person is a female and prefers product XYZ is

$$P(A_2 \text{ and } B_2) = \frac{400}{1{,}000} = 0.40$$

It is useful to construct a so-called **joint probability table** by dividing all entries in Table 2-1 by the total number of individuals (1,000). The resulting joint probability table is shown as Table 2-2. The figures in the table are examples of probabilities calculated as relative frequencies of occurrence, as discussed in section 2.1.

TABLE 2-1
1,000 persons classified by sex and by product preference

Sex	Prefers Product ABC B_1	Prefers Product XYZ B_2	Total
A_1: Male	200	300	500
A_2: Female	100	400	500
	300	700	1,000

TABLE 2-2
Joint probability table for 1,000 persons, classified by sex and product preference

Sex	Prefers Product ABC B_1	Prefers Product XYZ B_2	Marginal Probabilities
A_1: Male	0.20	0.30	0.50
A_2: Female	0.10	0.40	0.50
Marginal probabilities	0.30	0.70	1.00

Marginal Probabilities

In addition to the joint probabilities mentioned earlier, we can also separately obtain from Table 2-2 probabilities for each of the two classifications "sex" and "product preference." These probabilities, which are shown in the margins of the joint probability table, are referred to as **marginal probabilities**, or **unconditional probabilities**. For example, the marginal probability that a randomly chosen individual is a male is $P(A_1) = 0.50$, and the marginal probability that a person prefers product ABC is $P(B_1) = 0.30$.

The marginal probabilities for each classification are obtained by summing the appropriate joint probabilities. For example, the marginal probability that an individual prefers product ABC is 0.30. The event B_1, "prefers product ABC," consists of two mutually exclusive parts, "A_1 and B_1" ("male and prefers product ABC") and "A_2 and B_1" ("female and prefers product ABC"). Hence, we have

$$\begin{aligned} P(B_1) &= P[(A_1 \text{ and } B_1) \text{ or } (A_2 \text{ and } B_1)] \\ &= P(A_1 \text{ and } B_1) + P(A_2 \text{ and } B_1) \\ &= 0.20 + 0.10 = 0.30 \end{aligned}$$

Thus, to obtain marginal probabilities for product preference, we add the appropriate joint probabilities over the other classification (in this case, sex). This is an application of the addition rule for mutually exclusive events. All other marginal probabilities in Table 2-2 may be similarly calculated.

Conditional Probabilities

Often we are interested in how certain events are related to the occurrence of other events. In particular, we may be interested in the probability of the occurrence of an event given that another related event has occurred. Such probabilities are referred to as **conditional probabilities**. For instance, returning to the events discussed in Table 2-2, we may be interested in the probability that an individual prefers product ABC given that the individual is a male. This conditional probability is denoted $P(B_1|A_1)$ and is read "the probability of B_1 given A_1." The vertical line is read "given," and the event following the line, in this case A_1, is the one known to have occurred. We can now formally define a conditional probability.

Let A_1 and B_1 be two events in a sample space S. The conditional probability of B_1 given A_1, denoted $P(B_1|A_1)$, is

Conditional probability of B_1 given A_1

$$P(B_1|A_1) = \frac{P(A_1 \text{ and } B_1)}{P(A_1)} \quad \text{where } P(A_1) > 0 \tag{2.4}$$

The statement that $P(A_1) > 0$ is included in order to rule out the possibility of dividing by 0.

This definition may be illustrated by the example given in Table 2-2. The conditional probability that a person prefers product ABC given that the person is male is

$$P(B_1|A_1) = \frac{P(A_1 \text{ and } B_1)}{P(A_1)} = \frac{0.20}{0.50} = 0.40$$

We note that the conditional probability of event B_1 given A_1 is found by dividing the joint probability of A_1 and B_1 by the marginal probability of A_1. The rationale of this procedure becomes clear by returning to Table 2-1, where we see that the proportion of males who prefer product ABC is $\frac{200}{500} = 0.40$. Thus, the conditional probability of B_1 given A_1 is simply the proportion of times that B_1 occurs out of the total number of times that A_1 occurs.

Note that a chronological order is not necessarily implied in conditional probability. That is, in $P(B_1|A_1)$, event A_1 does not necessarily precede B_1 in time. In fact, using the same method of definition given in equation 2.4, we would have

$$P(A_1|B_1) = \frac{P(A_1 \text{ and } B_1)}{P(B_1)} \quad \text{where } P(B_1) > 0 \tag{2.5}$$

Thus, in Table 2-2 the conditional probability that an individual is a male given that the person prefers product ABC is

$$P(A_1|B_1) = \frac{P(A_1 \text{ and } B_1)}{P(B_1)} = \frac{0.20}{0.30} = 0.67$$

Conditional probabilities are important concepts in our everyday affairs and in managerial decision making. For instance, we may be interested in the probability that a friend will arrive at an appointment punctually given that he has said that he will appear at a certain time. A university is concerned about the probability that an applicant for admission will have a satisfactory academic performance at the university given that the applicant has certain aptitude test scores. A marketing executive would be interested in the probability that the sales volume for one of her company's products will increase given that she has made a commitment for an expensive sales promotion campaign for that product. In the illustration given in Table 2-2, the users of the market research survey data would be interested in the relationship between sex and product preference. Indeed, the following conditional probabilities in that illustration are revealing:

$$P(B_1|A_1) = \frac{0.20}{0.50} = 0.40 \quad \text{and} \quad P(B_1|A_2) = \frac{0.10}{0.50} = 0.20.$$

That is, the conditional probability of preference for product ABC given that the person is a male is 40%, whereas the corresponding conditional probability for a female is only 20%. Such sex differences in product preferences may be very important, for example, in choosing types of promotional effort to be used in selling the products.

Multiplication Rule

The multiplication rule for two events A_1 and B_1 follows immediately from the definition of conditional probability given in equation 2.4. Multiplying both sides of equation 2.4 by $P(A_1)$, we obtain the multiplication rule.

Multiplication rule for any two events A_1 and A_2

$$P(A_1 \text{ and } B_1) = P(A_1)P(B_1|A_1) \tag{2.6}$$

Equivalently, from equation 2.5, the multiplication rule may be stated as

$$P(A_1 \text{ and } B_1) = P(B_1)P(A_1|B_1) \tag{2.7}$$

Hence, $P(A_1 \text{ and } B_1)$ can be computed by either equation 2.6 or 2.7. As an example, let us apply these equations to the data of Table 2-2 to find the joint probability that a randomly selected person would be male and would prefer product ABC. By equation 2.6, we have

$$\begin{aligned} P(A_1 \text{ and } B_1) &= P(A_1)P(B_1|A_1) \\ &= (0.50)(0.40) = 0.20 \end{aligned}$$

and by equation 2.7, we obtain

$$\begin{aligned} P(A_1 \text{ and } B_1) &= P(B_1)P(A_1|B_1) \\ &= (0.30)(0.67) = 0.20 \end{aligned}$$

Of course, these calculations are given only to illustrate the principles involved in the multiplication rule. We would ordinarily not compute joint probabilities this way if the joint probabilities were already available as in Table 2-2.

The generalization of the multiplication rule to three or more events is straightforward.

Multiplication rule for any three events A_1, A_2, and A_3

(2.8) $$P(A_1 \text{ and } A_2 \text{ and } A_3) = P(A_1)P(A_2|A_1)P(A_3|A_2 \text{ and } A_1)$$

Multiplication rule for any n events $A_1, A_2, \ldots, A_n$

(2.9) $$P(A_1 \text{ and } A_2 \text{ and} \ldots \text{and } A_n) = P(A_1)P(A_2|A_1)P(A_3|A_2 \text{ and } A_1)\ldots P(A_n|A_{n-1} \text{ and} \ldots \text{and } A_1)$$

This notation means that the joint probability of the n events is given by the product of the probability that the first event A_1 has occurred, the conditional probability of the second event A_2 given that A_1 has occurred, the conditional probability of the third event A_3 given that both A_2 and A_1 have occurred, and so on. Of course, the n events can be numbered arbitrarily; any one of them may be the first event, any of the remaining $n - 1$ may be the second event, and so forth.

Let us consider a couple of examples of these ideas. Suppose the probability that a sales representative following up on a lead will make the sale is 0.2. Past experience indicates that 40% of such sales are for amounts in excess of $100. What is the probability that the representative will make a sale in excess of $100?

To solve this problem, let us use the following symbols:

$P(A_1)$: Probability that a sale is made

$P(A_2|A_1)$: Probability that the sale is in excess of $100 given that a sale is made

The required probability is given by

$$\begin{aligned} P(A_1 \text{ and } A_2) &= P(A_1)P(A_2|A_1) \\ &= (0.2)(0.4) = 0.08 \end{aligned}$$

It should be noted that $P(A_2|A_1)$ is a conditional probability because the 0.40 probability that a sale will exceed $100 depends on the sale actually being made.

Our second example is typical of many situations involving the sampling of human populations. Suppose we had a list of 10 individuals, 5 of whom reside in New York, 3 in Pennsylvania, and 2 in New Jersey. Suppose we select 3 names at random from the list, one at a time, so that at each drawing all remaining names have an equal chance of being selected. (The actual techniques involved in such sampling procedures are discussed in Chapter 4.) What is the probability that all 3 names are those of New York residents?

We designate the events of the first-drawn, second-drawn, and third-drawn names of New York residents as A_1, A_2, and A_3, respectively. The required joint probability is

$$\begin{aligned} P(A_1 \text{ and } A_2 \text{ and } A_3) &= P(A_1)P(A_2|A_1)P(A_3|A_2 \text{ and } A_1) \\ &= \left(\frac{5}{10}\right)\left(\frac{4}{9}\right)\left(\frac{3}{8}\right) = 0.083 \end{aligned}$$

Let us consider the factors on the right side of the equation. Because 5 of the 10 names are those of New York residents, $P(A_1) = \frac{5}{10}$. Given that a New York resident's name is selected, 4 of the 9 remaining names are of New York residents, so $P(A_2|A_1) = \frac{4}{9}$. Similarly, after the second New York name is chosen, there are 8 names remaining, 3 of which are of New York residents.

An important point concerning the use of the multiplication rule may be noted from this example. The probability calculation pertains to 3 names being drawn successively. What is the probability of obtaining 3 New York names if the 3 names are drawn *simultaneously* from the list? The answer to this question is exactly the same as in the preceding calculation. We can think of one of the 3 names as the "first," another as the "second," and another as the "third," although they have been drawn together. A_1 is again used to denote the event "the first name is of a resident of New York," and so on.

Because the *joint* probability of the *simultaneous* occurrence of events A_1, A_2, and A_3 is the same as their successive occurrence, we can state the following generalization about the use of the multiplication rule.

The multiplication rule may be used to obtain the joint probability of the successive or simultaneous occurrence of two or more events.

The example of sampling names from a list has interesting implications. In the example, we drew a sample of 3 names at random *without replacement* of the sampled elements. In human populations, sampling is usually carried out without replacement; that is, after the necessary data are obtained, an individual drawn into the sample is usually not replaced prior to the choosing of another individual. We have seen how the partial exhaustion of the population because of sampling had to be taken into account in the calculation of conditional probabilities. If the sampling had been with replacement, then the basic probabilities of selection remain unchanged after each item is replaced in the population. For example, if the sampling from the list had been with replacement, then the respective probabilities for New York residents would have been $P(A_1) = \frac{5}{10}$, $P(A_2) = \frac{5}{10}$, and $P(A_3) = \frac{5}{10}$. Hence the joint probability of the occurrence of A_1, A_2, and A_3 would have been

Sampling with replacement

$$P(A_1 \text{ and } A_2 \text{ and } A_3) = \left(\frac{5}{10}\right)\left(\frac{5}{10}\right)\left(\frac{5}{10}\right) = 0.125$$

In sampling *without replacement*, we are dealing with *dependent* events; in sampling *with replacement*, we are dealing with *independent* events.

The meaning of these concepts and the multiplication rules for independent events are discussed next.

Statistical Independence

In our discussion of Table 2-2, we saw that there was a relationship between product preference and sex. For example, we found that the probability that a male preferred product ABC was 0.40, whereas the corresponding probability for a female was 0.20. In other words, product preference *depends* on sex. Hence, the corresponding events involved (for example, "male" and "prefers product ABC") are said to be **dependent**. Since it is important in analysis and decision making to detect such relationships, it is correspondingly important to know when we are dealing with **independent** events.

We now turn to the concept of independence, usually referred to as **statistical independence**. If two events, say A_1 and B_1, are statistically independent, then knowing that one of them has occurred does not affect the probability that the other will occur; in such a case, $P(B_1|A_1) = P(B_1)$. For example, in tossing a fair coin twice, suppose we use the following notation for events:

A_1: Head on first toss
B_1: Head on second toss

The marginal, or unconditional, probability of obtaining a head on the second toss is $P(B_1) = \frac{1}{2}$. The conditional probability of obtaining a head on the second toss given that a head was obtained on the first toss is $P(B_1|A_1) = \frac{1}{2}$. Of course, the probability of obtaining a head on the second toss does not depend on whether a head or tail was obtained on the first toss. Thus, we note

$$P(B_1|A_1) = P(B_1) = \frac{1}{2}$$

We can now define statistical independence of two events. Two events A_1 and B_1 are *statistically independent* if

$$P(B_1|A_1) = P(B_1)$$

When this equality holds, it is also true[2] that

$$P(A_1|B_1) = P(A_1)$$

[2] If several events are *collectively independent* (or *mutually independent*), then every possible conditional probability for every combination of events must be equal to the corresponding unconditional probability. For example, for 3 events A_1, A_2, and A_3 to be collectively independent, the necessary and sufficient conditions are the following:

$$P(A_1) = P(A_1|A_2) \qquad P(A_1) = P(A_1|A_2 \text{ and } A_3)$$
$$P(A_2) = P(A_2|A_3) \qquad P(A_2) = P(A_2|A_1 \text{ and } A_3)$$
$$P(A_3) = P(A_3|A_1) \qquad P(A_3) = P(A_3|A_1 \text{ and } A_2)$$

When two events A_1 and B_1 are statistically independent, the multiplication rule given in equations 2.6 and 2.7 can be simplified.

Multiplication rule for two statistically independent events A_1 and B_1

$$P(A_1 \text{ and } B_1) = P(A_1)P(B_1) \quad (2.10)$$

Note that if two events A_1 and B_1 are statistically independent, equation 2.10 holds. The converse is also true. That is, if two events A_1 and B_1 are related according to equation 2.10, the two events are statistically independent; furthermore, $P(B_1|A_1) = P(B_1)$ and $P(A_1|B_1) = P(A_1)$.

The generalization of the multiplication rule for n collectively independent events (any one event is independent of any combination of the others) is given next.

Multiplication rule for n collectively independent events $A_1, A_2, \ldots, A_n$

$$P(A_1 \text{ and } A_2 \text{ and} \ldots \text{and } A_n) = P(A_1)P(A_2)P(A_3) \ldots P(A_n) \quad (2.11)$$

As Examples 2-5 through 2-7 show, it is possible to solve a variety of probability problems using only the addition and multiplication rules.

EXAMPLE 2-5

The probability that interest rates will rise has been assessed as 0.8. If they do rise, the probability that the stock market index will drop is estimated to be 0.9. If the interest rates do not rise, the probability that the stock market index will still drop is estimated as 0.4. What is the probability that the stock market index will drop?

Solution

$$P(A) = P\text{ (interest rates rise)} = 0.8$$

$$P(B) = P\text{ (stock market index drops)} = ?$$

Then, the probability of $\bar{A}$, the complement of A, "interest rates do not rise," is $P(\bar{A}) = 1 - 0.8 = 0.2$.

$$P(B|A) = P(\text{stock market index drops}|\text{interest rates rise}) = 0.9$$

$$P(B|\bar{A}) = P(\text{stock market index drops}|\text{interest rates do not rise}) = 0.4$$

By the multiplication rule

$$P(B \text{ and } A) = P(A)P(B|A) = 0.8 \times 0.9 = 0.72$$

and

$$P(B \text{ and } \bar{A}) = P(\bar{A})P(B|\bar{A}) = 0.2 \times 0.4 = 0.08$$

Hence,

$$\begin{aligned} P(B) &= P(\text{stock market index drops}) \\ &= P(B \text{ and } A) + P(B \text{ and } \bar{A}) = 0.72 + 0.08 = 0.80 \end{aligned}$$

EXAMPLE 2-6

The probability of a reduction in the cost of living index in any year compared with the previous year is estimated as 0.1. Assuming statistical independence, what is the probability that it will drop in at least one of the next three years?

Solution

Let D_1, D_2, and D_3 represent a "drop" in the cost of living index in the first, second, and third years, respectively. Correspondingly, let $\bar{D}_1$, $\bar{D}_2$, and $\bar{D}_3$ represent "no drop" in the cost of living index in the first, second, and third years, respectively. Then $P(D_1) = P(D_2) = P(D_3) = 0.1$ and $P(\bar{D}_1) = P(\bar{D}_2) = P(\bar{D}_3) = 0.9$. Let E denote the event "at least one drop in the cost of living index over the next three years." Then $\bar{E}$, the complement of E, denotes the event "no drop in the cost of living index over the next three years." Since the event $\bar{E}$ represents three successive years without a drop in the cost of living index, we obtain by the multiplication rule for independent events

$$P(\bar{E}) = P(\bar{D}_1 \text{ and } \bar{D}_2 \text{ and } \bar{D}_3) = (0.9)(0.9)(0.9) = 0.729$$

Therefore $P(E) = 1 - P(\bar{E}) = 1 - 0.729 = 0.271$.

EXAMPLE 2-7

The number of companies in which new investments were made are classified by country of investor and by industry.

a. Calculate the probability that a company selected from this group will be
   (1) A Japanese machinery company.
   (2) A transportation company.
   (3) A machinery company given that the investor is Japanese.
   (4) Either a German company or a machinery company.

b. Determine whether the investors' countries and the industries in which the investments are made are independent.

	Industry			
Country	Chemical	Transportation	Machinery	Total
Germany	19	13	2	34
Japan	8	6	2	16
Total	27	19	4	50

Solution

We use the following representation of events:

A_1: German $\quad B_1$: Chemical
A_2: Japanese $\quad B_2$: Transportation
B_3: Machinery

a. (1) $P(A_2 \text{ and } B_3) = \dfrac{2}{50} = 0.04$

(2) $P(B_2) = \dfrac{19}{50} = 0.38$

(3) $P(B_3|A_2) = \dfrac{P(B_3 \text{ and } A_2)}{P(A_2)} = \dfrac{2/50}{16/50} = \dfrac{2}{16} = 0.125$

(4) $P(A_1 \text{ or } B_3) = P(A_1) + P(B_3) - P(A_1 \text{ and } B_3)$

$$= \frac{34}{50} + \frac{4}{50} - \frac{2}{50}$$

$$= \frac{36}{50} = 0.72$$

b. In order for the country of investor and the industry of investment to be independent, the joint probability of each pair of A events and B events would have to equal the product of the respective unconditional probabilities. That is, the following equalities would have to hold:

$$P(A_1 \text{ and } B_1) = P(A_1)P(B_1) \qquad P(A_2 \text{ and } B_1) = P(A_2)P(B_1)$$
$$P(A_1 \text{ and } B_2) = P(A_1)P(B_2) \qquad P(A_2 \text{ and } B_2) = P(A_2)P(B_2)$$
$$P(A_1 \text{ and } B_3) = P(A_1)P(B_3) \qquad P(A_2 \text{ and } B_3) = P(A_2)P(B_3)$$

Clearly, these equalities do not hold, for example

$$P(A_1 \text{ and } B_1) \neq P(A_1)P(B_1)$$
$$\frac{19}{50} \neq \frac{34}{50} \times \frac{27}{50}$$

Another way of viewing the problem is that each conditional probability would have to equal the corresponding unconditional probability. Thus, the following equalities would have to hold:

$$P(A_1|B_1) = P(A_1) \qquad P(A_1|B_2) = P(A_1) \qquad P(A_1|B_3) = P(A_1)$$
$$P(A_2|B_1) = P(A_2) \qquad P(A_2|B_2) = P(A_2) \qquad P(A_2|B_3) = P(A_2)$$

These equalities do not hold, for example,

$$P(A_1|B_1) \neq P(A_1)$$
$$\frac{19}{27} \neq \frac{34}{50}$$

Exercises 2.2

1. A survey of families owning television sets in a remote area indicates that the probability of owning a color TV set is 0.21, the probability of owning a black and white TV is 0.74, and the probability of owning both a color TV and a black and white TV set is 0.13. Calculate the probability that a family from this area has a color TV set or a black and white TV set or both.

2. A company has 3 identical machines that produce the listed percentages of defective items. The production manager randomly chooses one article from the production of each machine. What is the probability that all 3 articles are defective? The production of the 3 machines may be considered independent.

Machine number	Percentage defective
1	5%
2	2%
3	3%

3. A new product has been advertised on television. Market researchers have estimated a probability of $\frac{1}{5}$ that a consumer in a certain metropolitan area will see this commercial. The researchers also believe that if such a consumer sees this commercial, the probability that he or she will purchase the product is $\frac{1}{3}$. Calculate the probability that a consumer in the metropolitan area will see the television commercial and will purchase the product.

4. Events A and B have the following probability structure:

$$P(A \text{ and } B) = \frac{1}{6}$$

$$P(A \text{ and } \bar{B}) = \frac{2}{9}$$

$$P(\bar{A} \text{ and } B) = \frac{1}{3}$$

   a. What is $P(\bar{A} \text{ and } \bar{B})$?
   b. Are A and B independent events?
   c. Are A and B mutually exclusive events?

5. In a lot of 100 TV sets, 3 are defective. If a sample of 2 of these sets is drawn without replacement, what is the probability that both are defective?

6. If the probability that Company A will buy Company B is 0.7, what are the odds that Company A will not buy Company B?

7. The table shows the distribution of students at a certain graduate school by sex and geographic area of residence.

   If a student is selected at random at this graduate school, what is the probability that
   a. the student is from the South?
   b. the student is a foreign female?
   c. the student is a male or from the East Coast?
   d. given that the student is from the Midwest, he is a male?

	East Coast	West	South	Midwest	Foreign
Male	155	17	16	15	7
Female	55	8	9	5	3

8. In roulette there are 38 slots in which a ball may land. There are numbers 0, 00, and 1 through 36. Odd numbers are red, even numbers are black and zeroes are green. A ball is thrown randomly into a slot.
   a. What is the probability that the slot is black?
   b. What is the probability that the slot is number 27?
   c. What is the probability that the slot is either red or number 27?
   d. What are the odds in favor of black?
   e. What are the odds in favor of number 27?
   f. If you repeatedly play black, in the long run what fraction of times will you win? What fraction of times will you lose?
9. The probability is 0.3 that a life insurance salesperson following up a magazine lead will make a sale. On a certain day, a salesperson has two leads. Assuming independence, what is the probability that the salesperson with sell
   a. both policies?
   b. exactly one policy?
   c. at least one policy?
10. A marketing vice-president is presenting new products to the finance committee for approval. The probability is 0.2 that the committee will approve a new product. Assuming independence, what is the probability that the vice-president will have to present more than five new products before one is approved by the committee?
11. A census of a district's 1,500 constituents with regard to opinion on import quotas for shoes showed 500 of the district's 600 Democrats favored the quotas and 550 constituents opposed the quotas. Assume that all the constituents can be classified as either Democrats or Republicans.
   a. What is the probability that a constituent selected at random will be a Republican opposed to the quotas?
   b. What is the probability that a constituent selected at random will favor the quotas?
   c. What is the probability that, if a Republican is chosen, he or she will favor the quotas?
   d. Are political affiliation (Democrat or Republican) and opinion on shoe import quotas independent? Prove your answer and give a short statement as to the implication of this finding.
12. A proposal concerning company contributions to the pension fund was presented by the leaders of Local 999, Sandbox Workers of America, to

the rank-and-file membership in all regions of the country. The results of an open ballot vote are listed by region and opinion.

a. What is the probability that a ballot selected at random is that of a midwestern union member opposed to the proposal?
b. What is the probability that a ballot selected at random is that of an eastern union member?
c. What is the probability that a ballot selected at random is that of a union member who is not opposed to the proposal?
d. If a ballot is selected at random from the Eastern group of union members, what is the probability that the union member is opposed to the proposal?
e. If a ballot is selected at random from the group of union members who responded favorably to the proposal, what is the probability that the union member comes from the Western region?
f. Are the union member's region and opinion on the proposal independent? If yes, prove it. If no, specify what the numbers in the table would have been had the two factors been independent.

Opinion	East	Region Midwest	West	Total
Opposed	255	220	125	600
Not opposed	145	130	125	400
Total	400	350	250	1,000

2.3 BAYES' THEOREM

The Reverend Thomas Bayes (1702–1761), an English Presbyterian minister and mathematician, considered the question of how to make inferences from observed sample data about the larger groups from which the data were drawn. His motivation was his desire to prove the existence of God by examining the sample evidence of the world about him. Mathematicians had previously concentrated on the problem of deducing the consequences of specified hypotheses. Bayes was interested in the inverse problem of drawing conclusions about hypotheses from observations of consequences. He derived a theorem that calculated probabilities of "causes" based on the observed "effects." The theorem may also be considered a means of revising probabilities of events based on additional information. In the period since World War II, a body of knowledge known as **Bayesian decision theory** has been developed to solve problems involving decision making under uncertainty.

Bayes' theorem is really nothing more than a statement of conditional probabilities. The following problem illustrates the nature of the theorem. Assume that 1% of the inhabitants of a country suffer from a certain disease. Let A_1 represent the event "has the disease" and A_2 denote "does not have the disease." Now suppose a person is selected at random from this population. What is the probability that this individual has the disease? Since 1% of the population has the disease and it is equally likely that any individual would be selected, we assign a probability of 0.01 to the event "has the disease." This probability, $P(A_1) = P(\text{has the disease}) = 0.01$, is referred to as a **prior probability** in the sense that it is assigned prior to the observation of any empirical information. Of course, $P(A_2) = 0.99$ is the corresponding prior probability that the individual does not have the disease.

Prior probability

Now let us assume that a new but imperfect diagnostic test has been developed, and let B denote the event "test indicates the disease is present." Suppose that through past experience it has been determined that the conditional probability that the test indicates the disease is present, given that the person has the disease, is

$$P(B|A_1) = 0.97$$

and the corresponding probability, given that the person does *not* have the disease, is

$$P(B|A_2) = 0.05$$

Suppose a person is selected at random and given the test and the test indicates that the disease is present. What is the probability that this person actually has the disease? In symbols, we want the conditional probability $P(A_1|B)$. It is important to note the nature of this question. The probability $P(\text{has the disease}|\text{test indicates the disease is present}) = P(A_1|B)$ is referred to as a **posterior probability**, or a **revised probability**, because it is assigned after the observation of empirical or additional information. The posterior probability is the type of probability computed with the help of Bayes' theorem. We will derive the value of $P(A_1|B)$ from probability principles developed in this chapter.

Posterior probability

By the definition of conditional probability given in equations 2.4 and 2.5, we have

$$P(A_1|B) = \frac{P(A_1 \text{ and } B)}{P(B)} \tag{2.12}$$

We compute the joint probability $P(A_1 \text{ and } B)$ in the numerator of equation 2.12 by the multiplication rule given in equation 2.6.

$$P(A_1 \text{ and } B) = P(A_1)P(B|A_1) \tag{2.13}$$

To compute the denominator of equation 2.12, we observe that

$$P(B) = P[(A_1 \text{ and } B) \text{ or } (A_2 \text{ and } B)]$$

Hence,

$$P(B) = P(A_1 \text{ and } B) + P(A_2 \text{ and } B) \tag{2.14}$$

since the two joint events (A_1 and B) and (A_2 and B) are mutually exclusive.

If we express the joint probabilities $P(A_1$ and $B)$ and $P(A_2$ and $B)$ according to the multiplication rule, equation 2.14 becomes

$$P(B) = P(A_1)P(B|A_1) + P(A_2)P(B|A_2) \tag{2.15}$$

Now, substituting into the numerator of equation 2.12 the expression for the joint probability $P(A_1$ and $B)$ given in equation 2.13, and substituting into the denominator the marginal probability $P(B)$ given in equation 2.15, we obtain the result known as Bayes' theorem.

Bayes' theorem for two basic events, A_1 and A_2

$$P(A_1|B) = \frac{P(A_1)P(B|A_1)}{P(A_1)P(B|A_1) + P(A_2)P(B|A_2)} \tag{2.16}$$

In terms of the disease example, by substituting the known values into the Bayes' theorem formula in equation 2.16, we find that

$$P(A_1|B) = \frac{(0.01)(0.97)}{(0.01)(0.97) + (0.99)(0.05)} = \frac{0.0097}{0.0592} = 0.16$$

Hence, the posterior probability that the individual has the disease given that the test indicated the presence of the disease is 0.16. In summary, if a person were selected at random from the population of this country, the prior probability that the individual has the disease is 0.01. On the other hand, after we have the empirical information that the test indicated the disease is present, we revise the probability that the individual has the disease upward to 0.16.

Although the posterior probability (0.16) is 16 times as large as the prior probability (0.01), 0.16 is still a surprisingly low probability that the individual has the disease given that the test indicated the disease was present. This results from the fact that there are a large number of persons in this population who do not have the disease but for whom the test would (falsely) indicate that the disease is present.

As we have seen, Bayes' theorem weights prior information with empirical evidence. The manner in which it does this may be seen by laying out the calculations in a form such as Table 2-3. The first column of Table 2-3 gives the basic events of interest, "has the disease" and "does not have the disease." The second column shows the prior probability assignments to these basic events. The third column shows the conditional probabilities of the additional information given the basic events. As noted in the column heading, such con-

TABLE 2-3
Bayes' theorem calculations for illustrative problem

Events A_i	Prior Probabilities $P(A_i)$	Likelihoods $P(B\|A_i)$	Joint Probabilities $P(A_i)P(B\|A_i)$	Revised Probabilities $P(A_i\|B)$
A_1: Has the disease	0.01	0.97	0.0097	$\frac{0.0097}{0.0592} = 0.16$
A_2: Does not have the disease	0.99	0.05	0.0495	$\frac{0.0495}{0.0592} = \frac{0.84}{1.00}$
	1.00		$P(B) = 0.0592$	

ditional probabilities are referred to as "likelihoods." In our illustration these are, respectively, $P(B|A_1) = P$(test indicates disease is present|has the disease) and $P(B|A_2) = P$(test indicates disease is present|does not have the disease). The fourth column gives the joint probabilities of the basic events and the additional information. Note that as indicated in equation 2.15, the sum of these joint probabilities is the marginal probability $P(B)$. When the joint probabilities are divided by their total, $P(B)$ (in this case, 0.0592), the results are the revised probabilities shown in the last column. The first probability shown in the last column ($0.0097/0.0592 = 0.16$) is the Bayes' theorem calculation required in the illustrative problem.

In the preceding problem, there were 2 basic events (A_i). If there are more than 2 basic events, then correspondingly, additional terms appear in the denominator of equation 2.16. Hence, we can make the following formal statement of Bayes' theorem for n basic events. Assume a set of complete and mutually exclusive events $A_1, A_2, \ldots, A_n$. The appearance of one of the A_i events is a necessary condition for the occurrence of another event B, which is observed. The probabilities $P(A_i)$ and $P(B|A_i)$ are known. The posterior probability of event A_1 given that B has occurred is given by Bayes' theorem.

Bayes' theorem for *n* basic events, $A_1, A_2, \ldots, A_n$

$$P(A_1|B) = \frac{P(A_1)P(B|A_1)}{P(A_1)P(B|A_1) + P(A_2)P(B|A_2) + \cdots + P(A_n)P(B|A_n)} \tag{2.17}$$

Although the theorem was stated in equations 2.16 and 2.17 for a particular A_i, namely A_1, it is a general statement, since any of the n events A_i can be designated A_1. In most applications of the theorem to decision problems, A_i represents events that precede the occurrence of the observed event B. In this connection, we can think of the theorem as answering the question "Given that event B has occurred, what is the probability that it was preceded by event A_1?" Or, as indicated earlier, $P(A_1|B)$ is the revised probability assigned to event A_1 after event B is observed.

In modern Bayesian decision theory, subjective prior probability assignments are made in many applications. It is argued that it is meaningful to assign prior probabilities concerning hypotheses based on degree of belief. Bayes' theorem is then a means of revising these probability assignments. In business applications, this means that executives' intuitions, subjective judgments, and current quantitative knowledge are captured in the form of prior probabilities; these figures undergo revision as relevant empirical data are collected. This procedure seems sensible and fruitful for a wide variety of applications. Examples 2-8 and 2-9 suggest some of the many possible types of applications of this interesting theorem.

EXAMPLE 2-8

A corporation uses a "selling aptitude test" to help it select its sales force. Past experience has shown that only 65% of all persons applying for a sales position achieved a classification of "satisfactory" in actual selling, whereas the remainder were classified "unsatisfactory." Of those classified as "satisfactory," 80% had scored a passing grade on the aptitude test. Only 30% of those classified "unsatisfactory" had passed the test. On the basis of this information, what is the probability that a candidate would be a "satisfactory" salesperson, given a passing grade on the aptitude test?

Solution

If A_1 stands for a "satisfactory" classification as a salesperson and B stands for "passes the test," then the probability that a candidate would be a "satisfactory" salesperson, given a passing grade on the aptitude test, is

$$P(A_1|B) = \frac{(0.65)(0.80)}{(0.65)(0.80) + (0.35)(0.30)} = 0.83$$

Thus, the tests are of some value in screening candidates. Assuming no change in the type of candidates applying for the selling positions, the probability that a random applicant would be satisfactory is 65%. On the other hand, if the company accepts only applicants who pass the test, this probability increases to 0.83.

EXAMPLE 2-9

A certain company is planning to market a new product. The company's marketing vice-president is particularly concerned about the product's superiority over the closest competitive product, which is sold by another company. The marketing vice-president assessed the probability of the new product's superiority to be 0.7. This executive then ordered a market survey, which indicated that the new product was superior to its competitor. Assume the market survey has the following reliability: If the product is really superior, the probability that the survey will indicate "superior" is 0.8. If the product is really worse than its competitor, the probability that the survey will indicate "superior" is 0.3. After completion of the market survey, what should be this executive's revised probability assignment to the event "new product is superior to its competitor"?

Solution

Let A_1 represent the event "new product is superior to its competitor" and B the event "market survey indicates that the new product is superior to its competitor." Then

$$P(A_1|B) = \frac{(0.7)(0.8)}{(0.7)(0.8) + (0.3)(0.3)} = 0.86$$

Exercises 2.3

1. A manufacturing firm produces its product with 3 machines: X, Y, and Z. Each machine produces finished articles ready for shipment with Machines X, Y, and Z accounting for 40%, 35%, and 25% of total production, respectively. The percentage of defectives produced by Machines X, Y, and Z are 4%, 3%, and 6%, respectively. An article was drawn at random from the merged production of the 3 machines. The article was defective. What is the probability that the article was produced by Machine X?

2. MAC Minerals, Inc., is involved in a mining exploration in the western United States. The chief engineer originally feels that there is a 50-50 chance that a significant mineral find will occur. A first test drilling is completed and the results are favorable. The probability that the test drilling would give misleading results is 0.3. What should be the engineer's revised probability that a significant mineral find will occur?

3. A man regularly plays darts in the recreation room of his home, observed by his 8-year-old son. The father has a history of making bull's-eyes 30% of the time. The son, who habitually sits injudiciously close to the dart board, reports to his father whether or not bull's-eyes have been made. However, as is often the case with 8-year-olds, the son is an imperfect observer. He reports correctly 90% of the time. On a particular occasion, the father throws a dart at the board. The son reports that a bull's-eye was made. What is the probability that a bull's-eye was indeed scored?

4. The probability that a correct tactical decision is made in a particular type of military operation has been found to be 0.80. Due to the length of time required to complete an operation of this type, the correctness of the decision cannot be immediately ascertained. However, by means of military intelligence, information can be quickly obtained as to whether or not the decision is correct and this intelligence is accurate 75% of the time. A tactical decision has recently been made and intelligence information has just been received indicating that the decision is correct. What probability would you assign for the correctness of the decision given this intelligence information?

5. If a machine is in correct adjustment, it will produce only 0.1% defective articles. However, if the machine is not in correct adjustment, 0.5% of the articles it produces will have defects. The probability that a worker adjusts the machine correctly is 0.90. As the quality control consultant, you want to reduce the uncertainty concerning whether the machine is in proper adjustment. Hence you select at random an article of the machine's output and you inspect it.
   a. If the article is found to be nondefective, what is the probability that the machine is actually in proper adjustment?
   b. If two articles are selected at random from the output of the machine and both are defective, what is the probability that the machine is in proper adjustment?

2.4

COUNTING PRINCIPLES AND TECHNIQUES

In the problems we have encountered so far, the pertinent sample spaces have been comparatively simple. However, in many situations the numbers of points in the appropriate sample spaces are so great that efficient methods are needed to count these points in order to arrive at required probabilities or answer other questions of interest. In this connection, it is useful to return to the concept of sequences of experimental trials to specify a simple but important fundamental principle.

The Multiplication Principle

If an experiment can result in n_1 distinct outcomes on the first trial, n_2 distinct outcomes on the second trial, and so forth for k sequential trials, then the total number of different sequences of outcomes in the k trials is $(n_1)(n_2) \cdots (n_k)$.

It is sometimes helpful to think in terms of the sequential performance of tasks rather than trials of an experiment. Thus, using somewhat different language than was employed in the context of experimental trials, if the first of a sequence of tasks can be performed in n_1 ways, the second in n_2 ways, and so forth for k tasks, then the sequence of k tasks can be carried out in $(n_1)(n_2) \cdots (n_k)$ ways.

Thus, if a coin is tossed and then a card is drawn at random from a standard deck of cards, there are $2 \times 52 = 104$ possible different sequences. For example, one such sequence of outcomes might be head, king of spades.

If a die is rolled 3 times, there are $6 \times 6 \times 6 = 216$ different sequences.

If it is possible to go from Philadelphia to Baltimore in 2 different ways and from Baltimore to Washington in 3 different ways, then there are $2 \times 3 = 6$ ways of going from Philadelphia to Washington via Baltimore.

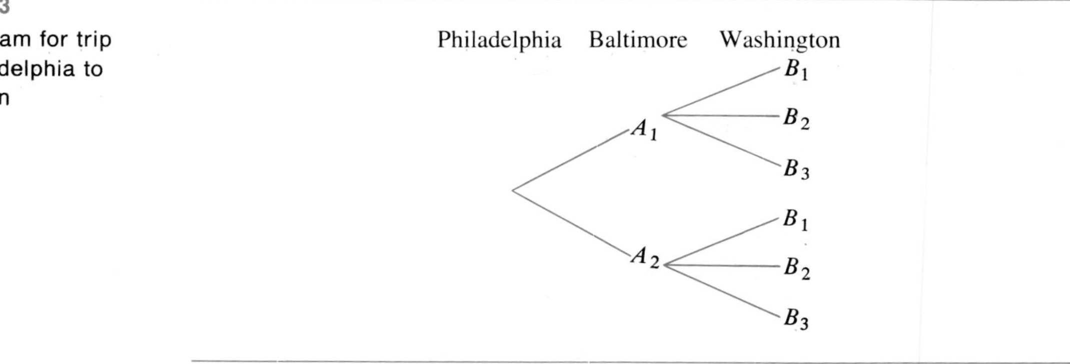

FIGURE 2-3 Tree diagram for trip from Philadelphia to Washington

A tree diagram is often helpful in demonstrating the total possible number of sequences. For example, in the case of the trip from Philadelphia to Washington, if A_1 and A_2 denote the two ways of going to Baltimore and B_1, B_2, and B_3 the 3 ways of proceeding from Baltimore to Washington, then the total number of possible sequences is indicated by the total number of different paths through the tree from left to right (see Figure 2-3).

In the following sections, the multiplication principle is used in many different types of problems.

Permutations

To handle the problem of counting points in complicated sample spaces, the counting techniques of *combination* and *permutation* are used. In this connection, it is helpful to think in terms of objects that occur in groups. These groups may be characterized by type of object, the number belonging to each type, and the way in which the objects are arranged. For example, consider the letters a, b, c, d, and e. There are 5 objects, one of each type. If we have the letters a, b, b, c, and c, there are 5 objects, one of type a, 2 of b, and 2 of c. Returning to the first group of objects, $a\ b\ c\ d\ e$, $b\ a\ c\ d\ e$, and $c\ d\ e\ a\ b$ differ in the *order* in which the 5 objects are arranged, but each of these groups contains the same number of objects of each type.

Suppose we have a group of n different objects. In how many ways can these n objects be arranged in order in a line? Applying the multiplication principle, we see that any one of the n objects can occupy the first position, any of the $n - 1$ remaining objects can occupy the second position, and so forth until we have only one possible object to occupy the nth position. Thus, the number of different possible arrangements of the n objects in a line consisting of n positions is

$$n! = (n)(n-1)\cdots(2)(1)$$

The symbol $n!$ is read "n factorial."

We shall be concerned only with cases for which n is a nonnegative integer. By definition, $0! = 1$.

Factorials

Some examples of **factorials** are

$$1! = 1$$

$$2! = 2 \times 1 = 2$$

$$3! = 3 \times 2 \times 1 = 6$$

$$\vdots$$

$$10! = 10 \times 9 \times \cdots \times 2 \times 1 = 3{,}628{,}800$$

It is useful to note that $n! = n(n-1)!$. Thus, $10! = 10 \times 9!$, and so on. We see from this relation that it makes sense to define $0! = 1$, since if we let $n = 1$, we have

$$1! = 1 \times 0!$$

and $0! = 1$. This enables us to maintain a consistent definition of factorials for all nonnegative integers.

Factorials obviously increase in size very rapidly. For example, how many different arrangements can be made of a deck of 52 cards if the cards are placed in a line? The answer, 52!, is a number that contains 68 digits.[3]

Permutation $_nP_x$

Frequently we are interested in choosing and arranging in order some subgroup of n different objects. If x of the objects ($x \leqslant n$) are to be selected and arranged in order, as in a line, then each such *ordered arrangement* is said to be a **permutation** of the n objects taken x at a time. The number of such permutations is denoted $_nP_x$. For example, suppose there are 50 persons competing in a contest for 3 rankings—first, second, and third. How many permutations of the 50 people taken 3 at a time are possible, that is, how many different rankings are possible? The answer is

$$_{50}P_3 = 50 \times 49 \times 48 = 117{,}600$$

This is true because any one of the 50 persons can occupy the first position, any of the remaining 49 the second position, and any of 48 could fill the third place. By the multiplication principle, the number of different sequences of first, second, and third rankings is obtained by the indicated product.

We can now generalize this procedure to obtain a convenient formula for the number of permutations of n different objects taken x at a time.

$$\begin{aligned} _nP_x &= (n)(n-1)\cdots[n-(x-1)] \\ &= (n)(n-1)\cdots(n-x+1) \\ &= \frac{(n)(n-1)\cdots(n-x+1)(n-x)!}{(n-x)!} \end{aligned}$$

$$_nP_x = \frac{n!}{(n-x)!} \qquad (2.18)$$

It can be seen, in general, that if there are x positions to be filled, the first position can be filled in n ways; after one object has been placed in the first

[3] Warren Weaver points out in *Lady Luck* (Garden City, NY: Doubleday, 1963), p. 88, about the number of possible arrangements in 52!, that, "If every human being on earth counted a million of these arrangements per second for twenty-four hours a day for lifetimes of eighty years each, they would have made only a negligible start in the job of counting all these arrangements—not a billionth of a billionth of one percent of them!"

position, $x - 1$ positions remain. The second can be filled in $n - 1$ ways, the third in $n - 2$ ways, and so forth down to the xth or last position, which can be filled in $n - (x - 1)$ ways. In writing down the factors that must be multiplied together, 0 is subtracted from n in the first position, 1 is subtracted from n in the second position, and so forth down to $(x - 1)$ subtracted from n in the xth position. The formula $n!/(n - x)!$ follows from the definition of a factorial, since $(n - x)!$ cancels out all factors after $(n - x + 1)$ in the numerator. Thus, in the contest problem, where $n = 50$ and $x = 3$, we have

$$ {}_{50}P_3 = \frac{50!}{(50-3)!} = \frac{50!}{47!} = \frac{50 \times 49 \times 48 \times 47!}{47!} = 50 \times 49 \times 48 $$

A special case of the formula for permutations occurs when all n objects are considered together. In this situation, we are concerned with the number of permutations of n different objects taken n at a time, which is

Permutation ${}_nP_n$

(2.19) $$ {}_nP_n = \frac{n!}{(n-n)!} = \frac{n!}{0!} = n! $$

For example, if a consumer were given one cup of coffee of each of five brands and asked to rank these according to preference, the total number of possible rankings (excluding the possibility of ties) would be

$$ {}_5P_5 = 5! = 120 $$

Combinations

In the case of permutations of objects, the order in which the objects are arranged is of importance.

> When order is not important, we are concerned with *combinations* of objects rather than permutations.

A simple example will illustrate the difference. Suppose the president of a company is interested in setting up a finance committee of two people and plans to select them from a group of 3 executives named Brown, Jones, and Smith. How many possible committees could be formed? Obviously, order is of no importance in this situation. That is, a committee consisting of Brown and Jones is no different from a committee of Jones and Brown. Using first letters to symbolize the 3 names, we can list 3 possible committees: BJ, BS, and JS.

This is an example of combinations of 3 objects taken 2 at a time. The terminology is similar to that used for permutations, so in general, we refer to the number of combinations that can be made of n different objects taken x at a time.

Using the same group of letters and treating them merely as symbols, if order of arrangement were important, the number of permutations of the 3 objects taken 2 at a time would exceed the number of combinations of 3 objects taken 2 at a time. The following 6 permutations can be made in this case:

BJ	JB
BS	SB
JS	SJ

To develop a formula for combinations, we need merely consider the relationship between numbers of combinations and numbers of permutations for the same group of n objects taken x at a time. Fixing attention for the moment on any particular combination, there are x objects filling x positions. How many permutations can be made of these x objects in the x positions? Clearly, any one of the x objects may fill the first position, $x - 1$ the second, and so forth down to one object for the xth position. Thus, $x!$ distinct permutations can be formed of the x objects in x positions. Therefore, the number of permutations that can be formed of n different objects taken x at a time is $x!$ times the number of combinations of these n objects taken x at a time.

The symbol for the number of combinations of n different objects taken x at a time is $\binom{n}{x}$.

Thus,

$$\binom{n}{x} x! = {}_nP_x$$

Solving for $\binom{n}{x}$ yields the following formula:

$$\binom{n}{x} = \frac{{}_nP_x}{x!} = \frac{n!}{x!(n-x)!} \tag{2.20}$$

Returning to the committee illustration, the number of combinations of the 3 people taken 2 at a time is

$$\binom{3}{2} = \frac{3!}{2!1!} = 3$$

which was the number previously listed. Similarly, the number of permutations of 3 objects taken 2 at a time is

$$_3P_2 = \frac{3!}{1!} = 6$$

which was the number of ordered arrangements listed earlier.

EXAMPLE 2-10

A woman is considering whether to invest her savings income in one or more of the following items: stocks, gold, and silver. How many options does she have?

Solution

There are 2 options to each investment. Therefore, by the multiplication principle, there are $2 \times 2 \times 2 = 8$ different possibilities.

EXAMPLE 2-11

An investor wishes to select a portfolio of 5 stocks from a list of 25 of the most active shares. How many different portfolios would he have to examine?

Solution

$$\binom{25}{5} = \frac{25!}{20!5!} = \frac{25 \times 24 \times 23 \times 22 \times 21}{5 \times 4 \times 3 \times 2 \times 1} = 53{,}130$$

EXAMPLE 2-12

There are 6 different operations in a manufacturing process. Let us refer to them as A, B, C, D, E, and F. Operation A must be performed first, and F must be performed last. All other operations may be performed in any order. How many different sequences of operations are possible?

Solution

Since A and F are in fixed positions (first and last), we need be concerned only with B, C, D, and E. Any of these 4 may be performed first after A, any of the remaining 3 may come second, and so on. Therefore, the number of possible different sequences is given by $_4P_4 = 24$.

EXAMPLE 2-13

An underwriting syndicate is to be formed from a group of investment banking firms, each of which is classified either as type A or as type B. There are 5 type A and 7 type B firms. In how many ways can a syndicate of 4 firms be formed if

a. It must consist of 2 firms of type A and 2 of type B?
b. It must consist of at least 2 firms of type A and at least one of type B?

Solution

a. We can think of the sequential selection of 2 firms from the 5 type A firms, followed by a selection of 2 firms from the 7 type B firms when the order of the selection is unimportant. Hence, the number of possible syndicates is

$$\binom{5}{2} \times \binom{7}{2} = 210$$

b. The following tabular arrangement gives the solution to part (b).

Different Methods of Forming the Syndicate	Number of Ways of Forming the Syndicate
2A, 2B	$\binom{5}{2} \times \binom{7}{2} = 210$
3A, 1B	$\binom{5}{3} \times \binom{7}{1} = 70$
	Total 280

Exercises 2.4

1. A sales manager wishes to place an advertisement in 2 journals. There are 5 feasible journals in which to advertise. In how many pairs of journals can the advertisement be placed?
2. A shopkeeper wishes to place each brand of detergent that he sells on a shelf.
   a. If he sells 4 brands of detergent, in how many ways can he arrange these brands on the shelf?
   b. If there was space available for displaying only 2 of the 4 brands, how many arrangements are possible?
3. A brief market research questionnaire requires the respondent to answer each of 10 successive questions with either a "yes" or a "no." The sequence of 10 "yes-no" responses is defined as the respondent's "profile." How many different possible profiles are there?
4. Consider a group of 5 persons, consisting of 3 men and 2 women, all of whom belong to an organization.
   a. How many committees of 3 persons can be formed from the group?
   b. In how many ways can the 2 positions, president and vice-president be formed?
   c. What is the probability that a committee of 2 persons chosen at random will consist of one man and one woman?
5. Lemon Motors orders 7 different upholstery colors for its cars and 12 different colors of body paint.
   a. How many different color combinations of body and upholstering are available to the customer?
   b. If Lemon Motors allows the customer to order a roof color different from the basic body color, how many additional different color coordinations of body, roof, and upholstering are available to the customer?
6. Mancuso's Furs, Inc., has just purchased a system to handle its accounts receivable. Each data card contains 80 columns in which either a number from 0 to 9 or a letter may be punched to represent information about an account. It is decided that each account will be assigned 4 identification

symbols that will be punched in the first 4 columns of the data card to identify each card with a particular account.

a. If only numbers are to be used, how many accounts can be handled by this method?
b. If the first column is to be a letter and the next 3 are to be numbers, how many accounts can be handled?
c. If either a letter or a number can be punched in each column, how many accounts can be handled?

7. A firm desires to build 6 new factories: 2 factories in the 13 Southern states, one factory in the 6 Middle Atlantic states, one factory in the 4 Far Western states, and 2 factories in the 8 Midwest states. If the firm wants to study the desirability of each possible combination of locations, how many combinations would the firm have to consider?
8. Susan McGurk, the financial vice-president of the Chemopetrol Company, is considering 5 similar investment proposals for the upcoming fiscal period. After analyzing the financial condition of the firm and estimating the firm's cash flow for the period, she decides that all portfolios consisting of 3 proposals are equally feasible and desirable in the long run.
   a. Assuming that the 3 investments in the adopted portfolio are made simultaneously, how many different portfolios are there from which to choose?
   b. If the 3 adopted proposals are implemented sequentially and it is desired that—due to differing cash payback periods—the order of implementation is a distinguishing factor in the comparison of otherwise identical portfolios, how many different portfolio arrangements are there?
   c. Under the assumptions of part (b) and assuming random selection, what is the probability that the portfolio selected will include proposals A, B, and C?
9. A motivational researcher shows a woman 15 projected colors for new fall clothes and asks her to pick her 5 favorite colors.
   a. Give a specific outcome of the experiment.
   b. How many such outcomes are there?
   c. How many outcomes will contain the color *russet*?
   d. What is the probability that russet will be one of the 5 favorites?
10. In a determination of preference of package design, a panel was given 4 different packaging designs and asked to rate them. How many different possible rankings could the panel have given (excluding ties)?
11. A committee consists of 5 union and 4 nonunion workers. In how many ways can a subcommittee of 5 be formed consisting of 3 union and 2 nonunion workers?
12. A committee consists of 10 people. It is decided to appoint a chairman, a vice-chairman, and a secretary-treasurer. How many different ways can this be done?

3

Discrete Random Variables and Probability Distributions

Managerial decisions are ordinarily made under conditions of uncertainty.

- A corporation treasurer makes investment decisions in the face of uncertainties concerning future movements of interest rates and stock market prices.
- A corporate executive committee may make a decision concerning expansion of manufacturing facilities despite uncertainty about future levels of demands for the company's products.
- An advertising manager makes decisions on advertising expenditures in various media without being certain of the sales that will be generated by these outlays.

In each of these cases, the outcomes of concern—such as interest rates, stock market prices, levels of demand, and sales that result from advertising—may assume a variety of values; we refer to the outcomes as **variables**. In statistical analysis, such variables are usually called random variables.

3.1 RANDOM VARIABLES

A **random variable** may be defined roughly as a variable that takes on different numerical values because of chance.[1]

[1] From a mathematical viewpoint, a random variable is a function consisting of the elements of a sample space and the numbers assigned to these sample elements. Ordinarily, a shortcut method of referring to a random variable is used. For example, we may refer to the random variables "the price of XYZ stock at some specified future time," "annual volume of sales of product ABC," or, in a coin-tossing example, "number of heads obtained in two tosses of a coin."

Probability distributions

In this chapter, we will be concerned primarily with **probability distributions** of random variables. This concept is central to all of statistics, and although we introduce it here in the context of the business decision problem, it is used in every field in which statistical methods are applied. Examples 3-1, 3-2, and 3-3 introduce the idea of a probability distribution.

EXAMPLE 3-1

Golden Harvest Fruit Mart stocks fresh fruit when it is in season. Raymond Hassel, the store manager, needs to know how many crates of grapes to order for each day during the season. Because the shelf life of grapes is only three days, he must be careful not to overorder. He decides to rely heavily on last year's daily demand experience to make his ordering decision. His records show the following daily demands in terms of numbers of crates sold on 25 days that he sold grapes:

Number of crates sold	3	4	5	6	7	8	9	10	11
Number of days	1	1	4	3	5	6	3	1	1

Hassel sets up the probability distribution based on relative frequencies shown in Table 3-1 for the number of crates sold. The two columns shown in Table 3-1 constitute the probability distribution for the random variable "number of crates sold." The following type of symbolism is conventionally used: If we let the symbol X stand for the random variable, then we can represent the values the random variable can assume by x. The probability that the random variable X will assume the value x is symbolized by $P(X = x)$, or simply $f(x)$. In Table 3-1, the values of the random variable are listed under the column headed x and the probabilities of these values are shown under $f(x)$. Thus,

$$P(X = 3) = f(3) = \frac{1}{25} = 0.04$$

TABLE 3-1
Probability distribution for the numbers of crates sold

Number of Crates Sold x	Probability $f(x)$
3	0.04
4	0.04
5	0.16
6	0.12
7	0.20
8	0.24
9	0.12
10	0.04
11	0.04
	1.00

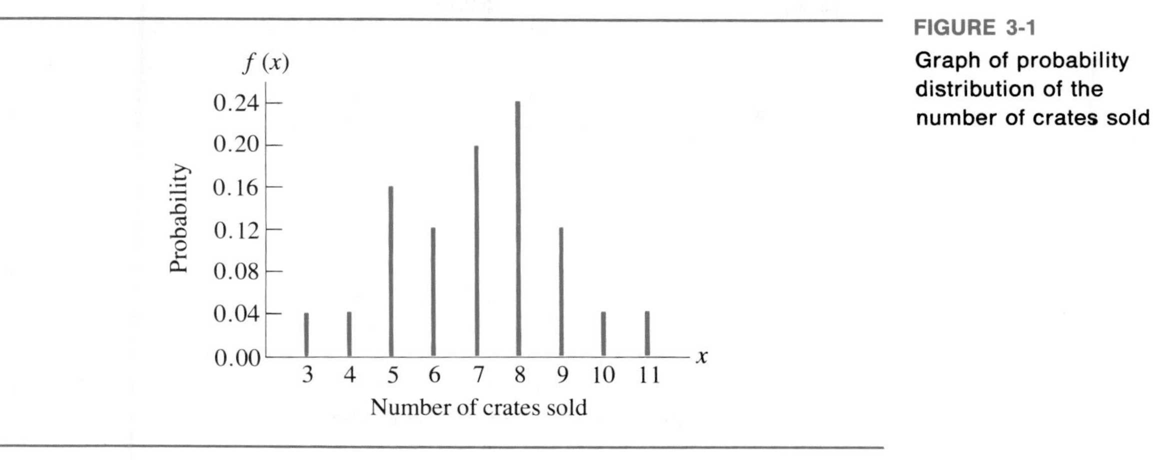

FIGURE 3-1
Graph of probability distribution of the number of crates sold

$$P(X = 4) = f(4) = \frac{1}{25} = 0.04$$

$$P(X = 5) = f(5) = \frac{4}{25} = 0.16$$

Note that the probabilities sum to one.

When a probability distribution is graphed, it is conventional to display the values of the random variable on the horizontal axis and their probabilities on the vertical scale. The graph of the probability distribution for the number of crates sold is shown in Figure 3-1.

EXAMPLE 3-2

A department store wants to conduct a rigorous screening of applicants for credit cards to minimize losses from uncollectibles. The store constructs a credit scoring formula based on such factors as a customer's income level, job experience, and prior credit record. With the use of a cutoff score, applicants are classified into two mutually exclusive categories: (1) good credit risks and (2) bad credit risks. Of all charge customers, 10% are classified as bad credit risks.

The store's credit department selects a sample of four customers at random from the list. What is the probability distribution of the random variable "number of good credit risks"? Assume that the list of charge customers is so large that even though the sample is drawn without replacement, there is only a negligible loss of accuracy if the computations are performed as though the sampling were carried out with replacement. That is, the partial exhaustion of the list when the items in the sample are drawn is so small that for practical purposes the probabilities of obtaining the two types of applicants remain unchanged.

Let A represent the occurrence of a good credit risk applicant and let $\bar{A}$ represent the occurrence of a bad credit risk applicant. The elements of the sample space for the experiment of drawing the sample of 4 customers are listed in Table 3-2.

TABLE 3-2
Elements of the sample space for the experiment of drawing a random sample of 4 customers

$\bar{A}\ \bar{A}\ \bar{A}\ \bar{A}$	$A\ A\ \bar{A}\ \bar{A}$	$A\ A\ A\ \bar{A}$
	$A\ \bar{A}\ A\ \bar{A}$	$A\ A\ \bar{A}\ A$
$A\ \bar{A}\ \bar{A}\ \bar{A}$	$A\ \bar{A}\ \bar{A}\ A$	$A\ \bar{A}\ A\ A$
$\bar{A}\ A\ \bar{A}\ \bar{A}$	$\bar{A}\ A\ A\ \bar{A}$	$\bar{A}\ A\ A\ A$
$\bar{A}\ \bar{A}\ A\ \bar{A}$	$\bar{A}\ \bar{A}\ A\ A$	
$\bar{A}\ \bar{A}\ \bar{A}\ A$	$\bar{A}\ A\ \bar{A}\ A$	$A\ A\ A\ A$

We denote by X the random variable "number of good credit risk applicants"; X can take on the values 0, 1, 2, 3, and 4. We can see from Table 3-2 that one sample element corresponds to the occurrence of *zero* good credit risk applicants; four elements to *one* good credit risk applicant; six elements to *two* good credit risk applicants; four elements to *three* good credit risk applicants; and one element to *four* good credit risk applicants. However, the sample elements are not equally likely. The probability of a good credit risk applicant is 0.9 and that of a bad credit risk applicant is 0.1. Considering one sample element each for zero, one, two, three, and four good credit risk applicants, we have the following probabilities:

$$P(\bar{A}\ \bar{A}\ \bar{A}\ \bar{A}) = (0.1)(0.1)(0.1)(0.1) = (0.1)^4$$

$$P(A\ \bar{A}\ \bar{A}\ \bar{A}) = (0.9)(0.1)(0.1)(0.1) = (0.9)(0.1)^3$$

$$P(A\ A\ \bar{A}\ \bar{A}) = (0.9)(0.9)(0.1)(0.1) = (0.9)^2(0.1)^2$$

$$P(A\ A\ A\ \bar{A}) = (0.9)(0.9)(0.9)(0.1) = (0.9)^3(0.1)$$

$$P(A\ A\ A\ A) = (0.9)(0.9)(0.9)(0.9) = (0.9)^4$$

To get the probability distribution of good credit risk applicants, we multiply the specified probabilities for elementary events by the number of such elements in the composite events "zero good credit risk applicants," "one good credit risk applicant," and so forth, as shown:

$$P(X = 0) = f(0) = 1(0.1)^4 = 0.0001$$

$$P(X = 1) = f(1) = 4(0.9)(0.1)^3 = 0.0036$$

$$P(X = 2) = f(2) = 6(0.9)^2(0.1)^2 = 0.0486$$

TABLE 3-3
Probability distribution of number of good credit risk customers

x	$f(x)$
0	0.0001
1	0.0036
2	0.0486
3	0.2916
4	0.6561

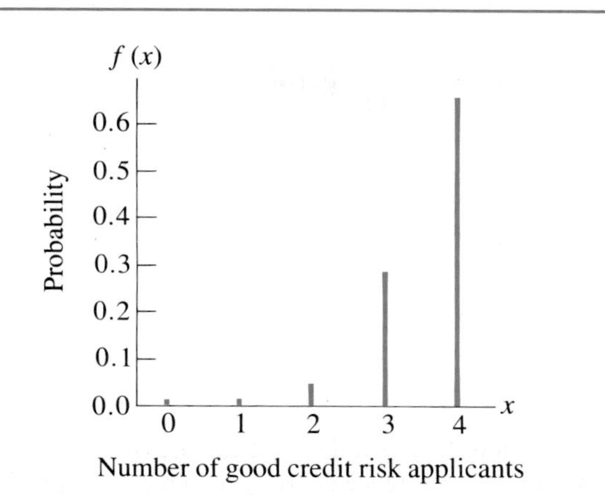

FIGURE 3-2
Graph of probability distribution of number of good credit risk applicants

$$P(X = 3) = f(3) = 4(0.9)^3(0.1) = 0.2916$$

$$P(X = 4) = f(4) = 1(0.9)^4 = 0.6561$$

These values are summarized in Table 3-3 as a probability distribution. Note that the probabilities add up to one.

Actually, it would not have been necessary to list all the elements in the sample space in order to derive this probability distribution. A much briefer method for computing the required probabilities is explained in section 3.4, where the binomial distribution is discussed.

A graph of the probability distribution is given in Figure 3-2.

EXAMPLE 3-3

Ace Food Company, a large grocery chain, is interested in diversifying its product line into the soft goods market. Karen Leslie, vice-president in charge of mergers and acquisitions, is negotiating the acquisition of Sandlers, a discount chain. To determine the price Ace Foods would have to pay per share for Sandlers, she sets up the probability distribution for the stock price shown in Table 3-4. A graph of the probability distribution appears in Figure 3-3.

TABLE 3-4
Probability distribution for the price of Sandlers common stock

Price of Sandlers Common Stock x	Probability $f(x)$
$33	0.10
34	0.25
35	0.50
36	0.10
37	0.05

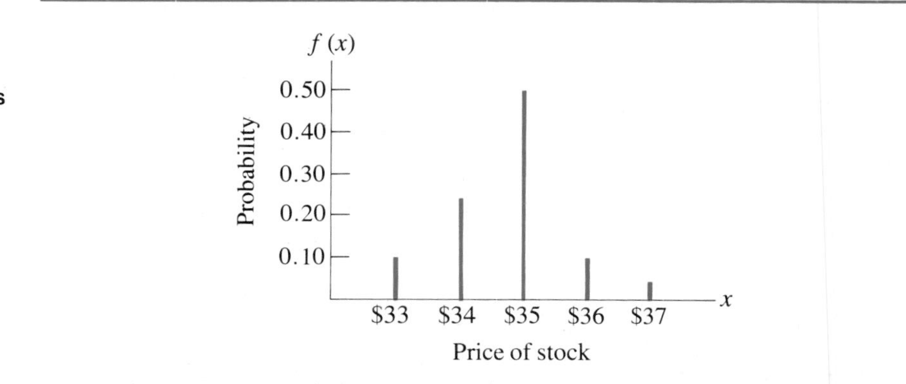

FIGURE 3-3
Graph of probability distribution of price of Sandlers common stocks

This is an example of a subjective probability distribution. By assigning a probability of 0.50 to \$35, Leslie indicates that she feels the odds are 50:50 that the stock price will be \$35 as opposed to any other figure. She feels that a price of \$35 is twice as likely to occur as a price of \$34, to which she assigns a probability of 0.25. She feels that prices of \$33, \$36, or \$37 are rather unlikely.

Types of Random Variables

Discrete random variables

Random variables are classified as either discrete or continuous. A **discrete random variable** is one that can take on only a finite or countable number of distinct values. Examples 3-1, 3-2, and 3-3 illustrate probability distributions of discrete random variables.

Continuous random variables

A random variable is said to be continuous in a given range if the variable can assume any value in that range. The term **continuous random variable** implies that variation takes place along a continuum. Examples of continuous variables include weight, length, velocity, rate of production, dosage of a drug, and the length of life of a given product. While discrete variables can be *counted*, continuous variables can be *measured* with some degree of accuracy.

> Although measured data are essentially discrete in the real world, the variable under measurement is often continuous.

It may be argued that, in the real world, all data are discrete. For example, if we measure weight with a measuring instrument that permits a determination only to the nearest thousandth of a pound, then the resulting data will be discrete in units of thousandths of a pound. Despite this discreteness of data caused by limitations of measuring instruments, it is nevertheless useful in many instances to use mathematical models that treat certain variables as continuous. We may conceive of a continuous mathematical model of heights of individuals—where the underlying data are measured and discrete—as a model of reality that is more accurate than the discrete data from which the model was derived rather than conceiving of such a model as merely a convenient approximation.

On the other hand, we often find it convenient to convert a variable that is conceptually continuous into a discrete one. Thus, in the case of heights of individuals, we may set up classifications such as tall, medium, and short rather than using measurements along a continuous scale of inches.

It is sometimes said that one indication of progress in science is the extent to which discrete variables can be converted into continuous variables.

- The physicist treats color in terms of the *continuous variable* of wavelengths rather than the *discrete* classification of names of colors.
- Measurement of temperature by means of a thermometer treats human body temperature as varying along a *continuous* scale (although the resulting measurements in degrees are discrete) rather than as *discrete* (as when temperature is judged to be "normal" or "high" by a hand placed on the forehead).

In applied problems, where a probability model is used to represent a real-world situation, we may work in terms of either discrete or continuous random variables, whichever system is most appropriate for the problem or decision-making situation in question. Only probability distributions of discrete random variables will be discussed in the remainder of this chapter.

Characteristics of Probability Distribution

In Examples 3-1, 3-2, and 3-3, we saw that the sum of the probabilities in each probability distribution was equal to one. It is possible to summarize the characteristics of probability distributions somewhat more formally.

A probability distribution of a discrete random variable X whose value at x is $f(x)$ possesses the following properties:[2]

1. $f(x) \geqslant 0$ for all real values of X
2. $\sum_{x} f(x) = 1$

[2] A somewhat simplified notation is used here. A mathematically more elegant notation would represent the values that the random variable X could assume as $x_1, x_2, \ldots, x_n$ with associated probabilities $f(x_1), f(x_2), \ldots, f(x_n)$. Then the two properties would appear as

1. $f(x_i) \geqslant 0$ for all i
2. $\sum_{i=1}^{n} f(x_i) = 1$

If X takes on an infinite number of values, then the second property would appear as

$$\sum_{i=1}^{\infty} f(x_i) = 1$$

Property 1 simply states that probabilities are greater than or equal to zero. The second property states that the sum of the probabilities in a probability distribution is equal to one. The notation

$$\sum_x f(x)$$

means "sum of the values of $f(x)$ for all values that x takes on."

While we will ordinarily use the term **probability distribution** to refer to both discrete and continuous variables, other terms are sometimes used to refer to probability distributions (also called probability functions).

- Probability distributions of discrete random variables are often referred to as **probability mass functions** or simply **mass functions** because the probabilities are *massed* at distinct points, for example, along the x axis.
- Probability distributions of continuous random variables are referred to as **probability density functions** or **density functions**.

Cumulative Distribution Functions

Frequently, we are interested in the probability that a random variable is less than, equal to, or greater than a given value. The **cumulative distribution function** is particularly useful in this connection. We may define this function as follows.

> Given a random variable X, the value of the cumulative distribution function at x, denoted $F(x)$, is the probability that X takes on values less than or equal to x.

Hence,

(3.1) $$F(x) = P(X \leqslant x)$$

In the case of a discrete random variable, it is clear that

(3.2) $$F(c) = \sum_{x \leqslant c} f(x)$$

The symbol

$$\sum_{x \leqslant c} f(x)$$

means "sum of the values of $f(x)$ for all values of x less than or equal to c."

EXAMPLE 3-4

We return to Example 3-3, involving the probability distribution for the price of Sandlers common stock. The probability that the price would be \$33 or less is $P(X \leqslant \$33) = F(\$33) = 0.10$; \$34 or less, $P(X \leqslant \$34) = F(\$34) = 0.35$; and so on. In Table 3-5, the

TABLE 3-5
Probability distribution and cumulative distribution function for the price of Sandlers common stock

Price of Stock x	Probability $f(x)$	Cumulative Probability $F(x)$
\$33	0.10	0.10
34	0.25	0.35
35	0.50	0.85
36	0.10	0.95
37	0.05	1.00

probability distribution and cumulative distribution function for the random variable "price of Sandlers common stock" are shown.

A graph of the cumulative distribution function is given in Figure 3-4. This graph is a *step function*; that is, the values change in discrete "steps" at the indicated integral values of the random variable, X. Thus, $F(x)$ takes the value 0 to the left of the point $x = \$33$, steps up to $F(x) = 0.10$ at $x = \$33$, and so on. The dot shown at the left of each horizontal line segment indicates the probability for the integral values of x. At these points, the values of the cumulative distribution function are read from the *upper* line segments.

We note the following relations in this problem, which follow from the definition of a cumulative distribution function:

$$F(\$33) = f(\$33) = 0.10$$

$$F(\$34) = f(\$33) + f(\$34) = 0.10 + 0.25 = 0.35$$

$$F(\$35) = f(\$33) + f(\$34) + f(\$35) = 0.10 + 0.25 + 0.50 = 0.85 \text{ and so on}$$

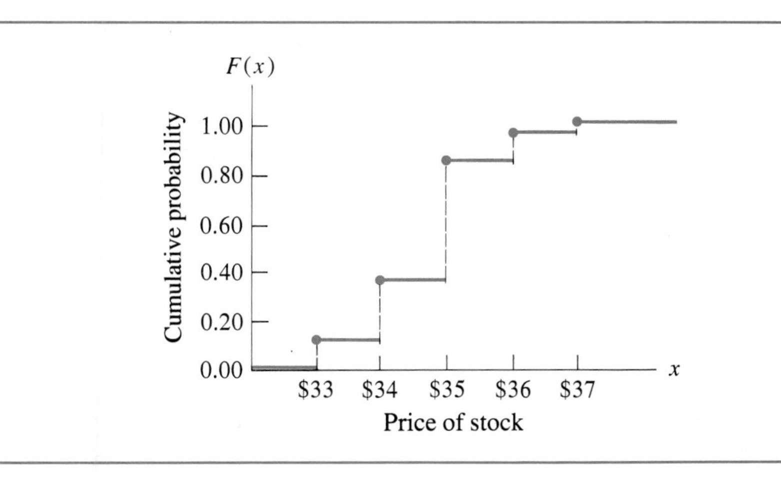

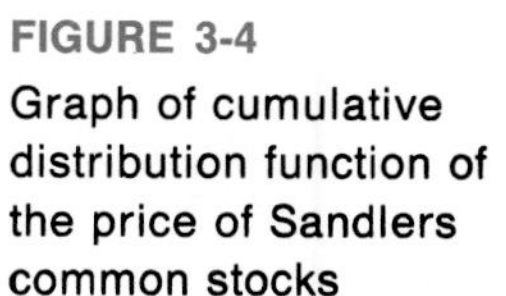

FIGURE 3-4
Graph of cumulative distribution function of the price of Sandlers common stocks

The probabilities that the price would be more than \$33, \$34, and \$35 are, respectively,

$$1 - F(\$33) = 0.90$$
$$1 - F(\$34) = 0.65$$
$$1 - F(\$35) = 0.15$$

EXAMPLE 3-5

Let us return to Example 3-2, which discussed the probabilities of obtaining a good risk in a sample of 4 credit applicants. A few questions will illustrate some uses of the cumulative distribution function.

a. What is the probability that no more than 3 applicants are good credit risks?

$$\begin{aligned} F(3) &= f(0) + f(1) + f(2) + f(3) \\ &= 0.0001 + 0.0036 + 0.0486 + 0.2916 = 0.3439 \end{aligned}$$

b. What is the probability that at least 2 applicants selected are good credit risks?

$$1 - F(1) = 1 - 0.0037 = 0.9963$$

c. Calculate the probability that more than 2 applicants selected are good credit risks.

$$1 - F(2) = 1 - 0.0523 = 0.9477$$

Exercises 3.1

1. Prove that the distribution of the random variable X has the properties of a probability distribution, given that

$$f(x) = \frac{x^2 + 2}{20} \quad \text{where } x = 1, 2, 3$$

2. The probability distribution of X, where X is the number of ships arriving at a port per day, is as follows: $f(0) = 0.4$, $f(1) = 0.3$, $f(2) = 0.2$, and $f(3) = 0.1$. Find the cumulative probability distribution and present it on a graph.

3. After investigating the income statement of a company, you find that the company operated at a \$1 million loss for 2 years, at \$1 million profit for 3 years, and at \$3 million profit for 3 years. Given that X is profit of the firm, write the probability distribution of X for the 8-year period.

4. Which of the following are valid probability functions?

a. $f(x) = \dfrac{x^2 + x}{2} \quad x = -1, 0, 1, 2$

b. $f(x) = \dfrac{x + 3}{12} \quad x = 0, 1, 2$

c. $f(x) = \dfrac{x}{2} \quad x = -1, 0, 1, 2$

d. $f(x) = \dfrac{x^2 + 2}{40} \quad x = 0, 1, \ldots, 4$

5. Find k such that the following are probability functions:

   a. $kx^2 \quad x = 0, 1, 2, 3$

   b. $\dfrac{k}{x^2} \quad x = 1, 2, 3, 4$

   c. $\dfrac{k}{x} \quad x = 1, 2, 3$

6. Consider the following random variable for the lifetime of a particular electronic component:

   $X = 1$ if the component lasts less than two years

   $X = 2$ if the component lasts at least two but less than three years

   $X = 3$ if the component lasts at least three but less than four years

   $X = 4$ if the component lasts at least four but less than five years

   Let

$$F(x) = \frac{x^2}{16}$$

   Find $f(x)$ in tabular form.

7. The marketing department of a large firm is making a survey of consumers in a large metropolitan area to determine interest in a new product the firm is developing. The consumers to be surveyed are drawn from names in a telephone book serving the area, so we can assume that the population sampled is large enough for us to treat the sampling as having been done with replacement. It is known that 40% of the consumers listed in the telephone book in this area already use Miracle Worker, a similar product now sold by the firm. Assume that a sample of 3 names is drawn.
   a. Show the elements of the sample space for the drawing. Let M represent the occurrence of a Miracle Worker customer and $\bar{M}$ represent a noncustomer.
   b. Determine the probability distribution of the random variable "number of Miracle Worker customers."
   c. Graph this probability distribution.

8. A large Japanese manufacturer of television sets is studying the number of defective sets per production run. The quality control manager selects a sample of 2 sets from a lot of 50 TV sets, 10 of which are known to be defective.
   a. Find the probability function of X.
   b. Find the probability that the sample contains 2 defective TV sets.

3.2

PROBABILITY DISTRIBUTIONS OF DISCRETE RANDOM VARIABLES

In many situations, it is useful to represent the probability distribution of a random variable by a general algebraic expression. Probability calculations can then be conveniently made by substituting appropriate values into the algebraic model. The mathematical expression is a compact summary of the process that has generated the probability distribution. Thus, the statement that a particular probability distribution is appropriate in a given situation contains a considerable amount of information concerning the nature of the underlying process. In the following sections, we discuss the uniform, binomial, multinomial, hypergeometric, and Poisson probability distributions of discrete random variables.

3.3

THE UNIFORM DISTRIBUTION

Sometimes, equal probabilities are assigned to all the possible values a random variable may assume. Such a probability distribution is referred to as a **uniform distribution**. For example, suppose a fair die is rolled once. The probability is $\frac{1}{6}$ that the die will show any given number on its uppermost face. The probability mass function in this case may be written as

$$f(x) = \tfrac{1}{6} \quad \text{for } x = 1, 2, \ldots, 6$$

A graph of this distribution is given in Figure 3-5.

As another illustration, let us consider the case of the Amgar Power Company, which produces electrical energy from geothermal steam fields. It takes about two years to build a production facility. In planning its production

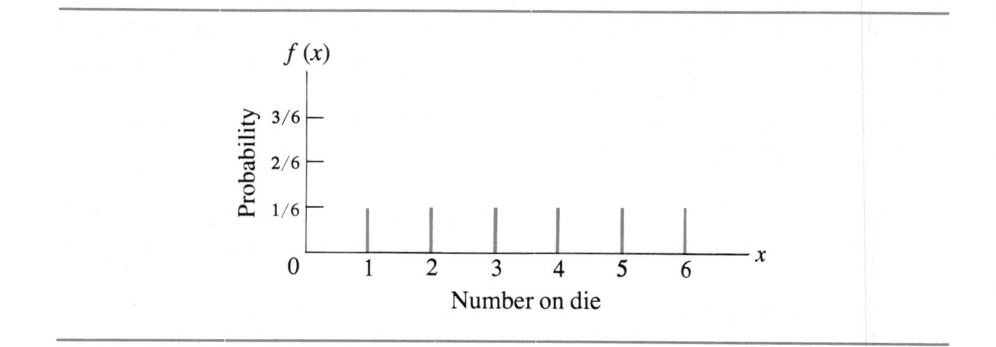

FIGURE 3-5
Graph of probability mass function of numbers obtained in a roll of a fair die

strategies, the company concludes that it is equally likely that demand two years hence will be 80,000, 90,000, 100,000, 110,000, or 120,000 kilowatts. Therefore, the probability distribution established by Amgar Power Company for this future demand is

$$f(x) = 0.20 \quad \text{for } x = 80{,}000, 90{,}000, \ldots, 120{,}000$$

Later, we will examine how such information is used in decision-making procedures.

Exercises 3.3

1. A roulette wheel has 38 equally spaced openings with numbers 00, 0, 1, 2, . . . , 36. Write the probability function of X, where X is any number appearing after a spin of the wheel.
2. An international airline flight is scheduled to arrive at Nairobi Airport at 7:30 A.M. A study has shown that the actual arrival time is uniformly distributed by minutes in the range 7:05 to 8:40 A.M. Let $X = 1$ represent arrival at 7:05 A.M., $X = 2$, arrival at 7:06 A.M., and so on.
   a. Write the mathematical expression for $f(x)$.
   b. What is the probability that the plane will be late?
   c. What is the probability that the plane will arrive after 8:00 A.M.?
   d. What is the probability that the plane will arrive at or after 8:00 A.M.?
   e. What is the probability that the plane will arrive before 7:30 A.M.?
3. Toss a coin 3 times and define the random variable X as the number of heads appearing on the 3 tosses. Is X uniformly distributed?

3.4 THE BINOMIAL DISTRIBUTION

The **binomial distribution**, in which there are two possible outcomes on each experimental trial, is undoubtedly the most widely applied probability distribution of a discrete random variable. It has been used to describe a large variety of processes in business and the social sciences as well as other areas. The process that gives rise to the binomial distribution is usually referred to as a **Bernoulli trial** or a **Bernoulli process**.[3] The mathematical model for a Bernoulli process is developed from a specific set of assumptions involving the concept of a series of experimental trials.

Bernoulli process

Let us envision a process or experiment characterized by repeated trials taking place under the following conditions or assumptions:

Assumptions

1. On each trial, there are two mutually exclusive possible outcomes, which are referred to as "success" and "failure." In somewhat different language,

[3] Named after James Bernoulli (1654–1705), a member of a family of Swiss mathematicians and scientists, who did some of the early significant work on the binomial distribution.

the sample space of possible outcomes on each experimental trial is $S = \{\text{failure, success}\}$.

2. The probability of a success, denoted p, remains constant from trial to trial. The probability of a failure, denoted q, is equal to $1 - p$.
3. The trials are independent. That is, the outcomes on any given trial or sequence of trials do not affect the outcomes on subsequent trials.

Random process

The outcome on any specific trial is determined by chance. Processes having this characteristic are referred to as **random processes**, or **stochastic processes**, and Bernoulli trials are one example of such processes.

Our aim is to develop a formula for the probability of x successes in n trials of a Bernoulli process. We start with a simple, specific case of a series of Bernoulli trials, 5 tosses of a coin. We calculate the probability of obtaining exactly 2 heads in 5 tosses; the resulting expression can then be generalized.

If we are tossing a fair coin 5 times, we may treat each toss as one Bernoulli trial. The possible outcomes on any particular trial are a head and a tail. Assume that the appearance of a head is a success. (Of course, the classification of one of the 2 possible outcomes as a "success" is completely arbitrary, and there is no necessary implication of desirability or goodness involved. For example, we may choose to refer to the appearance of a defective item in a production process as a success; or, if a series of births is treated as a Bernoulli process, the appearance of a female (male) may be classified as a success.) Suppose that the sequence of outcomes is

$$H\ T\ H\ T\ T$$

where H and T denote head and tail, as usual. We now introduce a convenient coding device for outcomes on Bernoulli trials. Let

$$x_i = 0 \quad \text{if the outcome on the } i\text{th trial is a failure}$$
$$x_i = 1 \quad \text{if the outcome on the } i\text{th trial is a success}$$

Then the outcomes of the previous sequence of tosses may be written as

$$1\,0\,1\,0\,0 \qquad (\text{representing} \quad H\ T\ H\ T\ T)$$

Since the probability of a success and a failure on a given trial are, respectively, p and q, the probability of this particular sequence of outcomes is, by the multiplication rule,

$$P(1, 0, 1, 0, 0) = pqpqq = q^3p^2$$

In this notation, for simplicity, commas have been used to separate the outcomes of the successive trials. Actually, though, this is the joint probability of the events that occurred on the 5 trials, that is, the probability of obtaining the specific sequences of successes and failures in the order in which they occurred. However, we are interested not in any specific order of results, but rather in the probability of obtaining a given number of successes in n trials. What then is the probability of obtaining exactly 2 successes in 5 Bernoulli trials? There are 9 other sequences that satisfy the condition of exactly 2 successes in 5 trials.

1 1 0 0 0	0 1 1 0 0	0 0 1 1 0
1 0 0 1 0	0 1 0 1 0	0 0 1 0 1
1 0 0 0 1	0 1 0 0 1	0 0 0 1 1

By the same reasoning used earlier, each of these sequences has the same probability, q^3p^2. We can obtain the number of such sequences from the formula for the number of combinations of n objects taken x at a time given in equation 2.20. Thus, the number of possible sequences in which two ones can occur is $\binom{5}{2}$. We indicated in equation 2.20 that

$$\binom{n}{x} = \frac{n!}{x!(n-x)!}$$

Thus,

$$\binom{5}{2} = \frac{5!}{2!3!} = 10$$

and we may write

$$P(\text{exactly 2 successes}) = \binom{5}{2} q^3 p^2$$

In the case of the fair coin example, we assign a probability of $\frac{1}{2}$ to p and $\frac{1}{2}$ to q. Hence,

$$P(\text{exactly 2 heads}) = \binom{5}{2}\left(\frac{1}{2}\right)^3\left(\frac{1}{2}\right)^2 = \frac{10}{32} = \frac{5}{16}$$

This result may be generalized to obtain the probability of (exactly) x successes in n trials of a Bernoulli process. Let us assume $n - x$ failures occurred followed by x successes, in that order. We may then represent this sequence as

$$\underbrace{0\,0\,0\cdots0}_{n-x\text{ failures}}\quad\underbrace{1\,1\,1\cdots1}_{x\text{ successes}}$$

The probability of this particular sequence is $q^{n-x}p^x$. The number of possible sequences of n trials resulting in exactly x successes is $\binom{n}{x}$.[4] Therefore, the

[4] Because the combination notation is universally used in connection with the binomial probability distribution, that convention is followed here. However, conceptually, we have here the number of distinct permutations that can be formed of n objects, $n - x$ of which are of one type and x of the other. Since the number of such permutations turns out to be equal to $\binom{n}{x}$, the combination notation may be used instead of that for permutations.

probability of obtaining x successes in n trials of a Bernoulli process is given by[5]

$$f(x) = \binom{n}{x} q^{n-x} p^x \quad \text{for } x = 0, 1, 2, \ldots, n \tag{3.3}$$

If we denote by X the random variable "number of successes in these n trials," then

$$f(x) = P(X = x)$$

The fact that this is a probability distribution is verified by noting the following conditions.

1. $f(x) \geqslant 0$ for all real values of x
2. $\sum_x f(x) = 1$

The first condition is verified by noting that since p and n are nonnegative numbers, $f(x)$ cannot be negative. The second condition is true because (as shown mathematically)

$$\sum_x \binom{n}{x} q^{n-x} p^x = (q + p)^n = 1^n = 1$$

Therefore, the term **binomial probability distribution**, or simply **binomial distribution**, is usually used to refer to the probability distribution resulting from a Bernoulli process.

> In problems where the assumptions of a Bernoulli process are met, we can obtain the probabilities of zero, one, or more successes in n trials from the respective terms of the binomial expansion of $(q + p)^n$, where q and p denote the probabilities of failure and success on a single trial and n is the number of trials.

The binomial distribution has two parameters, *n and p*.[6] Each pair of values for these parameters establishes a different distribution. Thus, the binomial

[5] The following method of writing the mathematical expression for such a probability distribution is often used:

$$f(x) = \binom{n}{x} q^{n-x} p^x \quad \text{for } x = 0, 1, 2, \ldots, n$$
$$= 0, \text{ elsewhere}$$

In this and other places where it is clear that $f(x)$ is equal to 0 for values of the random variable other than the specified ones, the notation on the last line will not be included.

[6] In this context, the term "parameters" refers to numerical quantities that are sufficient to specify a probability distribution. When particular values are assigned to the parameters of a probability function, a specific distribution in the family of possible distributions is defined. For example, $n = 10$, $p = \frac{1}{2}$ specifies a particular binomial distribution; $n = 20$, $p = \frac{1}{2}$ specifies another.

distribution is actually a family of probability distributions. Since computations become laborious for large values of n, it is advisable to use special tables. Selected values of the binomial cumulative distribution function are given in Table A-1 of Appendix A. The values of

$$F(c) = P(X \leqslant c) = \sum_{x \leqslant c} f(x) \quad \text{for } x = 0, 1, 2, \ldots, n$$

are shown in that table for $n = 2$ to $n = 20$ and $p = 0.05$ to $p = 0.50$ in multiples of 0.05. Values of $f(c)$, cumulative probabilities for p values greater than 0.50, and probabilities that x is greater than a given value or lies between two values can be obtained by appropriate manipulation of these tabulated values. Some of the examples that follow illustrate the use of the table. More extensive tables have been published by the National Bureau of Standards and Harvard University, but even these tables usually do not go beyond $n = 50$ or $n = 100$. For large values of n, approximations are available for the binomial distribution, and the exact values generally need not be determined.

For convenience in looking up individual terms of the binomial probability distribution, selected values of the binomial probability distribution are given in Table A-2 of Appendix A. The values of

$$f(x) = P(X = x) \quad \text{for } x = 0, 1, 2, \ldots, n$$

are shown in that table for $n = 1$ to $n = 20$ and $p = 0.05$ to $p = 0.50$ in multiples of 0.05. Of course, Table A-2 is particularly useful when values of individual terms of the binomial distribution are desired rather than sums of terms.

In the case of the binomial distribution, as with any other mathematical model, the correspondence between the real-world situation and the model must be carefully established.

In many cases, the underlying assumptions of a Bernoulli process are not met. For example, suppose that in a production process, items produced by a certain machine tool are tested as to whether they meet specifications. If the items are tested in the order in which they are produced, then the assumption of independence would doubtless be violated. That is, whether an item meets specifications would not be independent of whether the preceding item(s) did. If the machine tool had become subject to wear, it is quite likely that if it produced an item that did not meet specifications, the next item would fail to conform to specifications in a similar way. Thus, whether or not an item is defective would *depend* on the characteristics of preceding items. In the coin-tossing illustration, on the other hand, we imagined an experiment in which a head or tail on a particular toss did not affect the outcome on the next toss.

We can see from the assumptions underlying a Bernoulli process that the following is true.

> The binomial distribution is applicable to situations of *sampling from a finite population with replacement or sampling from an infinite population* with or without replacement.

In either of these cases, the probability of success may be viewed as remaining constant from trial to trial. If the population size is large relative to sample

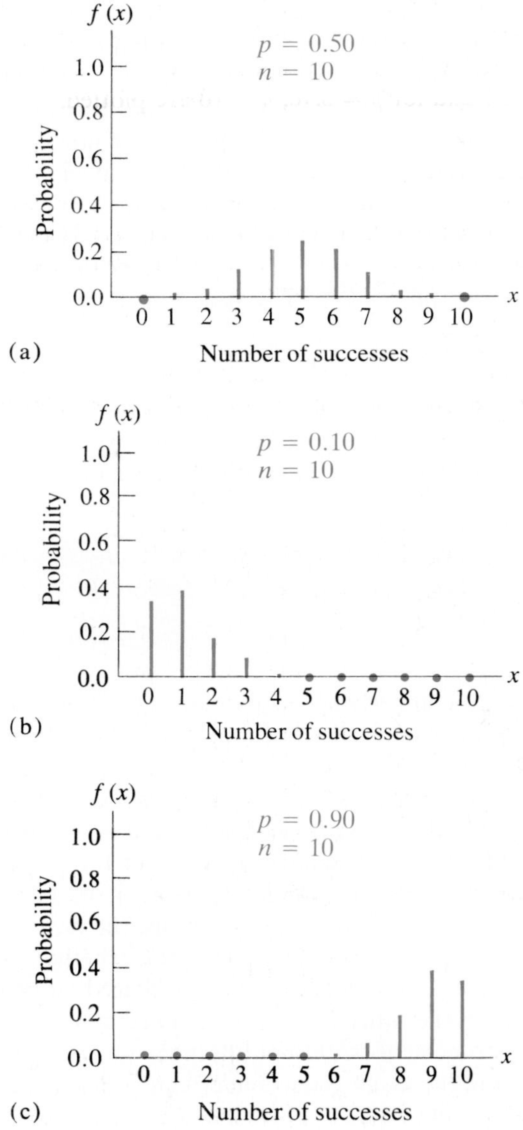

FIGURE 3-6
Graphs of binomial distributions. When $p = 0.50$, the distribution is symmetrical, as in (a); when $p \neq 0.50$, as shown in (b) and (c), the distribution is asymmetrical (skewed)

(a) $p = 0.50$, $n = 10$

(b) $p = 0.10$, $n = 10$

(c) $p = 0.90$, $n = 10$

size—that is, if the sample constitutes only a small fraction of the population—and if p is not very close in value to zero or to one, the binomial distribution is often sufficiently accurate, even though sampling may be carried out from a finite population without replacement. It is difficult to give universal rules of thumb on appropriate ratios of population size to sample size for this purpose. Some practitioners suggest a population size at least 10 times the sample size. However, the purpose of the calculations must determine the required degree of accuracy.

An important property of the binomial distribution is that when $p = 0.50$, the distribution is symmetrical. For example, see Figure 3.6(a), where $p = 0.50$ and $n = 10$. When $p \neq 0.50$, the distribution is asymmetrical (skewed). This property is illustrated in Figure 3.6(b) and (c), where the binomial distributions for $p = 0.10$, $n = 10$ and for $p = 0.90$, $n = 10$ are plotted.

EXAMPLE 3-6

The tossing of a fair coin 5 times was used earlier as an example of a Bernoulli process; the probability of obtaining 2 heads (successes) was calculated. Compute the probabilities of all possible numbers of heads and thus establish the particular binomial distribution that is appropriate in this case.

Solution

This problem is an application of the binomial distribution for $p = \frac{1}{2}$ and $n = 5$. Letting X represent the random variable "number of heads," the probability distribution is as follows:

x	$f(x)$
0	$\binom{5}{0}\left(\frac{1}{2}\right)^5\left(\frac{1}{2}\right)^0 = \frac{1}{32}$
1	$\binom{5}{1}\left(\frac{1}{2}\right)^4\left(\frac{1}{2}\right)^1 = \frac{5}{32}$
2	$\binom{5}{2}\left(\frac{1}{2}\right)^3\left(\frac{1}{2}\right)^2 = \frac{10}{32}$
3	$\binom{5}{3}\left(\frac{1}{2}\right)^2\left(\frac{1}{2}\right)^3 = \frac{10}{32}$
4	$\binom{5}{4}\left(\frac{1}{2}\right)^1\left(\frac{1}{2}\right)^4 = \frac{5}{32}$
5	$\binom{5}{5}\left(\frac{1}{2}\right)^0\left(\frac{1}{2}\right)^5 = \frac{1}{32}$
	1

EXAMPLE 3-7

Calculate the probability of obtaining at least one 6 in two rolls of a die (or in one roll of two dice) using the binomial distribution.

Solution

We view the two rolls of the die as Bernoulli trials. If we define the appearance of a 6 as a success, $p = \frac{1}{6}$, $q = \frac{5}{6}$, and $n = 2$. It is instructive to examine the entire probability distribution

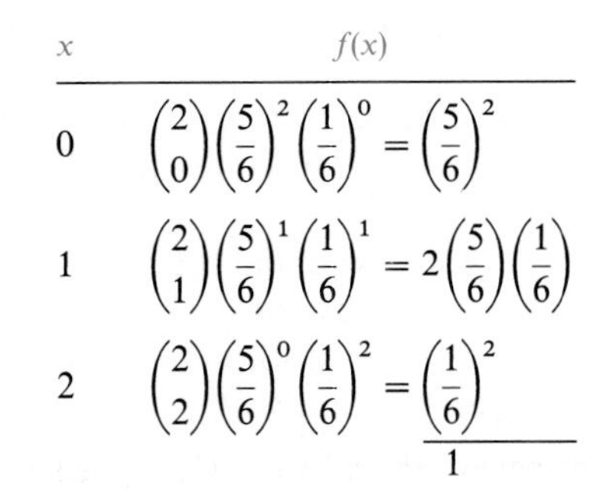

x	$f(x)$
0	$\binom{2}{0}\left(\frac{5}{6}\right)^2\left(\frac{1}{6}\right)^0 = \left(\frac{5}{6}\right)^2$
1	$\binom{2}{1}\left(\frac{5}{6}\right)^1\left(\frac{1}{6}\right)^1 = 2\left(\frac{5}{6}\right)\left(\frac{1}{6}\right)$
2	$\binom{2}{2}\left(\frac{5}{6}\right)^0\left(\frac{1}{6}\right)^2 = \left(\frac{1}{6}\right)^2$
	1

The expressions at the right side of the $f(x)$ column are in the form with which the student is probably most familiar for the terms in the expansion of $(\frac{5}{6} + \frac{1}{6})^2$.

The required probability is

$$P(\text{at least one 6}) = f(1) + f(2) = 2\left(\frac{5}{6}\right)\left(\frac{1}{6}\right) + \left(\frac{1}{6}\right)^2 = \frac{11}{36}$$

EXAMPLE 3-8

An interesting correspondence took place in 1693 between Samuel Pepys (author of the famous *Diary*) and Isaac Newton, in which Pepys posed a probability problem to the eminent mathematician. The question as originally stated by Pepys was:

> *A* has six dice in a box, with which he is to fling a 6.
> *B* has in another box twelve dice, with which he is to fling two 6s.
> *C* has in another box eighteen dice, with which he is to fling three 6s.
> (Question)—Whether *B* and *C* have not as easy a task as *A* at even luck?[7]

In rather flowery seventeenth century English, Newton replied and said, essentially, "Sam, I do not understand your question." Newton asked whether individuals *A*, *B*, and *C* were to throw independently and whether the question pertained to obtaining *exactly* one, two, or three 6s or *at least* one, two, or three 6s.

After an exchange of letters, in which Pepys supplied little help in answering these queries, Newton decided to frame the question himself. In modern language, Newton's wording would appear somewhat as follows:

> If *A*, *B*, and *C* toss dice independently, what are the probabilities that:
> *A* will obtain at least one 6 in a roll of six dice?
> *B* will obtain at least two 6s in a roll of twelve dice?
> *C* will obtain at least three 6s in a roll of eighteen dice?

Newton's reply to these questions involved some rather tortuous arithmetic. His work doubtless represented a respectable intellectual feat, considering the infantile state of probability theory at that time. Today, almost any beginning student of probability

[7] Schell, Emil D., "Samuel Pepys, Isaac Newton and Probability," *The American Statistician*, October 1960, pp. 27–30.

theory, standing on the shoulders of the giants who came before, would immediately see the application of the binomial distribution to the problem. Let us denote by $P(A)$, $P(B)$, and $P(C)$ the probabilities that A, B, and C would obtain the specified events. Then

$$P(A) = 1 - \binom{6}{0}\left(\frac{5}{6}\right)^{6}\left(\frac{1}{6}\right)^{0} \approx 0.67$$

$$P(B) = 1 - \binom{12}{0}\left(\frac{5}{6}\right)^{12}\left(\frac{1}{6}\right)^{0} - \binom{12}{1}\left(\frac{5}{6}\right)^{11}\left(\frac{1}{6}\right)^{1} \approx 0.62$$

$$P(C) = 1 - \binom{18}{0}\left(\frac{5}{6}\right)^{18}\left(\frac{1}{6}\right)^{0} - \binom{18}{1}\left(\frac{5}{6}\right)^{17}\left(\frac{1}{6}\right)^{1} - \binom{18}{2}\left(\frac{5}{6}\right)^{16}\left(\frac{1}{6}\right)^{2} \approx 0.60$$

Thus, $P(A) > P(B) > P(C)$.

Pepys admitted frankly that he did not understand Newton's calculations and furthermore that he did not believe the answer. He argued that since B throws twice as many dice as A, why can't he simply be considered two A's? Thus, he would have at least as great a probability of success as A. Of course, Pepys' question indicated that he was rather confused. There is no reason why the probability of at least two 6s in a roll of twelve dice should be twice the probability of at least one 6 in a roll of six dice, and as seen by the above calculations, indeed it is not.

EXAMPLE 3-9

The customer service manager of Courier Express is responsible for expediting late mail delivery. From past experience, she knows that prompt deliveries occur about 90% of the time. View this situation as a Bernoulli process and determine from Table A-1 of Appendix A the probabilities that in 10 deliveries

a. 3 or fewer deliveries will be late.
b. between 3 and 5 deliveries (inclusive) will be late.
c. 3 or more deliveries will be prompt.
d. at most, 8 deliveries will be prompt.
e. exactly 2 deliveries will be late.
f. 7 or more deliveries will be late.

Solution

Let $p = 0.10$ stand for the probability that a delivery will be late. (We define the probability of a success this way because Table A-1 gives p values only up to $p = 0.50$.) Therefore, $q = 0.90$ and $n = 10$. Let X represent the number of deliveries that do not arrive on time. Note that a failure to deliver mail on time is considered a "success" in this problem despite the undesirability of this outcome.

a. From Table A-1 of Appendix A, $F(3) = 0.9872$.

$$P(X \leqslant 3) = F(3) = \sum_{x=0}^{3}\binom{10}{x}(0.90)^{10-x}(0.10)^{x}$$

b. The probability of obtaining 3, 4, or 5 successes is the difference between "5 or fewer successes" and "2 or fewer successes." Thus,

$$P(3 \leqslant X \leqslant 5) = F(5) - F(2) = 0.9999 - 0.9298 = 0.0701$$

c. The event "3 or more failures" is the same as the event "7 or fewer successes." Hence,

$$P(X \leqslant 7) = F(7) = 1.0000$$

d. The event "at most, 8 failures" is the same as "8 or fewer failures" or "2 or more successes."

$$P(2 \leqslant X \leqslant 10) = F(10) - F(1) = 1.0000 - 0.7361 = 0.2639$$

e. The probability of "exactly 2 successes" is the difference between the probabilities "2 or fewer successes" and "one or fewer successes."

$$P(X = 2) = F(2) - F(1) = 0.9298 - 0.7361 = 0.1937$$

This probability may also be obtained directly by reference to Table A-2 where for $n = 10$, $x = 2$, we again find

$$f(x) = P(X = 2) = 0.1937$$

f. The event "7 or more successes" is the complement of the event "6 or fewer successes." Therefore,

$$P(X \geqslant 7) = 1 - P(X \leqslant 6) = 1 - F(6) = 1 - 1 = 0$$

Exercises 3.4

1. A certain type of plastic bag in the past has burst under a pressure of 15 pounds 10% of the time. If a prospective buyer tests 5 bags chosen at random under a pressure of 15 pounds, what is the probability that exactly one will burst?
2. A manufacturer of a certain type of film advertises that 95 out of 100 prints will develop. A person buys a roll of 10 prints and finds that 2 do not develop. If the manufacturer's claim is true, what is the probability that 2 or more prints will not develop?
3. Two employees of a company were arguing over whether a large barrel of articles had been produced by Machine 1. The first employee, R. C. Bradley, said she had taken a random sample of 20 articles from the barrel and found them all to be good. Since it is known that Machine 1 produces 10% defective, Bradley stated that she was almost certain that a random sample of 20 articles would contain at least one defective in view of the fact that a binomial distribution was applicable. Assuming the use of the binomial distribution, do you agree with Bradley? Use figures to support your position.
4. An oil exploration firm is formed with enough capital to finance 10 ventures. The probability of any exploration being successful is 0.1. What are the firm's chances of
   a. having exactly one successful exploration?
   b. having at least one successful exploration?
   c. going bankrupt in the 10 ventures?
5. A used car saleswoman claims that the odds are only 3:1 against her selling a car to any particular customer. If she attempts to sell automobiles

to 8 customers on a given day, what is the probability that she will make at least one sale?

6. The portfolio that an account executive has constructed for a discretionary account consists of 8 stocks. The broker feels that the probability that each stock will decrease in value over the next 6 months is 0.35 and that these changes in values are independent.
   a. What is the probability that exactly 4 will decrease in value?
   b. What is the probability that 4 or more will decrease in value?
   c. Does the assumption of independence here seem logical? If not, is the binomial distribution the appropriate distribution for this problem?

7. The director of placement at a liberal arts college is reviewing the file of a student who has just completed the process of interviewing for employment after graduation. Based on experience, the director estimates that an average candidate will get one job offer from 10 interviews, an above-average candidate will get one job offer from 5 interviews, and a below-average candidate will get one job offer from 20 interviews. The student under consideration received 3 offers from 12 interviews. What is the probability of this occurring if the student is
   a. an above-average candidate?
   b. an average candidate?
   c. a below-average candidate?

8. Graph the binomial distributions for $n = 11$ and $p = 0.2$, 0.5, and 0.8. What can be said about skewness of the binomial distribution as the value of p departs from 0.5?

9. Sales representative Maria Gonzales plans to visit 10 plant managers to try to sell an innovative product her company has developed. She estimates that her chance of landing a sale with any one plant manager is 0.6. Assuming independence, what is the probability that she will
   a. meet her quota of 7 sales?
   b. make no sales?
   c. make 10 sales?

10. A restaurant has determined that there is a probability of 20% that a customer will order Blidtz beer. If at a particular time there are 15 customers in the restaurant, what are the probabilities that
    a. at most, 2 customers will order Blidtz beer?
    b. at least one customer will order Blidtz beer?
    c. exactly one customer will order Blidtz beer?
    d. between 2 and 4 customers, inclusive, will order Blidtz beer?

11. A marketing vice-president for the Carlboro Cigarette Company is evaluating the impact of advertising the company's products in national news magazines. A study has shown that 20% of the readers who see the advertisement will be influenced to buy the products, 75% will have no reaction to the advertisement, and 5% will have a negative reaction to it. Assume independence.

a. If 20 readers of such a magazine see the advertisement, what is the probability that 5 of them will be influenced to buy the company's products?
b. What is the probability that 5 readers will have a negative reaction?

12. A corporate pension fund managed by the Innocuous Insurance Company holds positions in one bond in each of 15 municipalities. Due to the financial state of these municipalities, the manager has estimated the probability of the suspension of interest payments for each bond to be 0.25. Assuming independence, what is the probability of
    a. 3 municipalities suspending interest payments?
    b. fewer than 5 municipalities suspending interest payments?
    c. more than 10 municipalities suspending interest payments?
13. Ideal Corporation estimates that 3 out of 4 MBA graduates that join the firm become vice-presidents after at least 5 years. If the firm's claim is true, what is the probability that at most 13 out of 20 new MBA graduates that join the firm become vice-presidents after five years?

3.5 THE MULTINOMIAL DISTRIBUTION

In the case of the binomial distribution, there were two possible outcomes on each experimental trial. The **multinomial distribution** represents a straightforward generalization of the binomial distribution for the situation where there are more than two possible outcomes on each trial.

The assumptions underlying the multinomial distribution are completely analogous to those of the binomial distribution.

Assumptions

1. On each trial, there are k mutually exclusive possible outcomes, which may be referred to as $E_1, E_2, \ldots, E_k$. Therefore, the sample space of possible outcomes on each trial is $S = \{E_1, E_2, \ldots, E_k\}$.
2. The probabilities of outcomes $E_1, E_2, \ldots, E_k$, denoted $p_1, p_2, \ldots, p_k$, remain constant from trial to trial.
3. The trials are independent.

Under these assumptions, the probability that there will be x_1 occurrences of E_1, x_2 occurrences of $E_2, \ldots,$ and x_k occurrences of E_k in n trials is given by

(3.4) $$f(x_1, x_2, \ldots, x_k) = \frac{n!}{x_1!x_2! \cdots x_k!} p_1^{x_1} p_2^{x_2} \cdots p_k^{x_k}$$

where $x_1 + x_2 + \cdots + x_k = n$ and $p_1 + p_2 + \cdots + p_k = 1$.

The expression $f(x_1, x_2, \ldots, x_n)$ is the general term of the multinomial distribution

$$(p_1 + p_2 + \cdots + p_k)^n$$

Analogously, in the binomial distribution, the probability of x successes in n trials is given by

$$f(x) = \frac{n!}{x!(n-x)!} q^{n-x}p^x$$

which is the general term of $(q + p)^n$. In terminology similar to that of the multinomial distribution, $(q + p)^n$ may be written

$$(p_1 + p_2)^n$$

where $p_1 + p_2 = 1$.

EXAMPLE 3-10

The Bargain Center store is featuring a unique sale of ladies' stockings. For only 50¢, any customer may reach into a closed box containing a large number of pairs of stockings and remove a pair. The stockings have been classified into three boxes according to small, medium, or large sizes. Barbara Kane, a bargain hunter, draws 10 pairs from the "medium" box. Because of an error in size classification, about 15% of the pairs in the "medium" box are large, 5% are small, and the rest are medium sizes.

a. What is the probability that Ms. Kane will get exactly 5 pairs of medium size, one pair of large size, and 4 pairs of small size stockings?
b. What is the probability that she will get 7 medium size, 2 large size, and one small size?

Solution

a. Applying the multinomial distribution, we find the probability of exactly 5 pairs of medium size, one pair of large size, and 4 pairs of small size stockings is

$$f(5, 1, 4) = \frac{10!}{5!1!4!}(0.80)^5(0.15)(0.05)^4$$

$$= 0.0004$$

b. $f(7, 2, 1) = \frac{10!}{7!2!1!}(0.80)^7(0.15)^2(0.05)$

$= 0.0849$

EXAMPLE 3-11

The distribution of grades among MBAs in a well-known eastern school is as follows: 10% Distinguished (DS), 20% High Pass (HP), and 70% Pass (P). Suppose you select a sample of 3 student records at random from the records for the entire school. Give the probability distribution of the numbers of each grade level in the sample, assuming that the multinomial distribution is applicable.

Solution

If X_1, X_2, and X_3 stand for the numbers of DS, HP, and P grades, then the appropriate multinomial distribution is

$$f(x_1, x_2, x_3) = \frac{3!}{x_1!x_2!x_3!}(0.10)^{x_1}(0.20)^{x_2}(0.70)^{x_3}$$

$$x_1 + x_2 + x_3 = 3 \quad \text{and} \quad x_i = 0, 1, 2, 3$$

This probability distribution is given in Table 3-6.

TABLE 3-6
Multinomial distribution for numbers of DS, HP, and P grades: $p_1 = 0.10$, $p_2 = 0.20$, $p_3 = 0.70$, and $n = 3$

(x_1, x_2, x_3)	$f(x_1, x_2, x_3)$
3, 0, 0	0.001
0, 3, 0	0.008
0, 0, 3	0.343
2, 1, 0	0.006
2, 0, 1	0.021
1, 2, 0	0.012
1, 1, 1	0.084
1, 0, 2	0.147
0, 2, 1	0.084
0, 1, 2	0.294
	1.000

Exercises 3.5

1. From experience, a salesperson in a store that sells television sets has found that 45% of the people assigned to him buy a black and white TV set, 15% buy a color TV set, and the rest do not purchase. If 5 potential customers are assigned to the salesperson, what is the probability that he will sell two black and white TV sets and one color TV set?
2. A student who is taking 5 courses this semester estimates that in each course the chances of getting an A, B, and C are 0.1, 0.8, and 0.1 respectively. What is the probability of getting
   a. two A's and three B's?
   b. a B in every course?
3. A quality control system installed by Ace Computers, Inc., attempts to reduce loss due to production of defectives by categorizing components as good or defective and, if defective, identifying the degree of defectiveness. The classification resulted in the following probabilities:

Quality	Probability
Good	0.75
Defective, recheck	0.10
Defective, but repairable	0.03
Defective, but has salvage value	0.07
Defective, no value	0.05
	1.00

a. Assuming independence, if 8 parts are selected and tested on a given morning, what is the probability of obtaining 4 good components and one of each type of defective?
b. On a given morning, what is the probability that 4 out of the 8 tested items are good?
c. What is the name of the distribution used in part (a) and the distribution used in part (b)?

3.6 THE HYPERGEOMETRIC DISTRIBUTION

In section 3.4, the binomial distribution was discussed as the appropriate probability distribution for situations in which the assumptions of a Bernoulli process were met. A major application of the binomial distribution was the computation of probabilities for the sampling of finite populations *with replacement.* In most practical situations, sampling is carried out without replacement. In this section, we discuss the **hypergeometric distribution** as the appropriate model for sampling *without replacement.*

For example, if a sample of families is selected in a city in order to estimate the average income of all families in the city, sampling units are ordinarily not replaced prior to the selection of subsequent ones. That is, families are not replaced in the original population and thus given an opportunity to appear more than once in the sample. In fact, such samples are usually drawn in a single operation, without any possibility of drawing the same family twice. Moreover, in a sample drawn from a production process, articles are generally not replaced and given an opportunity to reappear in the sample. Thus, for both human populations and universes of physical objects, sampling is ordinarily carried out without replacement.

Suppose we have a list of 1,000 persons, 950 of whom are adults and 50 of whom are children. Numbers from 1 to 1,000 are assigned to these individuals. These numbers are printed on 1,000 identical discs, which are placed in a large bowl. We draw a sample of 5 chips at random from the bowl without replacement. These 5 chips may be drawn simultaneously or successively. For simplicity, we shall refer to this situation as the drawing of a random sample of 5 persons from the group of 1,000, although, of course, chips rather than persons are sampled. The process of numbering chips to correspond to persons and sampling the chips is simply a device to ensure randomness in sampling the population of 1,000 persons.

What is the probability that none of the 5 persons in the sample is a child? An alternative way of wording the question is, "What is the probability that all 5 persons in the sample are adults?" The sample space of possible outcomes in this experiment is the total number of samples of 5 persons that can be drawn from the population of 1,000 persons. This is the number of combinations that can be formed of 1,000 objects taken 5 at a time $\binom{1{,}000}{5}$. We can compute the required probability by obtaining the ratio of the number of sample points favorable to the event "none of the 5 persons is a child" to the total number of points in the sample space. The number of ways 5 adults can be drawn from the 950 adults is $\binom{950}{5}$. The number of ways no children can be selected from the 50 children is $\binom{50}{0}$. Therefore, the total number of ways of selecting 5 adults and no children from the population of 950 adults and 50 children is $\binom{950}{5}\binom{50}{0}$, by the multiplication principle.

Hence, the probability of obtaining no children (and 5 adults) in this sample of 5 persons is

$$\frac{\binom{950}{5}\binom{50}{0}}{\binom{1{,}000}{5}}$$

If we carry out the arithmetic, we see a very interesting fact.

$$\frac{\binom{950}{5}\binom{50}{0}}{\binom{1{,}000}{5}} = \frac{\dfrac{950!}{5!945!} \times \dfrac{50!}{0!50!}}{\dfrac{1{,}000!}{5!995!}} = \frac{950 \times 949 \times 948 \times 947 \times 946}{1{,}000 \times 999 \times 998 \times 997 \times 996} = 0.7734$$

Grouping the product obtained as a multiplication of 5 factors, we have

$$P(\text{no children}) = \left(\frac{950}{1{,}000}\right)\left(\frac{949}{999}\right)\left(\frac{948}{998}\right)\left(\frac{947}{997}\right)\left(\frac{946}{996}\right)$$

which is the result we would have arrived at if we had simply solved the original problem in terms of conditional probabilities. That is, the probability of obtaining an adult on the first draw is $\frac{950}{1{,}000}$; the probability of obtaining an adult on the second draw given that an adult was obtained on the first draw is $\frac{949}{999}$, and so on. Therefore, the joint probability of obtaining no children (5 adults) if the sampling is carried out without replacement is given by the multiplication of the 5 factors shown.

We can now state the general nature of this type of problem and the hypergeometric distribution as a solution to it. Suppose there is a population containing N elements, X of which are termed "successes," $N - X$ of which are denoted "failures." The corresponding terminology for a random sample of n elements drawn without replacement is that we require x successes and therefore

$n - x$ failures. The data of this general problem are tabulated below:

Population	Required Sample
X = Number of successes[8]	x = Number of successes
$N - X$ = Number of failures	$n - x$ = Number of failures
N = Total number in population	n = Total number in sample

The **hypergeometric distribution**, which gives the probability of x successes in a random sample of n elements drawn *without replacement*, is

$$f(x) = \frac{\binom{N-X}{n-x}\binom{X}{x}}{\binom{N}{n}} \quad \text{for } x = 0, 1, 2, \ldots, [n, X] \tag{3.5}$$

where the symbol $[n, X]$ means the smaller of n and X. For example, in the preceding illustration, if there had been only 10 children in the population (X) and the sample size had been 50 (n), the largest value that the number of children in the sample (x) could take on would be 10 (X). On the other hand, if X exceeded n, clearly x could be as large as n.

The hypergeometric distribution bears an interesting relationship to the binomial distribution.

Suppose, in the case of the population containing 950 adults and 50 children, we had been interested in the same probability (of obtaining no children in a random sample of 5 persons) but the sample was randomly drawn *with replacement*. Then, letting $q = 0.95$, $p = 0.05$, and $n = 5$ in the binomial distribution, we have

$$f(0) = \binom{5}{0}(0.95)^5(0.05)^0 = (0.95)^5 = 0.7738$$

Just as in the case of the hypergeometric distribution, where the required probability could have been computed by using the multiplication rule for *dependent* events, in the case of the binomial distribution, the probability could have been computed by simply using the multiplication rule for *independent* events

$$P(\text{no children}) = \left(\frac{950}{1{,}000}\right)\left(\frac{950}{1{,}000}\right)\left(\frac{950}{1{,}000}\right)\left(\frac{950}{1{,}000}\right)\left(\frac{950}{1{,}000}\right).$$

[8] Note that in order to maintain parallel notation for the population and sample in this case, the symbol X does *not* denote the *random variable* for number of successes in the sample, but is instead the total number of successes in the population being sampled.

Note that the hypergeometric and binomial probability values are extremely close in this illustration, agreeing exactly in the first three decimal places. We can show that when N increases without limit, the hypergeometric distribution approaches the binomial distribution.

The binomial probabilities may be used as approximations to hypergeometric probabilities when n/N is small.

A frequently used rule of thumb is that the population size should be at least 10 times the sample size ($N > 10n$) for the approximations to be used. However, the governing considerations, as usual, include such matters as the purpose of the calculations, whether a sum of terms rather than a single term is being approximated, and whether terms near the center or the extremes of the distribution are involved.

Just as the multinomial distribution represents the generalization of the binomial distribution when there are more than two possible classifications of outcomes, the hypergeometric distribution can be similarly extended.

No special name is given to the more general distribution; it also is referred to as the hypergeometric distribution. Assume a population that contains N elements, X_1 of type one, X_2 of type two, . . . , and X_k of type k. Suppose we require, in a sample of n elements drawn without replacement, that there be x_1 elements of type one, x_2 of type two, . . . , and x_k of type k. Tabulating the data in an analogous fashion to the two-outcome case, we have

Population (number of elements)	Required Sample (number of elements)
X_1 of type one	x_1 of type one
X_2 of type two	x_2 of type two
$\vdots$	$\vdots$
X_k of type k	x_k of type k

The **hypergeometric distribution**, which gives the probability of obtaining x_1 occurrences of type one, x_2 occurrences of type two, . . . , and x_k occurrences of type k in a random sample of n elements drawn without replacement, is

$$f(x_1, x_2, \ldots, x_k) = \frac{\binom{X_1}{x_1}\binom{X_2}{x_2}\cdots\binom{X_k}{x_k}}{\binom{N}{n}} \tag{3.6}$$

$$\text{for } x_i = 0, 1, 2, \ldots, [n, X_i]$$

where

$$\sum_{i=1}^{k} X_i = N \quad \text{and} \quad \sum_{i=1}^{k} x_i = n$$

EXAMPLE 3-12

The Humblest Oil Corporation has 100 service stations in a certain community; it has classified them according to merit of geographic location as follows:

Merit of Location	Number of Stations
Excellent	22
Good	38
Fair	27
Poor	10
Disastrous	3
	100

The corporation has a computer program for drawing random samples (without replacement) of its service stations. In a random sample of 20 of these stations, what is the joint probability of obtaining 6 excellent, 6 good, 4 fair, 3 poor, and one disastrous station?

Solution

$$f(6, 6, 4, 3, 1) = \frac{\binom{22}{6}\binom{38}{6}\binom{27}{4}\binom{10}{3}\binom{3}{1}}{\binom{100}{20}}$$

EXAMPLE 3-13

After taking a difficult exam, a certain student usually visits a bar, a restaurant, a theater, or a recreation center. In Center City, there are 30 bars, 50 restaurants, 10 theaters, and 10 recreation centers.

a. What is the probability that in the course of one semester, a student will have visited 6 bars, 10 restaurants, 2 theaters, and 2 recreation centers? Assume that the student does not go to the same place more than once, and assume that the visits are made randomly.
b. If the student visits the same place more than once, how would you modify your answer?

Solution

a. $$f(6, 10, 2, 2) = \frac{\binom{30}{6}\binom{50}{10}\binom{10}{2}\binom{10}{2}}{\binom{100}{20}}$$

b. $$f(6, 10, 2, 2) = \frac{20!}{6!10!2!2!}(0.30)^6(0.50)^{10}(0.10)^2(0.10)^2$$

EXAMPLE 3-14

What is the probability that a bridge hand, dealt from a well-shuffled deck of cards, will have either all hearts or all diamonds?

Solution

There are $\binom{52}{13}$ possible bridge hands. Using the hypergeometric distribution and the addition rule, the required probability is

$$\frac{\binom{13}{13}\binom{13}{0}\binom{13}{0}\binom{13}{0}}{\binom{52}{13}} + \frac{\binom{13}{0}\binom{13}{13}\binom{13}{0}\binom{13}{0}}{\binom{52}{13}} = \frac{1}{\binom{52}{13}} + \frac{1}{\binom{52}{13}} \approx \frac{2}{635 \text{ billion}}$$

If you are dealt such a hand, it is fair to say you have observed a rare event.

Exercises 3.6

1. Out of 200 workers, 160 belong to a union. If 4 workers are randomly selected, what is the probability that 2 of the workers belong to the union?
2. Plants A, B, and C produce identical electric fans. A number of fans from these plants were merged into a single lot. There were 5 fans from plant A, 4 fans from plant B, and 3 fans from plant C—for a total of 12 fans. If 3 fans were randomly selected from the lot, what is the probability that a fan from each plant appeared in the sample?
3. A hardware store has 10 accounts receivable with open balances. Of these accounts receivable, 6 have balances in excess of $1,000. An auditor selects 5 accounts receivable at random for audit. What is the probability that 3 of the accounts in the sample will have balances in excess of $1,000?
4. Six oil companies each send 4 executives to a conference on new developments in the energy field. Six representatives are chosen at random to lead discussion groups on various aspects of oil demand and supply in future years. What is the probability that
   a. exactly one representative from each company is a discussion leader?
   b. Company A has 4 discussion leaders?
   c. Company B has no discussion leaders?
5. In a hand (13 cards) in the game of bridge (52-card deck), what is the probability that you would have 2 diamonds, 3 hearts, 4 spades, and 4 clubs? Set up the mathematical expression. You need not evaluate it.
6. In a lot of 50 parts, 5 parts are defective. If 4 parts are randomly selected without replacement, what is the probability that
   a. exactly one part is defective?
   b. at most, 2 parts are defective?

3.7 THE POISSON DISTRIBUTION

Another useful probability function is the Poisson distribution, named for the Frenchman who developed it during the first half of the nineteenth century.[9] We will discuss the Poisson distribution first as a distribution in its own right, which is by far the most important use and which has many fruitful applications in a wide variety of fields. Then we will discuss the Poisson distribution as an approximation to the binomial distribution.

The Poisson Distribution Considered in Its Own Right

The Poisson distribution has been usefully employed to describe the probability functions of such phenomena as

- product demand
- demands for service
- numbers of telephone calls that come through a switchboard
- numbers of accidents
- numbers of traffic arrivals (such as trucks at terminals, airplanes at airports, ships at docks, and passenger cars at toll stations)
- numbers of defects observed in various types of lengths, surfaces, or objects.

All of the preceding illustrations have two elements in common.

Elements of a Poisson distribution

- The given occurrences can be described in terms of a discrete random variable, which takes on values 0, 1, 2, and so forth.
- There is some rate that characterizes the process producing the outcome. That **rate** is the number of occurrences per interval of *time* or *space*.

For example, product demand can be characterized by the number of units purchased in a specified period; the number of defects in a specified length of electrical cable can be counted. Product demand may be viewed as a process that produces random occurrences in continuous time; the observance of defects is a process that produces random occurrences in a continuum of space. In cases such as the defects example, the continuum may be one of *area* or *volume*

[9] Siméon Denis Poisson (1781–1840), was particularly noted for his applications of mathematics to the fields of electrostatics and magnetism. He wrote treatises in probability, calculus of variations, Fourier's series, and other areas.

as well as *length*. Thus, there may be a count of the number of blemishes in areas of sheetmetal used for aircraft or a count of the number of a certain type of microscopic particle in a unit of volume such as a cubic centimeter of a solution.

We can indicate the general nature of the process that produces a Poisson probability distribution by examining the occurrence of defects in a length of electrical cable. The length of cable has some rate of defects per interval: say, two defects per meter. If the entire length of cable is divided into subintervals of one millimeter each, then we might make the following assumptions:

Assumptions

1. The probability that exactly one defect occurs in each subinterval is a small number that is constant for each such subinterval.
2. The probability of two or more defects in a millimeter is so small that it may be considered to be zero.
3. The number of defects that occur in a millimeter does not depend on where that subinterval is located.
4. The number of defects that occur in a subinterval does not depend on the number of events in any other nonoverlapping subinterval.

Although the subinterval was a unit of *length* in the preceding example, analogous sets of assumptions would characterize examples in which the subinterval is a unit of *area*, *volume*, or *time*.

The Nature of the Poisson Distribution

As indicated previously,

> the Poisson distribution results from occurrences that can be described by a discrete random variable.

This random variable, denoted X, can take on values $x = 0, 1, 2, \ldots$ (where the three dots mean *ad infinitum*). That is, X can take on the values of all non-negative integers. The probability of exactly x occurrences in the Poisson distribution is

$$f(x) = \frac{\mu^x e^{-\mu}}{x!} \quad \text{for } x = 0, 1, 2, \ldots \tag{3.7}$$

where μ is the mean number of occurrences per interval and $e = 2.71828\ldots$ (the base of the Naperian or natural logarithm system).

We can see from equation 3.7 that the Poisson distribution has a single parameter symbolized by the Greek lower-case letter μ (mu). If we know the value of μ, we can write out the entire probability distribution.

The parameter μ can be interpreted as the average number of occurrences per interval of time or space that characterizes the process producing the Poisson distribution.

(The average referred to here is the arithmetic mean.) Thus, μ may represent an average of three units of demand per day, 5.3 demands for service per hour, 1.2 aircraft arrivals per five minutes, 1.5 defects per 10 feet of electrical cable, and so on.

In order to illustrate how probabilities are calculated in the Poisson distribution, we consider the following example. A study revealed that the number of telephone calls per minute coming through a certain switchboard between 10:00 A.M. and 11:00 A.M. on business days is distributed according to the Poisson probability function with an average μ of 0.4 calls per minute. What is the probability distribution of the number of telephone calls per minute during the specified time period?

Let X represent the random variable "number of telephone calls per minute" during the given time period. Then $\mu = 0.4$ calls per minute is the parameter of the Poisson probability distribution of this random variable. The probability that no calls will occur (come through the switchboard) in a given minute is obtained by substituting $x = 0$ in the Poisson probability function, equation 3.7. Hence,

$$P(X = 0) = f(0) = \frac{(0.4)^0 e^{-0.4}}{0!} \tag{3.8}$$

Since $(0.4)^0 = 1$ and $0! = 1$, equation 3.8 becomes simply

$$f(0) = e^{-0.4} = 0.670 \tag{3.9}$$

The value 0.670 for $f(0)$ can be found in Table A-10 of Appendix A, where exponential functions of the form e^x and e^{-x} are tabulated for values of x from 0.00 to 6.00 at intervals of 0.10.

Continuing with the calculation of the Poisson probability distribution, we find the probability of exactly one call in a given minute by substituting $x = 1$ in equation 3.7. Hence, $f(1)$ is given by

$$P(X = 1) = f(1) = \frac{(0.4)^1 e^{-0.4}}{1!} = (0.4)(0.670) = 0.268 \tag{3.10}$$

To find the other values of $f(x)$, we can use Table A-3 of Appendix A, which lists values of the cumulative distribution function for the Poisson distribution. That is, values of

$$F(c) = P(X \leqslant c) = \sum_{x=0}^{c} f(x)$$

or the probabilities of c or fewer occurrences, are provided for selected values of the parameter μ. As in Table A-1 for the binomial cumulative distribution, probabilities such as $1 - F(c)$ or $a \leqslant f(x) \leqslant b$ can be obtained by appropriate manipulation of the tabulated values.

TABLE 3-7
Poisson probability distribution of the number of telephone calls per minute coming through a certain switchboard between 10:00 A.M. and 11:00 A.M. on business days

Number of Calls x	Probability $f(x)$
0	0.670
1	0.268
2	0.054
3	0.007
4	0.001
	1.000

The use of Table A-3 will be illustrated in terms of our phone call example. To obtain the probability of no calls in a given minute, using $c = 0$ and $\mu = 0.4$ in Table A-3, we find the value of $F(0) = 0.670$. Of course, this is also the value of $f(0)$, since the probability of zero or fewer occurrences equals the probability of zero occurrences. Therefore, as before, $f(0) = 0.670$.

We find the probability of exactly one telephone call per minute by subtracting the probability of no calls from the probability of one or fewer calls, that is,

$$f(1) = F(1) - F(0) = 0.938 - 0.670 = 0.268$$

Similarly, using Table A-3, we find the values of $f(2)$, $f(3)$, and $f(4)$:

$$f(2) = F(2) - F(1) = 0.992 - 0.938 = 0.054$$

$$f(3) = F(3) - F(2) = 0.999 - 0.992 = 0.007$$

$$f(4) = F(4) - F(3) = 1.000 - 0.999 = 0.001$$

Although, as indicated earlier, the random variable X in the Poisson distribution takes on the values $0, 1, 2, \ldots,$ $F(4) = 1.00$ in this problem. This means that the probabilities of $5, 6, \ldots,$ occurrences are so small that they would appear as zero when rounded to three decimal places.

The required probability distribution for this problem is given in Table 3-7. Several other illustrations of the use of the Poisson distribution are given in Examples 3-15, 3-16, and 3-17.

EXAMPLE 3-15

On weekdays at a certain small airport, airplanes arrive at an average rate of three for the one-hour period 1:00 P.M. to 2:00 P.M. If these arrivals are distributed according to the Poisson probability distribution, what are the probabilities that

a. exactly zero airplanes will arrive between 1:00 P.M. and 2:00 P.M. next Monday?
b. either one or two airplanes will arrive between 1:00 P.M. and 2:00 P.M. next Monday?
c. a total of exactly two airplanes will arrive between 1:00 P.M. and 2:00 P.M. during the next three weekdays?

Solution

In this problem, we may use the parameter $\mu = 3$ arrivals per day for the period 1:00 P.M. to 2:00 P.M. Let X represent the random variable "number of arrivals during the specified time period." The mathematical solutions are given for parts (a), (b), and (c) to illustrate the theory involved. However, the answers may also be determined by looking up values in Table A-3 of Appendix A as indicated.

a. The random variable X follows the Poisson distribution with the parameter $\mu = 3$. Thus,

$$P(X = 0) = f(0) = \frac{3^0 e^{-3}}{0!} = 0.050$$

This value may be obtained from Table A-3 of Appendix A for $\mu = 3$, $c = 0$. We note that $f(0) = F(0)$.

b. Since exactly one arrival and exactly two arrivals are mutually exclusive events, we have, by the addition rule,

$$P(X = 1 \text{ or } X = 2) = f(1) + f(2) = \frac{3^1 e^{-3}}{1!} + \frac{3^2 e^{-3}}{2!} = 0.373$$

This value can be obtained from Table A-3 of Appendix A for $\mu = 3$. The required probability is $F(2) - F(0) = 0.423 - 0.050 = 0.373$.

c. A total of exactly two arrivals in three weekdays during the period 1:00 P.M.–2:00 P.M. can be obtained, for example, by having two arrivals on the first day, none on the second day, and none on the third day during the specified one-hour period. The total number of ways in which the event in question can occur is shown in Table 3-8.

Let P_2 represent the required probability. Using the multiplication and addition rules, and again using the parameter $\mu = 3$ arrivals per day during the period 1:00 P.M.–2:00 P.M., we have

$$\begin{aligned} P_2 &= 3[f(2)][f(0)]^2 + 3[f(1)]^2[f(0)] \\ &= 3\left(\frac{3^2 e^{-3}}{2!}\right)\left(\frac{3^0 e^{-3}}{0!}\right)^2 + 3\left(\frac{3^1 e^{-3}}{1!}\right)^2\left(\frac{3^0 e^{-3}}{0!}\right) \\ &= \frac{81}{2} e^{-9} = 0.005 \end{aligned}$$

The solution is greatly simplified if we change the time interval for which the parameter μ is stated. This has the effect of changing the random variable in the problem.

TABLE 3-8
Possible ways of obtaining a total of exactly 2 arrivals in 3 weekdays

Number of Arrivals		
Day 1	Day 2	Day 3
2	0	0
0	2	0
0	0	2
1	1	0
1	0	1
0	1	1

Thus, if μ = three arrivals *per day* during the period 1:00 P.M.–2:00 P.M., then μ = nine arrivals *per three days* during the same time period. The probability of exactly two arrivals in three weekdays during the given one-hour period can then be obtained by computing $P(X = 2)$, where X is a Poisson-distributed random variable denoting the number of arrivals *per three days*. The required probability is, therefore, obtained by simply computing $f(2)$ in a Poisson distribution with the parameter $\mu = 9$.

$$P_2 = f(2) = \frac{9^2 e^{-9}}{2!} = \frac{81}{2} e^{-9} = 0.005$$

This value can be obtained from Table A-3 of Appendix A for $\mu = 9$. The probability is given by $F(2) - F(1) = 0.006 - 0.001 = 0.005$.

This problem illustrates the point that considerable simplification of computations for Poisson processes can often be accomplished by convenient choice of parameters.

Appropriateness

We should note a few points concerning the appropriateness of the Poisson distribution in Example 3-15. We state at the beginning of the problem that the airplane arrivals were distributed according to the Poisson distribution. Whether it is appropriate to consider the past arrival distribution as a Poisson distribution during the specified time periods depends on the nature of the past data. Actual relative arrival frequencies can be tabulated and compared with the theoretical probabilities given by a Poisson distribution. Tests of "goodness of fit" for judging the closeness of actual and theoretical frequencies are discussed in Chapter 8.

In practice, the question often arises whether a given mathematical model is applicable in a certain situation. This requires careful examination of whether the underlying assumptions of the model are likely to be fulfilled by the real-world phenomena. For example, suppose certain cargo deliveries are made either on Mondays or Tuesdays between 1:00 A.M. and 2:00 P.M. Assuming that if a delivery is made on Monday, it will not be made on Tuesday, the fourth assumption of a Poisson process (given on page 124) is clearly violated. That is, the number of arrivals during the one-hour period on Tuesday *depends* on the number of arrivals during the corresponding period on Monday, and vice versa. Furthermore, if the nature of the aircraft arrivals is such that Monday and Tuesday always have more arrivals between 1:00 P.M. and 2:00 P.M. than do other weekdays, then the third assumption is violated. That is, if we were to count arrivals for the one-hour period for a given day (or two days, and so forth), then the number of occurrences obtained would depend on the day on which the count was begun.

Determining appropriateness of a distribution

The assumptions of a probability distribution rarely are met perfectly by a real-world process. Actual comparison of *the data generated by a process* with *the probabilities of the theoretical distribution* is the best way of determining the appropriateness of the distribution.

Of course, even if a given mathematical model (or other type of model) has provided a good description of past data, there is no guarantee that this state of affairs will continue. The analyst must be alert to changes in the environment that would make the model inapplicable. Experience in a given field aids considerably in judging whether so great a departure from assumptions has occurred that a model may no longer be applicable.[10]

[10] An appropriate thought here is perhaps contained in the anonymous bit of advice. "Good judgment comes from experience, and experience comes from poor judgment."

EXAMPLE 3-16

The Tai Ping Oriental Rug factory operates 10 looms for the manufacture of carpets. The oldest loom has been particularly troublesome, having to be stopped an average of 3 times per hour to repair broken strands of yarn. Every time the loom is stopped, the downtime is 5 minutes. Production efficiency has declined as a result of this problem. To solve the problem, production manager Max Baker is considering a capital budget allocation for a new loom. He decides that if the probability of the machine being down 15 minutes or more in an hour becomes greater than 0.50, he will make the capital budget request.

a. Determine the probability distribution of the number of repairs to be done on the old loom in one hour.
b. Should Baker request an allocation for a new loom?

Solution

Let X represent the random variable "number of repairs to be done per hour." Then, assuming that X is distributed according to the Poisson distribution, $\mu = 3$ repairs per hour.

a. The following table gives the probability distribution of X:

x	$f(x)$
0	0.050
1	0.149
2	0.224
3	0.224
4	0.168
5	0.101
6	0.050
7	0.022
8	0.008
9	0.003
10	0.001

The probabilities can be obtained from Table A-3 of Appendix A using the relationship $f(x) = F(x) - F(x - 1)$.

b. The probability that the machine will be down 15 minutes or more is the probability that X is equal to 3 repairs or more. Thus, from Table A-3 of Appendix A, we have

$$P(X \geqslant 3) = 1 - F(2) = 1 - 0.423 = 0.577$$

Therefore, Baker should request a new loom.

EXAMPLE 3-17

In connection with an auditing investigation, a certified public accountant discovers that the number of entries made in each of 6 accounts receivable is distributed according to the Poisson probability distribution with a parameter of $\mu = 2$ per day. Entries in the accounts may be assumed to be independent. What is the probability that on a specified day

a. none of the 6 accounts will receive any entries?
b. each of the 6 accounts will receive at least one entry?
c. exactly 3 accounts will receive no entries?

Solution

This problem illustrates a situation in which two different probability distributions must be used to provide a solution. In this case, the Poisson and binomial distributions are applicable.

a. Since the number of entries in a given account is Poisson distributed with an average of two entries per day ($\mu = 2$), the probability that a given account will receive no entries on a specified day is

$$P(\text{no entry in a given account}) = f(0) = \frac{2^0 e^{-2}}{0!} = e^{-2} = 0.135$$

Entries in different accounts are independent events. Therefore, by the multiplication rule we have

$$P(\text{no entry in all accounts}) = (e^{-2})^6 = e^{-12} \approx 0$$

b. The event that a given account will receive at least one entry is the complement of the event that the account receives no entry. Therefore,

$$P(\text{at least one entry in a given account}) = 1 - e^2$$

By the multiplication rule for independent events,

$$P(\text{at least one entry in each of the accounts}) = (1 - e^{-2})^6 \approx 0.419$$

c. Let p be the probability that a given account receives no entries on a specified day. From part (a), $p = e^{-2}$. If we think of p as the probability of success in a Bernoulli trial, then in this problem $q = 1 - e^{-2}$ and $n = 3$.

$$P(\text{no entries in exactly three accounts}) = \binom{6}{3}(1 - e^{-2})^3(e^{-2})^3 \approx 0.0318$$

The Poisson Distribution as an Approximation to the Binomial Distribution

The foregoing discussion concerned the use of the Poisson probability function as a distribution in its own right. We turn now to a consideration of the Poisson distribution as an approximation to the binomial distribution.

We saw that the Bernoulli process gives rise to a two-parameter probability function, the binomial distribution. Since computations involving the binomial distribution become quite tedious when n is large, it is useful to have a simple method of approximation.

The Poisson distribution is particularly suitable as an approximation to the binomial distribution when n is large and p is small.

Assume in the expression for $f(x)$ of the binomial distribution that, as n is permitted to increase without bound, p approaches zero in such a way that np remains constant. Let us denote this constant value for np as μ (which denotes the mean number of successes in n trials). Under these assumptions, it can be shown that the binomial expression for $f(x)$ approaches the value

$$f(x) = \frac{\mu^x e^{-\mu}}{x!}$$

where $\mu = np$ and e is the base of the natural logarithm system. Thus, we can see from equation 3.7 that the value approached by the binomial distribution under the given conditions is the value of the Poisson distribution. Hence, the Poisson distribution can be used as an approximation to the binomial probability function. In this context, the Poisson distribution is similar to the binomial distribution, because it gives the probability of observing x successes in n trials of an experiment, where p is the probability of success on a single trial. That is x, n, and p are interpreted in the same way as in the binomial distribution.

Because of the assumptions underlying the derivation of the Poisson distribution from the binomial distribution, the approximations to binomial probabilities are best when n is large and p is small.

A frequently used rule of thumb is that the Poisson approximation to the binomial distribution is appropriate when $p \leqslant 0.05$ and $n \geqslant 20$.

However, the Poisson distribution sometimes provides surprisingly close approximations even in cases when n is not large nor p very small. As an illustration of how these approximations may be carried out, we return to the problem of sampling the population of 950 adults and 50 children. The probability of observing no children in a random sample of five persons drawn with replacement was previously computed from the binomial distribution. We now compute the same probability using the Poisson distribution. Since n is only 5 in this problem, this is not an ideal situation for the use of the Poisson distribution for approximating binomial probabilities. Rather, it is an example of the surprisingly small errors observed in certain cases, even though n is small, and it is used here simply to carry out the arithmetic for a familiar illustration.

The binomial parameters in this problem were $p = 0.05$ and $n = 5$. Therefore,

$$\mu = np = 5 \times 0.05 = 0.25$$

Thus, in the Poisson distribution, the probability of no successes (children) is

$$f(0) = \frac{(0.25)^0 e^{-0.25}}{0!} = e^{-0.25}$$

From Table A-3 of Appendix A with $c = 0$ and $\mu = 0.25$, we find the value of $F(0)$, which in this case is equal to $f(0)$ (since the probability of zero or fewer successes equals the probability of zero successes). Therefore,

$$f(0) = 0.779$$

This figure is the same in the first two decimal places as the corresponding number (0.7738) obtained from the binomial probability (see section 3.6).

However, the percentage errors would be much larger for the other terms of the binomial distribution, representing probabilities of one, two, or more, successes. It is recommended, therefore, that the Poisson approximations not be used unless the conditions for n and p meet the limits outlined in the rule of thumb.

The parameter $\mu = np$ can be interpreted as the average number of successes per sample of size n.

Since p is the probability of success per trial and n is the number of trials, multiplication of n by p gives the average number of successes per n trials. In terms of the foregoing problem, we can interpret μ as a long-run relative frequency. The proportion of children in the population is $p = 0.05$, and a random sample of $n = 5$ persons was drawn with replacement from this population. Suppose samples of size $n = 5$ were repeatedly drawn with replacement from the same population and the number of children was recorded for each sample. It can be proven mathematically, and it seems intuitively reasonable, that the *average proportion* of children per sample of 5 persons is equal to $p = 0.05$. Furthermore, it follows that the *average number* of children per sample of 5 persons is equal to $np = 5(0.05) = 0.25$ children. The average referred to here is the arithmetic mean, obtained by totaling the proportions or numbers of children for all samples and dividing by the number of samples.

Example 3-18 represents a more justifiable use of the Poisson approximation to binomial probabilities than the preceding illustration, which involved a small sample size.

EXAMPLE 3-18

An oil exploration firm is formed with enough capital to finance 20 ventures. The probability of any exploration being successful is 0.10. What are the firm's chances of

a. exactly one successful exploration?
b. at least one successful exploration?
c. two or fewer successful explorations?
d. three or more successful explorations?

Assume that the population of possible explorations is sufficiently large to warrant binomial probability calculations.

Solution

a. Let $p = 0.10$ stand for the probability that an exploration will be successful and X for the number of successful explorations in 20 trials. Using the binomial distribution with parameters $p = 0.10$ and $n = 20$, we find that the probability of exactly one successful exploration is

$$P(X = 1) = f(1) = \binom{20}{1}(0.9)^{19}(0.1)^1$$

This probability may be determined from Table A-1 of Appendix A as

$$P(X = 1) = F(1) - F(0) = 0.3917 - 0.1216 = 0.2701$$

An approximation to this probability is given by the Poisson distribution with parameter

$$\mu = np = 20(0.10) = 2$$

The Poisson probability of exactly one success is

$$f(1) = \frac{2^1 e^{-2}}{1!}$$

which may be determined from Table A-3 of Appendix A as

$$P(X = 1) = F(1) - F(0) = 0.406 - 0.135 = 0.271$$

Thus, the percentage error is about 1 in 270, or about 0.4%.

Using Tables A-1 and A-3 of Appendix A for parts (b), (c), and (d), we have the following:

	Binomial	Poisson
b.	$P(X \geqslant 1) = 1 - F(0)$ $= 1 - 0.1216 = 0.8784$	$P(X \geqslant 1) = 1 - F(0)$ $= 1 - 0.135 = 0.865$
c.	$P(X \leqslant 2) = F(2) = 0.6769$	$P(X \leqslant 2) = F(2) = 0.677$
d.	$P(X \geqslant 3) = 1 - F(2)$ $= 1 - 0.6769 = 0.3231$	$P(X \geqslant 3) = 1 - F(2)$ $= 1 - 0.677 = 0.323$

Exercises 3.7

1. A department store has determined in connection with its inventory control system that the demand for a certain brand of portable radio was Poisson-distributed with a parameter $\mu = 4$ per day.

a. Determine the probability distribution of the daily demand for this item.
b. If the store stocks 5 of these items on a particular day, what is the probability that the demand will be greater than supply?

2. The telephone sales department of a certain department store receives an average of 24 calls per hour. It was found that the number of telephone calls followed the Poisson distribution. What is the probability that, between 10:00 A.M. and 10:05 A.M., there will be
   a. 3 calls?
   b. no calls?

3. The accident rate in a factory is 4 per month. It is believed that the accident rate is distributed according to the Poisson probability function. What is the probability that there are 6 accidents in a particular month at this factory?

4. On average, two customers per five minutes arrive at a gasoline station. What are the probabilities that
   a. more than one customer arrives in 5 minutes?
   b. there are more than 4 customers in 10 minutes?

5. There are 10% defectives in a batch of electrical resistors. If 10 resistors are randomly sampled, what is the probability that there will be two defectives?
   a. Use the binomial distribution.
   b. Use the Poisson distribution as an approximation to the binomial distribution.

6. A commercial bank has estimated an average of 60 customers per hour between 10:00 A.M. and 11:00 A.M. on Monday. What are the probabilities that, in one minute at that particular time and day, there are
   a. 2 customers?
   b. at most 2 customers?

7. Ships arrive at a West Coast unloading facility according to a Poisson distribution with an average rate of 4 per week.
   a. What is the probability that there will be exactly 5 arrivals in a given one-week period?
   b. What is the probability that there will be exactly 10 arrivals in a 2-week period?
   c. Explain why it is less probable to have 10 arrivals in 2 weeks than 5 arrivals in one week.

8. The Science Computer Consortium uses 20 computers to generate scientific information. The probability that one computer will not work on a given day is 0.05. Assume that the binomial distribution is the appropriate model.
   a. What is the probability that all computers will be working on a given day?
   b. What is the probability that at least 2 computers will not be working on a given day?

c. Now use the Poisson approximation to solve parts (a) and (b). Compare your results.

9. In a standard bolt of cloth produced by Berksquire Mills the average number of thread defects is 6. What is the probability that
   a. a bolt will have 8 or more thread defects?
   b. a bolt will have no thread defects?
   c. 5 bolts selected at random will all have 8 or more thread defects?

3.8 SUMMARY MEASURES FOR PROBABILITY DISTRIBUTIONS

In Chapter 1 we discussed summary, or descriptive, measures for empirical frequency distributions. We now turn to the corresponding measures for theoretical frequency distributions, that is, for probability distributions. These **summary measures for probability distributions** are essential components of modern quantitative techniques employed as aids for decision making under conditions of uncertainty.

Earlier in this chapter, we considered certain probability distributions for discrete random variables as appropriate mathematical models for real-world situations under specific sets of assumptions. Sometimes, from the nature of a problem, it is relatively easy to specify a suitable probability model. In other situations, the appropriate model is suggested only after substantial numbers of observations have been taken and empirical frequency distributions have been constructed. Whatever the method by which we arrive at probability distributions, we must be able to capture their salient properties in a few summary measures. These measures are the subject of the next two sections.

3.9 EXPECTED VALUE OF A RANDOM VARIABLE

Suppose the following game of chance were proposed to you: A fair coin is tossed. If it lands "heads," you win $10; if it lands "tails," you lose $5. What is the average amount that you would win per toss?

On any particular toss, you will either win $10 or lose $5. However, let us think in terms of a repeated experiment in which we toss the coin and play the game many times. Since the probability assigned to the event "heads" is $\frac{1}{2}$ and to the event "tails" $\frac{1}{2}$, in the long run, you would win $10 on half of the tosses and lose $5 on half. Therefore, the average (arithmetic mean) winnings per toss would be obtained by weighting both of the outcomes—$10 and −$5—by $\frac{1}{2}$ to yield a weighted mean of $2.50 per toss. In terms of equation 1.4, the weighted mean is

$$\bar{X}_w = \frac{\sum wX}{\sum w} = \frac{\$10(\frac{1}{2}) + (-\$5)(\frac{1}{2})}{\frac{1}{2} + \frac{1}{2}} = \$2.50 \text{ per toss}$$

Of course, when the weights are probabilities in a probability distribution, as in this case, the formula can be written without showing the division by the sum of the weights.

The average of $2.50 per toss is referred to as the **expected value** of the winnings. Note that on a single toss, only two outcomes are possible, namely, win $10 or lose $5. If these two possible outcomes are viewed as the possible values of a random variable, which occur with probabilities of $\frac{1}{2}$ each, then the expected value of the random variable is the mean of its probability distribution. More formally, if X is a discrete random variable that takes on the value x with probability $f(x)$, then the expected value of X, denoted $E(X)$, is

$$E(X) = \sum_x xf(x) \tag{3.11}$$

The expected value of a discrete random variable is obtained by multiplying (1) each value that the random variable can assume by (2) the probability of occurrence of that value; then all of these products are totaled.

Since the expected value is used frequently as a measure of central tendency for probability distributions, $E(X)$ is given as a simpler symbol, μ.

Hence, equation 3.11 can be rewritten as

$$\mu = E(X) = \sum_x xf(x)$$

TABLE 3-9
Calculation of the expected value for the coin-tossing problem

Value of Winnings x	Probability $f(x)$	Weighted Winnings $xf(x)$
$10	$\frac{1}{2}$	$5.00
−5	$\frac{1}{2}$	−2.50
	1	$2.50

$$\mu = E(X) = \sum_x xf(x) = \$2.50$$

The calculation of the expected value for the problem of tossing a coin is shown in Table 3-9. This calculation is similar to that for the mean of an empirical frequency distribution.

The expected value has a wide variety of applications in situations involving uncertain outcomes. Example 3-19 gives a simple illustration, in which the uncertainty is summarized in terms of a probability distribution of death for a particular type of insurance policyholder. Such probability distributions are ordinarily based on a large sample of observed experience, that is, on past relative frequencies of mortality. On the other hand, business problems often involve "one-time" decisions; with such problems, uncertainties can be summarized in terms of relevant subjective probability distributions. Example 3-20 illustrates such a situation.

EXAMPLE 3-19

An insurance company offers a 45-year-old man a $1,000 one-year term insurance policy for an annual premium of $12. Assume that the number of deaths per 1,000 is 5 for persons in this age group. What is the expected gain for the insurance company on a policy of this type?

Solution

We may think of this problem as representing a chance situation in which there are two possible outcomes, (1) the policy purchaser lives or (2) he dies during the year. Let X be a random variable denoting the dollar gain to the insurance company for these two outcomes. The probability that the man will live through the year is 0.995. In this case, the insurance company collects the premium of $12. The probability that the policy purchaser will die during the year is 0.005. In this case, the company has collected a premium of $12 but must pay the claim of $1,000, for a net gain of −$988. Thus, X takes on the values $12 and −$988 with respective probabilities 0.995 and 0.005. The calculation of expected gain for the insurance company is displayed in Table 3-10.

A couple of points may be noted. First, by insuring only one person, the company would not realize the expected gain but would have either the gain of $12 or the loss of $988. Hence, in order to realize expected gains, insurance companies "play the averages"

TABLE 3-10
Calculation of expected gain for an insurance company on a one-year term policy

Outcome	x	$f(x)$	$xf(x)$
Policyholder lives	$ 12	0.995	$11.94
Policyholder dies	−988	0.005	− 4.94
		1.000	$ 7.00

$$E(X) = \sum_x xf(x) = \$7.00$$

by insuring large numbers of individuals. Second, in setting a premium for this policy, the insurance company would take into account the usual expenses of doing business as well as the expected gain calculation.

EXAMPLE 3-20

A major retailing chain is considering bidding for space in a new suburban shopping mall that is still in the design stage. The chain has two basic alternatives. For each of these alternatives, it has carried out an analysis in which various net profit figures and subjective probabilities for the realization of these returns have been determined.

- Alternative A involves investing in a food outlet, a specialty shop group (cheese shop and delicatessen), and a garden and patio center. Net profits for this alternative are estimated as \$150,000, \$200,000, and \$300,000, with respective probabilities 0.4, 0.2, and 0.4.
- For alternative B, net profits from investing in a drugstore are estimated as \$120,000 and \$225,000, with a 50-50 chance of success.

Assuming that the net profit figures take into account the required capital investments in each project, which alternative is preferable from the standpoint of expected monetary return?

Solution

$$E(A) = (\$150{,}000)(0.4) + (\$200{,}000)(0.2) + (\$300{,}000)(0.4) = \$220{,}000$$

$$E(B) = (\$120{,}000)(0.5) + (\$225{,}000)(0.5) = \$172{,}500$$

Alternative A is estimated to yield higher expected net profit.

3.10 VARIANCE OF A RANDOM VARIABLE

As we have seen, the expected value of a random variable is analogous to the arithmetic mean of a frequency distribution of data. Similarly, the variability of a random variable may be measured in the same general way as the variability of a frequency distribution. In section 1.15 the variability of the observations in a population of data was referred to as the *variance*; the square root of the variance was called the *standard deviation*. The same terms are used for random variables. The **variance of a random variable** is the average (expected value) of the squared deviations from the expected value. The **standard deviation of a random variable** is the square root of the variance. The variance of a random variable may be written as $E(X - \mu)^2$ and may be calculated for discrete random variables by the following expression:

(3.12)
$$\sigma^2 = \sum_x (x - \mu)^2 f(x)$$

A mathematically equivalent formula that is easier for purposes of calculation is

$$\sigma^2 = E(X^2) - [E(X)]^2 \tag{3.13}$$

where

$$E(X^2) = \sum_x x^2 f(x)$$

and

$$E(X) = \sum_x x f(x)$$

In Examples 3-21 and 3-22, we illustrate the calculation of the variance and standard deviation for a few random variables. The calculation of the expected value as an intermediate step is also shown in Example 3-22.

EXAMPLE 3-21

Compute the mean and variance for the total obtained on the uppermost faces in a roll of two unbiased dice.

Solution

Let X denote the specified total on the two dice. Then

$$\begin{aligned}
\mu &= \sum_{x=2}^{12} x f(x) \\
&= 2(\tfrac{1}{36}) + 3(\tfrac{2}{36}) + 4(\tfrac{3}{36}) + 5(\tfrac{4}{36}) + 6(\tfrac{5}{36}) + 7(\tfrac{6}{36}) + 8(\tfrac{5}{36}) + 9(\tfrac{4}{36}) + 10(\tfrac{3}{36}) + 11(\tfrac{2}{36}) + 12(\tfrac{1}{36}) \\
&= 7
\end{aligned}$$

$$\begin{aligned}
\sigma^2 &= \sum_{x=2}^{12} x^2 f(x) - \mu^2 \\
&= 4(\tfrac{1}{36}) + 9(\tfrac{2}{36}) + 16(\tfrac{3}{36}) + 25(\tfrac{4}{36}) + 36(\tfrac{5}{36}) \\
&\quad + 49(\tfrac{6}{36}) + 64(\tfrac{5}{36}) + 81(\tfrac{4}{36}) + 100(\tfrac{3}{36}) + 121(\tfrac{2}{36}) + 144(\tfrac{1}{36}) - (7)^2 \\
&= 54\tfrac{5}{6} - 49 = 5\tfrac{5}{6}
\end{aligned}$$

EXAMPLE 3-22

A buyer for a large department store wants to project the required inventory for snow skis. Experience has shown that demand for skis has approximately the probability distribution shown in the first two columns of Table 3-11. Calculate the expected value and standard deviation of the number of snow skis sold.

Solution

The calculation of the expected value and standard deviation from equation 3.12 is given in Table 3-11. Also shown is the alternative calculation for the standard deviation, following from equation 3.13.

TABLE 3-11
Expected value and standard deviation of the number of snow skis sold

Pairs Sold x	Probability $f(x)$	$xf(x)$	$x - \mu$	$(x - \mu)^2$	$(x - \mu)^2 f(x)$
40	0.20	8	−29	841	168.20
60	0.30	18	−9	81	24.30
80	0.25	20	11	121	30.25
90	0.20	18	21	441	88.20
100	0.05	5	31	961	48.05
	1.00	69			359.00

$$\mu = E(X) = \sum_x xf(x) = 69 \text{ pairs}$$

$$\sigma^2 = E(X - \mu)^2 = \sum_x (x - \mu)^2 f(x) = 359$$

$$\sigma = \sqrt{359} = 18.95 \text{ pairs}$$

Alternative calculation of the standard deviation

$$\begin{aligned} E(X^2) &= \sum_x x^2 f(x) \\ &= 1{,}600(0.20) + 3{,}600(0.30) + 6{,}400(0.25) + 8{,}100(0.20) + 10{,}000(0.05) \\ &= 5{,}120 \end{aligned}$$

$$\sigma^2 = E(X^2) - [E(X)]^2 = 5{,}120 - (69)^2 = 359$$

$$\sigma = \sqrt{359} = 18.95 \text{ pairs}$$

EXAMPLE 3-23

Ventures Limited, Inc., is considering a proposal to develop a new calculator. The initial cash outlay would be $1 million, and development time would be one year. If successful, the firm anticipates that revenues over the 5-year life cycle of the product will be $1.5 million. If moderately successful, revenues will reach $1.2 million. If unsuccessful, the firm anticipates zero cash inflows. The firm assigns the following probabilities to the 5-year prospects for this product: successful, 0.60; moderately successful, 0.30; and unsuccessful, 0.10. What are the expected net profit and standard deviation of revenues? We will ignore the time value of money in this calculation.

Solution

Let X denote profit for the 5-year period, equal to total revenues for the 5-year period less development costs.

$$\begin{aligned} E(X) &= \$500{,}000(0.6) + (\$200{,}000)(0.3) + (-\$1{,}000{,}000)(0.1) \\ &= \$260{,}000 \end{aligned}$$

$$\begin{aligned} \sigma^2 &= (\$500{,}000 - \$260{,}000)^2(0.6) + (\$200{,}000 - \$260{,}000)^2(0.3) \\ &\quad + (-\$1{,}000{,}000 - \$260{,}000)^2(0.1) \\ &= 19.44 \times 10^{10} \end{aligned}$$

$$\sigma = \sqrt{19.44 \times 10^{10}} = \$440{,}908$$

3.11 EXPECTED VALUE AND VARIANCE OF SUMS OF RANDOM VARIABLES

We now discuss some of the important properties of expected values and variances of random variables. It can be shown that

$$E(X_1 + X_2 + \cdots + X_N) = E(X_1) + E(X_2) + \cdots + E(X_N) \quad (3.14)$$

The expected value of a sum of N random variables is equal to the sum of the expected values of these random variables.

A somewhat analogous relationship holds for variances of *independent* random variables. If $X_1, X_2, \ldots, X_N$ are N *independent* random variables, then

$$\sigma^2(X_1 + X_2 + \cdots + X_N) = \sigma^2(X_1) + \sigma^2(X_2) + \cdots + \sigma^2(X_N) \quad (3.15)$$

The variance of the sum of N *independent* random variables is equal to the sum of their variances.

Note that there is no restriction of independence in the case of equation 3.14. That is, equation 3.14 holds whether or not the variables are independent. Other properties of expected values and variances are summarized in Appendix C.

Example 3-24 (in the following section) illustrates the properties of expected values and variances discussed in this section.

3.12 TCHEBYCHEFF'S INEQUALITY

In this chapter we have discussed probability distributions of discrete random variables. In Chapter 5, we discuss probability distributions of continuous random variables, focusing on the normal distribution, probably the best-known and most widely applied probability function. In some situations, the underlying processes that produce data may not be well understood. Therefore, the appropriate form of probability distribution may be unknown. Tchebycheff's* inequality is useful in such situations because it sets bounds on probability statements regardless of the form of the underlying probability distribution and because it holds true for both discrete and continuous variables.

* The spelling of Tchebycheff varies considerably, depending on the way that the Russian version is transliterated. The following alternative spellings may be encountered in other sources: Chebycheff, Chebyshev, Tchebychev, Tchebycheff, and so forth.

Tchebycheff's inequality

If μ and σ are the mean and standard deviation of a probability distribution, then for any $k > 1$, at least $1 - (1/k^2)$ of the distribution is included within k standard deviations from the mean, that is, within the interval $\mu \pm k\sigma$.

Some examples of this distribution statement follow.

$$\text{If} \quad k = 2, \qquad \text{then} \quad 1 - \frac{1}{k^2} = 1 - \frac{1}{2^2} = \frac{3}{4}$$

$$\text{If} \quad k = 3, \qquad \text{then} \quad 1 - \frac{1}{k^2} = 1 - \frac{1}{3^2} = \frac{8}{9}$$

Therefore, we can make the following three statements.

- Regardless of the form of the probability distribution

  the range of $\mu \pm 2\sigma$ includes at least $\frac{3}{4}$ of the total probability

  the range of $\mu \pm 3\sigma$ includes at least $\frac{8}{9}$ of the total probability

- Tchebycheff's inequality is applicable to empirical data as well as to probability distributions. That is, if $\bar{x}$ and s are the mean and standard deviation of a set of observations, then we can say the following

  the range of $\bar{x} \pm 2s$ includes at least $\frac{3}{4}$ of the observations

  the range of $\bar{x} \pm 3s$ includes at least $\frac{8}{9}$ of the observations

- When we *know* the form of a probability distribution, we can make more specific statements about the distribution. For example, as we will see in Chapter 5, in a normal distribution with mean μ and standard deviation σ, we can state that:

  the range of $\mu \pm 2\sigma$ includes 95.5% of the total probability

  the range of $\mu \pm 3\sigma$ includes 99.7% of the total probability

However, Tchebycheff's inequality is a powerful tool because it enables us to make general statements about probability distributions and sets of data when we do not know the form of the distributions that are applicable to the random variables or data involved.

An example of the use of Tchebycheff's inequality is given in Example 3-24.

EXAMPLE 3-24

A client asks Joan Watson, an associate consultant at Consumer Pulse, Inc., to prepare a market segmentation study for a new product. Watson has scheduled the following activities to complete this project: (1) presentation of research design, (2) training of personnel, and (3) collating the data. The times required to complete these activities are independent of one another. Based on her experience with similar projects in the past, she estimates the means (expected times) and standard deviations of the completion times in weeks for each activity as follows:

Activity	Expected Time (in weeks)	Standard Deviation (in weeks)
1. Presentation of research	10	4
2. Training of personnel	5	1
3. Collating of data	6	3

Estimate the expected value and standard deviation of completion time for the entire project. Using Tchebycheff's inequality, give two probability statements concerning completion times for the entire project.

Solution

Let μ_1, μ_2, and μ_3 denote the expected values and let σ_1, σ_2, and σ_3 denote the standard deviations of the completion times of the three activities. Then the expected value and standard deviation of completion time for the entire project are

$$\mu = \mu_1 + \mu_2 + \mu_3 = 10 + 5 + 6 = 21 \quad \text{weeks}$$

$$\sigma = \sqrt{\sigma_1^2 + \sigma_2^2 + \sigma_3^2} = \sqrt{4^2 + 1^2 + 3^2} = 5.1 \quad \text{weeks}$$

The probability is at least 75% that the completion time of the entire project will lie in the range of

$$\begin{aligned} \mu \pm 2\sigma &= 21 \pm 2(5.1) \quad \text{weeks} \\ &= 10.8 \text{ to } 31.2 \quad \text{weeks} \end{aligned}$$

The probability is at least 89% that the completion time of the entire project will lie in the range of

$$\begin{aligned} \mu \pm 3\sigma &= 21 \pm 3(5.1) \quad \text{weeks} \\ &= 5.9 \text{ to } 36.1 \quad \text{weeks} \end{aligned}$$

Exercises 3.12

1. Let $X = -\$0.25$ if a customer does not purchase from a telephone sales promotion and $X = +\$10.89$ if a customer does purchase from a telephone sales promotion. If the probability that a customer will purchase is 10%, what is the expected value of X? Interpret the result.
2. The demand for a certain product is a discrete random variable and is uniformly distributed in the range of 10 to 20 items per day. State the probability distribution of this random variable and calculate the expected demand per day.
3. In the Fair Gain Bazaar, a booth offers you the chance to roll a fair die. You will receive or have to pay an amount in dollars equal to the uppermost face on a roll of the die. If a 2, 3, or 5 appears on the uppermost face, you lose. But if the die shows a 1, 4, or 6, you win. How much should you be willing to pay to participate in this game?

4. In investigating a company's financial reports, you find that the company operated at a $1 million loss for 3 years, $1 million profit for 5 years, and $5 million profit for 2 years. Calculate the expected profit and the standard deviation of the company's earnings.
5. You are considering buying one of two stocks, Volatile or Stable, both now priced at $46, for a one-month trading venture. The probability distribution for the closing prices of the two stocks (rounded to nearest $1) one month hence is estimated and tabulated.
   a. Find the expected values and risk of each stock. (Financial analysts often refer to variance as "risk.")
   b. Which stock would you purchase and why?

Volatile		Stable	
Price	f(Price)	Price	f(Price)
44	0.1	44	0.005
45	0.1	45	0.015
46	0.1	46	0.030
47	0.1	47	0.100
48	0.1	48	0.350
49	0.1	49	0.350
50	0.1	50	0.100
51	0.1	51	0.030
52	0.1	52	0.015
53	0.1	53	0.005

6. The probability distribution of the numbers of a weekly magazine in inventory is listed.

Number of Magazines in Inventory	Probability
0	0.03
1	0.06
2	0.07
3	0.11
4	0.17
5	0.14
6	0.14
7	0.10
8	0.09
9	0.07
10	0.02

a. Find the expected value and standard deviation of the number of magazines in inventory.
b. Using Tchebycheff's inequality, make two probability statements concerning the number of magazines in inventory.

7. As a marketing researcher of a toy manufacturing firm, you must decide on which of two new games to introduce this year. The probability distributions for the net profits of each product are listed. Find the expected net profit and standard deviation of net profit for each game.

Game A		Game B	
Net Profit (in thousands of dollars)	f(Net Profit)	Net Profit (in thousands of dollars)	f(Net Profit)
−5,000	0.1	0	0.2
0	0.2	+1,000	0.3
+3,000	0.4	+3,000	0.3
+6,000	0.3	+5,000	0.2

8. Suppose an insurance company offers a particular homeowner a $15,000 one-year term fire insurance policy on his home. Assume that the number of fires per 1,000 dwellings for this particular home safety classification is two. For simplicity, assume that any fire will cause damage of at least $15,000.
   a. If the insurance company charges a premium of $36 a year, what is its expected gain from this policy?
   b. Does this mean that the company will earn that many dollars on *this particular policy*? Explain.

9. A local retail clothing company has 10 stores. The net profit per week from each store has an expected value of $900 and a variance of $2,500. What is the expected total net profit per week and standard deviation for all 10 stores? Assume independence.

10. A life insurance sales office has 20 sales representatives. The arithmetic mean commission for these sales representatives is $260 with a standard deviation of $100. Assume independence.
    a. What are the mean and standard deviation for the *total commissions* paid in this sales office?
    b. If every sales representative is given a flat $50-per-week raise, what would be the new mean and standard deviation of weekly commissions for these sales representatives?
    c. If the commission rate is increased by 10%, what would be the new mean and standard deviation of weekly commissions for these sales representatives? (**Hint**: See rules 3 and 7 of Appendix C.)

3.13

JOINT PROBABILITY DISTRIBUTIONS

The discussion to this point has been concerned with probability distributions of discrete random variables considered one at a time. These are often referred to as *univariate probability distributions*, or *univariate probability mass functions.* In such distributions, probabilities are assigned to events pertaining to a single random variable.

In most realistic decision-making situations, however, more than one factor at a time must be taken into account. Frequently, the joint effects of several variables, some or all of which are interdependent, must be analyzed in terms of their impact on some objective the decision maker wishes to achieve.

In this section, we consider the joint probability distributions of discrete random variables, or, stated differently, probability distributions of two or more discrete random variables. Such functions are frequently referred to as **multivariate probability distributions**, the term **bivariate probability distribution** being used for the two-variable case.

We return to Table 2-2, the joint probability table for 1,000 persons classified by sex and product preference, for a simple example of a bivariate probability distribution. Table 2-2 is reproduced here for convenience as Table 3-12.

We now introduce some symbolism for the discussion of joint bivariate probability distributions. Instead of using the terminology of events, as we did earlier in discussing Table 2-2, we now use the language of random variables. Let X represent the random variable "product preference" and Y the random variable "sex". We let X and Y take on the following values:

$$X = 1 \text{ for "prefers product ABC"}$$
$$X = 2 \text{ for "prefers product XYZ"}$$

$$Y = 1 \text{ for "male"}$$
$$Y = 2 \text{ for "female"}$$

TABLE 3-12
Joint probability distribution for 1,000 persons classified by sex and product preference

	Product Preference		
Sex	Prefers Product ABC	Prefers Product XYZ	Marginal Probabilities
Male	0.20	0.30	0.50
Female	0.10	0.40	0.50
Marginal probabilities	0.30	0.70	1.00

TABLE 3-13
Bivariate probability distribution corresponding to Table 3-12

	x		
y	1	2	$P(Y = y)$
1	0.20	0.30	0.50
2	0.10	0.40	0.50
$P(X = x)$	0.30	0.70	1.00

Then $P(X = x \text{ and } Y = y)$ denotes the joint probability that X takes on the value x and Y takes on the value y.[11] For example, $P(X = 1 \text{ and } Y = 1)$ is the joint probability $P(\text{prefers product ABC and male}) = 0.20$, and so on. The bivariate probability distribution of X and Y is given in Table 3-13.

Marginal Probability Distributions

The values in the cells of Table 3-13 are the joint probabilities of the respective outcomes denoted by the column and row headings for X and Y. Also displayed in the table are the separate univariate probability distributions of X and Y. Earlier, we referred to the probabilities in the margins of the table as marginal probabilities. These probabilities form **marginal probability distributions**. The marginal distribution of X consists of the values of X shown in the column headings and the column totals at the bottom; the marginal distribution of Y consists of the values of Y shown in the row headings and the row totals at the right side of the table. These distributions are shown in Table 3-14. As indicated, the symbols $P(X = x)$ and $P(Y = y)$ are used to denote the probabilities for the marginal probability distributions of X and Y.

[11] The notation $P(X = x \text{ and } Y = y)$ should really read $P\{(X = x) \text{ and } (Y = y)\}$. The simplified symbolism is in common use and will be employed in this book.

TABLE 3-14
Marginal probability distributions of X and Y

x	Product Preference $P(X = x)$	y	Sex $P(Y = y)$
1	0.30	1	0.50
2	0.70	2	0.50
	1.00		1.00

Graph of Bivariate Probability Distribution

The probability distribution of a single discrete random variable is graphed by displaying the values of the random variable along the horizontal axis and the corresponding probabilities along the vertical axis. In the case of a bivariate distribution, two axes are required for the values of the random variables and a third for the measurement of probabilities. Usually, the joint values of the two variables are depicted on a plane (the x-y plane) and the associated probabilities are read along an axis perpendicular to the plane. A graph of the joint probability distribution of Table 3-13 is shown in Figure 3-7.

Conditional Probability Distributions

Another important type of probability distribution obtainable from a joint probability distribution is the **conditional probability distribution**. Using equations 2.4 and 2.5, we may calculate the conditional probability that X takes on a particular value (say, $X = 2$) given that Y takes on a particular value (say, $Y = 1$) by dividing the joint probability that $X = 2$ and $Y = 1$ by the marginal probability that $Y = 1$. For example, the conditional probability P(prefers

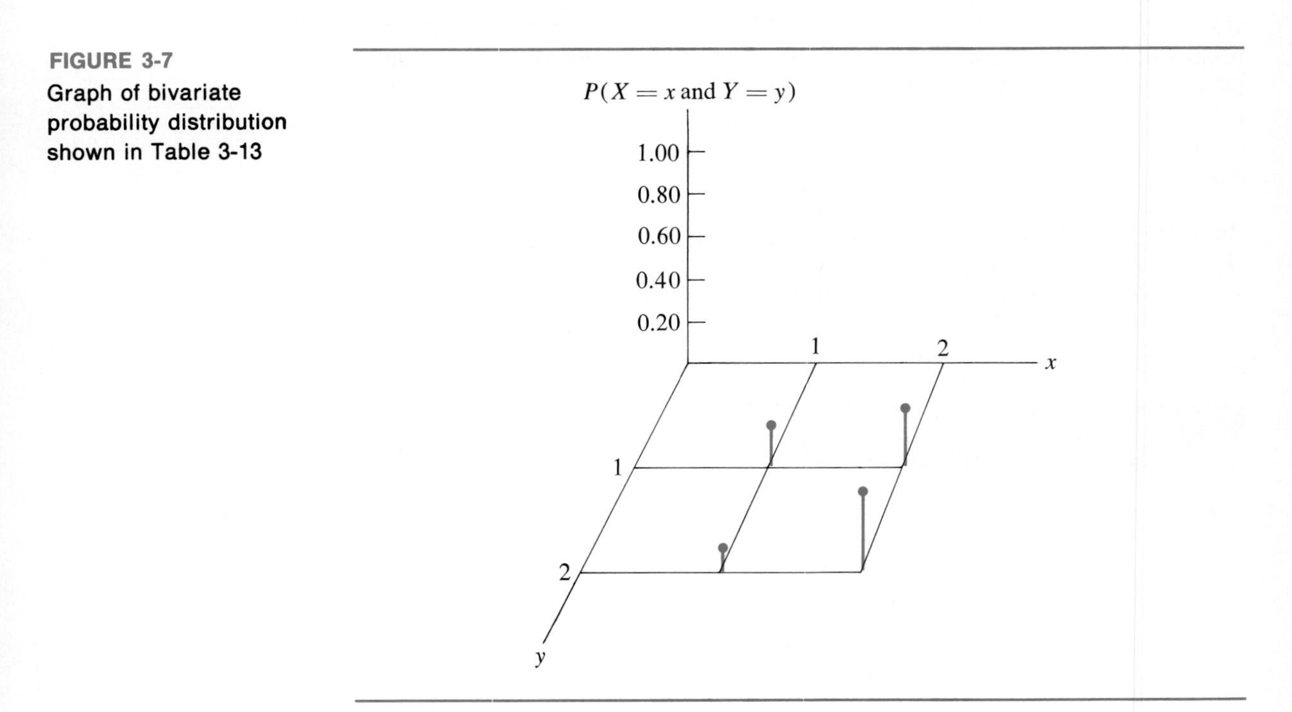

FIGURE 3-7
Graph of bivariate probability distribution shown in Table 3-13

TABLE 3-15
Conditional probability distribution for product preference (X) given that the individual is male ($Y = 1$)

x	$P(X = x \mid Y = 1)$
1	$\frac{0.20}{0.50} = 0.40$
2	$\frac{0.30}{0.50} = 0.60$
	1.00

product XYZ|male) = $P(X = 2 \mid Y = 1)$ is computed as follows:

$$P(X = 2 \mid Y = 1) = \frac{P(X = 2 \text{ and } Y = 1)}{P(Y = 1)} = \frac{0.30}{0.50} = 0.60$$

We can also find the conditional probability $P(X = 1 \mid Y = 1)$, thus forming a conditional probability distribution for X given $Y = 1$. This distribution is shown in Table 3-15.

Similarly, we can obtain the conditional probability distribution for X given $Y = 2$ by dividing the joint probabilities in the row $Y = 2$ by the row total, $P(Y = 2) = 0.50$.

Corresponding calculations give us the conditional probability distributions for Y given particular values of X. For example, the conditional probability distribution of sex (Y) given that the individual prefers product XYZ ($X = 2$) is shown in Table 3-16.

Independence

Returning to the terminology of events, we saw in section 2.2 that if two events A_1 and B_1 are statistically independent, $P(B_1 \mid A_1) = P(B_1)$ and $P(A_1 \mid B_1) = P(A_1)$. Using random variable notation, the analogous statement is that if X

TABLE 3-16
Conditional probability distribution for sex (Y) given that the individual prefers product XYZ ($X = 2$)

y	$P(Y = y \mid X = 2)$
1	$\frac{0.30}{0.70} = 0.43$
2	$\frac{0.40}{0.70} = 0.57$
	1.00

and Y are two *independent* random variables, then

$$P(X = x|Y = y) = P(X = x)$$

and

$$P(Y = y|X = x) = P(Y = y)$$

for all pairs of outcomes (x, y). For example, the first of these statements means that the conditional probability that X takes on a particular value x, given that Y takes on a particular value y, is equal to the marginal, or unconditional, probability that X takes on the particular value x.

Again returning to the terminology of events, we observed in section 2.2 that by the multiplication rule, if two events A_1 and B_1 are statistically independent, $P(A_1 \text{ and } B_1) = P(A_1)P(B_1)$. The analogous statement for random variables follows.

If X and Y are two independent random variables, then

$$P(X = x \text{ and } Y = y) = P(X = x)P(Y = y)$$

for all pairs of outcomes (x, y).

We illustrate this definition of independence by returning to Table 3-13. Suppose we consider the outcome pair (1, 1), that is, $X = 1$ and $Y = 1$. In this case,

$$P(X = 1|Y = 1) = \frac{0.20}{0.50} = 0.40$$

and

$$P(X = 1) = 0.30$$

Since $P(X = 1|Y = 1)$ is not equal to $P(X = 1)$, X and Y are not independent random variables. Alternatively, we observe

$$P(X = 1 \text{ and } Y = 1) = 0.20$$

$$P(X = 1) = 0.30$$

$$P(Y = 1) = 0.50$$

We note that

$$P(X = 1 \text{ and } Y = 1) = 0.20 \quad \neq \quad P(X = 1)P(Y = 1) = (0.30)(0.50) = 0.15$$

Again, we conclude that X and Y are not independent random variables. Our reasoning is based on the fact that for X and Y to be independent random variables, the independence conditions stated earlier must hold for *all* pairs of outcomes (x, y). Since the conditions do not hold for the pair (1, 1), that is, for $(X = 1, Y = 1)$, then X and Y are not independent.

If we retain the same marginal probability distributions shown in Table 3-13, what would the values of the joint probabilities have been if product preference (X) and sex (Y) had been *independent* random variables? To answer this question, we consider the marginal probability distributions in Table 3-13:

	x		
y	1	2	$P(Y = y)$
1			0.50
2			0.50
$P(X = x)$	0.30	0.70	1.00

We compute the required joint probabilities as follows:

$$P(X = 1 \text{ and } Y = 1) = P(X = 1)P(Y = 1) = (0.30)(0.50) = 0.15$$

$$P(X = 2 \text{ and } Y = 1) = P(X = 2)P(Y = 1) = (0.70)(0.50) = 0.35$$

and so on.

In Table 3-17, we show the resulting joint probability distribution if product preference and sex had been independent random variables. We now observe that $P(X = x|Y = y) = P(X = x)$ and $P(Y = y|X = x) = P(Y = y)$ for all pairs of outcomes (x, y). For example, for the outcome (1, 1), $P(X = 1|Y = 1) = \frac{0.15}{0.50} = 0.30 = P(X = 1)$. Also, we observe that $P(X = x \text{ and } Y = y) = P(X = x)P(Y = y)$ for all pairs of outcomes (x, y). For example, for the outcome (1, 1), $P(X = 1 \text{ and } Y = 1) = 0.15 = P(X = 1)P(Y = 1) = (0.30)(0.50) = 0.15$.

Note in Table 3-17 that 30% ($\frac{0.15}{0.50} = 0.30$) of the males prefer product ABC and 30% of the females prefer product ABC. Correspondingly, 70% of the males prefer product XYZ, and the same percentage applies for females as well. Hence, we can say that product preference is *independent* of sex.

TABLE 3-17
Joint probability distribution for 2 independent random variables X and Y

	x		
y	1	2	$P(Y = y)$
1	0.15	0.35	0.50
2	0.15	0.35	0.50
$P(X = x)$	0.30	0.70	1.00

TABLE 3-18
Sample of consumers in a certain community classified by income class and soft drink preference

Soft Drink Preference	Income Class			
	Upper	Middle	Lower	All Levels
Cola	650	830	820	2,300
Lime	300	260	290	850
Root beer	110	70	70	250
Total sample	1,060	1,160	1,180	3,400

EXAMPLE 3-25

Table 3-18 represents the result of a sample survey of consumers' preferences for soft drinks listed by income class in a certain community.

a. What is the probability that a person drawn at random from this sample belongs to the upper income class and is a cola drinker?
b. What is the probability that a person drawn at random from this sample belongs to the upper income class?
c. What is the probability of a person drawn at random from this sample being a cola drinker, given that the individual belongs to the upper income class?
d. What are the marginal probability distributions of income class and soft drink preference?
e. What is the conditional probability distribution of income class given that an individual in this sample is a cola drinker?
f. Is there evidence of dependence between soft drink preference and income class in this sample?
g. If the same marginal totals are retained, what would the numbers of persons in each cell of the table have to be for independence to exist between income class and soft drink preference?

TABLE 3-19
Relative frequencies of occurrence for a sample of consumers of a certain community classified by income class and soft drink preference

Soft Drink Preference	Income Class			
	Upper	Middle	Lower	All Levels
Cola	0.191	0.244	0.241	0.676
Lime	0.089	0.076	0.085	0.250
Root beer	0.032	0.021	0.021	0.074
Total sample	0.312	0.341	0.347	1.000

Solution

In order to answer these questions, we convert the table of numbers of occurrences (absolute frequencies) to probabilities (relative frequencies), by dividing each number in Table 3-18 by 3,400, the total number of persons in the sample. These relative frequencies are given in Table 3-19.

Let X and Y be random variables representing income class and soft drink preference, respectively, and taking on values as follows:

Income Class	x	Soft Drink Preference	y
Upper	0	Cola	0
Middle	1	Lime	1
Lower	2	Root beer	2

The joint probability distribution now appears in Table 3-20. The answers to the questions may now be given.

a. P(upper income class and cola drinker) $= P(X = 0 \text{ and } Y = 0) = 0.191$

b. P(upper income class) $= P(X = 0) = 0.312$

c. $P(\text{cola drinker} \mid \text{upper income class}) = P(Y = 0 \mid X = 0) = \dfrac{0.191}{0.312} = 0.612$

d. The marginal probability distributions given in the margins of the probability table are

x	$P(X = x)$	y	$P(Y = y)$
0	0.312	0	0.676
1	0.341	1	0.250
2	0.347	2	0.074
	1.000		1.000

TABLE 3-20
Joint probability distribution derived from Table 3-18 and 3-19

y \ x	0	1	2	$P(Y = y)$
0	0.191	0.244	0.241	0.676
1	0.089	0.076	0.085	0.250
2	0.032	0.021	0.021	0.074
$P(X = x)$	0.312	0.341	0.347	1.000

e. This conditional probability distribution is given by

$$P(X = x \mid Y = 0) = \frac{P(X = x \text{ and } Y = 0)}{P(Y = 0)} \qquad \text{where} \quad x = 0, 1, 2$$

Thus, each of the joint probabilities in the first row of Table 3-19 or 3-20 is divided by the marginal probability, or total, of that row. That is, the joint probabilities of being a cola drinker and in a specified income class are divided by the marginal probability of being a cola drinker. Therefore the required conditional probability distribution is

x	$P(x = x \mid Y = 0)$
0	$0.191 \mid 0.676 = 0.283$
1	$0.244 \mid 0.676 = 0.361$
2	$0.241 \mid 0.676 = 0.356$
	1.000

f. Income class and soft drink preference are *not independent* random variables. That is, it is not true that $P(X = x \text{ and } Y = y) = P(X = x)P(Y = y)$ for *all* values of x and y. For example,

$$P(X = 0 \text{ and } Y = 0) \neq P(X = 0)P(Y = 0)$$

Numerically,

$$0.191 \neq (0.312)(0.676)$$

Another way of indicating that income class and soft drink preference are not independent random variables is to note that the marginal distributions are not equal to the corresponding conditional distributions. Thus, for income class, the conditional and marginal probability distributions are

x	$P(X = x \mid Y = 0)$	$P(X = x \mid Y = 1)$	$P(X = x \mid Y = 2)$	$P(X = x)$
0	0.283	0.356	0.432	0.312
1	0.361	0.304	0.284	0.341
2	0.356	0.340	0.284	0.347
	1.000	1.000	1.000	1.000

g. As indicated in part (f), if income class and soft drink preference were independent, the marginal probability distributions would be equal to the corresponding conditional probability distributions. In terms of this problem, we observe that since 0.676 of persons in all income classes are cola drinkers [$P(Y = 0) = 0.676$], then under the assumption of independence, the same proportion of individuals in *each* income class would be cola drinkers. Thus, 67.6% of 1,060, 1,160, and 1,180 persons would be cola drinkers in the three respective income classes. In marginal totals,

TABLE 3-21
Classification of persons under the assumption of independence between income class and soft drink preference

Soft Drink Preference	Income Class			
	Upper	Middle	Lower	All Levels
Cola	717	785	798	2,300
Lime	265	290	295	850
Root beer	78	85	87	250
Total sample	1,060	1,160	1,180	3,400

the arithmetic for each group would be

$$\text{Upper income} \quad \frac{2{,}300}{3{,}400} \times 1{,}060 = 717$$

$$\text{Middle income} \quad \frac{2{,}300}{3{,}400} \times 1{,}160 = 785$$

$$\text{Lower income} \quad \frac{2{,}300}{3{,}400} \times 1{,}180 = 798$$

If we assume independence, we can give the numbers of persons in each cell of the table as in Table 3-21.

Exercises 3.13

1. The employees of Raskin Incorporated were surveyed and then classified by type of work and by sex. The results were tabulated.

Sex	Type of Work			
	Production	Office	Sales	Total
Female	12	25	13	50
Male	188	475	287	950
Total	200	500	300	1000

a. What is the joint probability that an employee who is randomly selected is a female working in production?
b. What is the marginal probability that an employee who is randomly selected works in production?
c. What is the conditional probability of an employee who is randomly selected being female, given that the employee works in production?

d. What are the marginal probability distributions of type of work and sex?
e. What is the conditional probability distribution of type of work given that an employee in this company is female?
f. Is there evidence of dependence between sex and type of work in this company?
g. If the same marginal totals are retained, what would the number of persons in each cell of the table have to be for independence to exist between type of work and sex?

2. A car dealer classified the previous month's car sales by type of car purchased and by method of payment. If a purchase is randomly selected, what is the probability of selling
a. a new car on credit?
b. any car on credit?
c. a new car, given that the purchase is for cash?

Type of car purchased	Method of Payment		Total
	Cash	Credit	
New car	4	46	50
Used Car	26	14	40
Total	30	60	90

3. The listed probability distributions pertain to the number of freight and passenger ships arriving on a given day at a certain port, where X is the number of freight ships and Y is the number of passenger ships. **Note**: In the notation used in this table, $f(x) = P(X = x)$ and $g(y) = P(Y = y)$. If the number of freight ships has no effect on the number of passenger ships arriving, what is the joint distribution of X and Y?

x	$f(x)$	y	$g(y)$
0	0.1	0	0.3
1	0.3	1	0.3
2	0.4	2	0.2
3	0.2	3	0.2

4. An analyst who was studying the operations of two dealerships owned by Barkley Holding, Inc., derived a bivariate probability distribution for sales per day of the two dealerships.

	X		
Y	0	1	2
0	0.1	0.1	0
1	0.1	0.5	0
2	0	0	0.2

X = the number of sales per day for Exotic Motors, Inc.
Y = the number of sales per day for Phlegmatic Motors, Inc.

a. Calculate $E(X + Y)$ and interpret your result.
b. Given that the sales of Exotic Motors, Inc., are one per day, what are the expected value and standard deviation of the number of sales per day for Phlegmatic Motors, Inc.?
c. What is the probability that the daily profit for the two dealerships combined would exceed \$4,000 if the sales prices were \$14,500 per unit at Exotic Motors and \$14,200 per unit at Phlegmatic Motors and cost per unit at both dealerships was \$13,000?

4

Statistical Investigations and Sampling

Throughout our lives, we are involved in answering questions and solving problems. For many of these questions and problems, careful, detailed investigations are simply inappropriate. For example, to answer the question, "What clothes should I wear today?" or the question, "What type of transportation should I take to get to a friend's house?" may not require painstaking, objective, scientific investigation.

This book is concerned with the investigation of problems that do require careful planning and an objective, scientific approach to arrive at meaningful solutions.

4.1 FORMULATION OF THE PROBLEM

A statistical investigation arises out of the need to solve some sort of a problem. Problems may be classified in a variety of ways.

Choosing among alternative courses of action is one sort of problem.

The problems of managerial decision making typically fall under this category. For example, an industrial corporation wishes to choose a particular plant site from several alternatives; a financial vice-president wishes to decide among alternative methods of financing planned increases in productive capacity; or an advertising manager must select advertising media from many possible choices.

Reporting information is another sort of problem.

Many illustrations can be given of statistical data collected for reporting purposes. For example, a trade association may report to its member firms on the characteristics of these companies. A research organization may publish data on the relationship between achievement of school children and the socioeconomic characteristics of their parents. An economist may report data on the frequency distribution of family incomes in a particular city. Even in the case of informational reporting, the data collected should have an ultimate decision-making purpose for someone.

Many problems originally arise as rather vague questions. These questions must be translated into a series of more specific questions, which then form the basis of the investigation. In most carefully planned investigations, the problem will be defined and redefined many times. The purposes and importance of an investigation will determine the type of study to be conducted.

There are instances when the objectives are particularly hard to define because of the many uses that will be made of the data and the large number of research consumers who will utilize the results of the study. For example, the U.S. Bureau of the Census publishes a wide variety of data on population, housing, manufacturers, and retail trade. It cannot specify in advance the many uses that will be made of these data.

In all studies it is critical to spell out as meticulously as possible the purposes and objectives of the investigation. All subsequent analysis and interpretation depends on these objectives, and only by stating very carefully what these objectives are can we know what questions have been answered by the inquiry.

4.2 DESIGN OF THE INVESTIGATION

Some investigations may be referred to as "controlled inquiries." We are all familiar with the scientist who controls variables in the laboratory. The scientist exercises control by manipulating the things and the events being investigated. For example, a chemist may hold the temperature of a gas constant while varying pressure and observing changes in volume. We begin by discussing observational or comparative studies, in which manipulation as in a laboratory is not possible.

Observational or Comparative Studies

In most statistical investigations in business and economics, it is not possible to manipulate people and events as directly as a physical scientist manipulates experimental materials. For example, if we want to investigate the effect of income on a person's expenditure pattern, it would not be feasible for us to vary this individual's income. On the other hand, we can observe the different expenditure patterns of people who fall in different income groups, and therefore we can make statistical generalizations about how expenditures vary with differences in income. This would be an example of a so-called **observational** study. In this type of study, the analyst essentially examines historical relationships among variables of interest. If one observes the important and relevant properties of the group under investigation, the study can be carried out in a controlled manner. For example, if we are interested in how family expenditures vary with family income and race, we can record data on family expenditures, family incomes, and race and then tabulate data on expenditures by income and racial classifications, such as white or black. Moreover, if we observe the differences in family expenditures for white and black families within the same income group, we have, in effect, "controlled" for the factor of income. That is, since the families observed are in the same income group, income cannot account for any differences in the expenditures observed.

If observational data represent historical relationships, it may be particularly difficult to ferret out causes and effects. For example, suppose we observe past data on the advertising expenses and sales of a particular company. Let us also assume that both of these series have been increasing over time. It may be quite incorrect to assume that the changes in advertising expenditures have caused sales to increase. If a company's practice in the past had been to budget 3% of last year's sales for advertising expense, one may state that advertising expenses depend on sales with a one-year lag. However, in this situation sales might be increasing quite independently of changes in advertising expenses. Therefore, one certainly would not be justified in concluding that changes in advertising expenses cause changes in sales. The point should also be made that many factors other than advertising may have influenced changes in sales. If data were not available on these other factors, it would not be possible to infer cause-and-effect relationships from these past observational data. The specific difficulty in attempting to derive cause-and-effect relationships in mathematical terms from historical data is that the various pertinent environmental factors will not ordinarily have been controlled or have remained stable.

Direct Experimentation or Controlled Studies

Direct experimentation studies are being increasingly used in fields other than the physical sciences, where they have traditionally been employed. In **direct experimentation** studies, the investigator directly controls or manipulates factors that affect a variable of interest. For example, a marketing experimenter may vary the amounts of direct mail exposure to a particular consumer audience. It is also possible to use different types of periodical advertising and observe the effects upon some experimental group. Various combinations of these direct mail exposures and periodical advertising may be used, as well as other types of promotional expenditures, such as a sales force. Thus, the investigator may be able to observe from the experiment that high levels of periodical advertising produce high levels of sales only if there is a high concentration of sales force activity. Such scientifically controlled experiments for generating statistical data, to which only brief reference is being made here, can be efficiently utilized to reduce the effect of uncontrolled variations. The real importance of this type of planning or design is that it gives greater assurance that the statistical investigation will yield valid and useful results.

Ideal Research Design

An important concept of a statistical inquiry is the **ideal research design**—meaning the investigators should think through at the design stage what the ideal research experiment would be, without reference either to the limitations of data available or to what data can feasibly be collected. Then if compromises must be made because of the practicalities of the real-world situation, the investigator will at least be completely aware of the specific compromises and expedients that have been employed. As an example, suppose we want to answer the question whether women or men are better automobile drivers. Clearly, it would be incorrect simply to obtain past data on the accident rates of men versus women. First of all, men may drive under quite different conditions than do women. For example, driving may constitute a large proportion of the work that many men do, whereas a larger proportion of women than men may drive primarily in connection with activities associated with the care of a home and family. The conditions of such driving differ considerably with respect to exposure to accident hazards. Many other reasons may be indicated for differences in accident rates between men and women apart from the essential driving ability of these two groups. Thus, as a first approximation to the ideal research design, perhaps we would like to have data for quite homogeneous groups of men and women, for example, women and men of essentially the same age, driving under essentially the same conditions, using the same types of automobiles. It may not be within the resources of a particular statistical investigation to gather data of this sort. However, once the ideal data required for a meaningful answer to the question have been thought through, the limitations of other, somewhat more practical sets of data become apparent.

Exercises 4.2

1. What is the difference between an observational study and a controlled study?
2. Why is it ordinarily very difficult to determine cause-and-effect relationships from historical data?
3. Why is it desirable to think through the ideal research design at the beginning stage of a statistical investigation?

4.3 CONSTRUCTION OF METHODOLOGY

An important phase of a statistical investigation is the construction of the conceptual or mathematical model to be used. A **model** is simply a representation of some aspect of the real world. Mechanical models are profitably used in industry as well as other fields of endeavor. For example, airplane models may be tested in a wind tunnel, or ship models may be tested in experimental water basins. Experiments may be carried out by varying certain factors and observing the effect of these variations on the mechanical models employed. Thus, we can manipulate and experiment with the models and draw corresponding inferences about their real-world counterparts. The advantages of this procedure are obvious compared with attempting to manipulate an experiment using the real-world counterparts, such as actual airplanes or ships, after they are constructed. In statistical investigations, mathematical models are often used to state in mathematical terms the relationships among the relevant variables.

Mechanical models

Mathematical models

Mathematical models are conceptual abstractions that attempt to describe, to predict, and often to control real-world phenomena.

For example, the law of gravity describes and predicts the relationship between the distance an object falls and the time elapsed. Such models can be tested by physical experimentation.

In well-designed statistical investigations, the nature of the model or models to be employed should be carefully thought through in the planning phases of the study.

Nature of models

In fact, the nature of these models provides the conceptual framework that dictates the type of statistical data to be collected. Let us consider a few simple examples. Suppose a market research group wants to investigate the relationship between expenditures for a particular product and income and several other socioeconomic variables. The investigators may want to use a mathematical model such as a **regression equation** (discussed later in this book), which states in mathematical form the relationship among the variables. When the investigators determine the variables that are most logically related to the expenditures for the product, they also determine the types of data that will have to be collected in order to construct their model.

Informational reporting models

Even in the case of relatively simple informational reporting, there is a conceptual model involved. For example, suppose an agency wishes to determine the unemployment rate in a given community. Also, assume that the agency must gather the data by means of a sample survey of the labor force in this community. The ratio "proportion unemployed" is itself a model. It states a mathematical relationship between the numerator (number of persons unemployed) and the denominator (total number of persons in the labor force). The agency may wish to go further and state the range within which it is highly confident that the true unemployment rate falls. In such a situation—as we shall see later when we study estimation of population values—there is an implicit model, namely the probability distribution of a sample proportion.

Suppose a company wishes to establish a systematic procedure for accepting or rejecting shipments from a particular supplier. Various types of models have been used to solve this sort of problem. The company may decide to accept or reject shipments on the basis of testing some hypothesis concerning the percentage of defective items observed. On the other hand, it may decide to base its acceptance procedure on the arithmetic mean value of some characteristic that is considered important. Other procedures are possible; for example, a **formal decision model** may be constructed. For these types of models, the probability distribution of the percentage of defective items produced by this company in the past may be required as well as data on the percentage of defective articles observed in a sample drawn from the particular incoming shipment in question. Obviously, the nature of the data to be observed and the nature of the analysis to be carried out will flow from the type of conceptual model used in the investigation.

Formal decision model

4.4 SOME FUNDAMENTAL CONCEPTS

Statistical Universe

In the problem formulation stage, we must define very carefully the relevant **statistical universe** of observations. The universe, or **population**, consists of the total collection of items or elements that fall within the scope of a statistical investigation.

> The purpose of defining a statistical population is to provide explicit limits for the data collection process and for the inferences and conclusions that may be drawn from the study.

The items or elements that form the population may be individuals, families, employees, schools, corporations, and so on. Time and space limitations must be specified, and it should be clear whether or not any particular element falls within or outside the universe.

In survey work, a listing of all the elements in the population is referred to as the **frame**, or **sampling frame**. A **census** is a survey that attempts to include every element in the universe. The word "attempts" is used here because complete coverage may not be effected in surveys of very large populations, despite every effort to do so. Thus, for example, the Bureau of the Census readily admits that its national "censuses" of population invariably result in underenumerations. Strictly speaking, any partial enumeration of a population constitutes a **sample**, but the term "census" is used as indicated here. In most practical applications, it is not even feasible to attempt complete enumerations of populations, and therefore, typically, only samples of items are drawn. If the population is well defined in space and time, the problem of selecting a sample of elements from it is considerably simplified.

Summary of terms

Let us illustrate some of these ideas by means of a simple example. Suppose we draw a *sample* of 1,000 families in a large city to estimate the arithmetic mean family income of all families in the city. The aggregate of all families in the city constitutes the *universe*, and each family is an *element* of the universe. The income of the family is a *characteristic* of the unit. A listing of all families in the city would comprise a *frame*. If, instead of drawing the sample of 1,000 families, an attempt had been made to include all families in the city, a *census* would have been conducted. The definition of the universe would have to be specific as to the geographic boundaries that constitute the city and also the period for which income would be observed. The terms "family" and "income" would also have to be rigidly defined. Of course, the precise definitions of all of these concepts would depend on the underlying purposes of the investigation.

Relative terms

The terms *universe* and *sample* are relative. An aggregate of elements that constitutes a population for one purpose may merely be a sample for another. Thus, if we want to determine the average weight of students in a particular classroom, the students in that room would represent the population. However, if we were to use the average weight of these students as an estimate of the corresponding average for all students in the school, then the students in the one room would be a sample of the larger population. The sample might not be a good one from a variety of viewpoints (such as representativeness), but nevertheless, it is a sample.

Finite populations

If the number of elements in the population is fixed, that is, if it is possible to count them and come to an end, the population is said to be **finite**. Such universes may range from a small to a large number of elements. For example, a small population might consist of the three vice-presidents of a corporation; a large population might be the retail transactions occurring in a large city during a one-year period. A point of interest concerning these two examples is that the vice-presidents represent a fixed and unchanging population, whereas the retail transactions illustrate a dynamic population that might differ considerably over time and space.

Infinite Populations

An **infinite population** is composed of an uncountable number of items. Usually such populations are conceptual constructs in which data are generated by processes that may be thought of as repeating indefinitely, such as the rolling of dice and the repeated measurement of the weight of an object. Sometimes, the population sampled is finite but so large that it makes little practical difference if it is considered to be infinite. For example, suppose that a population consisting of 1,000,000 manufactured articles contains 10,000 defectives. Thus, 1% of the articles is defective. If two articles are randomly drawn from the lot in succession, without replacing the first article after it is drawn, the probability of obtaining two defectives is

$$\left(\frac{10{,}000}{1{,}000{,}000}\right)\left(\frac{9{,}999}{999{,}999}\right)$$

For practical purposes, this product is equal to $(0.01)(0.01) = 0.0001$. If the population is considered infinite with 1% defective articles, the probability of obtaining two defectives is exactly 0.0001. Frequently, in situations when a finite population is extremely large relative to sample size, it is simpler to treat this population as infinite. Since a finite population is depleted by sampling without replacement whereas an infinite population is inexhaustible, and if the depletion causes the population to change only slightly, it may be simpler for computational purposes to consider the population infinite.

Sometimes, an infinite population may be a *process* that produces finite populations.

For example, a company may draw a sample from a lot from a particular supplier in order to decide whether to purchase from this supplier in the future. Thus, the purchaser makes a decision concerning the *manufacturing process* that produces future lots. The particular lot sampled for test purposes is a finite population. The process that produces the particular lot may be viewed as an infinite population. Care must be exercised in such situations to ensure that the manufacturing process is indeed a stable one and may validly be viewed as a single universe. Future testing may in fact reveal differences of such a magnitude that the conceptual universe should be viewed as having changed.

Target Populations

Another useful concept is the **target population**, or the universe about which inferences are desired. Sometimes in statistical work, it is impractical or perhaps impossible to draw a sample directly from this target population, but it is possible to obtain a sample from a closely related population. The list of elements that constitutes this sampled **frame** may be related to—but is definitely different from—the list of elements contained in the target population.

Sampling frame

For example, suppose we wish to predict the winner in a forthcoming municipal election by means of a polling technique. The target population is the collection of individuals who will cast votes on election day. However, it is not possible to draw a sample directly from this population, since the specific individuals who will show up at the polls on election day are unknown. It may be possible to draw a sample from a closely related population, such as the eligible voting population. In this case, the list of eligible voters constitutes the sampling frame. The percentage of eligible voters who would vote for a given candidate may differ from the corresponding figure for the election day population. Furthermore, the percentage of the eligible voter population who would vote for a given candidate will probably change as the election date approaches. Thus, we have a situation in which the population that can be sampled changes over time and is different from that about which inferences are to be made. In the case of election polling, the situation is further complicated by the fact that at the time the sample is taken, many individuals may not have made up their minds concerning the candidate for whom they will vote. Therefore, some assumption must be made about how these "undecideds" will break down as to voting preferences. It is common to use in-depth interviews, in which the undecided voters are questioned about the issues and individuals in the campaign, to help determine for whom the respondents will probably vote. In carefully run election polls, numerous sample surveys are taken, spaced through time, including some investigations near the election date. Then, trends can be determined in voting composition, and inferences can be made from populations that are defined close to election day. Some instances of incorrect predictions in national elections have resulted from failures to deal properly with the problem of undecided voters and from cessation of sampling too long before election day.

Adjustments

There is no easy answer to the question of how to adjust for the fact that the populations sampled are different from the election day population. For example, one approach to the problem of nonvoters is to conduct postelection surveys to determine the composition of the nonvoting group and to estimate their probable voting pattern had they shown up at the polls. Historical information of this sort could conceivably be used to adjust future polls of eligible voters. However, this is an expensive procedure, and the appropriate method of adjustment is fraught with problems.

In many statistical investigations, the target population coincides with a population that can be sampled. However, in any situation when one must sample a past statistical universe and yet make estimates for a future universe, the problem of inference about the target universe is present.

Control Groups

Probably the most familiar setting involving the concept of a control group is the situation in which an experimental group is given some type of treatment. In order to determine the effect of the treatment, another group is included in the experiment but is not given the treatment. These "no-treatment" cases are known as the **control group**. The effect of the treatment can then be determined by comparing the relevant measures for the "treatment" and "control" groups. For example, in testing the effectiveness of an inoculation against a particular disease, the inoculation may be administered to a group of school children (the treatment group) and not given to another group of school children (the control group).[1] The effectiveness of the inoculation can then be determined by comparing the incidences of the disease in the two groups. The experiment should be designed so that there is no systematic difference between the two groups at the outset that would make one group more susceptible to the disease than the other. Therefore, such experiments are sometimes designed with so-called "matched pairs," in which pairs of persons having similar characteristics, one from the treatment group and one from the control group, are drawn into the experiment. For example, if age and health are thought to have some effect on incidence of the disease, the experiment may require pairs of school children who are similar with respect to these characteristics so that the treatment can be given to one child of each pair and not to the other. Since the children are of similar ages and have the same health backgrounds, these factors cannot explain the fact that one child contracts the disease whereas the other does not. In the language of experimental design, age and general health conditions are said to have been **designed out** of the experiment. Numerous other techniques are employed in experimental design to ensure that treatment effects can be properly measured.

The concept of a control group is important in many statistical investigations in business and economics.

Forethought in selecting control groups

The results of an investigation may be uninterpretable unless one or more suitable control groups have been included in the study.

Sadly, it is often *after* statistical investigations are completed at considerable expense when it is found that because of faulty design and inadequate planning, the results cannot be meaningfully interpreted or the data collected are inappropriate for testing the hypotheses in question.

It is of paramount importance that during the planning stage, the investigators think ahead to the completion of the study.

They should ask, "If the collected data show thus-and-so, what conclusions can we reach?" This simple yet critical procedure will often highlight difficulties connected with the study design.

Example of use of control groups

The following example illustrates the use of control groups in statistical studies. Suppose a mail-order firm decides to conduct a study to determine the characteristics of its high-volume purchasers. Its purpose is to determine the dis-

[1] Difficult ethical questions arise in cases of this sort involving human experimentation. If the inoculation is indeed effective, its use clearly should not be withheld from anyone who wants it. In cases where the effectiveness of a new treatment is highly questionable, yet human experimentation appears necessary, the treatment group often is composed entirely of volunteers.

tinguishing characteristics of these customers in order to direct future campaigns to noncustomers who have similar attributes. Assume that the firm decides to do this by studying all its high-volume customers. At the conclusion of the investigation, it will be able to make statements such as, "The mean income of high-volume customers is so-many dollars." Or it may calculate that $X\%$ of these purchasers have a certain characteristic. Such population figures will be of virtually no use unless the company has an appropriate comparison group against which to assess them. The company wants to be able to isolate the distinguishing characteristics of high-volume purchasers. Thus, in studying its customers, the company should have separated them into two groups, "high-volume" and "non-high-volume." If it studied both groups, it would be in a position to determine those properties that are different between the two groups. Thus, if the company found that the high-volume and non-high-volume customers had the *same* mean incomes and that in *both* groups $X\%$ possessed a certain characteristic, it could not use these properties to distinguish between the two groups. The properties that differed most between the two groups would obviously be the ones most useful for spelling out the distinguishing characteristics of high-volume purchasers. In summary, the firm could have used the non-high-volume customers as a control group against which to compare the properties of the high-volume groups, which in the terminology used earlier would represent the "treatment group."

Selection of properties of control groups

Care must be used in the selection of the properties of the two groups to be observed. These properties should bear some logical relationships to the characteristic of high-volume versus non-high-volume purchasers. Otherwise, the properties may be spurious indicators of the distinguishing characteristics of the two groups. For example, income level would be logically related to purchasing volume; if the high-volume purchaser group had a substantially higher income than the non-high-volume group, then income would evidently be a reasonable distinguishing characteristic. On the other hand, suppose the high-volume purchaser group happened to have a higher percentage of persons who wore black shoes at the time of the survey than did the non-high-volume group. This characteristic of shoe color would *not* seem to be logically related to volume of purchases. Hence, we would not be surprised if the relationship between shoe color and volume of purchases disappeared in subsequent investigations or even reversed itself.

The construction of control groups

A couple of comments may be made on the construction of control groups. If the treatment group is symbolized as A and the control group as B, then an alternative control group to the one used would have been the treatment and control groups combined, or $A + B$. Thus, in the above example, if the relevant data had been available for the high-volume and non-high-volume customers combined (that is, for all customers), this group could have constituted the control. For example, let us assume for simplicity that there were equal numbers of high-volume and non-high-volume customers. Suppose that 90% of high-volume customers possessed characteristic X, whereas only 50% of non-high-volume customers had this characteristic. The same information would be given by stating that 90% of high-volume customers possessed characteristic X, whereas 70% of *all* customers possessed this characteristic. (The 70% figure, of course, is the weighted mean of 90% and 50%.) With the knowledge of equal numbers of persons in the high-volume and non-high-volume groups, we can infer that 50% of non-high-volume customers had the property in question. This point is important, because sometimes historical data may be available for an entire group $A + B$, whereas available resources may permit a study only of the treatment group A or (the more usual case) only a sample of this group. However, drawing conclusions about the present from a historical control group is dangerous, because systematic changes may have taken place in the treatment and control groups over time or in the surrounding conditions of the experiment. Therefore, the more scientifically desirable procedure is to design the treatment and control groups for the specific investigation in question.

Objectives determine nature of control groups

The general objectives of an investigation determine the control groups to be used. Thus, individuals who are not customers of a firm could constitute an appropriate control group for an experiment to determine the distinguishing characteristics of the firm's customers; or customers who have not purchased a specific product could constitute the control group in an experiment to find the particular traits of the customers who purchase the product. Note that time considerations and available resources usually permit only drawing of samples from the treatment and control populations rather than complete enumerations of these populations.

Examples of control groups

Some other brief examples of the use of control groups will be given here. A national commission wanted to investigate insurance conditions in cities in which civil disturbances in the form of riots had occurred. Specifically, the commission wished to study cancellation rates for burglary, fire, and theft policies in sections of these cities ("riot areas") primarily affected by the riots. The main purpose was to determine whether individuals and businesses in these areas were having difficulty retaining such policies because of cancellations by insurance companies. It became clear in the planning stages of the study that it would not be sufficient merely to measure cancellation rates in the cities in which riots had occurred, because it would not be possible in the absence of other information to judge whether these rates were low, average, or high. Therefore, a sample of individuals and businesses in cities that had not experienced riots was used as a control group. Another control group was established consisting of individuals and businesses in the "nonriot areas" of the cities that had experienced riots. Thus, the data on cancellation rates could be meaningfully interpreted. Comparisons were made between cities that had experienced riots and those that had not. Further comparisons were made between cancellation rates in riot areas and nonriot areas in cities where these disturbances had been present. The data disclosed that burglary, fire, and theft insurance cancellation rates were higher in cities in which riots had occurred. Furthermore, within cities in which these civil disturbances had been present, cancellation rates were higher in riot sections than in nonriot sections. It may be noted, parenthetically, that if the company policies on cancellations were known and data were available in suitable form, the same information could have been obtained from the company records. However, such information was not available, so the sample survey was required to obtain the indicated data.

Another illustration is the case of a company that wished to determine whether its labor costs were very different from such costs throughout the company's industry group. It obtained the ratio of labor costs to total operating costs for the company and compared these with a published distribution of such ratios for all firms in the industry for the same period. In this situation, all firms in the industry constituted the control group. This is an illustration of a rather obvious need for and choice of a control group. In many situations, the need and choice are somewhat more subtle.

Types of Measurement Errors

The concept of error is central throughout all statistical work. Wherever we have measurement, inference, or decision making, the possibility of error is present. In this section, we deal with **errors of measurement**. Errors of inference and decision making are treated in subsequent chapters.

Systematic errors

It is useful to distinguish the two types of errors that may be present in statistical measurements, namely, systematic errors and random errors. **Systematic errors** cause a measurement to be incorrect in some systematic way. Such errors are built into the procedures and they **bias** the results of a statistical investigation. These errors may occur in the planning stages or during or after the collection process. Examples of causes of systematic error are faulty design of a questionnaire (such as misleading or ambiguous questions), systematic mistakes in planning and carrying out the collection and processing of the data, nonresponse and refusals by respondents to provide information, and a great discrepancy between the sampling frame and the target universe. If observations have arisen from a sample drawn from a statistical universe, systematic errors are those that persist even when the sample size is increased. As a generalization, these errors may be viewed as arising primarily from inaccuracies or deficiencies in the measuring instrument.

Random errors

On the other hand, **random errors** or **sampling errors** or **experimental errors** may be viewed as arising from the operation of a large number of uncontrolled factors, conveniently described by the term "chance." As an example of this type of error, if repeated random samples of the same size are drawn from a statistical universe (with replacement of each sample after it is drawn), then a particular statistic, such as an arithmetic mean, will differ from sample to sample. These sample means tend to distribute themselves below and above the "true" population parameter (arithmetic mean), with small deviations between the statistic and the parameter occurring relatively frequently and large deviations occurring relatively infrequently. The word "true" has quotation marks around it because it refers to the figure that would have been obtained through complete coverage of the universe, that is, a complete census using the same definitions and procedures used in the samples. The difference between the mean of a particular sample and the population mean is said to be a *random error* (or a *sampling error*, as it is termed in later chapters). The complete collection of factors that could explain why the sample mean differed from the population mean is unknown, but we can conveniently lump them all together and refer to the difference as a random or chance error.

> **Random errors** are those that arise from differences found between the outcomes of trials (or samples) and the corresponding universe value using the same measurement procedures and instruments.

The sizes of the differences indicate reliability or precision.

> Random errors decrease on the average as sample size increases.

It is precisely for this reason that we prefer a large sample of observations to a small one, all other things being equal; that is, since sampling errors are on the average smaller for large samples, the results are more reliable or more precise.

Systematic and random errors may occur in experiments in which the variables are manipulated by the investigator or in survey work where observations are made on the elements of a population without any explicit attempt to manipulate directly the variables involved. A few examples will be given to show how bias, or systematic error, may be present in a statistical investigation. The problems of *how random errors are measured* and *what constitutes suitable models for the description of such errors* represent central topics of statistical methods and are discussed extensively later in this chapter and in Chapters 5 through 8.

Systematic Error: Biased Measurements

The possible presence of biased measurements in an experimental situation may be illustrated by a simple example. Suppose that a group of individuals measured the length of a 36-inch table top using the same yardstick. Let us further assume that the yardstick, although calibrated as though it were 36 inches long, was in fact 35 inches long and this fact was unknown to the individuals making the measurements. There would then be a systematic error of one inch present in each of the measurements, and a statistic such as an arithmetic mean of the readings would reflect this bias. In this situation, the systematic error could be detected if a correctly calibrated yardstick were used as a standard against which to test the incorrect yardstick. This is an important methodological point.

Systematic error often can be discovered through the use of an independent measuring instrument.

Even if the independent instrument is inaccurate, a comparison of the two measuring instruments may give clues about where the search for sources of bias should be made. The variation among the individual measurements made with the incorrect yardstick would be a measure of random error because the differences are not attributable to specific causes of variation. Note that the observations may have been very precise (although inaccurate), in the sense that each person's measurement was close to that of every other person. Thus, the random errors would be small, and there would be good *repeatability*, because in repeating the experiment, each measurement would be close to preceding measurements. These random or chance errors may be assumed to be compensating, in the sense that some observations would tend to be too large and some too small. Since the table top is 36 inches long, the measurements would tend to cluster around a value about one inch greater than the true length of the table top. In summary, we have a model in which each individual measurement may be viewed as the sum of three components: (1) the true value, (2) systematic error, and (3) random error. This relationship can be stated in equation form.

$$\text{Individual measurement} = \text{True value} + \text{Systematic error} + \text{Random error} \tag{4.1}$$

Systematic Error: Literary Digest Poll

A classic case of systematic error in survey sampling procedure is that of the *Literary Digest* prediction of the presidential election of 1936. During the election campaign between Franklin D. Roosevelt and Alf Landon, the *Literary Digest* magazine sent questionnaire ballots to a large list of persons whose names appeared in telephone directories and automobile registration lists. Over two million ballots—about one-fifth of the total number sent out—were returned by the respondents. On the basis of these replies, the *Literary Digest* erroneously predicted that Landon would be the next president of the United States. The reasons that the results of this survey were so severely biased are rather clear. In 1936, during the Great Depression, the presidential vote was cast largely along economic lines. The group of the electorate that did not own telephones or automobiles did not have an opportunity to be included in the sample. This group, which represented a lower economic level than owners of telephones and automobiles, voted predominantly for Roosevelt, the Democratic candidate. A second reason stemmed from the nonresponse group, which represented about four-fifths of those polled. Typically, individuals of high educational and high economic status are more apt to respond to voluntary questionnaires than those with low economic and educational status. Therefore, the nonresponse group doubtless contained a higher percentage of this low-status group than did the group that responded to the questionnaire. Again, this factor added a bias due to underrepresentation of Democratic votes. In summary, the sample used for prediction purposes contained a greater proportion of persons of high socioeconomic status than were present in the target population, namely, those who cast votes on election day. Since this factor of socioeconomic status was related to the way people voted, a systematic overstatement of the Republican vote was present in the sample data.

Two methodological lessons

Two methodological lessons can be derived from this example.

1. It is a dangerous procedure to sample a frame that differs considerably from the target population.
2. Procedures must be established to deal with the problem of nonresponse in statistical surveys.

Clearly, even if the proper target universe had been sampled in the previous case, the problem of nonresponse would still have to be properly handled.

Systematic Error: Method of Data Collection

Another example of bias will illustrate that the direction of systematic error may be associated with the nature of the agency that collects the data as well as the method by which the data are collected.

The alumni society of a large eastern university decided to gather information from the graduates of that institution to determine a number of characteristics, including their current economic status. One of the questions of interest was the amount of last year's gross income, suitably defined. A questionnaire was mailed to a random sample of graduates, and the results were tabulated from the returns. When frequency distributions were made and averages were calculated by year of graduation, it became clear that the income figures were unusually high compared with virtually any existing external data that could be examined; in other words, the income figures were clearly biased in an upward direction. It is fairly easy in this case to speculate on the causes of this upward systematic error. In this type of mail questionnaire, a higher nonresponse rate could be expected from those graduates whose incomes were relatively low than from those with higher incomes. That is, it appears reasonable that those with higher incomes would have a greater propensity to respond than others. Furthermore, if there were instances of misreporting of incomes, these probably tended to be overstatements rather than understatements, because of the desire to appear relatively economically successful.

On the other hand, let us consider the same sort of data as reported to the Internal Revenue Service on annual income tax returns. Doubtless, it is safe to say that relatively little overstatement of gross incomes occurs. Indeed, it seems reasonable to suppose that a downward bias exists in these data in the aggregate. Note that since responses to the Internal Revenue Service are mandatory, the effect of nonresponse may be considered negligible. Thus, the interesting situation is presented here of the same type of data being gathered by two different agencies, one set biased in an upward direction, the other in the opposite direction.

> In using secondary statistical information, we must exercise informed critical judgment to extract meaningful inferences.

This judgment must include practical considerations such as methods of data collection and auspices under which studies are conducted. Of course, false reporting is not easily overcome, particularly in situations in which no independent objective data are available against which the reported information may be checked.

Exercises 4.4

1. Explain the difference between a sample and a census.
2. Give an example of a situation in which the population of interest would be
   a. an infinite population.
   b. considered an infinite population, yet in reality is a finite population.
   c. considered an infinite population, because it is generated by a process that produces repeated observations.
   d. a target population.
3. Explain the way a control group should be used by criticizing the following situations:
   a. A study of the effectiveness of aspirin in relieving headaches showed that 90% of the adults in the study stated that they had obtained relief of pain 2 hours after taking 10 grains of aspirin.
   b. A stockbroker claims that everyone should invest with him because 85% of his recommendations last year performed better than market expectations (appropriately defined).
   c. A fertilizer company performed a study that showed that farmers who used Extra-Gro fertilizer last year had a 10% increase in crop yield over the preceding year. This company is now advertising that Extra-Gro will increase crop yields by 10%.
4. Determine an appropriate control group for each of the following studies, and briefly state why you feel it is appropriate.
   a. Investigators of the U.S. Olympic Committee wish to determine whether daily liquid protein supplements during training will produce stronger weight lifters for the United States.
   b. It is desired to try a new method of teaching sixth grade arithmetic in a certain city.
   c. It has been claimed that marathon runners do not suffer heart attacks. A medical investigator wishes to test this assertion.

5. A marketing research firm wanted to determine the effectiveness of a new advertising campaign. The firm presented the campaign to a group of individuals who already used the product and to a control group that was made up of individuals who did not use the product. Several weeks later, users reported an increase in use while the nonusers still did not use the product. It was concluded that the advertising campaign would be effective for increasing consumption by current users of the product.
   a. Explain why the control group was not adequate for the experiment conducted.
   b. Identify a more appropriate control group.
6. In each of the following situations, state whether a control group would be of use and, if so, what the control group would be.
   a. You are interested in investigating the accident rates in low-income urban areas, and you have data on numbers of accidents occurring in low-income urban areas, number of cars registered in low-income areas, actual area of low-income areas, and other similar information.
   b. You are interested in evaluating the effectiveness of a new safety lighting program to be installed in your plant.
7. Distinguish clearly between systematic error and random error. Explain which error will decrease as sample size increases and which error will not.
8. What is the population of interest if one is interested in the percentage of consumers in the New York City area who would buy a certain type of men's suit at varying prices?
   a. Is this a fixed and unchanging population or a dynamic population?
   b. Explain your answer.

4.5 FUNDAMENTALS OF SAMPLING

Purposes

Sampling is important in most applications of quantitative methods to managerial and other business problems for a number of reasons. In certain instances, sampling may represent the only possible or practicable method of obtaining the desired information. For example, in the case of processes, such as manufacturing, in which the universe is conceptually infinite (including all future as well as current production), a complete enumeration of the population is not possible. On the other hand, if sampling is a destructive process, a complete enumeration of the universe may be possible, but it would not be practical to do so. For example, if a military procurement agency wanted to test a shipment of bombs, it could detonate all of the bombs in a testing procedure and obtain complete information concerning the quality of the shipment. However, since there would be no usable product remaining, a sampling procedure is clearly the only practical way to assess the quality of the shipment.

Sampling procedures are often employed for overall effectiveness, cost, timeliness, and other reasons. A complete census, although it does not have sampling error introduced by a partial enumeration of the universe, nevertheless often contains greater total error than does a sample survey, because greater care can usually be exercised in a sample survey than in carrying out censuses. Errors in collection, classification, and processing of information may be considerably smaller in sample surveys, which can be carried out under far more carefully controlled conditions than large-scale censuses. For example, it may be possible to reduce response errors arising from lack of information, misunderstood questions, faulty recall, and other reasons only by intensive and expensive interviewing and measurement methods, which may be feasible in the case of a sample but prohibitively costly for a complete enumeration.

The employment of sampling rather than censuses for purposes of timeliness occurs in a variety of areas. A notable example is the wide array of government data on economic matters such as income, employment, and prices, which are collected on a sample basis at periodic intervals. Timeliness of publication of these results is of considerable importance. The more rapid collection and processing of data afforded by sampling procedures represents an important advantage over corresponding census methods.

Random and Nonrandom Selection

Items can be selected from statistical universes in a variety of ways. It is useful to distinguish random from nonrandom methods of selection. In this book, attention is focused on **random**, or **probability**, **sampling**, that is, sampling in which the probability of inclusion of every element in the universe is *known*. **Nonrandom sampling** methods are referred to as **judgment sampling** because judgment is exercised in deciding which elements of a universe to include in the sample. Such judgment samples may be drawn by choosing "typical" elements or groups of elements to represent the population. They may even involve random selection at one stage but allow the exercise of judgment in another. For example, areas may be selected at random in a given city, and interviewers may be instructed to obtain specified numbers of persons of given types within these areas but may be permitted to make their own decisions as to which individuals are brought into the sample.

Judgment Sampling

This book deals only with random, or probability, sampling methods (rather than judgment sampling) because of the clear-cut superiority of probability selection techniques.

> Random sampling is preferable to judgment sampling because of the precision with which estimates of population values can be made from the sample itself. In judgment selection there is no objective method of measuring the precision or reliability of estimates made from the sample.

This is an important advantage, since random sampling techniques provide an objective basis for measuring errors due to the sampling process and for stating the degree of confidence placed on estimates of population values.

Judgment samples can sometimes be usefully employed in the planning and design of probability samples. For example, when expert judgment is available, a pilot sample may be selected on a judgment basis in order to obtain information that will aid in the development of an appropriate sampling frame for a probability sample.

Simple Random Sampling

We have seen that a *random sample* or *probability sample* is a sample drawn in such a way that the probability of inclusion of every element in the population *is known*. A wide variety of types of such probability samples exist, particularly in the area of sample surveys. Experts in survey sampling have developed a large body of theory and practice aimed toward the optimal design of probability samples. This highly specialized area is often allocated an entire course or two in a graduate program in statistics. We will concentrate on the simplest and most fundamental probability method, namely, *simple random sampling*. The major body of statistical theory is based on this method of sampling.

Finite population sampling without replacement

We first define a simple random sample for the case of a finite population of N elements.

> A **simple random sample** of n elements is a sample drawn in such a way that *every* combination of n elements has an *equal* chance of being the sample selected.

Since most practical sampling situations involve sampling *without replacement*, it is useful to think of this type of sample as one in which each of the N population elements has an equal probability, $1/N$, of being the one selected on the first draw, each of the remaining $N - 1$ elements has an equal probability, $1/(N - 1)$, of being selected on the second draw, and so on until the nth sample item has been drawn. Since there are $\binom{N}{n}$ possible samples of n items, the probability that any sample of size n will be the one drawn is $1\Big/\binom{N}{n}$.

This concept of a simple random sample may be illustrated by the following example. Let the population consist of three elements, A, B, and C. Thus, $N = 3$. Suppose we wish to draw a simple random sample of two elements; then $n = 2$. Using equation 2.20 for the number of combinations that can be formed of N objects taken n at a time, we find that the number of possible samples is

$$\binom{3}{2} = \frac{3!}{2!1!} = 3$$

These three possible samples contain the following pairs of elements: (A, B), (A, C), and (B, C). The probability that any one of these three samples will be the one selected is $\frac{1}{3}$.

It is a property of simple random sampling that every element in the population has an equal probability of being included in the sample. However, many other sample designs possess this property as well, as for example, certain stratified sample and cluster sample procedures.

Finite population sampling with replacement

Simple random samples were defined above for the case of sampling a finite population without replacement. If a finite population is sampled *with replacement*, the same element could appear more than once in the sample. Since, for practical purposes, this type of sampling is virtually never employed, it will not be discussed any further here.

Infinite population sampling

On the other hand, simple random sampling of *infinite populations* is important, particularly in the context of sampling of processes. The following definition corresponds to the one given for finite populations.

> For an infinite population, a simple random sample is one in which on *every* selection, each element of the population has an *equal* probability of being the one drawn.

This is difficult to visualize in terms of actual sampling from a physical population. Therefore it is helpful to take a more formal approach and to use the language of random variables. Thus, we view the drawing of the sample as an experiment in which observations of values of a random variable are generated, and the successive sample observations or elements are the outcomes of trials of the experiment. Then a simple random sample of n observations is defined by the presence of two conditions: (1) the n successive trials of the experiment are independent, and (2) the probability of a particular outcome of a trial defined as a success remains constant from trial to trial. In terms of sampling a physical population, we may interpret these as meaning that (1) the n successive sample observations are independent and (2) the population composition remains constant from trial to trial.

To aid in the interpretation of the above definition, let us consider the case of drawing a simple random sample of n observations from a Bernoulli process. Assume a situation in which a fair coin is tossed. A simple random sample of n observations would be the sample consisting of the outcomes on *n independent* tosses of the coin. Thus, if the number 0 denotes the appearance of a tail and 1 the appearance of a head, the following notation might designate a particular simple random sample of five observations, in which two tails and three heads were obtained in the indicated order (0, 1, 0, 1, 1). The random variable in this illustration may be designated as "number of heads (or tails) obtained in five tosses of a fair coin." In summary, we note that (1) the tosses of the coin were statistically independent and (2) the probability of obtaining a head (or tail) remained constant from trial to trial. It is conventional to use an abbreviated method of referring to such a sample as "a sample of five independent observations from a Bernoulli process," or "a sample of five independent observations from a binomial distribution."

The term "random sample," although it properly refers to a sample drawn with known probabilities, is often used to mean "simple random sample." The student should be aware of this alternative usage.

Methods of Simple Random Sampling

Although it is easy to state the definition of a simple random sample, it is not always obvious how such a sample is to be drawn from an actual population. The following are two useful methods:

Drawing chips from a bowl

We first restrict our attention to the most straightforward situation, in which the population is finite and the elements are easily identified and can be numbered. For example, suppose there are 100 students in a college freshman class and we wish to draw a simple random sample of 10 of these students without replacement. We could assign numbers from 1 to 100 to each of the students and place these numbers on physically similar disks (or balls, slips of paper, and so on), which could then be placed in a bowl. We shake the bowl to mix the disks thoroughly and then proceed to draw the sample. The first disk is drawn, and we record the number written on it. We then shake the bowl again, draw the second disk, and record the result. The process is repeated until we have drawn 10 numbers. The students corresponding to these 10 numbers constitute the required simple random sample.

Tables of random numbers

If the population size is very large, the above procedure can become quite unwieldy and time-consuming. Furthermore, it may introduce biases if the disks are not thoroughly mixed. Therefore, in recent years, there has been a marked tendency to use tables of random digits for the purpose of drawing such samples. These tables are useful for the selection of other types of probability samples, as well.

A **table of random digits** is simply a table of digits that has been generated by a random process. For ease of use, the digits are usually combined into groups, for example, of five digits each. Thus, a table of random digits could be generated by the process of drawing chips from a bowl similar to the one just described. The digits 0, 1, 2, . . . , 9 could be written on disks and the disks placed in a bowl and then drawn, one at a time, *replacing the selected disk after each drawing*. Thus, on each selection, the population would consist of the 10 digits. The recorded digits would constitute a particular sequence of random digits. These tables are now usually produced by a computer that has been programmed to generate random sequences of digits.

Using random digits

We now illustrate the use of random digits using Table 4-1. Suppose there were 9,241 undergraduates at a large university and we wished to draw a simple random sample of 300 of these students. Each of the 9,241 students could be assigned a four-digit number, say, from 0001 to 9241. This list of names and numbers would constitute the sampling frame. We now turn to a table of

TABLE 4-1
Random digits

98389	95130	36323	33381	98930	60278	33338	45778	86643	78214
17245	58145	89635	19473	61690	33549	70476	35153	41736	96170
01289	68740	70432	43824	98577	50959	36855	79112	01047	33005
98182	43535	79938	72575	13602	44115	11316	55879	78224	96740
59266	39490	21582	09389	93679	26320	51754	42930	93809	06815
42162	43375	78976	89654	71446	77779	95460	41250	01551	42552
50357	15046	27813	34984	32297	57063	65418	79579	23870	00982
11326	67204	56708	28022	80243	51848	06119	59285	86325	02877
55636	06783	60962	12436	75218	38374	43797	65961	52366	83357
31149	06588	27838	17511	02935	69747	88322	70380	77368	04222
25055	23402	60275	81173	21950	63463	09389	83095	90744	44178
35150	34706	08126	35809	57489	51799	01665	13834	97714	55167
61486	33467	28352	58951	70174	21360	99318	69504	65556	02724
44444	86623	28371	23287	36548	30503	76550	24593	27517	63304
14825	81523	62729	36417	67047	16506	76410	42372	55040	27431
59079	46755	72348	69595	53408	92708	67110	68260	79820	91123
48391	76486	60421	69414	37271	89276	07577	43880	08133	09898
67072	33693	81976	68018	89363	39340	93294	82290	95922	96329
86050	07331	89994	36265	62934	47361	25352	61467	51683	43833
84426	40439	57595	37715	16639	06343	00144	98294	64512	19201
41048	26126	02664	23909	50517	65201	07369	79308	79981	40286
30335	84930	99485	68202	79272	91220	76515	23902	29430	42049
33524	27659	20526	52412	86213	60767	70235	36975	28660	90993
26764	20591	20308	75604	49285	46100	13120	18694	63017	85112
85741	22843	16202	48470	97412	65416	36996	52391	81122	95157

Source: The Rand Corporation, *A Million Random Digits with 100,000 Normal Deviates* (New York: Free Press, 1955), excerpt from page 387. Copyright 1955 by The Rand Corporation. Used by permission.

random digits in order to select a simple random sample of 300 such four-digit numbers. We may begin on any page in the table and proceed in any systematic manner to draw the sample. Assume we decide to use the first four columns of each group of five digits, beginning at the upper left and reading downward. Starting with the first group of digits, we find the sequence 98389. Since we are using the first four digits, we have the number 9838. This exceeds the largest number in our population, 9241, so we ignore this number and read down to pick up the next four-digit number, 1724. This is the number of the first student in the sample. Reading down consecutively, we find 0128, 9818 (which we ignore), 5926, and so on until 300 four-digit numbers between 0001 and 9241 have been specified. If any previously selected number is repeated, we simply ignore the repeated appearance and continue. In this illustration, we read downward on the page, but we could have read laterally, diagonally, or in any other systematic fashion. The important point is that each four-digit number has an equal probability of selection, regardless of what systematic method of drawing is used, and regardless of what numbers have aleady preceded.

Methods are available for drawing types of samples other than simple random samples and even for situations in which the elements have not been listed beforehand. Many of the tables include instructions for their use, but we will not pursue the subject further here.

Computer-generated sequences

Computers are often used these days to generate sequences of random numbers. However, even when a computer is used, it is useful to refer back to the idea of tables of random numbers to understand how a universe of elements is mixed in the abstract for sampling purposes.

Exercises 4.5

1. Give at least three possible reasons why sampling procedures may be preferable to a census in certain situations.
2. State whether each of the following statements is true or false and explain your answer.
   a. Judgment sampling is good, since we can get an objective measure of the random error.
   b. When costs permit a census to be taken, a census is usually preferable to sampling.
   c. Systematic errors often can be reduced by better procedures, while random errors can be reduced only by larger sample sizes.
3. The school board of Upper Bayport wished to determine voter opinion concerning a special assessment to permit the expansion of school services. Upper Bayport is an industrial community on the fringe of a metropolitan area and has a population of 25,000. There are 5,000 students enrolled in the public schools of the community. The board selected a random sample of these students and sent questionnaires to their parents. If you had been asked to assist the board, would you have approved the universe it studied? Defend your position.

5

Sampling Distributions

In Chapter 1, we examined how to compute the arithmetic mean and standard deviation from the data contained in a sample. We now consider how such statistics differ from sample to sample if repeated simple random samples of the same size are drawn from a statistical population. The probability distribution of such a statistic is referred to as its **sampling distribution**. Thus, we may have a sampling distribution of a proportion, a sampling distribution of a mean, and so on.

> Sampling distributions are the foundations of statistical inference and are of considerable importance in modern statistical decision theory as well.

We begin our discussion by considering the sampling distributions of numbers of occurrences and proportions of occurrences.

5.1 SAMPLING DISTRIBUTION OF A NUMBER OF OCCURRENCES

We can illustrate the meaning and properties of the sampling distribution of a number of occurrences by means of a typical example. Let us assume a 10% probability that any given customer who enters Johnson's Supermarket will purchase ice cream. We conceive of the purchasing process as an infinite population. Thus, we may view the successive purchases and failures to purchase as outcomes of a series of Bernoulli trials. That is, in terms of this problem, the three requirements of a Bernoulli process are as follows.

Bernoulli trials

1. There are two possible outcomes on each trial: customer purchases ice cream or customer fails to purchase ice cream.
2. The probability of a purchase of ice cream remains constant from trial to trial.
3. The trials are independent.

Suppose we draw a simple random sample of five customers who enter the supermarket on a particular day and note the number of purchasers of ice cream in the sample. This number is a random variable that can take on the values 0, 1, 2, 3, 4, or 5. Since we are dealing with a Bernoulli process, the probabilities of obtaining these numbers of purchasers may be computed by means of a binomial distribution with $p = 0.10$, $q = 0.90$, and $n = 5$. Therefore, the respective probabilities are given by the expansion of the binomial $(0.9 + 0.1)^5$. This probability distribution is shown in Table 5-1, using the same notation as in Chapter 3.

We can also interpret the probability distribution given in Table 5-1 as a sampling distribution. Since the number of ice cream purchasers observed in a sample of five customers is a sample statistic, Table 5-1 displays the probability distribution of this sample statistic. If we took repreated simple random

TABLE 5-1
Probability distribution of the number of ice cream purchasers in a simple random sample of 5 customers: $p = P(\text{ice cream purchaser}) = 0.10$

Number of Ice Cream Purchasers x	Probability $f(x)$
0	$\binom{5}{0}(0.9)^5(0.1)^0 = 0.59$
1	$\binom{5}{1}(0.9)^4(0.1)^1 = 0.33$
2	$\binom{5}{2}(0.9)^3(0.1)^2 = 0.07$
3	$\binom{5}{3}(0.9)^2(0.1)^3 = 0.01$
4	$\binom{5}{4}(0.9)^1(0.1)^4 \approx 0.00$
5	$\binom{5}{5}(0.9)^0(0.1)^5 \approx 0.00$
	1.00

samples of five customers each and if the probability that any customer would purchase ice cream was 10%, then we would observe no ice cream purchasers in 59% of these samples, one ice cream purchaser in 33% of the samples, and so forth.

The probability distribution may now be called a *sampling distribution of number of occurrences.*

5.2 SAMPLING DISTRIBUTION OF A PROPORTION

Frequently, it is convenient to consider proportion of occurrences rather than number of occurrences. We can convert the numbers of occurrences to proportions by dividing by the sample size. The sample proportion, denoted $\bar{p}$ (pronounced "p-bar"), may be calculated from

$$\bar{p} = \frac{x}{n} \tag{5.1}$$

where x is the number of occurrences of interest and n is the sample size. In the above example, $\bar{p}$ takes on the possible values $\frac{0}{5} = 0.00$, $\frac{1}{5} = 0.20, \ldots, \frac{5}{5} = 1.00$ with the same probabilities as the corresponding numbers of ice cream purchasers. (Note that the number of occurrences in a sample of size n is given by $x = n\bar{p}$. This may be seen by multiplying both sides of equation 5.1 by n.) The sampling distribution of $\bar{p}$ is given in Table 5-2. In keeping with the usual convention, the probabilities are denoted $f(\bar{p})$.

TABLE 5-2
Sampling distribution of the proportion of ice cream purchasers in a simple random sample of 5 customers: $p = P(\text{ice cream purchaser}) = 0.10$

Proportion of Ice Cream Purchasers $\bar{p}$	Probability $f(\bar{p})$
0.00	0.59
0.20	0.33
0.40	0.07
0.60	0.01
0.80	0.00
1.00	0.00
	1.00

TABLE 5-3
Calculation of the mean and standard deviation of the sampling distribution of the number of ice cream purchasers in a simple random sample of 5 customers: $p = P(\text{ice cream purchaser}) = 0.10$

$x = n\bar{p}$	$f(x)$	$xf(x)$	$x - \mu_x$	$(x - \mu_x)^2$	$(x - \mu_x)^2 f(x)$
0	0.59	0.00	−0.5	0.25	0.1475
1	0.33	0.33	+0.5	0.25	0.0825
2	0.07	0.14	+1.5	2.25	0.1575
3	0.01	0.03	+2.5	6.25	0.0625
4	0.00	0.00	+3.5	12.25	0.0000
5	0.00	0.00	+4.5	20.25	0.0000
	1.00	0.50			0.4500

$$\mu_{n\bar{p}} = \mu_x = \sum xf(x) = 0.50 \text{ ice cream purchasers}$$

$$\sigma_{n\bar{p}} = \sigma_x = \sqrt{\sum(x - \mu)^2 f(x)} = \sqrt{0.4500} = 0.67 \text{ ice cream purchasers}$$

Properties of the Sampling Distributions of $\bar{p}$ and $n\bar{p}$

We turn now to the properties of the sampling distributions of number of occurrences $n\bar{p}$ and proportion of occurrences $\bar{p}$. The means and standard deviations of these distributions are of particular interest in statistical inference. The calculation of these two measures is given in Table 5-3 for number of ice

TABLE 5-4
Calculation of the mean and standard deviation of the sampling distribution of the proportion of ice cream purchasers in a simple random sample of 5 customers: $p = P(\text{ice cream purchaser}) = 0.10$

$\frac{x}{n} = \bar{p}$	$f(\bar{p})$	$\bar{p}f(\bar{p})$	$\bar{p} - \mu_{\bar{p}}$	$(\bar{p} - \mu_{\bar{p}})^2$	$(\bar{p} - \mu_{\bar{p}})^2 f(\bar{p})$
0.00	0.59	0.000	−0.10	0.01	0.0059
0.20	0.33	0.066	+0.10	0.01	0.0033
0.40	0.07	0.028	+0.30	0.09	0.0063
0.60	0.01	0.006	+0.50	0.25	0.0025
0.80	0.00	0.000	+0.70	0.49	0.0000
1.00	0.00	0.000	+0.90	0.81	0.0000
		0.100			0.0180

$$\mu_{\bar{p}} = \mu_{x/n} = \sum \bar{p}f(\bar{p}) = 0.10 = 10\% \text{ ice cream purchasers}$$

$$\sigma_{\bar{p}} = \sigma_{x/n} = \sqrt{\sum(\bar{p} - \mu_{\bar{p}})^2 f(\bar{p})} = \sqrt{0.0180} = 0.134 = 13.4\% \text{ ice cream purchasers}$$

cream purchasers and in Table 5-4 for proportion of ice cream purchasers. Subscripts are used to indicate the random variable for which these measures are computed. For example, $\mu_{n\bar{p}}$ denotes the mean of the random variable $n\bar{p}$. The definitional formulas (3.11) and (3.12) were used for these computations. In actual applications, calculations such as those in Tables 5-3 and 5-4 are never made to obtain the mean and standard deviation of a binomial distribution because convenient computational formulas are available. The calculations are given here only to aid in understanding the meaning of sampling distributions.

General formulas for the mean and standard deviation can be derived by substituting $\binom{n}{x} q^{n-x}p^x$ for $f(x)$ and $f(\bar{p})$ in the definitional formulas and performing appropriate manipulations. The results of these derivations are summarized in Table 5-5.

Let us illustrate the use of the formulas given in Table 5-5 for the preceding distributions of number and proportion of ice cream puchasers. Substituting $p = 0.10$, $q = 0.90$, and $n = 5$, we obtain

Number of Ice Cream Purchasers

Mean	$\mu_{n\bar{p}} = np = 5 \times 0.1 = 0.50$ ice cream purchasers
Standard deviation	$\sigma_{n\bar{p}} = \sqrt{npq} = \sqrt{5 \times 0.1 \times 0.9} = \sqrt{0.45}$ $= 0.67$ ice cream purchasers

Proportion of Ice Cream Purchasers

Mean	$\mu_{\bar{p}} = p = 0.10 = 10\%$ ice cream purchasers
Standard deviation	$\sigma_{\bar{p}} = \sqrt{\frac{pq}{n}} = \sqrt{\frac{0.10 \times 0.90}{5}} = \sqrt{0.0180} = 0.134$ $= 13.4\%$ ice cream purchasers

TABLE 5-5
Formulas for the mean and standard deviation of a binomial distribution

Random Variable	Mean	Standard Deviations
Number of occurrences ($n\bar{p}$)	$\mu_{n\bar{p}} = np$	$\sigma_{n\bar{p}} = \sqrt{npq}$
Proportion of occurrences ($\bar{p}$)	$\mu_{\bar{p}} = p$	$\sigma_{\bar{p}} = \sqrt{\frac{pq}{n}}$

Of course, these are the same results obtained in the longer calculations shown in Tables 5-3 and 5-4. Let us interpret these results in terms of the appropriate sampling distributions. In this example, where the sample statistic is *number* of ice cream purchasers $n\bar{p}$, the mean of the binomial distribution is $\mu_{n\bar{p}} = np = (5)(0.10) = 0.50$ ice cream purchasers. This says that if we take repeated simple random samples of five customers each and if the probability is 10% that any customer would purchase ice cream, then there will be, on the average, one-half an ice cream purchaser per sample. (If the sample size were $n = 200$ with $p = 0.10$, then we would expect to obtain, on the average, $(200)(0.10) = 20$ ice cream purchasers per sample.) The standard deviation, $\sigma_{n\bar{p}} = \sqrt{npq} = 0.67$, is a measure of the variation in *number* of ice cream purchasers attributable to the chance effects of random sampling.

Number

Proportion

When the sample statistic is *proportion* of ice cream purchasers $\bar{p}$, the mean $\mu_{\bar{p}} = p = 0.10$, or 10% purchasers. This means that if we draw simple random samples of five customers each, we will observe, on the average, 10% ice cream purchasers in the samples. The standard deviation, $\sigma_{\bar{p}} = \sqrt{pq/n} = 0.134$ or 13.4%, is a measure of variation attributable to the chance effects of sampling, expressed in terms of *proportion* of ice cream purchasers.

The binomial distribution in this example is skewed, as shown in Figure 5-1. Two horizontal scales are shown in this graph to depict corresponding values of $n\bar{p}$ and $\bar{p}$.

As we saw in section 3.4, if p is less than 0.5, the distribution tails off to the right as in Figure 5-1. On the other hand, if p exceeds 0.5, the skewness is to the left. If p is held fixed and if the sample size n is made larger, the sampling distributions of $n\bar{p}$ and $\bar{p}$ become more symmetrical. This is an important property of the binomial distribution from the standpoint of sampling theory and practice, and we examine it further in section 5.3. An illustration of this property is given in Figure 5-2, in which p is held fixed at 0.20. The sampling distribution of $\bar{p}$ is shown for sample sizes of $n = 5$, 10, and 100.

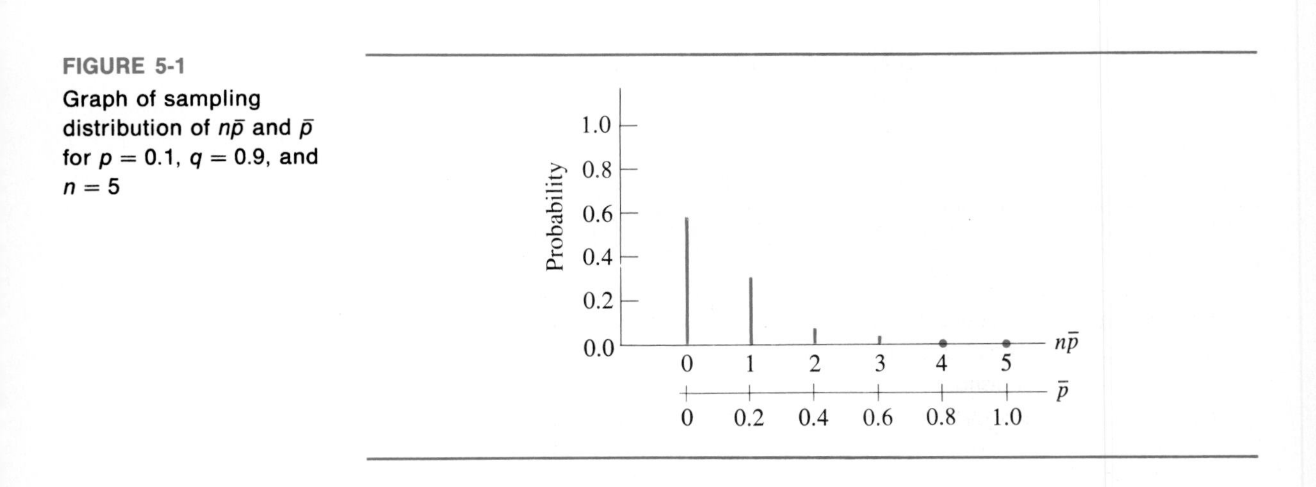

FIGURE 5-1
Graph of sampling distribution of $n\bar{p}$ and $\bar{p}$ for $p = 0.1$, $q = 0.9$, and $n = 5$

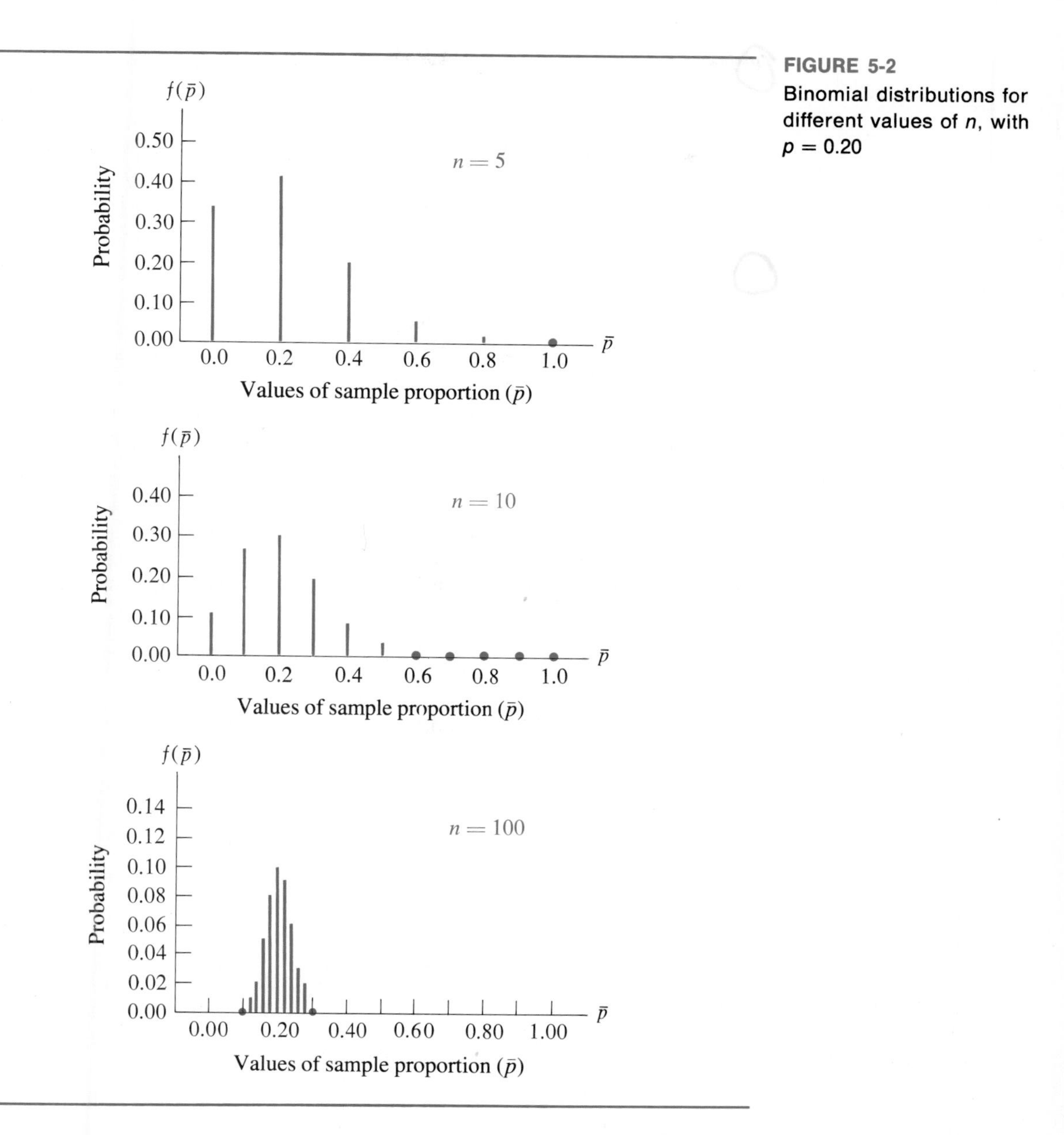

FIGURE 5-2
Binomial distributions for different values of n, with $p = 0.20$

Exercises 5.2

1. An inspector of a bottling company's assembly line draws 10 filled bottles at random for inspection. If a bottle contains within half an ounce of the proper amount, it is classified as good; otherwise, it is classified as defective.
   a. List the possible numbers of defective bottles that the inspector could obtain in a sample.
   b. Calculate the mean and standard deviation for the number of defectives and the proportion of defectives, assuming the probability of a defective is 0.10.

2. Assume that 40% of a large population favors a certain type of weapons-control legislation. If people selected at random are questioned in regard to the legislation, what are the mean and the standard deviation of the sampling distribution of the number favoring the legislation? Assume independence.
   a. If 20 people are questioned?
   b. If 40 people are questioned?
   c. If 80 people are questioned?
3. Last year, 5% of the companies headquartered in a certain industrialized nation lost money from normal operations. What are the mean and the standard deviation of the sampling distribution of the proportion of money losers in a random sample of companies drawn from this large population? Assume independence.
   a. Calculate first for a random sample of 20 companies.
   b. Perform the same calculation for a random sample of 40 companies.
4. At a certain large manufacturing company, 80% of the employees want to establish a trade union. For a random sample of 100 employees, calculate the mean and the standard deviation of the sampling distribution of the proportion of employees who want a trade union.
5. An accountant makes mistakes on 15% of the tax returns she prepares. During the recent tax season the accountant prepared a large number of returns. If the IRS samples 10 of the returns prepared by this accountant, calculate
   a. the probability that none of the returns are prepared incorrectly.
   b. the mean and the standard deviation of the sampling distribution of incorrectly prepared tax returns (of the 10 sampled).

5.3 CONTINUOUS DISTRIBUTIONS

Thus far, we have dealt solely with probability distributions of *discrete* random variables. Probability distributions of *continuous* random variables are also of considerable importance in statistical theory. We turn now to an examination of such distributions, with particular emphasis on the meaning of their graphs. You may want to review the definitions of discrete and continuous variables given in section 3.1.

The binomial distribution that we have been discussing in this chapter is an example of a probability distribution of a *discrete* random variable. We have graphed such distributions by erecting **ordinates** (vertical lines) at distinct values along the horizontal axis. To gain better insight into the meaning of a graph of the probability distribution of a continuous random variable, let us graph a binomial distribution as a **histogram** (bar graph). We assume a situation in which a fair coin is tossed twice and the random variable of interest is the number of heads obtained. The probabilities of zero, one, and two heads are,

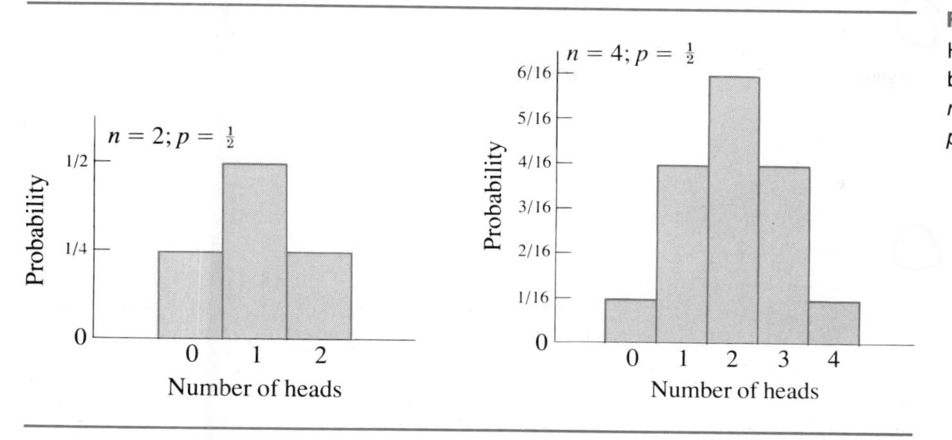

FIGURE 5-3 Histograms of the binomial distribution for $n = 2$, $p = \frac{1}{2}$, and $n = 4$, $p = \frac{1}{2}$

respectively, $\frac{1}{4}$, $\frac{1}{2}$, and $\frac{1}{4}$. This is an illustration of a binomial distribution in which $p = \frac{1}{2}$ and $n = 2$. If the coin is tossed four times, the probabilities of zero, one, two, three, and four heads are respectively, $\frac{1}{16}$, $\frac{4}{16}$, $\frac{6}{16}$, $\frac{4}{16}$, and $\frac{1}{16}$. This is a binomial distribution in which $p = \frac{1}{2}$ and $n = 4$. Graphs of these distributions in the form of histograms are given in Figure 5-3. Using these histograms, let us now interpret zero, one, two, three, and four heads not as discrete values, but rather as midpoints of classes whose respective limits are $-\frac{1}{2}$ to $\frac{1}{2}$, $\frac{1}{2}$ to $1\frac{1}{2}$, $1\frac{1}{2}$ to $2\frac{1}{2}$, and so on. The probabilities or relative frequencies associated with these classes are represented on the graphs by the areas of the rectangles or bars. Thus, in the graph for $n = 4$, since the rectangle for the class interval $2\frac{1}{2}$ to $3\frac{1}{2}$ has four times the area of that from $3\frac{1}{2}$ to $4\frac{1}{2}$, it represents four times the probability. If we were to represent the histogram for $n = 4$ with a smooth continuous curve, the curve would pass through the rectangle for three heads as shown in Figure 5-4(a). In Figure 5-4(b), the curve is simplified to a straight

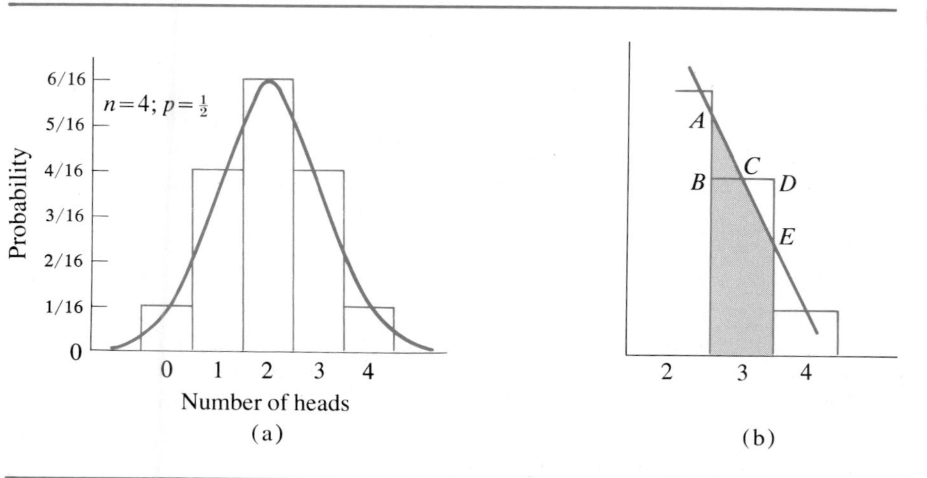

FIGURE 5-4 Approximation of a histogram by a continuous curve

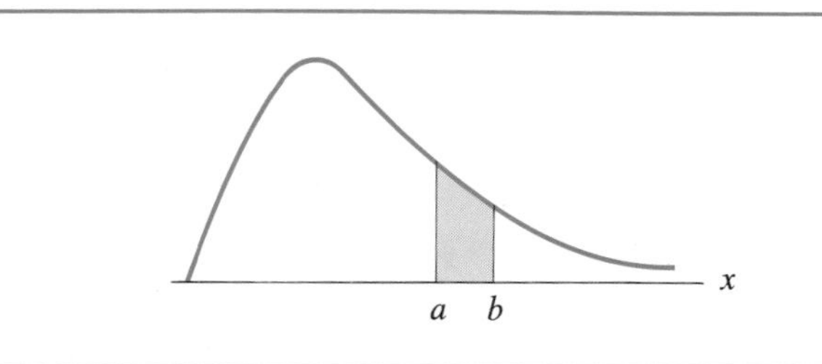

FIGURE 5-5
Graph of a continuous distribution: the shaded area represents the probability that the random variable X lies between a and b

line, and it is clear that the shaded area under the curve for the class interval $2\frac{1}{2}$ to $3\frac{1}{2}$ is approximately equal to the area of the rectangle representing the probability of three heads, because the included area *ABC* is about equal to the excluded area *CDE*.

Summary

> In the approximation of a histogram by a smooth curve, the area under the curve bounded by the class limits for any given class represents the probability of occurrence of that class.

In the foregoing illustration, if we had increased n greatly, say to 50 or 100, and decreased the width of the rectangles, the corresponding shape of the histogram would approach that of a continuous curve more closely.

> Since the total area of the rectangles in a histogram representing a probability distribution of a discrete random variable is equal to *one*, the total area under a continuous curve representing the probability distribution of a continuous random variable is correspondingly equal to *one*.

Furthermore, the area under the curve lying between the two vertical lines erected at points a and b on the x axis represents the probability that the random variable X takes on values in the interval a to b.[1] This is depicted in Figure 5-5.

[1] Let the value of the probability distribution of a random variable X at x be denoted $f(x)$. If X is discrete, the probability that X lies between a and b inclusive [in the closed interval (a, b)] is

$$P(a \leqslant X \leqslant b) = \sum_{x=a}^{b} f(x)$$

If X is continuous, the probability that X lies between a and b is

$$P(a \leqslant X \leqslant b) = \int_a^b f(x)\,dx$$

The reader acquainted with integral calculus can see that this definition in the continuous case is the counterpart of the summation in the discrete case. Also, it can be seen that the graphic interpretation of the probability in the continuous case is the area bounded by the curve whose value at x is $f(x)$, the x axis, and the ordinates at a and b. If the probability distribution is continuous at a and b, it makes no difference whether we consider $P(a \leqslant X \leqslant b)$ or $P(a < X < b)$, because the probability is zero that X is exactly equal to a or exactly equal to b.

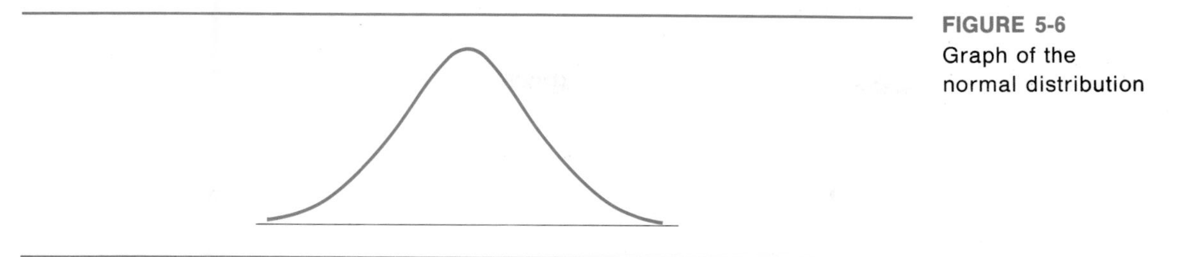

FIGURE 5-6
Graph of the normal distribution

In the continuous case, since there are an infinite number of points between a and b, the probability that X lies between a and b may be viewed as the sum of an infinite number of ordinates erected from a to b. Intuitively, we see this sum as identical with the area bounded by the curve, the horizontal axis, and the ordinates at a and b.

In the discrete case, $f(x)$ denotes the probability that a random variable X takes on the value x. In the continuous case, $f(x)$ cannot be interpreted as the probability of an event x, since there are an infinite number of x values and the probability of any one of them must be considered zero.[2]

For continuous random variables, probabilities can be interpreted graphically only in terms of *areas between* two values.

By mathematics beyond the scope of this text, it can be shown that if p is held fixed while n is increased without limit, in the binomial distribution, the distribution approaches a particular continuous distribution referred to as the **normal distribution, normal curve**, or **Gaussian distribution.**[3] Although our illustration has been for the case $p = \frac{1}{2}$, this is not a necessary condition for the proof. Even if the binomial distribution is not symmetrical (that is, $p \neq \frac{1}{2}$), it still approaches the normal distribution as n increases. The shape of the normal curve is shown in Figure 5-6. Actually, in the early mathematical derivations, the binomial variable x was expressed in **standard units**, that is, as

$$\frac{x - \mu_{n\bar{p}}}{\sigma_{n\bar{p}}} = \frac{x - np}{\sqrt{npq}}$$

and n was assumed to increase without limit. Modern proofs use other approaches to arrive at the same result.

A brief comment on standard units is useful at this point, because such units are widely employed, particularly in sampling theory and statistical inference. Standard units are merely an example of the previously mentioned standard score (see section 1.15).[4]

[2] For example, $P(X = a) = \int_a^a f(x)\,dx = 0$.

[3] After the German mathematician and astronomer Karl Friedrick Gauss.

[4] Other terms used to refer to standard scores or standard units include *standardized unit*, *standardized form*, and *standard form*.

The **standard score** is the deviation of a value from the mean stated in units of the standard deviation.

In general, it is of the form $(x - \mu)/\sigma$, where x denotes the value of the item and μ and σ are the mean and standard deviation of the distribution. In the case of a binomially distributed random variable, as indicated in Table 5-5, the mean and standard deviation of X (the number of successes in n trials) are, respectively, np and $\sqrt{npq}$. Hence, the standard score or standard unit is $(x - np)/\sqrt{npq}$.

5.4 THE NORMAL DISTRIBUTION

The normal distribution plays a central role in statistical theory and practice, particularly in the area of statistical inference. Because of the relationship we mentioned in section 5.3, the normal distribution is useful as an approximation to the binomial distribution in many instances when the latter is the theoretically correct one. We shall see that calculations involving the normal curve are generally much easier than those involving the binomial distribution because of the simple, compact form of tables of areas under the normal curve.

In addition to its use as an approximation to the binomial distribution, the normal distribution is important in its own right in sampling applications. Before we consider such applications, let us examine the basic properties of the normal distribution.

Properties of the Normal Curve

Probability distributions of continuous random variables can be described by the same types of measures (such as means, medians, and standard deviations) as are used for discrete random variables.

An important characteristic of the normal curve is that we need know only the mean and standard deviation to compute the entire distribution.

The normal probability distribution is defined by the equation

$$f(x) = \frac{1}{\sqrt{2\pi}\sigma} e^{-(1/2)[(x-\mu)/\sigma]^2} \tag{5.2}$$

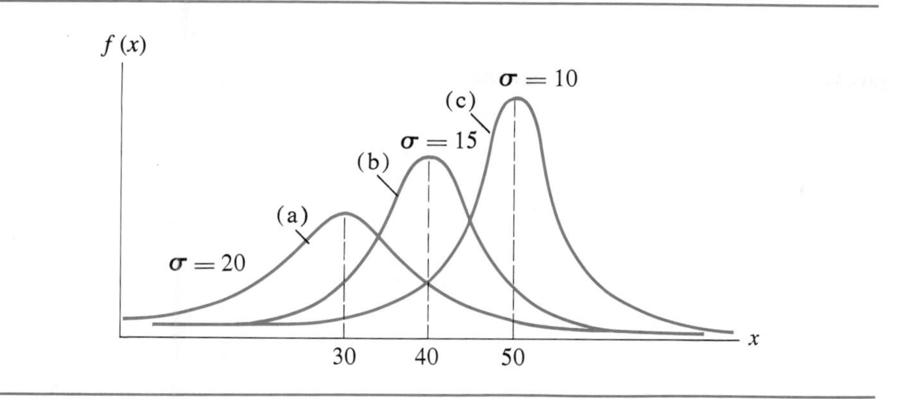

FIGURE 5-7
Three normal probability distributions

In this equation, the mean μ and the standard deviation σ, which determine the location and spread of the distribution, are said to be the two parameters of the normal distribution. This is analogous to the situation for the binomial distribution, in which the parameters are n and p.[5] Thus, for given values of μ and σ, if we substitute a value of x into equation 5.2, we can compute the corresponding value of $f(x)$. Following the usual convention, the values x of the random variable of interest are plotted along the horizontal axis and the corresponding ordinates $f(x)$ along the vertical axis. Figure 5-7 shows three normal probability distributions that differ in their locations and spreads. Distribution (c) has the largest mean (50), and distribution (a) has the smallest mean (30). On the other hand, the standard deviation of (c), which is 10, is the smallest of the three, while the standard deviation of (a), which is 20, is the largest. Thus, the normal distribution defined by equation 5.2 represents a family of distributions, with each specific member of that family determined by particular values of the parameters μ and σ.

Graphically, a normal curve is bell-shaped and symmetrical around the ordinate erected at the mean, which lies at the center of the distribution. Recall that the total area under the graph of a continuous probability distribution is equal to one; since one-half of the area (representing probability) lies to the left (right) of the mean, the probability is 0.5 that a value of x will fall below (above) the mean. The values of x range from minus infinity to plus infinity. As we move farther away from the mean, either to the right or to the left, the ordinates $f(x)$ get smaller. Thus, moving in either direction from the mean, the curve is **asymptotic** to the x axis; that is, the curve gets closer to the horizontal axis but never reaches it. However, for practical purposes, we rarely need to consider

[5] The numbers π and e are simply constants that arise in the mathematical derivation; their approximate values are $\pi = 3.1416$ and $e = 2.7183$. The number π is the familiar quantity that appears in numerous mathematical formulas, such as the expression for the area of a circle, $A = \pi r^2$, where A denotes the area and r the radius. The constant e is the base of the natural logarithm system, as indicated in the discussion of the Poisson distribution in section 3.7.

x values lying beyond three or four standard deviations from the mean, since nearly the entire area is included within this range. Stated differently, there is virtually no area in the tails of a normal distribution beyond three or four standard deviations from the mean.

Areas under the Normal Curve

We now turn to the use of the areas under the normal curve. Although it was important to define the distribution as in equation 5.2 in order to observe the relationship between x values and $f(x)$ values, in most applications in statistical inference we are not interested in the ordinates of the curve. Rather, since the normal curve is a continuous distribution and since it is a useful probability distribution, we are interested in the areas under the curve.

Normally distributed variables

It is convenient to use the term **normally distributed** for variables that have normal probability distributions, and we shall do so here. The term "normal" merely refers to probability distributions described by equation 5.2; it does not imply that other distributions are in some sense "abnormal." Normally distributed variables occur in a variety of units, such as dollars, pounds, inches, and hours. Any normally distributed variable can be transformed into a form applicable to a single table of areas under the normal curve, regardless of the units of the original data. The transformation used for this purpose is that of the standard unit or, as it is often called in the case of a normal distribution, the **standard score**. As we noted earlier, to express an observation of a variable in standard units, we obtain the deviation of this observation from the mean of the distribution and then state this deviation in multiples of the standard deviation. For example, suppose a variable is normally distributed with mean 100 pounds and standard deviation 10 pounds. If one observed value of this variable is 120 pounds, what is this number in standardized units? The deviation of 120 pounds from 100 pounds is +20 pounds, in units of the original data. Dividing +20 pounds by 10 pounds, we obtain +2. Thus, if one standard deviation equals 10 pounds, a deviation of +20 pounds from the mean lies *two standard deviations* above the mean.

Standard score

Let us state this notion in general form. The standard score, that is, the number of standard units z for an observation x from a probability distribution is defined by

$$z = \frac{\text{Value} - \text{Mean}}{\text{Standard deviation}} = \frac{x - \mu}{\sigma} \tag{5.3}$$

where x = the value of the observation
μ = the mean of the distribution
σ = the standard deviation of the distribution

In the illustration, $z = \frac{120-100}{10} = \frac{20}{10} = +2$. The "+2" indicates a value lying two standard deviations *above* the mean. If the observation had been 80, we would have $z = \frac{80-100}{10} = \frac{-20}{10} = -2$. The "$-2$" denotes a value lying two standard deviations *below* the mean.

We now turn to another example to illustrate the use of a table of areas under the normal curve. The Watts Renewal Corporation has a manufacturing process that produces light bulbs whose lifetimes are normally distributed with an arithmetic mean of 1,000 hours and a standard deviation of 200 hours. Figure 5-8 shows the relationship between values of the original variable (x values) and values in standard units (z values).

Suppose we wish to determine the proportion of light bulbs produced by this process with lifetimes between 1,000 and 1,400 hours, indicated by the shaded area in Figure 5-8. We can obtain this value from Table A-5 of Appendix A, which gives areas under the normal curve lying between vertical lines erected at the mean and at specified points above the mean stated in multiples of standard deviations (z values).

- The left column of Table A-5 gives z values to one decimal place.
- The column headings give the second decimal place of the z value.
- The entries in the body of Table A-5 represent the area included between the vertical line at the mean and the line at the specified z value.

Thus, returning to our example, the z value for 1,400 hours is $z = (1{,}400 - 1{,}000)/200 = +2$. In Table A-5, we find the value 0.4772; hence 47.72% of the area in a normal distribution lies between the mean and a value two standard deviations above the mean. We conclude that 0.4772 is the proportion of light bulbs produced by this process with lifetimes between 1,000 and 1,400 hours.

We now note a general point about the distribution of z values. Comparing the x scales and z scales in Figure 5-8, we see that for a value at the

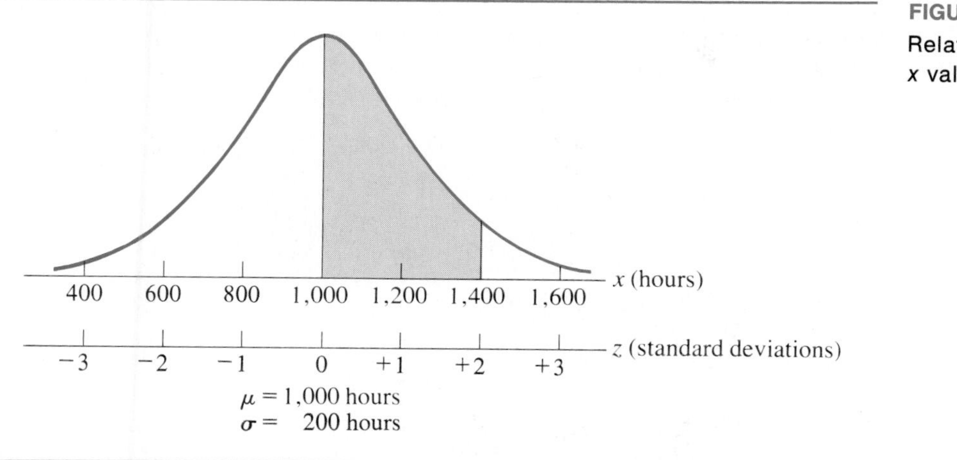

FIGURE 5-8
Relationship between x values and z values

mean in the distribution of x, $z = 0$. If an x value is at $\mu + \sigma$ (that is, one standard deviation above the mean), $z = +1$, and so on. Therefore the probability distribution of z values, referred to as the **standard normal distribution**, is simply a normal distribution with mean zero and standard deviation one.[6]

EXAMPLE 5-1

What is the proportion of light bulbs produced by the Watts Renewal Corporation with lifetimes between 800 and 1,200 hours?

Solution

First, we transform these values to deviations from the mean in units of the standard deviation.

$$\text{If } x = 1{,}200, \quad z = \frac{1{,}200 - 1{,}000}{200} = +1$$

$$\text{If } x = 800, \quad z = \frac{800 - 1{,}000}{200} = -1$$

Thus, we want to determine the area in a normal distribution that lies within one standard deviation of the mean. Table A-5 of Appendix A gives entries only for positive z values. However, since the normal distribution is symmetrical, the area between the mean and a value one standard deviation *below* the mean is the same as the area between the mean and a value one standard deviation *above* the mean. From Table A-5 we find that 34.13% of the area lies between the mean and a value one standard deviation above the mean. Hence, we double this area to find that about 68.3% of the light bulbs produced by this process have lifetimes between 800 and 1,200 hours. The required area is shown in Figure 5-9(a). Note that in general, about 68.3% of the area in a normal distribution lies within one standard deviation of the mean.

EXAMPLE 5-2

What is the proportion of light bulbs produced by the Watts Renewal Corporation with lifetimes between 1,150 and 1,450 hours?

Solution

Both 1,150 and 1,450 lie above the mean of 1,000 hours. We can determine the required probability by obtaining (1) the area between the mean and 1,450 and (2) the area between the mean and 1,150 and then subtracting (2) from (1).

$$\text{If } x = 1{,}450 \quad z = \frac{1{,}450 - 1{,}000}{200} = \frac{450}{200} = 2.25$$

$$\text{If } x = 1{,}150, \quad z = \frac{1{,}150 - 1{,}000}{200} = \frac{150}{200} = 0.75$$

Table A-5 gives 0.4878 as the area corresponding to $z = 2.25$ and 0.2734 for $z = 0.75$.

[6] We note that Table A-5 gives values of the integral

$$\int_0^{z_0} f(z)\,dz \quad \text{where } z_0 = \frac{x_0 - \mu}{\sigma} \quad \text{and } f(z) = \frac{1}{\sqrt{2\pi}} e^{-z^2/2}$$

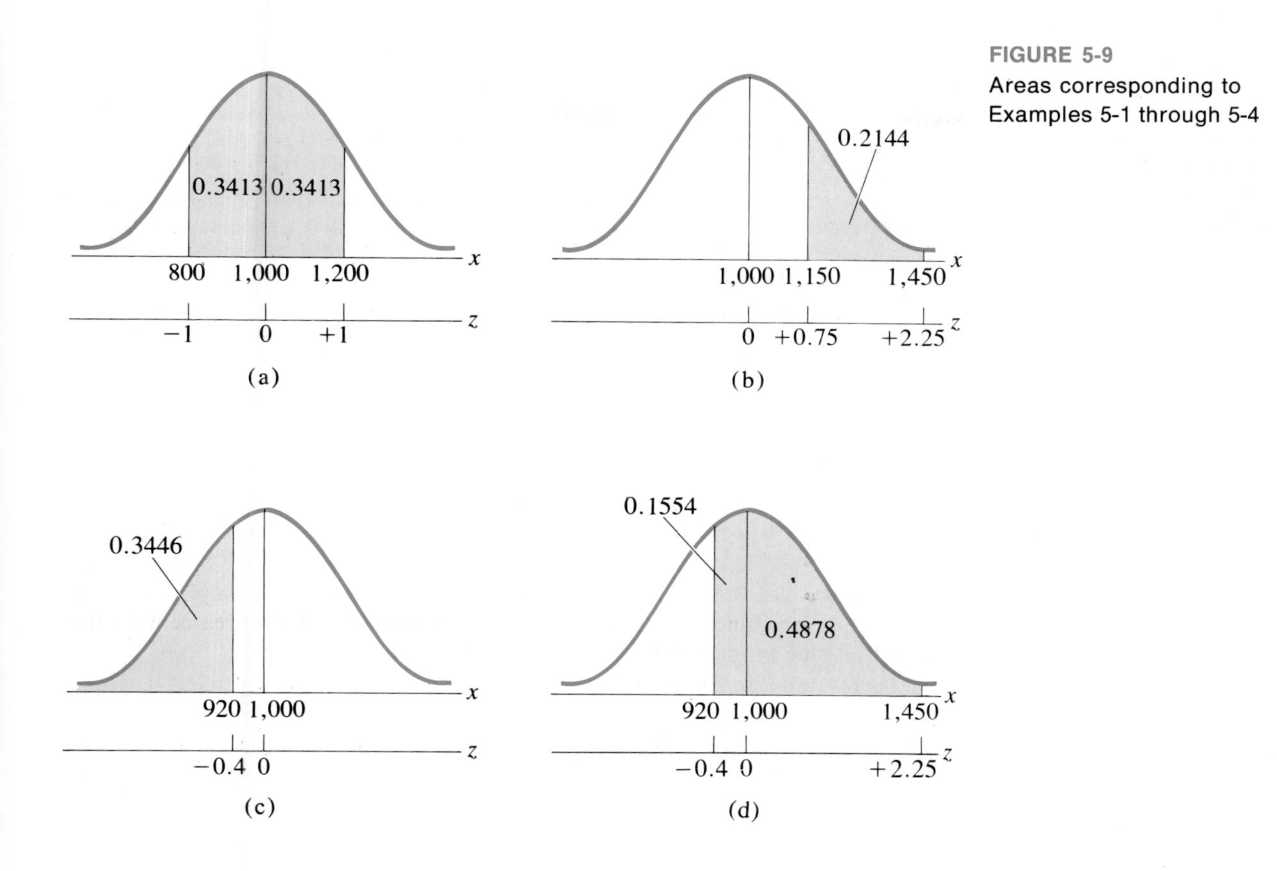

FIGURE 5-9 Areas corresponding to Examples 5-1 through 5-4

Subtracting 0.2734 from 0.4878 yields 0.2144, or 21.44%, as the result. This area is shown in Figure 5-9(b).

EXAMPLE 5-3

What is the proportion of light bulbs produced by the Watts Renewal Corporation with lifetimes less than 920 hours?

Solution

The observation "920 hours" lies below the mean. We solve this problem by determining the area between the mean and 920 and subtracting this value from 0.500, which is the entire area to the left of the mean.

$$\text{If } x = 920, \quad z = \frac{920 - 1{,}000}{200} = -0.40$$

Only positive z values are shown in Table A-5, but we look up $z = 0.40$ and find 0.1554, which is also the area between the mean and $z = -0.40$. Subtracting 0.1554 from 0.5000 gives the desired result, 0.3446. The area corresponding to this probability is shown in Figure 5-9(c).

EXAMPLE 5-4

What is the proportion of light bulbs produced by this process with lifetimes between 920 and 1,450 hours?

Solution

Since 920 lies below the mean and 1,450 lies above the mean, we determine (1) the area lying between 920 and the mean and (2) the area lying between 1,450 and the mean, and we add (1) to (2). The respective z values for 920 and 1,450 were previously determined as -0.40 and $+2.25$, with corresponding areas of 0.1554 and 0.4878. Adding these two figures, we obtain 0.6432 as the proportion of light bulbs with lifetimes between 920 and 1,450 hours. The corresponding area is shown in Figure 5-9(d).

We stated that in the normal distribution, the range of the x variable extends from minus infinity to plus infinity. Yet, in Examples 5-1, 5-2, 5-3, and 5-4, negative lifetimes are not possible. This illustrates the point that a variable may be said to be normally distributed provided that the normal curve constitutes a good fit to its empirical frequency distribution within a range of about three standard deviations from the mean. Since virtually all the area is included in this range, the situation in the tails of the distribution is considered negligible.

It is useful to note the percentages of area that lie within integral numbers of standard deviations from the mean of a normal distribution. These values have been tabulated in Table 5-6. Hence, as we observed in Example 5-1, about 68.3% of the area in a normal distribution lies within plus or minus one standard deviation from the mean. The reader should verify the other figures from Table A-5. Let us restate these probability figures in terms of rough statements of odds.

Since about two-thirds of the area in a normal distribution lies within one standard deviation, the odds are about two to one that an observation will fall within that range.

Correspondingly, the odds are about 95:5, or 19:1, for the two standard deviation range and 997:3, or about 332:1, for three standard deviations.

TABLE 5-6

Percentages of area that lie within specified intervals around the mean in a normal distribution

Interval	Percentage of Area
$\mu \pm \sigma$	68.3
$\mu \pm 2\sigma$	95.5
$\mu \pm 3\sigma$	99.7

The Normal Curve as an Approximation to the Binomial Distribution

In section 5.3, we indicated that the normal curve can be used as an approximation to the binomial distribution for the calculation of probabilities for which the binomial distribution is the theoretically correct distribution. This approximation is possible because the binomial distribution approaches the normal distribution when n becomes large. In general, the approximations are better when the value of p in the binomial distribution is close to $\frac{1}{2}$ than when p is close to zero or one, because for $p = \frac{1}{2}$ the binomial distribution is symmetrical, and as we have seen, the normal curve is a symmetrical distribution. However, the normal distribution often provides surprisingly good approximations even when $p \neq \frac{1}{2}$ and even when n is not very large.

A popular rule states that the normal distribution is an appropriate approximation to the binomial distribution when both $np \geqslant 5$ and $n(1 - p) \geqslant 5$.

Under these conditions, the binomial distribution can be closely approximated by a normal curve with the same mean and standard deviation. We illustrate the use of the normal curve as an approximation to the binomial distribution by two examples. Example 5-5 illustrates the approximation of the probability of a single term in the binomial distribution by a normal curve calculation. Example 5-6 illustrates a corresponding calculation for a sum of terms in the binomial distribution.

EXAMPLE 5-5

Assume that 20% of a large population smoke at least one pack of cigarettes a day. What is the probability that a randomly drawn sample of 20 individuals will contain exactly 4 who smoke at least one pack a day?

Solution

Using equation 3.3 for the binomial distribution with $n = 20$, $p = 0.20$, and $q = 0.80$, we have

$$P(X = 4) = f(4) = \binom{20}{4}(0.80)^{16}(0.20)^4$$

This probability is evaluated from Table A-1 of Appendix A as

$$P(X = 4) = F(4) - F(3) = 0.6296 - 0.4114 = 0.2182$$

In order to obtain the normal curve approximation to this probability of exactly 4 such smokers, we set up a normal curve with the same mean and standard deviation as the given binomial distribution and find the area between 3.5 and 4.5, as shown in Figure 5-10. We obtain the area between 3.5 and 4.5 because the random variable in the binomial distribution is discrete, whereas in the case of the normal curve, it is continuous. Hence,

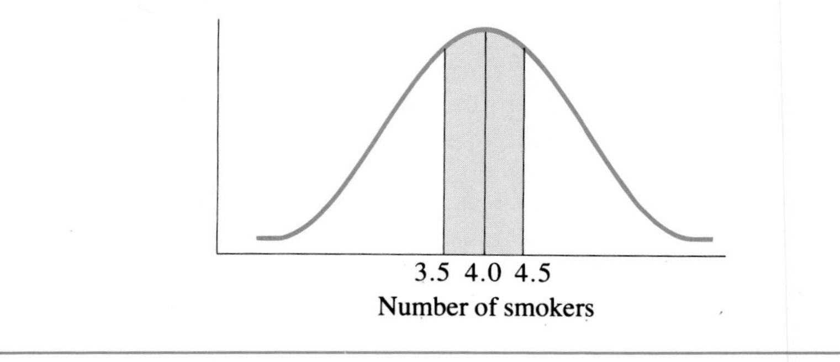

FIGURE 5-10
Area under the normal curve for the probability of obtaining exactly 4 individuals who smoke at least one pack of cigarettes a day in a randomly drawn sample of 20 individuals: $p = 0.20$

as shown in Figure 5-11, if the binomial probabilities are depicted graphically as a histogram, the true probability of 4 occurrences is given by the area of the rectangles centered at 4.0. To approximate this area by a corresponding area under the normal curve, we treat 4 smokers as the value at the midpoint of a class whose limits are 3.5 and 4.5. The mean and standard deviation of the binomial distribution in this problem are.

$$\mu = np = (20)(0.20) = 4$$

$$\sigma = \sqrt{npq} = \sqrt{(20)(0.20)(0.80)} = 1.79$$

Using these numbers as the mean and standard deviation of the approximating normal curve, we calculate the z values for 3.5 and 4.5 as follows:

$$z_1 = \frac{3.5 - 4}{1.79} = -0.28 \qquad z_2 = \frac{4.5 - 4}{1.79} = 0.28$$

The area corresponding to a z value of 0.28 is 0.1103, and doubling this yields the desired approximation, $2(0.1103) = 0.2206$. Hence, 0.2206 is the normal curve approximation to the true binomial probability, which is 0.2182.

EXAMPLE 5-6

Referring to Example 5-5, what is the probability that a randomly drawn sample of 20 individuals will contain 3 or more persons who smoke at least one pack of cigarettes a day?

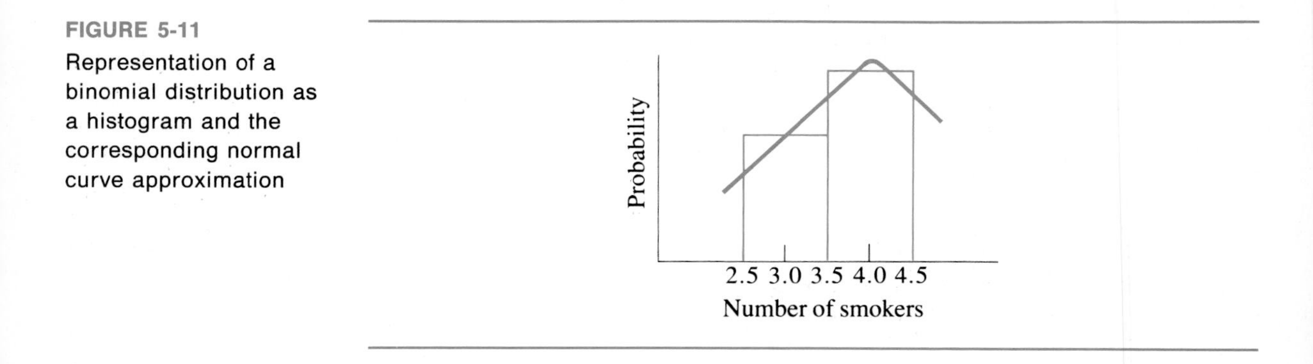

FIGURE 5-11
Representation of a binomial distribution as a histogram and the corresponding normal curve approximation

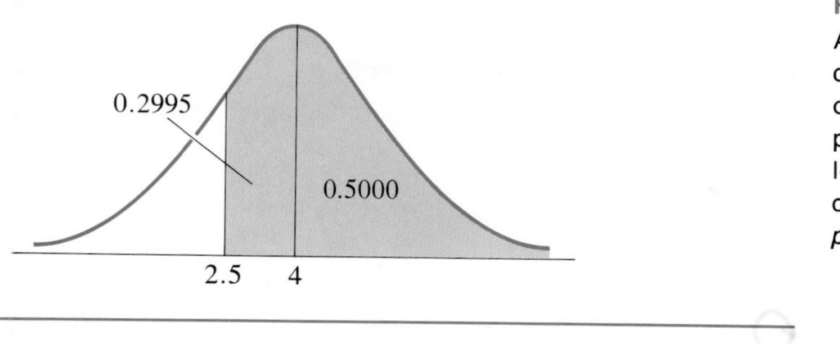

FIGURE 5-12
Area under the normal curve for the probability of obtaining 3 or more persons who smoke at least one pack of cigarettes a day: $p = 0.20$

Solution

Summing the appropriate terms in equation 3.3, we find

$$P(X \geqslant 3) = \sum_{x=3}^{20} f(x) = \sum_{x=3}^{20} \binom{20}{x} (0.80)^{20-x}(0.20)^{x}$$

This probability is evaluated from Table A-1 of Appendix A as

$$P(X \geqslant 3) = 1 - F(2) = 1 - 0.2061 = 0.7939$$

The corresponding normal curve approximation is shown graphically in Figure 5-12. The z value for 2.5 is calculated as follows:

$$z = \frac{2.5 - 4}{1.79} = -0.84$$

Therefore, the desired area is $0.2995 + 0.5000 = 0.7995$. The closeness of this approximation to the true binomial probability of 0.7939 illustrates that normal curve approximations involving sums of terms usually are closer to the true probabilities than are approximations for individual terms in the binomial distribution.

When the probability of a *single term* in the binomial distribution is desired, as in Example 5-5, it is necessary to use an interval *one unit wide* centered on that term, because the binomial distribution is discrete whereas the normal curve is continuous. On the other hand, this correction is often dispensed with when *sums of terms* in the binomial distribution are desired and the *sample size is large.* Thus, in Example 5-6, if the sample size had been, say, 100, we could have used the z value for 3 rather than for 2.5 in calculating the probability of 3 or more persons who smoke at least one pack of cigarettes a day. Problems at the end of this section that require the use of this "continuity correction" are identified.

Exercises 5.4

1. State whether the following random variables are discrete or continuous.
   a. The weight of an adult female.
   b. The number of errors made in one week of a computer billing operation.

   c. $X = 0$ if the family owns its home and $X = 1$ if the family rents.
   d. Tensile strength of an aluminum bar in pounds per square inch.

2. The weights of bags of sugar have a mean of 50 pounds and a standard deviation of 10 pounds. Assuming a normal distribution, what is the probability that a bag weighs
   a. between 45 and 62 pounds?
   b. less than 50 pounds?
   c. more than 67 pounds?
   d. less than 38 pounds?
3. The scores on an achievement test given to high school students throughout the United States are normally distributed with a mean score of 500 and a standard deviation of 100.
   a. What is the probability that a test score is less than 600?
   b. What is the probability that a test score is between 450 and 650?
   c. What is the probability that a test score is between 600 and 650?
   d. The probability is 0.85 that a test score is more than what value?
4. The average length of life of a certain brand of car is 15 years with a standard deviation of 2.5 years. Assuming a normal distribution, find the probability that this type of car has fewer than 10 years of life.
5. The weight of Greenbay Brand cookies packed in a box is normally distributed with a standard deviation of 1.6 ounces. If 4% of these boxes of cookies weigh more than 8 ounces, calculate the average weight of Greenbay Brand cookies.
6. The lifetime of a particular model of a stereo cartridge is normally distributed with a mean of 1,520 hours and a standard deviation of 125 hours. Find the probability that one of these cartridges will last
   a. more than 1,750 hours
   b. at most, 1,480 hours
   c. between 1,420 and 1,480 hours
   d. between 1,420 and 1,750 hours
7. For a 4-week period, 80 turkeys were fed a new type of diet. The weight of the turkeys increased by an average of 2.9 pounds with a 0.2-pound standard deviation. Assuming that the increases in weight were normally distributed, estimate the number of turkeys whose weight gain was between 3.0 and 3.2 pounds.

8. In answering parts (a) and (b), use the normal approximation to the binomial distribution with the continuity correction. The production manager at a tire manufacturing company reports that 40 tires are classified as defective in an experimental run of 100 tires produced yesterday. If 15 tires are randomly chosen from the run of 100 tires with replacement, what is the probability that
   a. 5 tires are defective?
   b. between 4 and 7 tires are defective?

9. The probability is 0.2 that any customer who enters a certain store will purchase a carton of Alpine Milk. If 2,500 customers enter the store, what is the minimum number of cartons of Alpine Milk the store must have on hand if the probability that it will run out of this product is to be at most 1%? Assume independence.
10. According to Pulp Paper Company, 5% of all trees cut down cannot be used in the processing of paper. Assume independence, and suppose that 20 trees are cut down in a given period.
    a. Find the exact probability that 2 or more trees will not be usable for paper production.
    b. Use the normal approximation to the binomial distribution with the continuity correction factor to obtain an approximation to the exact probability.
11. The Crude Oil Company has two main oil fields, one in the Alaskan tundra region and one in the Texas Panhandle. The company has enjoyed 25% success in finding retrievable oil deposits with its drillings in the Alaskan field, whereas its successful drillings in Texas have been only 20% of the total drillings. Next year, the company plans 400 new drillings in each field. If the success percentages stay the same, what are the probabilities of the company establishing new successful oil wells in Alaska and in Texas? Assume independence.
    a. What is the probability of establishing 90 or more in Alaska?
    b. What is the probability of establishing 90 or more in Texas?

5.5 SAMPLING DISTRIBUTION OF THE MEAN

In sections 5.1 and 5.2, we discussed sampling distributions of numbers of occurrences and percentages of occurrences. We now turn to another important probability distribution, namely, the sampling distribution of the arithmetic mean. For brevity, we shall use the term **the sampling distribution of the mean,** or simply **the sampling distribution of $\bar{x}$.**[7] To illustrate the nature of this distribution, let us return to the Watts Renewal Corporation manufacturing process that produces light bulbs whose lifetimes are normally distributed with an arithmetic mean of 1,000 hours and a standard deviation of 200 hours. We now interpret this distribution as an infinite population from which simple random samples can be drawn. It is possible for us to draw a large number of such samples of a given size, say $n = 5$, and compute the arithmetic mean lifetime of the five light bulbs in each sample. In accordance with our usual terminology,

[7] Although we used the symbol $\bar{X}$ in Chapter 3 to denote a sample mean, we will henceforth use instead the symbol $\bar{x}$ in sampling theory and statistical inference. The lower-case notation is more convenient because of the use of $\bar{x}$ as a subscript.

each such sample mean is referred to as a statistic. Since these statistics will usually differ from one another, we can consider them values of a random variable for which we can construct a frequency distribution. The universe mean of 1,000 hours is the parameter around which these sample statistics will be distributed, with some sample means lying below 1,000 and some lying above it. If we draw any finite number of samples, the sampling distribution is referred to as an **empirical sampling distribution**. On the other hand, if we conceive of drawing all possible samples of the given size, the resulting sampling distribution is a **theoretical sampling distribution**. Statistical inference is based on these theoretical sampling distributions, which are nothing more than probability distributions of the relevant statistics.

> In most practical situations, only one sample is drawn from a statistical population in order to test a hypothesis or to estimate the value of a parameter.

The work implied in generating a sampling distribution by drawing repeated samples of the same size is virtually never carried out, except perhaps as a learning experience. However, the reader must realize that the sampling distribution provides the underlying theoretical structure for decisions based on single samples.

Sampling from Normal Populations

What are the salient characteristics of the sampling distribution of the mean, if samples of the same size are drawn from a population in which values are normally distributed? To answer this question, we begin by assuming that the sample size is 5. In terms of our problem, this means that a random sample of 5 light bulbs is drawn from the above-mentioned population, and the mean lifetime of these 5 bulbs, denoted $\bar{x}_1$, is determined. Then, another sample of 5 bulbs is drawn, and the mean $\bar{x}_2$ is determined. Let us assume that the first mean is 990 hours, which falls below the population mean, and that the second mean is 1,022 hours, which lies above the population mean. The theoretical frequency distribution of $\bar{x}$ values of all such simple random samples of 5 bulbs would constitute the sampling distribution of the mean for samples of size 5.

Characteristics of the sampling distribution

Intuitively, we can see what some of the characteristics of such a distribution might be. A sample mean would be just as likely to lie above the population mean of 1,000 hours as below it. Small deviations from 1,000 hours would occur more frequently than large deviations. Furthermore, because of the effect of averaging, we would expect less dispersion or spread among these sample means than among the values of the individual items in the original population; that is, the standard deviation of the sampling distribution of the mean should be less than the standard deviation of the values of individual items in the population.

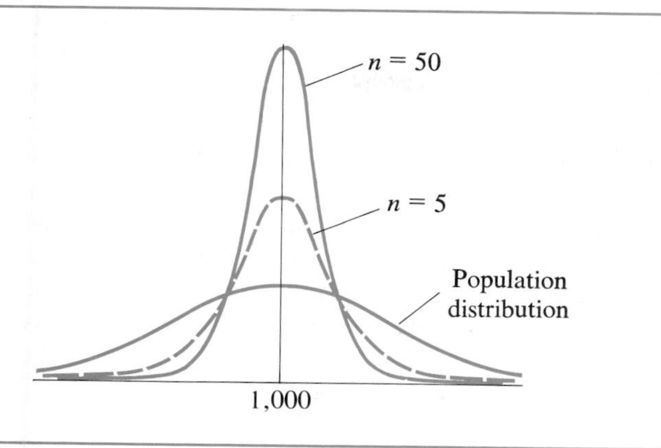

FIGURE 5-13 Relationship between a normal population distribution and normal sampling distributions of the mean for $n = 5$ and $n = 50$

Other characteristics of sampling distributions of the mean might be noted. If samples of size 50 rather than 5 had been drawn, another sampling distribution of the mean would be generated. Again, we would expect the means of these samples to cluster around the population mean of 1,000 hours. However, we would expect to find even less dispersion among these sample means than in the case of samples of size 5, because the larger the sample, the closer the sample mean is likely to be to the population mean. Thus, the standard deviation of the sampling distribution, which measures chance error inherent in the process of using samples to approximate population values, would decrease with increasing sample size. Another characteristic of these sampling distributions, which is not at all intuitively obvious but can be proved mathematically, is that if the original population distribution is normal, sampling distributions of the mean will also be normal. Figure 5-13 displays the relationships we have just discussed for the case of a normal population. For the population distribution, the horizontal axis represents values of individual items (x values). For the sampling distributions, the horizontal axis represents the means of samples of size 5 and 50. Since all three of the distributions are probability distributions of continuous random variables, the vertical axis pertains to probability densities.

The foregoing material introduces the following theorem:

Theorem 5.1

If a random variable X is normally distributed with mean μ and standard deviation σ, then the random variable "the mean $\bar{x}$ of a simple random sample of size n"[8] is also normally distributed with mean $\mu_{\bar{x}} = \mu$ and standard deviation

$$\sigma_{\bar{x}} = \frac{\sigma}{\sqrt{n}}$$

[8] See footnote 7.

In this statement of the theorem, we have used somewhat more formal language than in the preceding discussion. Instead of saying that the values of individual items in a population are normally distributed, we refer to normal distributions of the random variables X and $\bar{x}$.

An interesting aspect of the theorem is that the expected value (mean) of the sampling distribution of the mean, symbolized $\mu_{\bar{x}}$, is equal to the original population mean μ. This relationship is proved in rule 13 of Appendix C for the more general case of simple random samples of size n from *any* infinite population. The standard deviation of the sampling distribution of the mean—usually referred to as the **standard error of the mean** and denoted $\sigma_{\bar{x}}$—is given by

$$\sigma_{\bar{x}} = \frac{\sigma}{\sqrt{n}} \tag{5.4}$$

This relationship is proved in rule 12 of Appendix C, again for the more general case of sampling from any infinite population.

Important implications

Important implications follow from equation 5.4. We can think of any sample mean $\bar{x}$ as an estimate of the population mean μ. The difference between the statistic $\bar{x}$ and the parameter μ, $\bar{x} - \mu$, is referred to as a **sampling error**. (For example, if $\bar{x}$ were exactly equal to μ and were used as an estimate of μ, there would be no sampling error.) Therefore $\sigma_{\bar{x}}$, which is a measure of the spread of the $\bar{x}$ values around μ, is a measure of **average sampling error**; that is, it measures the amount by which $\bar{x}$ can be expected to vary from sample to sample. Another interpretation is that $\sigma_{\bar{x}}$ is a measure of the *precision* with which μ can be estimated using $\bar{x}$. Referring to equation 5.4, we see that $\sigma_{\bar{x}}$ varies directly with the dispersion in the original population σ and inversely with the square root of the sample size n. Thus, the greater the dispersion among the items in the original population, the greater the expected sampling error in using $\bar{x}$ as an estimate of μ; similarly, the smaller the population dispersion, the smaller the expected sampling error. In the limiting case in which every item in the population has the same value, the population standard deviation is zero; therefore, the standard error of the mean is also zero. This indicates that the mean of a sample from such a population would be a perfect estimate of the corresponding population mean, since there could be no sampling error. For example, if every item in the population weighed 100 pounds, the population mean weight would be 100 pounds. All the items in any sample would weigh 100 pounds, and the sample mean would be 100 pounds; thus, the sample mean would estimate the population mean with perfect precision. As the population dispersion increases, estimation precision decreases.

The fact that the standard error of the mean varies inversely with the square root of sample size means that there is a diminishing return in sampling effort.

Quadrupling sample size only halves the standard error of the mean; multiplying sample size by nine cuts the standard error only to one-third its previous value.

Sampling from Nonnormal Populations

We mentioned in the foregoing discussion that if a population is normally distributed, the sampling distribution of $\bar{x}$ is also normal. However, many population distributions of business and economic data are not normally distributed. What then is the nature of the sampling distribution of $\bar{x}$?

It is a remarkable fact that for almost all types of population distributions, the sampling distribution of $\bar{x}$ is approximately normal for sufficiently large samples.

This relationship between the shapes of the population distribution and of the sampling distribution of the mean has been summarized in what could be the most important theorem of statistical inference, the **Central Limit Theorem**. The theorem is stated in terms of the z variable for the sampling distribution of the mean and the approach of the distribution of this variable to the standard normal distribution. The formal statement of the theorem follows.

Central Limit Theorem

Central Limit Theorem

Theorem 5.2

If a random variable X, either discrete or continuous, has a mean μ and a finite standard deviation σ, then the probability distribution of $z = (\bar{x} - \mu)/\sigma_{\bar{x}}$ approaches the standard normal distribution as n increases without limit.

This theorem is a general one, since it makes no restrictions on the shape of the original population distribution. The requirement of a finite standard deviation is not a practical restriction at all, since virtually all distributions involved in real-world problems satisfy this condition. Note also that the z variable in this theorem is a transformation in terms of deviations of the sample mean $\bar{x}$ from the mean of the sampling distribution of means μ stated in multiples of the standard deviation of that distribution $\sigma_{\bar{x}}$. Thus, it is exactly the same type of transformation as was used in the illustrative examples given earlier in this chapter.

$$z = \frac{\text{Value} - \text{Mean}}{\text{Standard deviation}}$$

Here, however, all the values pertain to the sampling distribution of the mean.

It is useful at this point to restate Theorems 5.1 annd 5.2 in the less formal context of sampling applications and to summarize the importance of the concepts involved.

Theorem 5.1 restated

If a population distribution is normal, the sampling distribution of $\bar{x}$ is also normal for samples of all sizes.

Theorem 5.2 restated

If a population distribution is nonnormal, the sampling distribution of $\bar{x}$ may be considered approximately normal for a large sample.

The first of these theorems states that the sampling distribution of $\bar{x}$ will be *exactly* normal if the population is normal. The second, the Central Limit Theorem, assures us that no matter what the shape of the population distribution, the sampling distribution of $\bar{x}$ approaches normality as the sample size increases. For a wide variety of population types, samples do not even have to be very large for the sampling distribution of $\bar{x}$ to be approximately normal. For example, only in the case of highly skewed populations would the sampling distribution of $\bar{x}$ be appreciably skewed for samples larger than about 20. For most types of populations, the approach to normality is quite rapid as n increases.

Finite Population Multiplier

In our discussion of sampling distributions, we have dealt with infinite populations. However, many of the populations in practical problems are finite, as for example, the employees in a given industry, the households in a city, and the counties in the United States. As a practical matter, formulas already obtained for infinite populations can be applied in most cases to finite populations as well. In those cases when the results for infinite populations are not directly applicable, a simple correction factor is applied to the formula for the standard deviation of the relevant sampling distribution.

In simple random sampling from an *infinite population*, we have seen that the sampling distribution of $\bar{x}$ has a mean $\mu_{\bar{x}}$ equal to the population mean μ and a standard deviation $\sigma_{\bar{x}}$ equal to $\sigma/\sqrt{n}$.

Standard error of the mean for finite populations

In sampling from a *finite population*, the mean $\mu_{\bar{x}}$ of the sampling distribution of $\bar{x}$ again is equal to the population mean μ but the standard deviation (standard error of the mean) is given by the following formula.

$$\sigma_{\bar{x}} = \sqrt{\frac{N-n}{N-1}}\frac{\sigma}{\sqrt{n}} \tag{5.5}$$

Here N is the number of elements in the population and n is the number in the sample. The quantity $\sqrt{(N-n)/(N-1)}$ is usually referred to as the **finite population correction** or the **finite correction factor**. Thus, we see that in the case of a finite population, the standard error of the mean is equal to the finite population correction multiplied by $\sigma/\sqrt{n}$, which is the standard error of the mean in the infinite case. The finite population correction is approximately equal to one when the population size N is large relative to the sample size n; therefore, when we are choosing samples of size n from a much larger (but finite) population, the standard error of the mean $\sigma_{\bar{x}}$ is for practical purposes equal to $\sigma/\sqrt{n}$, as was the case when we sampled an infinite population.

To see why the finite correction factor is approximately one when population size is large relative to sample size, note that the factor $\sqrt{(N-n)/(N-1)}$ is approximately equal to $\sqrt{(N-n)/N}$ for large populations, since the subtraction of one in the denominator is negligible. We can now write

$$\sqrt{\frac{N-n}{N}} = \sqrt{1-\frac{n}{N}} = \sqrt{1-f}$$

where $f = n/N$ is referred to as the **sampling fraction** because it measures the fraction of the population contained in the sample. Thus, if the population size is $N = 1{,}000$ and the sample size is $n = 10$, then $f = \frac{10}{1{,}000} = \frac{1}{100}$. In such a case, the finite population correction is very close to one. Here, for example, $\sqrt{1-\frac{1}{100}} \approx 1$ (the symbol $\approx$ means "is approximately equal to"). In summary, in this case, $\sigma_{\bar{x}}$ is practically equal to $\sigma/\sqrt{n}$

A generally employed rule of thumb is that the formula $\sigma_x = \sigma/\sqrt{n}$ may be used whenever the size of the population is at least 20 times that of the sample, or in other words, whenever the sample represents 5% or less of the population.

A striking implication of equation 5.5 is that as long as the population is large relative to the sample, sampling precision becomes a function of sample size alone and does not depend on the proportion of the population sampled. Of course, we assume in this statement that the population standard deviation is constant. For example, let us assume a situation in which we draw a simple random sample of size $n = 100$ from each of two populations. Each population has a standard deviation equal to 200 units ($\sigma = 200$). In order to observe the effect of increasing the number of elements in the population, we further assume the population are of different sizes, namely, $N = 10{,}000$ and $N = 1{,}000{,}000$. The standard error of the mean for the population of 10,000 elements is, by equation 5.5,

$$\sigma_{\bar{x}} = \sqrt{\frac{10{,}000-100}{10{,}000-1}}\left(\frac{200}{\sqrt{100}}\right) \approx \sqrt{1}\left(\frac{200}{10}\right) \approx 20$$

For the population of 1,000,000 elements, we have

$$\sigma_{\bar{x}} = \sqrt{\frac{1,000,000 - 100}{1,000,000 - 1}}\left(\frac{200}{\sqrt{100}}\right) \approx \sqrt{1}\left(\frac{200}{10}\right) \approx 20$$

Thus, increasing the population size from 10,000 to 1,000,000 has virtually no effect on the standard error of the mean, since the finite population correction is approximately equal to one in both instances. Indeed, if the population size were increased to infinity, the same result would again be obtained for the standard error.

The finding that it is the absolute size of the sample, and not the proportion of the population sampled, that basically determines sampling precision is difficult for many people to accept intuitively. In fact, prior to the introduction of statistical quality control procedures in American industry, arbitrary methods such as sampling 10% of the items of incoming shipments, regardless of shipment size were quite common. Managers had vague feelings in these cases that approximately the same sampling precision was obtained by maintaining a constant sampling fraction. However, widely different standard errors resulted from large variations in the absolute sizes of the samples. The interesting principle that emerges from this discussion is this:

> For cases in which the populations are large relative to the samples, the absolute amount of work done (sample size), not the amount of work that might conceivably have been done (population size), is important in determining sampling precision.

We leave it to the reader's judgment whether this finding can be applied to other areas of human activity as well.

In our subsequent discussion of statistical inference, we will be concerned with measures of sampling error for proportions as well as for means. Therefore, we note the corresponding formula for the standard error of a proportion in sampling a finite population. In section 5.2, we indicated that the standard deviation of the sampling distribution of a proportion, which we now refer to as the **standard error of a proportion**, is given by $\sigma_{\bar{p}} = \sqrt{pq/n}$. Since our discussion referred to sampling as a **Bernoulli process**, it pertained to the sampling of an infinite population. The corresponding formula for the standard error of a proportion for a simple random sample of size n from a finite population is as follows.

Standard error of a proportion for finite populations

(5.6)
$$\sigma_{\bar{p}} = \sqrt{\frac{N-n}{N-1}}\sqrt{\frac{pq}{n}}$$

The same sorts of approximation considerations discussed in the case of the mean are pertinent here as well. Hence, if the size of the population is at least 20 times that of the sample, the formula for infinite populations may be used.

Other Sampling Distributions

In this chapter, we have discussed sampling distributions of numbers and proportions of occurrences and sampling distributions of the mean. Just as we were able to use the binomial distribution as a sampling distribution of numbers of occurrences under the appropriate conditions, other distributions may similarly be used under other sets of conditions. However, it is frequently far simpler to use normal curve methods based on the operation of the Central Limit Theorem. Two other continuous sampling distributions, which we have not yet examined, are important in elementary statistical methods: the student t distribution and the chi-square distribution. They will be discussed at the appropriate places in connection with statistical inference.

Exercises 5.5

1. Management of a wholesale outlet is considering a new sales campaign. It is known that the average purchase per customer is $200 and the standard deviation is $15. If 36 customers are randomly chosen, calculate the probability that the average (mean) purchase is more than $204.
2. A production manager is considering the purchase of new card-printing machines. On average, these machines can print 1,900 cards per day and the standard deviation is 200. Assume that the number of cards printed per day is normally distributed.
   a. What is the probability that a machine can print more than 2,100 cards per day?
   b. If 30 machines are randomly chosen from the manufacturing company, what is the probability that the mean number of cards printed per day is fewer than 1,850?
3. The personnel department of a manufacturing company found that 80% of the employees want to join a trade union. If a random sample of 100 employees is drawn, what is the probability that the sample will contain at least 75% who want to join a trade union?
4. It is known from national statistics that the mean family income in a certain suburban area is $3,160 per month with a standard deviation of $800 per month. In a random sample of 50 families drawn from this suburban area, what is the probability that the mean family income will be
   a. greater than $3,000?
   b. less than $3,000?
   c. between $3,200 and $3,300?
5. The family income distribution in a certain large city is characterized by skewness to the right. A census reveals that the mean family income is $26,000 and the standard deviation is $2,000. If a simple random sample of 100 families is drawn, what is the probability that the sample mean family income will differ from the city mean income of $26,000 by more than $200?

6. A specification calls for a drug to have a therapeutic effectiveness for a mean period of 50 hours. The standard deviation of the distribution of the period of effectiveness is known to be 16 hours. Shipments of the drug are to be accepted if the mean period exceeds 48 hours in a sample of 64 items drawn at random. Suppose the actual mean period of effectiveness of the drug in a given shipment is 44 hours. What is the probability that the shipment will be accepted when, in fact, it should not be?
7. The following is known about the clients of a successful stockbroker:
   (1) Rather large commissions are generated by 15% of the clients because they are "heavy traders."
   (2) The dollar sizes of the clients' accounts are normally distributed, with an arithmetic mean of $20,000 and a standard deviation of $1,500.
   a. If 10 accounts are randomly selected, what is the probability of obtaining exactly one "heavy trader"?
   b. If one account is selected, what is the probability that its size will be between $18,500 and $23,000?
   c. If a random sample of 36 accounts is selected, what is the probability that the sample mean will be between $20,200 and $20,400?
   d. If one account is randomly selected, the probability is 0.209 that its size will exceed a certain dollar amount. What is this dollar amount?
8. Do you agree or disagree with the following statements? Explain your answer.
   a. The probability that there are 10 defective items in a sample of 200 is the same as the probability that 95% of the items in a sample of 200 are not defective.
   b. You could be reasonably certain that you would get exactly 5,000 billion "heads" if you flipped a fair coin 10,000 billion times—because the probability that the proportion of "heads" equals $\frac{1}{2}$ converges to 1 as the number of tosses increases to infinity.
   c. According to the Central Limit Theorem, if the mean and variance of a variable are known, we can use the normal distribution to approximate the probability that the variable will exceed some number.
   d. If X is normally distributed, the only information we need know about X to answer probability statements about it is its mean and standard deviation.
   e. The mean of a sample is always exactly normally distributed.
9. The mean salary earned by civil servants in 5 salary levels in a large municipality is $30,500 with a standard deviation of $1,138.
   a. Would it be correct to say that approximately 99.7% of all civil servants in these 5 salary levels earn between $27,086 and $33,914? Why or why not?
   b. Would it be correct to say that if simple random samples of 100 civil servants each were repeatedly drawn from these classifications, the average salary of the group would be between $30,158.60 and $30,841.40 approximately 99.7% of the time?

c. Would it be correct to say that if simple random samples of 10,000 civil servants each were repeatedly drawn from these classifications, the average salary of the group would be between $30,272.40 and $30,727.60 approximately 99.7% of the time? Why or why not?
d. As an approximation, would part (b) or part (c) be more likely to be correct?

10. The mean salary of the presidents of 100 different small electronic controls companies is $68,900 with a standard deviation of $4,210. A certain business magazine decides to make a study of the presidents of these 100 firms.
    a. Suppose 25 presidents are selected at random. What are the mean and standard deviation of the distribution of average salaries for all possible samples of size 25?
    b. Suppose 50 presidents are selected at random. What are the mean and standard deviation of the distribution of average salaries for all possible samples of size 50?
    c. Suppose 100 presidents are selected at random. What are the mean and standard deviation of the distribution of average salaries for all possible samples of size 100?

11. The average amount of time spent per week in meetings by the 15 top executives of a certain company is 15 hours with a standard deviation of 2 hours. At random, 5 executives are selected and asked various questions, one of which is the amount of time per week spent in meetings. The answers from the executives will be used for the orientation of newly hired management aspirants. The mean for the 5 executives questioned was 14.8 hours with a standard deviation of 1.5 hours.
    a. Are 14.8 hours and 1.5 hours the mean and the standard deviation of the sampling distribution of samples of 5 executives selected from the population of 15?
    b. Define the sampling distribution of the mean time spent per week in meetings for samples of 5 executives selected from 15. Calculate the mean and standard deviation of the sampling distribution.

Review Exercises for Chapters 1 through 5

1. There were 100 production runs, of 100 articles each, made on Machine 1. The percentages of defective articles for the 100 runs were recorded as shown in the frequency distribution.

Percentage Defective	Number of Production Runs
0 and under 2	8
2 and under 4	24
4 and under 6	30
6 and under 8	22
8 and under 10	16
Total	100

   a. Compute the mean and standard deviation of the percentages of defective articles produced in the 100 production runs on Machine 1.
   b. On a similar set of 100 production runs made on Machine 2, the mean and the standard deviation of the percentages of defective articles were 10.0% and 3.2%, respectively. On which of the two machines was there greater relative variability in the percentage of defective articles produced?
   c. Recall that 100 articles were produced in each of the 100 runs on Machine 1. Assume that the 10,000 articles from the 100 runs have been merged. Runs containing fewer than 2% defectives have been classified as grade A, runs containing from 2% to 8% defectives are classified as grade B, and runs containing 8% to 10% defectives are classified as grade C. If a random sample of 10 articles is drawn without replacement from the lot of 10,000 articles referred to in the frequency distribution, what is the probability that this random sample of 10 articles will contain exactly 3 articles from grade A lots and 6 or more articles from grade C lots? Set up a mathematical expression from which the result can be calculated, but do not carry out the arithmetic.

2. In a large industrial complex, the average number of accidents per month is 2. A month is here defined as a 4-week period.
   a. What is the probability that at least one accident will occur during the next 2-week period?
   b. Assuming independence, what is the probability that in the next 5 months, at least one accident will be observed in 4 of these months?
3. Let X and Y, respectively, be the number of homes owned and the number of automobiles owned by families in a high-income area. Assume the following joint distribution for X and Y.

$$f(x, y) = \frac{1}{15} xy^2 \qquad x = 1, 2 \quad y = 1, 2$$

Note: $f(x, y)$ is an alternative notation for $P(X = x \text{ and } Y = y)$
   a. What is the probability that a randomly selected family will own 2 homes?
   b. Find the mean and standard deviation for Y, the number of automobiles owned.
4. Six people—4 men and 2 women—are ranked by the personnel director of a company. Assume there is no difference in the quality of the applicants so that each ordering is equally likely.
   a. What is the probability that the 4 men will be ranked highest (that is, 1, 2, 3, 4)?
   b. If a sample of 3 persons is drawn from these 6 persons without replacement, what is the probability that the sample would contain fewer than 2 women? Carry out your solution to obtain a numerical result.
5. The number of automobile accidents per day, X, on a certain expressway over the last 40 days are listed.

Number of Accidents x	Frequency (number of days)
0	9
1	12
2	11
3	6
4	2

   a. On a second highway in the area, the sample standard deviation for the number of accidents per day, Y, was computed to be 1.4. On the basis of the sample results, what can be said about the relative variability in accidents per day on the two highways? Assume that Y is Poisson distributed. (**Hint**: In the Poisson distribution, $\mu = \sigma^2$.)
   b. What is the median value of X, the number of accidents per day on the first expressway?

6. A manufacturer submits bids on 10 different government contracts. In the past, this manufacturer's bid has been the low one and the manufacturer obtained the contract 10% of the time. Assume this percentage holds and assume independence.
   a. What is the probability that the firm will obtain at least one of the 10 contracts?
   b. What is the probability that the firm will obtain fewer than 3 contracts?
   c. If the firm bids on 100 contracts, what is the approximate probability of obtaining at least one contract? Use the Poisson distribution.

7. The accounts receivable of a large corporation are being examined. Each account will be classified as either "active" or "bad debt." In the past, 5% of all accounts proved to be bad.
   a. If a random sample of 20 accounts is drawn, what is the probability of finding at least one bad account in the sample?
   b. If 5 random samples are drawn, what is the probability of finding at least one bad account in every one of the 5 samples?
   c. In the total sample of 100 accounts, 11 bad accounts were observed. Since the expected number would be 5 = (0.05)(100), one examiner expressed doubts about the correctness of the 5% bad debt rate. If the rate actually were 5%, what percentage of the time would 11 or more bad accounts be found in a sample of 100?

8. A manufacturer of cranes carried out a study that showed that the average number of sales per month can be accurately modeled by a Poisson distribution whose mean is

$$\mu = 5 + \frac{x}{\$10{,}000}$$

   where x represents the monthly dollar amount spent on advertising.
   a. If the firm's monthly advertising budget is $20,000, what is the probability that sales will be at most 6 in a given month?
   b. What is the probability that sales will exceed 12 in a particular 2-month period when the monthly budget for advertising is $20,000?
   c. If the firm spends $20,000 for advertising in each particular month, what is the probability that sales will be at most 6 in 6 out of 12 months?

9. Assume that the Department of Labor has estimated that 30% of the U.S. labor force are dissatisfied with their jobs. At Wharton Industries, Inc., 20 presently employed candidates are being interviewed for a position. Assume perfect randomness in choosing these 20 candidates from the labor force.
   a. What is the probability that at least 4 candidates are dissatisfied with their present jobs?
   b. Using the Poisson approximation, find the probability in part (a).
   c. Using the normal approximation, find the same probability in part (a).
   d. Are the approximations in parts (b) and (c) justified? Why or why not?

10. Suppose that a random sample is drawn from a population of women known to have a mean age of 30, with a standard deviation of 3 years. The population is normally distributed.
    a. What is the probability that a randomly selected woman will be over 35 years of age?
    b. What is the probability, in the sample of 36 women, that the mean age will be less than 31?
    c. Do your answers to parts (a) and (b) depend on the assumption that the population is normally distributed? Explain.
11. The A. P. Tomlinson Company produces articles that are components of an electronic product. The process produces articles whose lengths are normally distributed with a mean of 6.00 inches and a standard deviation of 0.30 inch. Specifications for individual articles are 5.60 inches to 6.50 inches.
    a. What proportion of articles are too short?
    b. What proportion of articles are too long?
    c. Articles that are too short must be discarded. Articles that are too long can be reworked for a small additional cost. If the loss incurred in discarding articles is much more than rework cost per article, what suggestions would you make concerning possible changes in the process?
12. A certain manufacturing process produces 20% defective items.
    a. If a simple random sample of 20 items is inspected, what is the probability that exactly one is defective? Use the normal curve approximation to the binomial and compare your answer to the binomial probability.
    b. If a simple random sample of 100 articles is drawn, what is the probability of observing between 8% and 13% defective?
    c. Find an upper limit which would be exceeded only 2 times in 100, on the average, by the percentage defective in a simple random sample of 100 items drawn from this process.
13. In reviewing its accounts receivable, a firm determined a probability of 0.15 that an account should be classified as "overdue." Assume this probability is constant from account to account and assume independence.
    a. In a random sample of 10 such accounts, what is the probability that more than 3 would be overdue?
    b. In 5 such groups of 10 accounts, what is the probability that more than 3 overdue accounts would be observed exactly twice?
    c. Assume that, in a group of 1,000 accounts receivable that were being worked on, 100 were overdue. If a sample of 20 accounts were drawn from the 1,000 without replacement, what is the probability that there would be fewer than 2 overdue accounts? Set up an expression which, when evaluated, would yield the answer. You need not carry out the arithmetic.

14. A power station uses diesel engines as the motive power for generating electricity. It has been observed that two types of breakdowns occur in the system: engine failure and generator failure. Based on past records, a joint probability distribution of the annual number of defects was derived.

		Generator defects Y		
		0	1	2
Engine defects X	0	0.20	0.15	0.05
	1	0.15	0.10	0.05
	2	0.10	0.15	0
	3	0.05	0	0

a. Given that exactly one engine defect will occur, what is the probability that one or more generator defects will occur?
b. Given that one or two engine defects will occur, what is the probability that one or fewer generator defects will occur?
c. Are X and Y independent random variables? Justify your answer.
d. Compute the mean and variance of number of engine defects.

6

Estimation

We said at the beginning of Chapter 3 that decisions must often be made when only incomplete information is available and there is uncertainty concerning the outcomes that must be considered by the decision maker. The remainder of this book deals with methods by which rational decisions can be made under such circumstances. In our brief introduction to probability theory, we have begun to see how probability concepts can be used to cope with problems of uncertainty. **Statistical inference** uses this theory as a basis for making reasonable decisions from incomplete data. Statistical inference treats two different classes of problems: (1) estimation, which is discussed in this chapter; and (2) hypothesis testing, which is examined in Chapters 7 and 8. In both cases, inferences are made about population characteristics from information contained in samples.

6.1

POINT AND INTERVAL ESTIMATION

The need to estimate population parameters from sample data stems from the fact that it is usually too expensive or not feasible to enumerate complete populations to obtain the required information. The cost of complete censuses of finite populations may be prohibitive; complete enumerations of infinite populations are impossible.

> **Statistical estimation** procedures provide us with the means to estimate population parameters with desired degrees of precision.

Numerous examples can be given of the need to estimate pertinent population parameters in business and economics. A marketing organization may be interested in estimates of average income and other socioeconomic characteristics of the consumers in a metropolitan area; a retail chain may want an estimate of the average number of pedestrians per day who pass a certain corner; or a bank may want an estimate of average interest rates on mortgages in a certain section of the country. Undoubtedly, in all of these cases, exact accuracy is not required, and estimates derived from sample data would probably provide appropriate information to meet the demands of the practical situation.

Two different types of estimates of population parameters are of interest: point estimates and interval estimates.

Point estimate

A **point estimate** is a single number used as an estimate of the unknown population parameter.

For example, the arithmetic mean income of a sample of families in a metropolitan area may be used as a point estimate of the corresponding population mean for all families in that metropolitan area; or the percentage of a sample of eligible voters in a political opinion poll who state that they would vote for a particular candidate may be used as an estimate of the corresponding unknown percentage in the relevant population.

A distinction can be made between an *estimate* and an *estimator*. Consider the illustration of estimating the population figure for arithmetic mean income of all families in a metropolitan area from the corresponding sample mean.

Estimate

- The numerical value of the sample mean is said to be an **estimate** of the population mean figure.

Estimator

- The statistical measure used (that is, the *method* of estimation) is referred to as an **estimator**.

For example, the sample mean $\bar{x}$ is an *estimator* of the population mean. When a specific number is calculated for the sample mean, say $8,000, that number is an *estimate* of the population mean figure.

Choosing an estimate

Whether we use a point estimate rather than an interval estimate depends on the purpose of the investigation. For example, for planning purposes, a marketing department may estimate a single figure for annual sales of one of its company's products and may then break that figure down into monthly sales estimates. These figures may be passed on to the production department for the planning of production requirements. The production department may in turn convert its production requirements into materials purchasing plans for the purchasing department. If the marketing department estimated annual sales as a range, say from $10 million to $12 million, rather than as a single figure, this could unduly complicate the subsequent steps of obtaining monthly breakdowns and planning production and purchasing requirements. However, for many practical purposes, having only a single point estimate of a population parameter is not sufficient. Any single point estimate will be either right or wrong. It would certainly seem useful, and perhaps even necessary, to have in addition to a point estimate, some notion of the degree of error that might be involved in using this estimate. Interval estimation is useful in this connection.

Interval estimate

An **interval estimate** of a population parameter is a statement of two values between which we have some confidence that the parameter lies.

Thus, an interval estimate in the example of the population arithmetic mean income of families in a metropolitan area might be $24,100 to $25,900. An interval estimate for the percentage of defectives in a shipment might be 3% to 5%.

We may have a great deal of confidence or very little confidence that the population parameter is included in the range of the interval estimate, so we must attach some sort of probabilistic statement to the interval. The procedure used to create such a probabilistic statement is **confidence interval estimation.** The confidence interval is an interval estimate of the population parameter. A confidence coefficient such as 90% or 95% is attached to this interval to indicate the degree of confidence or credibility placed on the estimated interval.

Confidence interval estimation

6.2 CRITERIA OF GOODNESS OF ESTIMATION

Numerous criteria have been developed by which to judge the goodness of point estimators of population parameters. A rigorous discussion of these criteria requires some complex mathematics that falls outside the scope of this text. However, it is possible to gain an appreciation of the nature of these criteria in an intuitive, nonrigorous way.

Let us return to our illustration of estimating the arithmetic mean income of families in a metropolitan area. This arithmetic mean—assuming suitable definitions of income, family, and metropolitan area—is an unknown population parameter that we designate as μ. Suppose we took a simple random sample of families from this population and calculated the arithmetic mean $\bar{x}$, the median Md, and the mid-range $(x_{\max} + x_{\min})/2$, where $x_{\max}$ and $x_{\min}$ are the largest and smallest sample observations. Which method would be the best estimator of the population mean? Probably you would answer that the sample mean $\bar{x}$ is the best estimator. In fact, if this question had not been raised, it might not even have occurred to you to use any statistic other than the sample mean as an estimator of the population mean. However, why do you think the sample mean represents the best estimator? It may not be easy to articulate your answer to that question. The sample mean is preferable to the other estimators by the generally utilized criteria of goodness of estimation of classical statistical inference. Let us briefly examine the nature of a few of these criteria: *unbiasedness*, *consistency*, and *efficiency*.

Unbiasedness

An estimator—such as a sample arithmetic mean—is a random variable, because it may take on different values, depending on which population elements are drawn into the sample. Rule 13 of Appendix C proves that the expected value of this random variable is the population mean. In symbols, we have

$$E(\bar{x}) = \mu \tag{6.1}$$

where $\bar{x}$ = the sample mean
μ = the population mean

> If the expected value of a sample statistic is equal to the population parameter for which the statistic is an estimator, the statistic or estimator is said to be **unbiased**.

In symbols, if θ is a parameter to be estimated and $\hat{\theta}$ is a sample statistic used to estimate θ, then $\hat{\theta}$ is said to be an *unbiased estimator* of θ if

$$E(\hat{\theta}) = \theta \tag{6.2}$$

If we say that a given estimator is unbiased, we are simply saying that this method of estimation is correct *on the average*. That is, if the method is employed repeatedly, the average of all estimates obtained from this estimator is equal to the value of the population parameter. Clearly, an unbiased estimator does not guarantee useful individual estimates. The differences of these individual estimates from the value of the population parameter may represent large errors. The simple fact that the bias, or long-run average of these errors, is zero may be of little practical importance. Furthermore, if we have two unbiased estimators, we require additional criteria in order to make a choice between them.

Just as the sample mean is an unbiased estimator of the population mean, the sample variance as defined in equation 1.8 is an unbiased estimator of the population variance.[1] In symbols, as proven in rule 14 of Appendix C, we have

$$E(s^2) = E\left[\frac{\Sigma(x - \bar{x})^2}{n - 1}\right] = \sigma^2 \tag{6.3}$$

Equation 6.3 is a mathematical restatement of the point made in section 1.15 that when the sample variance is defined with divisor $n - 1$, it is an unbiased estimator of the population variance.

[1] We can see from the nature of the proof in rule 14 of Appendix C, $\Sigma(x - \bar{x})^2/(n - 1)$ is not an unbiased estimator of the variance of a *finite* population. Perhaps surprisingly, both $\sqrt{\Sigma(x - \bar{x})^2/n}$ and $\sqrt{\Sigma(x - \bar{x})^2/(n - 1)}$ are biased estimators of the population standard deviation.

Consistency

It is clear from the preceding discussion that knowing only that an estimator is unbiased gives us insufficient information as to the goodness of that method of estimation. Closeness of the estimator to the parameter seems to be of primary importance. Both the concepts of *consistency* and *efficiency* deal with this property of closeness. Consider the sample mean $\bar{x}$ as an estimator of the population parameter μ for an infinite population. What happens to the possible values of $\bar{x}$ as the sample size n increases? On an intuitive basis, we would certainly expect $\bar{x}$ to lie closer to μ as n becomes larger. Generally, if an estimator, say $\hat{\theta}$, approaches closer to the parameter θ as the sample size n increases, $\hat{\theta}$ is said to be a **consistent estimator** of θ.

In terms of sampling, this idea of consistency means that the sampling distribution of the estimator becomes more and more "tightly packed" around the population parameter as the sample size increases. Figure 6-1(a) illustrates this concept for the sample mean as an estimator of μ, the mean of an infinite population. The graph represents the respective sampling distributions of the sample mean $\bar{x}$ for samples of size n_1, of size n_2, and of size n_3 drawn from the same population; n_3 is larger than n_2, which is larger than n_1. We know from

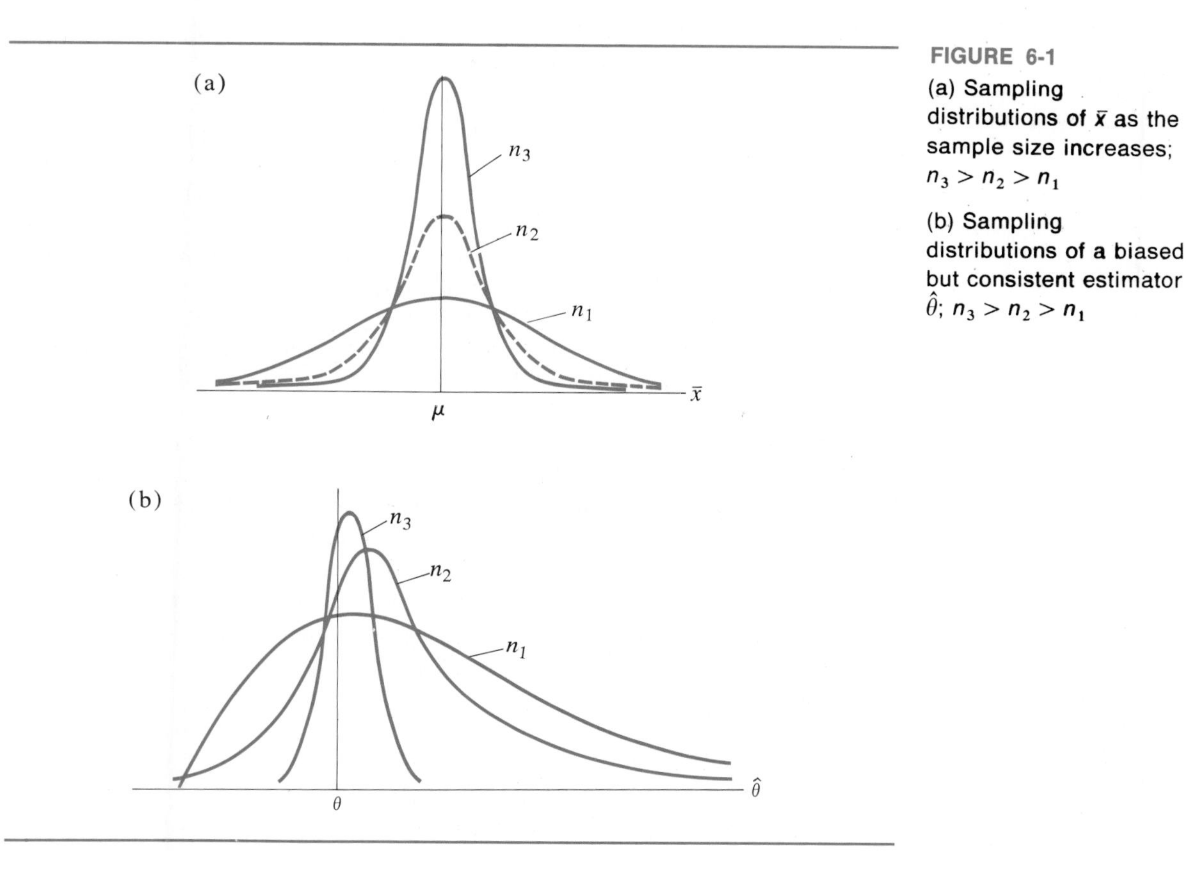

FIGURE 6-1

(a) Sampling distributions of $\bar{x}$ as the sample size increases; $n_3 > n_2 > n_1$

(b) Sampling distributions of a biased but consistent estimator $\hat{\theta}$; $n_3 > n_2 > n_1$

the relationship $\sigma_{\bar{x}}^2 = \sigma^2/n$ that the sampling variance $\sigma_{\bar{x}}^2$ decreases as n increases. Note that all three sampling distributions center on the population parameter μ, since $\bar{x}$ is an unbiased estimator of μ.

Figure 6-1(b) illustrates the concept of consistency for $\hat{\theta}$, a **biased estimator** of a parameter θ. As in Figure 6-1(a), the graph represents the respective sampling distributions of an estimator, in this case $\hat{\theta}$, for samples of size n_1, of size n_2, and of size n_3 drawn from the same population, with $n_3 > n_2 > n_1$. None of the distributions are centered on the population parameter θ. Since $E(\hat{\theta})$ is not equal to θ, $\hat{\theta}$ is a biased estimator of θ. However, we can see from the graph that as the sample size increases, the sampling distribution becomes more "tightly packed" around θ.

Efficiency

The concept of efficiency refers to the sampling variability of an estimator. If two competing estimators are both unbiased, the one with the smaller variance (for a given sample size) is said to be relatively more efficient. More formally, if $\hat{\theta}_1$ and $\hat{\theta}_2$ are two unbiased estimators of θ, their relative efficiency is defined by the ratio

$$\frac{\sigma_{\hat{\theta}_2}^2}{\sigma_{\hat{\theta}_1}^2} \tag{6.4}$$

where $\sigma_{\hat{\theta}_1}^2$ is the smaller variance.

Let us consider as an example a simple random sample of size n drawn from a normal population with mean μ and variance σ^2. Suppose we want to consider the relative efficiency of the sample mean $\bar{x}$ and the sample median Md as estimators of the population mean μ. Both estimators are unbiased. We know that the variance of the sample mean $\bar{x}$ is $\sigma_{\bar{x}}^2 = \sigma^2/n$. It can be shown that the variance of the sample median Md is approximately $\sigma_{Md}^2 = 1.57\ \sigma^2/n$. Therefore, the relative efficiency of $\bar{x}$ with respect to Md is

$$\frac{\sigma_{Md}^2}{\sigma_{\bar{x}}^2} = \frac{1.57\sigma^2/n}{\sigma^2/n}$$
$$= 1.57$$

We can interpret this result in terms of sample sizes. If the sample median rather than the sample mean were used as an estimator of the mean μ of a normal population, then in order to obtain the same precision provided by the sample mean, a sample size 57% larger would be required. Stated differently, the required sample size for the sample median would be 157% of that for the sample mean. Figure 6-2 shows the sampling distributions for these two estimators.

Other criteria for goodness of estimation

Other criteria for goodness of estimation may be found in standard texts on mathematical statistics. As we have seen, sampling error decreases with increasing sample size, and biased but consistent estimators approach the population parameter as sample size increases. However, greater cost is incurred

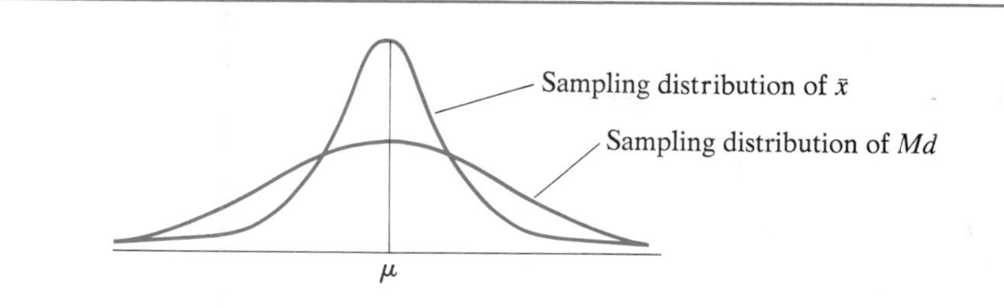

FIGURE 6-2
Sampling distributions of the mean $\bar{x}$ and median *Md* for the same sample size

with larger sample sizes. Therefore, most practical estimation situations involve trade-offs between these considerations.

Two problems of interest in this chapter are the estimation of a population proportion and a population mean. In the case of estimating a population proportion, if we assume a Bernoulli process generating n sample observations, it can be shown that the observed sample proportion of successes $\bar{p}$ = (number of successes)/(number of sample observations) is an unbiased, efficient, and consistent estimator. Similarly, the mean $\bar{x}$ of a simple random sample of n observations is an unbiased, efficient, and consistent estimator of the population mean. It is not surprising that in many applications of statistical methods, sample proportions or sample means are used as the "best" point estimators of the corresponding population parameters. Perhaps the reader has had occasion to calculate a sample proportion or mean and has found it intuitively appealing to use such a figure as an estimator of the corresponding population parameter. If so, this intuitive approach was supported by sound statistical theory.

Exercises 6.2

1. Determine which of the following estimates of sales for next year you would like to receive from the marketing department.
   a. Sales will be $1,000,000.
   b. Sales will fall between $500,000 and $2,500,000 with 100% probability.
   c. Sales will fall between $900,000 and $1,100,000 with 95% probability.
2. Sampling error decreases with increasing sample size, and biased but consistent estimators approach the population parameter as sample size increases. In view of the benefits of increasing sample sizes, why don't researchers try to take as large a sample as possible?
3. Is each of the following statements about an estimator or an estimate? If the statement is about an estimate, is it a point or an interval estimate?
   a. The sales manager feels that February's sales will be between $40,000 and $50,000.
   b. In certain situations, the median of a class may be a better measure of central tendency than the mean. For example, in a research study on the savings habits of American families, the large savings held by the wealthy may distort the mean.

c. A bank auditor, after examining a sample of the bank's loan portfolio, says that 4% of all loans in the portfolio will have to be written off as bad debts.

d. A security analyst predicts that the Dow Jones average will be between 900 and 1,100 by the end of the year.

4. State whether each of the following statements are true or false, and explain your answer.

a. An unbiased estimator is always better than a consistent but biased estimator.

b. If an unbiased estimator of σ^2 is desired, it is best to calculate the sample variance as

$$\frac{\sum(x - \bar{x})^2}{n - 1}$$

c. If $\hat{\theta}_1$ is a biased estimator of θ and if $\hat{\theta}_2$ is an unbiased estimator of θ and if the ratio $\sigma^2_{\hat{\theta}_2}/\sigma^2_{\hat{\theta}_1}$ is greater than 1, then $\hat{\theta}_1$ is relatively more efficient than $\hat{\theta}_2$.

6.3 CONFIDENCE INTERVAL ESTIMATION (LARGE SAMPLES)

As indicated in section 6.1, for many practical purposes, it is not sufficient merely to have a single point estimate of a population parameter. It is usually necessary to have an estimation procedure that measures the degree of precision involved. In classical statistical inference, the standard procedure for this purpose is confidence interval estimation.

Confidence Interval Estimation

We will explain the rationale of confidence interval estimation in terms of an example in which a population mean is the parameter to be estimated. Suppose a manufacturer has a very large production run of a certain brand of tire and wants to obtain an estimate of their arithmetic mean lifetime by drawing a simple random sample of 100 tires and subjecting them to a forced life test. Let us assume that, from long experience in manufacturing this brand of tire, the manufacturer knows that the population standard deviation for a production run is $\sigma = 3{,}000$ miles. (Of course, ordinarily the standard deviation of a population is not known exactly and must be estimated from a sample, just as are the mean and other parameters. However, let us assume in this case that the population standard deviation is indeed known.) When the sample of 100 tires

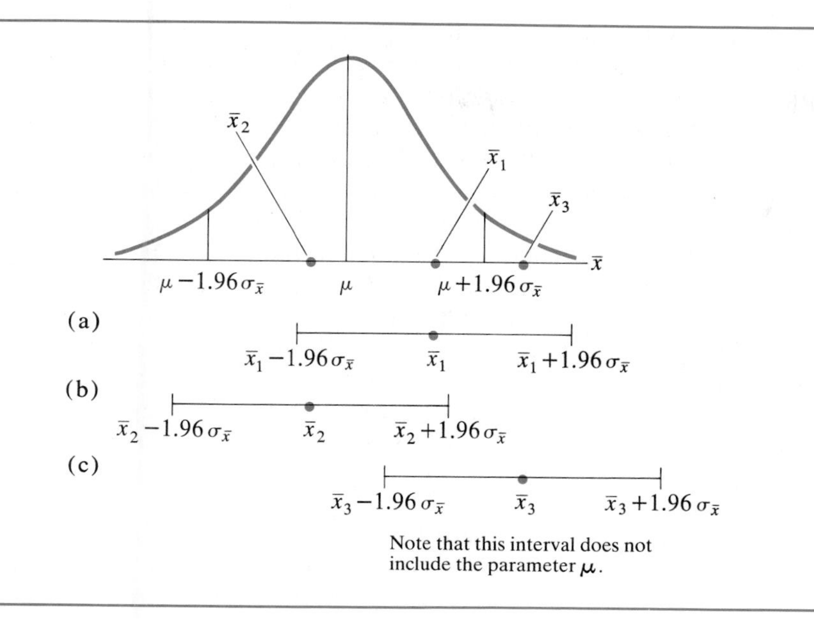

FIGURE 6-3
Sampling distribution of the mean and confidence interval estimates for three illustrative samples

is drawn, a mean lifetime of 32,500 miles is observed. Thus, we denote $\bar{x} =$ 32,500 miles. This sample mean is our best *point* estimate of the population mean lifetime, that is, of the mean lifetime of all tires in the production run. Additionally, we would like an *interval* estimate of the population mean lifetime. That is, we would like to be able to state that the population mean is between two limits, say $\bar{x} - 2\sigma_{\bar{x}}$ and $\bar{x} + 2\sigma_{\bar{x}}$ where $\bar{x} - 2\sigma_{\bar{x}}$ is the lower limit of the interval and $\bar{x} + 2\sigma_{\bar{x}}$ is the upper limit. Furthermore, we would like to have a high degree of confidence that the true population mean is included in this interval.

The procedure in confidence interval estimation is based on the concept of the sampling distribution. In this example, since we are dealing with the estimation of a mean, the appropriate distribution is the sampling distribution of the mean. We will review some fundamentals of this distribution to lay the foundation for confidence interval estimation. Figure 6-3 shows the sampling distribution of the mean for simple random samples of size $n = 100$ from a population with an unknown mean, denoted μ, and a standard deviation $\sigma =$ 3,000. We assume that the sample is large enough so that by the Central Limit Theorem, stated in section 5.5, the sampling distribution may be assumed to be normal, even if the population is nonnormal. The standard error of the mean,[2] which is the standard deviation of this sampling distribution, equals $\sigma_{\bar{x}} = \sigma/\sqrt{n}$.

[2] Strictly speaking, the finite population correction should be shown in this formula, but we will assume the population is so large relative to sample size that for practical purposes the correction factor is equal to one. The mean of the sampling distribution $\mu_{\bar{x}}$ is equal to the population mean μ.

In our work with the normal sampling distribution of the mean, we have learned how to make probability statements about sample means, given the value of the population mean. Thus, in terms of the data of this problem, we can state that if repeated simple random samples of 100 tires each were drawn from the production run, 95% of the sample means $\bar{x}$ would lie within 1.96 standard error units of the mean of the sampling distribution (the population mean), or between $\mu - 1.96\sigma_{\bar{x}}$ and $\mu + 1.96\sigma_{\bar{x}}$. This range is indicated on the horizontal axis of the sampling distribution in Figure 6-3. For emphasis, vertical lines show the endpoints of this range. As usual, we determine the 1.96 figure from Table A-5 of Appendix A, where we find that 47.5% of the area in a normal distribution is included between the mean of a normal distribution and a value 1.96 standard deviations to the right of the mean; thus, by symmetry, 95% of the area is included in a range of ± 1.96 standard deviations from the mean. In terms of relative frequency, 95% of the $\bar{x}$ values of samples of size 100 would lie in this range if repeated samples were drawn from the given population.

How then might we construct the desired interval estimate for the population parameter? Let us consider again the repeated simple random samples of size 100 from the population of the production run of tires. Suppose our first sample yields a mean that exceeds μ but falls between $\mu + 1.96\sigma_{\bar{x}}$. The position of this sample mean, denoted $\bar{x}_1$, is shown on the horizontal axis of Figure 6-3. Suppose now that we set up an interval from $\bar{x}_1 - 1.96\sigma_{\bar{x}}$ to $\bar{x}_1 + 1.96\sigma_{\bar{x}}$. This interval is shown immediately below the graph in Figure 6-3(a). As seen in the figure, this interval, which may be written as $\bar{x}_1 \pm 1.96\sigma_{\bar{x}}$, includes the population parameter μ. This follows from the fact that $\bar{x}_1$ fell less than $1.96\sigma_{\bar{x}}$ from the mean of the sampling distribution μ.

Now, let us assume our second sample from the same population yields the mean $\bar{x}_2$, which lies on the horizontal axis to the left of μ, but again at a distance less than $1.96\sigma_{\bar{x}}$ from μ. Again we set up an interval of the sample mean $\pm 1.96\sigma_{\bar{x}}$, or from $\bar{x}_2 - 1.96\sigma_{\bar{x}}$ to $\bar{x}_2 + 1.96\sigma_{\bar{x}}$. This interval, shown below the graph in Figure 6-3(b), includes the population mean μ.

Finally, suppose a third sample is drawn from the same population, with the mean $\bar{x}_3$ shown on the horizontal axis of Figure 6-3. This sample mean lies to the right of μ, but at a distance *greater than* $1.96\sigma_{\bar{x}}$ above μ. When we set up the range $\bar{x}_3 - 1.96\sigma_{\bar{x}}$ to $\bar{x}_3 + 1.96\sigma_{\bar{x}}$, this interval shown below the graph in Figure 6-3(c), does not include μ.

We can imagine a continuation of this sampling procedure. Since 95% of the sample means fall within $1.96\sigma_x$ of μ, we can assert that 95% of the intervals of the type $\bar{x} \pm 1.96\sigma_{\bar{x}}$ include the population parameter μ. Now, we can get to the crux of confidence interval estimation. In the problem originally posed, as in most practical applications, only one sample was drawn from the population, not repeated samples. On the basis of the single sample, we were required to estimate the population parameter. The procedure simply establishes the interval $\bar{x} \pm 1.96\sigma_{\bar{x}}$ and attaches a suitable statement to it. The interval itself is referred to as a **confidence interval**. Thus, in our original problem the required confidence interval is

Confidence interval

$$\bar{x} \pm 1.96\sigma_{\bar{x}} = \bar{x} \pm 1.96\frac{\sigma}{\sqrt{n}} = 32{,}500 \pm 1.96\frac{3{,}000}{\sqrt{100}}$$

$$= 32{,}500 \pm 588 = 31{,}912 \text{ to } 33{,}088 \text{ miles}$$

We must be very careful how we interpret this confidence interval. It is incorrect to make a probability statement about this *specific* interval. For example, it is incorrect to state that the probability is 95% that the mean lifetime μ of all tires falls in this interval. *The population mean is not a random variable*; hence, probability statements cannot be made about it. The unknown population mean μ either lies in the interval or it does not. We must return to the line of argument used in explaining the method and indicate that the values of the random variable are the intervals of $\bar{x} \pm 1.96\sigma_{\bar{x}}$, not μ. Thus, if repeated simple random samples of the same size were drawn from this population and the interval $\bar{x} \pm 1.96\sigma_{\bar{x}}$ were constructed from each of them, then 95% of the statements that the interval contains the population mean μ would be correct. Another way of putting it is that in 95 samples out of 100, the mean μ would lie within intervals constructed by this procedure. The 95% figure is referred to as a **confidence coefficient** to distinguish it from the type of probability calculated when deductive statements are made about sample values from known population parameters.[3]

Confidence coefficient

Interpretation and use of interval estimation

Despite the above interpretation of the meaning of a confidence interval, where the probability pertains to the estimation procedure rather than to the specific interval constructed from a single sample, the fact remains that we ordinarily must make an inference on the basis of the single sample drawn. We will not draw the repeated samples implied by the interpretational statement. For example, in the tire illustration, an inference was required about the production run based on the particular sample of 100 tires in hand, and an interval estimate of a mean lifetime of 31,912 to 33,088 miles was obtained. If the confidence coefficient attached to the interval estimate is high, then the investigator will assume that the interval estimate is correct. This interval may or may not encompass the actual value of the population parameter μ. However, since 95% of intervals so constructed would include the value of the mean lifetime μ of all tires in the production run, we will behave as though this particular interval does include the actual value.

[3] This paragraph presents the standard interpretation of confidence intervals provided by classical statistics. Bayesian statistics (discussed in Chapters 14 through 17) disputes this interpretation, arguing that if μ is unknown, it may be treated as a random variable. Hence, in Bayesian statistics, "prior probability" statements may be made about μ based on subjective assessments, and "posterior probability" statements may be made that combine the prior probabilities and information obtained from sampling.

> It is desirable to obtain a relatively narrow interval with a high confidence coefficient associated with it. One without the other is not particularly useful.

Thus, for example, in estimating a proportion (say, the proportion of persons in the labor force who are unemployed), we can assert even without sample data that the percentage lies somewhere between zero and 100% with a confidence coefficient of 100%. Obviously, this statement is neither profound nor useful, because the interval is too wide. On the other hand, if the interval is narrow but has a low associated confidence coefficient, say 10%, the statement would again have little practical utility.

Confidence coefficients such as 0.90, 0.95, and 0.99 and limits of two or three standard errors (such as 0.955 or 0.997) are conventionally used. Limits of two or three standard errors are those obtained by making an estimate of a population parameter and adding and subtracting two or three standard errors to establish confidence intervals. For a fixed confidence coefficient and population standard deviation, the only way to narrow a confidence interval and increase the precision of the statement is to increase the sample size. This is readily apparent from the way the confidence interval was constructed in the tire example. We computed $\bar{x} \pm 1.96\sigma_{\bar{x}}$, where $\sigma_{\bar{x}} = \sigma/\sqrt{n}$. If the 1.96 figure and σ are fixed, we can decrease the width of the interval only by increasing the sample size n since $\sigma_{\bar{x}}$ is inversely related to $\sqrt{n}$. Thus, the marginal benefit of increased precision must be measured against the increased cost of sampling. Later in this section, we discuss a method of determining the sample size required for a specified degree of precision.

One final point may be made before turning to confidence interval estimation of different types of population parameters. Ordinarily, as indicated in the tire example, the standard deviation of the population σ is unknown. Therefore, it is not possible to calculate $\sigma_{\bar{x}}$, the standard error of the mean. However, we can estimate the standard deviation of the population from a sample and use this figure to calculate an estimated standard error of the mean. We use this estimation technique in the examples that follow.

Interval Estimation of a Mean (Large Samples)

We will use examples to discuss confidence interval estimation and will concentrate first on situations in which the sample size is large. Our discussion will then focus, in turn, on interval estimation of a mean, a proportion, the difference between means, and the difference between proportions. Finally, we will briefly treat corresponding estimation procedures for small samples.

Let us look at Example 6-1, an illustration of interval estimation of a mean from a large sample.

EXAMPLE 6-1

National Motors selected a random sample of 120 of its cars of the same model in order to determine the mean gas mileage of this model. The sample results were $\bar{x} = 33.2$ miles per gallon, $s = 4.6$, and $n = 120$. Determine the 99% confidence interval for the mean gas mileage of this model of car.

Solution

This problem differs from the illustration of the mean lifetime of tires only because the population standard deviation is unknown. The usual procedure for large samples ($n > 30$) is simply to use the sample standard deviation as an estimate of the corresponding population standard deviation. Using s as an estimator of σ, we can compute an estimated standard error of the mean $s_{\bar{x}}$. We have

$$s_{\bar{x}} = \frac{s}{\sqrt{n}} = \frac{4.6}{\sqrt{120}} = 0.42$$

Hence, we may use $s_{\bar{x}}$ as an estimator of $\sigma_{\bar{x}}$, and because n is large we invoke the Central Limit Theorem to argue that the sampling distribution of $\bar{x}$ is approximately normal. We have assumed that the finite population correction equals one. The confidence interval, in general, is given by

$$\bar{x} \pm z s_{\bar{x}} \tag{6.5}$$

where z is the multiple of standard errors and $s_{\bar{x}}$ now replaces $\sigma_{\bar{x}}$, which was used when the population standard deviation was known. For a 99% confidence coefficient, $z = 2.58$. Therefore, the required interval is

$$33.2 \pm 2.58(0.42) = 33.2 \pm 1.08$$

So, the population mean is roughly between 32.12 and 34.28 miles per gallon with a 99% confidence coefficient. The same interpretation given earlier for confidence intervals applies here.

Interval Estimation of a Proportion (Large Samples)

In many situations, it is important to estimate a proportion of occurrences in a population from sample observations. For example, it may be of interest to estimate the proportion of unemployed persons in a certain city, the proportion of eligible voters who intend to vote for a particular political candidate, or the proportion of students at a university who favor changing the grading system. In all these cases, the corresponding proportions observed in simple random samples may be used to estimate the population proportions. Before turning to a description of how this estimation is accomplished, we will briefly establish the conceptual underpinnings of the procedure.

In Chapter 5, we saw that under certain conditions, the binomial distribution is the appropriate sampling distribution for the number of successes x in a simple random sample of size n. Furthermore, we noted in section 5.5 that if p, the proportion to be estimated, is not too close to zero or one, the binomial distribution can be closely approximated by a normal curve with the same mean and standard deviation, that is, $\mu = np$ and $\sigma = \sqrt{npq}$.

It is a simple matter to convert from a sampling distribution of *number of successes* to the corresponding distribution of *proportion of successes.* If x is the number of successes in a sample of n observations, then the proportion of successes in the sample is $\bar{p} = x/n$. Hence, dividing the formulas in the preceding paragraph by n, we find that the mean and standard deviation of the sample proportion become

(6.6) $$\mu_{\bar{p}} = p$$

and

(6.7) $$\sigma_{\bar{p}} = \sqrt{\frac{pq}{n}}$$

Note that the subscript $\bar{p}$ has been used on the left sides of equations 6.6 and 6.7 in keeping with the symbolism used for the corresponding values for the sampling distribution of the mean, namely, $\mu_{\bar{x}}$ and $\sigma_{\bar{x}}$. Therefore, in summary, if p is not too close to zero or one, the sampling distribution of $\bar{p}$ can be closely approximated by a normal curve with the mean and standard deviation given in equations 6.6 and 6.7. It is important to observe in this connection that the Central Limit Theorem holds for sample proportions as well as for sample means.

Note that the above discussion pertains to cases in which the population size is large compared with the sample size. Otherwise, equation 6.7 should be multiplied by the finite population correction $\sqrt{(N-n)/(N-1)}$. As we mentioned in Chapter 5, $\sigma_{\bar{p}}$ is referred to as the **standard error of a proportion.**

To illustrate confidence interval estimation for a proportion, we shall make assumptions similar to those in the preceding example. In Example 6-2, we assume a large simple random sample drawn from a population that is very large compared with the sample size.

EXAMPLE 6-2

In an urban area, a simple random sample of 800 voters revealed that 560 opposed the reelection of their mayor. What are the 95.5% confidence limits for the proportion of all voters in this city who would not like the current mayor elected to another term?

Solution

In this problem, we want a confidence interval estimate for p, a population proportion. We have obtained the sample statistic $\bar{p} = \frac{560}{800} = 0.70$, which is the sample proportion of the voters who do not want the present mayor reelected. As noted earlier, for large sample sizes and for p values not too close to zero or one, the sampling distribution of $\bar{p}$ may be approximated by a normal distribution with $\mu_{\bar{p}} = p$ and $\sigma_{\bar{p}} = \sqrt{pq/n}$. Here we encounter the same type of problem as in interval estimation of the mean. The formula for the exact standard error of a proportion, $\sigma_{\bar{p}} = \sqrt{pq/n}$, requires the values of the unknown population parameters p and q. Hence, we use an estimation procedure similar to that used for the mean. Just as we used s to approximate σ, we can substitute the corresponding sample statistics $\bar{p}$ and $\bar{q}$ for the parameters p and q in the formula for $\sigma_{\bar{p}}$ in order to calculate an estimated standard error of a proportion, $s_{\bar{p}} = \sqrt{\bar{p}\bar{q}/n}$.

Using the same reasoning as that for interval estimation of the mean, we can state a two-sided confidence interval estimate for a population proportion as

(6.8)
$$\bar{p} \pm zs_{\bar{p}}$$

In this problem, $z = 2$ since the confidence coefficient is 95.5%. Hence, substituting into equation 6.8, we obtain the following interval estimate of all voters in this city who would not like the mayor to seek another term:

$$0.70 \pm 2\sqrt{\frac{0.70 \times 0.30}{800}} = 0.70 \pm 0.0324$$

Thus, the population proportion is estimated to be included in the interval 0.6676 to 0.7324, or roughly between 66.8% and 73.2% with a 95.5% confidence coefficient.

Interval Estimation of the Difference between Two Means (Large Samples)

The foregoing examples of *estimation* of a population mean and proportions are based on single samples. We now examine interval estimation of the difference between means and the difference between proportions based on data obtained from two independent large samples. First, we examine an example of confidence interval estimation of the difference between two population means.

EXAMPLE 6-3

A large department store chain was interested in analyzing the difference between the average dollar amount of its delinquent charge accounts in the northeastern and western regions of the country for a certain year. The store took two independent simple random samples of these delinquent charge accounts, one from each region. The mean and standard deviation of the dollar amounts of these delinquent accounts were calculated to the nearest dollar. The northeastern region is denoted as 1 and the western region as 2.

Sample 1	Sample 2
$\bar{x}_1 = \$76$	$\bar{x}_2 = \$65$
$s_1 = \$25$	$s_2 = \$22$
$n_1 = 100$	$n_2 = 100$

The analysts decided to establish 99.7% confidence limits for $\mu_1 - \mu_2$, where μ_1 and μ_2 denote the respective population mean sizes of delinquent accounts. Of course, a point estimate of $\mu_1 - \mu_2$ is given by $\bar{x}_1 - \bar{x}_2$. The required theory for the interval estimate is based on the fact that the sampling distribution of $\bar{x}_1 - \bar{x}_2$ for two large independent samples is exactly normal, if the population of differences is normal, with mean and standard deviation

(6.9)
$$\mu_{\bar{x}_1 - \bar{x}_2} = \mu_1 - \mu_2$$

and

(6.10)
$$\sigma_{\bar{x}_1 - \bar{x}_2} = \sqrt{\frac{\sigma_1^2}{n_1} + \frac{\sigma_2^2}{n_2}}$$

where σ_1 and σ_2 represent the respective population standard deviations of sizes of delinquent accounts.[4]

Since the population standard deviations σ_1 and σ_2 are unknown, and since the sample sizes are large, the sample standard deviations may be substituted into the formula for $\sigma_{\bar{x}_1-\bar{x}_2}$ to give an estimated standard error of the difference between two means,

$$s_{\bar{x}_1-\bar{x}_2} = \sqrt{\frac{s_1^2}{n_1} + \frac{s_2^2}{n_2}}$$

As usual with problems of this type, the population of differences may not be normal, and the population standard deviations are unknown. However, since the samples are large, we can use the Central Limit Theorem to assert that the sampling distribution of $\bar{x}_1 - \bar{x}_2$ is approximately normal. The required confidence limits are given by

$$(\bar{x}_1 - \bar{x}_2) \pm z s_{\bar{x}_1-\bar{x}_2} \tag{6.11}$$

The calculation for $s_{\bar{x}_1-\bar{x}_2}$ in this problem is

$$s_{\bar{x}_1-\bar{x}_2} = \sqrt{\frac{(25)^2}{100} + \frac{(22)^2}{100}} = \$3.33$$

Since a 99.7% confidence interval is desired, the value of z is three. Therefore, substituting into equation 6.11 gives

$$(\$76 - \$65) \pm 3(\$3.33) = \$11 \pm \$9.99$$

Hence, to the nearest dollar, confidence limits for $\bar{x}_1 - \bar{x}_2$ are \$1 and \$21. It is a worthwhile exercise for the reader to attempt to express in words specifically what this confidence interval means.

Interval Estimation of the Difference between Two Proportions (Large Samples)

The procedure for constructing a confidence interval estimate for the difference between two proportions is analogous to the technique used in constructing a confidence interval estimate for means.

EXAMPLE 6-4

A credit reference service investigated two simple random samples of customers who applied for charge accounts in two different department stores. The service was interested in the proportion of applicants in each store who had annual incomes exceeding \$20,000. Confidence limits of 90% were established for the difference $p_1 - p_2$, where p_1 and p_2 represent the population proportions of applicants in each store whose incomes exceeded \$20,000.

[4] Equation 6.9 follows from rule 4 of Appendix C. By rule 11 of Appendix C, the variance of the difference $\bar{x}_1 - \bar{x}_2$ is $\sigma^2_{\bar{x}_1-\bar{x}_2} = \sigma^2_{\bar{x}_1} + \sigma^2_{\bar{x}_2}$. This is an application of the principle that the variance of the difference between two *independent* random variables is equal to the sum of the variances of these variables. Taking the square root of both sides of this equation, we obtain $\sigma_{\bar{x}_1-\bar{x}_2} = \sqrt{\sigma^2_{\bar{x}_1} + \sigma^2_{\bar{x}_2}}$, where $\sigma^2_{\bar{x}_1}$ and $\sigma^2_{\bar{x}_2}$ are simply the variances of the sampling distributions of $\bar{x}_1$ and $\bar{x}_2$. Substituting $\sigma^2_{\bar{x}_1} = \sigma_1^2/n_1$ and $\sigma^2_{\bar{x}_2} = \sigma_2^2/n_2$, we get equation 6.10.

The sample data were

Store 1	Store 2
$\bar{p}_1 = 0.50$	$\bar{p}_2 = 0.18$
$\bar{q}_1 = 0.50$	$\bar{q}_2 = 0.82$
$n_1 = 150$	$n_2 = 160$

where these symbols have their conventional meanings. As in the preceding example, we can start with a point estimate. The number $\bar{p}_1 - \bar{p}_2$ is the obvious point estimate of $p_1 - p_2$, and we can assume that the sampling distribution of $\bar{p}_1 - \bar{p}_2$ is approximately normal with mean

$$\mu_{\bar{p}_1 - \bar{p}_2} = p_1 - p_2$$

and standard deviation

$$\sigma_{\bar{p}_1 - \bar{p}_2} = \sqrt{\frac{p_1 q_1}{n_1} + \frac{p_2 q_2}{n_2}}$$

Since the population proportions p_1 and p_2 are unknown and the sample sizes are large, $\bar{p}_1$ and $\bar{p}_2$ may be substituted for p_1 and p_2 to obtain the estimated standard error of the difference between percentages.

(6.12) $$s_{\bar{p}_1 - \bar{p}_2} = \sqrt{\frac{\bar{p}_1 \bar{q}_1}{n_1} + \frac{\bar{p}_2 \bar{q}_2}{n_2}}$$

As in the procedure for differences between means, we use the Central Limit Theorem to argue that the sampling distribution of $\bar{p}_1 - \bar{p}_2$ is approximately normal, and we establish confidence limits of

(6.13) $$(\bar{p}_1 - \bar{p}_2) \pm z s_{\bar{p}_1 - \bar{p}_2}$$

In this problem, the value of $s_{\bar{p}_1 - \bar{p}_2}$ is

$$s_{\bar{p}_1 - \bar{p}_2} = \sqrt{\frac{(0.50)(0.50)}{150} + \frac{(0.18)(0.82)}{160}} = 0.051$$

and since a 90% confidence coefficient is desired, $z = 1.65$. Therefore, the required confidence interval for the difference in the proportion of the applicants in the two stores whose incomes exceeded $20,000 is

$$(0.50 - 0.18) \pm 1.65(0.051) = 0.32 \pm 0.084$$

The confidence limits are 0.236 and 0.404.

Exercises 6.3

1. A department store studied the size of purchases made by a random sample of 100 customers. The mean and standard deviation were $24.75 and $5.50 respectively. Construct a 90% confidence interval for the population mean.
2. A random sample of 100 special extra-large bricks has a mean weight of 110 pounds and a standard deviation of 20 pounds. Calculate a 95% confidence interval for the mean weight of all such special bricks.

3. In a survey of the coffee-drinking habits of American people by the manufacturer of a leading brand, 108 consumers out of 196 responded that they preferred the leading brand. Establish a 99% confidence interval for the actual proportion of consumers who preferred the leading brand.
4. A simple random sample of 100 alumni of a certain university yielded a modal income of $35,000, a median income of $38,500, and a mean income of $40,100. The standard deviation was $4,100. If you wanted to estimate the mean income of the alumni of this university with 90% confidence, what would your interval estimate be?
5. In a simple random sample of 30 firms within a large industry, the arithmetic mean number of employees per firm was 780, with a standard deviation of 40. Establish a 96% confidence interval for the population mean.
6. A market research firm was engaged to study characteristics of the City Centre Hotel and Suburban Square Inn. A random sample of 35 guests was drawn at the City Centre Hotel. The mean annual income and standard deviation were $36,000 and $2,800, respectively. At Suburban Square Inn, 40 guests were sampled. The mean income and standard deviation were $32,000 and $2,500, respectively. Construct a 95% confidence interval for the difference in mean annual incomes of guests of these two hotels.
7. In a random sample of 200 male shoppers, 132 responded "yes" to a certain question. Out of 150 female shoppers, 90 responded "yes" to that question. Construct a 99% confidence interval for the difference in the proportions.
8. A random sample of 40 type A machines revealed an arithmetic mean use age of 12.8 years, with a standard deviation of 2.6 years. In a random sample of 50 type B machines, an arithmetic mean of 10.1 years, with a standard deviation of 2.2 years, was observed. Construct a 95.5% confidence interval for the difference in the mean use age of the two types of machines.
9. On the basis of a sample of 100 people, pollster A estimates that the percentage of people who are going to vote "yes" on a bond issue is 50%. Pollster B estimates the "yes" percentage as 55% on the basis of a sample of 100 different randomly selected people. Is there reason to say that the difference between the results is due to the different methods the pollsters use?
10. In a random sample of 900 residents of city A, an arithmetic mean income of $23,500, with a standard deviation of $5,700, was observed. In a random sample of 400 residents of city B, an arithmetic mean income of $23,000, with a standard deviation of $4,200, was observed for the same period.
    a. Construct a 95% confidence interval for the mean income of all residents in city A.
    b. Construct a 95% confidence interval for the mean income of all residents in city B.
    c. Construct a 95% confidence interval for the difference in mean income between the residents of the two cities.

6.4 CONFIDENCE INTERVAL ESTIMATION (SMALL SAMPLES)

The estimation methods discussed thus far are appropriate when the sample size is large. The distinction between large and small sample sizes is important when the population standard deviation is *unknown* and therefore must be estimated from sample observations. The main point is as follows. We have seen that the ratio $z = (\bar{x} - \mu)/\sigma_{\bar{x}}$ (where $\sigma_{\bar{x}} = \sigma/\sqrt{n}$) is normally distributed for all sample sizes if the population is normal and approximately normally distributed for large samples if the population is not normally distributed. In words, this ratio is z = (sample mean − population mean)/*known* standard error. Furthermore, in section 6.3 we observed that for *large samples*, even if an *estimated* standard error is used in the denominator of this ratio, the sampling distribution may be assumed to be a standard normal distribution for practical purposes. However, the ratio (sample mean − population mean)/*estimated* standard error is not approximately normally distributed for *small* samples, so the theoretically correct distribution, known as the t distribution, must be used instead.* Although the underlying mathematics involved in the derivation of the t distribution is complex and beyond the scope of our book, we can get an intuitive understanding of the nature of that distribution and its relationship to the normal curve.

The ratio $(\bar{x} - \mu)/(s/\sqrt{n})$ is referred to as the t statistic. That is,

(6.14)
$$t = \frac{\bar{x} - \mu}{s/\sqrt{n}}$$

where, as defined in equation 1.9, the sample standard deviation $s = \sqrt{\Sigma(x - \bar{x})^2/(n - 1)}$ is an estimator of the unknown population standard deviation σ.

Let us examine the t statistic and its relationship to the standard normal statistic, $z = (\bar{x} - \mu)/\sigma_{\bar{x}}$. We noted that the denominator of the z ratio represents a *known* standard error, because it is based on a known population standard deviation. On the other hand, the denominator of the t statistic represents an *estimated* standard error, because s is an estimator of the population standard deviation.

The number $n - 1$ in the formula for s is referred to as the **number of degrees of freedom**, which we will denote by the Greek lower-case letter ν

* Early work on the ***t* distribution** was carried out in the early 1900s by W. S. Gossett, an employee of Guinness Brewery in Dublin. Since the brewery did not permit publication of research findings by its employees under their own names, Gossett adopted "Student" as a pen name. Consequently, in addition to the term "t distribution" used here, the distribution has come to be known as "Student's distribution" or "Student's t distribution" and is so referred to in many books and journals.

(pronounced "nu"). It is not feasible to give a simple verbal explanation of this concept. From a purely mathematical point of view, the number of degrees of freedom ν is simply a parameter that appears in the formula of the t distribution. However, in the present discussion, in which s is used as an estimator of the population standard deviation σ, the $n - 1$ may be interpreted as the number of independent deviations of the form $x - \bar{x}$ present in the calculation of s. Since the total of the deviations $\Sigma(x - \bar{x})$ for n observations equals zero, only $n - 1$ of them are independent. This means that if we were free to specify the deviations $x - \bar{x}$, we could designate only $n - 1$ of them independently. The nth one would be determined by the condition that the n deviations must add up to zero. Therefore, in the estimation of a population standard deviation or a population variance, if the divisor $n - 1$ is used in the estimator, then $n - 1$ degrees of freedom are present.

The t distribution

The t distribution has been derived mathematically under the assumption of a normally distributed population.[5] As with the standard normal distribution, the t distribution is symmetrical and has a mean of zero. However, the standard deviation of the t distribution is greater than that of the normal distribution but approaches the latter figure as the number of degrees of freedom (and, therefore, the sample size) becomes large. It can be demonstrated mathematically that for an infinite number of degrees of freedom, the t distribution and normal distribution are exactly equal. The approach to this limit is quite rapid. Hence, a widely applied rule of thumb considers samples of size $n > 30$ "large," and for such samples, the standard normal distribution may appropriately be used as an approximation to the t distribution, even though the latter is the theoretically correct functional form. Figure 6-4 shows the graphs of several t curves for different numbers of degrees of freedom. We can see from these graphs that the t curves are lower at the mean and higher in the tails than the standard normal distribution. As the number of degrees of freedom increases, the t distribution rises at the mean and lowers at the tails until, for an infinite number of degrees of freedom, it coincides with the normal distribution. The use of tables of areas for the t distribution is explained in Example 6-5.

[5] The t distribution has the form

$$f(t) = c\left(1 + \frac{t^2}{\nu}\right)^{-(\nu+1)/2}$$

where $t = \dfrac{\bar{x} - \mu}{s_{\bar{x}}}$ (as previously defined)

c = a constant required to make the area under the curve equal to one

$\nu = n - 1$, the number of degrees of freedom

The variable t ranges from minus infinity to plus infinity. The constant c is a function of ν, so that for a particular value of ν, the distribution of $f(t)$ is completely specified. Thus, $f(t)$ is a family of functions, one for each value of ν.

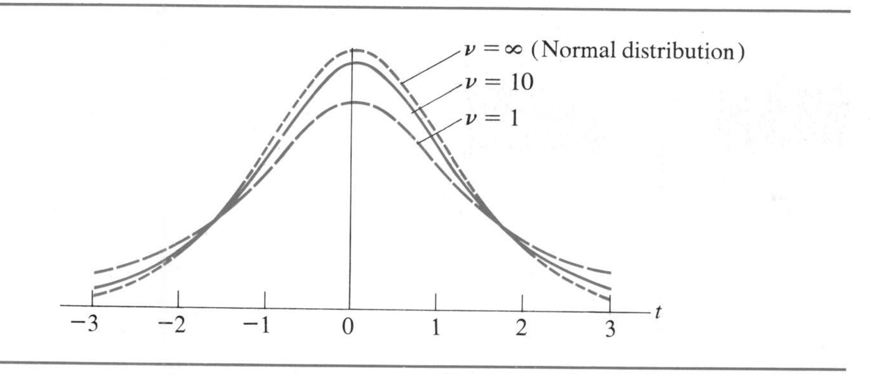

FIGURE 6-4
The t distributions for $\nu = 1$ and $\nu = 10$ compared with the normal distribution ($\nu = \infty$)

EXAMPLE 6-5

Assume that a simple random sample of 9 automobile tires was drawn from a large production run of a certain brand of tire. The mean lifetime of the tires in the sample was $\bar{x} = 32{,}010$ miles. This sample mean is the best single estimate of the corresponding population mean. The population standard deviation is unknown. Hence, an estimate of the population standard deviation is calculated by the formula $s = \sqrt{\Sigma(x - \bar{x})^2/(n - 1)}$. The result is $s = 2{,}520$ miles. What is the interval estimate for the population mean at a 95% level of confidence?

Solution

Reasoning as we did in the case of the normal sampling distribution for means, we find that confidence limits for the population mean, using the t distribution, are given by

(6.15)
$$\bar{x} \pm t \frac{s}{\sqrt{n}}$$

where t is determined for $n - 1$ degrees of freedom. The number of degrees of freedom is one less than the sample size, that is, $\nu = n - 1 = 9 - 1 = 8$.

Just as the z values in Examples 6-1 and 6-2 represented multiples of standard errors, the t value in equation 6.15 represents a multiple of estimated standard errors. We find the t value in Table A-6 of Appendix A.

A brief explanation of Table A-6 is required. In the table of areas under the normal curve, areas lying between the mean and specified z values were given. However, in the case of the t distribution, since there is a different t curve for each sample size, no single table of areas can be given for all these distributions. Therefore, for compactness, a t table shows the relationship between areas and t values for only a few "percentage points" in different t distributions. Specifically, the entries in the body of the table are t values for areas of 0.01, 0.02, 0.05, and 0.10 in the two tails of the distribution combined.

In this problem, we refer to Table A-6 of Appendix A under column 0.05 for 8 degrees of freedom, and we find $t = 2.306$. This means that, as shown in Figure 6-5, for 8 degrees of freedom, a total of 0.05 of the area in the t distribution lies below $t = -2.306$ or above $t = 2.306$. Correspondingly, the probability is 0.95 that for 8 degrees of freedom, the t value lies between -2.306 and 2.306.

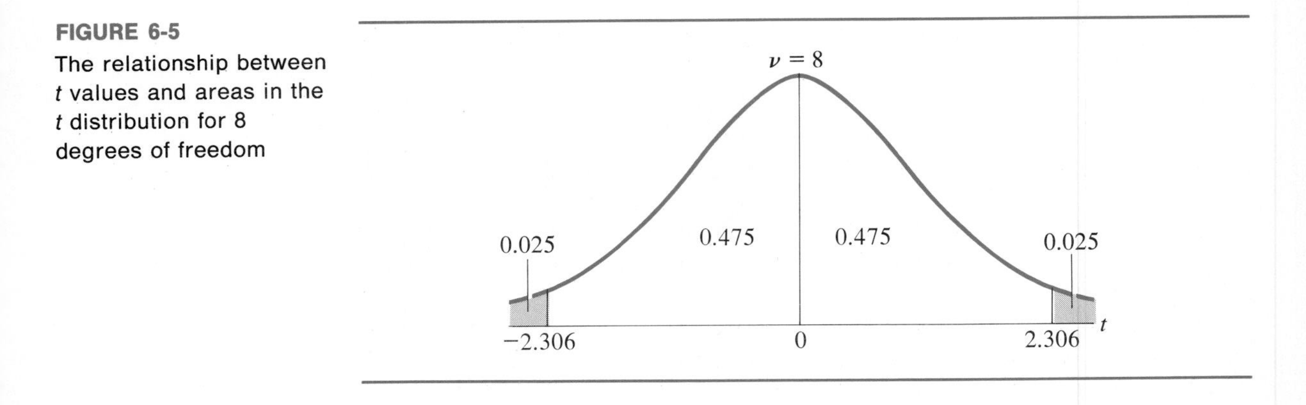

FIGURE 6-5 The relationship between t values and areas in the t distribution for 8 degrees of freedom

In this problem, substituting $t = 2.306$ into equation 6.15, we obtain the following 95% confidence limits:

$$32{,}010 \pm 2.306(840) = 32{,}010 \pm 1{,}937.04 \text{ miles}$$

Hence, to the nearest mile, the confidence limits for the estimate of the mean lifetime of all tires in the production run are 30,073 and 33,947 miles.

The interpretation of this interval and the associated confidence coefficient is the same as in the case of large samples and a normal distribution. Comparing this procedure with the corresponding method for large samples for 95% confidence limits, we note that the t value of 2.306 replaces the 1.96 figure that is appropriate for the normal curve. Thus, with small samples we have a wider confidence interval, leading to a vaguer result. This is to be expected, because σ is estimated by s using a small sample size n.

Note that since Table A-6 shows areas in the combined tails of the t distribution, we had to look under the column headed 0.05 for the t value corresponding to a 95% confidence interval. Correspondingly, we would find t values for 90%, 98%, and 99% confidence intervals under the columns headed 0.10, 0.02, and 0.01, respectively.

Exercises 6.4

1. A manufacturer of electronic components has a contract with NASA to provide components for weather satellites. The contract specifies that the components must have a mean life of 2 months when operating under extreme temperature conditions. To determine whether its components meet the requirements specified, the company decides to test several of the components, selected randomly. Since the testing process consists of subjecting the components (which are very costly) to extreme conditions that destroy the components, the company uses a small random sample. The results obtained from this sample are $n = 9$ components, $\bar{x} = 9$ weeks and $s = 4$ weeks.

a. State a 95% confidence interval for μ, the population mean life of the components.
b. Can the company be sure that there is no more than a 5% chance that the population mean μ lies outside this interval? Explain.

2. A company ran a test to determine the length of time required to complete service calls. The following times, in minutes, were obtained for a simple random sample of 9 service calls: 48, 51, 28, 66, 81, 36, 40, 59, and 50.
   a. Construct a 99% confidence interval for the mean time for completion of service calls.
   b. If times to complete service calls followed a highly skewed distribution, would the range you set up in part (a) really be a 99% confidence interval?
3. A random sample of 25 accounts receivable is randomly selected. The sample mean is found to be $7,850, with a standard deviation of $200. Set up a 90% confidence interval for the population mean.
4. A simple random sample of 12 small firms yielded a mean amount of property insurance of $750,000 with a standard deviation of $100,000. Construct a 99% confidence interval for the mean amount of property insurance carried by the relevant population of small firms. What would your answer be if the same results had been obtained with a simple random sample of 24 firms?
5. The time required by a worker to finish a task is normally distributed. Six observations were recorded in hours as follows: 8, 12, 10, 9, 5, and 16. Estimate the mean time required to finish the task; use an interval estimate with a 90% confidence coefficient.

6.5 DETERMINATION OF SAMPLE SIZE

In all of the examples thus far, the sample size n was given. However, we could ask the following question: How large should a sample be in a specific situation? If a sample larger than necessary is used, resources are wasted; if the sample is too small, the objectives of the analysis may not be achieved.

Sample Size for Estimation of a Proportion

Statistical inference provides the following answer to the question of sample size. Let us assume an investigator desires to estimate a certain population parameter and wants to know how large a simple random sample is required. We assume that the population is very large relative to the prospective sample size.

To specify the required sample size any investigator must answer two questions.

1. What degree of precision is desired?
2. What probability is attached to obtaining the desired precision?

Clearly, the greater the degree of desired precision, the larger will be the necessary sample size; similarly, the greater the probability specified for obtaining the desired precision, the larger will be the required sample size. We will use examples to indicate the technique of determining sample size for estimation of a population proportion and a population mean.

EXAMPLE 6-6

Suppose we would like to conduct a poll among eligible voters in a city in order to determine the percentage who intend to vote for the Democratic candidate in an upcoming election. We want a 95.5% probability that we will estimate the percentage that will vote Democratic within ± 1 percentage point. What is the required sample size?

Solution

We answer the question by first indicating the rationale of the procedure and then condensing this rationale into a simple formula. The statement of the question gives a relationship between the sampling error that we are willing to tolerate and the probability of obtaining this level of precision. In this problem, $2\sigma_{\bar{p}}$ must equal 0.01. This means that we are willing to have a probability of 95.5% that our sample percentage $\bar{p}$ will fall within 0.01 of the true but unknown population proportion p (see Figure 6-6). We may now write

$$2\sigma_{\bar{p}} = 0.01$$

or

$$2\sqrt{\frac{pq}{n}} = 0.01$$

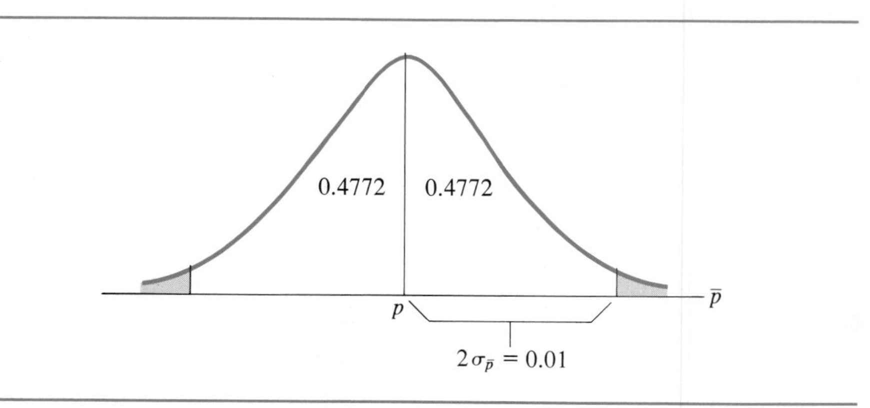

FIGURE 6-6 Sampling distribution of a proportion showing the relationship between the sampling error and the probability of obtaining this degree of precision

and

$$\sqrt{\frac{pq}{n}} = 0.005$$

In all our previous problems, the sample size n was known, but here n is the unknown for which we must solve. However, it appears that there are too many unknowns, namely, the population parameters p and q as well as n. Therefore, we must estimate values for p and q, and then we can solve for n. Suppose we wanted to make a conservative estimate for n. What should we guess as a value for p? In this context, a conservative estimate is an estimate made in such a way as to ensure that the sample size will be large enough to deliver the desired precision. In this problem, the "most conservative" estimate for n is given by assuming $p = 0.50$ and $q = 0.50$. This follows from the fact that the product $pq = 0.25$ is larger for $p = 0.50$ and $q = 0.50$ than for any other possible values of p and q where $p + q = 1$. For example, if $p = 0.70$ and $q = 0.30$, then $pq = 0.21$, which is less than 0.25. The relationships between possible values of p and the corresponding values of pq are shown in Figure 6-7. Thus, the largest, or "most conservative," value of n is determined by substituting $p = q = 0.50$ as follows:

$$0.005 = \sqrt{\frac{0.50 \times 0.50}{n}}$$

Squaring both sides gives

$$0.000025 = \frac{0.50 \times 0.50}{n}$$

and

$$n = \frac{0.50 \times 0.50}{0.000025} = 10{,}000$$

Hence, to achieve the desired degree of precision, a simple random sample of 10,000 eligible voters would be required. Of course, the large size of this sample is attributable to the high degree of precision specified. If $\sigma_{\bar{p}}$ were doubled from 0.005 to

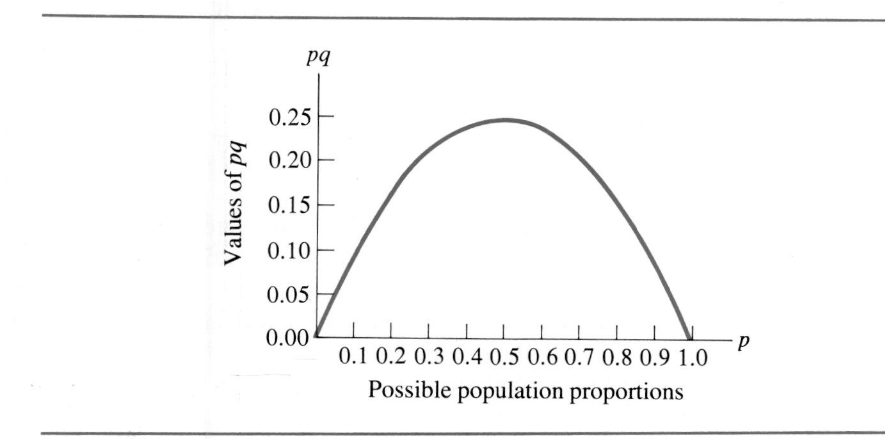

FIGURE 6-7 Relationship between possible values of p and the corresponding products pq (where $q = 1 - p$)

0.01, the required sample size would be cut down to one-fourth of 10,000, or 2,500. This follows from the fact that the standard error varies inversely with the square root of sample size.

In this election problem, we assumed $p = q = 0.50$, although less conservative estimates are possible if we believe that $p \neq 0.50$. In problems involving proportions, we would use whatever past knowledge we have to estimate p. For example, suppose we wanted to determine the sample size to estimate an unemployment rate and we knew from past experience that for the community of interest the proportion of the labor force that was unemployed was somewhere between 0.05 and 0.10. We then would assume $p = 0.10$, since this would give us a more conservative estimate (larger sample size) than assuming $p = 0.05$ or any value between 0.05 and 0.10. (For example, if we assumed $p = 0.07$ and in fact the true value of p was 0.09, the sample size we determined from a calculation involving $p = 0.07$ would not be large enough to give us the specified precision.) Assuming $p = 0.10$ assures us of obtaining the desired degree of precision regardless of what the true value of p is, as long as it is in the range 0.05 to 0.10.

We can summarize this calculation for sample size by noting that we start with the following statement:

$$D = z\sigma_{\bar{p}}$$

where D is the tolerated deviation (in percentage points) and z is the multiple of standard errors corresponding to the specified probability of obtaining this precision. Then for infinite populations or large populations relative to sample size, we have

$$n = \frac{z^2(p)(1 - p)}{D^2} \tag{6.16}$$

Hence, in the preceding voting problem, application of equation 6.16 yields

$$n = \frac{(2)^2(0.50)(0.50)}{(0.01)^2} = 10{,}000$$

Sample Size for Estimation of a Mean

The required sample size for estimation of a mean can be determined by an analogous calculation. Suppose we wanted to estimate the arithmetic mean hourly wage for a group of skilled workers in a certain industry. Let us further assume that from prior studies we estimate that the population standard deviation of the hourly wages of these workers is about \$.15. How large a sample size would be required to yield a probability of 99.7% that we will estimate the mean hourly wage of these workers within ±\$.03?

Since the 99.7% probability corresponds to a level of three standard errors, we can write

$$3\sigma_{\bar{x}} = \$.03$$

or

$$\frac{3\sigma}{\sqrt{n}} = \$.03$$

and

$$\frac{\sigma}{\sqrt{n}} = \$.01$$

The population standard deviation σ is known from past experience. Hence, substituting $\sigma = \$.15$ gives

$$\frac{\$.15}{\sqrt{n}} = \$.01$$

and

$$\sqrt{n} = \frac{\$.15}{\$.01} = 15$$

Squaring both sides yields the solution

$$n = (15)^2 = 225$$

Therefore, a simple random sample of 225 of these workers would be required. In summary, if we calculate $\bar{x}$ for the hourly wages of a simple random sample of 225 of these workers, we can estimate the mean wage rate of all skilled workers in this industry within \$.03 with a probability of 99.7%.

In this problem, we assumed that an estimate of the population standard deviation was available from prior studies; this situation may exist for governmental agencies that conduct repeated surveys of wage rates, population, and the like. If the population standard deviations (or estimated population standard deviations) in these past studies are not erratic or excessively unstable, they provide useful bases for estimating σ values in the above procedures for computing sample size.

Of course, an estimate of the population standard deviation may not be available from past experience. It may be possible, however, to get a rough estimate of σ if there is at least some knowledge of the total range of the basic random variable in the population. For example, suppose we know that the difference between the wages of the highest- and lowest-paid workers is about \$1.20. In a normal distribution, a range of three standard deviations on either side of the mean includes virtually the entire distribution. Thus, a range of 6σ includes almost all the frequencies, and we may state

$$6\sigma \approx \$1.20$$

or

$$\sigma \approx \$.20$$

Of course, the population distribution is probably nonnormal and \$1.20 may not be exact. Consequently, the estimate of σ may be quite rough. Nevertheless, we may be able to obtain a reasonably good estimate of the required sample size in a situation where, in the absence of this "guestimating" procedure, we may be at a loss for any notion of a suitable sample size.

If we express in the form of an equation the technique for calculating a required sample size when estimating a population mean, we have

$$n = \frac{z^2\sigma^2}{D^2} \tag{6.17}$$

where D is the tolerated deviation. For the problem involving hourly wages, this gives

$$n = \frac{(3)^2(\$.15)^2}{(\$.03)^2} = 225$$

In equations 6.16 and 6.17, we assume that the n value determined is sufficiently large for the assumption of a normal sampling distribution as appropriate and that the populations are large relative to this sample size. If the populations are not large relative to sample size, appropriate formulas may be derived for n by taking into account the finite population correction. The formulas for the sample sizes required for estimation of a proportion and mean, respectively, are

$$n = \frac{p(1-p)}{\dfrac{D^2}{z^2} + \dfrac{p(1-p)}{N}} \tag{6.18}$$

$$n = \frac{\sigma^2}{\dfrac{D^2}{z^2} + \dfrac{\sigma^2}{N}} \tag{6.19}$$

In equations 6.18 and 6.19, N is the number of elements in the population (population size), and all other symbols are as previously defined. Equations 6.18 and 6.19 correspond, respectively, to equations 6.16 and 6.17, in which we assumed infinite populations.

For example, in the voting problem discussed earlier in this section, if the population size had been 50,000, then the required sample size computed from equation 6.18 would be

$$n = \frac{(0.50)(0.50)}{\dfrac{(0.01)^2}{(2)^2} + \dfrac{(0.50)(0.50)}{50{,}000}} = 8{,}333$$

This result may be compared with the sample size of 10,000 computed earlier from equation 6.16 assuming an infinite population size.

Analogously, in the problem involving the required sample size for estimating the population mean hourly wage for a group of skilled workers in a certain industry, let us assume a population size of 1,000 such skilled workers. Then the required sample size computed from equation 6.19 would be

$$n = \frac{(0.15)^2}{\dfrac{(\$.03)^2}{(3)^2} + \dfrac{(0.15)^2}{1{,}000}} = 184$$

This figure may be compared with the sample size of 225 computed previously from equation 6.17 assuming an infinite population size. It would be instructive for you to verify that if a population size of 1,000,000 had been assumed, the sample size computed from equation 6.19 would be 224.9, which rounds off to 225—the same figure obtained by assuming an infinite sample size.

Exercises 6.5

1. A general estimate is needed of the proportion of households that own electric fans in a certain community. There is no information available, so a sample survey will be taken. The market researchers indicate that they wish to estimate the required proportion within 2 percentage points.
   a. How large must a simple random sample be for a 90% confidence coefficient?
   b. If the researchers make a preliminary estimate that the proportion is 0.30, how large must the sample be for the same confidence coefficient?
2. For a study of required numbers of salespersons, a department store wants to determine the mean time a salesperson spends with a customer. The analyst wants to be 99% confident that the estimate will be within 2 minutes of the true figure. If it is guessed that the standard deviation of the time a salesperson spends with each customer is 4 minutes, how large should the selected sample be to estimate the mean time?
3. A major oil company intends to conduct a survey of people who live in the New England region to determine their views on the issue of offshore drilling on the continental shelf. The analyst in charge of the project wishes to be 98% confident that the estimate of the proportion of New Englanders who favor offshore drilling lies within 4 percentage points of the true percentage. How large a sample should be drawn?
4. A government official working on economic policy desires to ascertain the proportion of businesspersons who feel that inflation is a more serious problem than unemployment. The official is certain that no more than 20% hold that view. If the objective is to estimate this proportion within 5 percentage points of the true figure, how large a simple random sample should be taken? Use a 95.5% confidence coefficient.

5. The vice-president in charge of customer service for the Ambivalent Automobile Corporation wants to authorize a survey to determine the percentage of customers who are completely satisfied with the present warranty on the company's products. The optimistic executive believes that 90% of the customers are so satisfied. If the estimate is to lie within 3 percentage points of the true proportion, with 97% confidence, how large a simple random sample should be drawn?
6. A statistician wishes to determine the average hourly earnings for employees in a given occupation in a particular state. A pilot study obtains point estimates of $6.20 for the mean and $.50 for the standard deviation. The statistician then specifies that when he takes his random sample, he wants to be 95.5% confident that the maximum error of estimate will not exceed $.05.
   a. Discuss the sense in which he will have 95.5% confidence in the estimate.
   b. What size should the sample be?
7. A public accountant wishes to estimate the percentage of companies in the United States that use the LIFO method of pricing inventory. The accountant intends to do this on the basis of a random sample and wishes to be 95% confident that the estimate lies within 3 percentage points of the true percentage of companies using LIFO. The accountant is quite certain that no more than 25% of the companies in the United States use this method of pricing inventory. How large should the simple random sample be?
8. One can always decrease the width of a confidence interval by increasing the sample size. Why not determine the desired width of interval first and then sample accordingly?
9. What are two ways of decreasing the width of a confidence interval for μ, given that the best point estimator of a sample is $\bar{x}$?
10. A popular U.S. senator—who strongly encourages constituents to express their views on issues through letters and telegrams—chairs a subcommittee that is attempting to formulate legislation to limit the amount of money that organizations can contribute to the campaign of a presidential candidate. As a guideline, the senator has decided to estimate the mean of the many limitations (in dollars) that have been suggested in the thousands of letters received from the voters favoring such a limit. Instead of actually calculating the mean, the senator's staff will take a random sample of the letters and telegrams that have mentioned dollar amounts for the limitation.
    a. Estimate how large the sample should be in order for the appropriate interval estimate of the mean limitation to be no wider than $100. A 90% confidence coefficient is desired, and the standard deviation is taken to be $500.
    b. If the confidence coefficient was set at 99%, how large would the sample have to be?

Computer Exercises 6.5

Assume that the 75 families in Appendix E constitute a simple random sample of all families in a suburban area.

1. Compute the 95% confidence intervals for annual income, family size, and age of the highest income earner.
2. Compute the 98% confidence intervals for annual income, family size, and age of the highest income earner.
3. Compute the 99% confidence interval for the proportion of families that are home owners.

7

Hypothesis Testing

We will now focus on hypothesis testing—the second basic subdivision of statistical inference. **Hypothesis testing** addresses the important question of how to choose among alternative propositions or courses of action, while controlling or minimizing the risks of making wrong decisions. We will now briefly and informally summarize the rationale involved in testing hypotheses and then explain the details of these testing procedures by means of examples.

7.1 THE RATIONALE OF HYPOTHESIS TESTING

To gain some insight into the reasoning involved in statistical hypothesis testing, we will consider a nonstatistical hypothesis-testing procedure with which we are all familiar. The basic process of inference involved is strikingly similar to that employed in statistical methodology.

Consider the process by which an accused individual is judged in a court of law. Under Anglo-Saxon law, the man before the bar is assumed to be innocent. The burden of proof of his guilt rests on the prosecution. Using the language of hypothesis testing, let us say that we want to test the **hypothesis**, which we denote by H_0, that the man before the bar is innocent. The **alternative hypothesis** H_1 is that the defendant is guilty. The jury examines the evidence to determine whether the prosecution has demonstrated that this evidence is inconsistent with the basic hypothesis H_0 of innocence. If the jurors decide the evidence is inconsistent with H_0, they reject that hypothesis and accept its alternative H_1 that the defendant is guilty.

If we analyze the situation that results when the jury makes a decision, we find that four possibilities exist in terms of the basic hypothesis H_0:

1. The defendant is innocent (H_0 is true), and the jury finds that he is innocent (accepts H_0); hence, the correct decision is made.
2. The defendant is innocent (H_0 is true), but the jury finds him guilty (rejects H_0); hence, an error is made.
3. The defendant is guilty (H_0 is false), and the jury finds that he is guilty (rejects H_0); hence, the correct decision is made.
4. The defendant is guilty (H_0 is false), but the jury finds him innocent (accepts H_0); hence, an error is made.

In possibilities (1) and (3), the jury reaches the correct decision; in possibilities (2) and (4), it makes an error. Let us consider these errors in terms of conventional statistical terminology. In possibility (2), hypothesis H_0 is erroneously rejected. The basic hypothesis H_0 that is being tested for possible rejection is generally referred to as the **null hypothesis**. Hypothesis H_1 is designated the **alternative hypothesis**.

Type I error

To reject the null hypothesis when in fact it is true is referred to as a **Type I error**.

In possibility (4), hypothesis H_0 is accepted in error.

Type II error

To accept the null hypothesis when it is false is termed a **Type II error**.

Under our legal system, the commission of a Type I error is considered far more serious than the commission of a Type II error. Thus, we feel that it is a more grievous mistake to convict an innocent person than to let a guilty person go free.

Had we made H_0 the hypothesis that the defendant is guilty, the meaning of Type I and Type II errors would have been the reverse of the first formulation; what had previously been a Type I error would become a Type II error, and a Type II would become a Type I error.

In the statistical formulation of hypotheses, how we choose to exercise control over Type I and Type II errors serves as a basic guide in stating the hypotheses to be treated.

In this chapter, we will see how errors are controlled in hypothesis testing.

TABLE 7-1
The relationship between actions concerning a null hypothesis and the truth or falsity of the hypothesis

Action Concerning Hypothesis H_0	State of Nature	
	H_0 Is True (innocent)	H_0 Is False (guilty)
Accept H_0	Correct decision	Type II error
Reject H_0	Type I error	Correct decision

The four possible jury decisions in our example are summarized in Table 7-1. Here, the headings are stated in modern decision-theory terminology and require brief explanations. When hypothesis testing is viewed as a problem in decision making, two alternative actions can be taken: "accept H_0" or "reject H_0." The two alternatives, truth or falsity of hypothesis H_0, are viewed as "states of nature" or "states of the world" that affect the consequences, or "payoff," of the decision. The payoffs are indicated in the table in terms of the correctness of the decision or the type of error incurred. We can see from the framework of this hypothesis-testing problem that what we need is some criterion on which to base the decision to accept or reject the null hypothesis H_0. Classical hypothesis testing attacks this problem by establishing **decision rules** based on data derived from simple random samples. The sample data are analogous to the evidence investigated by the jury. The decision procedure attempts to assess the risks of making incorrect decisions, and, in a sense, which we will examine, to minimize them.

Terminology

Decision rules

The Hypothesis-Testing Procedure

Hypothesis-testing procedures can be used to solve two basic types of decision problems. In the first type of problem, we want to know whether a population parameter has changed from or differs from a particular value. Here, we are interested in detecting whether the population parameter is *either* larger than or smaller than a particular value. For example, suppose that the mean family income in a certain city was determined from a census to be \$26,500 for a particular year, and two years later we want to discover whether the mean income has *changed*. If it is not feasible to take another census, we may draw a simple random sample of families and try to reach a conclusion based on this sample. As we did with the judicial decisions, we can set up two competing hypotheses and choose between them.

- The null hypothesis H_0 would simply be an assertion that the mean family income was unchanged from the \$26,500 figure; in statistical language, we write this hypothesis H_0: $\mu = \$26{,}500$, and μ is the mean family income in the city.

- The alternative hypothesis is that the mean family income *has* changed or, in statistical terminology, H_1: $\mu \neq \$26{,}500$.

In this example, we would observe the mean family income in a simple random sample of, say, 1,000 families. If the value of the sample mean $\bar{x}$ differs from the population mean $\mu = \$26{,}500$ by more than we would be willing to attribute to chance sampling error, we will reject the null hypothesis H_0 and accept its alternative H_1. On the other hand, if the difference between the sample mean and the population mean assumed under H_0 is small enough to be attributed to chance sampling error, we will accept H_0. How do we know for what values of the sample statistic to reject H_0 and for what values to accept H_0? The answer to this question is the essence of hypothesis testing.

> The hypothesis-testing procedure is simply a decision rule that specifies whether the null hypothesis H_0 should be accepted or rejected for every possible value of a statistic observable in a simple random sample of size *n*.

The set of possible values of the sample statistic is referred to as the **sample space**. Therefore, the test procedure divides the sample space into mutually exclusive parts called the **acceptance region** and the **rejection region** or **critical region**.

The nature of the division of the sample space for the example we have been discussing is illustrated in Figure 7-1. From the sampling theory developed in Chapter 5, we know that, given a population with a mean of $26,500 and a known standard deviation, there would be sampling variation among the means of samples of the same size drawn from that population. According to the central limit theorem, the sampling distribution of the mean for large sample sizes may be assumed to be normal, regardless of the shape of the population

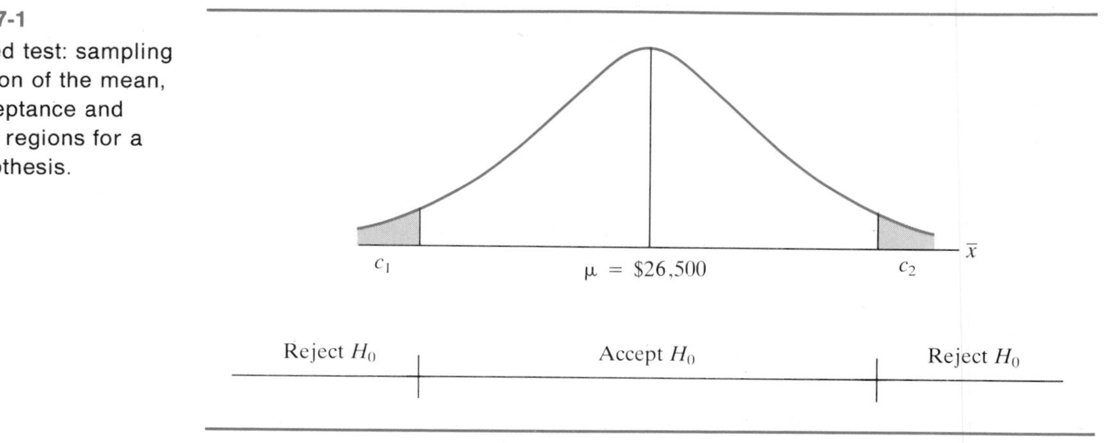

FIGURE 7-1
Two-tailed test: sampling distribution of the mean, with acceptance and rejection regions for a null hypothesis.

distribution. To decide whether family mean income has changed from \$26,500, we determine two values, denoted by c_1 and c_2 in Figure 7-1, that set limits on the amount of sampling variation we feel is consistent with the null hypothesis. The decision rule in this case would be: (1) If the mean income of the sample of 1,000 families lies below c_1 or above c_2, we will reject the null hypothesis and conclude that the mean family income of the city has changed from \$26,500; (2) if the sample mean lies between c_1 and c_2, we cannot reject H_0 and we will not be able to conclude that the city's mean family income has changed.

A test in which we want to determine whether a population parameter has *changed*—regardless of the direction of change—is referred to as a **two-tailed test**, because the null hypothesis can be rejected by observing a statistic that falls in either of the two tails of the appropriate sampling distribution.

Two-tailed test

In the second type of hypothesis test, we wish to find out whether a sample comes from a population that has a parameter *less* than or *more* than a hypothesized value.

Decision problems in which attention is focused on the direction of change give rise to **one-tailed tests.**

One-tailed test

The following example illustrates such a test. Suppose that in the past, under carefully specified driving conditions, the gas mileage of a compact car has averaged 30 miles per gallon (mpg) or less. The manufacturer of this car has redesigned the engine and wishes to test its claim that the average gas mileage is now greater than 30 miles per gallon. The company's engineers decide to test the claim using a simple random sample of 50 new cars. The null hypothesis H_0 to be tested is "the true mean is equal to or less than 30 miles per gallon"; the alternative hypothesis H_1 is "the true mean is greater than 30 miles per gallon." Mathematically, these two hypotheses may be expressed

$$H_0: \mu \leqslant 30 \text{ mpg}$$

$$H_1: \mu > 30 \text{ mpg}$$

where μ denotes the mean number of miles per gallon.

From long experience, the automobile manufacturer has found that the population standard deviation is 5 miles per gallon. In this problem, we will assume that it is *known* from experience that the population standard deviation $\sigma = 5$ miles per gallon.

In the random sample of 50 automobiles included in the test, the mean number of miles per gallon observed is 32 (that is, $\bar{x} = 32$ mpg). Should we conclude that the automobile manufacturer's claim is valid—that the average gas mileage of this car is now greater than 30 mpg? Or, should we conclude

FIGURE 7-2
One-tailed test: sampling distribution of the mean, with acceptance and rejection regions for a null hypothesis.

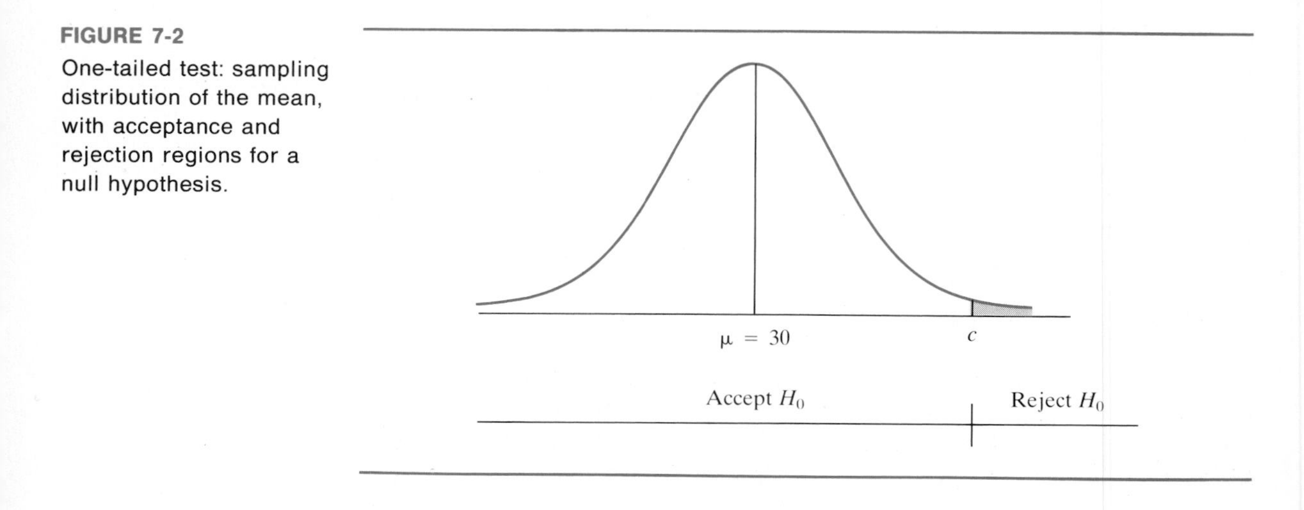

that $\bar{x} = 32$ does not differ significantly from 30 mpg and that the difference may be attributable to chance sampling error?

In this case, the decision rule would be to reject H_0 if the mean of the sample $\bar{x}$ is more than some appropriate number c, and to accept H_0 if $\bar{x}$ is equal to or less than c. The decision rule is diagrammed in Figure 7-2. Again, the number c represents the limit on the amount of sampling variation that we feel is consistent with the null hypothesis.

Concept of the Null Hypothesis

A null hypothesis is a statement about a population parameter, as in our first example (H_0: $\mu = \$26{,}500$) and in our second example (H_0: $\mu \leqslant 30$ mpg). Note that the equality sign is included in the null hypotheses, which is standard practice. By having the null hypothesis assert that the population parameter is equal to some specific value, we are able to decide for which values of the observed sample statistic we will reject or accept that hypothesis. For example, when we hypothesized that $\mu = \$26{,}500$ in the first example and included in the second hypothesis that $\mu = 30$ mpg, we were able to establish sampling distributions of the sample mean $\bar{x}$ and to decide which values of $\bar{x}$ would cause us to reject the null hypothesis and which values of $\bar{x}$ would cause us to accept it. Furthermore, we can specify how much risk of making Type I errors we are willing to tolerate. For example, if we hypothesize that $\mu = 30$ mpg, we can compute the probability of making a Type I error (that is, the probability of erroneously rejecting that hypothesis). This concept will be applied later in the chapter. These points explain why null hypotheses are set up in the form stated earlier, rather than in the form H_0: $\mu \neq \$26{,}500$ or H_0: $\mu > 30$ miles per gallon.

One-sample tests

We will now turn to the application of some of these ideas. First, we will consider **one-sample tests**, which are tests of hypotheses based on data contained in a single sample. Tests involving means and proportions will be studied in that order.

7.2 ONE-SAMPLE TESTS (LARGE SAMPLES)

In the one-sample tests discussed in this section, we will assume three things.

Assumptions

1. The sample size is large ($n > 30$).
2. The sample size is determined before the test is conducted.
3. The population standard deviation is known.

The assumption of a large sample size ($n > 30$) is made so that we can assume that the sampling distributions used in the testing procedure are normal. This is a useful simplification compared to the small-sample case, as we will see at the end of this chapter.

To construct a more detailed illustration of how the hypothesis-testing procedure can be used, we will return to the example of testing an automobile manufacturer's claim that after redesigning the engine of a compact car, the car's gasoline mileage is now in excess of 30 mpg.

We proceed to convert the testing of the automobile manufacturer's claim to a hypothesis-testing framework. This case is an example of a one-tailed test, or a **one-sided alternative**.

There are six steps involved in a test of statistical hypothesis.

Six steps

1. Determine the null and alternative hypotheses.
2. Select a level of significance for the test.
3. Choose a test statistic.
4. Select a sample size.
5. Determine the decision rule.
6. Reach a conclusion based on the sample drawn. This involves the rejection or retention of the null hypothesis.

We will now conduct the test, taking these steps in order. For ease of discussion, steps 3 and 4 and steps 5 and 6 have been combined.

State the null and alternative hypotheses (1)

We begin by stating the null and alternative hypotheses in statistical terms. As we have seen earlier, if we let μ represent the average number of miles per gallon of a compact car with the redesigned engine, the null and alternative hypotheses can be stated

$$H_0: \mu \leqslant 30 \text{ mpg}$$

$$H_1: \mu > 30 \text{ mpg}$$

Specify the level of significance (2)

Our decision will be based on the data observed in the simple random sample of compact cars produced by the automobile manufacturer. The question we want to answer is "Are the sample data so inconsistent with the null hypothesis that we must reject that hypothesis?" In designing the test, we must specify the risk that we are willing to run of rejecting the null hypothesis when it is true. In other words, we must specify the probability of committing a Type I error—or, as it is commonly designated, the **level of significance** of the test. Levels of significance such as 0.05 or 0.01 are conventionally used. Of course, such figures are rather arbitrary, but low levels are ordinarily used so that the probability of committing a Type I error will be quite low. In this problem, we will assume that we do not want the risk of erroneously rejecting the null hypothesis to be greater than 0.05. The level of significance is denoted by the Greek letter α. Hence, in this problem, $\alpha = 0.05$. In a one-tailed test, α represents the *maximum* probability of a Type I error. In a two-tailed test in which the null hypothesis consists of only one value of a population parameter, α represents *the* probability of a Type I error.

Table 7-2 summarizes the alternative hypotheses tested in our gasoline mileage problem and the possible actions related to these hypotheses. After a brief discussion of the meanings of Type I and Type II errors in this problem, we will identify the remaining steps in the hypothesis-testing procedure and carry them out quantitatively.

From Table 7-2, we can observe that a Type I error in this problem (the incorrect rejection of the null hypothesis) takes the form of concluding that the compact car delivers more than 30 mpg on the average when the true average is 30 mpg or less. In this problem, a Type II error takes the form of concluding that the compact car delivers 30 mpg or less on the average when actually the true average is greater than 30 mpg.

Choose the test statistic and sample size (3, 4)

Since the hypotheses in this problem are related to the *mean* number of miles per gallon that the compact car yields, the **test statistic** we want to observe is the mean number of miles per gallon delivered by the sample of cars. In this problem, we are assuming that a *simple random sample of 50 cars* is used in the test. Of course, as the sample size increases, the precision of the test increases (the expected amount of sampling error decreases). The choice of sample size

TABLE 7-2
Gasoline mileage problem: the relationship between possible actions and hypotheses related to the number of miles per gallon delivered by a certain compact car

Action Concerning Hypothesis H_0	State of Nature	
	H_0: $\mu \leqslant 30$ mpg	H_1: $\mu > 30$ mpg
Accept H_0	No error	Type II error (Accept $H_0 \mid H_0$ False)
Reject H_0	Type I error (Reject $H_0 \mid H_0$ True)	No error

is essentially an economic decision, because as the sample size increases, the cost involved in performing the test increases.

Determine the decision rule and conclusion (5, 6)

We will now turn to the question of how to establish a decision rule on which to base our acceptance or rejection of the null hypothesis. As indicated earlier, a hypothesis-testing rule is simply a procedure that specifies the action to be taken for each possible sample outcome. Thus, we are interested in partitioning the sample space into a region in which we will reject the null hypothesis and a region in which we will accept it. In every hypothesis-testing problem, the partitioning of the appropriate sample space is accomplished by assuming that the null hypothesis is true and considering the appropriate sampling distribution. This follows from the fact that specifying the probability of making a Type I error determines how the sample space will be partitioned.

In this particular problem, the question concerns the *mean* number of miles per gallon, so that the sampling distribution of the mean is the appropriate distribution. Because our sample size is large ($n > 30$), we use the central limit theorem and assume that the sampling distribution of means is normal. The normal distribution of samples of size 50 from a population in which $\mu = 30$ mpg (the upper limit of the hypothesis the manufacturer must reject to demonstrate the claim that $\mu > 30$) and $\sigma = 5$ mpg is shown in Figure 7-3. (Recall that we are assuming we *know* from past experience that the population standard deviation σ is equal to 5 mpg.) As indicated in the graph, the shaded region represents 5% of the area under the normal curve. Referring to Table A-5 in Appendix A, we see that 5% of the area in a normal distribution lies to the right of $z = 1.65$ (45% of the area lies between $z = 0$ and $z = 1.65$). Therefore, in this problem, the point above which we would reject the null hypothesis H_0 and conclude that the automobile manufacturer's claim has been verified is a

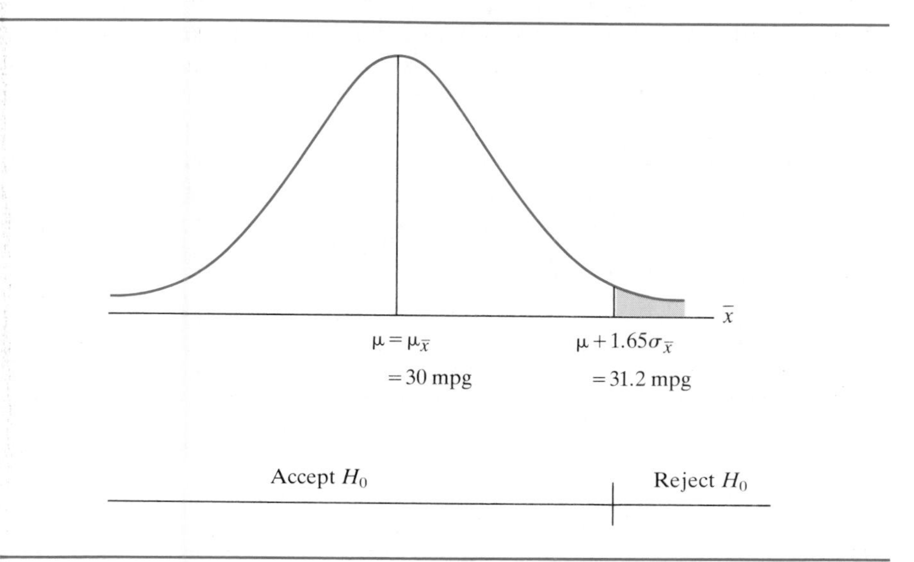

FIGURE 7-3
Sampling distribution of the mean, showing regions of acceptance and rejection of H_0 (population parameters $\mu = 30$ mpg and $\sigma = 5$ mpg; sample size $n = 50$).

sample mean with a value greater than $\mu + 1.65\sigma_{\bar{x}}$. The standard error of the mean $\sigma_{\bar{x}}$ is

$$\sigma_{\bar{x}} = \frac{\sigma}{\sqrt{n}} = \frac{5}{\sqrt{50}} = 0.71 \text{ mpg}$$

Thus, the critical value above which we would reject H_0 is

$$\mu + 1.65\sigma_{\bar{x}} = 30 + 1.65(0.71) = 31.2 \text{ mpg}$$

We can now see that this type of hypothesis-testing situation is referred to as a "one-tailed test" or a "one-sided alternative" because rejection of the null hypothesis takes place in only *one tail* of the sampling distribution.

Summary

In summary, on testing a simple random sample of 50 automobiles and observing $\bar{x}$, the sample mean number of miles per gallon, the automobile manufacturer should proceed according to the following decision rule:

DECISION RULE

1. If $\bar{x} > 31.2$ mpg, reject H_0 (claim is valid).
2. If $\bar{x} \leqslant 31.2$ mpg, accept H_0 (claim is invalid).[1]

We can now answer our original question. Should we conclude that the automobile manufacturer's claim is valid—that the average gasoline mileage of the compact car is now greater than 30 mpg? Since the sample yields a mean of $\mu = 32$ mpg, which exceeds the critical value of 31.2 mpg, the null hypothesis H_0 should be rejected, and we can conclude that the manufacturer's claim is valid. Another way of stating this conclusion is to say that "the sample mean of 32 is significantly greater than 30"; that is, it is unlikely that the difference between those two figures can be attributed merely to chance sampling error. Hence, again we conclude that the manufacturer's claim is valid.

Alternative method

It is instructive to examine an alternative method of stating the decision rule. Instead of working with the original units (in this case, miles per gallon), we could calculate the z value in a standard normal distribution that corresponds to $\bar{x} = 31.2$ mpg. If the sample z value lies to the right of the critical value of 1.65, H_0 will be rejected; if the sample z value lies to the left of 1.65, H_0 will be accepted. Thus, the decision rule can be rephrased as follows:

DECISION RULE

1. If $z > 1.65$, reject H_0 (claim is valid).
2. If $z \leqslant 1.65$, accept H_0 (claim is invalid).

[1] There is some ambiguity about whether the equal sign should appear in the rejection portion or the acceptance portion of the decision rule. From the theoretical point of view, this positioning is inconsequential. The normal curve is a continuous probability distribution. Thus, the probability of observing exactly $\bar{x} = 31.2$ mpg is 0.

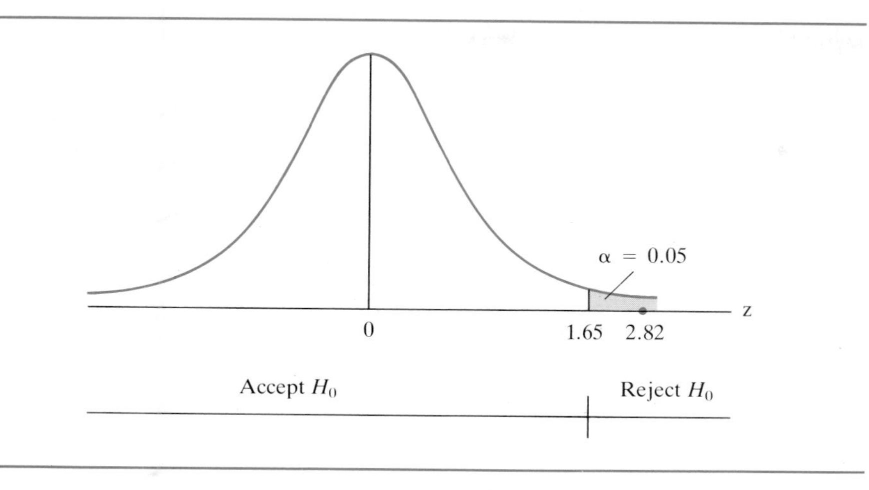

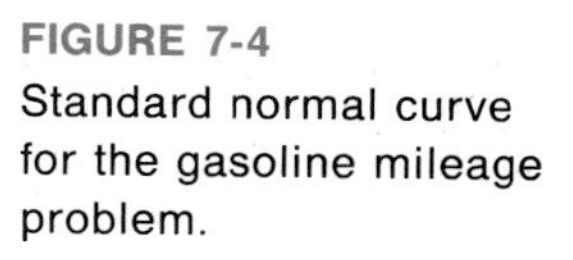
FIGURE 7-4
Standard normal curve for the gasoline mileage problem.

The $\bar{x}$ value of 32 mpg corresponds to

$$z = \frac{32 - 30}{0.71} = \frac{2.0}{0.71} = 2.82$$

Therefore, because a mean of 32 falls 2.82 standard error units above the mean of the sampling distribution and the dividing line is at 1.65 units, the null hypothesis H_0 is rejected. This situation is graphed in Figure 7-4.

The form of the decision rule is inconsequential. However, it is instructive when considering the rationale of the test to observe the implications of the computed z value. In this case, for example, an $\bar{x}$ value of 32 mpg corresponds to a z value of 2.82. We can observe from Table A-5 in Appendix A that about 0.0018 of the area in a normal distribution lies to the right of a z value of 2.82. Thus, interpreting the z value of 2.82 in terms of the gasoline mileage problem, if a sample of 50 cars is drawn at random from a population of automobiles that delivers an average gasoline mileage of 30 mpg with a standard deviation of 5 mpg, the probability of observing a sample mean of 32 mpg or greater is 0.0018. Because this sample result is so unlikely given the hypothesis that $\mu =$ 30 mpg, we reject that hypothesis.

In modern applications of hypothesis testing, probabilities such as 0.0018 are termed *p values*.

The ***p* values** represent the probability that if H_0 is true, we will observe a statistic ($\bar{x}$, in this case) that deviates by chance from the parameter being tested ($\mu = 30$ mpg, in this case) by a greater degree than is observed.

If a significance level of $\alpha = 0.05$ is used, as in the present problem, a p value of less than 0.05 would represent a significant result—that is, a rejection of the null hypothesis. The use of p values reflects the desire to report the results obtained and not to rely too heavily on arbitrarily significance levels such as 0.05 and 0.01, so that users of the results can draw their own conclusions.

For values of $\mu < 30$ mpg in our problem, the probability of a Type I error is less than 0.05. We can see this from Figure 7-3. If $\mu < 30$ mpg (that is, if the sampling distribution shifts to the left), then less than 5% of the area will lie in the rejection region above 31.2 mpg. As the value of μ becomes smaller, the probability of committing a Type I error becomes lower. This makes sense in terms of the gasoline mileage problem. If the true number of miles per gallon delivered by the cars with redesigned engines is lower, then the probability that the null hypothesis H_0: $\mu \leqslant 30$ mpg will be erroneously rejected is lower. For $\mu = 30$ mpg, the probability of committing a Type I error is 0.05. Now the meaning of $\alpha = 0.05$, the significance level in this problem, becomes clear: it is the maximum probability of committing a Type I error. This sort of interpretation is typical in one-tailed tests.

Until this point, we have considered only situations in which the acceptance and the rejection of the null hypothesis result in just two possible actions. Furthermore, we have concentrated on the determination of decision rules stemming from control of Type I errors without reference to the corresponding implications for Type II errors. We will deal with these matters subsequently, and it is advisable not to clutter the present discussion with too many details. However, the following four points summarize the hypothesis-testing procedure discussed thus far.

Summary of the procedure

1. A null hypothesis and its alternative are drawn up. The null hypothesis is framed in such a way that we can compute the probability of committing a Type I error.
2. A level of significance α is determined. This controls the risk of committing a Type I error.
3. A decision rule is established by partitioning the relevant sample space into regions of acceptance and rejection of the null hypothesis. This partitioning is accomplished by considering the relevant sampling distribution. The nature of the null hypothesis and the choice of α determine the partition.
4. The decision rule is applied to a sample of size n. The null hypothesis is accepted or rejected. Rejection of the null hypothesis implies acceptance of the alternative hypothesis.

Remarks

A number of points can be made concerning the statistical theory involved in the preceding problem. First, the normal curve is used as the appropriate sampling distribution of the mean. If the population distribution of mean gasoline mileage is normal, the normal curve is the theoretically correct sampling distribution. Note that no statement at all is made in the gasoline mileage problem about the population distribution. The normal curve is used for the sampling distribution of $\bar{x}$, based on the specification of the central limit theorem that no matter what the shape of the population, the sampling distribution of $\bar{x}$ will be approximately normal for a sample as large as $n = 50$.

Second, no finite population correction factor is used in the calculation of the standard error of the mean, despite the fact that the sample of 50 automobiles has been drawn without replacement from a finite population. However, the population size may be assumed to be very large relative to the sample size. Therefore, the finite population correction factor can be assumed to be approximately equal to 1 in this case.

Third, the population standard deviation σ is assumed to be known. If the population standard deviation is unknown and the sample size is large (say $n > 30$), then the sample standard deviation s may be substituted for σ. Hence, instead of calculating $\sigma_{\bar{x}} = \sigma/\sqrt{n}$, an estimated standard error of the mean is computed $s/\sqrt{n}$. In all other respects, the decision procedure remains the same. (In section 7.4, we will discuss how to deal with the situation in which the sample size is small and the population standard deviation is unknown.)

Fourth, the nature of the z value computed in the problem is worth noting. In Chapter 5, the concept of a standard score was discussed in the context of a normally distributed *population.* In that case, $z = (x - \mu)/\sigma$ represents a deviation of the value of an individual item from the mean of the population, expressed as a multiple of the population standard deviation. In our hypothesis-testing problem, the z values are of the form $z = (\bar{x} - \mu)/\sigma_{\bar{x}}$. Such a z value represents a deviation of a sample mean from the mean of the sampling distribution of $\bar{x}$, stated in multiples of the standard deviation of that distribution $\sigma_{\bar{x}}$. As we noted previously, the mean of the sampling distribution of $\bar{x}$, denoted by $\mu_{\bar{x}}$, is equal to the population mean μ. Thus, we use μ and $\mu_{\bar{x}}$ interchangeably. As a generalization, in hypothesis-testing problems, z values take the form

$$z = \frac{\text{Statistic} - \text{Parameter}}{\text{Standard error}}$$

For example, in the hypothesis-testing problem just discussed, the $\bar{x}$ value is the sample statistic, the population mean μ is the parameter, and the standard error of the mean $\sigma_{\bar{x}}$ is the appropriate standard error.

Fifth, we note that the size of the sample, $n = 50$, has been predetermined in our illustration. Thus, the sample is large and predetermined, and the construction of the decision rule is based on the control of only one type of incorrect decision (Type I errors). The next section dealing with the power curve discusses the measurement of Type II errors for such a test.

Finally, it is important to realize that we cannot *prove* that a null hypothesis is false or that a null hypothesis is true. All that we can do is discredit a null hypothesis or fail to discredit it on the basis of sample data.

Actually, a single sample statistic such as $\bar{x}$ is consistent with an infinite number of hypotheses concerning μ.[2] From the standpoint of decision making and subsequent behavior, if sample data do not discredit a null hypothesis, we will act as though that hypothesis is true.

The hypothesis-testing procedure outlined thus far has concentrated on the control of Type I errors. The question of how well this test controls Type II errors naturally arises. When the null hypothesis is false, how frequently does the decision rule lead us to accept it erroneously? This question is answered by means of the **power curve**, also called the **power function**, which can be computed from the information given in the problem and the decision rule. The Greek letter β is used to denote the probability of committing a Type II error; thus, β represents the probability of accepting the null hypothesis when it is false. In the gasoline mileage problem, the null hypothesis H_0 is false for each value of μ satisfying the alternative hypothesis H_1: $\mu > 30$ mpg. Therefore, for each particular value of μ greater than 30 mpg, we can determine a β value. Actually, by convention, the power curve gives the complementary probability to β (that is, $1 - \beta$) for each value of the alternative hypothesis. Thus, it indicates the probability of rejecting the null hypothesis for each value for which the null hypothesis is false, which, of course, represents the probability of selecting the correct course of action in each case. $1 - \beta$ is referred to as the "power of the test" for each particular value of the alternative hypothesis. For completeness, in a power curve, the probabilities of rejection are also shown for each value for which the null hypothesis is true.

The power curve

Summary

A power curve is a function that gives the probabilities of rejecting the null hypothesis H_0 for all possible values of the parameter tested. Therefore, it measures the ability of the decision rule to discriminate between true and false hypotheses.

The power curve is useful in assessing the risks of making both Type I and Type II errors when a decision rule is employed.

The power curve for the gasoline mileage problem (shown in Figure 7-5) has the typical S shape of a power curve in a one-tailed test, with the rejection region in the right tail. For a one-tailed test with the rejection region in the left tail, the curve would be reverse S-shaped, dropping from the upper left corner to the lower right corner on the graph. Rejection probabilities for the null hypothesis are shown on the vertical axis, and possible values of the population parameter μ appear on the horizontal axis. Specifically, the figures plotted on the vertical axis are conditional probabilities of the form $P(\text{rejection of } H_0|\mu) = P(\bar{x} > 31.2 \text{ mpg}|\mu)$.

[2] You have probably had the disconcerting experience of watching a number of experts in disagreement after they have observed ostensibly the same basic set of data. Perhaps you share the experience of finding it easier to accept and reject hypotheses when no data are available at all.

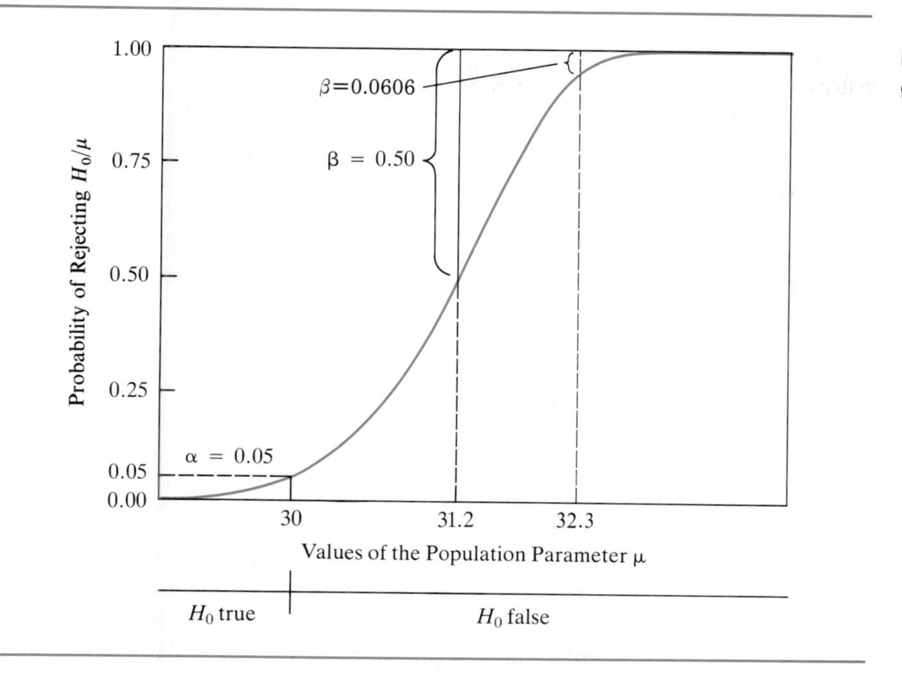

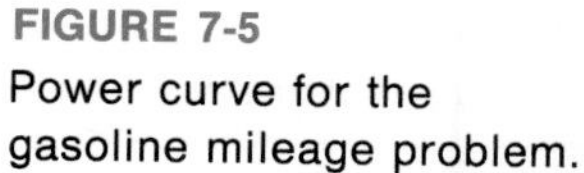
FIGURE 7-5
Power curve for the gasoline mileage problem.

The nature of the power curve in Figure 7-5 can be determined by considering a couple of the plotted values. The value of $\alpha = 0.05$ is shown for $\mu = 30$ mpg, indicating the significance level of the test. We can see that this is the maximum probability of erroneously rejecting the null hypothesis H_0, because the ordinates of the curve drop off to the left as μ decreases in the region where H_0 is true. The heights of the ordinates of the power curve to the left of $\mu = 30$ mpg represent the probabilities of making Type I errors. The heights of the ordinates to the right of $\mu = 30$ mpg represent the values of $1 - \beta$, or the probabilities of rejecting the null hypothesis when it is false. Therefore, the complementary distances from points on the curve to 1.0 are values of β, or probabilities of making Type II errors. One such value is displayed in the graph for $\mu = 31.2$ mpg. Recall that our decision rule requires the rejection of H_0 if the sample mean $\bar{x}$ is greater than 31.2 mpg. If the population mean is 31.2 mpg, the probability of observing $\bar{x}$ values less than 31.2 mpg and therefore of rejecting H_0 is obviously 0.50. This situation is graphed in Figure 7-6.

Some computation is required to obtain the β value for $\mu = 32.3$ mpg. To compute this figure, we must refer to the sampling distribution of $\bar{x}$, given that $\mu = 32.3$, and calculate the probability that a sample mean would lie in the acceptance region $\bar{x} \leqslant 31.2$ mpg. The z value for 32.3 mpg is

$$z = \frac{31.2 - 32.3}{0.71} = -1.55$$

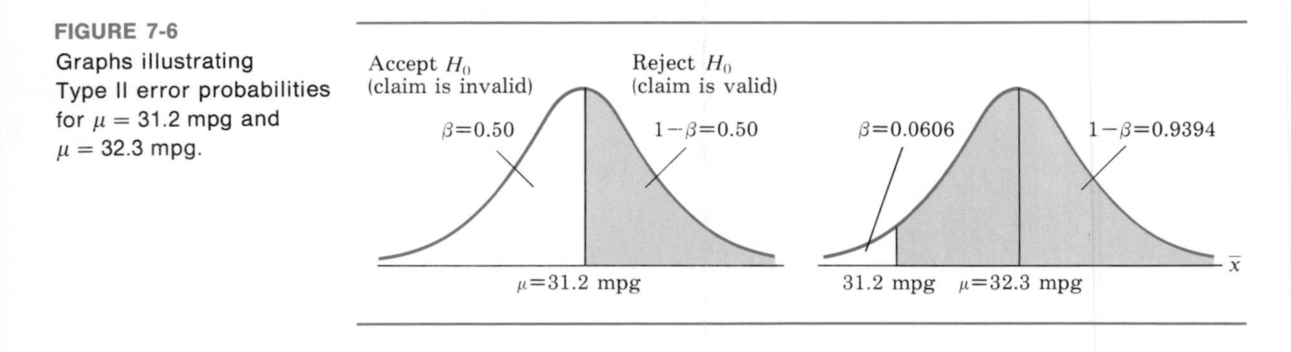

FIGURE 7-6 Graphs illustrating Type II error probabilities for $\mu = 31.2$ mpg and $\mu = 32.3$ mpg.

Thus, if $\mu = 32.3$ mpg, the critical $\bar{x}$ value of 31.2 lies 1.55 standard errors of the mean below 32.3 mpg. Referring to Table A-5 of Appendix A, we find a figure of 0.4394, which when subtracted from 0.5000 (the area to the left of $\mu = 32.3$ mpg) gives 0.0606 for β, the probability of erroneously accepting H_0.

For values of μ slightly more than 30 mpg, the probability of a Type II error is very high. In fact, we can see in Figure 7-5 that these probabilities exceed 0.50 for μ values between 30 mpg and 31.2 mpg. This simply indicates that the power of the test is low when the value of μ satisfies the alternative hypothesis (H_1: $\mu > 30$ mpg) but is close to values of μ satisfying the null hypothesis (H_0: $\mu \leqslant 30$ mpg). For a fixed sample of size n, β can be decreased only by increasing α and vice versa. If α is fixed, then as sample size is increased, β is reduced for all values of μ in the region where H_0 is false. In this type of one-tailed test, the ideal power curve would be ⌠ shaped, with the vertical line occurring at $\mu = 30$ mpg. Thus, the probability of rejecting H_0 would always be equal to 0.0 when H_0 is true and to 1.0 when H_0 is false. However, this ideal curve is clearly unattainable when sample data are used to test hypotheses, because sampling error will always be present.

Deciding the level of significance

The trade-off relationship between Type I and Type II errors for a sample of fixed size is such that the level of significance should be decided *by considering the relative seriousness of the two types of errors* before conducting hypothesis tests.

Using the power curve

How does the decision maker use power curves in setting up an appropriate hypothesis-testing procedure? In the preceding discussion, the decision rule was determined for a fixed sample size n; α was specified, and the critical value of $\bar{x} = 31.2$ mpg was computed. Suppose that on examination of the resulting power curve, the decision maker feels that the β values are too high. For instance, in the present example, we determined that if an automobile actually delivers 31.2 miles per gallon, then $\beta = 0.50$. To reduce this risk, the decision maker (the automobile manufacturer) could decrease the critical value of $\bar{x}$, thereby raising the level of significance, which had been previously set at $\alpha = 0.05$. This would shift the entire power curve to the left. Thus, we see that

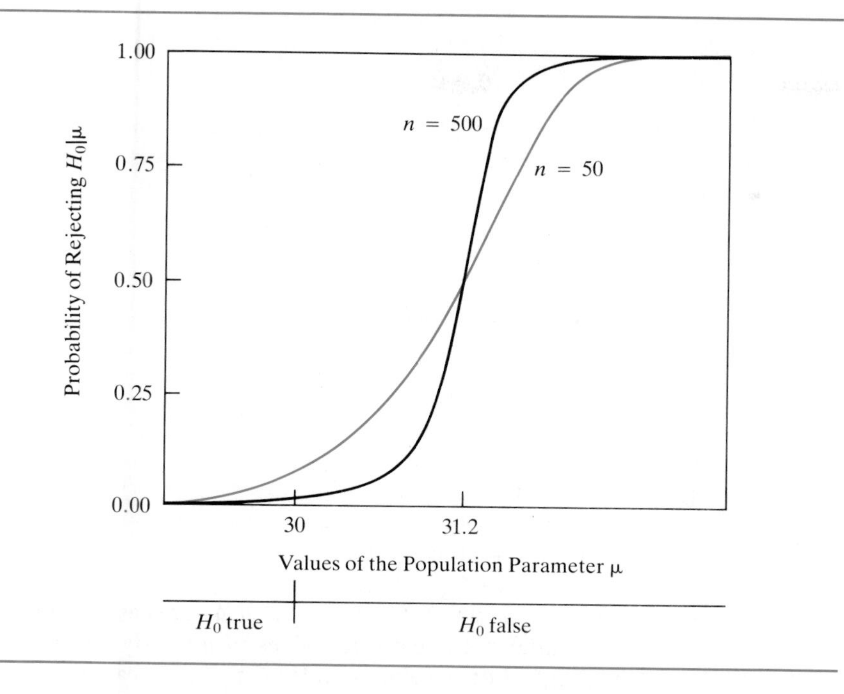

FIGURE 7-7 Comparison of power curves for two sample sizes ($n = 50$ and $n = 500$) for the gasoline mileage problem.

if the sample size n is unchanged, the only way to reduce β levels is to increase α. On the other hand, if the decision maker is unwilling to increase the significance level α, then the only way to reduce β values is to increase the sample size n. Of course, larger sample sizes involve increased costs.

To illustrate that larger sample sizes provide more discriminating tests, let us consider the effect of increasing the sample size in the gasoline mileage example from $n = 50$ to $n = 500$, while retaining the critical value at 31.2 mpg. The corresponding curves for the tests provided by these two sample sizes appear in Figure 7-7. Note from the figure that the larger sample size provides a more discriminating test because its power curve implies lower probabilities both of accepting false hypotheses and of rejecting true hypotheses. Thus, the power curve for the larger sample size lies much closer to the ideal power curve.

Operating characteristic curves

It is usual practice in industrial quality control to use **operating characteristic curves** (succinctly referred to as **O-C curves**), rather than power curves, to evaluate the discriminating power of a test. The O-C curve is simply the complement of the power curve. The probability of acceptance, rather than the probability of rejection, of the null hypothesis is plotted on the vertical axis of the O-C curve.

Test Involving a Proportion: Two-Tailed Test

The preceding discussion dealt with a test of a hypothesis about a mean. We now turn to hypothesis testing for a proportion.

EXAMPLE 7-1

Let us consider the case of an advertising agency that developed a general theme for the commercials on a certain TV show based on the assumption that 50% of the show's viewers were over 30 years of age. The agency was interested in determining whether the percentage had changed in either an upward or downward direction. If we use the symbol p to denote the proportion of all viewers of the show, we can state the null and alternative hypotheses as follows:

$$H_0: p = 0.50 \text{ viewers over 30 years of age}$$

$$H_1: p \neq 0.50 \text{ viewers over 30 years of age}$$

Let us assume that the agency wished to run a 5% risk of erroneously rejecting the null hypothesis of "no change," or, H_0: $p = 0.50$. That is, the agency decided to test the null hypothesis at the 5% significance level ($\alpha = 0.05$).

In order to test the hypothesis, the agency conducted a survey of a simple random sample of 400 viewers of the TV show. Of the 400 viewers, 210 were over 30 years of age and 190 were 30 years of age or less. What conclusion should the agency reach?

In this problem, as contrasted with the previous one, the null hypothesis concerns a single value of p, which is a hypothetical population parameter of 0.50. The alternative hypothesis includes all other possible values of p. We can understand the reason for setting up the hypotheses this way by reflecting on how the test will be conducted. The hypothesized parameter under the null hypothesis is $p = 0.50$. We have observed in a sample a certain proportion, denoted $\bar{p}$, who were over 30 years of age. The testing procedure involves a comparison of $\bar{p}$ with the hypothesized value of p to determine whether a significant difference exists between them. If $\bar{p}$ does not differ significantly from p, and we accept the null hypothesis that $p = 0.50$, what we really mean is that the sample is consistent with a hypothesis that half the viewers of the TV show are over 30. On the other hand, if $\bar{p}$ is greater than 0.50 and a significant difference between $\bar{p}$ and p is observed, we will conclude that more than half the viewers are over 30. If the observed $\bar{p}$ is less than 0.50 and a significant difference from $p = 0.50$ is observed, we will conclude that fewer than half the viewers are over 30.

It is important to note that in hypothesis-testing procedures, the two hypotheses and the significance level of the test must be selected before the data are examined.

We can easily see the difficulty with a procedure that would permit the investigator to select α after examination of the sample data. It would always be possible to accept a null hypothesis simply by choosing a sufficiently small significance level, thereby setting up a large enough region of acceptance. Thus, the first step in our problem is setting up the competing hypotheses, with the null hypothesis stated in such a way that the probability of a Type I error can be calculated. We have accomplished this by a single-valued null hypothesis, H_0: $p = 0.50$. Our next step is to set the significance level, which we have taken as $\alpha = 0.05$.

We proceed with the test. The simple random sample of size 400 is drawn, the statistic $\bar{p}$ is observed, and we can now establish the appropriate decision rule. Since the sample size is large, the theory developed in Chapter 5 allows

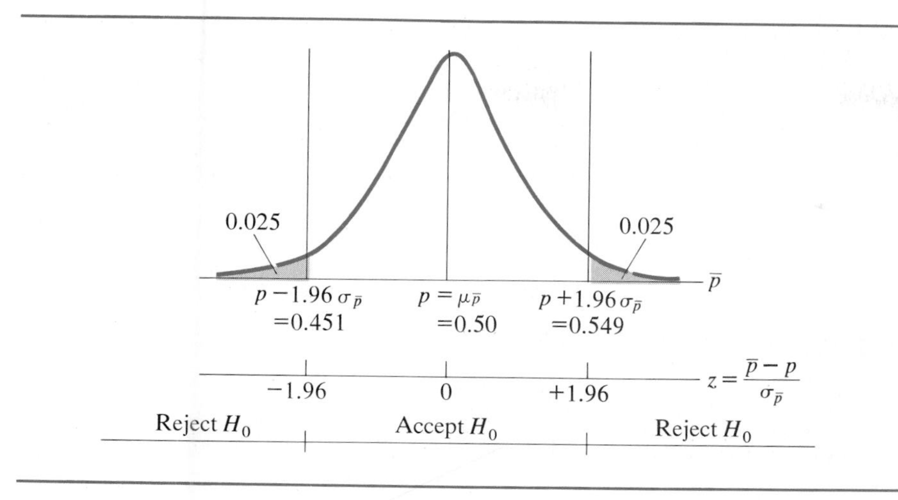

FIGURE 7-8
Sampling distribution of a proportion, with $p = 0.50$ and $n = 400$. This is a two-tailed test with $\alpha = 0.05$

us to use the normal curve as an appropriate approximation for the sampling distribution of the percentage $\bar{p}$. As in the preceding problem, for illustrative purposes, we will establish the decision rule in two different forms, first in terms of $\bar{p}$ values, then in terms of the corresponding z values in a standard normal distribution. Under the assumption that the null hypothesis is true, the sampling distribution of $\bar{p}$ has a mean of p and a standard deviation $\sigma_{\bar{p}} = \sqrt{pq/n}$. Again, we ignore the finite population correction, because the population is large relative to the sample size. The sampling distribution of $\bar{p}$ for the present problem, in which $p = 0.50$, is shown in Figure 7-8. Also shown is the horizontal axis of the corresponding standard normal distribution, scaled in z values.

Since the null hypothesis will be rejected by an observation of a $\bar{p}$ value that lies significantly below or significantly above $p = 0.50$, we clearly must use a two-tailed test. The critical regions (rejection regions) are displayed in Figure 7-8. The arithmetic involved in establishing regions of acceptance and rejection of H_0 is as follows. The standard error of $\bar{p}$ is

$$\sigma_{\bar{p}} = \sqrt{\frac{(0.50)(0.50)}{400}} = 0.025$$

Referring to Table A-5 of Appendix A, we find that 2.5% of the area in a normal distribution lies to the right of $z = +1.96$, and therefore 2.5% also lies to the left of $z = -1.96$. Thus, we establish a significance level of 5% by marking off an acceptance range for H_0 of $p \pm 1.96\sigma_{\bar{p}}$. The calculation is

$$p + 1.96\sigma_{\bar{p}} = 0.50 + (1.96)(0.025) = 0.50 + 0.049 = 0.549$$

$$p - 1.96\sigma_{\bar{p}} = 0.50 - (1.96)(0.025) = 0.50 - 0.049 = 0.451$$

We can now state the decision rule. The agency draws a simple random sample of 400 viewers of the TV show, observes $\bar{p}$ (the proportion of persons in the sample who are over 30 years of age), and then applies the following decision rule.

DECISION RULE

1. If $\bar{p} < 0.451$ or $\bar{p} > 0.549$, reject H_0
2. If $0.451 \leqslant \bar{p} \leqslant 0.549$, accept H_0

Again, for illustrative purposes, we restate the decision rule in terms of z values.

DECISION RULE

1. If $z < -1.96$ or $z > +1.96$, reject H_0
2. If $-1.96 \leqslant z \leqslant +1.96$, accept H_0

Let us apply this decision rule to the present problem. The observed sample $\bar{p}$ was

$$\bar{p} = \frac{210}{400} = 0.525$$

and

$$z = \frac{\bar{p} - p}{\sigma_{\bar{p}}} = \frac{0.525 - 0.500}{0.025} = +1.0$$

Therefore, the null hypothesis H_0 is accepted. This leads us to a rather vague conclusion. If $\bar{p}$ fell in the rejection region in the right tail of the sampling distribution in Figure 7-8 (that is, if $\bar{p}$ were greater than 0.549), we would conclude that more than half the viewers of the TV show are over 30 years of age. If $\bar{p}$ had fallen in the left tail of the rejection region, we would conclude that less than half are over 30. However, $\bar{p}$ lies in the acceptance region, so we *cannot* conclude that more than half of the viewers are over 30 or that less than half are over 30. The sample evidence is *consistent with the hypothesis of a 50-50 split.* Thus, acceptance of the null hypothesis means that on the basis of the available evidence, we simply are not in a position to conclude that more than half of the viewers are over 30 years of age or that less than half of them are. In some instances, the best course is to reserve judgment.

Further Remarks

Power curves can be computed for two-tailed tests in an analogous way to that for one-tailed tests. However, such calculations will not be illustrated here. The power curve for the two-tailed test in the TV viewer problem is shown in Figure 7-9. The curve has the characteristic U shape of a power function for

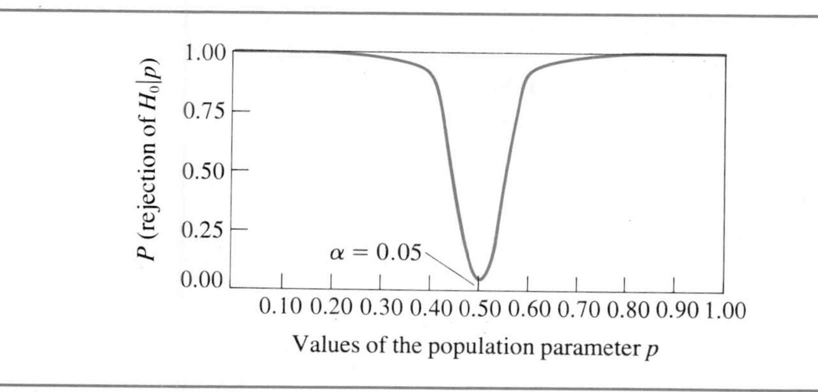

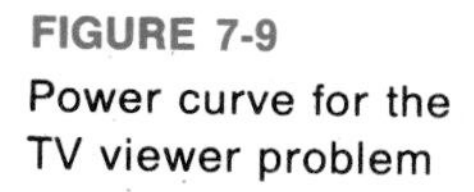

FIGURE 7-9
Power curve for the TV viewer problem

a two-tailed test. The possible values of the parameter p are shown on the horizontal axis. The height of the ordinate at $p = 0.50$ is 0.05, which is the value of α. Since the only value of p for which the null hypothesis is true is 0.50, the ordinates at all other p values denote probabilities of rejecting the null hypothesis when it is false. The complements of these ordinates are equal to the values of β, the probabilities of making Type II errors.

The hypothesized proportion according to the null hypothesis in this problem was 0.50, because the agency was interested in whether or not more than 50% of the viewers were over 30 years of age. On the other hand, if we wanted to test the assertion that the population proportion was 0.55, 0.60, or some other number, we would have used these figures as the respective hypothesized parameters. Also in this problem, the null hypothesis was single-valued ($p = 0.50$), whereas the alternative hypothesis was many-valued ($p \neq 0.50$). This resulted in a two-tailed test.

> Whether a test of a hypothesis is one-tailed or two-tailed depends on the question to be answered.

For example, suppose an assertion had been made that less than 50% of the viewers of the TV show were over 30. In hypothesis testing, we ordinarily place the burden of proof on the person(s) who has made the assertion. Hence, we will conclude the assertion is correct only if the sample proportion $\bar{p}$ is significantly less than 0.50. Assume a test at the 5% significance level. The null and alternative hypotheses in this instance would be

$$H_0\colon p \geqslant 0.50$$

$$H_1\colon p < 0.50$$

This would involve a one-tailed test with a rejection region lying in the left tail and containing 5% of the area under the normal curve. The critical region would be in the left tail, since only a significant difference for a $\bar{p}$ value lying *below* 0.50 could result in the rejection of the stated null hypothesis. The decision rule in terms of z values follows.

DECISION RULE

1. If $z < -1.65$, reject H_0
2. If $z \geq -1.65$, accept H_0

On the other hand, if the assertion had been that more than 50% of the viewers were over 30 and if we performed a test at the 5% significance level, the rejection region would be the 5% area in the right tail. The corresponding hypotheses and decision rule would be

$$H_0: p \leq 0.50$$
$$H_1: p > 0.50$$

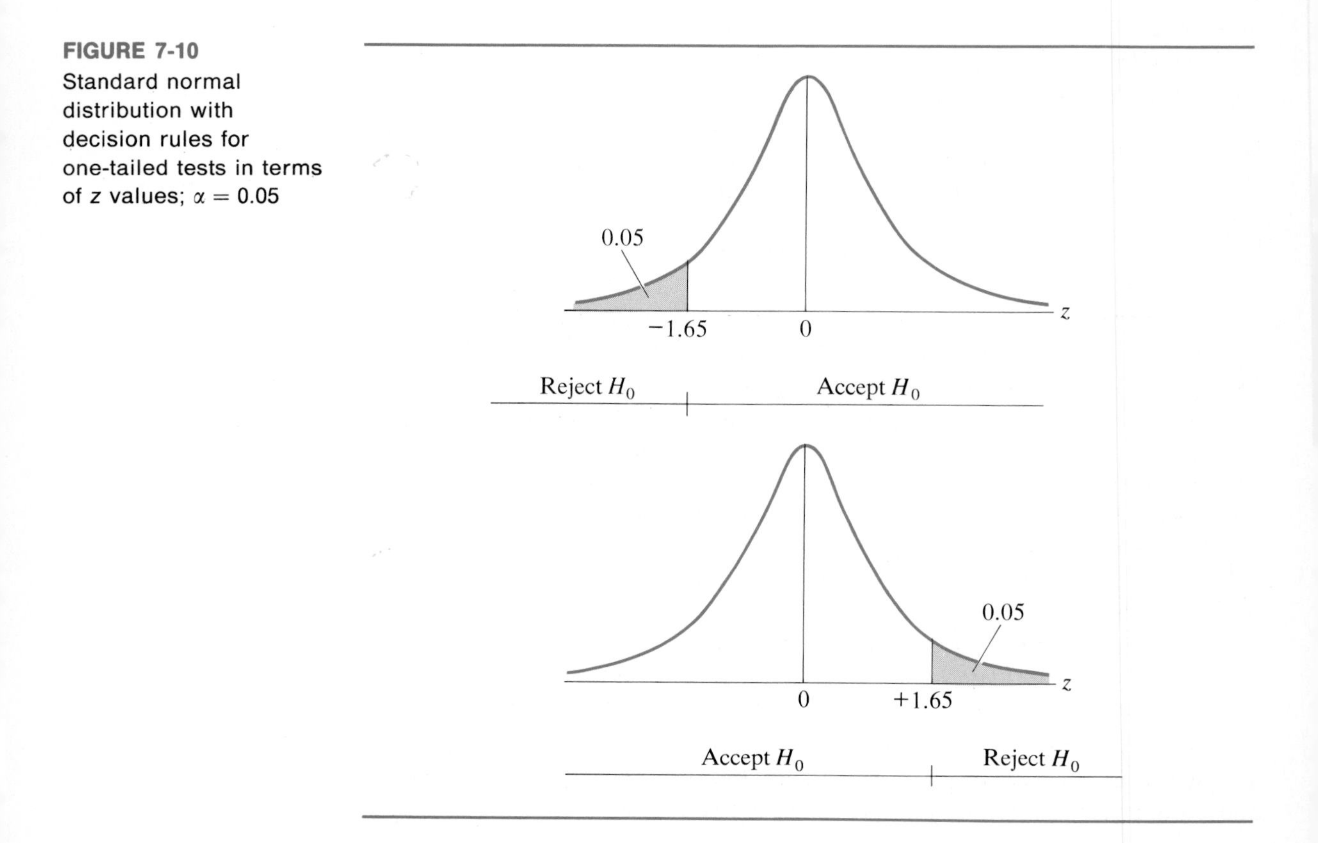

FIGURE 7-10 Standard normal distribution with decision rules for one-tailed tests in terms of z values; $\alpha = 0.05$

DECISION RULE

1. If $z \geq +1.65$, reject H_0
2. If $z < +1.65$, accept H_0

Standard normal distributions with the decision rules for these one-tailed tests are depicted in Figure 7-10. When tests are conducted for means or other statistical measures, they may also be either one- or two-tailed, depending on the context of the problem.

Exercises 7.2

1. Distinguish between a parameter and a statistic.
2. Support or criticize the following statement: "You will make very few errors in hypothesis testing if you set the significance level (α) very low, say 0.001. Therefore, you should always do so."
3. Distinguish briefly between
   a. One-tailed test and two-tailed test
   b. Type I and Type II errors
4. What is meant by "the power of a test"? How is the power related to the Type II error?
5. Agree with or criticize the following statements:
   a. If you have a choice of applying two different tests to decide between a certain null hypothesis and an alternative hypothesis, both tests having the same α level, then you should use the one with the smaller β error.
   b. Beta (β) represents the probability that one will reject the null hypothesis incorrectly.
   c. The power curve has the probability of rejecting (1) the null hypothesis on one axis and (2) the possible values of the statistic on the other.
6. United States Pharmacopoeia (U.S.P.) specifications for a certain drug call for an average therapeutically effective period of at least 37 hours. The standard deviation for the period of effectiveness is known to be 11 hours. A shipment of this drug will be accepted or rejected on the basis of a simple random sample of 100 items drawn from the shipment.
   a. What decision rule should be used if the maximum probability of erroneously rejecting the incoming shipment is to be 0.10, that is, $\alpha = 0.10$?
   b. State clearly the null and alternative hypotheses.
   c. Indicate the decision to be reached if a simple random sample shows a mean period of effectiveness of 35 hours. Show your reasoning.
7. The company you work for sells consumer packaged goods. Since the managers of the company are uncertain about the consumer acceptance of one particular package design, they decide to test acceptance. A simple

random sample of 400 consumers is drawn and each person is asked his or her opinion of the design. Management feels that more than 85% of consumers should like the package design in order for the company to continue producing it. Unfamiliar with statistics, the managers decide that if 340 or fewer of those interviewed like the design, the company will discontinue using it. Determine the probability of a Type 1 error for H_0: $p \leqslant 0.85$ and explain whether this cutoff level should be used.

8. A manufacturer of commemorative coins wishes to determine whether the proportion of gold in these coins has changed. His advertising claims an 80% content of pure gold. He is willing to run a risk of 0.01 of erroneously deciding the content has changed. Assuming a random sample of 625 coins is drawn, set up the decision rule for the test.

9. A major department store chain reports that in the past, 8% of all its credit charges were never collected and were written off as bad debts. Recently a central computer was installed in the home office with on-line access to all branches. For any credit purchase over $50, the cashier must call the home office for a credit check before allowing the sale. Customers who are in arrears are asked to see the branch credit manager for an approval of the sale. To test whether this system is working, 10,000 accepted credit charges were chosen at random. It was found that 725 of these charges were uncollected and written off as bad debts.
   a. At a 5% significance level, do you think the system represents an improvement over past experience?
   b. What is the level of significance or p-value for 725 uncollected charges?

10. Over the past 10 years, the Snow Mountain Ski Resort has averaged 120 skiers per day during the winter season (130 days) with a standard deviation of 10 skiers. In a random sample of 50 days during the most recent ski season, the mean number of customers was 118.
   a. Assuming $\alpha = 0.05$, would you conclude that the average number of skiers per day at this resort has changed? Use the finite population correction.
   b. Suppose you decided not to use $\alpha = 0.05$ as in part (a). Instead you used the following decision rule:
      1. Accept H_0 if $117 \leqslant \bar{x} \leqslant 123$
      2. Reject H_0 otherwise

      What is the level of significance or p-value for this test?

11. It has been determined that the arithmetic mean length of life of safety tires produced by the Firerock process is 30,000 miles with a standard deviation of 4,000 miles. A new process, Morelife, is developed and used on a simple random sample of 100 tires.
   a. How long must this sample of 100 tires last on the average in order for the president of Safety Tire Company to conclude that tires produced by the new process will have a higher mean than tires prepared by the old process? The president is willing to run a risk of no more than 0.02 of drawing such a conclusion in error.

b. Suppose the universe mean for the new process is 30,200 miles and you employ the critical value established in part (a). What is the probability of reaching an incorrect conclusion?
c. If you reach an incorrect conclusion in part (b), will it be a Type I or Type II error? Why?

12. Assume that the Food and Drug Administration limits the caffein content of cola drinks to 1.2 grains per 12-ounce bottle and that the actual caffein content of cola drinks is normally distributed and varies from 0.55 to 0.85 grains per 12-ounce bottle. Let $\mu = 0.70$ grains and $\sigma = 0.05$ grains. Suppose the FDA adopts an inspection plan that calls for rejection of a production lot if the population mean caffein content exceeds 0.7 grains per bottle. Assume that the sample size is 225 and that mistaken rejection of lots should not occur more than 1% of the time. Then what values of sample means will lead to rejection of a lot?

13. Over the past 10 years, students receiving their MBA degrees from the Branford Business School received an average of 6 job offers, with a standard deviation of 2. In a random sample of 100 members of the most recent class, the mean number of job offers was 5.5. The school graduates 500 students each year. Assuming $\alpha = 0.05$, would you conclude that there has been a change in the average number of job offers received by Branford graduates? Use the finite population correction.

14. Last year, a wholesale distributor found that the mean sales per invoice was \$60, with standard deviation \$20. This year, a random sample of 400 invoices is to be drawn in order to test the hypothesis that the mean sale per invoice has not changed. It is assumed that α will not change. The acceptance region of the test is agreed to be

$$\text{if } \$58.72 < \bar{x} < \$61.28, \text{ accept } H_0$$

a. Suppose that in fact this year $\mu = \$61$. What is the probability of accepting H_0?
b. Calculate and explain the meaning of the power of the test when $\mu = \$61$.
c. What level of significance was used for the test?

15. Suppose you are responsible for the quality control of a certain part bought from a supplier. Inspection tests destroy the part, so you must use sampling. A 5% defective rate is tolerable, but in your sample of 100 from a lot of 10,000, you discover 8 parts are defective. Is this sufficient evidence that the lot of parts has too many defectives?

16. Two hundred purchasers of a new product were selected at random and were asked whether they liked the product. The company conducting the survey felt that at least 20% of all purchasers must like the product in order for the firm to continue marketing it. Therefore, the firm decided that if 30 or fewer people responded favorably, it would stop marketing the product.
a. State in words the nature of the Type I error involved here.
b. How large is the risk of such an error?

17. Given the following hypothesis test:

    H_0: The percentage p of losing stocks selected by a stockbroker is 20%
    H_1: The percentage p of losing stocks selected by a stockbroker is greater than 20%

    The sample is 50, and the significance level is 5%. The decision rule is

    If the percentage of losers is greater than 29.34%, reject H_0 and switch to another stockbroker. Otherwise, accept H_0 and keep the present stockbroker.

    Comment on the following statements:
    a. If p is really greater than 20%, the probability that we will reject H_0 is 0.05.
    b. If 32% of the sample stocks are losers, this proves that p is greater than 20%.
    c. This hypothesis test is a one-tailed test; the decision rule would be the same for H_0: $p \leqslant 20\%$.
    d. If 33% of the stocks are losers, the probability that we would make a Type II error is

$$P\left(z < \frac{0.33 - 0.20}{\sqrt{\frac{(0.20)(0.80)}{50}}}\right)$$

18. A financial analyst has determined that the net present value of a certain investment proposal is a random variable with a mean of \$10,000 and a standard deviation of \$750. A second investment is possible; a sample of 30 investments in the past similar to this alternative produced a sample mean value of \$9,800 and a standard deviation of \$1,000. The financial analyst's assistant has noted, "Since the sample mean is not significantly less than \$10,000 at $\alpha = 0.05$, and the risk of this type of investment is less than that of the first proposal, we have strong evidence that we should choose the second alternative." Do you agree with the assistant's statement? Why or why not?

Computer Exercises 7.2

Assume that the 75 families in Appendix E constitute a simple random sample of all families in a suburban area.

1. Is the hypothesis tenable that the mean annual family of all families in this suburban area is \$39,000? Use $\alpha = 0.05$.
2. A public official has claimed that the mean annual income of families in this suburban area is in excess of \$34,000. Would you consider this claim valid at the 0.01 level of significance?

3. The same public official has stated that the percentage of renters has declined significantly from 56%, which was the true figure for this suburban area in a previous year. Would you consider this official's statement to be correct? Use a significance level of 0.05.

7.3 TWO-SAMPLE TESTS (LARGE SAMPLES)

The discussion thus far has involved testing of hypotheses using data from a single random sample. Another important class of problems involves the question of whether statistics observed in two simple random samples differ significantly. Recalling that all statistical hypotheses are statements concerning population parameters, we see that this question implies a corresponding question about the underlying parameters in the populations from which the samples were drawn. For example, if the statistics observed in the two samples are arithmetic means (say, $\bar{x}_1$ and $\bar{x}_2$), the question is whether we are willing to attribute the difference between these two sample means to chance sampling errors or whether we will conclude that the populations from which the samples were drawn have **unequal means**. We shall illustrate these tests, first for differences between means and then for differences between proportions. In both cases, we assume (as we did in section 7.2) that we are using large samples.

Considering unequal means

Test for Difference between Means: Two-Tailed Test

A consulting firm conducting research for a client was asked to test whether the wage levels of unskilled workers in a certain industry were the same in two different geographical areas, referred to as Area A and Area B. The firm took simple random samples of the unskilled workers in the two areas and obtained the following sample data for weekly wages:

Area	Mean	Standard Deviation	Size of Sample
A	$\bar{x}_1 = \$250.01$	$s_1 = \$4.00$	$n_1 = 100$
B	$\bar{x}_2 = 245.21$	$s_2 = 4.50$	$n_2 = 200$

If the client wished to run a risk of 0.02 of incorrectly rejecting the hypothesis that the population means in these two areas were the same, what conclusion should be reached? Note that the samples need not be the same size; different sample sizes have been assumed in this problem in order to keep the example completely general.

Let us refer to the means and standard deviations of *all* unskilled workers in this industry in Areas A and B, respectively, as μ_1 and μ_2, and σ_1 and σ_2.

These are the population parameters corresponding to the sample statistics $\bar{x}_1$, $\bar{x}_2$, s_1, and s_2. The hypotheses to be tested are

$$H_0: \mu_1 - \mu_2 = 0$$

$$H_1: \mu_1 - \mu_2 \neq 0$$

That is, the null hypothesis asserts that the population parameters μ_1 and μ_2 are equal. We form the statistic $\bar{x}_1 - \bar{x}_2$, the difference between the sample means. If $\bar{x}_1 - \bar{x}_2$ differs significantly from zero, the hypothesized value for $\mu_1 - \mu_2$, we will reject the null hypothesis and conclude that the population parameters μ_1 and μ_2 are indeed different.

Since the risk of a Type I error has been set, we turn now to determining the decision rule based on the appropriate random sampling distribution. Let us examine some of the important characteristics of this distribution. The two random samples are independent; that is, the probabilities of selection of the elements in one sample are not affected by the selection of the other sample. Hence, $\bar{x}_1$ and $\bar{x}_2$ are independent random variables. It has been shown that the mean and standard deviation of the sampling distribution of $\bar{x}_1 - \bar{x}_2$ are, respectively,

$$\mu_{\bar{x}_1 - \bar{x}_2} = 0 \tag{7.1}$$

and

$$\sigma_{\bar{x}_1 - \bar{x}_2} = \sqrt{\frac{\sigma_1^2}{n_1} + \frac{\sigma_2^2}{n_2}} \tag{7.2}$$

For large, independent samples, the sampling distribution of $\bar{x}_1 - \bar{x}_2$ is approximately normal by the Central Limit Theorem.

Summary

If $\bar{x}_1$ and $\bar{x}_2$ are the means of two large independent samples from populations with means μ_1 and μ_2 and standard deviations σ_1 and σ_2, and if we hypothesize that μ_1 and μ_2 are equal, then the sampling distribution of $\bar{x}_1 - \bar{x}_2$ may be approximated by a normal curve with mean $\mu_{\bar{x}_1 - \bar{x}_2} = 0$ and standard deviation $\sigma_{\bar{x}_1 - \bar{x}_2} = \sqrt{\sigma_1^2/n_1 + \sigma_2^2/n_2}$.

It is helpful to think of this sampling distribution as the frequency distribution that would be obtained by grouping the $\bar{x}_1 - \bar{x}_2$ values observed in repeated pairs of samples drawn independently from two populations with the same means.

The standard deviation $\sigma_{\bar{x}_1 - \bar{x}_2}$ is referred to as the **standard error of the difference between two means**. We see from equation 7.2 that we must know the population standard deviations in order to calculate this standard error. However, for *large samples*, we can approximate σ_1 and σ_2 using the sample standard

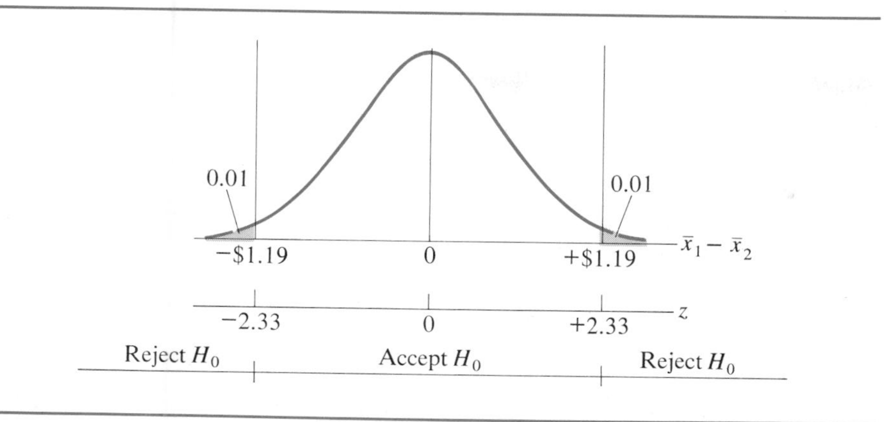

FIGURE 7-11
Sampling distribution of the difference between two means. This is a two-tailed test with $\alpha = 0.02$

deviations s_1 and s_2. The resulting estimated (or approximate) standard error is symbolized $s_{\bar{x}_1 - \bar{x}_2}$ and may be written

(7.3)
$$s_{\bar{x}_1 - \bar{x}_2} = \sqrt{\frac{s_1^2}{n_1} + \frac{s_2^2}{n_2}}$$

We can now establish the decision rule for the problem. The test is clearly two-tailed, because the hypothesis of equal population means would be rejected if $\bar{x}_1 - \bar{x}_2$ differed significantly from zero in either the positive or the negative direction. The sampling distribution of $\bar{x}_1 - \bar{x}_2$ is shown in Figure 7-11. The horizontal scale of the distribution shows the difference between the sample means $\bar{x}_1 - \bar{x}_2$. As indicated, the mean of the distribution is equal to zero; in other words, under the null hypothesis, the expected value of $\bar{x}_1 - \bar{x}_2$ is zero. Another way of interpreting the zero is that under the null hypothesis H_0: $\mu_1 - \mu_2 = 0$, we have assumed that the mean wages of the populations of unskilled workers are the same in Area A and Area B for the industry in question. Since the significance level is 0.02, 1% of the area under the normal curve is shown in each tail. From Table A-5 of Appendix A we find that in a normal distribution 1% of the area lies to the right of $z = +2.33$ and (by symmetry) 1% to the left of $z = -2.33$. Thus, we would reject the null hypothesis if the sample difference $\bar{x}_1 - \bar{x}_2$ fell more than 2.33 standard errors from the expected value of zero. The estimated standard error of the difference between means, $s_{\bar{x}_1 - \bar{x}_2}$, is, by equation 7.3,

$$s_{\bar{x}_1 - \bar{x}_2} = \sqrt{\frac{(\$4.00)^2}{100} + \frac{(\$4.50)^2}{200}} = \$.51$$

and

$$2.33 s_{\bar{x}_1 - \bar{x}_2} = (2.33)(\$.51) = \$1.19$$

Thus, the decision rule may be stated as follows:

DECISION RULE

1. If $\bar{x}_1 - \bar{x}_2 < -\$1.19$ or $\bar{x}_1 - \bar{x}_2 > \$1.19$, reject H_0
2. If $-\$1.19 \leqslant \bar{x}_1 - \bar{x}_2 \leqslant \$1.19$, accept H_0

In terms of z values, we have

DECISION RULE

1. If $z < -2.33$ or $z > +2.33$, reject H_0
2. If $-2.33 \leqslant z \leqslant +2.33$, accept H_0

where

$$z = \frac{\bar{x}_1 - \bar{x}_2}{s_{\bar{x}_1 - \bar{x}_2}}$$

Note that this z value is in the usual form of the ratio [(Statistic − Parameter)/Standard error]. The difference $\bar{x}_1 - \bar{x}_2$ is the statistic. The parameter under test is $\mu_1 - \mu_2$, which by the null hypothesis is zero and thus need not be shown in the numerator of the ratio. As previously indicated, we have substituted an approximate standard error for the true standard error in the denominator.

Applying this decision rule to the problem, we have

$$\bar{x}_1 - \bar{x}_2 = \$250.01 - \$245.21 = \$4.80$$

and

$$z = \frac{\bar{x}_1 - \bar{x}_2}{s_{\bar{x}_1 - \bar{x}_2}} = \frac{\$4.80}{\$.51} = 9.4$$

Since \$4.80 far exceeds \$1.19 (and correspondingly, 9.4 far exceeds 2.33), the null hypothesis is rejected. Hence, it is extremely unlikely that these two samples were drawn from populations having the same mean. We conclude that the sample mean wages of unskilled workers in this industry *differed significantly* between Areas A and B and thus that the population means *differ* between Areas A and B. Note that it is incorrect to use the term "significant difference" when referring to the relationship between two population parameters (in this case, the population means). The student should also keep in mind that, in this as in all other hypothesis-testing situations, we are assuming random sampling. Obviously, if the samples were not randomly drawn from the two populations, the foregoing procedure and conclusion would be invalid.

Test for Difference between Proportions: Two-Tailed Test

Another important case of two-sample hypothesis testing is one in which the observed statistics are proportions. The decision procedure is conceptually the same as when the sample statistics are means; only the computational details differ. In order to illustrate the technique, let us consider the following example. Workers in the Stanley Marino Company and Rock Hayden Company, two firms in the same industry, were asked whether they preferred to receive a specified package of increased fringe benefits or a specified increase in base pay. For brevity, we will refer to the companies as the S.M. Company and the R.H. Company and the proposed increases as "increased fringe benefits" and "increased base pay." In a simple random sample of 150 workers in the S.M. Company, 75 indicated that they preferred increased base pay. In the R.H. Company, 103 out of a simple random sample of 200 preferred increased base pay. In each company, the sample was less than 5% of the total number of workers. It was desirable to have a low probability of erroneously rejecting the hypothesis of equal proportions of workers in the two companies who preferred increased base pay. Therefore, a 1% level of significance was used for the test. Can it be concluded at the 1% level of significance that these two companies differed in the proportion of workers who preferred increased base pay?

Using the subscripts 1 and 2 to refer to the S.M. Company and R.H. Company, respectively, we can organize the sample data in a table, where $\bar{p}_1$ and $\bar{q}_1$ refer to the sample proportions in the S.M. Company in favor of and opposed to increased base pay, respectively. The sample size in the S.M. Company is denoted n_1. Corresponding notation is used for the R.H. Company.

S.M. Company	R.H. Company
$\bar{p}_1 = \frac{75}{150} = 0.50$	$\bar{p}_2 = \frac{103}{200} = 0.515$
$\bar{q}_1 = \frac{75}{150} = 0.50$	$\bar{q}_2 = \frac{97}{200} = 0.485$
$n_1 = 150$	$n_2 = 200$

If we designate the population proportions in favor of increased pay in the two companies as p_1 and p_2, then in a manner analogous to that of the preceding problem, we set up the two hypotheses

$$H_0: p_1 - p_2 = 0$$

$$H_1: p_1 - p_2 \neq 0$$

The underlying theory for the test is similar to that in the two-sample test for the difference between two means. If $\bar{p}_1$ and $\bar{p}_2$ are the observed sample proportions in large simple random samples drawn from populations with parameters p_1 and p_2, then the sampling distribution of the statistic $\bar{p}_1 - \bar{p}_2$ has a mean

(7.4)
$$\mu_{\bar{p}_1 - \bar{p}_2} = p_1 - p_2$$

and a standard deviation

(7.5)
$$\sigma_{\bar{p}_1 - \bar{p}_2} = \sqrt{\sigma^2_{\bar{p}_1} + \sigma^2_{\bar{p}_2}}$$

where $\sigma^2_{\bar{p}_1}$ and $\sigma^2_{\bar{p}_2}$ are the variances of the sampling distributions of $\bar{p}_1$ and $\bar{p}_2$. Assuming a binomial distribution, $\sigma^2_{\bar{p}_1} = p_1q_1/n_1$ and $\sigma^2_{\bar{p}_2} = p_2q_2/n_2$. Although the sampling was conducted without replacement, each of the samples constituted only a small percentage of the corresponding population (less than 5%), and the binomial distribution assumption appears reasonable. Thus, equation 7.5 becomes

$$\sigma_{\bar{p}_1-\bar{p}_2} = \sqrt{\frac{p_1q_1}{n_1} + \frac{p_2q_2}{n_2}} \tag{7.6}$$

If we hypothesize that $p_1 = p_2$ (the null hypothesis in this problem) and refer to the common value of p_1 and p_2 as p, equations 7.4 and 7.6 become

$$\mu_{\bar{p}_1-\bar{p}_2} = p - p = 0 \tag{7.7}$$

and

$$\sigma_{\bar{p}_1-\bar{p}_2} = \sqrt{\frac{pq}{n_1} + \frac{pq}{n_2}} = \sqrt{pq\left(\frac{1}{n_1} + \frac{1}{n_2}\right)} \tag{7.8}$$

Since the common proportion p hypothesized under the null hypothesis is unknown, we estimate it for the hypothesis test by taking a weighted mean of the observed sample percentages. If we refer to this "pooled estimator" as $\hat{p}$, we have

$$\hat{p} = \frac{n_1\bar{p}_1 + n_2\bar{p}_2}{n_1 + n_2} \tag{7.9}$$

The numerator of equation 7.9 is simply the total number of "successes" in the two samples combined, and the denominator is the total number of observations in the two samples. The standard deviation in equation 7.8, $\sigma_{\bar{p}_1-\bar{p}_2}$, is referred to as the **standard error of the difference between two proportions**. Substituting the "pooled estimator" $\hat{p}$ for p in equation 7.8, we have the following formula for the *estimated* or *approximate* standard error $s_{\bar{p}_1-\bar{p}_2}$

$$s_{\bar{p}_1-\bar{p}_2} = \sqrt{\hat{p}\hat{q}\left(\frac{1}{n_1} + \frac{1}{n_2}\right)} \tag{7.10}$$

Summary

We can now summarize these results. Let $\bar{p}_1$ and $\bar{p}_2$ be proportions of successes observed in two large, independent samples from populations with parameters p_1 and p_2. If we hypothesize that $p_1 = p_2 = p$, we obtain a pooled estimator $\hat{p}$ for p, where $\hat{p} = (n_1\bar{p}_1 + n_2\bar{p}_2)/(n_1 + n_2)$.

Then, the sampling distribution of $\bar{p}_1 - \bar{p}_2$ may be approximated by a normal curve with mean $\mu_{\bar{p}_1-\bar{p}_2} = 0$ and estimated standard deviation

$$s_{\bar{p}_1-\bar{p}_2} = \sqrt{\hat{p}\hat{q}\left(\frac{1}{n_1} + \frac{1}{n_2}\right)}$$

We may think of this sampling distribution as the frequency distribution of $\bar{p}_1 - \bar{p}_2$ values observed in repeated pairs of samples drawn independently from two populations having the same proportions.

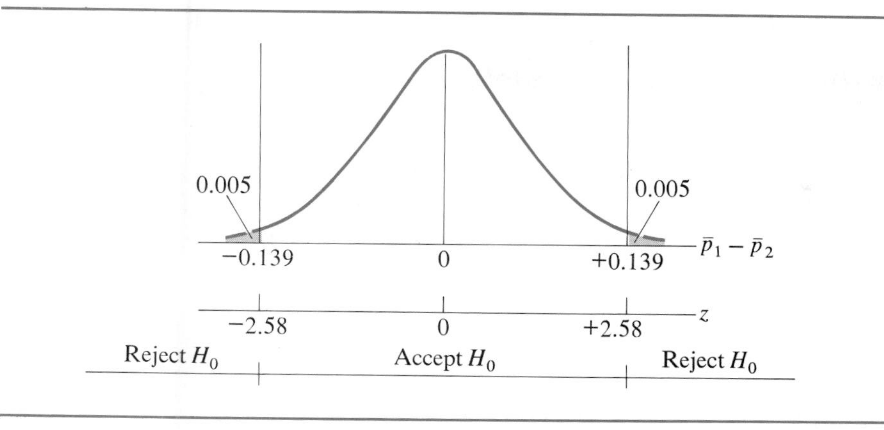

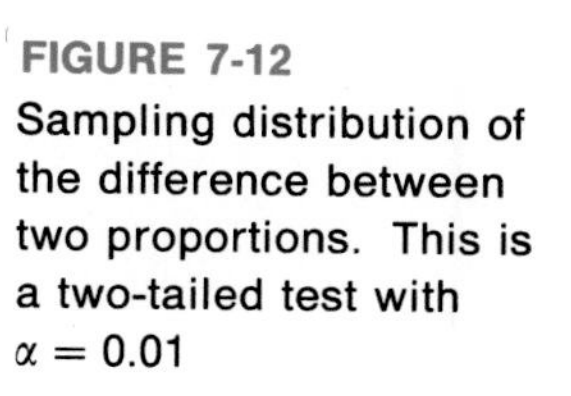
FIGURE 7-12
Sampling distribution of the difference between two proportions. This is a two-tailed test with $\alpha = 0.01$

Proceeding to the decision rule, we see again that the test is two-tailed, because the hypothesis of equal population proportions would be rejected for $\bar{p}_1 - \bar{p}_2$ values significantly above or below zero. The sampling distribution of $\bar{p}_1 - \bar{p}_2$ for the present problem is shown in Figure 7-12. Since the significance level is $\alpha = 0.01$, the area under the normal distribution curve shown in each tail is $\alpha/2$ or $\frac{1}{2}(1\%)$. Referring to Table A-5 of Appendix A, we find that 0.005 of the area under a normal curve lies above a z value of $+2.58$, and thus the same percentage lies below $z = -2.58$. Hence, rejection of the null hypothesis $H_0\colon p_1 - p_2 = 0$ occurs if the sample difference $\bar{p}_1 - \bar{p}_2$ falls more than 2.58 standard error units from zero. By equation 7.10, the standard error of the difference between proportions is

$$s_{\bar{p}_1 - \bar{p}_2} = \sqrt{\hat{p}\hat{q}\left(\frac{1}{n_1} + \frac{1}{n_2}\right)} = \sqrt{(0.51)(0.49)\left(\frac{1}{150} + \frac{1}{200}\right)}$$
$$= 0.054$$

where

$$\hat{p} = \frac{n_1\bar{p}_1 + n_2\bar{p}_2}{n_1 + n_2} = \frac{(150)(0.50) + (200)(0.515)}{150 + 200} = \frac{75 + 103}{150 + 200} = 0.51$$

Hence,

$$2.58 s_{\bar{p}_1 - \bar{p}_2} = (2.58)(0.054) = 0.139$$

Therefore, the decision rule is

DECISION RULE

1. If $\bar{p}_1 - \bar{p}_2 < -0.139$ or $\bar{p}_1 - \bar{p}_2 > +0.139$, reject H_0
2. If $-0.139 \leqslant \bar{p}_1 - \bar{p}_2 \leqslant 0.139$, accept H_0

In terms of z values, the rule is

DECISION RULE

1. If $z < -2.58$ or $z > +2.58$, reject H_0
2. If $-2.58 \leqslant z \leqslant +2.58$, accept H_0

where

$$z = \frac{\bar{p}_1 - \bar{p}_2}{s_{\bar{p}_1 - \bar{p}_2}}$$

Applying this decision rule yields

$$\bar{p}_1 - \bar{p}_2 = 0.500 - 0.515 = -0.015$$

and

$$z = \frac{\bar{p}_1 - \bar{p}_2}{s_{\bar{p}_1 - \bar{p}_2}} = \frac{-0.015}{0.054} = -0.28$$

Summary

Thus, the null hypothesis is accepted. In summary, the sample proportions $\bar{p}_1$ and $\bar{p}_2$ did not differ significantly, and therefore we cannot conclude that the two companies differed with respect to the proportion of workers who preferred increased base pay. This conclusion may be useful, for example, in testing a claim of a participant in labor negotiations that such a difference exists. Our reasoning is based on the finding that if the population proportions were equal, a difference between the sample proportions as large as the one observed could not at all be considered unusual.

Test for Differences between Proportions: One-Tailed Test

The two preceding examples illustrated two-tailed tests for cases in which data are available for samples from two populations. Just as in the one-sample case, the question we wish to answer may give rise to a one-tailed test. In order to illustrate this point, let us examine the following problem.

Two competing drugs are available for treating a certain physical ailment. There are no apparent side effects from administration of the first drug, whereas there are some definite side effects (nausea and mild headaches) from use of the second. A group of medical researchers has decided that it would nevertheless be willing to recommend use of the second drug in preference to the first if the proportion of cures effected by the second were higher than those by the first drug. The group felt that the potential benefits of achieving increased cures of the ailment would far outweigh the disadvantages of the possible side effects. On the other hand, if the proportion of cures effected by the second drug was equal to or less than that of the first drug, the group would recommend use of the first one. In terms of hypothesis testing, we can state the alternatives and consequent actions as

$$H_0: p_2 \leqslant p_1 \text{ (use the first drug)}$$

$$H_1: p_2 > p_1 \text{ (use the second drug)}$$

where p_1 and p_2 denote the population proportions of cures effected by the first and second drugs. Another way we may write these alternatives is

$$H_0: p_2 - p_1 \leqslant 0 \text{ (use the first drug)}$$

$$H_1: p_2 - p_1 > 0 \text{ (use the second drug)}$$

For purposes of comparison, note that the hypotheses in the preceding problem, a two-tailed testing situation, were

$$H_0: p_1 = p_2$$

$$H_1: p_1 \neq p_2$$

or in the alternative form (in terms of differences)

$$H_0: p_1 - p_2 = 0$$

$$H_1: p_1 - p_2 \neq 0$$

Clearly, the present problem involves a one-tailed test, in which we would reject the null hypothesis only if the sample difference $\bar{p}_2 - \bar{p}_1$ differed significantly from zero and was a positive number.

The medical researchers used the drugs experimentally on two random samples of persons suffering from the ailment, administering the first drug to a group of 80 patients and the second drug to a group of 90 patients. By the end of the experimental period, 52 of those treated with the first drug were classified as "cured," whereas 63 of those treated with the second drug were so classified. The sample results may be summarized in a table, with p denoting "proportion cured" and q denoting "proportion not cured."

First Drug	Second Drug
$\bar{p}_1 = \frac{52}{80} = 0.65$ cured	$\bar{p}_2 = \frac{63}{90} = 0.70$ cured
$\bar{q}_1 = \frac{28}{80} = 0.35$ not cured	$\bar{q}_2 = \frac{27}{90} = 0.30$ not cured
$n_1 = 80$	$n_2 = 90$

The pooled sample proportion cured is

$$\hat{p} = \frac{52 + 63}{80 + 90} = \frac{115}{170} = 0.676$$

and the estimated standard error of the difference between proportions is

$$s_{\bar{p}_2-\bar{p}_1} = \sqrt{(0.676)(0.324)\left(\frac{1}{80} + \frac{1}{90}\right)} = 0.0719$$

Since the medical group wished to maintain a low probability of erroneously adopting the second drug, it selected a 1% significance level for the test. This means that 1% of the area under the normal curve lies to the right of $z = +2.33$. Therefore, the null hypothesis would be rejected if $\bar{p}_2 - \bar{p}_1$ falls at least 2.33 standard error units above zero. In terms of proportions,

$$2.33s_{\bar{p}_2-\bar{p}_1} = 2.33(0.0719) = 0.168$$

Hence, the decision rule is

DECISION RULE

1. If $\bar{p}_2 - \bar{p}_1 > 0.168$, reject H_0
2. If $\bar{p}_2 - \bar{p}_1 \leqslant 0.168$, accept H_0

In terms of z values, the rule is

DECISION RULE

1. If $z > +2.33$, reject H_0
2. If $z \leqslant +2.33$, accept H_0

where

$$z = \frac{\bar{p}_2 - \bar{p}_1}{s_{\bar{p}_2-\bar{p}_1}}$$

In the present problem,

$$\bar{p}_2 - \bar{p}_1 = 0.70 - 0.65 = 0.05$$

so

$$z = \frac{0.70 - 0.65}{0.0719} = 0.70$$

Thus, the null hypothesis is accepted. On the basis of the sample data, we cannot conclude that the second drug accomplishes a greater proportion of cures than the first. The sampling distribution of $\bar{p}_2 - \bar{p}_1$ is given in Figure 7-13. Note that it is immaterial whether we state the difference between proportions

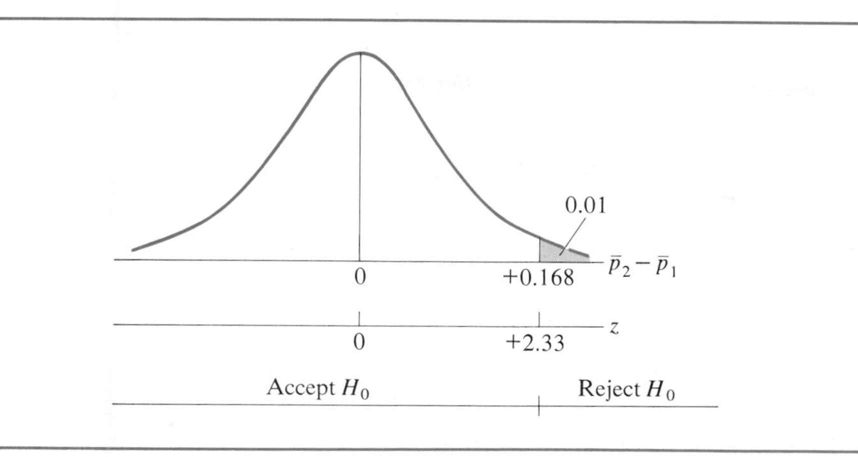

FIGURE 7-13
Sampling distribution of the difference between two proportions. This is a one-tailed test with $\alpha = 0.01$

as $\bar{p}_1 - \bar{p}_2$ or $\bar{p}_2 - \bar{p}_1$, but we must exercise care concerning the correspondence between the way the hypothesis is stated, the sign of the difference between the sample proportions, and the tail of the sampling distribution in which rejection of the null hypothesis takes place.

Exercises 7.3

1. Match the correct test statistics with the following four null hypotheses.

   (1) $\bar{x}$ (2) $\bar{p}$ (3) $\bar{x}_1 - \bar{x}_2$ (4) $\bar{p}_1 + \bar{p}_2$ (5) $n\bar{p}$ (6) $\bar{p}_1 - \bar{p}_2$ (7) $\bar{x}/\bar{p}$

   a. H_0: $\mu_1 = \mu_2$
   b. H_0: $p = p_0$
   c. H_0: $p_1 = p_2$
   d. H_0: $\mu = \mu_0$

2. A railroad company installed 100 red oak ties in two sets of 50. The two sets were treated with creosote by two different processes. After a number of years of service, it was found that 22 ties in the first set and 18 ties in the second set were still in good condition. Are you justified in claiming that there is a real difference between the preserving properties of the two processes if you are willing to be wrong no more than 5% of the time?

3. A bank is considering opening a new branch in one of two neighborhoods. One of the factors considered by the bank is whether the average family incomes in the two neighborhoods differ. From census records, the bank draws two simple random samples of 200 families each. Formulate an appropriate null hypothesis and test it at the 5% significance level, based on the following information.

$$\bar{x}_1 = \$25{,}600 \qquad \bar{x}_2 = \$25{,}490$$
$$s_1 = \$500 \qquad s_2 = \$700$$
$$n_1 = 200 \qquad n_2 = 200$$

4. The Colson Brewing Company felt that two of its market areas exhibited equivalent sales patterns. The company wished to reduce its advertising budget, but it decided to test the hypothesis about sales patterns prior to any advertising change. Random sample data for daily consumption are listed by area.
   a. Would you conclude that areas 1 and 2 have equivalent mean daily consumption rates? Justify your answer using a 2% significance level.

   In answering part (a), you had to locate a sample statistic (the arithmetic mean) on a random sampling distribution. Draw a rough sketch of this distribution, showing
   b. the value and location of the hypothetical parameter
   c. the value and location of the statistic
   d. the horizontal scale description

Area	Mean Daily Consumption	Standard Deviation	Sample Size
1	1500	140	100
2	1450	120	150

5. Suppose that you took a census of the incomes of all attorneys in two towns and found that the mean income for attorneys was $50,800 in Wells Pines and was $52,600 in Oak Run.

   Can you conclude statistically that the average income of attorneys in Oak Run exceeded that of Wells Pines attorneys? Would you test a hypothesis? If yes, what hypothesis? If no, why not?
6. In a simple random sample of 200 employees of the Orange Computer Corporation, 120 favored a merger with the MBI Corporation. In a simple random sample of 300 employees of the MBI corporation, 231 were in favor of a merger with the Orange Computer Corporation.
   a. Do you believe that there is a real difference in the two sets of employees regarding the desirability of merger? Justify your answer statistically and indicate which significance level you used.
   b. Explain specifically the meaning of a Type I error in this particular problem.
7. A company specializing in wood products owns a large tract of timber land. The trees on this tract are sprayed periodically in order to reduce damage to the trees by insects. A new spray has become available on the market, and company executives have asked their research department to determine whether the new spray is more effective than the old spray in reducing the number of damaged trees. The investigators sprayed 200 trees with the old spray and 200 trees with the new spray. After a period of time, they

observed damage in 86 trees that had been sprayed with the old spray, while only 74 trees that had been sprayed with the new spray were damaged. At a 0.01 significance level, would you be willing to conclude that the new spray is more effective? Your answer should include a clear statement of the statistical hypotheses and of the decision rule.

7.4 THE t DISTRIBUTION: SMALL SAMPLES WITH UNKNOWN POPULATION STANDARD DEVIATION(S)

The hypothesis-testing methods discussed in the preceding sections are appropriate for large samples. In this section, we concern ourselves with the case when the sample size is small. The underlying theory is exactly the same as that given in section 6.3, in which confidence interval estimation for small samples was discussed.

In hypothesis testing, as in confidence interval estimation, the distinction between large and small sample tests becomes important when the population standard deviation is *unknown* and therefore must be *estimated* from the sample observations.

The main principles will be reviewed here. The statistic $(\bar{x} - \mu)/s_{\bar{x}}$, where $s_{\bar{x}}$ denotes an estimated standard error, is not approximately normally distributed for all sample sizes. As we have noted earlier, $s_{\bar{x}}$ is computed by the formula $s_{\bar{x}} = s/\sqrt{n}$ where s represents an estimate of the true population standard deviation. For large samples, the ratio $(\bar{x} - \mu)/s_{\bar{x}}$ is approximately normally distributed, and we may use the methods discussed in sections 7.2 and 7.3. However, since this statistic is not approximately normally distributed for small samples, the t distribution should be used instead. The use of the t distribution for testing a hypothesis concerning a population mean is demonstrated in Example 7-2. A small sample ($n \leqslant 30$) is assumed, and the population standard deviation is unknown.

EXAMPLE 7-2

One-sample Test of a Hypothesis about the Mean: Two-tailed Test

The personnel department of a company developed an aptitude test for a certain type of semiskilled worker. The individual test scores were assumed to be normally distributed. The developers of the test asserted a tentative hypothesis that the arithmetic mean grade obtained by this type of semiskilled worker would be 100. It was agreed that this hypothesis would be subjected to a two-tailed test at the 5% level of significance. The aptitude test was given to a simple random sample of 16 semiskilled workers with the following results:

$$\bar{x} = 94$$
$$s = 5$$
$$n = 16$$

The competing hypotheses are

$$H_0: \mu = 100$$
$$H_1: \mu \neq 100$$

Solution

To carry out the test, the following quantities were calculated:

$$s_{\bar{x}} = \frac{s}{\sqrt{n}} = \frac{5}{\sqrt{16}} = 1.25$$

and

$$t = \frac{\bar{x} - \mu}{s_{\bar{x}}} = \frac{94 - 100}{1.25} = -4.80$$

The significance of this t value is judged from Table A-6 of Appendix A. The meaning of the table was discussed in section 6.3. We will explain the use of the table for this hypothesis-testing problem. Let us set up the areas of acceptance and rejection of the hypothesis. Since the sample size is 16, the number of degrees of freedom is $\nu = 16 - 1 = 15$. Looking along the row of Table A-6 for 15 under the column 0.05, we find the t value, 2.131. This means that in a t distribution for $\nu = 15$, the probability is 5% that t is greater than 2.131 or is less than -2.131. Thus, in the present problem, at the 5% level of significance, the null hypothesis $H_0: \mu = 100$ is rejected if a t value exceeding 2.131 or less than -2.131 is observed. Since the computed t value in this problem is -4.80, we reject the null hypothesis. In other words, we are unwilling to attribute the difference between our sample mean of 94 and the hypothesized population mean of 100 merely to chance errors of sampling. The t distribution for this problem is shown in Figure 7-14.

A few remarks can be made about this problem. Since the computed t value of -4.80 was so much less than zero, the null hypothesis would have been rejected even at

FIGURE 7-14
The t distribution for $\nu = 15$

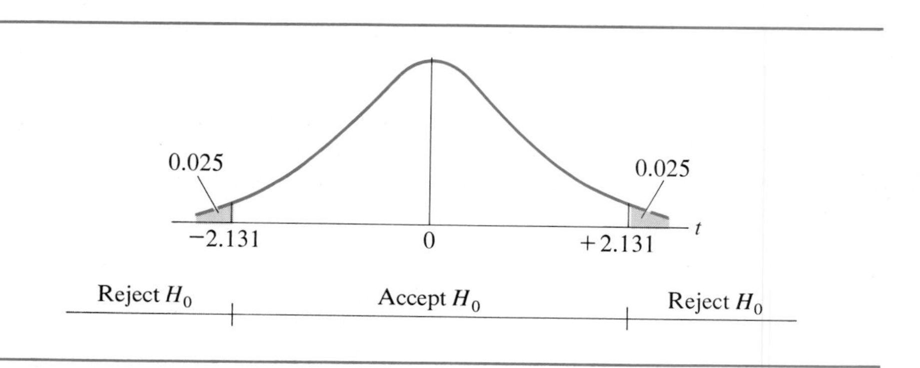

the 2% or 1% level of significance (see Table A-6 of Appendix A). Had the test been one-tailed at the 5% level of significance, we would have had to obtain the critical t value by looking under 0.10 in Table A-6, since the 0.10 figure is the combined area in both tails. Thus, for a one-tailed test at the 5% level of significance and a rejection region in the lower tail, the critical t value would have been -1.753.

It is interesting to compare these critical t values with analogous critical z values for the normal curve. From Table A-5 of Appendix A, we find that the critical z values at the 5% level of significance are -1.96 and 1.96 for a two-tailed test and -1.65 for a one-tailed test with a rejection region in the lower tail. As we have just seen, the corresponding figures for the critical t values in a test involving 15 degrees of freedom are -2.131 and 2.131 for a two-tailed test and -1.753 for a one-tailed test.

An underlying assumption in applying the t test is that the population is closely approximated by a normal distribution. Since the population standard deviation σ is unknown, the t distribution is the theoretically correct sampling distribution. However, if the sample size had been large, even with an unknown population standard deviation, the normal curve could have been used as an approximation to the t distribution. As we saw in this problem, for $\nu = 15$, a total of 5% of the area in the t distribution falls to the right of $t = +2.131$ and to the left of $t = -2.131$. The corresponding z values in the normal distribution are $+1.96$ and -1.96. From Table A-6, we see that the t value for $\nu = 30$ is 2.042. The closeness of this figure to $+1.96$ gives rise to the usual rule of thumb, which uses $n > 30$ as the arbitrary dividing line between large-sample and small-sample methods. We have used this convenient rule in Chapters 6 and 7. However, what constitutes a suitable approximation really depends on the context of the particular problem. Furthermore, if the population is highly skewed, a sample size as large as 100 may be required for the assumption of a normal sampling distribution of $\bar{x}$ to be appropriate.

EXAMPLE 7-3

Two-sample Test for Means: Two-tailed Test

A marketing research effort has been launched to determine how different names for a particular new diet food product could possibly affect sales. The food products in the market research study were identical, but they were given two different names, Slimall and Supersate.

In an initial study, 10 randomly selected potential consumers were asked to taste the product named Slimall and to rate it on a scale of 1 (very poor) to 10 (outstanding). Similarly, another 10 potential consumers were asked to taste the product named Supersate and to rate it on a scale of 1 to 10. The statistics were calculated:

Slimall	Supersate
$\bar{x}_1 = 8.6$	$\bar{x}_2 = 6.8$
$s_1 = 1.1$	$s_2 = 1.3$
$n_1 = 10$	$n_2 = 10$

Do the names produce different ratings? That is, should the difference observed between the two sample means be considered significant? Use $\alpha = 0.01$.

Solution The alternative hypotheses are

$$H_0: \mu_1 - \mu_2 = 0$$

$$H_1: \mu_1 - \mu_2 \neq 0$$

To test the null hypothesis, we have the t statistic

$$t = \frac{(\bar{x}_1 - \bar{x}_2) - 0}{s_{\bar{x}_1 - \bar{x}_2}} = \frac{\bar{x}_1 - \bar{x}_2}{s_{\bar{x}_1 - \bar{x}_2}}$$

where $s_{\bar{x}_1 - \bar{x}_2}$ is the estimated standard error of the difference between the two means.

Unlike the case with large samples, here we must assume equal population variances. An estimate of this common variance is obtained by pooling the two sample variances into a weighted average, using the numbers of degrees of freedom, $n_1 - 1$ and $n_2 - 1$, as weights. This pooled estimate of the common variance, which we denote as $\hat{s}^2$, is given by

$$\hat{s}^2 = \frac{(n_1 - 1)s_1^2 + (n_2 - 1)s_2^2}{n_1 + n_2 - 2} \tag{7.11}$$

The estimated standard error of the difference between two means is then

$$s_{\bar{x}_1 - \bar{x}_2} = \sqrt{\frac{\hat{s}^2}{n_1} + \frac{\hat{s}^2}{n_2}} = \hat{s}\sqrt{\frac{1}{n_1} + \frac{1}{n_2}} \tag{7.12}$$

A number of alternative mathematical expressions are possible for equation 7.12, but because of its similarity in appearance to previously used standard error formulas, we shall use it in this form.

We now work out the present problem. Substitution into equation 7.11 gives

$$\hat{s}^2 = \frac{(10 - 1)(1.1)^2 + (10 - 1)(1.3)^2}{10 + 10 - 2} = 1.45$$

and

$$\hat{s} = \sqrt{1.45} = 1.204$$

Thus, the estimated standard error is

$$s_{\bar{x}_1 - \bar{x}_2} = (1.204)\sqrt{\frac{1}{10} + \frac{1}{10}} = 0.538$$

and the t value is

$$t = \frac{8.6 - 6.8}{0.538} = 3.35$$

The number of degrees of freedom in this problem is $n_1 + n_2 - 2$, that is, $10 + 10 - 2 = 18$. We can explain the number of degrees of freedom in this case as follows. In the one-sample case, when the sample standard deviation is used as an estimate of the population standard deviation, there is a loss of one degree of freedom; hence, the number of degrees of freedom is $n - 1$. In the two-sample case, each of the sample variances is used in the pooled estimate of the population variance; hence, two degrees of freedom are lost, and the number of degrees of freedom is $n_1 + n_2 - 2$.

The critical t value at the 1% significance level for 18 degrees of freedom is 2.878 (see Table A-6 of Appendix A). Since the observed t value, 3.35, exceeds this critical t value, the null hypothesis is rejected, and we conclude on the basis of the sample data that the population means are indeed different. In terms of the problem, we are unwilling to attribute the difference in the Slimall and Supersate average taste ratings to chance errors of sampling.

Computer Output for Tests of Hypotheses

There are many computer packages for carrying out statistical calculations and tests. Figure 7-15 is an example of computer output for a one-sample test of an hypothesis. Note that both z tests and t tests can be done. In the last line of this output, a p value of 0.00369 is shown. As indicated in section 7.2, this p value means that a z value greater than the one obtained (2.681) would occur with a frequency of only 0.00369, or fewer than 4 times in 1,000. Hence, the sample mean differs significantly from the null hypothesis mean at such conventional significance levels as 0.05 and 0.01.

FIGURE 7-15
Computer output for a one-sample test of an hypothesis

```
MEAN TEST

ENTER NULL HYPOTHESIS MEAN
•      30

DO YOU WANT A Z TEST FOR A T TEST? ENTER Z OR T
•      Z

ENTER ASSUMED POPULATION SIGMA
•      6

ONE-TAILED OR TWO-TAILED TEST? ENTER 1 OR 2
•      1

ENTER DATA
•      28.6  28.8  29.2  29.7  30.3  30.5  30.8  31.1  31.3  31.4
       31.6  32.0  32.2  32.3  32.5  32.8  32.9  33.1  33.2  33.4
       33.4  33.6  33.7  33.8  33.9  33.9  34.1  34.2  34.3  35.5
       37.9  38.0  38.4

SAMPLE MEAN IS 32.8
Z STATISTIC EQUALS 2.681
P-VALUE IS 0.00369
```

Exercises 7.4

1. A fast-food restaurant chain wished to test the effects of a new advertising campaign on sales. Five stores in the chain were randomly selected to use the new campaign for 10 weeks. During the study the average total weekly sales for the 5 stores combined was $198,000 with a standard deviation of $20,000.
   a. If the prestudy weekly sales average of these stores was $188,000, can you conclude that there has been a change in the average total weekly sales? Use a two-tailed test with a 0.05 level of significance.
   b. If the sample period had been one year instead of 10 weeks (assuming everything else remained unchanged), would you reach the same conclusion as in part (a)?
2. (In this problem, a statistical procedure has been misused. Describe the incorrect procedure.) NASA wants to test the reliability of a certain magneto relay used in the Surveyor spacecraft. According to specifications, these relays should have an average life of 180 hours before failure. A sample of 20 switches is tested, and the average switch life is 177.5 hours with a standard deviation of 4. Since the z statistic

$$z = \frac{177.5 - 180}{4/\sqrt{20}}$$

is less than -1.65, it is concluded (on the basis of the normal distribution) at a 0.05 significance level that the switches do not meet specifications.
3. In a study of use of bar soap by a simple random sample of 15 suburban families, the consumption of such soap was found to have an arithmetic mean of 60 ounces per family per month with a standard deviation of 10 ounces. In another similar study of 11 urban families, consumption was found to average 55 ounces with a standard deviation of 12 ounces.
   a. At the 10% level of significance, would you conclude that there was a statistically significant difference in the sample averages of consumption of bar soap?
   b. For this problem, state the null and alternative hypotheses and the decision rule employed.
4. An industrial psychologist obtained a productivity score of 7 work teams under normal noise conditions and found an average of 35 and a sample variance of 54. She also measured the productivity score of 4 other work teams under reduced noise conditions and found an average of 44 and a sample variance of 60. She concluded: "Since the difference in means is not significant at $\alpha = 0.01$, we can safely conclude that noise level has no effect on average productivity."
   a. Do you agree that the difference is not significant at $\alpha = 0.01$? Justify your answer.
   b. Do you agree with the conclusion regarding noise level and productivity? What does the concept of power of the test have to do with your answer?

5. A successful tax advisor is analyzing a motion picture tax shelter for possible investment. The prospectus claims that movies in the same general category as the one under consideration have grossed an average of $10 million each in the past. This is an optimal level of receipts for the current film (that is, gross receipts less than this would result in an inadequate return on investment or a loss of investment, and gross receipts more than this would result in taxable income, which is not the primary objective of a tax shelter). The advisor has taken a random sample of similar films and has compiled statistics to test the claim of the prospectus.

Film	Gross Receipts (in $Millions)
1	11.5
2	7
3	9
4	10
5	10
6	12
7	8.5
8	9
9	8
10	6
11	7.5
12	9.5

Formulate an appropriate test using $\alpha = 0.02$.

7.5 THE DESIGN OF A TEST TO CONTROL BOTH TYPE I AND TYPE II ERRORS°

Relatively simple hypothesis-testing situations have been considered in this chapter in order to convey the basic principles of classical hypothesis testing. These tests have assumed that sample size was predetermined and that only the risks of Type I errors were to be controlled formally by the decision procedure. Of course, we assumed that a power curve would be constructed in all cases, and therefore, we could make an examination of the risks of Type II errors for

° This section is optional. It may be omitted without interfering with the continuity of the discussion.

parameter values not included in the null hypothesis. However, nothing was included in the formal testing procedure to control the level of risk of a Type II error for any specific parameter value.

In this section, we consider a method of controlling the levels of both Type I and Type II errors simultaneously in the same test. In the previous tests, which controlled only Type I errors—that is, where only α was specified—one point on the power curve was determined. In a test designed to control both Type I and Type II errors, two specific points on the power curve are determined.

As an illustration, we consider a one-tailed test problem similar to the one discussed in section 7.2. This problem involves an acceptance sampling procedure in which $\alpha = 0.05$ for an incoming shipment whose parts have a mean heat resistance of 2250°F. We call this the "producer's risk"; that is, the producer runs a 5% risk of having a "good" shipment with $\mu = 2250$°F rejected in error. Because H_0 is $\mu \geqslant 2250$°F, the maximum risk of rejecting a good shipment is 0.05. The probability of erroneous rejection drops below 5% for shipments with means in excess of 2250°F. Suppose now it is agreed to fix the "consumer's risk" or β at 0.03 for a shipment whose mean was 2150°F. That is, a shipment whose parts have a mean heat resistance of 2150°F does not meet specifications, and we want the probability of erroneously accepting such a shipment to be 0.03. We make the assumption that the population standard deviation is $\sigma =$ 300°F. The solution to this problem involves the determination of the *sample size* required to give the desired levels of control of both types of errors. After the required sample size has been calculated, the appropriate decision rule can be specified. We assume the sample size will be large enough to use normal sampling distributions for $\bar{x}$.

Figure 7-16 gives a graphic representation of the error controls specified in the preceding paragraph. In a standard solution to this problem involving control of only Type I errors, the critical value for acceptance or rejection of the null hypothesis H_0: $\mu \geqslant 2250$°F is 2200°F. The critical value is now un-

FIGURE 7-16
Acceptance sampling problem: Control of Type I and Type II errors

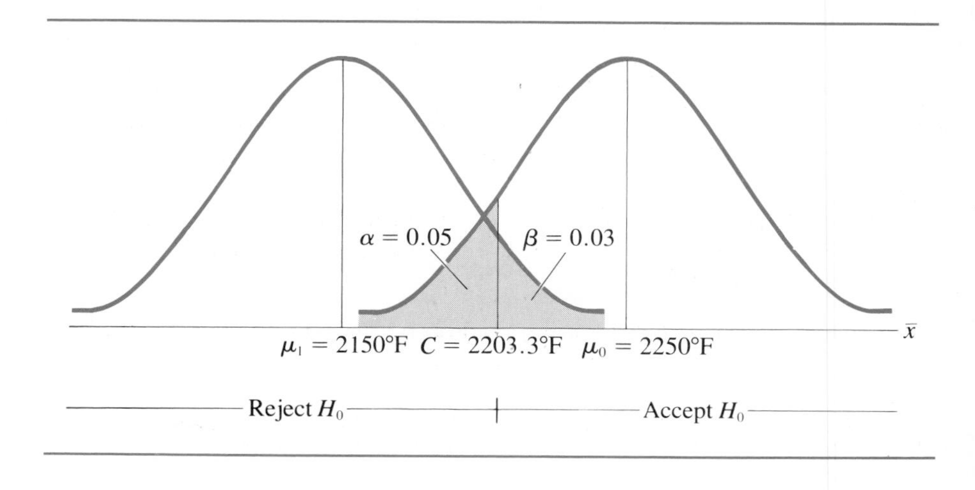

known and will have to be evaluated. Let us denote the new critical value as C, the mean of 2250°F under the null hypothesis as μ_0, and the mean of 2150°F under the alternative hypothesis as μ_1. The area in the left tail of the sampling distribution of $\bar{x}$ when $\mu_0 = 2250°\text{F}$ is shown as 0.05, denoting the Type I error. As is readily determined, the critical point C lies 1.65 standard error units to the left of the mean $\mu_0 = 2250°\text{F}$. Therefore, $z_0 = 1.65$. Under the alternative $\mu_1 = 2150°\text{F}$, we want the probability that an $\bar{x}$ value lies in the acceptance region to be 0.03. From Table A-5, we ascertain that 0.03 of the area in a normal sampling distribution lies to the right of a value 1.88 standard error units above the mean. Hence, $z_1 = 1.88$. Therefore, we can write the following relationships for the critical point C:

(7.13)
$$C = \mu_0 - z_0\sigma_{\bar{x}} = \mu_0 - z_0\frac{\sigma}{\sqrt{n}}$$

(7.14)
$$C = \mu_1 + z_1\sigma_{\bar{x}} = \mu_1 + z_1\frac{\sigma}{\sqrt{n}}$$

Substituting the numerical values for this problem, we obtain

$$C = 2250°\text{F} - 1.65\frac{(300°\text{F})}{\sqrt{n}}$$

$$C = 2150°\text{F} + 1.88\frac{(300°\text{F})}{\sqrt{n}}$$

Setting the right-hand sides of these two equations equal to one another yields the solution $n = 112$ (to the nearest integer). Therefore, a simple random sample of 112 parts from the incoming shipment would be required in order to obtain the desired levels of error control.

We can express this required sample size by solving the simultaneous equations 7.13 and 7.14. The solution is

(7.15)
$$n = \left[\frac{(z_0 + z_1)\sigma}{(\mu_0 - \mu_1)}\right]^2$$

Of course, substitution of the numerical values into this formula again yields $n = 112$. Note that both z_0 and z_1 are taken as positive, since the matter of whether C lies above or below μ_0 and μ_1 is taken care of by the signs in equations 7.13 and 7.14.

Now the decision rule can be stated. The critical value C can be obtained by substituting into either of the two simultaneous equations. Substituting into the first equation yields[4]

$$C = 2250°\text{F} - 1.65\frac{(300°\text{F})}{\sqrt{112}} = 2203.3°\text{F}$$

[4] Actually the values $C = 2203.2°\text{F}$ and $2203.4°\text{F}$ are obtained from the first and second equations, respectively. The discrepancy is due to rounding off n to an integral value.

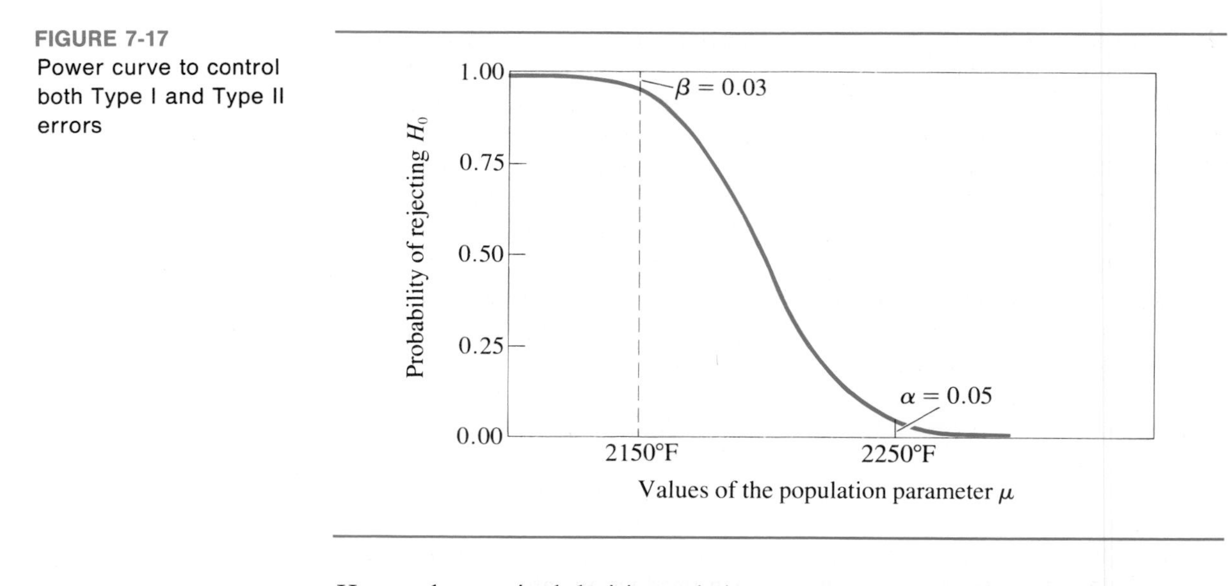

FIGURE 7-17
Power curve to control both Type I and Type II errors

Hence, the required decision rule is

DECISION RULE

1. If $\bar{x} < 2203.3$°F, reject H_0 (reject the shipment)
2. If $\bar{x} \geqslant 2203.3$°F, accept H_0 (accept the shipment)

The power curve for this test is displayed in Figure 7-17.

This one-tailed test for a mean illustrates the general method for controlling the levels of both Type I and Type II errors. Analogous tests can be constructed for proportions and for two-tailed tests as well.

In the type of quality control problem described here, the producer and the consumer might negotiate to determine the levels of Type I and Type II risks that they would agree to tolerate. These tolerable risks in turn would primarily depend on the costs involved to the producer and consumer of these two types of errors.

Exercises 7.5

1. A particular drug is supposed to have a mean period of effectiveness of at least 50 hours. Assume that the standard deviation of the period of effectiveness was known to be 10 hours. In an acceptance sampling procedure, a significance level of $\alpha = 0.01$ was adopted for an incoming shipment in which the mean period of effectiveness was 50 hours. Suppose it was agreed to fix the "consumer's risk" or β at 0.05 for a shipment whose mean period of effectiveness was only 47 hours of effectiveness. Determine the sample size required to give the desired levels of control for both types of errors, and state the decision rule in terms of the sample mean period of effectiveness.

2. The owner of a small business has determined that in the past the mean amount of sales per sales invoice has been \$30 and that the standard deviation was \$9. She wishes to run a test to find out whether the mean has decreased from \$30, using a Type I error probability of 0.05. In this test, she wants to maintain a Type II error probability of 0.01 of reaching an incorrect conclusion if the mean sales per sales invoice has actually dropped to \$28. Determine the number of invoices to be sampled and the decision rule in terms of the sample mean.
3. Management wants to determine whether production will be increased by instituting a bonus plan. A mean production rate of 50 units per day and a standard deviation of 8 units per day had been established before the proposed bonus plan. A sample test was agreed upon in which a random sample of workers would be given a bonus plan to test the null hypothesis H_0: $\mu \leqslant 50$ versus the alternative hypothesis H_1: $\mu > 50$ using an $\alpha = 0.03$. It was desired to maintain a Type II error probability of 0.02 if the mean production rate had risen to 54 units per day under the bonus plan. What size sample should be used in this situation? Specify the decision rule to be followed in terms of the sample mean production rate.

7.6 THE t TEST FOR PAIRED OBSERVATIONS

In the two-sample tests considered so far, the two samples had to be independent. That is, the values of observations in one sample had to be *independent* of the values in the other. Situations arise in practice in which this condition does not hold. In fact, the two samples may consist of pairs of observations made on the same individual, the same object, or more generally, the same selected population elements. Clearly, the independence condition is violated in these cases.

As a concrete example, let us consider a case in which a group of 10 men was given a special diet, and it was desired to test weight loss in pounds at the end of a two-week period. The observed data are shown in Table 7-3.

As indicated in Table 7-3, X_1 denotes the weight before the diet and X_2 the weight after the diet. It would be incorrect to run a t test to determine whether there is a significant difference between the mean of the X_1 values and the mean of the X_2 values, because of the nonindependence of the two samples. An individual's weight after the test is certainly not independent of his weight before the test. Each of the d values, $d = X_2 - X_1$, represents a difference between two observations on the same individual. The assumption is made that the subtraction of one value from the other removes the effect of factors other than that of the diet; that is, we assume that these other factors affect each member of any pair of X_1 and X_2 values in the same way.

TABLE 7-3
Weights before and after a special diet for a simple random sample of 10 men

Man	Weight before Diet X_1	Weight after Diet X_2	Difference in Weight $d = X_2 - X_1$	$d - \bar{d}$	$(d - \bar{d})^2$
1	181	178	−3	+1	1
2	172	172	0	+4	16
3	190	185	−5	−1	1
4	187	184	−3	+1	1
5	210	201	−9	−5	25
6	202	201	−1	+3	9
7	166	160	−6	−2	4
8	173	168	−5	−1	1
9	183	180	−3	+1	1
10	184	179	−5	−1	1
			−40		60

$$\bar{d} = \frac{-40}{10} = -4 \text{ pounds}$$

$$s_d = \sqrt{\frac{\Sigma(d - \bar{d})^2}{n - 1}} = \sqrt{\frac{60}{10 - 1}} = 2.58 \text{ pounds}$$

$$s_{\bar{d}} = \frac{s_d}{\sqrt{n}} = \frac{2.58}{\sqrt{10}} = 0.82 \text{ pounds}$$

We can state the hypotheses to be tested as

$$H_0\colon \mu_2 - \mu_1 = 0$$

$$H_1\colon \mu_2 - \mu_1 < 0$$

where μ_1 and μ_2 are the population mean weights before and after the diet, respectively. Let us assume that the test is to be carried out with $\alpha = 0.05$. The null hypothesis states that there is no difference between mean weight after the diet and mean weight before the diet, whereas the alternative hypothesis says that mean weight after the diet is less than mean weight before. We can visualize the situation as one in which, if the null hypothesis is true, a population of numbers represents differences in the weight after the diet and before the diet, and the mean of these numbers is zero. Hence, we wish to test the hypothesis that our simple random sample of $d = X_2 - X_1$ values comes from this universe. The procedure used in these cases is to obtain the mean of the sample differences and to test whether this average $\bar{d}$ differs significantly from zero. The estimated standard error of $\bar{d}$, denoted $s_{\bar{d}}$, is given by

$$s_{\bar{d}} = \frac{s_d}{\sqrt{n}}$$

where

$$s_d = \sqrt{\frac{\Sigma(d - \bar{d})^2}{n - 1}}$$

In this problem, $s_{\bar{d}} = 0.82$ pounds, as shown by the calculations in Table 7-3. Assuming that the population of differences (d values) is normally distributed, the ratio $(\bar{d} - 0)/s_{\bar{d}}$ is distributed according to the t distribution.

$$t = \frac{\bar{d} - 0}{s_{\bar{d}}} = \frac{-4}{0.82} = -4.88$$

The number of degrees of freedom is $n - 1$, where n is the number of d values. Hence, in this problem, $n - 1 = 10 - 1 = 9$. The test is one-tailed, because only a $\bar{d}$ value that is negative and significantly different from zero could result in acceptance of the alternative hypothesis, H_1: $\mu_2 - \mu_1 < 0$. Since $\alpha = 0.05$ and the test is one-tailed, we look in Table A-6 under the heading 0.10. The critical t value for $v = 9$ is 1.833, which for our purposes is interpreted as -1.833. Since the observed t value of -4.88 is less than (lies to the left of) this critical point, the null hypothesis is rejected and we accept its alternative. Therefore, on the basis of this experiment, we conclude that the special diet does result in an average weight loss over a two-week period.

This method of pairing observations is also used to reduce the effect of extraneous factors that could cause a significant difference in means, whereas the factor whose effect we are really interested in may not have resulted in such a difference if the extraneous factors had not been present.

For example, if medical experimenters wanted to test two different treatments to judge which was better, they might administer one treatment to one group of persons and the other treatment to a second group. Suppose on the basis of the usual significance test for means, it is concluded that one treatment is better than the other. Let us also assume that the group receiving the supposedly better treatment was much younger and much healthier at the beginning of the experiment than the other group and that these factors could have an effect on the reaction to the treatments. Then clearly, the relative effectiveness of the two treatments would be obscured.

On the other hand, assume that individuals were selected in pairs in which both members were about the same age and in about the same health condition. If the first treatment is given to one member of a pair and the second treatment to the other, and then a difference measure is calculated for the effect of treatment, neither age nor health condition would affect this measurement. Ideally, we would like to select pairs that are identical in all characteristics other than the factor whose effects we are attempting to measure. Obviously, as a practical matter this is impossible, but the guiding principle is clear. Once differences are taken between members of each pair, the t test proceeds exactly as in the preceding example. Note that in the weight example, the differences were measured on the same individual, whereas in the present illustration the differences are derived from the two members of each pair.

The method of paired observations is a useful technique. Compared with the standard two-sample t test, in addition to the advantage that we do not have to assume that the two samples are independent, we also need not assume that the variances of the two samples are equal.

7.7 SUMMARY AND A LOOK AHEAD

In this chapter, we have considered some classical hypothesis-testing techniques. These tests represent only a few of the simplest methods. All the cases discussed thus far have involved only one or two samples, but methods are available for testing hypotheses concerning three or more samples. The cases we have dealt with also have tested only one parameter of a probability distribution. However, techniques are available for testing whether an entire frequency distribution is in conformity with a theoretical model, such as a specified probability distribution. Finally, the tests we have considered involved a final decision on the basis of the sample evidence; that is, a decision concerning acceptance or rejection of hypotheses was reached on the basis of the evidence contained in one or two samples. Some of the broader decision procedures are discussed in subsequent chapters.

Although classical hypothesis-testing techniques of the type discussed in this chapter have been widely applied in a great many fields, it would be incorrect to infer that their use is noncontroversial and that they can simply be employed in a mechanical way. At this point, it suffices to indicate that the methods discussed are admittedly incomplete and that Bayesian decision theory addresses itself to the required completion. Thus, for example, in hypothesis testing, establishing significance levels such as 0.05 or 0.01 inevitably appears to be a rather arbitrary procedure, despite the fact that the relative seriousness of Type I and Type II errors is supposed to be considered in designing a test. Although costs of Type I and Type II errors can theoretically be considered in the classical formulation, as a matter of practice, they are rarely included explicitly in the analysis. In Bayesian decision theory, the costs of Type I and Type II errors, as well as the payoffs of correct decisions, are an explicit part of the formal analysis.

Moreover, in classical hypothesis testing, decisions are reached solely on the basis of current sample information without reference to any prior knowledge concerning the hypothesis under test. On the other hand, Bayesian decision theory provides a method for combining prior knowledge with current sample information for decision-making purposes. These Bayesian decision-theory methods are discussed in subsequent chapters.

Exercises 7.7

1. The marketing research department of the Pacific Petroleum Company wished to test the effect of increased radio advertising on the sales of one of the firm's major products, a gasoline additive that increases the efficiency

of automobile engines. The manager in charge of the project decided to increase substantially the radio advertising of the product in 13 metropolitan areas in the Western sales region for a 5-week period. The average weekly sales (in units) of the product for the 5-week period before the project began and for the 5-week trial period in the 13 metropolitan areas are given below:

Metropolitan Area	Average Weekly Product Sales (in thousands of units) Before Trial Period	During Trial Period
Seattle	55	60
Olympia	55	58
Portland	47	48
San Francisco	38	42
Sacramento	30	32
Los Angeles	68	65
San Diego	16	20
Reno	50	53
Las Vegas	23	28
Boise	36	34
Tucson	25	27
Phoenix	23	22
Salt Lake City	23	26

Using the paired t test, would you conclude that increased radio advertising increases sales? Assume that other factors affecting sales are unchanged over the 10-week period, and use a 5% significance level.

2. Ten people were placed on a special diet. The results after 10 weeks were as follows:

Person	Weight in Pounds before Diet	Weight in Pounds after Diet
A	210	209
B	200	194
C	185	184
D	174	174
E	193	191
F	190	190
G	184	180
H	225	225
I	240	237
J	215	212

Using the paired t-test, would you say the diet was successful? Use $\alpha = 0.05$.

3. Ten middle managers from the Mammoth Manufacturing Company attended a 3-week session of group meetings and workshops to increase their managerial effectiveness and sensitivity. The top management of the firm evaluates its middle managers using the "grid" methodology; each manager is given a rating consisting of two numbers between 1 (worst rating) and 9 (best rating). The first number indicates the evaluation of the manager's concern for production, and the second reflects the manager's concern for people. The table shows the ratings given to each manager before and after attending the session:

Manager	Rating before Session	Rating after Session
1	7, 5	8, 6
2	4, 5	6, 8
3	8, 2	8, 4
4	8, 7	7, 7
5	6, 6	6, 9
6	4, 2	7, 4
7	7, 3	8, 7
8	3, 6	6, 7
9	4, 9	7, 8
10	6, 5	4, 5

Assuming consistency in the top management's rating determinations, would you conclude that the session has had a positive effect on the managers' ratings (a) on concern for production? (b) on concern for people? Use a 2.5% significance level.

8

Chi-Square Tests and Analysis of Variance

In Chapter 7, procedures were discussed for testing hypotheses with data obtained from a single simple random sample or from two such samples. For example, we considered tests of whether two population proportions or two population means were equal. Obvious generalizations of such techniques are tests for the equality of more than two proportions or more than two means. The two topics discussed in this chapter supply these generalizations. **Chi-square tests**[1] provide the basis for judging whether more than two population proportions may be considered to be equal; **analysis of variance** techniques provide ways to test whether more than two population means may be considered to be equal.

We discuss χ^2 tests first, considering the topics of (1) goodness of fit and (2) independence. **Tests of goodness of fit** provide a means for deciding whether a particular theoretical probability distribution—such as the binomial distribution—is a close enough approximation to a sample frequency distribution for the population from which the sample was drawn to be described by the theoretical distribution. **Tests of independence** constitute a method for deciding whether the hypothesis of independence between different variables is tenable. This procedure provides a test for the equality of more than two population proportions. Both types of χ^2 tests furnish a conclusion on whether a set of *observed* frequencies differs so greatly from a set of *theoretical* frequencies that the hypothesis under which the theoretical frequencies were derived should be rejected.

Tests of goodness of fit

Tests of Independence

[1] The procedures are referred to as "χ^2 tests" or "chi-square tests," where the symbol χ is the Greek lower-case letter "chi" (pronounced "kye").

8.1

TESTS OF GOODNESS OF FIT

One of the major problems in the application of probability theory, statistics, and mathematical models in general is that the real-world phenomena to which they are applied usually depart somewhat from the assumptions embodied in the theory or models. For example, let us consider use of the binomial probability distribution in a particular problem. As indicated in section 3.4, two of the assumptions involved in the derivation of the binomial distribution are

Assumptions

1. The probability of a success p remains constant from trial to trial.
2. The trials are independent.

We consider whether these assumptions are met in the following problem.

A firm bills its accounts at a 2% discount for payment within 10 days and for the full amount due for payment after 10 days. In the past, 40% of all invoices have been paid within 10 days. In a particular week, the firm sends out 20 invoices. Is the binomial distribution appropriate for computing the probabilities that 0, 1, 2, . . . , 20 firms will receive the discount for payment within 10 days?

Considering the possible use of the binomial distribution, we can let $p = 0.40$ represent the probability that a firm will receive the discount and $n = 20$ firms represent the number of trials. Does it seem reasonable to assume that p, the probability of receiving the discount, is 0.40 for each firm? Past relative frequency data for *each* firm could be brought to bear on this question. In most practical situations, we would probably find that the practices of individual firms vary widely, with some firms nearly always receiving discounts, some firms rarely receiving discounts, and most firms falling somewhere between these two extremes.

Does the assumption of independence seem tenable in this problem? That is, does it seem reasonable that whether one firm receives the discount is independent of whether another firm does? Probably not, since general monetary conditions affect many firms in a similar way. For example, when money is "tight" and it is difficult to acquire adequate amounts of working capital, the fact that one firm does not receive the discount is related to, rather than *independent* of, whether other firms have done so. Moreover, there may be traditional practices in certain industries concerning whether or not discounts are taken. Other factors that would interfere with the independence assumption may also be present.

How great a departure from the assumptions underlying a probability distribution—or, more generally, from the assumptions embodied in any theory or mathematical model—can be tolerated before we should conclude that the distribution, theory, or model is no longer applicable? This complex question cannot be readily answered by any simple universally applicable rule.

The purpose of χ^2 "goodness of fit" tests is to provide one type of answer to this question by comparing *observed frequencies* with *theoretical*, or *expected, frequencies* derived under specified probability distributions or hypotheses.

The sequence of steps in performing goodness of fit tests is similar to previously discussed hypothesis-testing procedures.

Goodness of fit tests

1. Null and alternative hypotheses are established, and a significance level is selected for rejection of the null hypothesis.
2. A random sample of observations is drawn from a relevant statistical population or process.
3. A set of expected, or theoretical, frequencies is derived under the assumption that the null hypothesis is true. This is generally an assumption that a particular probability distribution is applicable to the statistical population under consideration.
4. The observed frequencies are compared with the expected, or theoretical, frequencies.
5. If the aggregate discrepancy between the observed and theoretical frequencies is too great to attribute to chance fluctuations at the selected significance level, the null hypothesis is rejected.

We now illustrate goodness of fits tests and discuss some of the underlying theory for an example involving a uniform probability distribution (see section 3.3).

Coffee-tasting problem

Suppose a consumer research firm wished to determine whether any of 5 brands of coffee was preferred by coffee drinkers in a certain metropolitan area. The firm took a simple random sample of 1,000 coffee drinkers in the area and conducted the following experiment. Each consumer was given 5 cups of coffee, one of each brand (A, B, C, D, and E), without identification of the individual brands. The cups were presented to each consumer in a random order determined by sequential selection from 5 paper slips, each containing one of the letters A, B, C, D, and E. Table 8-1 shows the numbers of coffee drinkers who stated that they liked the indicated brands best.

Denoting the true proportions of preference for each brand as p_A, p_B, p_C, p_D, and p_E, we can state the null and alternative hypotheses as follows:

$$H_0\colon p_A = p_B = p_C = p_D = p_E = 0.20$$

$$H_1\colon \text{The } p \text{ values are not all equal}$$

TABLE 8-1
Number of coffee drinkers in a certain metropolitan area who most preferred the specified brand of coffee

Brand Preference	Number of Consumers
A	210
B	312
C	170
D	85
E	223
	1,000

That is, if in the population from which the sample was drawn there were no differences in preference among the 5 brands, 20% of coffee drinkers would prefer each brand. An equivalent way of stating these hypotheses is

H_0: The probability distribution is uniform

H_1: The probability distribution is not uniform

In other words, we want to know whether the sample of 1,000 coffee drinkers can be considered a random sample from a population in which the proportions who prefer each of the 5 brands are equal. Of course, this hypothesis is only one of many that could conceivably be formulated.

> One of the strengths of the goodness of fit test is that it permits a variety of different hypotheses to be raised and tested.

If the null hypothesis of no difference in preference were true, the *expected* or *theoretical* number of the 1,000 coffee drinkers in the sample who would prefer each brand would be $0.20 \times 1{,}000 = 200$. Hence, the expected frequency corresponding to each of the observed frequencies in Table 8-1 is 200. We can now compare the set of observed frequencies with the set of theoretical frequencies derived under the assumption that the null hypothesis is true. The test statistic we compute to make this comparison is known as chi-square, denoted χ^2. The computed value of χ^2 is

$$\chi^2 = \sum \frac{(f_o - f_t)^2}{f_t} \tag{8.1}$$

where f_o = an **o**bserved frequency
f_t = a **t**heoretical (or expected) frequency

As we can see from equation 8.1, if every observed frequency is exactly equal to the corresponding theoretical frequency, the computed value of χ^2 is

TABLE 8-2
Calculation of the χ^2 statistic for the coffee-testing problem

Brand Preference	(1) Observed Frequency f_o	(2) Theoretical (expected) Frequency f_t	(3) $(f_o - f_t)$	(4) $(f_o - f_t)^2$	(5) $\frac{\text{Column 4}}{\text{Column 2}}$ $\frac{(f_o - f_t)^2}{f_t}$
A	210	200	10	100	0.5
B	312	200	112	12,544	62.7
C	170	200	−30	900	4.5
D	85	200	−115	13,225	66.1
E	223	200	23	529	2.6
Total	1,000	1,000			136.4

$$\chi^2 = \sum \frac{(f_o - f_t)^2}{f_t} = 136.4$$

zero. This is the smallest value χ^2 can have. The larger the discrepancies between the observed and theoretical frequencies, the larger is χ^2.

The computed value of χ^2 is a random variable that takes on different values from sample to sample. That is, χ^2 has a sampling distribution just as do the test statistics discussed in Chapter 7. We wish to answer the following question: Is the computed value of χ^2 so large that we must reject the null hypothesis? In other words, are the aggregate discrepancies between the observed frequencies f_o and theoretical frequencies f_t so large that we are unwilling to attribute them to chance, and have to reject the null hypothesis? The calculation of χ^2 for the present problem is shown in Table 8-2.

The χ^2 Distribution

Before we can answer the questions raised in the preceding paragraph, we must digress for a discussion of the appropriate sampling distribution. Then we will complete the solution to the coffee-tasting problem. It can be shown that for large sample sizes the sampling (probability) distribution of χ^2 can be closely approximated by the χ^2 distribution whose probability function is

$$f(\chi^2) = c(\chi^2)^{(v/2)-1}e^{-\chi^2/2} \tag{8.2}$$

where $e = 2.71828\ldots$
v = number of degrees of freedom
c = a constant depending only on v

The χ^2 distribution has only one parameter v, the number of degrees of freedom. This is similar to the case of the t distribution, discussed in section 6.4.

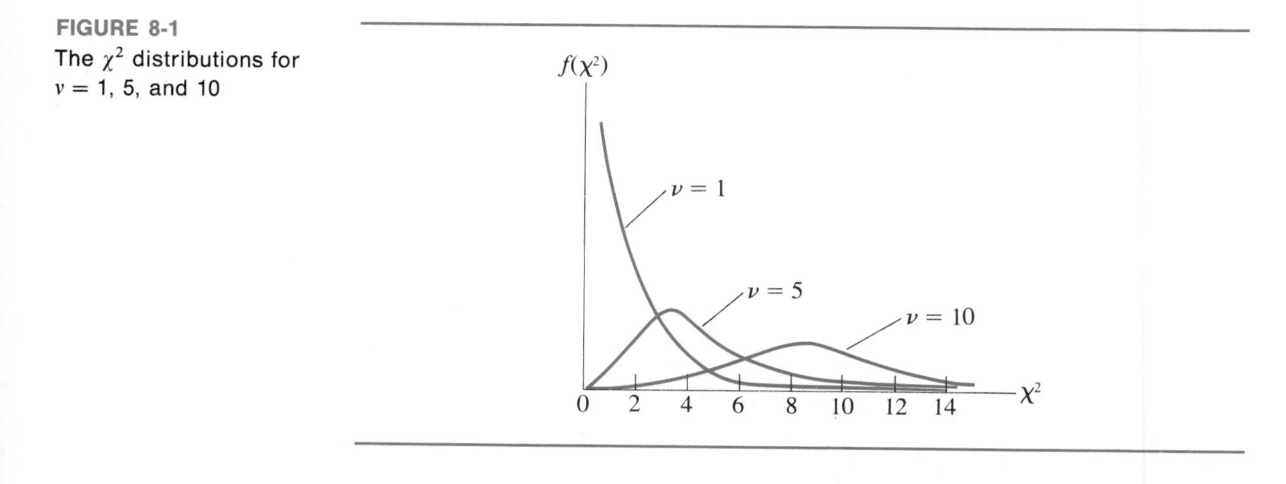

FIGURE 8-1
The χ^2 distributions for $\nu = 1$, 5, and 10

Hence, $f(\chi^2)$ is a family of distributions, one for each value of ν. χ^2 is a continuous random variable greater than or equal to zero. For small values of ν, the distribution is skewed to the right. As ν increases, the distribution rapidly becomes symmetrical. In fact, for large values of ν, the χ^2 distribution is closely approximated by the normal curve. Figure 8-1 depicts the χ^2 distributions for $\nu = 1$, 5, and 10.

Since the χ^2 distribution is a probability distribution, the area under the curve for each value of ν equals one. Because there is a separate distribution for each value of ν, it is not practical to construct a detailed table of areas. Therefore, for compactness, a χ^2 table generally shows the relationship between areas and values of χ^2 for only a few levels of significance in different χ^2 distributions.

Table A-7 of Appendix A shows χ^2 values corresponding to selected areas in the right tail of the χ^2 distribution. These tabulations are shown separately for the number of degrees of freedom listed in the left column. The χ^2 values are shown in the body of the table, and the corresponding areas are shown in the column headings.

As an illustration of the use of the χ^2 table, let us assume a random variable having a χ^2 distribution with $\nu = 8$. In Table A-7, we find a χ^2 value of 15.507 corresponding to an area of 0.05 in the right tail. The relationships described in this illustrative problem are shown in Figure 8-2. Hence, if the random variable has a χ^2 distribution with $\nu = 8$, the probability that χ^2 is greater than 15.507 is 0.05. Let us give the corresponding interpretation in a hypothesis-testing context. If the null hypothesis being tested is true, the probability of observing a χ^2 figure greater than 15.507 because of chance variation is equal to 0.05. Therefore, for example, if the null hypothesis were tested at the 0.05 level of significance and we calculated $\chi^2 = 16$, we would reject the null hypothesis, because so large a χ^2 value would occur less than 5% of the time if the null hypothesis is true. We now turn to a discussion of the rules for determining the number of degrees of freedom involved in a χ^2 test.

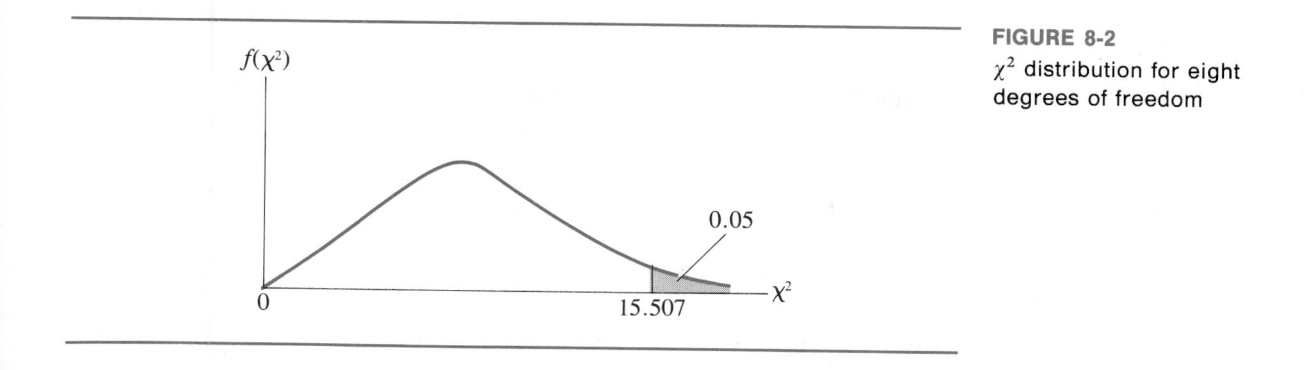

FIGURE 8-2
χ^2 distribution for eight degrees of freedom

Number of Degrees of Freedom

We have seen that a χ^2 goodness of fit test involves a comparison of a set of observed frequencies, denoted f_o, with a set of theoretical frequencies, denoted f_t. Let k be the number of classes for which these comparisons are made. For example, in the coffee-tasting problem $k = 5$, because there are 5 classes for which we computed relative deviations of the form $(f_o - f_t)^2/f_t$. To determine the number of degrees of freedom, we must reduce k by one for each restriction imposed. In the coffee-tasting example, the number of degrees of freedom is equal to $\nu = k - 1 = 5 - 1 = 4$. The rationale for this computation follows.

In the calculations shown in Table 8-2, there are 5 classes for which f_o and f_t values are to be compared. Hence, we start with $k = 5$ degrees of freedom. However, we have forced the total of the theoretical frequencies, Σf_t, to be equal to the total of the observed frequencies, Σf_o; that is, 1,000. Therefore, we have reduced the number of degrees of freedom by one and there are now only 4 degrees of freedom. That is, once the total of the theoretical frequencies is fixed, only 4 of the f_t values may be freely assigned to the classes; when these 4 have been assigned, the fifth class is immediately determined, because the theoretical frequencies must total 1,000.

The number of degrees of freedom is reduced by one for each restriction imposed in the calculation of the theoretical frequencies. An additional degree of freedom is lost for each parameter value that must be estimated from the sample. For instance, if the mean μ of a Poisson distribution must be estimated from the mean $\bar{x}$ of a sample, there would be a reduction of one degree of freedom. In summary, we can state the following rules for determining ν, the number of degrees of freedom in a χ^2 test in which k classes of observed and theoretical frequencies are compared:

Rules for determining ν, the number of degrees of freedom

1. If the only restriction is $\Sigma f_t = \Sigma f_o$, the number of degrees of freedom is $\nu = k - 1$.
2. If, in addition to the previous restriction, m parameters are replaced by sample estimates, the number of degrees of freedom is $\nu = k - 1 - m$.

Decision Procedure

We can now return to the coffee-tasting example to perform the goodness of fit test. Let us assume we wish to test the null hypothesis at the 0.05 level of significance. Since the number of degrees of freedom is 4, we find the critical value of χ^2, which we denote as $\chi^2_{0.05}$, to be 9.488 (Table A-7 of Appendix A). This means that if the null hypothesis is true, the probability of observing a χ^2 value greater than 9.488 is 0.05. Specifically in terms of the coffee problem, this means that an aggregate discrepancy between the observed and theoretical frequencies larger than 9.488 will occur only 5% of the time if there is no difference in preference among brands. We can state the decision rule for this problem in which $\chi^2_{0.05} = 9.488$ as follows:

DECISION RULE

1. If $\chi^2 > 9.488$, reject H_0
2. If $\chi^2 \leqslant 9.488$, accept H_0

Since the computed χ^2 value in this problem is 136.4 and is thus much larger than the critical $\chi^2_{0.05}$ value of 9.488, we reject the null hypothesis. Therefore, we conclude that real differences exist in consumer preference among the brands of coffee involved in the experiment. In statistical terms, we cannot consider the 1,000 coffee drinkers in the experiment to be a simple random sample from a population whose members prefer each of the 5 brands in equal proportions. In terms of goodness of fit, we reject the null hypothesis that the probability distribution is uniform. Hence, we conclude that the uniform distribution is decidedly not a "good fit" to the sample data.

We now turn to other examples of χ^2 goodness of fit tests.

In Example 8-1, the hypothesis that the population probability distribution is Poisson is tested and accepted. A similar procedure could be used to test whether a shift away from a previously established Poisson distribution has occurred. The number of degrees of freedom would be $k - 1$ if μ was estimated from previous results or $k - 1 - 1$ if μ was estimated from the mean $\bar{x}$ of the present sample data.

EXAMPLE 8-1

Jonathan Falk, a management scientist, was developing an inventory control system for a manufacturer of a diversified product line. He wanted to determine whether the Poisson distribution was an appropriate model for the demand for a particular product. He obtained the frequency distribution of the number of units of this product demanded per day for the past 200 business days. That distribution is shown in columns (1) and (2) of Table 8-3(a). The mean number of units per day is shown at the bottom of the table; it is obtained by dividing the total of column (3) by the total of column (2).

TABLE 8-3

(a) Number of units of a particular product demanded per day for the past 200 business days

(1) Number of Units Demanded per Day x	(2) Observed Number of Days f_o	(3) Column 1 × Column 2 $f_o x$
0	11	0
1	28	28
2	43	86
3	47	141
4	32	128
5	28	140
6	7	42
7	0	0
8	2	16
9	1	9
10	1	10
Total	200	600

$$\bar{x} = \frac{600}{200} = 3 \text{ units per day}$$

(b) Theoretical distribution of demand assuming a Poisson distribution

(1) Number of Units Demanded per Day x	(2) Probability $f(x)$	(3) Column 2 × 200 Expected Number of Days f_t
0	0.050	10.0
1	0.149	29.8
2	0.224	44.8
3	0.224	44.8
4	0.168	33.6
5	0.101	20.2
6	0.050	10.0
7	0.022	4.4
8	0.008	1.6
9	0.003	0.6
10	0.001	0.2
Total	1.000	200.0

Using the mean of this sample of observations, $\bar{x} = 3$ units demanded per day, as an estimate of the parameter μ of the corresponding theoretical Poisson distribution, Falk calculated the Poisson probability distribution shown in the first two columns of Table 8-3(b). Multiplying the probabilities in column (2) by 200 days, he obtained the theoretical, or expected, frequencies if the demand were distributed according to the Poisson distribution. These theoretical frequencies are shown in column (3). For example, if the *probability* that zero units would be demanded is 0.050, then in 200 days the *expected number* of days in which zero units would be demanded is $0.050 \times 200 = 10.0$ days, the first entry in column (3). The analyst was now able to apply a χ^2 goodness of fit test using the actual number of days in column (2) of Table 8-3(a) as the observed frequencies f_o and the expected number of days in column (3) of Table 8-3(b) as the theoretical frequencies f_t. These two sets of frequencies are shown in columns (2) and (3) of Table 8-4, where the calculation of the χ^2 value is carried out. The hypothesis under test in this problem may be stated as follows:

H_0: The population probability distribution is Poisson with $\mu = 3$

Assume that we wish to test the null hypothesis at the 0.05 level of significance.

We can see in Table 8-4 that the last 4 classes of Table 8-3 for 7, 8, 9, and 10 units of demand have been combined into one class titled "7 or more." Both f_o and f_t values have been cumulated for the 4 classes, and a single relative deviation of the form $(f_o - f_t)^2/f_t$ has been calculated for the combined class. There are now 8 classes, $k = 8$, in Table 8-4 for which the χ^2 value has been computed and from which the number of

TABLE 8-4
Calculation of the χ^2 statistic for the demand distribution problem

(1) Number of Units Demanded per Day	(2) Observed Number of Days f_o	(3) Theoretical Number of Days f_t	(4) $f_o - f_t$	(5) $(f_o - f_t)^2$	(6) $\frac{(f_o - f_t)^2}{f_t}$
0	11	10.0	1.0	1.00	0.10
1	28	29.8	−1.8	3.24	0.11
2	43	44.8	−1.8	3.24	0.07
3	47	44.8	2.2	4.84	0.11
4	32	33.6	−1.6	2.56	0.08
5	28	20.2	7.8	60.84	3.01
6	7	10.0	−3.0	9.00	0.90
7 or more	4	6.8	−2.8	7.84	1.15
Total	200	200.0	0		5.53

$$\chi^2 = \sum \frac{(f_o - f_t)^2}{f_t} = 5.53$$

degrees of freedom will be determined. The reason for this combination of classes will be explained at the completion of the problem.

Let us now compute the number of degrees of freedom in the test. As indicated in the earlier discussion, the number of degrees of freedom is given by $v = k - 1 - m$, where m is the number of parameters that have been replaced by sample estimates. Since the sample mean $\bar{x}$ was used as the estimate of the parameter μ in the Poisson distribution, $m = 1$. Hence, the number of degrees of freedom is $v = 8 - 1 - 1 = 6$. For 6 degrees of freedom, the critical value of χ^2 at the 0.05 level of significance is $\chi^2_{0.05} = 12.592$ (Table A-7 of Appendix A). Therefore, since the observed χ^2 value of 5.53 is less than 12.592, we accept the null hypothesis.

In other words, the aggregate discrepancy between the observed and theoretical frequencies is sufficiently small for us to *conclude that the Poisson distribution with $\mu = 3$ is a good fit*. Based on this result, the Poisson distribution can reasonably be used as a model for demand for the product.

Rule concerning size of theoretical frequencies

As indicated earlier, for large sample sizes, the probability function of the computed χ^2 values can be closely approximated by the χ^2 distribution given in equation 8.2, which is the distribution of a *continuous* random variable. However, there are only a finite number of possible combinations of f_t values, and hence only a finite number of computed χ^2 values. Thus, a computed χ^2 figure is one value of a *discrete* random variable. If the sample size is large, the approximation of the probability distribution of this discrete random variable by the continuous chi-square distribution will be a good one. This is analogous to approximating the binomial distribution, which is discrete, by the normal curve, which is continuous (see section 5.3).

When the expected frequencies (the f_t values) are small, the approximation discussed in the preceding paragraph is inadequate. A frequently used rule is that each f_t value should be equal to or greater than 5. This is why the classes for 7, 8, 9, and 10 units of demand were combined in Example 8-1 in the computation of the χ^2 value. As shown in Table 8-4, the computed f_t value for the combined class is equal to 6.8, which satisfies the commonly used rule of thumb for a minimum expected frequency.

In Example 8-2, the investigator tested a hypothesis using the χ^2 goodness of fit procedure, and she tested another related hypothesis using a standard one-tailed test for a proportion. The conclusions from these two tests were consistent. However, Example 8-2 illustrates the tentative nature of conclusions drawn from hypothesis-testing procedures. It is conceivable that a different sample could have led to the acceptance of a hypothesis involving $p = 0.522$ when used in the χ^2 test, but if that same sample were used in the hypothesis test of a proportion, it could have led to the acceptance of the hypothesis that $p = 0.50$, even at the same significance level.

EXAMPLE 8-2

Researcher Lana Mauro hypothesized that the determination of sex in human births could be considered a Bernoulli process. However, she suspected that male and female births were not equally likely. Specifically, she believed that large families tended to have more male children than female. She had data for a simple random sample of 320 families with 5 children each, which had been drawn for another purpose, so she decided to conduct a partial test of her theory using these data. The frequency distribution of male children in these 320 families is shown in Table 8-5.

Since Mauro hypothesized that human births could be treated as a Bernoulli process, she decided to fit a binomial distribution to the data. She further decided to

TABLE 8-5
Calculation of $\bar{p}$, the proportion of male children in the sample of 320 families

(1) Number of Male Children x	(2) Observed Number of Families f_o	(3) Column 1 × Column 2 $f_o x$
0	12	0
1	42	42
2	92	184
3	108	324
4	46	184
5	20	100
Total	320	834

$\bar{x} = \frac{834}{320} = 2.61$ male children per family

$\bar{p} = \frac{2.61}{5} = 0.522$

estimate p, the probability of a male birth, by using $\bar{p}$, the proportion of male births in the sample. As indicated in Table 8-5, she calculated $\bar{x}$, the mean number of male children per family, and divided that figure by 5. These computations are shown in columns 1, 2, and 3 and at the bottom of Table 8-5. As shown, the sample proportion of male children was $\bar{p} = 0.522$ for the sample of 320 families. Therefore, the investigator stated the null hypothesis as follows:

H_0: The population probability distribution is binomial with $p = 0.522$

Since 0.05 was the conventional level of significance ordinarily used by other researchers in her field, Mauro decided to use that level of significance in testing the hypothesis. Letting

$$p = \text{the probability of a male birth} = 0.522$$

$$q = \text{the probability of a female birth} = 0.478$$

and

$$n = \text{number of children per family} = 5$$

she computed the binomial probability distribution given in columns (1) and (2) of Table 8-6(a). Multiplying the probabilities in column (2) by 320 families, she determined the expected number of families with 0, 1, 2, 3, 4, and 5 male children each. These theoretical

TABLE 8-6

(a) Calculation of expected frequencies of number of male children: Assumed binomial distribution with $p = 0.522$ and $n = 5$

(1) Number of Male Children x	(2) Probability[a] $f(x)$	(3) Column 2 × 320 Expected Number of Families f_t
0	0.025	8.00
1	0.136	43.52
2	0.298	95.36
3	0.324	103.68
4	0.178	56.96
5	0.039	12.48
	1.000	320.00

(b) Calculation of the χ^2 statistic for the number of male children problem: Fit of binomial distribution with $p = 0.522$ and $n = 5$

(1) Number of Male Children	(2) Observed Number of Families f_o	(3) Theoretical Number of Families f_t	(4) $f_o - f_t$	(5) $(f_o - f_t)^2$	(6) $\frac{(f_o - f_t)^2}{f_t}$
0	12	8.00	4.00	16.00	2.00
1	42	43.52	−1.52	2.31	0.05
2	92	95.36	−3.36	11.29	0.12
3	108	103.68	4.32	18.66	0.18
4	46	56.96	−10.96	120.12	2.11
5	20	12.48	7.52	56.55	4.53
Total	320	320.00	0		8.99

$$\chi^2 = \sum \frac{(f_o - f_t)^2}{f_t} = 8.99$$

[a] These probabilities can only be approximated by using Tables A-1 and A-2 in Appendix A because those tables give probabilities only for p values of 0.05, 0.10, 0.15, and so on.

frequencies are shown in column (3) of Table 8-6(a). She then proceeded with the χ^2 test of goodness of fit using the observed frequencies in column (2) of Table 8-5 as the f_o values and the expected frequencies in column (3) of Table 8-6(a) as f_t. Table 8-6(b) shows the calculation of the χ^2 value, 8.99.

The number of classes is $k = 6$, and $m = 1$ because one sample estimate $\bar{p}$ was used to replace a population parameter p. Therefore, the number of degrees of freedom is $v = k - 1 - m = 6 - 1 - 1 = 4$. For $v = 4$, the critical χ^2 value is $\chi^2_{0.05} = 9.488$. Since the computed χ^2 value was only 8.99, the null hypothesis that the population probability distribution is binomial with $p = 0.522$ was accepted. Hence, Mauro concluded that the evidence represented by the observed frequency distribution on the number of male children in the sample families was consistent with the hypothesis of the operation of a Bernoulli process with $p = 0.522$ for births.

As a further test of her belief that large families have more male than female children, the investigator carried out a one-tailed test of the null hypothesis H_0: $p = 0.50$ males against the alternative hypothesis H_1: $p > 0.50$ males at the 0.05 level of significance. The sample statistic $\bar{p} = 0.522$ males for a sample of 1,600 births (5 children per family times 320 families) differed significantly from $p = 0.50$. She was pleased to observe this additional evidence in favor of her belief that male births tended to occur more frequently than female births in large families.

There is a subtle point involved in the example.

Under classical hypothesis-testing procedures, the hypothesis should be set up before the data are gathered.

However, note that in the null hypothesis of the χ^2 test (H_0: The population probability distribution is binomial with $p = 0.522$), the hypothesized value of p was derived from the sample statistic $\bar{p}$.

In practice, many hypotheses are tested after sample evidence is obtained, sometimes by persons who controlled the collection and tabulation of the data, and sometimes by others.

Classical statistics does not provide separate techniques for testing hypotheses before and after sample evidence has been collected. However, Bayesian decision theory, discussed in Chapters 14 through 17, provides different techniques for decision making prior to obtaining sample data and for the incorporation of sample information with prior knowledge.

Exercises 8.1

1. The manager of a large industrial plant, wishing to determine whether the number of employees absent from work is related to the day of the week, has obtained data for one year of absence days.

Suppose that the manager wishes to test the hypothesis that the number of employees absent does not depend on the day of the week. At $\alpha = 0.05$, what conclusion should he reach?

Day of week	M	T	W	Th	F
Total absences for the year	500	285	255	323	602

2. In an investigation of trust departments of commercial banks in the United States, a simple random sample of portfolios containing investments in 6 different stocks was taken. The portfolios were examined to determine how many of the stocks declined in price during the past year. The results are listed in a table.

   Determine whether the binomial distribution is a "good fit" to these data. Assume that the probability of a particular stock's declining in price during the past year is 0.50. Assume a 0.05 level of significance.

Number of stocks declining	0	1	2	3	4	5	6
Number of portfolios	25	205	483	583	500	168	36
Total number of portfolios = 2,000							

3. An automobile manufacturer wishes to test for consumer preferences among 5 new models. By a simple random selection process, 500 consumers were selected. The table shows the numbers of consumers who preferred each model.

   Using a chi-square test, determine whether the null hypothesis "H_0: The probability distribution is uniform" should be rejected. Test at both the 0.05 and 0.01 levels of significance.

Preferred model	A	B	C	D	E
Number of consumers	225	185	230	187	173

4. Over a 100-day period, the number of telephone calls coming into the switchboard of a company during a certain minute between 2:00 P.M. and 4:00 P.M. were tabulated.

   Fit a Poisson distribution to the data, and use a chi-square test to determine the "goodness of fit." Use a 0.01 level of significance.

Number of calls	0	1	2	3	4	5	6	7	8
Observed number of days	5	7	30	40	7	4	5	1	1

5. A set of 5 coins was tossed 1,000 times. The number of times that 0, 1, 2, 3, 4, and 5 heads were obtained is shown in the table.

   Determine whether the binomial distribution is a good fit to these data. Assume the probability of a head is $\frac{1}{2}$ and use a 0.02 level of significance.

Number of heads	0	1	2	3	4	5
Number of tosses	36	138	348	287	165	26
Total number of tosses: 1,000						

6. A consumer research firm wished to determine whether drinkers of iced tea in a certain metropolitan area had a real difference in taste preference among 5 brands of iced tea. The firm took a simple random sample of 100 iced-tea drinkers in the area and conducted the following experiment: Five glasses of iced tea, each containing one of the 5 brands, were marked A, B, C, D, and E. The 5 glasses were presented to each consumer in a random order determined by sequential selection from 5 paper slips, each slip containing one of the letters A, B, C, D, and E. The table shows the number of iced-tea customers who stated that they most preferred each indicated brand.

   Use a χ^2 test to determine whether the null hypothesis

   $$H_0\text{: the probability distribution is uniform}$$

   should be rejected. Do the test using both a 0.05 and a 0.01 level of significance.

Preferred brand	A	B	C	D	E
Number of drinkers	27	16	22	18	17

8.2 TESTS OF INDEPENDENCE

Another important application of the χ^2 distribution is in testing for the independence of two variables on the basis of sample data. The general nature of the test is best explained with a specific example.

TABLE 8-7
A simple random sample of 10,000 families classified by number of automobiles and telephones owned

Number of Telephones Owned	Number of Automobiles Owned (A_1) Zero	(A_2) One	(A_3) Two	Total
(B_1) Zero	1,000	900	100	2,000
(B_2) One	1,500	2,600	500	4,600
(B_3) Two or More	500	2,500	400	3,400
Total	3,000	6,000	1,000	10,000

Automobile-ownership problem

In an investigation of the socioeconomic characteristics of the families in a certain city, a market research firm wished to determine whether the number of telephones owned was independent of the number of automobiles owned. The firm obtained this ownership information from a simple random sample of 10,000 families who lived in the city.

$r \times c$ contingency table

The results are shown in Table 8-7. This type of table, which has one basis of classification vertically across the rows (in this case, number of telephones owned) and another basis of classification horizontally across the columns (in this case, number of automobiles owned), is known as a **contingency table**. If the table has three rows and three columns, as Table 8-7 has, it is called a **three-by-three** (often written 3×3) **contingency table**. In general, in an $r \times c$ contingency table, where r denotes the number of rows and c denotes the number of columns, there are $r \times c$ **cells**. For example, in the 3×3 table under discussion, there are $3 \times 3 = 9$ cells with observed frequencies. In a 3×2 table, there are $3 \times 2 = 6$ cells, and so on.

The χ^2 test consists of calculating expected frequencies under the hypothesis of independence and comparing the observed and expected frequencies.

The competing hypotheses under test in this problem may be stated as follows:

H_0: The number of automobiles owned is independent of the number of telephones owned

H_1: The number of automobiles owned is not independent of the number of telephones owned

Calculation of Theoretical (Expected) Frequencies

Since we are interested in determining whether the hypothesis of independence is tenable, we calculate the theoretical, or expected, frequencies by assuming that the null hypothesis is true. We observe from the marginal totals in the last column of Table 8-7 that $\frac{2,000}{10,000}$, or 20%, of the families do not own telephones. If the null hypothesis H_0 is true—that is, if ownership of automobiles is independent of ownership of telephones—then 20% of the 3,000 families owning no automobiles, 20% of the 6,000 families owning one automobile, and 20% of the 1,000 families owning 2 automobiles would be expected to have no telephones.

Thus, the expected number of "no-car" families who do not own telephones is

$$\frac{2,000}{10,000} \times 3,000 = 600$$

This *expected* frequency corresponds to 1,000, the *observed* number of "no-car" families who do not own telephones.

Similarly, the expected number of "one-car" families who do not own telephones is

$$\frac{2,000}{10,000} \times 6,000 = 1,200$$

This figure corresponds to the 900 shown in the first row.

In general, the theoretical or expected frequency for a cell in the ith row and jth column is calculated as follows:

$$(f_t)_{ij} = \frac{(\Sigma \text{ row } i)(\Sigma \text{ column } j)}{\text{Grand total}} \tag{8.3}$$

where $(f_t)_{ij}$ = the theoretical (expected) frequency for a cell in the ith row and jth column

Σ row i = the total of the frequencies in the ith row

Σ column j = the total of the frequencies in the jth column

grand total = the total of all of the frequencies in the table

For example, the theoretical frequency in the first row and first column of Table 8-8 (whose rationale of calculation was just explained) is computed by

TABLE 8-8
Expected frequencies for the problem on the relationship between telephone and automobile ownership

Number of Telephones Owned	Number of Automobiles Owned			
	Zero	One	Two	Total
Zero	600	1,200	200	2,000
One	1,380	2,760	460	4,600
Two or More	1,020	2,040	340	3,400
Total	3,000	6,000	1,000	10,000

equation 8.3 as

$$(f_t)_{11} = \frac{(2{,}000)(3{,}000)}{10{,}000} = 600$$

In order to keep the notation uncluttered, we will drop the subscripts denoting rows and columns for f_t values in the subsequent discussion.

The expected frequencies for the present problem are shown in Table 8-8. Because of the method of calculating the expected frequencies, the totals in the margins of the table are the same as the totals in the margins of the table of observed frequencies (Table 8-7). Note that the method of computing the expected frequencies under the null hypothesis of independence is simply an application of the multiplication rule for independent events given in equation 2.10. For example, in Table 8-7, the "no-car" and "no-telephone" categories have been denoted A_1 and B_1, respectively. Under independence, $P(A_1 \text{ and } B_1) = P(A_1)P(B_1)$. The marginal probabilities $P(A_1)$ and $P(B_1)$ are given by

$$P(A_1) = \frac{3{,}000}{10{,}000} = 0.30$$

$$P(B_1) = \frac{2{,}000}{10{,}000} = 0.20$$

$$P(A_1 \text{ and } B_1) = P(A_1)P(B_1) = (0.30)(0.20) = 0.06$$

Multiplying this joint probability by the total frequency (10,000), we obtain the expected frequency previously derived for the upper left cell.

$$0.06 \times 10{,}000 = 600$$

The χ^2 Test

How great a departure from the theoretical frequencies under the assumption of independence can be tolerated before we reject the hypothesis of independence? The purpose of the χ^2 test is to provide an answer to this question by comparing observed frequencies with the theoretical, or expected, frequencies derived under the hypothesis of independence. The test statistic used to make this comparison is the chi-square statistic, $\chi^2 = \Sigma(f_o - f_t)^2/f_t$, as defined in equation 8.1, where f_o is an observed frequency and f_t is a theoretical frequency.

Number of Degrees of Freedom

The number of degrees of freedom in the contingency table must be determined in order to apply the χ^2 test. The number of degrees of freedom in a 3×3 contingency table is 4, calculated as follows. In determining the expected frequencies, we used the marginal row and column totals. With 3 rows, only 2 row totals are "free," since the row totals must sum to Σf_o, which is 10,000 in the present illustration. Correspondingly, with 3 columns, 2 column totals are "free." This gives us the freedom to specify 4 cell totals, where the "free" columns and "free" rows intersect.

In general, in a contingency table containing r rows and c columns, there are $(r - 1)(c - 1)$ degrees of freedom.

Thus, in a 2×2 table, $v = (2 - 1)(2 - 1) = 1$; in a 3×2 table, $v = (3 - 1)(2 - 1) = 2$; in a 3×3 table, $v = (3 - 1)(3 - 1) = 4$.

We now return to our example to perform the χ^2 test of independence. Again denoting the observed frequencies as f_o and the expected frequencies as f_t, we have shown the calculation of the χ^2 statistic in Table 8-9. No cell designations are indicated, but of course every f_o value is compared with the corresponding f_t figure. As shown at the bottom of the table, the computed value of χ^2 is equal to 794.3. The number of degrees of freedom is $(r - 1)(c - 1)$ or $(3 - 1)(3 - 1) = 4$. In Table A-7 of Appendix A, we find a critical value at the 0.01 level of significance of $\chi^2_{0.01} = 13.277$. This means that if the null hypothesis is true, the probability of observing a χ^2 value greater than 13.277 is 0.01. Specifically in terms of the problem, this means that if ownership of telephones was independent of ownership of automobiles, an aggregate discrepancy between the observed and theoretical frequencies larger than a χ^2 value of 13.277 would occur only 1% of the time. We can state the decision rule for this problem

TABLE 8-9
Calculation of the χ^2 statistic for the telephone and automobile ownership problem

	Observed Number of Families f_o	Expected Number of Families f_t	$f_o - f_t$	$(f_o - f_t)^2$	$\frac{(f_o - f_t)^2}{f_t}$
	1,000	600	400	160,000	266.7
	1,500	1,380	120	14,400	10.4
	500	1,020	−520	270,400	265.1
	900	1,200	−300	90,000	75.0
	2,600	2,760	−160	25,600	9.3
	2,500	2,040	460	211,600	103.7
	100	200	−100	10,000	50.0
	500	460	40	1,600	3.5
	400	340	60	3,600	10.6
Total	10,000	10,000	0		$\chi^2 = 794.3$

as follows:

DECISION RULE

1. If $\chi^2 > 13.277$, reject H_0
2. If $\chi^2 \leqslant 13.277$, accept H_0

Since the computed χ^2 value of 794.3 so greatly exceeds this critical value, the null hypothesis of independence between telephone and automobile ownership is rejected.

Further Comments

We have seen how the χ^2 test for independence in contingency tables is a means of determining whether a relationship exists between two bases of classification, or in other words, whether a relationship exists between two variables. Although this type of tabulation provides a basis for testing whether there is a dependence between the two classificatory variables, it does not yield a method for estimating the values of one variable from known values or assumed values of the other. In the next chapter, which deals with regression and correlation analysis, methods for providing such estimates are discussed. For example, **regression analysis** provides a method for estimating or predicting the number of telephones owned by a family with a specific number of automobiles. Regression analysis, in particular, provides a powerful tool for stating in explicit mathematical form the nature of the relationship that exists between two or more variables.

However, we may obtain at least some indication of the nature of the relationship between the two variables in a contingency table. Equivalently to the null hypothesis of independence rejected in our example, we have rejected the null hypothesis H_0: $p_1 = p_2 = p_3$, where p_1, p_2, and p_3 denote the population proportions of zero-, one-, and two-car families who do not have telephones. Reference to Table 8-8 makes it obvious why the null hypothesis was rejected. Of the 3,000 families who did not own automobiles, $\frac{1,000}{3,000} = 0.33$ did not own a telephone. Let $\bar{p}_1 = 0.33$. The corresponding proportions of one- and two-car families who did not own telephones were $\bar{p}_2 = \frac{900}{6,000} = 0.15$ and $\bar{p}_3 = \frac{100}{1,000} = 0.10$. Hence, we have concluded that it is highly unlikely that these three statistics represent samples drawn from populations that have the same proportions $(p_1 = p_2 = p_3)$. Clearly, the proportion of no-telephone families declines as automobile ownership increases. The data suggest a strong relationship between the ownership of telephones and automobiles for the families studied.

A powerful generalization develops from the preceding discussion. It can be shown that a χ^2 test applied to a 2×2 contingency table is algebraically identical to the two-sample test for difference between proportions by the methods of section 7.3 using equation 7.10 to calculate the estimated standard error of the difference. This means that the test of the hypothesis of independence carried out in a χ^2 test for a 2×2 contingency table is identical to the testing of the following hypotheses:

$$H_0: p_1 = p_2$$

$$H_1: p_1 \neq p_2$$

As we have seen, in our illustrative problem involving a 3×3 contingency table, we tested the null hypothesis

$$H_0: p_1 = p_2 = p_3$$

against the alternative that the p values were not all equal. The analogous test can be applied in general to c categories, where $c \geqslant 2$.

Additional Comments

Since the sampling distribution of the χ^2 statistic, $\chi^2 = \Sigma(f_o - f_t)^2/f_t$, is only an approximation to the theoretical distribution defined in equation 8.2, the sample size must be large to yield a good approximation. As in the goodness of fit tests, in contingency tables, cells with expected frequencies of less than five should be combined.

Furthermore, in 2×2 tables (that is, when there is one degree of freedom), an adjustment known as **Yates' correction for continuity** may be used. We introduce this correction because the theoretical χ^2 distribution is continuous, whereas the tabulated values in Table A-7 of Appendix A are based on the distribution of the discrete χ^2 statistic of equation 8.1. We apply the correction by computing the following χ^2 statistic:

(8.4)
$$\chi^2 = \sum \frac{(|f_o - f_t| - \frac{1}{2})^2}{f_t}$$

In this correction, $\frac{1}{2}$ is subtracted from the absolute value of the difference between f_o and f_t before squaring. The effect is to reduce the calculated value of χ^2 compared with the corresponding calculation by equation 8.1 without the correction.[2] In an example such as the one just discussed, where the expected frequencies are large, the effect of this correction is clearly unimportant, but it may be of greater significance for smaller samples.

We have seen that in both χ^2 goodness of fit tests and tests of independence, the null hypothesis is rejected when large enough values of χ^2 are observed. Some investigators have raised the question whether the null hypothesis

[2] For a more complete discussion of Yates' correction, see F. Yates, "Contingency Tables Involving Small Numbers and the χ^2 Test." Suppl. *J. Royal Stat. Soc.*, 1, 1934, 217–235, and Snedecor, George W. and William G. Cochran, *Statistical Methods*, 6th ed., Iowa State University Press, 1967.

should also be rejected when the computed value of χ^2 is too low, that is, too close to zero. This is a situation in which the observed frequencies f_o appear to *agree too well* with the theoretical frequencies f_t. The recommended course of action is to examine the data closely to see whether errors have been made in recording them. Perhaps the data rather than the null hypothesis should be rejected. One researcher's experience is relevant to this point. He was analyzing some data on oral temperatures and found that a disturbingly large number of the recorded temperatures were equal to the "normal" figure of 98.6°F. He suspected that these data were "too good to be true." Upon investigation, he found that the temperatures were recorded by relatively untrained nurses' aides. Several of them had misread temperatures by recording the number to which the arrow on the thermometer pointed, namely, 98.6°! Clearly, this was a case in which the data, rather than an investigator's null hypothesis, should be rejected.

8.3 SAMPLE CHI-SQUARE PROBLEM AND COMPUTER OUTPUT

Since the calculations for chi-square problems are repetitive, such problems are particularly appropriate for computer solution. Many canned programs designed to be used at computer terminals calculate the value of χ^2. One such program in the BASIC programming language is illustrated in Example 8-3.

EXAMPLE 8-3

A subscription service stated that preferences for different national magazines were independent of geographical location. A survey was taken in which 300 persons randomly chosen from 3 areas were given a choice among 3 different magazines. Each person expressed a preference and the results were tabulated by region. On the basis of these data, would you agree with the subscription service's assertion? Use a 0.05 level of significance.

	Magazine			
Region	X	Y	Z	Total
New England	75	50	175	300
Northeastern	120	85	95	300
Southern	105	110	85	300
Total	300	245	355	900

Figure 8-3 shows the computer output for this problem.

FIGURE 8-3
Computer output for Example 8-3

```
901 DATA 3,3,75,50,175,120,85,95,105,110,85
XEQ

75
50
175

120
85
95

105
110
85

CHI SQUARE EQUALS        73.871659 ON        4 DEGREES OF FREEDOM.
**        331 OUT OF DATA.
```

Interpretation of Computer Output

The computer output in Figure 8-3 begins with the following line

901 DATA 3,3,75,50,175,120,85,95,105,110,85

- The "_" is the computer prompt that asks for instructions. The analyst supplies the remainder of this line.
- The "901" is a line number referring to the program itself. In this particular program the number "901" must precede a data statement.
- The first two numbers following the word "DATA" (in this case, "3,3") tell the computer the number of rows and columns in the table containing the data.
- The numbers that follow (75,50, . . . ,110,85) are the data, entered by rows, omitting totals.
- The computer then responds by asking "_" (indicating a need for further instructions), to which the analyst answers "X EQ" (telling the computer to execute the chi-square program).

This particular version of the program prints the data once again as a check before computing the chi-square value. With 3 rows and 3 columns there are $(3 - 1)(3 - 1) = 4$ degrees of freedom. For 4 degrees of freedom and a 5% risk of a Type I error, the critical value of chi-square is 9.488 (see Appendix Table A-7).

Since 73.871659, the calculated value of chi-square, is greater than 9.488, the null hypothesis of independence must be rejected. We would disagree with the subscription service's statement that preferences for the 3 different national magazines were independent of geographical location. The "** 331 OUT OF DATA" indicates the end of the computer output.

Exercises 8.3

1. Components are supplied to a television manufacturer by 2 subcontractors. Each component is tested with respect to 5 characteristics before it is accepted by the manufacturer. Records have been kept for one month on the number of different types of defects for each of the 2 suppliers. From these records, a table has been constructed. Based on the recorded data, would you conclude that type of defect and supplier are independent? Use $\alpha = 0.01$.

	Type of Defect					
Supplier	A	B	C	D	E	Total
1	70	10	10	30	0	120
2	10	10	20	20	20	80
Total	80	20	30	50	20	200

2. A federal investigator from the Justice Department has been advised by the personnel director of a large corporation that the firm does not discriminate against women or minority men in its hiring practices. Using a random sample from the firm's personnel records, the investigator compiled a table. The figures reflect the number of applicants for employment during the past year who were hired or rejected in each classification shown.

   Do these data support the hypothesis that the proportions hired were the same for the 3 categories of applicants? Use a 0.05 level of significance.

	Males			
	Caucasian	Minority	Females	Total
Number hired	175	25	100	300
Number rejected	275	125	300	700
Total	450	150	400	1,000

3. The subcommittee on social welfare in a state legislature conducted research to determine whether income level was related to opinion on welfare expenditures. An aide took a poll of 600 people from each of 3 income classes and noted their opinions on the issue of increased welfare expenditures.

   Based on the results of the poll, test the hypothesis that income level and opinion on welfare expenditures are independent. Use a 0.05 level of significance.

Income Level	Number Who Thought Spending Was: Too Low	Satisfactory	Too High	Total
Above $30,000	180	210	210	600
$15,000 to $30,000	200	220	180	600
Below $15,000	195	245	160	600
Total	575	675	550	1,800

4. The director of placement at a prestigious private university wished to know if the starting salaries of recent graduates were independent of their grades in school. An assistant obtained the grades of all graduates with degrees for a recent year.

   Using the data listed, test the hypothesis that starting salary and grade point average are independent. Use a 0.01 and a 0.05 significance level.

Range of Starting Salaries	Range of Grade Point Averages Under 2.0	2.0 to 3.0	Over 3.0	Total
$26,000 and above	20	45	35	100
$23,000 and under $26,000	40	90	70	200
Below $23,000	15	65	70	150
Total	75	200	175	450

5. The PTS Corporation has test-marketed a new automobile gasoline additive that is supposed to increase gas mileage. Simple random samples of automobile owners were taken in 3 cities. Consumers were asked to try the product for one month and then were asked, "Would you purchase this product?" (See tabulated data on next page.)

   From the data based on the answers of 600 automobile owners in 3 cities, would you conclude that respondents' preferences are independent of city location? Use $\alpha = 0.02$.

6

Test Results	Boston	Atlanta	Minneapolis	Total
Will purchase	100	160	190	450
Will not purchase	50	40	60	150
Total	150	200	250	600

6. A firm is testing to determine if a buyer's preference for flavor of ice cream is independent of that person's sex, in each of the following 2 situations, where the sample results have been summarized. Find the degrees of freedom and the critical value of the test statistic. Then indicate whether the null hypothesis of independence should be accepted or rejected.
   a. Vanilla, strawberry, and chocolate are considered ($\chi^2 = 3.72$ and $\alpha = 0.05$).
   b. Pistachio, banana, caramel pecan, and chocolate mint chip are considered ($\chi^2 = 20.582$ and $\alpha = 0.01$).
7. A research organization obtained data concerning a sample of 200 stock investments made by each of 4 unusually successful mutual funds in a year when the stock market rose sharply. In the 4 funds, the number of stocks that suffered losses were 6, 8, 9, and 12. Test whether the proportions of stocks suffering losses are the same for each of these 4 mutual funds. Use a 0.05 significance level.

8.4 ANALYSIS OF VARIANCE: TESTS FOR EQUALITY OF SEVERAL MEANS

In section 8.2, we saw that

> the χ^2 test provides a generalization of the two-sample test for proportions

and enables us to test for the significance of the difference among c ($c > 2$) sample *proportions.* Conceptually, this represents a test of whether the c samples can be treated as having been drawn from the same population or, in other words, from populations having the same proportions.

In this section, we consider an ingenious technique known as the analysis of variance.

> The analysis of variance is a generalization of the two-sample test for means

and enables us to test for the significance of the difference among c $(c > 2)$ sample *means*. As with the χ^2 test, this technique represents a test of whether the c samples can be treated as having been drawn from the same population or, more precisely, from populations having the same means.

Analysis of variance

The **analysis of variance** technique uses sample information to determine whether three or more treatments yield different results. The term "*treatment*" (conventionally used in the analysis of variance) derives from work in the field of agricultural experimentation—in which the treatments may be different types of fertilizer applied to plots of farm land, different types of feeding methods for animals, and so forth. We will use the conventional terminology. In the example that follows, the treatments are different teaching methods.

A central point is that the analysis of variance—literally a technique that analyzes or tests variances—provides us with a test for the significance of the difference among *means*. The rationale by which a test of variances is, in fact, a test for means will be explained shortly.

Teaching methods problem

As an example, suppose we wish to test whether 3 methods of teaching a basic statistics course differ in effectiveness.

Method 1. The lecturer neither works out nor assigns problems.

Method 2. The lecturer works out and assigns problems.

Method 3. The lecturer works out and assigns problems. Students are also required to construct and solve their own problems.

The same professor teaches 3 different sections of students, using one of the 3 methods in each class. All of the students are sophomores at the same university and are randomly assigned to the 3 sections. There are only 12 students in the experiment, 4 in each of the 3 different sections.

In practice, a substantially larger number of observations would be required to furnish convincing results; however, limiting the group to 12 permits us to examine the principles of the analysis without cumbersome computational detail.

It was agreed that student grades on a final examination covering the work of the entire course would be used as the measure of effectiveness. The final examination was graded using 25 as the maximum score and zero as the minimum score. The final examination grades of the 12 students in the 3 sections are given in Table 8-10. As shown in the table, the mean grades for students taught by methods (1), (2), and (3) were 17, 20, and 23, respectively, and the overall average of the 12 students, referred to as the "grand mean," was 20. (Note that the grand mean of 20 is the same figure that would be obtained by adding up all 12 grades and dividing by 12.)

TABLE 8-10

(a) Final examination grades of 12 students taught by 3 different methods

(b) Notation corresponding to the data listed in part (a)

Student	Teaching Method 1	2	3	i	X_{i1}	X_{i2}	X_{i3}
1	16	19	24	1	X_{11}	X_{12}	X_{13}
2	21	20	21	2	X_{21}	X_{22}	X_{23}
3	18	21	22	3	X_{31}	X_{32}	X_{33}
4	13	20	25	4	X_{41}	X_{42}	X_{43}
Total	68	80	92	Total	$\sum_i X_{i1}$	$\sum_i X_{i2}$	$\sum_i X_{i3}$
Mean	17	20	23	Mean	$\bar{X}_1$	$\bar{X}_2$	$\bar{X}_3$

Grand mean $= \dfrac{17 + 20 + 23}{3} = 20$ $\qquad \bar{\bar{X}} = \dfrac{\bar{X}_1 + \bar{X}_2 + \bar{X}_3}{3}$

Notation

At this point, we introduce some useful notation. In Table 8-10(a), there are 4 rows and 3 columns. As in the discussion of χ^2 tests for contingency tables, r represents the number of rows and c represents the number of columns. There is a total of $r \times c$ observations in the table, in this case $4 \times 3 = 12$. Let X_{ij} be the score of the ith student taught by the jth method (treatment), where $i = 1$, 2, 3, 4 and $j = 1, 2, 3$. (Thus, for example, X_{12} denotes the score of student 1 taught by method 2 and is equal to 19; $X_{23} = 21$, and so on.) In this problem, the 3 different methods of instruction are indicated in the columns of the table, and interest centers on the differences among the scores in the 3 columns. This is typical of the so-called **one-factor** (or **one-way**) **analysis of variance**, in which an attempt is made to assess the effect of only one factor (in this case, instructional method) on the observations. In the present problem, there are 3 columns. Hence, we denote the *values* in the columns as X_{i1}, X_{i2}, and X_{i3}, and the *totals* of these columns are denoted as

$$\sum_i X_{i1}, \quad \sum_i X_{i2}, \quad \text{and} \quad \sum_i X_{i3}.$$

The subscript i under the summation signs indicates that the total of each of the columns is obtained by summing the entries over the row. Adopting a simplified notation, we will refer to the means of the 3 columns as $\bar{X}_1$, $\bar{X}_2$, and $\bar{X}_3$, or in general, $\bar{X}_j$. Finally, we denote the *grand mean* as $\bar{\bar{X}}$ (pronounced "X double-bar"), where $\bar{\bar{X}}$ is the mean of all $r \times c$ observations. Since each column in our example contains the same number of observations, $\bar{\bar{X}}$ can be obtained by taking the mean of the 3 sample means $\bar{X}_1$, $\bar{X}_2$, and $\bar{X}_3$. This notation is summarized in Table 8-10(b) so that you can compare the notation with the corresponding entries in Table 8-10(a).

The Hypothesis to Be Tested

As indicated earlier, we want to test whether the 3 methods of teaching a basic statistics course differ in effectiveness. We have calculated the following mean final examination scores of students taught by the 3 methods, $\bar{X}_1 = 17$, $\bar{X}_2 = 20$, and $\bar{X}_3 = 23$. The statistical question is: Can the three samples represented by these 3 means be considered as having been drawn from populations having the same mean? Denoting the population means corresponding to $\bar{X}_1$, $\bar{X}_2$, and $\bar{X}_3$ as μ_1, μ_2, and μ_3, respectively, we can state the null hypothesis as

$$H_0: \mu_1 = \mu_2 = \mu_3$$

This hypothesis is to be tested against the alternative hypothesis

$$H_1: \text{The means } \mu_1, \mu_2, \text{ and } \mu_3 \text{ are not all equal.}$$

What we wish to determine is whether the differences among the sample means $\bar{X}_1$, $\bar{X}_2$, and $\bar{X}_3$ are too great to be attributed to the chance errors of drawing samples from populations having the same means. If we decide that the sample means differ significantly, our substantive conclusion is that the teaching methods differ in effectiveness.

Although we will specify the assumptions underlying the test procedure at the end of the problem, we indicate one of them of this point, namely, the assumption that the variances of the 3 populations are all equal.

Decomposition of Total Variation

Before discussing the procedures involved in the analysis of variance, we consider the general rationale underlying the test.

> If the null hypothesis that the 3 population means (μ_1, μ_2, and μ_3) are equal is true, then both the variation among the sample means ($\bar{X}_1$, $\bar{X}_2$, and $\bar{X}_3$) and the variation within the 3 groups reflect chance errors of the sampling process.

- The first of these types of variation is conventionally referred to as between-treatment variation (the word "between" rather than "among" is used even when there are more than two groups present). **Between-treatment variation** is variation of the sample means $\bar{X}_1$, $\bar{X}_2$, and $\bar{X}_3$ around the grand mean $\bar{\bar{X}}$. This variation is sometimes referred to as variation between the c means, between-group variation, and between-column variation.

- The second type of variation is referred to as within-treatment variation (also known as within-group variation and within-column variation). **Within-treatment variation** is variation of the individual observations within each column from their respective means $\bar{X}_1$, $\bar{X}_2$, and $\bar{X}_3$.

Under the null hypothesis that the population means are equal, the between-treatment variation and the within-treatment variation would be expected not to differ significantly from one another after adjustment for degrees of freedom, since they both reflect the same type of chance sampling errors. If the null hypothesis is false and the population column means are indeed different, then the between-treatment variation should significantly exceed the within-treatment variation. This follows from the fact that the between-treatment variation would now be produced by the inherent differences among the treatment means as well as by chance sampling error. On the other hand, the within-treatment variation would still reflect chance sampling errors only.

A comparison of between-treatment variation and within-treatment variation yields information concerning differences among the treatment means.

This is the central insight provided by the analysis of variance technique.

Terms

The term **variation** is used in statistics in a specific way to refer to a sum of squared deviations and is often referred to simply as a **sum of squares**. When a measure of variation is divided by an appropriate number of degrees of freedom (as we have seen earlier in this text), it is referred to as a *variance*, and in the analysis of variance, a variance is referred to as a **mean square**. For example, the variation of a set of sample observations, denoted X, around their mean $\bar{X}$ is $\Sigma(X - \bar{X})^2$. Dividing this sum of squares by the number of degrees of freedom $n - 1$ (where n is the number of observations), we obtain $\Sigma(X - \bar{X})^2/(n - 1)$, the sample variance, which is an unbiased estimator of the variance of an infinite population as indicated in section 1.15. This sample variance can also be referred to as a mean square.

We now proceed with the analysis of variance by calculating the between-treatment variation and within-treatment variation for our problem.

Between-Treatment Variation

As indicated earlier, the between-treatment variation, or **between-treatment sum of squares**, measures the variation among the sample column means. It is calculated as follows:

Between-treatment sum of squares

(8.5)
$$\sum_j r(\bar{X}_j - \bar{\bar{X}})^2$$

TABLE 8-11
Calculation of the between-treatment sum of squares for the teaching methods problem

$$(\bar{X}_1 - \bar{\bar{X}})^2 = (17 - 20)^2$$
$$= 9$$
$$(\bar{X}_2 - \bar{\bar{X}})^2 = (20 - 20)^2$$
$$= 0$$
$$(\bar{X}_3 - \bar{\bar{X}})^2 = (23 - 20)^2$$
$$= 9$$
$$\sum_j r(\bar{X}_j - \bar{\bar{X}})^2 = 4(9) + 4(0) + 4(9)$$
$$= 72$$

where r = number of rows (sample size involved in the calculation of each column mean)[3]
$\bar{X}_j$ = the mean of the jth column (treatment)
$\bar{\bar{X}}$ = the grand mean
$\sum_j$ = summation taken over all columns

As indicated in equation 8.5, the between-treatment sum of squares is calculated by the following steps:

1. Compute the deviation of each treatment mean from the grand mean.
2. Square the deviations obtained in step 1.
3. Weight each deviation by the sample size involved in calculating the respective mean. In our example, all sample sizes are the same and are equal to the number of rows, $r = 4$.
4. Sum over all columns the products obtained in step 3.

The calculation of the between-treatment sum of squares for the example involving 3 different teaching methods is given in Table 8-11. As indicated in the table, the between-treatment variation is 72.

Within-Treatment Variation

The **within-treatment sum of squares** is a summary measure of the random errors of the individual observations around their column (treatment) means. The

[3] In this example, equal sample sizes (equal numbers of rows) are assumed. Subsequently, we generalize this approach to allow for different sample sizes.

TABLE 8-12
Calculation of the within-treatment sum of squares for the teaching methods problem

i	$(X_{i1} - \bar{X}_1)$	$(X_{i1} - \bar{X}_1)^2$	$(X_{i2} - \bar{X}_2)$	$(X_{i2} - \bar{X}_2)^2$	$(X_{i3} - \bar{X}_3)$	$(X_{i3} - \bar{X}_3)^2$
1	$(16 - 17) = -1$	1	$(19 - 20) = -1$	1	$(24 - 23) = 1$	1
2	$(21 - 17) = 4$	16	$(20 - 20) = 0$	0	$(21 - 23) = -2$	4
3	$(18 - 17) = 1$	1	$(21 - 20) = 1$	1	$(22 - 23) = -1$	1
4	$(13 - 17) = -4$	16	$(20 - 20) = 0$	0	$(25 - 23) = 2$	4
		34		2		10

$$\sum_j \sum_i (X_{ij} - \bar{X}_j)^2 = 34 + 2 + 10 = 46$$

formula for its computation is

Within-treatment sum of squares

$$\sum_j \sum_i (X_{ij} - \bar{X}_j)^2 \qquad (8.6)$$

where X_{ij} = the value of the observation in the ith row and jth column
$\bar{X}_j$ = the mean of the jth column
$\sum_j \sum_i$ = means that the squared deviations are first summed over all sample observations within a given column, then summed over all columns

As indicated in equation 8.6, the within-treatment sum of squares is calculated as follows:

1. Calculate the deviation of each observation from its column mean.
2. Square the deviations obtained in step 1.
3. Add the squared deviations within each column.
4. Sum over all columns the figures obtained in step 3.

The computation of the within-treatment variation for the teaching methods problem is given in Table 8-12.

Total Variation

The between-treatment variation and within-treatment variation represent the two components of the total variation in the overall set of experimental data. The total variation, or **total sum of squares**, is calculated by adding the squared deviations of all of the individual observations from the grand mean $\bar{\bar{X}}$. Hence, the formula for the total sum of squares is

(8.7)
$$\sum_j \sum_i (X_{ij} - \bar{\bar{X}})^2$$
Total sum of squares

The total sum of squares is computed by the following steps:

1. Calculate the deviation of each observation from the grand mean.
2. Square the deviations obtained in step 1.
3. Add the squared deviations over all rows and columns.

The total sum of squares, or total variation of the 12 observations in the teaching methods problem, is $(16 - 20)^2 + (21 - 20)^2 + \cdots + (25 - 20)^2 = 118$. Referring to the results obtained in Tables 8-11 and 8-12, we see that the total sum of squares, 118, is equal to the sum of the between-treatment sum of squares, 72, and the within-treatment sum of squares, 46. In general, the following relationship holds:

(8.8)
$$\begin{array}{c}\text{Total}\\ \text{variation}\end{array} = \begin{array}{c}\text{Between-treatment}\\ \text{variation}\end{array} + \begin{array}{c}\text{Within-treatment}\\ \text{variation}\end{array}$$

Although, as we have indicated earlier, the test of the null hypothesis in a one-factor analysis of variance involves only the between-treatment variation and the within-treatment variation, it is useful to calculate the total variation as well. This computation is helpful as a check procedure and is instructive in indicating the relationship between total variation and its components.

Shortcut Computational Formulas

The formulas we have given for calculating the between-treatment sum of squares (8.5), the within-treatment sum of squares (8.6), and the total sum of squares (8.7) clearly reveal the rationale of the analysis of variance procedure. However, the following shortcut computation formulas are often used to calculate these sums of squares.

(8.9)
$$\frac{\sum_j T_j^2}{r} - C$$
Between-treatment sum of squares

(8.10)
$$\sum_j \sum_i X_{ij}^2 - \sum_j \frac{T_j^2}{r}$$
Within-treatment sum of squares

(8.11)
$$\sum_j \sum_i X_{ij}^2 - C$$
Total sum of squares

where C, the so-called **correction term**, is given by

Correction term

(8.12)
$$C = \frac{T^2}{rc}$$

T_j is the total of the r observations in the jth column, and T is the grand total of all rc observations, that is,

Grand total of all rc observations

(8.13)
$$T = \sum_j \sum_i X_{ij}$$

All other terms are as previously defined.

These formulas are especially useful when the column means and grand mean are not integers. The shortcut formulas not only save time and computational labor, but also are more accurate because of avoidance of rounding problems, which usually occur with the use of equations 8.5, 8.6, and 8.7. The shortcut computations for the teaching methods example are as follows:

$$C = \frac{(240)^2}{(3)(4)} = 4{,}800$$

$$\text{Between-treatment sum of squares} = \frac{(68)^2 + (80)^2 + (92)^2}{4} - 4{,}800 = 72$$

$$\text{Within-treatment sum of squares} = (16)^2 + (21)^2 + \cdots + (25)^2 - \frac{(68)^2 + (80)^2 + (92)^2}{4} = 46$$

$$\text{Total sum of squares} = (16)^2 + (21)^2 + \cdots + (25)^2 - 4{,}800 = 118$$

It is recommended that the shortcut formulas be used, particularly when carrying out computations by hand.

Number of Degrees of Freedom

Although the preceding discussion was in terms of *variation* or *sums of squares* rather than *variance*,

> the actual test of the null hypothesis in the analysis of variance involves a comparison of the *between-treatment variance* with the *within-treatment variance*

or, in equivalent terminology, a comparison of the *between-treatment mean square* with the *within-treatment mean square*. Hence, the next step in our procedure is to determine the number of degrees of freedom associated with each of the measures of variation. As stated earlier in this section, if a measure of variation—that is, a sum of squares—is divided by the appropriate number of degrees of freedom, the resulting measure is a variance—that is, a mean square.

Between-treatment sum of squares

The number of degrees of freedom associated with the between-treatment sum of squares is $c - 1$. We can see the reason for this by applying the same general principles indicated earlier for determining number of degrees of freedom in t tests and χ^2 tests. Since there are c columns, or c group means, there are c sums of squares involved in measuring the variation of these column means around the grand mean. Because the sample grand mean is only an estimate of the unknown population mean, we lose one degree of freedom. Hence, there are $c - 1$ degrees of freedom present. The number of degrees of freedom in our example, which has 3 different teaching methods (that is, 3 treatments), is $c - 1 = 3 - 1 = 2$.

Within-treatment variation

The number of degrees of freedom associated with the within-treatment variation is $rc - c = c(r - 1)$. This may be reasoned as follows. There are a total of rc observations. In determining the within-treatment variation, the squared deviations within each treatment were taken around the treatment (column) mean. There are c treatment means, each of which is an estimate of the true unknown population treatment mean. Hence, there is a loss of c degrees of freedom, and c must be subtracted from rc, the total number of observations.

Alternatively, there are r squared deviations in each treatment taken around the treatment mean and a total sum of squares for the treatment. We can assign $r - 1$ of the sums of squares arbitrarily, and the last becomes fixed in order for the sum to equal the column sum. Since there are c treatments, we have $c(r - 1)$ degrees of freedom. In the present problem, the number of degrees of freedom associated with the within-treatment sum of squares is $c(r - 1) = 3(4 - 1) = 9$.

Within-treatment sum of squares

Within-treatment variance

A simpler way to denote the number of degrees of freedom associated with the within-treatment variance is $n - c$, where n is the total number of observations. We observe that $n = rc$. For example, in the present illustration, $n = (4)(3) = 12$. Thus, we see that the number of degrees of freedom associated with the within-treatment variance may be written

$$c(r - 1) = cr - c = n - c$$

Total variation

The number of degrees of freedom associated with the total variation is $rc - 1 = n - 1$. There are rc squared deviations taken from the sample grand mean $\bar{\bar{X}}$. Since $\bar{\bar{X}}$ is an estimate of the true unknown population mean, there is a loss of one degree of freedom. Alternatively, in the determination of the total sum of squares, there are rc squared deviations. We may assign $rc - 1$ squared deviations arbitrarily, but the last one is constrained in order for the sum to be equal to the total sum of squares. In the example, the number of degrees of freedom associated with the total variation is $rc - 1 = (4)(3) - 1 = 11$, or $n - 1 = 12 - 1 = 11$.

Just as the between-treatment and the within-treatment variations sum to the total variation, the numbers of degrees of freedom associated with the between-treatment and within-treatment variations add to the number associated with the total variation. In symbols,

$$rc - 1 = (c - 1) + (rc - c) \tag{8.14}$$

and

$$n - 1 = (c - 1) + (n - c)$$

In the teaching methods problems, the numerical values corresponding to equation 8.14 are $11 = 2 + 9$.

The Analysis of Variance Table

An analysis of variance table for the teaching methods problem is given in Table 8-13. The table uses the standard form to summarize the results of an analysis of variance. In columns (1), (2), and (3) are listed the possible sources of variation, the sum of squares for each of these sources, and the number of degrees of freedom associated with each of the sums of squares. The term **error** has been included in parentheses after "within treatments" to indicate that such variation is attributed to chance sampling error. "Error" is a frequently used term in computer printouts of analyses of variance. We again note that both sums of squares and numbers of degrees of freedom are additive; that is, these figures for between-treatment and within-treatment sources of variation add to the corresponding figure for total variation. Dividing the sums of squares in column (2) by the numbers of degrees of freedom in column (3) yields the between-treatment and within-treatment variances shown in column (4). As indicated earlier, **mean square** is another name for a sum of squares divided by

TABLE 8-13
Analysis of variance table for the teaching methods problem

(1) Source of Variation	(2) Sum of Squares	(3) Degrees of Freedom	(4) Mean Square
Between treatments	72	2	36
Within treatments (error)	46	9	5.11
Total	118	11	

$$F(2, 9) = \frac{36}{5.11} = 7.05$$

$$F_{0.05}(2, 9) = 4.26$$

Since $7.05 > 4.26$, reject H_0

TABLE 8-14
General format of a one-factor analysis of variance table

(1) Source of Variation	(2) Sum of Squares	(3) Degrees of Freedom	(4) Mean Square
Between treatments	SSA	$v_1 = c - 1$	$SSA/(c - 1)$
Within treatments (error)	SSE	$v_2 = n - c$	$SSE/(n - c)$
Total	SST	$n - 1$	

$$F(v_1, v_2) = \frac{SSA/(c - 1)}{SSE/(n - c)}$$

the appropriate number of degrees of freedom, and it is conventional to use this term in an analysis of variance table. Thus, in our problem, the between-treatment mean square is equal to $\frac{72}{2} = 36$. The within-treatment mean square is equal to $\frac{46}{9} = 5.11$. The test of the null hypothesis that the population treatment means are equal is carried out by a comparison of the between-treatment mean square with the within-treatment mean square.

Table 8-14 gives the general format of a one-factor analysis of variance table. The sums of squares are denoted as follows:

SSA = between-treatment sum of squares

SSE = within-treatment sum of squares

SST = total sum of squares

The rationale for the SSA symbol for the treatment sum of squares is as follows: The SS denotes "sum of squares," and the A indicates the first treatment. Hence, if there were additional treatments, the notation for their sums of squares would be SSB, SSC, and so forth. In the symbol SSE for the within-treatment sum of squares, the E denotes error while the SS again stands for sum of squares. The SSE symbol, referred to as the "error sum of squares," is a widely used notation. Other notation is as previously defined or as given in the next subsection.

The F Test and F Distribution

The comparison of the between-treatment mean square (variance) with the within-treatment mean square (variance) is made by computing their ratio, referred to as F. Hence, F is given by

$$F = \frac{\text{Between-treatment variance}}{\text{Within-treatment variance}} = \frac{SSA/(c - 1)}{SSE/(n - c)} \tag{8.15}$$

F ratio

In the F ratio, the between-treatment variance is always placed in the numerator and the within-treatment variance in the denominator. Under the null hypothesis that the population treatment means are equal, the F ratio would tend to equal one. If the population treatment means do indeed differ, then the between-treatment mean square will tend to exceed the within-treatment mean square, and the F ratio will be greater than one. In terms of our problem concerning different teaching methods, if F is large, we will reject the null hypothesis that the population mean examination scores are all equal; that is, we will reject $H_0\colon \mu_1 = \mu_2 = \mu_3$. If F is close to one, we will accept the null hypothesis. We can determine how large the test statistic F must be in order to reject the null hypothesis by referring to the probability distribution of the F random variable. This distribution is complex, and its mathematical expression is given here for reference only. Fortunately, critical values of the F ratio have been tabulated for frequently used significance levels analogous to the case of the χ^2 distribution. The probability density function of F is

F test

$$f(F) = kF^{(\nu_1/2)-1}\left(1 + \frac{\nu_1 F}{\nu_2}\right)^{-(\nu_1+\nu_2)/2} \tag{8.16}$$

where ν_1 = the number of degrees of freedom of the numerator of F
ν_2 = the number of degrees of freedom of the denominator of F
k = a constant depending only on ν_1 and ν_2

The underlying assumptions[4] are that two independent random samples are drawn from normally distributed populations with equal variances σ_1^2 and σ_2^2. The term **homoscedasticity** is used in this and other statistical tests for the assumption of equal variances. Unbiased estimators $\hat{\sigma}_1^2$ and $\hat{\sigma}_2^2$ of the population variances are constructed from the sample, and

Homoscedasticity

$$F = \frac{\hat{\sigma}_1^2}{\hat{\sigma}_2^2}$$

F distribution

The F distribution is similar to the distributions of t and χ^2 in that it is actually a family of distributions. Each pair of values of ν_1 and ν_2 specifies a different distribution. F is a continuous random variable that ranges from zero to infinity. Since the variances in both the numerator and denominator of the F ratio are squared quantities, F cannot take on negative values. The F distribution has a single mode, and although the specific distribution depends on the values of ν_1 and ν_2, its shape is generally asymmetrical and skewed to the right. The distribution tends towards symmetry as ν_1 and ν_2 increase. We will use the notation $F(\nu_1, \nu_2)$ to denote the F ratio defined in equation 8.15,

[4] Research in recent years on "robustness" has shown that minor departures from these assumptions, particularly that of normality, do not materially affect the results of the analysis.

where the numerator and denominator are between-treatment mean squares and within-treatment mean squares with v_1 and v_2 degrees of freedom, respectively. Table A-8 of Appendix A presents the critical values of the F distribution for two selected significance levels, $\alpha = 0.05$ and $\alpha = 0.01$. In this table, v_1 values are listed across the columns and v_2 values are listed down the rows. There are two entries in the table corresponding to every pair of v_1 and v_2 values. The upper figure (in lightface type) is an F value that corresponds to an area of 0.05 in the right tail of the F distribution with v_1 and v_2 degrees of freedom. That is, it is an F value that would be exceeded only 5 times in 100 if the null hypothesis under consideration were true. The lower figure (in boldface type) is an F value corresponding to a 0.01 area in the right tail.

***F* table**

We will illustrate the use of the F table in terms of the teaching methods problem.

***F* table used for teaching methods problem**

Assuming that we wish to test the null hypothesis $H_0: \mu_1 = \mu_2 = \mu_3$ at the 0.05 level of significance, we find in Table A-8 of Appendix A that for $v_1 = 2$ and $v_2 = 9$ degrees of freedom, an F value of 4.26 would be exceeded 5% of the time if the null hypothesis were true. As indicated at the bottom of Table 8-13, we denote this critical value as $F_{0.05}(2, 9) = 4.26$. This relationship is depicted in Figure 8-4. Again referring to Table 8-13, since the computed value of the F ratio (the between-treatment mean square over the within-treatment mean square) is 7.05, and therefore greater than the critical value of 4.26, we reject the null hypothesis. Hence, we conclude that the treatment means (that is, the sample means of final examination scores in classes taught by the 3 teaching methods) differ significantly. The inference about the corresponding population means is that they are not all the same. Referring to Table 8-10(a), we see that average grades under method (3) exceed those under method (2), which are higher than those under method (1).

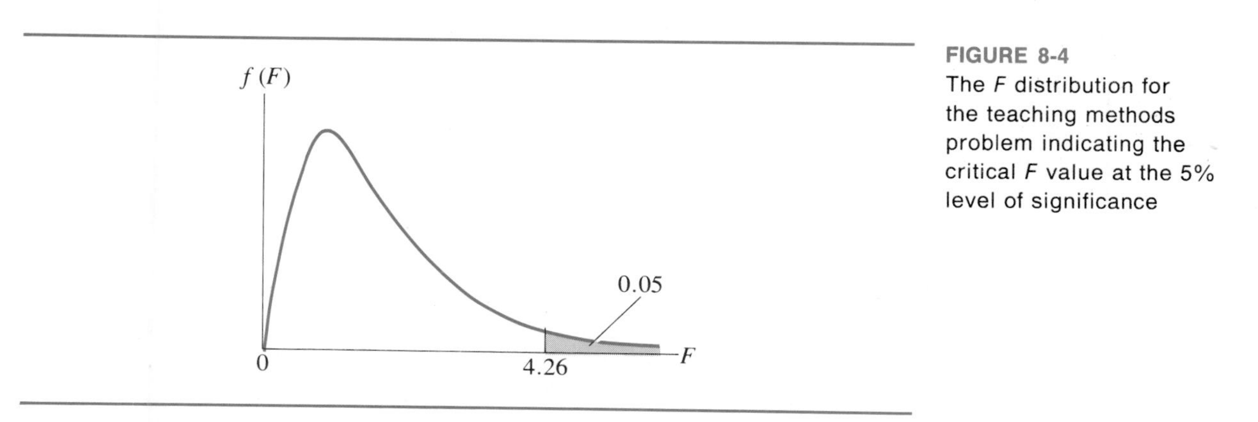

FIGURE 8-4
The F distribution for the teaching methods problem indicating the critical F value at the 5% level of significance

Conclusion on teaching methods

Based on these data, our inference is that the 3 teaching methods are not equally effective, and there is evidence that method (3) is the most effective and method (1) is the least effective.

Caveat

The foregoing example was used to illustrate the rationale involved in the analysis of variance, the statistical technique employed and the nature of the conclusions that can be drawn. However, we repeat the warning that the sample sizes in this illustration are too small for safe conclusions to be drawn about differences in effectiveness of the 3 teaching methods. After all, only 4 observations were made under each teaching method. Although the risk of a Type I error was controlled at 0.05 in this problem, the risk of Type II errors may be intolerably high. Suffice it to say that larger sample sizes are generally required. Also, the example was in terms of a treatment (teaching methods) applied to groups of equal size. In the next subsection, we discuss the use of a computer program for carrying out an analysis of variance.

Analysis of Variance and the Computer

The computer output for a canned BASIC analysis of variance program is shown in Figure 8-5. The output pertains to the teaching methods problem discussed in this section.

FIGURE 8-5
Computer output for the teaching methods problem

```
1202 DATA 12,3
1212 DATA 4,4,4
1222 DATA 16,21,18,13,19,20,21,20,24,21,22,25
XEQ
ANOVA TABLE:

    SOURCE              SS                  DF              MS%%

GRAND TOTAL         4918.000000          12%%
GRAND MEAN          4800.000000           1%%
TREATMENTS            72.000000           2               36.0000000000%%
ERROR                 46.000000           9                5.1111111111%%

F = 7.0434782608696 ON 2  AND  9  DEGREES OF FREEDOM

EXACT PROB. OF F= 7.0434782608696 WITH (  2 ;  9  ) D.F . IS 0.01444

**       812 XEQ "END".
```

An explanation of the printout follows.

1202 DATA 12,3	The computer asks for instructions and the operator replies that the data consists of 12 observations in 3 columns. (As in the chi-square program, the number preceding the word DATA corresponds to a line number in the program and has no significance for the output.)
1212 DATA 4,4,4	This tells the computer that the 3 columns consist of 4 observations each.
1222 DATA 16,21, . . . ,22,25	This enters the data one column at a time.
X EQ	This is the instruction to the computer to execute the program.

The analysis of variance (ANOVA) table is the same as the one computed "by hand" and shown in Table 8-13. The only changes are that in this particular printout the value of

$$\sum_j \sum_i X_{ij}^2 = 4918$$

is shown as the GRAND TOTAL and $C = T^2/rc = 4800$ is labeled GRAND MEAN. The value of F and the degrees of freedom are computed as shown with the change in labeling—what was called "between-columns sum of squares" is now called TREATMENTS and what was called "within-treatment sum of squares" is now called ERROR.

In addition, the exact probability of an F value as large or larger than the F obtained, given the degrees of freedom, is also shown. The advantage of this calculation is that it eliminates the need to refer to the F table. In this problem, the exact probability of an F value as large as 7.043 . . . or larger (given 2 and 9 degrees of freedom) is 0.01444, indicating that an F value as large as 7.043 . . . would be exceeded 1.444% of the time due to chance. This is consistent with the results obtained in Figure 8-5—the null hypothesis can be rejected with a 5% risk of a Type I error but not with a 1% risk of a Type I error.

One-Factor Analysis of Variance: Unequal Sample Sizes

In the teaching methods example, the samples were all the same size; the same number of students (4) were taught by each method. Although it is often simplest in the collection of data to work with samples of the same size, this is not always feasible. The analysis of variance computational procedure given earlier is easily adjusted for differing sample sizes. The general format for a one-factor analysis of variance for unequal sample sizes is the same as for equal sample sizes, as given in Table 8-14. However, the shortcut computational formulas differ slightly.

Let us now consider the teaching methods problem as an example of unequal sample sizes. However, we shall assume that there were only nine students in the experiment: two in the first section, three in the second section, and four in the third section.

The final examination grades of these students are given in Table 8-15(a), and the analysis of variance is carried out in Table 8-15(b).

Columns (1), (2), and (3) of Table 8-15(b) list the possible sources of variation, the sums of squares for each of these sources, and the number of degrees of freedom associated with the sums of squares. Note that both sums of squares and degrees of freedom are additive; that is, these figures for between-treatment and within-treatment sources of variation add to the corresponding total variation figure.

The sums of squares have been computed by the shortcut formulas given in equations 8.17, 8.18, and 8.19, which are similar to the shortcut formulas of equations 8.9, 8.10, and 8.11 for equal-sized samples.

Between-treatment sum of squares

$$\sum_j \frac{T_j^2}{r_j} - C = \frac{(37)^2}{2} + \frac{(60)^2}{3} + \frac{(92)^2}{4} - \frac{(189)^2}{9} = 31.5 \tag{8.17}$$

Within-treatment sum of squares

$$\sum_j \sum_i X_{ij}^2 - \sum_j \frac{T_j^2}{r_j} = (16)^2 + (21)^2 + (19)^2 + \cdots + (25)^2 - \frac{(37)^2}{2} - \frac{(60)^2}{3} - \frac{(92)^2}{4} = 24.5 \tag{8.18}$$

TABLE 8-15

(a) Final examination grades of 9 students taught by 3 different methods

Teacher	Teaching Method 1	2	3
1	16	19	24
2	21	20	21
3	—	21	22
4	—	—	25
Total	37	60	92
Mean	18.5	20.0	23.0

(b) Analysis of variance table for the teaching methods problem with unequal sample sizes

(1) Source of Variation	(2) Sum of Squares	(3) Degrees of Freedom	(4) Mean Square
Between treatments	31.5	2	15.75
Within treatments (error)	24.5	6	4.08
Total	56.0	8	

$$F(2, 6) = \frac{15.75}{4.08} = 3.86$$

$$F_{0.05}(2, 6) = 5.14$$

Since $3.86 < 5.14$, accept H_0

Total sum of squares

(8.19)
$$\sum_j \sum_i X_{ij}^2 - C = SSA + SSE = 31.5 + 24.5$$
$$= 56$$

where C, the correction term, is given by

(8.20)
$$C = \frac{T^2}{\sum_j r_j} = \frac{(189)^2}{9}$$

where T_j and r_j are the total and number, respectively, of observations in the jth column, and T is the grand total of all observations as given in equation 8.13.

The number of degrees of freedom associated with the between-treatment sum of squares is $c - 1 = 3 - 1 = 2$. The number of degrees of freedom associated with the within-treatment variation is $n - c = 9 - 3 = 6$. As shown in Table 8-15(b), the analysis of variance computations are carried out in similar fashion to the equal-sized sample case. Since the F test results in the acceptance of the null hypothesis, we cannot conclude that there is any difference in the effectiveness of the three teaching methods as measured by final examination grades.

The one-factor teaching methods illustration used in this chapter has another characteristic: This type of example is conventionally designated as a **completely randomized experimental design** and it involves the random selection of independent samples from c (the number of columns) normal populations. This implies that the students taught by each of the three methods were randomly selected from the sophomore class.

Two-Factor Analysis of Variance

Let us look more critically at the possible interpretations of our findings. We assumed that the same teacher taught three different sections of the basic statistics course using three specified methods—1, 2, and 3. Final examination scores seemed to indicate that method 3 was the most effective and method 1 the least effective teaching method.

New data for the teaching-methods problem

Suppose the method 3 class is given early in the morning, when the teacher (and students, too?) is fresh, wide awake, and enthusiastic. On the other hand, suppose the classes taught by methods 2 and 1 are in the middle of the day and the late afternoon, respectively. Let us assume that by late afternoon the instructor is tired, sleepy, and rather unenthusiastic. In this case, the differences in teaching effectiveness may not be attributable solely to the different teaching methods but rather to some unknown mixture of the difference in teaching methods and the aforementioned time-of-day factors.

Time of day may be thought of as an extraneous factor that influences the variable of interest, teaching methods. A **two-factor** or **two-way analysis of variance** isolates the influence of this extraneous factor so the effects of the main variable may be more accurately judged.

Randomized block design

There are two basic types of the two-factor analysis of variance: the *randomized block design* and the *completely randomized design.* The **randomized block design** is appropriate for the situation just described; we are interested in the influence of a certain factor (for example, teaching methods), but we also wish to isolate the effects of a second extraneous factor, such as time of day.

> The term "block" derives from experimental design work in agriculture, in which parcels of land are referred to as blocks. In a randomized block design, treatments are randomly assigned to units within each block. In testing the yield of different fertilizers, for example, this design ensures that the best fertilizer is applied to all types of soil, not just the best soil.

In the teaching methods illustration, the times of day could be treated as the blocks, and the teaching methods (treatments) would be randomly assigned within the blocks. Thus, methods 1, 2, and 3 would each be equally represented in the early morning, middle of the day, and late afternoon. The null hypothesis is the same as in the one-factor analysis of variance, namely H_0: $\mu_1 = \mu_2 = \mu_3$. However, the design removes the variation in time of day from the comparison of the three teaching methods. No inference is attempted about the effect of time of day, since time of day is viewed as an extraneous factor. The extraneous blocking factor usually represents time, location, or experimental material. Just as the one-factor analysis of variance represents a generalization of the t test for means of two independent samples, the randomized block design represents a generalization of the t test for paired observations discussed in section 7.5.

The second type of two-factor analysis of variance is used when inferences about both factors are desired.

> Suppose that 4 different instructors were teaching the statistics course and that inferences were desired about differences in effectiveness among the 3 methods of teaching *and* among the 4 different teachers.

Completely randomized design

Of course, the sample sizes in this experiment and in the randomized block design would have to be much larger than the 12 students used earlier. This two-factor study would use a completely randomized design, rather than the randomized block design. In the **completely randomized design**, the sample units (students, in this case; sections of students in a more realistic example) would be randomly assigned to each factor combination. For example, teacher 1 using

TABLE 8-16
Final examination grades of 12 students for 3 different teaching methods by 4 different teachers

Teacher	Teaching Method 1	2	3	Row Sum T_i
1	24	16	19	59
2	21	21	20	62
3	22	18	21	61
4	25	13	20	58
Column sum (T_j)	92	68	80	240

method 1, teacher 1 using method 2, and so on, would represent factor combinations. This experimental design then permits the testing of two null hypotheses:

1. H_0: No difference in population mean final examination scores among the different teaching methods
2. H_0: No difference in population mean final examination scores among the different teachers

Table 8-16 shows final examination scores for the three teaching methods by four different teachers.

To test the two hypotheses, we set up a general format of a two-factor analysis of variance table as shown in Table 8-17. As before,

$$SSA = \sum_j \frac{T_j^2}{r} - \frac{T^2}{rc}$$

$$= \frac{(92)^2}{4} + \frac{(68)^2}{4} + \frac{(80)^2}{4} - \frac{(240)^2}{(4)(3)}$$

$$= 72.00$$

TABLE 8-17
General format of a two-factor analysis of variance table

(1) Source of Variation	(2) Sum of Squares	(3) Degrees of Freedom	(4) Mean Square
Teaching methods	SSA	$c - 1$	$SSA/(c-1)$
Teachers	SSB	$r - 1$	$SSB/(r-1)$
Error	SSE	$(r-1)(c-1)$	$SSE/(r-1)(c-1)$
Total	SST	$rc - 1$	

$$SST = \sum_j \sum_i X_{ij}^2 - \frac{T^2}{rc}$$

$$= (24)^2 + (21)^2 + (22)^2 + \cdots + (20)^2 - \frac{(240)^2}{(4)(3)}$$

$$= 118.00$$

And analogous to SSA, the sum of squares within columns is

$$SSB = \sum_i \frac{T_i^2}{c} - \frac{T^2}{rc}$$

$$= \frac{(59)^2}{3} + \frac{(62)^2}{3} + \frac{(61)^2}{3} + \frac{(58)^2}{3} - \frac{(240)^2}{(4)(3)}$$

$$= 3.33$$

Since $SST = SSA + SSB + SSE$, the sum of squares of the error term, SSE, can be found as a residual by subtraction. $SSE = 118.00 - 72.00 - 3.33 = 42.67$.

An analysis of variance table for the two-factor teaching methods and teacher problem is given in Table 8-18, and the two hypotheses are tested as shown.

Hypothesis test for 3 teaching methods

$$F(2, 6) = \frac{36.00}{7.11} = 5.06$$

$$F_{0.05}(2, 6) = 5.14$$

Since $5.06 < 5.14$, accept H_0

Hypothesis test for 4 teachers

$$F(3, 6) = \frac{1.11}{7.11} = 0.16$$

$$F_{0.05}(3, 6) = 4.76$$

Since $0.16 < 4.76$, accept H_0

TABLE 8-18
Analysis of variance table for the two-factor teaching problem

(1) Source of Variation	(2) Sum of Squares	(3) Degrees of Freedom	(4) Mean Squares
Teaching methods	72.00	$3 - 1 = 2$	36.00
Teachers	3.33	$4 - 1 = 3$	1.11
Error	42.67	$(3 - 1)(4 - 1) = 6$	7.11
Total	118.00	11	

Summary

Since neither of the two null hypotheses was rejected, we cannot conclude from these data on examination grades that there were real differences in effectiveness among the 3 teaching methods or among the 4 teachers.

Advantages of two-factor analyses

Two-factor analyses of variance have some distinct advantages. Note from the example that we were able to test two separate null hypotheses from the same set of experimental data. We did not need to run two one-factor experiments to get information about two factors. Furthermore, certain types of questions can be answered by two-factor designs but cannot be treated in one-factor analyses. For example, the interaction, or joint effects, of the two factors may be examined as well as their separate effects.

Only a brief introduction to the analysis of variance has been given in this chapter. More elaborate designs than those considered in this book are available; they attempt to control and test for the effects of more factors, both qualitative and quantitative.

Further Remarks

Importance of careful experimental design

As we saw in the one-factor teaching methods example, observed differences in teaching effectiveness may not have been attributable solely to the different teaching methods, but rather to some unknown mixture of teaching methods and factors associated with time of day, and perhaps other factors as well. Hence, unless careful thought is given to the experimental design from which data are to be collected, erroneous inferences may be drawn. This applies equally to the hypothesis-testing methods considered earlier, for example in Chapter 7, because we might have had only two teaching methods to compare rather than three. Thus, we must guard against mechanical or rote application of statistical techniques such as hypothesis-testing methods. In this book, we consider the general principles involved in some of the simpler, basic procedures. More refined and sophisticated techniques may very well be required in particular instances.

Alternative interpretations

One of the points we have attempted to convey in the preceding discussion is that statistical results are virtually always consistent with more than one interpretation. The researcher must avoid naively leaping to conclusions and must give careful consideration to alternative interpretations and explanations. We conclude this chapter with two anonymous humorous stories that are relevant to the point that alternative interpretations and explanations of experimental results are often possible.

An investigator wished to determine the differential effects involved in drinking various types of mixed drinks. Therefore, he had subjects drink substantial quantities of scotch and water, bourbon and water, and rye and water. All of the subjects became intoxicated. The investigator concluded that since water was the one factor common to all of these drinks, the imbibing of water makes people drunk.

The heroine of our second story is a grammar school teacher, who wished to explain the harmful effects of drinking liquor to her class of 8-year-olds. She placed two glass jars of worms on her desk. Into the first jar, she poured some water. The worms continued to move about, and did not appear to have been adversely affected at all by the contact with the water. Then she poured a bottle of whiskey into the second jar. The worms became still and appeared to have been mortally stricken.

The teacher then called on a student and asked, "Johnny, what is the lesson to be learned from this experiment?" Johnny, looking thoughtful, replied, "I guess it proves that it is good to drink whiskey, because it will kill any worms you may have in your body."

Exercises 8.4

1. As head of a department of a consumer research organization, you have the responsibility for testing and comparing lifetimes of an electrical product for 4 different brands. Suppose you test the lifetime of 3 of these products for each of the 4 brands. Your test data entries each represent the lifetime of one of these products (measured in hundreds of hours). Would you conclude that the mean lifetimes of the 4 brands are equal?

A	B	C	D
20	25	24	23
19	23	20	20
21	21	22	20

2. A manufacturer has a choice of 3 subcontractors from whom to buy parts. The manufacturer, before deciding from whom he will buy, purchases 5 batches from each subcontractor. There are the same number of parts in each batch. Based on the number of defectives per batch (as given in the table), would you conclude that there is no real difference among these 3 subcontractors in the average number of defectives produced per batch? Use a 0.01 level of significance.

	Subcontractors		
Batch	A	B	C
1	35	15	25
2	25	20	40
3	30	25	40
4	35	15	35
5	20	30	30

3. A manufacturer wishes to select the best advertising display for a new product from among 5 different displays. The manufacturer randomly selects 25 different stores and places each type of display in 5 stores. For the first 6 months, the manufacturer records the following average amounts sold per store and the store-to-store variance for each display (expressed in dozens). Based on the listed data, can the manufacturer assume that the displays are equally effective? Use a 0.01 significance level.

Type of Display	Mean	Variance
1	78	9
2	76	7
3	77	8
4	74	8
5	76	10

4. The merger specialist for an investment banking firm conducted a study of past mergers and acquisitions for a client considering an acquisition. The specialist compiled data on 100 acquisition deals, which fall into 5 categories. For each type of deal, the percentage of stockholders tendering their stock was calculated, and the results were averaged over the sample of 20 in each category. The resulting means and variances are listed here.

   The client is interested in maximizing the percentage of shareholders tendering their stock. Would you conclude that there has been no real difference among these 5 types of acquisition deals in the average percentage of shareholders offering their stock for exchange? Use a 0.05 significance level.

Type of Deal	Mean	Variance
All cash	90	20
Common stock	85	18
Debt	80	19
Preferred stock	80	17
Convertible preferred	85	15

5. The marketing director for the Superficial Cosmetics Company is trying to determine which of 5 marketing strategies would be most effective for a major new product, Flawless Facial Cream. For an 8-week period, he selects 30 cities randomly and initiates each strategy in 6 of the communities. The table indicates the average sales level and variance for each strategy;

all sales figures are in thousands of units. Can the director conclude from the listed data that the strategies are equally effective? Use a significance level of 0.01.

Type of Strategy	Mean	Variance
1	60	15
2	60	17
3	63	16
4	55	15
5	57	16

Review Exercises for Chapters 6 through 8

1. The following two facts are known about the clients of a large stockbrokerage firm: (1) Rather large commissions are generated by 20% of the clients (who are what is known as "heavy traders") and (2) the dollar sizes of the clients' accounts (defined in terms of annual value of transactions) are normally distributed with an arithmetic mean of \$20,000 and a standard deviation of \$1500.
   a. If 50 accounts are randomly selected, what is the probability that 25% or more "heavy traders" would be included in the sample?
   b. If one account is randomly selected, what is the probability that its size will be between \$18,500 and \$23,000?
   c. If one account is randomly selected, the probability is 0.209 that its size will exceed a certain dollar amount. What is this dollar amount?
2. A cigarette manufacturer wishes to estimate the average nicotine content of the company's cigarettes. A random sample of 100 cigarettes is chosen and the nicotine content of each is determined. A sample arithmetic mean of 3.20 milligrams per cigarette is obtained. Based on past experience, it appears that 0.4 milligrams is a reasonable estimate of the population standard deviation.
   a. Determine a 99% confidence interval for the population mean nicotine level.
   b. It is decided that the interval estimate determined in part (a) is not sufficiently precise. An estimate is required which is within 0.05 milligrams of the true population mean value. Using the same confidence level, determine how many cigarettes must be sampled to satisfy this requirement.
3. A previous study showed that it is only worthwhile to market a certain product in an area where the mean yearly income per family is at least \$25,000. In order to determine the mean yearly income in Feltonville, 100 families were chosen by randomly selecting residential phone numbers and asking the respondent for the family's yearly income.
   a. Give reasons why the average family income obtained from the sample might not agree with the corresponding average in the population.

b. Instead of asking the respondent for the actual yearly income, you ask for the group in which the income falls. The range of possible incomes is divided into the following groups, and the number of families falling in each group is recorded. Compute the mean income using the data given.

Family Income	Number of Families
\$0 to under \$10,000	40
\$10,000 to under \$20,000	28
\$20,000 to under \$40,000	17
\$40,000 to under \$60,000	15
\$60,000 and above	0

4. In a survey of the coffee-drinking habits of the American people, 400 adults were randomly sampled in each region. The following means and standard deviations were obtained from these two samples:

Pounds of Coffee Used per Adult per Year	
New England Region	North Central Region
$\bar{x}_1 = 20.0$	$\bar{x}_2 = 24.0$
$s_1 = 9.0$	$s_2 = 12.0$

a. Test the hypothesis that there is no difference in coffee consumption between the two regions at the 0.05 level of significance.

Decide whether the following statements are true or false. Give a concise explanation for each decision. These 3 questions are based on the test performed in part (a).

b. The probability of a Type II error is the same when $\mu_1 - \mu_2 = 23$ lbs. $-$ 21 lbs. as when $\mu_1 - \mu_2 = 20$ lbs. $-$ 22 lbs.

c. The size of the critical (i.e., rejection) region increases when we decrease the level of significance, holding everything else constant.

d. Since $\alpha + \beta = 1$, increasing α brings about decreases in the probabilities of making Type II errors.

5. A consumer group was studying the sales of a standard model of a certain automobile with a specified set of accessories. The group obtained data on prices from a random sample of 3 dealers in each of 3 cities.

a. Test the hypothesis that the mean price of these automobiles was the same in each of the 3 cities. Use $\alpha = 0.05$.

b. What assumption underlying the analysis of variance procedure do you consider to be the most dubious one in the context of this problem? Explain.

Sales of a Standard Model Automobile with Specified Accessories (in hundreds of dollars)

	Atlanta	Philadelphia	New York
	72	69	72
	68	70	75
	70	71	72
Mean	70	70	73
Variances	4	1	3

(Variances were computed using $n - 1$ in the denominators.)

6. You wish to advertise your product in *Youth* magazine. Your product is specifically designed for the teenage market (ages 15–19).
   a. To test *Youth* magazine's statement that at least 90% of its readers are in this age group, you randomly sample 100 of its readers and find that 88 of them are in this age group. If you are willing to run a maximum risk of 0.05 of rejecting *Youth* magazine's statement erroneously, would you reject the statement? Demonstrate why or why not.
   b. Using the decision rule implied by part (a) and assuming the true proportion of teenage readers of *Youth* magazine was 87%, what is the probability that *Youth* magazine's statement would be erroneously accepted in a simple random sample of 100 readers?
   c. In another project, you wish to estimate the percentage of *Youth* magazine's readers who are Black. You are virtually certain that this percentage is 20% or less. You wish to have a 90% probability of estimating the percentage of Black readers within two percentage points. How large a simple random sample would be necessary?
7. A company asked 400 purchasers of a new product whether they liked the product. The company conducting the survey felt that at least 20% of all purchasers must like the product in order for the firm to continue marketing it. The firm wished to run a low risk of incorrectly discontinuing the marketing of the product. Therefore, the firm decided that if fewer than 68 people responded favorably, it would stop marketing the product.
   a. State briefly in your own words the nature of the Type I error involved here. Calculate the level of significance (α) associated with the company's decision procedure.
   b. If the true (but unknown) percentage of purchasers who liked this new product was 15%, what is the probability that the firm's decision procedure would erroneously lead to continuance of marketing of the product?
8. Over the Counter, Inc., currently pays its new stockbrokers a commission of 2% above the base salary of $20,000 per year. The firm wants to determine whether or not it is preferable for them to pay a fixed salary of $30,000

per year. The president of the firm finds that the 4 newest employees who have been with the company for at least one year averaged (arithmetic mean) \$11,000 per year for their first year's commission with a standard deviation of \$1000.

a. Find a 95% two-sided confidence interval for the average yearly commissions earned by new employees, using the figures for the 4 newest employees. What assumptions did you make?
b. What is the minimum number of employees that need to be sampled so that the "precision" of the 95% confidence interval (i.e., half of the length of the interval) is within $\frac{1}{3}\sigma$?
c. Instead of using $\bar{X}$ as an estimator, the president wants to weight the 4 newest employees as follows.

$$e = \frac{X_1 + 2X_2 + 2X_3 + X_4}{6}$$

If σ is assumed to be equal to \$1000 and $e =$ \$11,200 from our sample, what would be the 95% confidence interval for the average yearly commissions earned by new employees? Carry out the calculations using $e =$ \$11,200 and the standard error of e.

9. Assume that the Department of Labor has estimated that 30% of the U.S. labor force are dissatisfied with their jobs. At Wharton Industries, Inc., 20 presently employed candidates are being interviewed for a position. Assume perfect randomness in choosing these 20 candidates from the labor force.
   a. What is the probability that at least 4 are dissatisfied with their present jobs?
   b. Using the Poisson approximation, find the probability in part (a).
   c. Using the normal approximation, find the same probability.
   d. Are the approximations in parts (b) and (c) justified? Why or why not?
10. A computer manufacturer produces large numbers of microprocessor chips with great care, but only a certain proportion p meets acceptable standards for use in the company's computers. Let us make the rather strong assumption that the production of each chip is a Bernoulli trial, with probability p of making a good chip, and probability $1 - p$ of making an unacceptable one. The n trials under consideration will be treated as independent events.
    a. Suppose that $n = 100$ chips were produced, and suppose $x = 32$ met the acceptance standard. What is your estimated value of p?
    b. The proportion of acceptable chips in the sample is approximately normally distributed when n is large. What are your estimates of the mean and variance of that distribution based on the sample data in part (a)?
    c. Find the confidence interval that will include the unknown parameter p 98% of the time for many repeated random samples of $n = 100$ chips from the population. Use the large-sample normal distribution spec-

ified in part (b) and the estimate of its variance that you calculated in part (b).

d. At a later date, the computer manufacturer wishes to make another estimate of p. He is convinced that p is equal to 40% or less. How large a simple random sample would be required to estimate p with 90% confidence that the estimate would be within 2 percentage points of the true population figure?

11. In a recent investigation at a major U.S. airport, researchers studied the arrivals of aircraft scheduled to land between 7:00 P.M. and 7:15 P.M. Over a 5-day span with ideal weather conditions, there were 81 landings with an average delay time of 12 minutes. The sample variance was found to be 20.25 (minutes)^2.
    a. Construct a 97% confidence interval estimate of the average delay time for aircraft under the stated weather conditions.
    b. Assume that the standard deviation of the delay time for landings under the stated weather conditions is in fact 5 minutes. How large a sample size is required to estimate average delay time to within 2 minutes with probability 0.99?
    c. The research team reported a confidence interval estimate of (11.359, 12.641) after analyzing the data. What confidence level did the research team use?

12. Statebank is a large institution that does an extensive amount of financial consulting. The executive vice president in charge of the consulting division wants to determine whether it is worthwhile to have MBAs who are employed by the firm attend a two-week mini-course prior to engaging in any consulting activities. As a pilot study, the personnel department sets up a scoring system (higher scores are better) to measure performance.

    A year after the mini-course, the personnel department tested the performance of 16 MBAs who attended the mini-course and 16 MBAs who did not attend the mini-course. The data on performance are summarized.

Performance Scores	
With Mini-Course	Without Mini-Course
$\bar{x}_1 = 74$	$\bar{x}_2 = 68$
$s_1^2 = 78$	$s_2^2 = 84$

    a. Are these data statistically significant for the conclusion that the mini-course is worthwhile, using $\alpha = 0.05$? Show your work, including the alternative hypotheses and decision rule that you have used.
    b. Since the mini-course is expensive, it is decided to conduct the mini-course only if it does better by more than two performance points on the average. What are the null and alternative hypotheses now? Would you conduct the mini-course using $\alpha = 0.05$? Show your work.

Are the following statements true or false? These questions refer to the test constructed in part (a).

c. If $\mu_1 = 70$, you can make only a type II error.

d. If the acceptance region was $\bar{x}_1 - \bar{x}_2 \leqslant 2$, then α would be greater than 0.05 (i.e., $\alpha > 0.05$)

e. If we increase the sample sizes for $\alpha = 0.05$, we would have smaller values of β.

13. *Financial Week* magazine recently conducted a survey of its readers in selected metropolitan areas. One of the questions asked for annual income. A rough summary of the data for 3 of the areas includes the following number of respondents.

Annual Income in 3 Metropolitan Areas

	Area A	Area B	Area C
Less than $30,000	10	25	25
$30,000–$60,000	40	45	25
More than $60,000	10	10	10

a. *Financial Week's* director of advertising believes that the readers in area A in the stated income ranges, from lowest to highest, are in the ratio 1:2:1. Do the data support this hypothesis? Use $\alpha = 0.05$.

b. Are the 3 metropolitan areas alike in their income profiles for the readers of *Financial Week*? Test using $\alpha = 0.01$.

14. The scores on an aptitude test used by a large company are normally distributed with a mean of 500 and a standard deviation of 50. The top 10% of the scores have been classified as "top management potential."

a. What is the probability that an aptitude test score lies between 400 and 450?

b. If a random sample of 100 test scores is selected, what is the probability that the mean score lies between 495 and 505?

c. If a random sample of 100 test scores is selected, what is the probability that the percentage which are classified as having "top management potential" lies in the range of 13% to 16%?

15. A candy wholesaler has found from extensive experience that his monthly sales to drugstores of a certain size has a mean of $120 and a standard deviation σ of $24. Recently the wholesaler acquired $n = 9$ additional drugstores of the same sort when a competitor went out of business. During the first complete month of business the wholesaler's average sales to the new stores was $\bar{x} = \$108$.

a. Test the hypothesis at the 5% level of significance that the new stores came from a population with mean $120 as opposed to an alternative of a lower mean. You may assume that monthly sales is a normal random variable, and that the 9 stores constitute a simple random sample. Your answer should include a statement of the hypotheses and a clear indication of the decision rule used.

b. Now suppose that the population standard deviation $\sigma = \$24$ was not known, but the wholesaler calculated the unbiased sample variance to be $s^2 = 324$. Test the same hypotheses on mean sales under these assumptions at the 5% level.

c. Assume that the *true population mean* of monthly sales for the new stores was \$98.80. With the decision rule you derived in part (a), what is the probability that you would erroneously accept the hypothesis that the population mean was \$120?

16. You are approached by an entrepreneur who offers you a limited partnership in a real estate development he is organizing. From past experience you are willing to assume that the risk of the venture can be measured by $\sigma = 6\%$ and that the rate of return of the venture follows a normal distribution. You investigate 36 previous projects of this developer in order to estimate the mean percentage return on investment, μ. The average return from these 36 projects x was 10%.
    a. Find a 95% confidence interval for μ.
    b. If you want to be 95% confident that the maximum error of your estimate will not exceed 1%, what sample size should you take?
    c. This venture will earn less than what level of return only 10% of the time if, in fact, the true mean $\mu = 10\%$?

17. It has been claimed that about 20% of all offers to prospective executives include short-run stock options (options good for 3 years or less) and another 20% include long-run options (good for more than 3 years). It is also claimed that these options have no effect on the likelihood of an executive accepting the offer. A headhunter checked her recent files and found a total of 200 offers with data broken down as shown in the table.

	Type of Stock Option			
	No Options	Short-run Options	Long-run Options	Total
Accepted	64	24	32	120
Rejected	46	16	18	80
Total	110	40	50	200

a. Are the data from the 200 offers significantly different from the 60% no options, 20% short-run, 20% long-run claim? Justify your answer using $\alpha = 0.05$.

b. Would you conclude that the type of option has an effect on acceptance or rejection of offers? Justify your answer using $\alpha = 0.05$.

18. A soft drink company tests 4 television advertisements of a soft drink by inviting 25 individuals into a studio to watch the 4 advertisements and then to rate each of the advertisements. The first 2 advertisements are similar in that they focus on thirst quenching. The last 2 advertisements are similar in that they compare the soft drink to the main competitor's product. The data are tabulated:

Thirst Quenching Ads		Product Comparison Ads	
Ad 1	Ad 2	Ad 3	Ad 4
$\bar{x}_1 = 62$	$\bar{x}_2 = 56$	$\bar{x}_3 = 54$	$\bar{x}_4 = 68$
$s_1^2 = 466.0$	$s_2^2 = 420.0$	$s_3^2 = 422.0$	$s_4^2 = 508.0$

a. An advertisement is acceptable if it is fairly certain that its mean rating exceeds 50. In particular, is the claim that the mean rating for the Advertisement 1 exceeds 50 supported by the data? Use $\alpha = 0.05$.
b. It is also of interest to see whether the first 2 advertisements differ significantly and if the last 2 advertisements differ significantly. In particular, do the data support the claim that the mean ratings of the third and fourth advertisements are different? Use $\alpha = 0.05$.
c. Do the data support the claim that the means of the 4 advertisements are not all the same? Use $\alpha = 0.05$.

19. A consumer lobby wishes to determine the proportion of voters currently in favor of an anti-pollution proposition on the ballot in an upcoming election. They want an interval estimate with a confidence level of 99%. A polling organization offers to do the survey for a charge of $5 per voter sampled. Assume that on this issue the number of undecided voters is so small as to be negligible.
a. If the lobby has $25,000 budgeted for the project, determine the degree of precision obtainable in the survey estimate.
b. What would be the cost of a survey that would provide an estimate accurate to within ± 0.01 of the true proportion?

20. Ventures Unlimited is a firm that buys ideas for new products from inventors, designs the products, and then markets them. One inventor sends in his idea for a new computer terminal that is faster than any terminal on the market but is expensive to manufacture. Ventures Unlimited feels that if it were likely that more than 20% of large firms are willing to consider buying the product, it would be worthwhile to design and manufacture a prototype.
a. A survey is sent out to a simple random sample of 100 of these firms, and 30 firms show an interest in buying the terminal. Is this a statistically significant result indicating that the company should proceed with the manufacturing of the prototype assuming $\alpha = 0.05$? Show your work, stating the alternative hypotheses and decision rule that you have used.

b. The prototype is manufactured and the sales in the test market are measured for 25 weeks. The arithmetic mean for weekly sales was 105 terminals. We assume that the standard deviation for weekly sales is $\sigma = 20$ terminals. Ventures Unlimited has decided that if it is conclusive that the mean for weekly sales exceeds 100 terminals, then manufacturing should commence. Can this claim be made using $\alpha = 0.01$? Show your work.

c. If, in fact, the true mean for weekly sales is 110 terminals, what is the probability that Ventures Unlimited would erroneously decide not to manufacture the product using the test constructed in part (b)?

21. The Burger Chief fast food chain has a number of restaurants located in the Philadelphia area. The chain claims that it dispenses hamburgers of virtually identical content, quality, and size at all locations. A total of 7 hamburgers purchased at the chain's Main Line location contained beef weighing an average of 3.8 ounces per hamburger, with a sample variance of 0.070. At the North Philadelphia restaurant, however, a sample of 11 hamburgers yielded corresponding values equal to 3.5 and 0.068.

a. Set up and test a hypothesis versus a suitable alternative to check the chain's claim. Use $\alpha = 0.01$.

b. Test the null hypothesis that the average weight of the beef in the hamburgers at the North Philadelphia location is at least that of the hamburgers at the Main Line location, versus the alternative that it is not. Use $\alpha = 0.05$.

c. A confidence interval estimate of (3.4857–4.1143) ounces was reported for the average weight of beef in the hamburgers at the Main Line location based on the data in part (a). What confidence level was used? Show your work.

22. A dean found a 10-year-old study that showed that 30% of the MBAs then had found the program difficult, while only 20% had found it easy. The remainder thought that the level was just right. From the current matriculants, 200 students are interviewed at random. The backgrounds of the 200 current students are also recorded.

a. If 40 of the current students feel that the program is difficult and 80 feel it is easy, should it be concluded that the reactions now are different from 10 years ago? Use $\alpha = 0.01$ and show your work.

b. From the data, would you conclude that students' backgrounds affect how they feel about the degree of difficulty of the program? Use $\alpha = 0.05$.

	Backgrounds of Current Students		
	Technical	Business	Other
Easy	30	20	30
Just right	15	10	55
Difficult	5	20	15

23. Dietrich Investments is a large firm which specializes in developing portfolios for nonprofit institutions. Each portfolio is assigned to one of its analysts. In 1980 fifteen analysts were hired: Five of these attended a special program, but otherwise had only a high school diploma; 5 other analysts had a college degree but no graduate training; the remaining 5 had MBA degrees. One portfolio is randomly selected from each analyst. The 15 percentage returns are combined into the 3 groups according to educational background.

High School	College	MBA
$\bar{x}_1 = 6$	$\bar{x}_2 = 7$	$\bar{x}_3 = 8$
$s_1^2 = 0.36$	$s_2^2 = 0.64$	$s_3^2 = 0.25$

a. Dietrich Investments wants to be confident that the mean percentage return for each group is greater than 5. Is it safe to hire high school graduates? Show your work. Use $\alpha = 0.05$.
b. Can it be shown that there is a significant difference in performance between high school graduates and MBAs? Show your work. Use $\alpha = 0.05$.
c. Test the hypothesis of no difference in performance among the 3 groups. Show your work. Use $\alpha = 0.05$.

9

Regression Analysis and Correlation Analysis

Prediction is required in virtually every aspect of the management of enterprises. Indeed, business planning and decision making are inseparable from prediction. Forecasting of sales, earnings, costs, production, personnel requirements, inventories, purchases, and capital requirements is the foundation of company planning and control. We could give many other illustrations of the need for prediction in the management of private and public organizations.

9.1 EXPRESSING RELATIONSHIPS AMONG VARIABLES

In this chapter, we concern ourselves with a broad class of techniques for prediction, namely, *regression* and *correlation analysis*. This type of analysis is undoubtedly one of the most widely used statistical methods. **Regression analysis** provides the basis for predicting the values of a variable from the values of one or more other variables; **correlation analysis** enables us to assess the strength of the relationships (correlations) among the variables.

Equations are used in mathematics to express the relationships among variables. In fields such as geometry or trigonometry, these mathematical equations, or functions, express the **deterministic (exact) relationships** among the variables of interest.

Deterministic relationships

- The equation $A = s^2$ describes the relationship between s (the length of the side of a square) and A (the area of the square).

- The equation $A = ab/2$ expresses the relationship between b (the length of any side of a triangle), a (the altitude or perpendicular distance to that side from the angle opposite it, and A (the area of the triangle).

By substituting numerical values for the variables on the right-hand sides of these equations, we can *determine* the *exact values* of the quantities on the left-hand sides.

Statistical relationships

In the social sciences and in fields such as business and government administration, exact relationships are not generally observed among variables, but rather **statistical relationships** prevail. That is, certain average relationships may be observed among variables, but these average relationships do not provide a basis for perfect predictions.

- If we know how much money a corporation spends on television advertising, we cannot make an exact prediction of the amount of sales this promotional expenditure will generate.
- If we know a family's net income, we cannot make an exact forecast of the amount of money that family saves, but we can measure statistically how family savings vary, on the average, with differences in income. On the other hand, we can measure statistically how sales vary, on the average, with differences in television advertising.

We can also determine to what extent actual figures vary from these average relationships. On the basis of these relationships, we may be able to estimate the values of the variables of interest closely enough for decision-making purposes. The techniques of regression and correlation analysis are important statistical tools in this measurement and estimation process.

Techniques of Analysis

The term **regression analysis** refers to the methods by which estimates are made of the values of a variable from a knowledge of the values of one or more other variables, and to the measurement of the errors involved in this estimation process. The term **correlation analysis** refers to methods for measuring the degree of association among these variables.

Two-variable linear regression and correlation analysis

We begin by discussing **two-variable linear regression and correlation analysis**. The term **linear** means that an equation of a straight line of the form $Y = A + BX$, where A and B are fixed numbers, is used to describe the average relationship between the two variables and to carry out the estimation process. The factor whose values we wish to estimate is referred to as the **dependent variable** and is denoted by the symbol Y. The factor from which these estimates are made is called the **independent variable** and is denoted by X.

In the formulation of a regression analysis, the investigator should use prior knowledge and the results of past research to select independent variables that are potentially helpful in predicting the values of the dependent variable. Hence, the investigator should use any available information concerning the direction of cause and effect in the selection of relevant variables. As indicated earlier, correlation analysis can be used to measure the strength or degree of correlation among the variables of interest.

Let us consider some illustrative cases of variables that it is reasonable to assume are related to one another, that is, correlated. If suitable data were available, we might attempt to construct an equation that would permit us to estimate the values of one variable from the values of the other. The first named factor in each pair is the variable to be estimated, that is, the dependent variable, and the second one is independent. Consumption expenditures might be estimated from a knowledge of income; investment in telephone equipment from expenditures on new construction; personal net savings from disposable income; commercial bank interest rates from Federal Reserve Bank discount rates; and success in college from Scholastic Aptitude Test scores.

Of course, additional definitions are required to attach meaning to the estimation problems listed above. Thus, in the illustration of consumption expenditures and income, we would have to specify whose expenditures and whose income are involved. If we wanted to estimate family consumption expenditures from family income, the family would be the "unit of association." The estimating equation would be constructed from data representing observations of these two variables for individual families. We would have to define the variables more specifically. For example, we might be interested in estimates of annual family consumption expenditures from annual family net income, where again these terms would require precise definitions.

In each of these examples, we can specify other independent variables that might be included to obtain good estimates of the dependent variable. Hence, in estimating a family's consumption expenditures, we might wish to use knowledge of the size of the family in addition to information on the family's income. This would be an illustration of **multiple regression analysis**, where two independent variables (family income and family size) are used to obtain estimates of a dependent variable. In this chapter, we shall consider two-variable problems involving a dependent factor and only one independent factor. Chapter 10 covers multiple regression problems involving more than one independent variable.

Multiple regression analysis

The Simple Two-Variable Linear Regression Model

The use of a variable to predict the values of another variable may be viewed as a problem of statistical inference. The population consists of all relevant pairs of observations of the dependent and independent variables. Generally, estimates or predictions must be made from only a sample of that population. For example, suppose we are interested in estimating a family's annual food expenditures (the dependent variable) from a knowledge of the family's annual net income (the independent variable). Furthermore, let us assume that we are

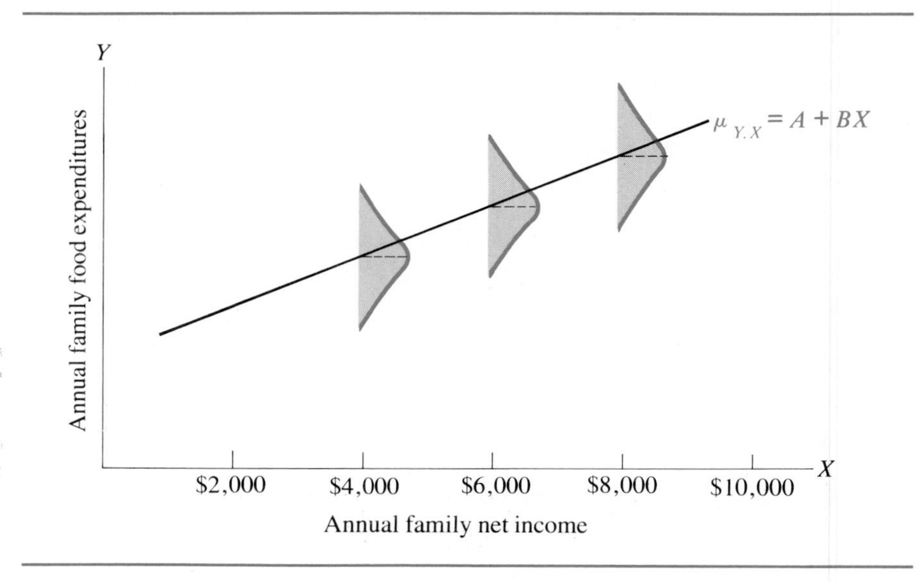

FIGURE 9-1
The linear regression model and conditional probability distributions of annual family food expenditures for selected annual family net incomes

interested in these estimates for 1986 for low-income families with $10,000 or less of annual net income in a particular metropolitan area. Figure 9-1 shows what the population relationship between these two variables might look like for families in the given income range if we assume that this relationship is linear (that is, in the form of a straight line). The fact that the line runs from the lower left to the upper right side of the graph indicates that as family annual net income increases, family food expenditures also increase. We denote this **population regression line** or **true regression line** as

Population regression line

$$\mu_{Y.X} = A + BX \tag{9.1}$$

where the meaning of $\mu_{Y.X}$ is explained in the next paragraph.

In order to understand the assumptions involved in the linear regression model, let us consider families with certain specified incomes (say, $4,000, $6,000, and $8,000), as indicated in Figure 9-1. If we now focus on families at a given income level (say, $6,000), the probability distribution of the Y variable, food expenditures, is a **conditional probability distribution** of Y given $X = \$6,000$. Such a conditional probability distribution may be symbolized in the usual way as $f(Y|X = \$6,000)$ or, in general, $f(Y|X)$. This distribution has a mean, which may be denoted $\mu_{Y.X}$, and a standard deviation, which may be denoted $\sigma_{Y.X}$. In the linear regression model, the assumed relationship between $\mu_{Y.X}$ and X can be graphed as a straight line. That is, the means of the conditional probability distributions are assumed to lie along a straight line. In this model of a linear regression function, A and B are population parameters that must be *estimated* from sample data.

The conditional probability distributions describe the variability of the Y values. Thus, for families with $6,000 annual net income in the given metropolitan area, the conditional probability distribution indicates the variability of family annual food expenditures around the conditional mean food expenditures, $\mu_{Y.X}$ (in this case $\mu_{Y.\$6,000}$). The symbol $\mu_{Y.X}$ represents the mean of the Y variable for a given X value.

Assumptions of the linear regression model

In addition to the assumption of a linear relationship, the following assumptions are involved in the use of linear regression model:

1. The Y values are independent of one another.
2. The conditional probability distributions of Y given X are normal.
3. The conditional standard deviations $\sigma_{Y.X}$ are equal for all values of X.

A few comments about these assumptions are in order. The first assumption means that there is independence between successive observations. This means, for example, that a low value for Y on the first observation does not imply that the second Y value will also be low. In certain situations, such an assumption may not be particularly valid. We see this clearly in time-series data that move in cycles around a fitted trend line. Observations in the expansion phase of a business cycle will tend to lie above the trend line and to have relatively high values; the opposite is true for observations in a contraction or recession phase. Hence, successive observations in this case tend to be related rather than independent.

The second assumption means that for each value of X, we are assuming that the Y values are normally distributed around $\mu_{Y.X}$. As we will see, this assumption is useful for making probability statements about estimates of the dependent variable Y.

Homoscedasticity

The third assumption implies that there is the same amount of variability around the regression line at each value of the independent variable, X. This characteristic is referred to as **homoscedasticity**. Note that in regression analysis, according to the second assumption, only Y is considered a random variable. X is considered fixed. Hence, if we attempt to predict a Y value (for example, a family's food expenditures) from a knowledge of X (for example, a family's income), the predicted Y value is subject to error. X is assumed to be known without error. On the other hand, in correlation analysis, both X and Y are treated as normally distributed random variables.

Clearly these three assumptions are never perfectly met in the real world. In many situations, however, these assumptions are approximately true and the model is useful. In this chapter, we develop the standard linear regression model under these assumptions.

Formulation of the problem

Before turning to the methodology of regression and correlation analysis beginning in the next section, it is important to emphasize that the *formulation of the problem* is of critical importance. In the formulation stage, the analyst must

- identify the dependent and independent variables
- specify the relationships among these variables (for example, linear or nonlinear)

- decide what the relevant population is (for example, low-income families in a particular metropolitan area)
- supply the data on the variables for a sample or for the entire relevant population.

9.2 SCATTER DIAGRAMS

Plotting data on a graph is useful in studying the relationship between two variables. A graph allows visual examination of the extent to which the variables are related and aids in choosing the appropriate type of model for estimation. The chart used for this purpose is known as a **scatter diagram**, which is a graph on which each plotted point represents an observed pair of values of the dependent and independent variables. We will illustrate this by plotting a scatter diagram for the data given in Table 9-1. These figures represent observations for a sample of 10 low-income families of annual expenditures on food, which we shall treat as the dependent variable, Y (the factor to be estimated), and annual net income, X, which is the independent variable (the factor from which the estimates are to be made). As in the illustration given in section 9.1, we assume that the 10 families constitute a simple random sample of families with \$10,000 or less of annual net income in a metropolitan area in 1986. Although the sample size is too small to draw useful conclusions that would apply to all such families in a metropolitan area, we shall use such a small

TABLE 9-1
Annual food expenditures and annual net income of a sample of 10 low-income families in a metropolitan area in 1986

Family	Annual Food Expenditures (\$00) Y	Annual Income (\$000) X
A	22	8
B	23	10
C	18	7
D	9	2
E	14	4
F	20	6
G	21	7
H	18	6
I	16	4
J	19	6

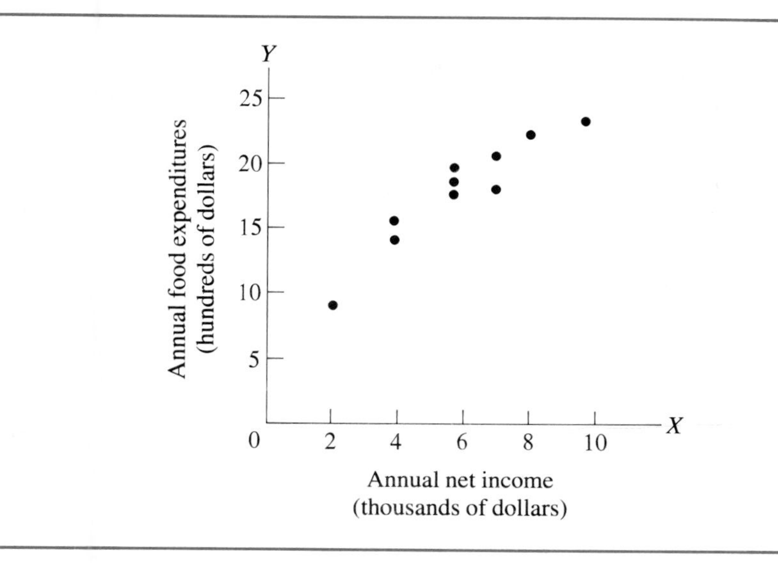

FIGURE 9-2
A scatter diagram of annual food expenditures and annual net income of a sample of 10 low-income families in a metropolitan area in 1986.

sample in order to limit the amount of arithmetic needed. Furthermore, we have assumed relatively low incomes in order to simplify the numerical work.

Scatter diagram

Figure 9-2 presents the data of Table 9-1 plotted as a scatter diagram. On the Y axis are plotted the figures on food expenditures and on the X axis annual net income. This follows the standard convention of plotting the dependent variable along the Y axis and the independent variable along the X axis. The pair of observations for each family determines one point on the scatter diagram. For example, for family A, a point is plotted corresponding to $X = 8$ along the horizontal axis and $Y = 22$ along the vertical axis; for family B, a point is plotted corresponding to $X = 10$ and $Y = 23$. An examination of the scatter diagram gives some useful indications of the nature and strength of the relationship between the two variables. For example, depending on whether the Y values tend to increase or to decrease as the values of X increase, there is a *direct* or *inverse* relationship, respectively, between the two variables. The configuration in Figure 9-2 indicates a general tendency for the points to run from the lower left to the upper right side of the graph. Hence, as noted in section 9.1, as income increases, food expenditures tend to increase. This is an example of a **direct relationship** between the two variables. On the other hand, if the scatter of points runs from the upper left to the lower right (that is, if the Y variable tends to decrease as X increases), there is an **inverse relationship** between the variables. Also, an examination of the scatter diagram gives an indication of whether a straight line appears to be an adequate description of the average relationship between the two variables. If a straight line is used to describe the average relationship between Y and X, a **linear relationship** is present. However if the points on the scatter diagram appear to fall along a

Direct relationship

Inverse relationship

Linear relationship

FIGURE 9-3
Scatter diagrams

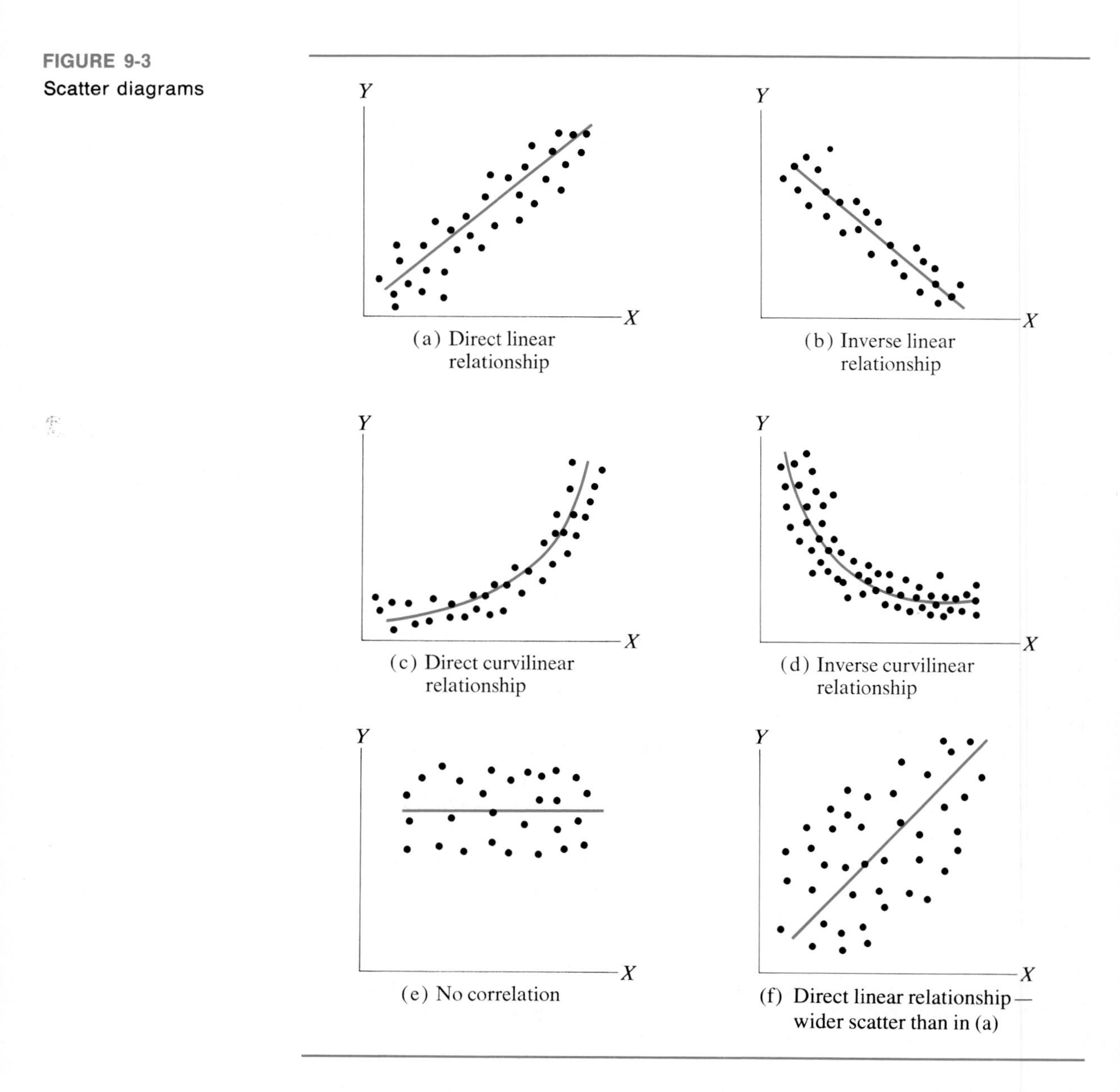

Curvilinear relationship

curved line rather than a straight line, a **curvilinear relationship** exists. Figure 9-3 presents illustrative combinations of the foregoing types of relationships. Parts (a), (b), (c), and (d) of Figure 9-3 show, respectively, direct linear, inverse linear, direct curvilinear, and inverse curvilinear relationships. The points tend to follow a straight line sloping upward in (a), a straight line sloping downward in (b), a curved line sloping upward in (c), and a curved line sloping downward in (d). Of course, the relationships are not always so obvious. In (e), the points appear to follow a horizontal straight line. Such a case depicts "no correlation" between the X and Y variables, or no evident relationship, since the horizontal line implies no change in Y, on the average, as X increases. In (f), the points follow a straight line sloping upward as in (a), but there is a much wider scatter of points around the line than in (a).

In our present problem, we assume that our prior expectation was for a direct linear relationship between the two variables for low-income families and that a visual examination of Figure 9-2 confirmed this expectation. In the next section, we discuss the procedures used in regression and correlation analysis. We begin by considering what we hope to accomplish through the use of such techniques.

9.3 PURPOSES OF REGRESSION AND CORRELATION ANALYSIS

What does a regression and correlation analysis attempt to accomplish in studying the relationship between two variables, such as expenditures on food and net annual income of families? We will concentrate on these basic goals, which emphasize the relationships contained in the particular sample under study. Later, we will consider other objectives involving statistical inference, that is, inferences concerning the population from which the sample was drawn.

The first two objectives and the statistical procedures involved in their accomplishment fall under the heading of *regression analysis*, whereas the third objective and related procedures are classified as *correlation analysis*. These objectives are stated below, and the statistical measures used to achieve the objectives are named. However, the mathematical definitions of these measures are postponed until the discussion of their use in the problem involving family expenditures on food and family income.

Regression analysis

The first purpose of regression analysis is to provide estimates of values of the dependent variable from values of the independent variable.

The device used to accomplish this estimation procedure is the *sample regression line*, which is a line fitted to the data by a method we shall describe in the next section. The **sample regression line** describes the average relationship between the X and Y variables in the sample data. The equation of this line, known as the **sample regression equation**, provides estimates of the mean value of Y for each value of X. Hence, in the illustration given in section 9.1, we could obtain estimates from the sample regression equation of the mean family food expenditures for each level of family income.

A second goal of regression analysis is to obtain measures of the error involved in using the regression line as a basis of estimation.

For this purpose, the *standard error of estimate* and related measures are calculated. The **standard error of estimate** measures the scatter, or spread, of the observed values of Y around the corresponding values estimated from the fitted regression line. **Measures of forecast error** take into account this scatter as well as the probable difference between the regression line fitted to sample data and the true but unknown population regression line.

Correlation analysis

The third objective, which we have classified as correlation analysis, is to obtain a measure of the degree of association or correlation between the two variables.

The **coefficient of correlation** and the **coefficient of determination**, calculated for this purpose, measure the strength of the relationship between the two variables.

9.4 ESTIMATION USING THE REGRESSION LINE

As indicated in the preceding section, to accomplish the first objective of a regression analysis, we must obtain the mathematical equation of a line that describes the average relationship between the dependent and independent variables. We can then use this line to estimate values of the dependent variable. Since the present discussion is limited to *linear* regression analysis, we are referring to a straight line. Ideally, we would like to obtain the equation of the straight line that best fits the data. Let us defer for the moment what we mean by "best fits" and review the concept of the equation of a straight line.

The equation of a straight line is $Y = a + bX$, where a is the so-called "Y intercept," or the computed value of Y when $X = 0$, and b is the slope of the line, or the amount by which the computed value of Y changes with each unit change in X.

Let us review by means of a simple illustration the relationship between the equation $Y = a + bX$ and the straight line that represents the graph of the equation. Suppose the equation is

$$Y = 2 + 3X \tag{9.2}$$

Thus, $a = 2$ and $b = 3$. If we substitute a value of X into this equation, we can obtain the corresponding computed value of Y. Each pair of X and Y values represents a single point. Although only two points are required to determine a straight line, several pairs of X and Y values for the line $Y = 2 + 3X$ are shown next to the graph corresponding to the line shown in Figure 9-4. On this graph, since the a value in the equation of the line is two, the line intersects the Y axis at a height of two units. Also, since the b value, or slope of the line, is three, we note that the Y values increase by three units each time X increases by one unit. This is shown graphically in Figure 9-4 as a rise of three units in the line when X increases by one unit .

The terms "regression line" and "regression equation" for the estimating line and equation stem from the pioneer work in regression and correlation analysis of the British biologist Sir Francis Galton in the nineteenth century. The lines that he fitted to scatter diagrams of data on heights of fathers and sons in this early work came to be known as "regression lines" and the equations of these lines as "regression equations," because Galton found that the heights of the sons "regressed" toward an average height. Unfortunately, the terminology has persisted. Thus, these terms for the estimating line and estimating equation are used in the wide variety of fields in which regression analysis is applied, despite the fact that the original implication of a regression toward an average is not necessarily present for the phenomena under investigation.

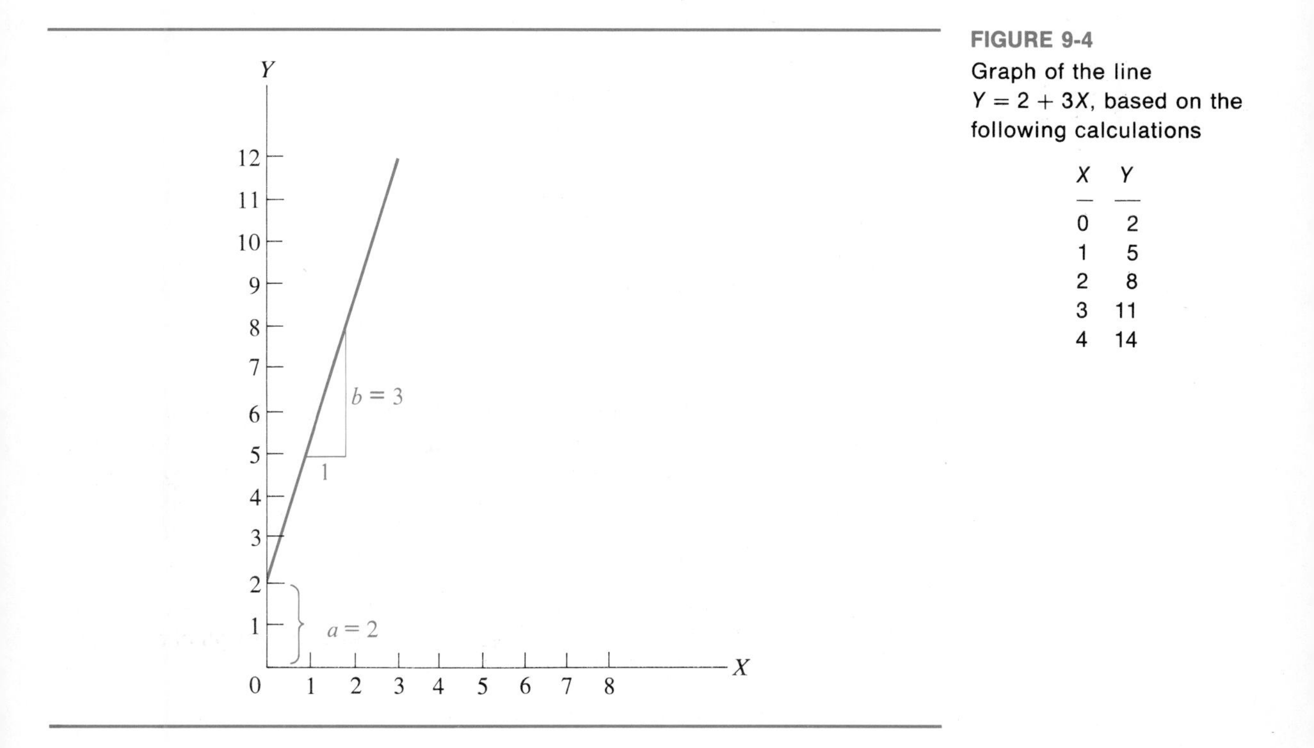

FIGURE 9-4
Graph of the line $Y = 2 + 3X$, based on the following calculations

X	Y
0	2
1	5
2	8
3	11
4	14

We now turn to the question of obtaining a best-fitting line to the data plotted on a scatter diagram in a two-variable linear regression problem. The fitting procedure discussed is the method of least squares, undoubtedly the most widely applied curve-fitting technique in statistics.

The Method of Least Squares

In discussing the linear regression model in section 9.1, we denoted the population or true regression line as $\mu_{Y.X} = A + BX$. Correspondingly, the sample regression line, which is the best-fitting line to the sample data, is denoted as

Sample regression line

(9.3) $$\hat{Y} = a + bX$$

where a and b represent estimates of A and B in the population regression line.

Goodness of fit

In order to establish a best-fitting line to a set of data on a scatter diagram, we must have criteria concerning what constitutes **goodness of fit**. A number of criteria that might at first seem reasonable turn out to be unsuitable. For example, we might entertain the idea of fitting a straight line to the data in such a way that half of the points fall above the line and half below. However, such a line may represent a quite poor fit to the data if, say, the points that fall above the line lie very close to it whereas the points below deviate considerably from it.

Method of least squares

Let us now consider the most generally applied curve-fitting technique in regression analysis, namely, the **method of least squares**. This method imposes the requirement that the *sum of the squares* of the deviations of the observed values of the dependent variable from the corresponding computed values on the regression line must be a minimum. Thus, if a straight line is fitted to a set of data by the method of least squares, it is a "best fit" in the sense that the sum of the squared deviations, $\Sigma(Y - \hat{Y})^2$, is less than it would be for any other possible straight line. Another useful characteristic of the least squares straight line is that it passes through the point of means $(\bar{X}, \bar{Y})$, and therefore makes the total of the positive and negative deviations equal to zero. In summary, the least squares straight line possesses the following mathematical properties:

(9.4) $$\Sigma(Y - \hat{Y})^2 \text{ is a minimum}$$

(9.5) $$\Sigma(Y - \hat{Y}) = 0$$

Figure 9-5 presents graphically the nature of the least squares property. A scatter plot is shown around a least squares sample regression line, denoted $\hat{Y} = a + bX$. Let us consider the first point, whose coordinates are $X = 2$ and $Y = 10$. The vertical distance of the point from the X axis is its Y value, in this case, 10. The $\hat{Y}$ value for the same point is the vertical distance from the X axis to the regression line. In this case, $\hat{Y} = 7$. The vertical distance of the point from the corresponding value on the regression line is the deviation, or residual, $Y - \hat{Y}$. In this illustration, $Y - \hat{Y} = 10 - 7 = 3$. Since the Y value

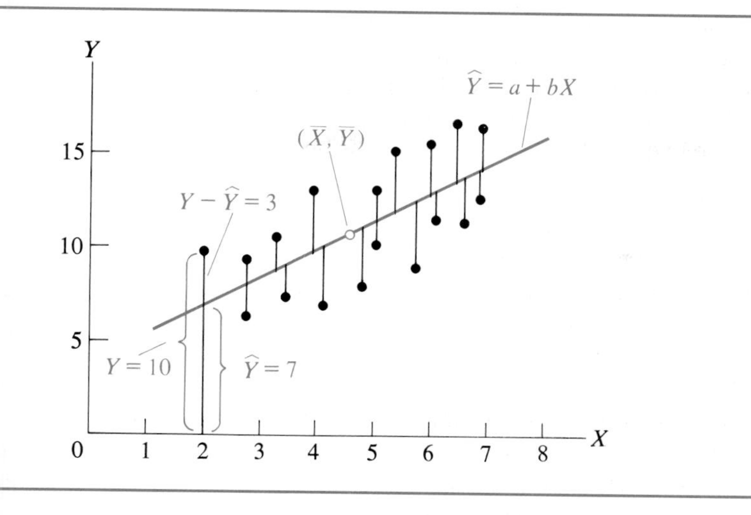

FIGURE 9-5
Scatter diagram and graphic representation of deviations from a fitted least squares regression line

is above the regression line, the deviation is positive; for Y values below the line, the deviations are negative. The square of the deviation for the first point is $3^2 = 9$. As mentioned above, the line fitted by this method has the property that the total of all squared deviations is less than the corresponding totals for any other straight line that could have been fitted to the same data.

The regression line is depicted as passing through the point established by the means of the X and Y variables, $(\bar{X}, \bar{Y})$, which in this case is (4.5, 11). It can be shown that a least squares straight line must include this point of means and that this property makes the algebraic sum of the deviations above and below the line equal to zero. That is,

> the sum of the deviations of the points lying above the regression line, which are positive, and the deviations of the points lying below the regression line, which are negative, is zero.

It can be shown that the a and b values derived by the method of least squares are unbiased, efficient, and consistent estimators of the corresponding parameters A and B in the regression line. Considering the "best fit" properties mentioned earlier and these other desirable properties of this estimation method, it is perhaps not surprising that the method of least squares is the standard method of curve-fitting in regression analysis.

Computational Procedure

By using calculus methods to apply the condition that the sum of the squared deviations from a straight line must be a minimum, two equations known as the **normal equations** are derived.[1] These equations can be solved for the values of a and b in the regression equation $\hat{Y} = a + bX$. The values of a and b can then be determined from the following general solutions for a and b:

Normal equations

$$a = \bar{Y} - b\bar{X} \tag{9.6}$$

$$b = \frac{\Sigma XY - n\bar{X}\bar{Y}}{\Sigma X^2 - n\bar{X}^2} \tag{9.7}$$

where $\bar{X}$ and $\bar{Y}$ are the arithmetic means of the X and Y variables.

Let us return to the problem involving the sample of 10 low-income families and assume that we have decided to fit a *straight line* to the data. From the original observations, we can determine the various quantities (n, ΣY, ΣX, ΣXY, and ΣX^2) required in equations 9.6 and 9.7, where n is the number of pairs of X and Y values (in this case, 10). For our illustration, the computation of the required totals is shown in Table 9-2. Although ΣY^2 is not needed for the calculation of a and b, its computation is also shown. This figure is useful for calculating the standard error of estimate, to be discussed shortly.

From Table 9-2, we compute the means of X and Y as

$$\bar{X} = \frac{\Sigma X}{n} = \frac{60}{10} = 6 \text{ (thousands of dollars)}$$

$$\bar{Y} = \frac{\Sigma Y}{n} = \frac{180}{10} = 18 \text{ (hundreds of dollars)}$$

Substituting the additional quantities $n = 10$, $\Sigma XY = 1{,}159$, and $\Sigma X^2 = 406$ from Table 9-2 into equations 9.6 and 9.7, we obtain the following values for a and b:

$$b = \frac{1{,}159 - (10)(6)(18)}{406 - 10(6)^2} = \frac{79}{46} = 1.717$$

$$a = 18 - 1.717(6) = 18 - 10.302 = 7.698$$

Hence, the least squares regression line is

$$\hat{Y} = 7.698 + 1.717X \tag{9.8}$$

If a family drawn from the same population had an annual net income of \$8,000 in 1986, its estimated annual food expenditures from equation 9.8 would be

$$\hat{Y} = 7.698 + 1.717(8) = 21.434 \text{ (hundreds of dollars)}$$

[1] The derivation of the equations is as follows. We denote the sum of squared deviations that must be minimized as some function of the unknown quantities a and b. Thus, let

$$F(a, b) = \Sigma(Y - \hat{Y})^2$$

Substituting $a + bX$ for $\hat{Y}$ into the above equation gives

$$F(a, b) = \Sigma(Y - a - bX)^2$$

We impose the condition of a minimum value for $F(a, b)$ by obtaining its partial derivatives with

TABLE 9-2
Computions for a regression and correlation analysis for the data shown in Table 9-1

Family	Y	X	XY	X^2	Y^2
A	22	8	176	64	484
B	23	10	230	100	529
C	18	7	126	49	324
D	9	2	18	4	81
E	14	4	56	16	196
F	20	6	120	36	400
G	21	7	147	49	441
H	18	6	108	36	324
I	16	4	64	16	256
J	19	6	114	36	361
	180	60	1,159	406	3,396

Because the best-fitting line to the sample data also estimates the population regression line $\mu_{Y.X} = A + BX$, where $\mu_{Y.X}$ is a conditional mean, we can state that the $\hat{Y}$ value of \$2,143.40 is the best estimate of *mean* annual family food expenditures for an $X = 8$ (\$8,000) annual family income. Thus, the sample regression line, or prediction line as it is often called, may be used to predict a future value of Y or to estimate a mean value of Y for a given value of X.

Terms

Incidentally, when the context is clear, the terms **regression line** and **regression equation** are often used for **sample regression line** and **sample regression equation**. That practice is used in this chapter and in Chapter 10.

By plotting the point thus determined ($X = 8$, $Y = 21.434$) and one other point, or by plotting any two points derived from the regression equation, we can graph the regression line. The line is shown in Figure 9-6, along with the original data. Hence, in this case $a = 7.698$ means that the estimated annual food expenditures for a family whose income is zero dollars in 1986 is 7.698 (hundreds of dollars), or \$769.80. Since no families in the original sample had

respect to a and b and setting them equal to zero. Thus,

$$\frac{\partial F(a, b)}{\partial a} = -2\Sigma(Y - a - bX) = 0$$

$$\frac{\partial F(a, b)}{\partial b} = -2\Sigma(Y - a - bX)(X) = 0$$

Solving these equations yields

$$\Sigma Y = na + b\Sigma X$$
$$\Sigma XY = a\Sigma X + b\Sigma X^2$$

from which equations 9.6 and 9.7 can be derived. A check reveals that the second derivatives of $F(a, b)$ are positive, so a minimum has been found.

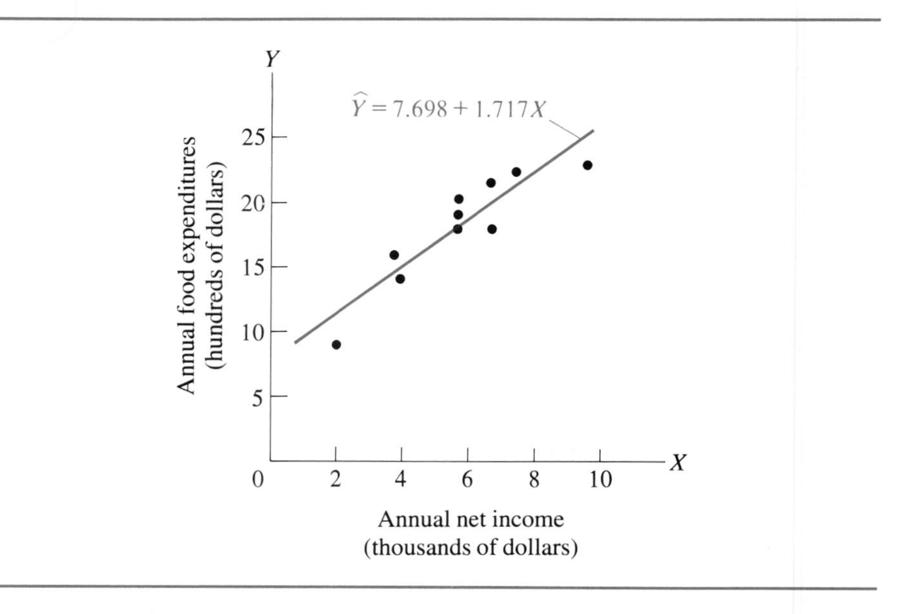

FIGURE 9-6
Least squares regression line for food expenditures and annual net income of a sample of 10 low-income families in a metropolitan area in 1986

incomes less than $2,000, it would be extremely hazardous to make predictions for families with incomes less than the $2,000 figure. Prediction outside the range of the original observations is discussed in a later section of this chapter.

Sample regression coefficient

The b value in the regression equation is often referred to as the **sample regression coefficient** or **sample slope coefficient**. The figure of $b = 1.717$ indicates first of all that the slope of the regression line is positive. Thus, as income increases, estimated food expenditures increase. Taking into account the units in which the X and Y variables are stated, $b = 1.717$ means that for two families whose annual net incomes differ by $1,000, the *estimated difference* in their annual food expenditures is $171.70. This is an interpretation in terms of the *regression line*. If we think of the figure $b = 1.717$ in terms of the *sample studied*, we can say that for two families whose annual net incomes differ by $1,000, their annual food expenditures differ, *on the average*, by $171.70.

Exercises 9.4

1. List 3 main objectives of simple two-variable regression and correlation analysis. What statistical measures are used to achieve these objectives?
2. a. What is the meaning of the term, *method of least squares*?
   b. Describe the main assumptions of the linear regression model.
3. An insurance company wished to examine the relationship between income and amount of life insurance held by heads of families with incomes of $25,000 or less. The company drew a simple random sample of 12 family heads and listed the results shown in Table 9-3.

   a. Determine the linear regression equation using the method of least squares with income as the independent variable. Calculate $\hat{Y}$ for each income level given above.

TABLE 9-3
Relationship between income and amount of life insurance

Family	Amount of Life Insurance ($000)	Income ($000)
A	32	14
B	40	19
C	50	23
D	20	12
E	22	9
F	35	15
G	55	22
H	45	25
I	28	15
J	22	10
K	24	12
L	35	16

b. What is the meaning of the regression coefficient b in this case?
c. What is your estimate of the amount of life insurance carried by a family head from the same population whose income is $20,000?

4. The scatter diagram and regression line in Figure 9-7 show the relationship between hot dog sales and beer sales at a college football stadium during the last 5 seasons. On the basis of this chart, estimate the values of a and b (using specific numbers) in the equation of the regression line $\hat{Y} = a + bX$.

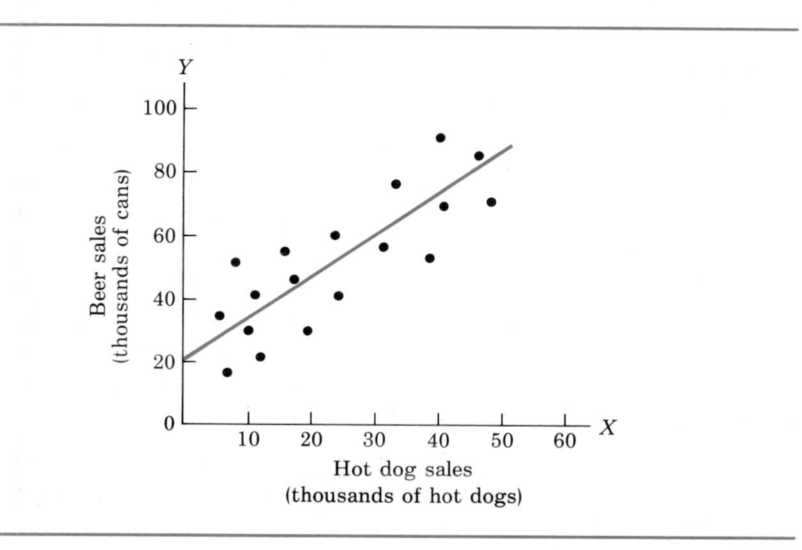

FIGURE 9-7
Relationship between sales of hot dogs and beer at a college football stadium.

5. A national education organization is trying to build a strong case for increased expenditures on public education below the college level. The organization has taken a random sample of towns and cities across the United States and has collected the following information:

$X =$ average annual expenditure per student by town or city (in dollars), ranges from \$475 to \$700

$Y =$ average performance of college-oriented student on standard college entrance examination

$n = 200$

$\Sigma X = 125{,}000$

$\Sigma Y = 120{,}000$

$\Sigma XY = 77{,}625{,}000$

$\Sigma X^2 = 80{,}000{,}000$

a. Using the least squares method, estimate an appropriate regression equation.
b. Give two interpretations of the regression coefficient b.
c. The range of possible scores on the standard examinations is 200 to 800. Calculate the estimated average performance for college-oriented students in a city that spends an average of \$800 per student on public education. Is there any problem with this estimate?

6. A research analyst for a federal reserve bank has compiled information on loan activity of a simple random sample of 45 banks in the reserve bank's district. The analyst has calculated the following statistics on net return on loans as a percentage of total loans from each bank (Y) and total deposits, in tens of millions of dollars, for each bank (X):

$X =$ total deposits (in tens of millions of dollars)

$Y =$ net return on loans as a percentage of total loans

$n = 45$

$\Sigma X = 787.5$

$\Sigma Y = 225$

$\Sigma XY = 3800$

$\Sigma X^2 = 15{,}150$

a. Estimate a least squares regression equation, obtaining estimates for a and b.
b. The Naugatuck Valley National Bank has total deposits of approximately \$225 million. Estimate the net return on loans, using the above equation.

9.5 CONFIDENCE INTERVALS AND PREDICTION INTERVALS IN REGRESSION ANALYSIS

Standard Error of Estimate

Now that we have seen that the regression equation is used for estimation, we can turn to the second objective in section 9.3—obtaining measures of the error involved in using the regression line for estimation. If there is a great deal of scatter of the observed Y values around the fitted line, estimates of Y values based on computed values from the regression line will not be close to the observed Y values. On the other hand, if every point falls on the regression line, insofar as the **sample observations** are concerned, perfect estimates of the Y values can be made from the fitted regression line. Just as the standard deviation was used as a measure of the scatter of a set of observations about the mean, an analogous measure of dispersion of observed Y values around the regression line is desirable. The measure of dispersion, referred to as the **standard error of estimate** (or the **standard deviation of the residuals**), is obtained by solving the equation

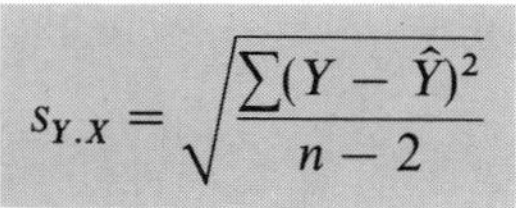

(9.9) $$s_{Y.X} = \sqrt{\frac{\sum(Y - \hat{Y})^2}{n - 2}}$$

where, as before, n is the sample size.

Although $s_{Y.X}$ is called the *standard error* of estimate, it is not a standard error in the sense that the term was used in Chapters 6 and 7, that is, as a measure of sampling error. The standard error of estimate simply measures the scatter of the observed values of Y around the corresponding computed $\hat{Y}$ values on the regression line. The sum of the squared deviations is divided by $n - 2$ because this divisor makes $s^2_{Y.X}$ an unbiased estimator of the conditional variance around the population regression line,[2] denoted as $\sigma^2_{Y.X}$. The $n - 2$ represents the number of degrees of freedom around the fitted regression line. In general, the denominator is $n - k$ where k is the number of constants in the regression equation. In the case of a straight line, the denominator is $n - 2$ because two degrees of freedom are lost when a and b are used as estimates of the corresponding constants in the population regression line.

Notation $s_{Y.X}$

It is useful to consider the nature of the notation for the standard error of estimate $s_{Y.X}$. In the discussion of dispersion in section 1.15, we used the symbol s to denote the standard deviation of a sample of observations. The use of the letter s in $s_{Y.X}$ is analogous, since, as explained in the preceding

[2] In using $s_{Y.X}$ to estimate $\sigma_{Y.X}$, the deviation of each Y value is taken around its own estimated conditional mean $\hat{Y}$. We can combine the squares of these deviations, which are obtained from different conditional probability distributions, because of the assumption of the linear regression model that the conditional standard deviations $\sigma_{Y.X}$ are all equal.

paragraph, $s_{Y.X}$ is also a measure of dispersion computed from a sample. However, since both the variables Y and X are present in a two-variable regression and correlation analysis, subscript notation is required to distinguish among the various possible dispersion measures. Hence, the notation for the standard error of estimate, where Y and X are, respectively, the dependent and independent variables, is $s_{Y.X}$. The letter to the left of the period in the subscript is the dependent variable, and the letter to the right denotes the independent variable. Subscripts are also required to distinguish standard deviations around the means of the two variables. Thus, s_Y denotes the standard deviation of the Y values of a sample around the mean $\bar{Y}$, and s_X denotes the standard deviation of the X values around their mean $\bar{X}$.

In a problem containing large numbers of observations, the computation of the standard error of estimate using equation 9.9 clearly involves a great deal of arithmetic. Calculation of Y for each X value in the sample is required, and then the arithmetic implied by the formula must be carried out. A useful shortcut formula involving only quantities already computed is given by

$$s_{Y.X} = \sqrt{\frac{\Sigma Y^2 - a\Sigma Y - b\Sigma XY}{n-2}} \tag{9.10}$$

All quantities required by equation 9.10 were calculated for our illustrative problem in Table 9-3, or were computed in obtaining the constants of the regression line. Hence, the standard error of estimate for these data is

$$s_{Y.X} = \sqrt{\frac{3{,}396 - (7.698)(180) - (1.717)(1{,}159)}{10-2}} = 1.595 \text{ (hundreds of dollars)}$$

Since the standard error of estimate $s_{Y.X}$ is an estimate of $\sigma_{Y.X}$ (the standard deviation around the true but unknown population regression line), $s_{Y.X}$ may be used and interpreted as a standard deviation. If every sample point falls on the regression line—that is, if there is no scatter around the line—then $s_{Y.X} = 0$. This indicates that the regression line is a perfect fit to the sample data.

The larger the value of $s_{Y.X}$, the greater is the scatter around the regression line.

As indicated earlier, in the case of a perfect linear relationship between X and Y for a *sample* of data, given a value of X, we could estimate or predict the corresponding *sample* Y value perfectly. However, to make predictions about values in the population not included in our sample, we would have to take account of the fact that the sample regression line we have fit to the data plotted on a scatter diagram may differ from the unknown regression line because of chance errors of sampling. That is, because of chance sampling errors, the a and b values in the sample regression line, $\hat{Y} = a + bX$, may differ from the A

and B values in the true population regression line, $\mu_{Y.X} = A + BX$. This means that the height and slope of the sample regression line may differ from the height and slope of the population regression line. Therefore, in making predictions about items in the population that are not included in the sample, we cannot simply use $s_{Y.X}$ in establishing confidence intervals. We must use appropriate standard errors that take the aforementioned chance errors into account. We now turn to the discussion of such confidence intervals.

Types of Interval Estimates

Two types of estimates, or predictions, of values of the dependent variable are ordinarily made in regression analysis. Interval estimates are generally made for both types of estimates, with the *width of the intervals* indicating the precision of the estimation procedure.

Interval estimates for a *conditional mean* are usually referred to as **confidence intervals**. For example, in our food expenditures illustration, we may be interested in estimating *average* food expenditures for families with annual net incomes of \$8,000; that is, we may want an estimate of the mean of the conditional probability distribution of food expenditures for families with annual net incomes of \$8,000.

Confidence intervals

Estimates that involve predicting an *individual value* of the dependent variable Y are referred to as **prediction intervals**. Sometimes we wish to predict a *single value* of the dependent variable Y, rather than a conditional average value as in the first type of estimate. Thus, for example, we may wish to predict food expenditures for a *particular* family whose income is \$8,000. Such single values cannot be predicted with as much precision as for conditional means.

Prediction intervals

We consider these two types of estimates in the following subsections.

Confidence Interval Estimate of a Conditional Mean

In Chapter 6, we discussed methods for estimating a population parameter from a statistic observed in a simple random sample. Thus, for example, we can establish a *confidence interval* for the population mean food expenditures for the sample of 10 families for whom data were shown in Table 9-1. Assuming that the dependent variable Y is normally distributed, and using equation 6.15, we can calculate the confidence limits for population mean food expenditures as

Confidence limits of population mean

$$\bar{Y} \pm ts_{\bar{Y}} = \bar{Y} \pm t\frac{s_Y}{\sqrt{n}}$$

where $\bar{Y}$ is the sample mean for the desired confidence level, t is determined for $n - 1$ degrees of freedom for the desired confidence level, $s_{\bar{Y}}$ is the estimated standard error of the mean, s_Y is the sample standard deviation of the Y data, and n is the sample size (in this case, 10). The symbol Y is used for the variable of interest here rather than the X used in equation 6.15. Carrying out the arithmetic in this illustration for a 95% confidence interval, we have:

$$\bar{Y} = 18 \text{ (hundreds of dollars)}$$

$$s_Y = \sqrt{\frac{\Sigma(Y - \bar{Y})^2}{n - 1}} = \sqrt{\frac{156}{9}} = 4.163 \text{ (hundreds of dollars)}$$

We have $t = 2.262$ for $v = 10 - 1 = 9$ degrees of freedom (obtained from Table A-6 of Appendix A for a 95% confidence coefficient, that is, 5% area in both tails, or 2.5% in each tail), and

$$s_{\bar{Y}} = \frac{s_Y}{\sqrt{n}} = \frac{4.163}{\sqrt{10}} = 1.317$$

Therefore, $\bar{Y} \pm 2.262 s_{\bar{Y}} = 18 \pm 2.262(1.317)$ (hundreds of dollars).

Hence, the 95% confidence limits are 15.021 and 20.979 (hundreds of dollars). That is, population mean food expenditures would be included in 95% of the intervals so constructed.

Interval estimates can be determined from a regression equation in an essentially similar way. Let us consider the construction of an interval estimate for a conditional mean $\mu_{Y.X}$. Returning to our illustration, suppose we wanted to estimate the *mean* or expected food expenditures for a given family with an annual net income of $8,000. As we have seen, another way of stating this is that we wish to estimate the mean of the conditional probability distribution of food expenditures for $X = 8$ (thousands of dollars). An unbiased estimate of this conditional mean, denoted $\mu_{Y.8}$, is given by $\hat{Y}$ from our sample regression line for $X = 8$ (thousands of dollars). We previously calculated $\hat{Y}$ in section 9.4 as

$$\hat{Y} = 7.698 + 1.717(8) = 21.434 \text{ (hundreds of dollars)}$$

This figure of $\hat{Y} = 21.434$ (hundreds of dollars) is the point on the sample regression line that corresponds to $X = 8$ (thousands of dollars), and it is our best estimate of the conditional mean of Y, the vertical coordinate of the point on the population regression line when $X = 8$. This situation is depicted in Figure 9-8. The confidence interval estimate for the conditional mean (also called the confidence interval for the regression line) is given by

Confidence interval estimate for the conditional mean

(9.11) $$\hat{Y} \pm t s_{\hat{Y}}$$

where $s_{\hat{Y}}$ is the estimated standard error of the conditional mean and t is the t multiple determined for $n - 2$ degrees of freedom for the desired confidence level. The number of degrees of freedom is $n - 2$ because two degrees of freedom are lost by using a and b from the sample regression equation, $\hat{Y} = a + bX$, to estimate A and B in the population regression line, $\mu_{Y.X} = A + BX$.

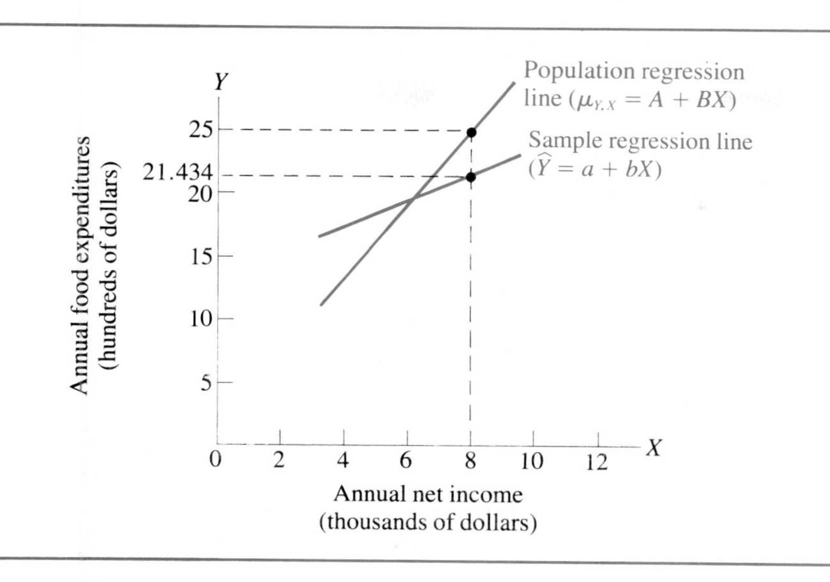

FIGURE 9-8
Estimated and population mean food expenditures when income equals $8,000

The estimated standard error of the conditional mean is given by

Estimated standard error of the conditional mean

(9.12)
$$s_{\hat{Y}} = s_{Y.X}\sqrt{\frac{1}{n} + \frac{(X_0 - \bar{X})^2}{\Sigma(X - \bar{X})^2}}$$

This expression for $s_{\hat{Y}}$ can be used to obtain a confidence interval estimate for the mean or expected value of annual food expenditures for a given value of X, say X_0.

Continuing with our illustration, for a family with an annual net income of $8,000, we compute the estimated standard error of the conditional mean. A more convenient form of equation 9.12 for computation purposes is

A more convenient form of equation 9.12

(9.13)
$$s_{\hat{Y}} = s_{Y.X}\sqrt{\frac{1}{n} + \frac{(X_0 - \bar{X})^2}{\Sigma X^2 - \frac{(\Sigma X)^2}{n}}}$$

Substituting into equation 9.13 for $X_0 = 8$ (thousands of dollars), we get

$$s_{\hat{Y}} = 1.595\sqrt{\frac{1}{10} + \frac{(8 - 6)^2}{406 - \frac{(60)^2}{10}}}$$

$$s_{\hat{Y}} = 0.6897 \text{ (hundreds of dollars)}$$

Hence, a 95% confidence interval for the conditional mean obtained by substitution into equation 9.11 is

$$21.434 \pm 2.306(0.6897) \quad \text{or} \quad (19.844, 23.024) \text{ in hundreds of dollars}$$

FIGURE 9-9
95% confidence interval for the conditional mean

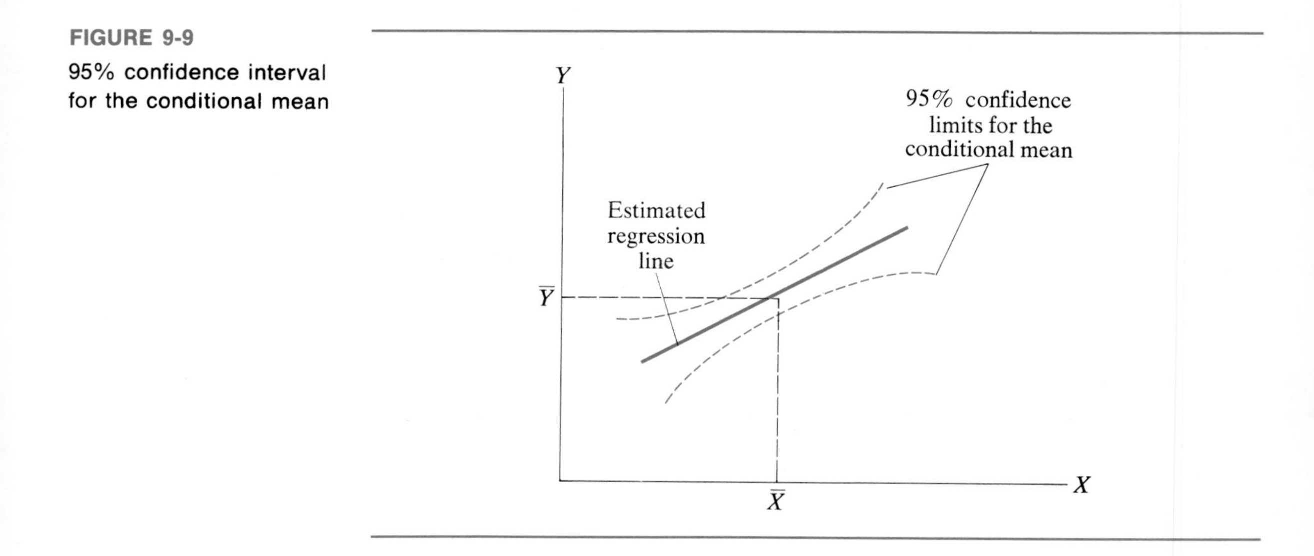

where the t value of 2.306 for a 95% confidence coefficient was obtained from Table A-6 of Appendix A for $n - 2 = 10 - 2 = 8$ degrees of freedom. Therefore, the desired 95% confidence limits are about \$1,980 to \$2,300 for the mean food expenditure for a family with an annual net income of \$8,000.

Note from equation 9.12, the formula for $s_{\hat{Y}}$, that the farther an X value lies from the mean, the larger is $(X_0 - \bar{X})^2$ and the wider is the confidence interval at that value of X. This concept is intuitively plausible because the greater the distance from the mean, the greater would be the effect of an error in the estimated slope of the regression line. This situation is depicted in Figure 9-9.

Prediction Interval for an Individual Value of Y

Just as in the case of estimating a conditional mean for a given X value, we obtain an unbiased estimate in predicting an *individual Y value* for a given X from the point on the estimated regression line corresponding to X. That is, the best point prediction for the individual Y value is obtained from the regression $\hat{Y} = a + bX$ by substituting the given X value.

However, for a given X value, we cannot predict an individual Y value with as much precision as we can estimate a conditional mean. This follows from the fact that the sampling error in predicting an individual value of Y is greater than the sampling error involved in estimating a conditional mean value of Y.

It can be shown that the standard deviation of the error $(Y - \hat{Y})$ of predicting an individual value of Y when $X = X_0$ is

Standard error of forecast (of an individual value of *Y*)

$$s_{\text{IND}} = s_{Y.X}\sqrt{1 + \frac{1}{n} + \frac{(X_0 - \bar{X})^2}{\Sigma(X - \bar{X})^2}} \tag{9.14}$$

As in the case of the standard error of the conditional mean, a more convenient form of equation 9.14 for purposes of calculation is

(9.15)

$$s_{\text{IND}} = s_{Y.X}\sqrt{1 + \frac{1}{n} + \frac{(X_0 - \bar{X})^2}{\Sigma X^2 - \frac{(\Sigma X)^2}{n}}}$$

A more convenient form of equation 9.14

The prediction of an individual value of Y is usually referred to as an **individual forecast**, and s_{IND} is called the **standard error of forecast.**[3] The standard error of forecast can be used to set up a prediction interval for an individual Y value in a manner analogous to setting up confidence intervals for the conditional mean.

Returning to the food expenditures example, assume that we want to predict food expenditures for a particular family with an annual net income of $8,000. The prediction interval for an individual Y value is given by

(9.16)

$$\hat{Y} \pm ts_{\text{IND}}$$

Prediction interval of an individual Y value

where $\hat{Y}$ is the individual forecast as determined from the sample regression line, s_{IND} is the standard error of forecast, and t is the t multiple determined for $n - 2$ degrees of freedom for the desired confidence level.

Substituting into formula 9.15 for $X_0 = 8$ (thousands of dollars), we obtain for the standard error of forecast

$$s_{\text{IND}} = 1.595\sqrt{1 + \frac{1}{10} + \frac{(8 - 6)^2}{406 - \frac{(60)^2}{10}}}$$

$$s_{\text{IND}} = 1.738 \text{ (hundreds of dollars)}$$

Therefore, by formula 9.16, a 95% prediction interval for the individual forecast of the Y value for a family with an $8,000 annual net income is

$$21.434 \pm 2.306(1.738)$$

where the t value of 2.306 for a 95% confidence coefficient was obtained earlier from Table A-6 for $n - 2 = 10 - 2 = 8$ degrees of freedom.

Hence, the prediction interval is from 17.426 to 25.442 (hundreds of dollars), or approximately $1,740 to $2,540 of food expenditures. We note that, as expected, these limits are wider than those previously computed for the conditional mean. Also, as was true for confidence limits for the conditional mean, the farther the X value is from $\bar{X}$, the wider is the prediction interval. This characteristic of prediction intervals for individual forecasts is depicted in Figure 9-10.

[3] Actually, s_{IND} should be referred to as the "estimated standard error of forecast" just as $s_{\hat{Y}}$ is the "estimated standard error of the conditional mean." In both of these cases, "estimated" is appropriate because $s_{Y.X}$ is an estimate of the population $\sigma_{Y.X}$. However, it is conventional to refer to the "standard error of forecast" without the use of the word "estimated."

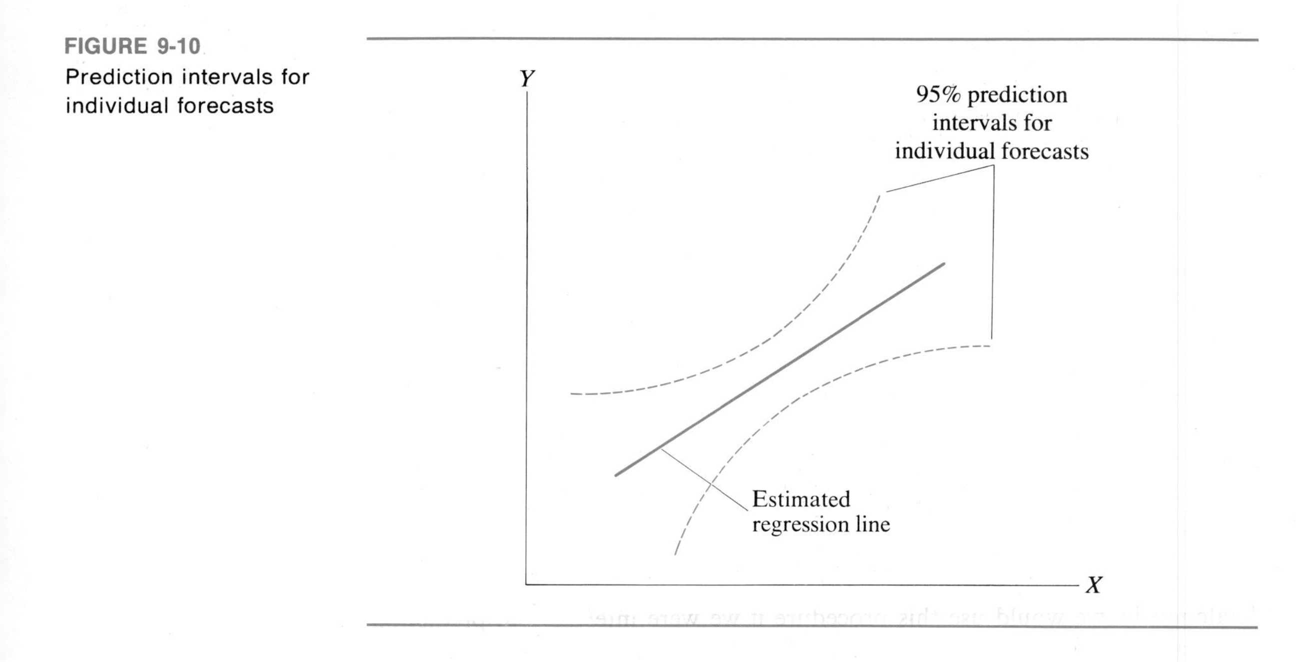

FIGURE 9-10
Prediction intervals for individual forecasts

Width of Confidence and Prediction Intervals

We can see from equations 9.12 and 9.14 that a number of factors affect the sizes of the estimated standard error of the conditional mean and the standard error of forecast and, hence, the width of confidence and prediction intervals. The larger the sample size n, the smaller are these standard errors, and therefore the narrower are the widths of the intervals. This agrees with our intuitive notion that larger sample sizes provide greater precision of estimation.

Second, as noted earlier, the greater the deviation of X from $\bar{X}$, the greater is the standard error and the wider are the intervals for the given X value. This means that confidence and prediction intervals may be quite wide for very small or very large X values compared with the analogous intervals for average-sized X values.

Third, the larger the estimated standard error of estimate $s_{Y.X}$, the larger are the confidence and prediction intervals. This agrees with the intuitive idea that the more variable (less uniform) the data, the less precise would be predictions from these data.

Finally, note that the more variability there is in the sample of X values (in our illustrative problem, the more variability there is in family incomes), the larger is $\Sigma(X - \bar{X})^2$, and hence the smaller are the standard errors and the narrower are the confidence and prediction intervals. This is consistent with the idea that, for example, if we observe a relationship for families with large variation in incomes, we would expect to make a better estimate of food expenditures than if the variation in incomes was small.

Further Remarks

We have discussed estimation procedures for confidence intervals for conditional means and for prediction intervals for individual forecasts. Setting up confidence intervals for conditional means is appropriate whenever we are interested in estimating the *average value* of Y for a given X. In this chapter, we illustrated the estimation of *average food expenditures* for families with $8,000 of income. The same procedure would be used if in other regression analyses we were interested, for example, in confidence intervals for *average* or *expected sales* for an expenditure of $1 million on advertising, for *average number of accidents* for a company with a certain number of workers exposed to some hazard, or *average weight* for black males in the United States who are 70 inches tall.

On the other hand, prediction intervals for individual forecasts are appropriate when we are interested in predicting an *individual Y value* for a given X. Conceptually, we wish to predict the Y value for a new observation drawn from the same population as the sample from which we have constructed the regression line. Hence, in this chapter, we illustrated the prediction interval estimation of *actual food expenditures* of a *particular* family with an $8,000 income. Analogously, we would use this procedure if we were interested in prediction intervals for the *specific amount of sales* for the *next company* that spends $1 million on advertising, for the *specific number* of *accidents* for a *particular company* with 150 employees exposed to some hazard, or the *actual weight* of a *particular* black male in the United States who is 70 inches tall.

In terms of statistical theory, when we estimate a conditional mean, we estimate the value of a *parameter*; when we predict an individual value of Y, we predict the value of a *random variable.* Stated differently, in the former case we are trying to estimate a point on the population regression line; in the latter case we are trying to predict an individual value of Y.

The usefulness of confidence intervals or prediction intervals depends on the purposes for which they are used. For example, for long-range planning purposes, relatively wide limits may be appropriate and useful. On the other hand, for short-term operational decision making, narrower and therefore more precise intervals may be required. In a two-variable regression analysis, the standard error of estimate $s_{Y.X}$ may be so large as to yield confidence and prediction intervals that are too wide for the investigator's purposes. In such a case, the investigator may introduce additional independent variables to obtain greater precision of estimation and prediction. We discuss the use of two or more independent variables in Chapter 10, which deals with *multiple* regression and correlation analysis.

Exercises 9.5

1. Texecon Products, Inc., a major producer of household products, has introduced 14 new products during the past two years. The marketing research unit needs an estimate of the relationship between first-year sales and an appropriate independent variable; this relationship will be used in future

TABLE 9-4
Texecon Products, Inc., marketing research data

Product	First-Year Sales ($ millions)	Customer Awareness (%)	Advertising Expenditures ($ millions)
E-Z-Kleen	82	50	$1.8
Sud-Z-Est	46	45	1.2
Alumofoil	17	15	0.4
Backscratcher	21	15	0.5
Pest Killer	112	70	2.5
Liquid Lush	105	75	2.5
Bubbly Bath	65	60	1.5
Wipe Away	55	40	1.2
Whirlwind	80	60	1.6
Magic Mop	43	25	1.0
Cobweb Cure	79	50	1.5
Oven Eater	24	20	0.7
Dirt-B-Gone	30	30	1.0
Dustbowl	11	5	0.8

planning in marketing and advertising. The researchers have constructed a variable called "customer awareness," measured by the proportion of consumers who had heard of the product by the third month after its introduction. The data are shown in Table 9-4.

a. Find the least squares regression equation, with customer awareness as the independent variable.
b. Calculate the standard error of estimate, $s_{Y.X}$. What assumption is necessary in using $s_{Y.X}$ to estimate $\sigma_{Y.X}$, the conditional standard deviation around the population regression line?
c. Calculate a 95% confidence interval estimate for conditional mean sales for a customer awareness level of 35%.
d. Obtain a 95% prediction interval for new product sales for the same customer awareness level as in part (c).
e. Which interval was wider: that obtained in part (c) or that obtained in part (d)? Why?

2. The marketing research department of Texecon Products, Inc. decided that the customer awareness variable in exercise 1 suffered from too many shortcomings. You have been instructed to perform the same analysis as in exercise 1, using advertising expenditures (listed in Table 9-4) as the independent variable.
   a. Estimate the linear regression equation, using the method of least squares.

b. Calculate the standard error of estimate, $s_{Y.X}$.
c. Obtain a 95% confidence interval estimate for conditional mean sales for an advertising expenditure level of $1 million.
d. Obtain a 95% prediction interval for new product sales for the same advertising expenditure level as in part (c).
e. What factors affect the width of confidence and prediction interval estimates?

3. A federal agency is completing a study on the feasibility of national health insurance and has requested information from its research department on family medical expenditures. The researchers have collected data on 122 families and present these results:

$$X = \text{annual family income (thousands of dollars)}$$

$$Y = \text{annual family medical expenditures (hundreds of dollars)}$$

$$n = 122$$

$$\Sigma X = 1{,}464$$

$$\Sigma X^2 = 18{,}000$$

$$\Sigma Y = 1{,}220$$

$$\Sigma Y^2 = 12{,}475$$

$$\Sigma XY = 14{,}900$$

a. Estimate the linear regression equation, using the method of least squares, with annual family income as the independent variable.
b. Calculate the standard error of estimate, $s_{Y.X}$.
c. Estimate the annual family medical expenditures for a family with an annual income of $8,000. Do the same for a family with an income of $25,000. Compare the two results as percentages of annual family income.
d. Obtain 95% confidence interval estimates for the conditional mean family medical expenditures for the two families in part (c).

9.6 CORRELATION ANALYSIS: MEASURES OF ASSOCIATION

In the preceding two sections, regression analysis was discussed, with emphasis on estimation and measures of error in the estimation process. We now turn to correlation analysis, in which the basic objective is to obtain a measure of the degree of association between two variables. In this analysis, interest centers on the strength of the relationship between the variables, that is, on how well the variables are correlated. The assumptions of the two-variable correlation model are as follows:

1. Both X and Y are random variables.
2. Both X and Y are normally distributed. The two distributions need not be independent.
3. The standard deviations of the Ys are assumed to be equal for all values of X, and the standard deviations of the Xs are assumed to be equal for all values of Y.

Note that in the correlation model, both X and Y are assumed to be random variables. On the other hand, in the regression model, only Y is a random variable, and the Y observations are treated as a random sample from the conditional distribution of Y for a given X.

The Coefficient of Determination

A measure of the amount of correlation between Y and X can be explained in terms of the relative variation of the Y values around the regression line and the corresponding variation around the mean of the Y variable. The term **variation**, as used in statistics, conventionally refers to a sum of squared deviations.

The variation of Y values around the regression line is measured by

Variation of Y values around regression line

$$\Sigma(Y - \hat{Y})^2 \tag{9.17}$$

The variation of Y values around the mean of the Y variable is measured by

Variation of Y values around the mean

$$\Sigma(Y - \bar{Y})^2 \tag{9.18}$$

Equation 9.17 is the sum of the squared vertical deviations of the Y values from the regression line. Equation 9.18 is the sum of the squared vertical deviations of the Y values from the horizontal line $Y = \bar{Y}$. The relationship between the variations around the regression line and the mean can be summarized in a single measure to indicate the degree of association between X and Y. The measure used for this purpose is the **sample coefficient of determination**, defined as follows:

Sample coefficient of determination

$$r^2 = 1 - \frac{\Sigma(Y - \hat{Y})^2}{\Sigma(Y - \bar{Y})^2} \tag{9.19}$$

As we shall see from the subsequent discussion, r^2 may be interpreted as the proportion of variation in the dependent variable Y that has been accounted for, or "explained," by the relationship between Y and X expressed in the regression line. Hence, it is a measure of the degree of association or correlation between Y and X.

To present the rationale of this measure of strength of the relationship between Y and X, we will consider two extreme cases, zero linear correlation and perfect direct linear correlation. The term *linear* indicates that a straight line has been fitted to the X and Y values, and the term *direct* indicates that the line has a positive slope.

Two sets of data are labeled (a) and (b) and are shown in the scatter diagrams in Figure 9-11. The data in (a) and (b) illustrate the cases of zero linear correlation and perfect direct linear correlation, respectively. In the discussion that follows, we will assume that the observations shown in (a) and (b) represent simple random samples from their respective universes. Therefore, we employ notation appropriate to samples. If we assumed that the observations represent population data, the notation would change correspondingly. The calculations given below the scatter diagrams in Figure 9-11 will be explained in terms of the data displayed in the charts.

Case (a) represents a situation in which $\bar{Y}$, the mean of the Y values, coincides with a least squares regression line fitted to these data. Even without doing the arithmetic, we can see why this is so. The slope of the regression line is zero, because the same Y values are observed for $X = 1, 2, 3,$ and 4. Thus, the regression line would coincide with the mean of the Y values, balancing deviations above and below the regression line. Another way of observing this

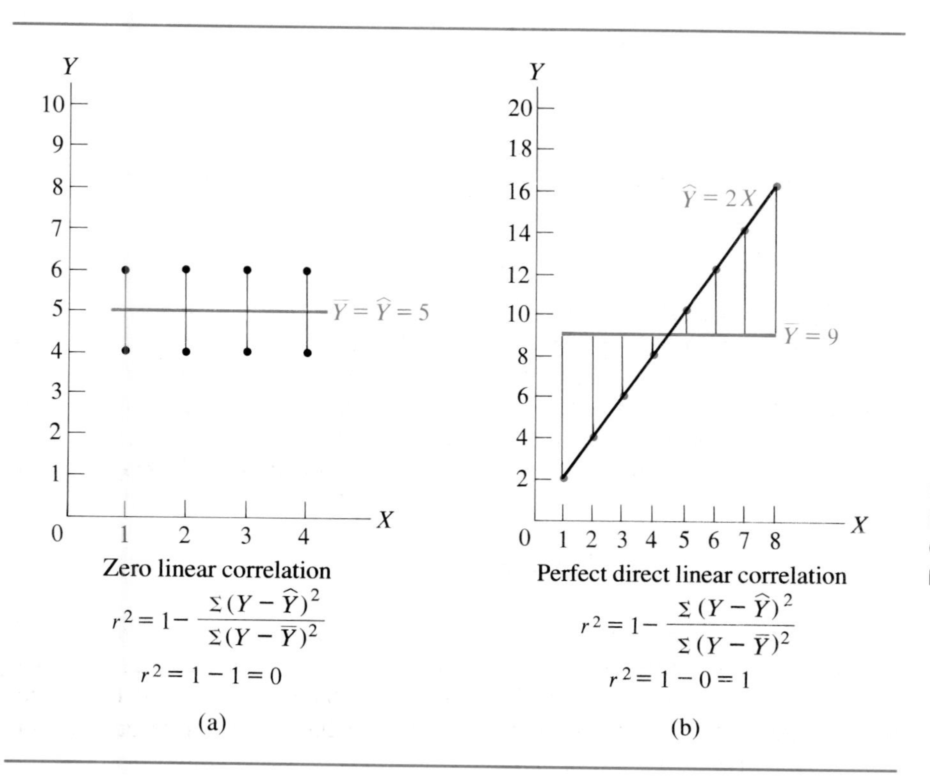

FIGURE 9-11
Scatter diagrams representing zero linear correlation and perfect direct linear correlation based on the following sets of data for observations A through H.

	(a)		(b)	
	X	Y	X	Y
A	1	4	1	2
B	1	6	2	4
C	2	4	3	6
D	2	6	4	8
E	3	4	5	10
F	3	6	6	12
G	4	4	7	14
H	4	6	8	16
		$\bar{Y} = 5$		$\bar{Y} = 9$

relationship is in terms of the first of the two equations used to solve for a and b. In equation 9.6, $a = \bar{Y} - b\bar{X}$, and since $b = 0$, then $a = \bar{Y}$. Hence, the regression line has a Y intercept equal to $\bar{Y}$. Since it is also a horizontal line, the regression line coincides with $\bar{Y}$. From the point of view of estimation of the Y variable, the regression line represents no improvement over the mean of the Y values. This can be shown by a comparison of $\Sigma(Y - \hat{Y})^2$, the variation around the regression line, and $\Sigma(Y - \bar{Y})^2$, the variation around the mean of the Y values. In this case, the two variations are equal. These variations may be interpreted graphically as the sum of the squares of the vertical distances between the points on the scatter diagram and $\bar{Y}$, shown in Figure 9-11(a).

Now, let us consider case (b). In this situation, the regression line is a perfect fit to the data. The regression equation is a simple one, which can be determined by inspection. The Y intercept is zero, since the line passes through the origin (0, 0). The slope is two, because for every unit increase in X, Y increases by two units. Hence, the regression equation is $\hat{Y} = 2X$, and all the data points lie on the regression line. As far as the data in the sample are concerned, perfect predictions are provided by this regression line. Given a value of X, the corresponding value of Y can be correctly estimated from the regression equation, indicating a perfect linear relationship between the two variables. Again, a comparison can be made of $\Sigma(Y - \hat{Y})^2$ and $\Sigma(Y - \bar{Y})^2$. Since all points lie on the regression line, the variation around the line $\Sigma(Y - \hat{Y})^2$ is zero. On the other hand, the variation around the mean, $\Sigma(Y - \bar{Y})^2$, is some positive number, in this case, 168.

As indicated in Figure 9-11(a), when there is no linear correlation between X and Y, the sample coefficient of determination, r^2, is zero. This follows from the fact that since $\Sigma(Y - \hat{Y})^2$ and $\Sigma(Y - \bar{Y})^2$ are equal, the ratio $\Sigma(Y - \hat{Y})^2/\Sigma(Y - \bar{Y})^2$ equals one. Hence, from equation 9.19, $r^2 = 0$, because the computation of the coefficient of determination requires subtraction of this ratio from one.

On the other hand, as indicated in Figure 9-11(b), when there is perfect linear correlation between X and Y, the sample coefficient of determination, r^2, is one. In this case, the variation around the regression line is zero, while the variation around the mean is some positive number. Thus, the ratio $\Sigma(Y - \hat{Y})^2/\Sigma(Y - \bar{Y})^2$ is zero. Hence, from equation 9.19, $r^2 = 1$ when the value of this ratio is subtracted from one.

In realistic problems, r^2 falls somewhere between the two limits zero and one. An r^2 value close to zero suggests not much linear correlation between X and Y; an r^2 value close to one connotes a strong linear relationship between X and Y.

Population Coefficient of Determination

The measure r^2, called the **sample coefficient of determination**, pertains only to the sample of n observations studied. The regression line computed from the sample may be viewed as an estimate of the true population regression line, which may be denoted

The true population regression line

(9.20)
$$\mu_{Y.X} = A + BX$$

The corresponding population coefficient of determination is defined as

Population coefficient of determination

(9.21)
$$\rho^2 = 1 - \frac{\sigma^2_{Y.X}}{\sigma^2_Y}$$

The use of the symbol ρ^2 (rho squared) adheres to the usual convention of employing a Greek letter for a population parameter corresponding to the same letter in our alphabet that denotes a sample statistic. In the definition of ρ^2, $\sigma^2_{Y.X}$ is the variance around the population regression line $\mu_{Y.X} = A + BX$, and σ^2_Y is the variance around the population mean of the Ys, denoted μ_Y. Both the sample and population coefficients of determination are equal to one minus a ratio of the variability around the regression line to the variability around the mean of the Y values.

A slightly different form of the *sample* coefficient of determination, which is directly parallel to equation 9.21, is

(9.22)
$$r_c^2 = 1 - \frac{s^2_{Y.X}}{s^2_Y} = 1 - \frac{\Sigma(Y - \hat{Y})^2/(n-2)}{\Sigma(Y - \bar{Y})^2/(n-1)}$$

The quantity r_c^2 is the **corrected** (or adjusted) **sample coefficient of determination.** This terminology is used because $s^2_{Y.X}$ and s^2_Y are estimators of $\sigma^2_{Y.X}$ and σ^2_Y that make the appropriate corrections or adjustments for degrees of freedom.[4] Since $s^2_{Y.X}$ and s^2_Y are unbiased estimators of $\sigma^2_{Y.X}$ and σ^2_Y, the adjusted sample coefficient of determination r_c^2, rather than the unadjusted coefficient r^2, is ordinarily used in estimating the population coefficient of determination, ρ^2. However, in the discussion that follows we use only the unadjusted value r^2, because of the complication associated with the divisors $n - 2$ and $n - 1$ in the adjusted measure.

Interpretation of the Coefficient of Determination

Let us consider in more detail the specific interpretations that may be made of coefficients of determination. For convenience, only the sample coefficient r^2 will be discussed, but the corresponding meanings for ρ^2 are obvious.

An important interpretation of r^2 may be made in terms of variation in the dependent variable Y, which has been explained by the regression line. We conceive of the problem of estimation in terms of "explaining" or accounting

[4] The relationship between r_c^2 and r^2 is given by $r_c^2 = 1 - (1 - r^2)\left(\frac{n-1}{n-2}\right)$. For large sample sizes $\left(\frac{n-1}{n-2}\right)$ is close to 1 and r_c^2 and r^2 are approximately equal.

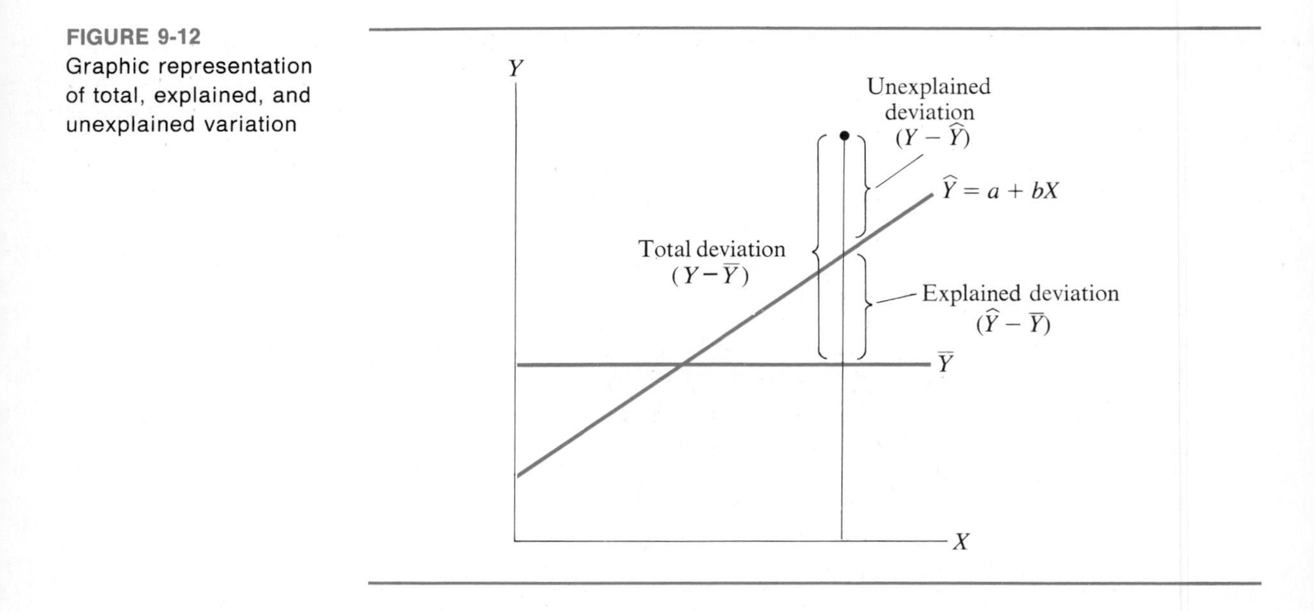

FIGURE 9-12
Graphic representation of total, explained, and unexplained variation

for the variation in the dependent variable Y. Figure 9-12, on which a single point is shown, gives a graphic interpretation of the situation. In this context, if $\bar{Y}$, the mean of the Y values, were used to estimate the value of Y, the total deviation would be $Y - \bar{Y}$. We can think of total deviation as being composed of the following elements:

Total deviation

Total deviation = Explained deviation + Unexplained deviation

$$(Y - \bar{Y}) = (\hat{Y} - \bar{Y}) + (Y - \hat{Y})$$

If the regression line were used to estimate the value of Y, we would now have a closer estimate. As shown in the figure, there is still an "unexplained deviation" of $(Y - \hat{Y})$, but we have explained $(\hat{Y} - \bar{Y})$ out of the total deviation by assuming that the relationship between X and Y is given by the regression line.

In an analogous manner, we can partition the total variation of the dependent variable (or total sum of squares), $\Sigma(Y - \bar{Y})^2$, as follows:

Total variation

Total variation = Explained variation + Unexplained variation

$$\Sigma(Y - \bar{Y})^2 = \Sigma(\hat{Y} - \bar{Y})^2 + \Sigma(Y - \hat{Y})^2$$

The ratio $\Sigma(Y - \hat{Y})^2/\Sigma(Y - \bar{Y})^2$ is the proportion of total variation that remains unexplained by the regression equation; correspondingly, $1 - [\Sigma(Y - \hat{Y})^2/\Sigma(Y - \bar{Y})^2]$ represents the *proportion of total variation in Y* that has been explained by the regression equation.

These ideas may be summarized as follows:

(9.23)

$$r^2 = 1 - \frac{\Sigma(Y - \hat{Y})^2}{\Sigma(Y - \bar{Y})^2} = 1 - \frac{\text{Unexplained variation}}{\text{Total variation}}$$

$$r^2 = \frac{\text{Explained variation}}{\text{Total variation}}$$

A simple numerical example helps to clarify these relationships. Let $\Sigma(Y - \bar{Y})^2 = 10$ and $\Sigma(Y - \hat{Y})^2 = 4$. Thus, $r^2 = 1 - \frac{4}{10} = \frac{6}{10} = 60\%$. In this problem, 10 units of total variation in Y have to be accounted for. After we fit the regression line, the residual variation or unexplained variation amounts to four units. Hence, 60% of the total variation in the dependent variable is explained by the relationship between Y and X expressed in the regression line.

Calculation of the Sample Coefficient of Determination

The computation of r^2 from the definitional formula, equation 9.19, becomes quite tedious, particularly with a large sample. Just as in the case of the standard error of estimate, shorter methods of calculation are ordinarily used. These shortcut formulas are particularly helpful when computations are carried out by hand or on a calculator; even when computers are used, they represent more efficient methods of computation. Such a formula, which involves only quantities already calculated, is

(9.24)

$$r^2 = \frac{a\Sigma Y + b\Sigma XY - n\bar{Y}^2}{\Sigma Y^2 - n\bar{Y}^2}$$

Substituting into equation 9.24, we obtain

$$r^2 = \frac{(7.698)(180) + (1.717)(1{,}159) - 10(18)^2}{(3{,}396) - 10(18)^2} = 0.870$$

Thus, for our sample of 10 low-income families, about 87% of the variation in annual food expenditures was explained by the regression equation, which related such expenditures to annual net income.

The Coefficient of Correlation

A widely used measure of the degree of association between two variables is the coefficient of correlation, which is simply the square root of the coefficient of determination. Thus, the population and sample coefficients of correlation are

(9.25)

$$\rho = \pm\sqrt{\rho^2}$$

Population coefficient of correlation

FIGURE 9-13
Scatter diagram representing perfect inverse linear correlation, $r = -1$

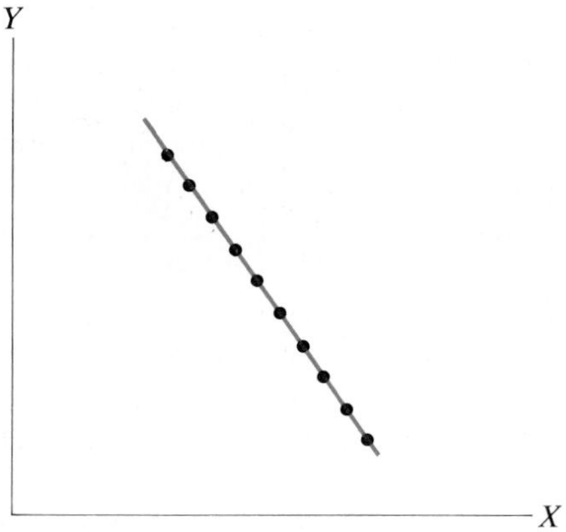

and

Sample coefficient of correlation

(9.26) $$r = \pm\sqrt{r^2}$$

Again, for convenience, our discussion will relate only to the sample value.

The algebraic sign attached to r is the same as that of the regression coefficient, b. Thus, if the slope of the regression line b is positive, then r is also positive; if b is negative, r is negative.[5] Hence r ranges in value from -1 to $+1$. A figure of $r = -1$ indicates a perfect inverse linear relationship, $r = +1$ indicates a perfect direct linear relationship, and $r = 0$ indicates no linear relationship.

Scatter diagrams for the cases of $r = +1$ ($r^2 = 1$) and $r = 0$ ($r^2 = 0$) were given in Figure 9-11. A corresponding scatter diagram for $r = -1$, the case of perfect inverse linear correlation, is shown in Figure 9-13. As indicated in the graph, the slope of the regression line is negative and every point falls on the line. Thus, for example, if the slope of the regression line, b, were equal to -2, with each increase of one unit in X, Y would decrease by two units. Since all points fall on the regression line in the case of perfect inverse correlation, $\Sigma(Y - \hat{Y})^2 = 0$. Therefore, substituting into equation 9.19 to compute the sample coefficient of determination, we have $r^2 = 1 - 0 = 1$. Taking the square root, we obtain $r = \pm\sqrt{1} = \pm 1$. However, since the b value is negative (that is, X and Y are inversely correlated), we assign the negative sign to r, and $r = -1$.

In our problem,

$$r = \sqrt{0.870} = 0.933$$

[5] An interesting relationship between r and b is given by $b = r[\Sigma(Y - \bar{Y})^2/\Sigma(X - \bar{X})^2]$. Since $\Sigma(Y - \bar{Y})^2$ and $\Sigma(X - \bar{X})^2$ are positive numbers, b has the same sign as r. We also note that when the regression line is horizontal, $b = 0$ and $r = 0$.

The sign is positive because b was positive, indicating a direct relationship between food expenditures and net income.

Despite the rather common use of the coefficient of correlation, it is preferable for interpretation purposes to use the coefficient of determination. As we have seen, r^2 can be interpreted as a proportion or a percentage figure. When the square root of a percentage is taken, the specific meaning becomes obscure. Furthermore, since r^2 is a decimal value (unless it is equal to zero or one), its square root, or r, is a larger number. Thus, the use of r values to indicate the degree of correlation between two variables tends to give the impression of a stronger relationship than is actually present. For example, an r value of $+0.7$ or -0.7 seems to represent a reasonably high degree of association. However, since $r^2 = 0.49$, less than half of the total variance in Y has been explained by the regression equation.

Observe that the values of r and r^2 do not depend on the units in which X and Y are stated nor on which of these variables is selected as the dependent or independent variable. Whether a value of r or r^2 is considered high depends somewhat on the specific field of application. With some types of data, r values in excess of about 0.80 are relatively unusual. On the other hand, particularly in the case of time-series data, r values in excess of 0.90 are quite common. In the following section, we consider the matter of determining whether the observed degree of correlation in a sample is sufficiently large to justify a conclusion that correlation between X and Y actually exists in the population.

9.7 INFERENCE ABOUT POPULATION PARAMETERS IN REGRESSION AND CORRELATION

In the procedures discussed to this point, computation and interpretation of *sample* measures have been emphasized. However, from our study of statistical inference, we know that sample statistics ordinarily differ from corresponding population parameters because of chance errors of sampling. Therefore, it is useful to have a protective procedure against the possible error of concluding from a sample that an association exists between two variables, while actually no such relationship exists in the population from which the sample was drawn. Hypothesis-testing techniques, such as those discussed in Chapter 7, can be employed for this purpose.

Inference about the Population Correlation Coefficient, ρ

Let us assume a situation in which we take a simple random sample of n units from a population and make paired observations of X and Y for each unit. The sample correlation coefficient r, as defined in equation 9.26, is calculated.

The procedure tests the hypothesis that the population correlation coefficient, ρ, is zero in the universe from which the sample was drawn. In keeping with the language used in Chapter 7, we wish to test the null hypothesis that $\rho = 0$ versus the alternative that $\rho \neq 0$. Symbolically, we may write

$$H_0: \rho = 0$$

$$H_1: \rho \neq 0$$

If the computed r values in successive samples of the same size from the population were distributed normally around $\rho = 0$, we would only have to know the standard error of r, σ_r, to perform the usual test involving the normal distribution. Although r values are not normally distributed, a similar procedure is provided by the following statistic:

$$t = \frac{r - \rho}{s_r} = \frac{r}{\sqrt{(1 - r^2)/(n - 2)}} \tag{9.27}$$

which has a t distribution for $n - 2$ degrees of freedom. The estimated standard error of r is $s_r = \sqrt{(1 - r^2)/(n - 2)}$, and we have $\rho = 0$ by the null hypothesis. Note that despite the previous explanation that r^2 is easier to interpret than r, the hypothesis-testing procedure is in terms of r rather than r^2. The reason is that under the null hypothesis, $H_0: \rho = 0$, the sampling distribution of r leads to the t statistic, a well-known distribution that is relatively easy to work with. On the other hand, under the same hypothesis of no correlation in the universe, r^2 values, which range from zero to one, would not even be symmetrically distributed, and the sampling distribution would be more difficult to deal with. Suppose we wish to test the hypothesis that $\rho = 0$ at the 5% level of significance for our problem involving 10 low-income families. Since $r = 0.933$ and $n = 10$, substitution into equation 9.27 yields

$$t = \frac{0.933}{\sqrt{\dfrac{1 - 0.870}{10 - 2}}} = 7.32$$

Referring to Table A-6, we find a critical t value of 2.306 at the 5% level of significance for eight degrees of freedom. Therefore, the decision rule is

DECISION RULE

1. If $-2.306 \leqslant t \leqslant 2.306$, accept H_0
2. If $t < -2.306$ or $t > 2.306$, reject H_0

Since our computed t value is 7.32, far in excess of the critical value, we conclude that the sample r value differs significantly from zero. We reject the hypothesis that $\rho = 0$ and conclude that a positive relationship exists between annual food expenditures and annual net income in the population from which our sample was drawn. Since the critical t value is 3.355 at the 1% level of significance (the smallest level shown in Table A-6 of Appendix A), it is extremely unlikely that an r value as high as 0.93 would have been observed in a sample of 10 items drawn from a population in which X and Y were uncorrelated.

Comments

A few comments may be made concerning this hypothesis-testing procedure. First of all, this technique is valid only for a hypothesized universe value of $\rho = 0$. Other procedures must be used for assumed universe correlation coefficients other than zero.[6]

Second, only Type I errors are controlled by this testing procedure. That is, when the significance level is set at, say, 5%, the test provides a 5% risk of incorrectly rejecting the null hypothesis of no correlation. No attempt is made to fix the risks of Type II errors (that is, the risk of accepting $H_0\colon \rho = 0$ when $\rho \neq 0$) at specific levels.

Third, even though the sample r value is significant according to this test, in some instances the amount of correlation may not be considered substantively important. For example, in a large sample, a low r value may be found to differ significantly from zero. However, since relatively little correlation has been found between the two variables, we may be unwilling to use the relationship observed between X and Y for decision-making purposes. Furthermore, prediction intervals based on the use of the applicable standard errors of estimate may be too wide to be of practical use.

Fourth, the distributions of t values computed by equation 9.27, approach the normal distribution as sample size increases. Hence, for large sample sizes, the t value is approximately equal to z in the standard normal distribution, and critical values applicable to the normal distribution may be used instead. For example, in the preceding illustration, in which the critical t value was 2.306 for 8 degrees of freedom at the 5% level of significance, the corresponding critical z value would be 1.96 at the same significance level. For large sample sizes, these values would be much closer.

Fifth, as noted earlier, in correlation analysis we assume that both X and Y are normally distributed *random variables*. Hence, in order to use a hypothesis-testing procedure about the value of ρ, such as the one illustrated above, X and Y should both be random variables. On the other hand, in the regression model, the independent variable X is not a random variable. In a particular analysis, if the X values are indeed fixed or predetermined, it is not proper to use the sample correlation coefficient r to test hypotheses about ρ. However, in that case, as we have seen earlier, r or r^2 may be used to measure the effectiveness of the regression equation in explaining variation in the dependent variable Y.

[6] Fisher's z transformation may be used when ρ is hypothesized to be nonzero. In this procedure, a change of variable is made from the sample r to a statistic z, defined as $z = \frac{1}{2} \log_e[(1 + r)/(1 - r)]$. This statistic is approximately normally distributed with mean $\mu_z = \frac{1}{2} \log_e[(1 + \rho)/(1 - \rho)]$ and standard deviation $\sigma_z = 1/\sqrt{n - 3}$.

Inference about the Population Regression Coefficient, B

In many cases, a great deal of interest is centered on the value of b, the slope of the regression line computed from a sample. Statistical inference procedures involving either hypothesis testing or confidence interval estimation are often useful for answering questions concerning the size of the population regression coefficient B in the population regression equation $\mu_{Y.X} = A + BX$.

In order to illustrate the hypothesis-testing procedure for a regression coefficient, let us return to the data in our problem, in which $b = 1.717$. We interpreted this figure to mean that the estimated difference in annual food expenditures for two families whose annual net income in 1986 differed by \$1,000 was \$172. Suppose that on the basis of similar studies in the same metropolitan area, it had been concluded that in previous years, a valid assumption for the true population regression coefficient was $B = 2$. Can we conclude that the population regression coefficient has changed?

To answer this question, we use a familiar hypothesis-testing procedure. We establish the following null and alternative hypotheses:

$$H_0: B = 2$$

$$H_1: B \neq 2$$

Assume that we were willing to run a 5% risk of erroneously rejecting the null hypothesis that $B = 2$. The procedure involves a two-tailed t test in which the estimated standard error of the regression coefficient, denoted s_b, is given by

Estimated standard error of the regression coefficient

$$s_b = \frac{s_{Y.X}}{\sqrt{\Sigma(X - \bar{X})^2}} \tag{9.28}$$

Hence, s_b, the estimated standard deviation of the sampling distribution of b values, is a function of the scatter of points around the regression line and the dispersion of the X values around their mean. The t statistic, computed in the usual way, is given by

***t* distribution**

$$t = \frac{b - B}{s_b} \tag{9.29}$$

We calculate s_b according to equation 9.28 as follows:

$$s_b = \frac{1.595}{\sqrt{46}} = 0.235$$

Substituting this value for s_b into equation 9.29 gives

$$t = \frac{1.717 - 2}{0.235} = -1.204$$

Since the same level of significance and the same number of degrees of freedom ($n - 2$) are involved as in the preceding test for the significance of r, the decision rule is identical. With critical t values of ± 2.306 at the 5% level of significance, we cannot reject the null hypothesis that $B = 2$. Hence, we cannot conclude that the regression coefficient has changed from $B = 2$ for families in the given metropolitan area. That is, we retain the hypothesis that if a family had an annual net income \$1,000 higher than that of a second family, then on the average, the first family's food expenditures would be \$200 higher.

The corresponding confidence interval procedure involves setting up the interval

$$b \pm ts_b \tag{9.30}$$

In this problem, the 95% confidence interval for B is $1.717 \pm (2.306)(0.235) = 1.717 \pm 0.542$. Therefore, we can assert that the population B figure is included in the interval 1.175 to 2.259 with an associated confidence coefficient of 95%.

Note that we used the t distribution in both the hypothesis-testing and confidence interval procedures just discussed. As in previous examples, normal curve procedures can be used for large sample sizes. Hence, for two-tailed hypothesis testing at the 5% level of significance and for 95% confidence interval estimation, the 2.306 t value given in the preceding examples would be replaced by a normal curve z value of 1.96.

A frequently used hypothesis test concerning the parameter B is for the null hypothesis, H_0: $B = 0$. This test determines whether the slope of the sample regression line differs significantly from a hypothesized value of zero. A slope of zero for the population regression coefficient B implies that there is no linear relationship between the variables X and Y and that the population regression line is horizontal. In other words, a B value of zero in the linear model implies that all of the conditional probability distributions have the same mean. Hence, regardless of the value of X, all conditional probability distributions of Y are identical, so the same value of Y would be predicted for all values of X.

If B is assumed to be zero, equation 9.29 reduces to

$$t = \frac{b}{s_b} \tag{9.31}$$

Substitution into equation 9.31 in the present example yields

$$t = \frac{1.717}{0.235} = 7.31$$

Of course, the same number of degrees of freedom, $n - 2 = 10 - 2 = 8$, is involved in this test as in the preceding test that $B = 2$, and at the 5% level of significance, the critical t values are again ± 2.306. Hence, with a t value of 7.31, we reject the hypothesis that $B = 0$. We conclude that the slope of the population regression line is not zero. Based on our simple random sample, our best estimate of the slope of the population regression line is 1.717. Note that the null hypotheses H_0: $\rho = 0$ and H_0: $B = 0$ are equivalent assumptions. In packaged computer programs for regression and correlation analysis, the printouts generally do not show the t test for H_0: $\rho = 0$. However, the printouts for simple two-variable and multiple regression analysis either display the t values for all regression coefficients (b values) or display all regression coefficients and their corresponding standard errors, so that t values may easily be computed.

Caution

A final caution is appropriate. The tests discussed here pertain to the simple two-variable linear regression model. Although we might conclude on the basis of such tests that there is no linear relationship between X and Y, some other statistically significant relationship (such as curvilinear or logarithmic) may exist: moreover, if additional variables are present but unnoticed, they may obscure the relationship (linear or otherwise) between X and Y.

9.8 CAVEATS AND LIMITATIONS

Regression analysis and correlation analysis are useful and widely applied techniques. However, it is important to understand the limitations of these methods and to interpret the results with care.

Cause and Effect Relationships

In correlation analysis, the value of the coefficient of determination r^2 is calculated. This statistic measures the degree of association between two variables. Neither this quantity nor any other statistical technique that measures or expresses the relationship among variables can prove that one variable is the *cause* and one or more other variables are the *effects*. Indeed, through the centuries philosophical speculation and debate have considered the meaning of cause and effect and whether such a relationship can ever be demonstrated by experimental methods. In any event, a measure such as r^2 does not prove the existence of a cause and effect relationship between two variables X and Y.

When a high value of r^2 is obtained, X may be producing variations in Y, or third and fourth variables W and Z may be producing variations in both X and Y. Numerous examples, frequently humorous in nature, have been given to demonstrate the pitfalls in attempting to draw conclusions about cause and effect in such cases. For example, if the average salaries of ministers are associated with the average price of a bottle of Scotch whiskey over time, a high degree of correlation between these two variables will probably be observed. Doubtless, we would be reluctant to conclude that the fluctuations in ministers' salaries cause the variations in the price of a bottle of Scotch, or vice versa.

In this case, a third variable, which we may conveniently designate as the general level of economic activity, produces variations in both of the variables. From the economic standpoint, salaries of ministers represent the price paid for a particular type of labor; the cost of a bottle of Scotch is also a price. When the general level of economic activity is high, both of these prices tend to be high. When the general level of economic activity is low, as in periods of recession or depression, both of these prices tend to be lower than during more prosperous times. Thus, the high degree of correlation between the two variables of interest is produced by a third variable (and possibly others); certainly, neither variable is *causing* the variations in the other.

Furthermore, it is important to keep in mind the problem of sampling error. As we have seen, in a particular sample a high degree of correlation, either direct or inverse, may be observed, when in fact no correlation (or very little correlation) exists between the two variables in the population.

Finally, in applying critical judgment to the evaluation of observed relationships, we must be on guard against "nonsense correlations" in which no meaningful unit of association is present. For example, suppose we record in a column labeled X the distance from the ground of the skirt hemlines of the first 100 women who pass a particular street corner. In a column labeled Y, we record 100 observations of the heights of the Himalaya mountains along a certain latitude at 5-mile intervals. It is possible that a high r^2 value might be obtained for these data. Clearly, the result is nonsensical, because there is no meaningful unit or entity through which these data are related. In the illustrative example in this chapter, expenditures and income were observed for the same family. Hence, the family may be referred to as the *unit of association.* A **unit of association** might be a time period or some other entity, but it must provide a reasonable link between the variables studied.

Extrapolation beyond the Range of Observed Data

In regression analysis, an estimating equation is established on the basis of a particular set of observations. A great deal of care must be exercised in predicting values of the dependent variable based on values of the independent variable outside the range of the observed data. Such predictions are referred to as **extrapolations**. For example, in the problem considered in this chapter, a regression line was computed for low-income families whose annual net incomes ranged from $2,000 to $10,000. It would be extremely unwise to make a prediction of food expenditures for a family with an annual net income of $25,000 using the computed regression line. To do so would imply that the straight-line relationship could be projected up to a value of $25,000 for the independent variable. Clearly, in the absence of other information, we simply do not know whether the same functional form of the estimating equation is valid outside the range of the observed data. In fact, in certain cases, unreasonable or even impossible values may result from such extrapolations. For example, suppose a regression line with a negative slope had been computed relating the percentage of defective articles produced (Y) with the number of weeks of on-the-job training received (X) by a group of workers. An extrapolation for a large enough number of weeks of training would produce a negative value for the percentage of articles produced, which is an impossible result. Clearly in this case, although the computed estimating equation may be a good description of the relation between X and Y within the range of the observed data, an equation with different parameters or even a completely different functional form is required outside this range. Without a specific investigation or a pertinent theory, we simply do not know what the appropriate estimating device is outside the range of observed data.

However, sometimes the exigencies of a situation require an estimate, even though it is impractical or impossible to obtain additional data. Extrapolations and alternative methods of prediction have to be used, but the limitations and risks involved must be kept constantly in mind.

Other Regression Models

So far, we have considered only one form of the regression model, namely, a straight-line equation relating the dependent variable Y to the independent variable X. Sometimes, theoretical considerations indicate that this is the required model. On the other hand, a linear model is often used either because the theoretical form of the relationship is unknown and a linear equation appears to be adequate or because the theoretical form is known but rather complex and a linear equation may provide a sufficiently good approximation. In all cases, the determination of the most appropriate regression model should result from a combination of theoretical reasoning, practical considerations, and careful scrutiny of the available data.

Often, the straight-line model $\hat{Y} = a + bX$ is not an adequate description of the relationship between the two variables. In some situations, models involving transformations of one or both of the variables may provide better fits to the data. For example, if the dependent variable Y is transformed to a new variable, log Y, a regression equaton of the form

$$\log \hat{Y} = a + bX \tag{9.32}$$

may yield a better fit. Insofar as arithmetic is concerned, log Y is substituted for Y everywhere that Y appeared previously in the formulas. However, care must be used in the interpretation. The antilogarithm of log $\hat{Y}$ must be taken to provide an estimate of the dependent variable Y for a given value of X. Note that different assumptions are involved in this model than in the model $\hat{Y} = a + bX$. We now assume that log Y rather than Y is a normally distributed random variable.

Furthermore, the logarithmic model implies that there is constant *percentage* change in $\hat{Y}$ per unit change in X, whereas the $\hat{Y} = a + bX$ model implies a constant *amount* of change in $\hat{Y}$ per unit change in X.

Possible transformations include the use of square roots, reciprocals, and logarithms of one or both of the variables. As an example of one such useful transformation, if a straight-line equation is fitted to the logarithms of both variables, the model takes the form

$$\log \hat{Y} = a + b \log X \tag{9.33}$$

The regression coefficient b in this model has an interesting interpretation, if as in the illustrative example used in this chapter, Y is a consumption variable and X is income. For such variables, the regression coefficient b in the model $\hat{Y} = a + bX$ can be interpreted as a marginal propensity-to-consume coefficient; that is, it estimates the dollar change in consumption per dollar change in income. Analogously, in the model $\log \hat{Y} = a + b \log X$, the coefficient b can be interpreted as an income elasticity of consumption coefficient; that is, it estimates the *percentage change* in consumption per 1% *change* in income. Of course, fitting an equation of the form $\log \hat{Y} = a + b \log X$ in the illustration under discussion implies that the income elasticity of consumption is constant over the range of income observed. Similarly, a model of the form $\hat{Y} = a + bX$ implies that the marginal propensity to consume is constant over the range of observed income. In fact, according to Keynesian economic theory, the marginal propensity to consume (for total consumption expenditures) decreases with increasing income. Fitting such regression models clearly cannot be merely a mechanistic procedure, but must involve a combination of knowledge of the field of application, good judgment, and experimentation.

In some applications, a curvilinear regression function may be more appropriate than a linear one. Polynomial functions are particularly convenient to fit by the method of least squares. The straight-line regression equation $\hat{Y} = a + bX$ is a polynomial of the first degree, since X is raised to the first power. A second-degree polynomial would involve a regression function of the form

$$\hat{Y} = a + bX + cX^2 \tag{9.34}$$

in which the highest power to which X is raised is 2. This is the equation of a second-degree parabola, which is characterized by *one change in direction* in $\hat{Y}$ as X increases, whereas in the case of a straight line, no changes in direction can take place. A third-degree polynomial permits *two changes in direction*, and so on. In the straight-line function, the amount of change in $\hat{Y}$ is constant per unit change in X. In the second-degree parabola, the amounts of change in $\hat{Y}$ may decrease or increase per unit change in X, depending on the shape of the function. Figure 9-14 shows two scatter diagrams for situations in which a

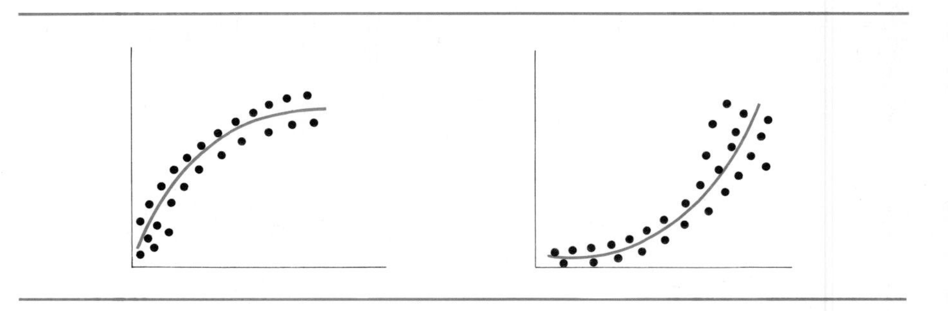

FIGURE 9-14
Scatter diagrams for 2 situations in which a second-degree polynomial regression function might be appropriate

second-degree regression function of the form of equation 9.34 may provide a good fit. The probable shape of the regression function has been indicated. Analogous situations could be portrayed for cases of inverse relationships between X and Y.

In the case of the straight-line regression function, the application of the method of least squares leads to 2 normal equations that must be solved for a and b. Analogously, to obtain the values of a, b, and c in the second-degree polynomial function, the following 3 simultaneous equations must be solved:

$$
\begin{aligned}
\Sigma Y &= na + b\Sigma X + c\Sigma X^2 \\
\Sigma XY &= a\Sigma X + b\Sigma X^2 + c\Sigma X^3 \\
\Sigma X^2 Y &= a\Sigma X^2 + b\Sigma X^3 + c\Sigma X^4
\end{aligned}
\tag{9.35}
$$

Various computer programs have been developed to solve such normal equation systems, provide for transformations of the variables, and calculate all the regression and correlation measures we have discussed (as well as others). In applied problems in which large quantities of data are present and considerable experimentation with the form of the regression model is required, or in which more complex models than those so far discussed are appropriate, the use of computers may be the only feasible method of implementation.

The discussion in this chapter has been limited to two-variable regression and correlation analysis. In many problems, the inclusion of more than one independent variable in a regression model may be required to provide useful estimates of the dependent variable. Suppose, for example, that in the illustration involving family food expenditures and family income, poor predictions were made based on the single independent variable "income." Other factors such as family size, age of the head of the family, and number of employed persons in the family might be considered as additional independent variables to aid in the estimation of food expenditures. When two or more independent variables are utilized, the problem is referred to as a *multiple regression and correlation analysis.* A description of the technique is included in the next chapter.

Exercises 9.8

Note: For simplicity in these exercises, substitute r_c for r in the t test of the null hypothesis H_0: $\rho = 0$. That is, use the t statistic

$$t = \frac{r_c}{\sqrt{(1 - r_c^2)/(n - 2)}}$$

1. As the manager of a local bank, you are given the following information, collected from a simple random sample of checking account customers:

$\hat{Y} = 6 - 0.01X$

X = balance in a customer's checking account

Y = number of bad checks written per month by the customer

$X > -\$10$

$X < \$500$

$n = 200$

$s_Y^2 = 25$

$s_{Y.X}^2 = 8$

   a. Calculate r_c^2 and interpret your answer.
   b. Explain the meaning of the negative regression coefficient b = −0.01.
   c. Do you think that there is any correlation (for all checking accounts) between the balance and the number of bad checks written per month? Justify your answer statistically.

2. The Continental Corporation derived the following data from a random sample of 96 of its salespeople:

$\hat{Y} = 3 + 25X$

X = number of calls on prospects by a salesperson

Y = sales ($ hundreds) made by the salesperson

$n = 96$

$s_Y = 85$

$s_{Y.X} = 25$

   a. Explain the meaning of the 3 and 25 in the regression equation.
   b. Calculate and interpret r_c^2 in terms of this problem.
   c. Distinguish among the measures s_Y, $s_{Y.X}$, and the standard error of r. Place particular emphasis on the distributions involved.
   d. Test the significance of the correlation coefficient. Indicate the hypothesis tested and the meaning of your conclusion in terms of the problem.

3. A market research firm wishes to develop a model to predict purchases of tennis balls by city, based on the number of tennis courts in a city. A simple random sample of 50 cities developed the following data:

$$X = \text{number of courts in a city}$$

$$Y = \text{thousands of tennis balls sold in the city}$$

$$\bar{X} = 235$$

$$\bar{Y} = 375$$

$$\Sigma XY = 4{,}435{,}650$$

$$\Sigma X^2 = 2{,}780{,}850$$

$$n = 50$$

$$s_X^2 = 400$$

$$s_Y^2 = 625$$

$$s_{Y.X}^2 = 249.64$$

a. What is the equation of the estimated regression line that you would use to predict Y from X?
b. Calculate a 95% confidence interval for B.
c. Calculate r_c^2 and test to determine the significance of the correlation coefficient.

4. The Leland Cranston School of Finance placement office has initiated a study of the determinants of a graduate's starting salary level. The independent variable selected for consideration in one part of the study was a rating of all graduates developed by the school based on quality of undergraduate school, academic performance, work experience, and achievement in nonacademic activities. The following statistics were computed:

$$Y = \text{starting salary (thousands of dollars)}$$

$$X = \text{rating}$$

$$n = 62$$

$$X \geqslant 0$$

$$X \leqslant 100$$

$$\hat{Y} = 10 + 0.3X$$

$$s_{Y.X} = 5$$

$$s_Y = 12$$

$$\Sigma(X - \bar{X})^2 = 10{,}000$$

a. Calculate the adjusted coefficient of determination and the adjusted coefficient of correlation. Why is it preferable for interpretation purposes to use r_c^2 rather than r_c?
b. Is the estimated regression coefficient b significantly different from zero? Use a 1% significance level.

5. The BCF Freight System, Inc., was considering the use of a model to predict company sales. As a first step in the development of such a model, the researchers decided to test the usefulness of industry sales as an indicator for predicting company sales. They used the data listed in Table 9-5.
   a. Estimate the linear regression equation, using the method of least squares.
   b. Calculate the adjusted and unadjusted coefficients of determination.
   c. Obtain a prediction interval estimate of company sales for a level of industry sales of $3,100 million. Do you perceive any difficulty in connection with this estimate of company sales?

TABLE 9-5
BCF Freight System, Inc., test of industry sales as an indicator of company sales

Year	Company Sales ($ millions)	Industry Sales ($ millions)
1975	60	615
1976	71	760
1977	76	800
1978	92	885
1979	101	950
1980	115	1020
1981	130	1255
1982	148	1600
1983	193	1855
1984	204	2075

6. A production manager computed the following regression equation between weekly production and cost per unit:

$$\hat{Y} = 12.5 - 0.1X$$

$$X = \text{weekly production (in thousands of units)}$$

$$Y = \text{cost per unit (in dollars)}$$

$$s_Y = 1.5$$

$$s_{Y.X} = 0.8$$

$$n = 102 \text{ weeks}$$

Weekly production range: 0 to 20,000 units

$$\bar{X} = 12.5$$

$$\Sigma(X - \bar{X})^2 = 1{,}650$$

a. Interpret the regression coefficient b specifically in terms of this problem.
b. Calculate and interpret the adjusted coefficient of determination.
c. An executive asserted that there is no linear relationship between unit cost and production level. Do you agree?
d. Of what utility is the Y intercept ($a = 12.5$)?
e. Would a cost per unit of $13.30 be considered an unusually high cost for a production level of 10,000 units? Why or why not?

10

Multiple Regression and Correlation Analysis

Multiple regression analysis represents a logical extension of two-variable regression analysis. Instead of a single independent variable, two or more independent variables are used to estimate the values of a dependent variable. However, the fundamental concepts in the analysis remain the same.

10.1 PURPOSES

Just as in the analysis involving the dependent and only one independent variable, the following three general purposes apply to multiple regression and correlation analysis:

1. To derive an equation that provides estimates of the dependent variable from values of two or more independent variables.
2. To obtain measures of the error involved in using this regression equation as a basis for estimation.
3. To obtain a measure of the proportion of variance in the dependent variable accounted for, or "explained by," the independent variables.

Means by which purposes are accomplished

The first purpose is accomplished by deriving an appropriate regression equation by the method of least squares. The second purpose is achieved through the calculation of a standard error of estimate and related measures. The third purpose is accomplished by computing the multiple coefficient of determination, which is analogous to the coefficient of determination in the two-variable case and, as indicated in (3) above, measures the proportion of variation in the dependent variable explained by the independent variables.

As an example, let us return to our Chapter 9 problem of estimating family food expenditures from family net income for low-income families, both variables being stated on an annual basis. As indicated at the end of that chapter, the use of additional variables to income might improve prediction of the dependent variable. We select family size as a second independent variable. Estimates of food expenditures may now be made from the following linear multiple regression equation:

(10.1)

$$\hat{Y} = a + b_1X_1 + b_2X_2$$

where $\hat{Y}$ = family food expenditures (estimated)
X_1 = family net income
X_2 = family size

and a, b_1, and b_2 are numerical constants that must be determined from the data in a manner analogous to that of the two-variable case. For simplicity, we have assumed a linear regression function.

We carry out a multiple regression and correlation analysis by fitting the linear regression equation 10.1 to data for the indicated variables. The basic data for family food expenditures, family income, and family size are shown in

TABLE 10-1

Computations for linear multiple regression analysis: family food expenditures (Y), family income (X_1), and family size (X_2)

Family	Annual Food Expenditures ($00) Y	Annual Net Income ($000) X_1	Family Size (number in family) X_2	X_1Y	X_2Y	X_1X_2	Y^2	X_1^2	X_2^2
A	22	8	6	176	132	48	484	64	36
B	23	10	7	230	161	70	529	100	49
C	18	7	5	126	90	35	324	49	25
D	9	2	2	18	18	4	81	4	4
E	14	4	3	56	42	12	196	16	9
F	20	6	4	120	80	24	400	36	16
G	21	7	4	147	84	28	441	49	16
H	18	6	3	108	54	18	324	36	9
I	16	4	3	64	48	12	256	16	9
J	19	6	3	114	57	18	361	36	9
Total	180	60	40	1,159	766	269	3,396	406	182
Mean	18	6	4						

the first three columns of Table 10-1. The data for the first two of these variables are the same as those given in Table 9-1 for the two-variable problem solved in Chapter 9. The data on family size represent the total number of persons in each of the families in the sample.

10.2 THE MULTIPLE REGRESSION EQUATION

We begin the analysis by using the method of least squares to obtain the best-fitting three-variable linear regression equation of the form given in equation 10.1. In the two-variable regression problem, the method of least squares was used to obtain the best-fitting straight line. Analogously in the present problem, the method of least squares is used to obtain the best-fitting plane. In a three-variable regression problem, the points can be plotted in three dimensions, along the X_1, X_2, and Y axes (analogous to the case of a two-variable problem, in which the points are plotted in two dimensions along an X and Y axis). The best-fitting plane would pass through the points as shown in Figure 10-1, with

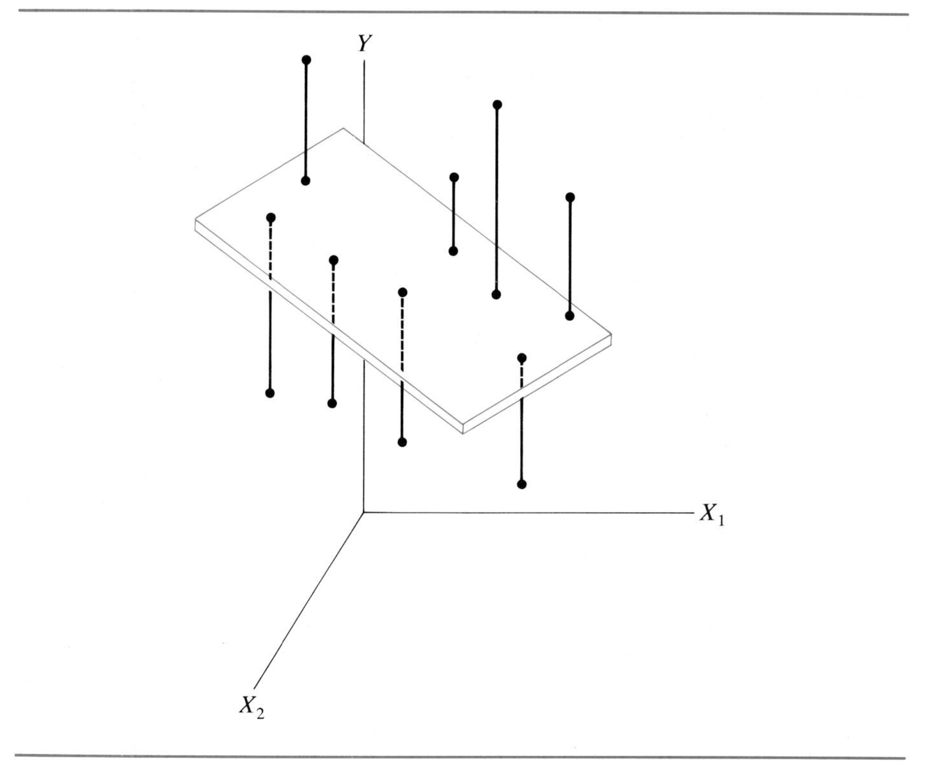

FIGURE 10-1
Graphs of a multiple regression plane for data on the variables Y, X_1, and X_2

some falling above and some below the plane in such a way that $\Sigma(Y - \hat{Y})^2$ is a minimum. Whereas in our previous illustration (involving two variables), two normal equations resulted from the minimization procedure, now three normal equations must be solved to determine the values of a, b_1, and b_2:[1]

Normal equations

(10.2)
$$\begin{aligned} \Sigma Y &= na + b_1\Sigma X_1 + b_2\Sigma X_2 \\ \Sigma X_1 Y &= a\Sigma X_1 + b_1\Sigma X_1^2 + b_2\Sigma X_1 X_2 \\ \Sigma X_2 Y &= a\Sigma X_2 + b_1\Sigma X_1 X_2 + b_2\Sigma X_2^2 \end{aligned}$$

The calculations of the required sums are shown in Table 10-1. Substitution into the normal equations 10.2 yields

$$\begin{aligned} 180 &= 10a + 60b_1 + 40b_2 \\ 1{,}159 &= 60a + 406b_1 + 269b_2 \\ 766 &= 40a + 269b_1 + 182b_2 \end{aligned}$$

Solving these three equations simultaneously, we obtain the following values for a, b_1, and b_2:

$$\begin{aligned} a &= 7.918 \\ b_1 &= 2.363 \\ b_2 &= -1.024 \end{aligned}$$

Although the calculations to obtain the values of a, b_1, and b_2 can be simplified somewhat, solving the normal equations is clearly a laborious mathematical procedure. In the present example, three equations in three unknowns had to be solved. If we needed to determine four constants, four equations would have to be solved. Fortunately, since the advent of electronic computers, numerous packaged programs for multiple regression and correlation analysis have become available. Therefore, the computer solves the normal equations and carries out the other calculations. Since the use of electronic computing equipment is usually the only feasible alternative for carrying out multiple regression and correlation analyses, we will focus on the concepts involved in such analyses and on the interpretation of computer output rather than the calculations by which we obtain the output.

Returning to our problem, we find that the multiple regression equation may now be written as

(10.3)
$$\hat{Y} = 7.918 + 2.363X_1 - 1.024X_2$$

[1] In a manner similar to that of the two-variable case, a function of the form

$$F(a, b_1, b_2) = \Sigma(Y - \hat{Y})^2 = \Sigma(Y - a - b_1X_1 - b_2X_2)^2$$

is set up. This function is minimized by the standard calculus method of taking its partial derivatives with respect to a, b_1, and b_2 and equating these derivatives to zero. This procedure results in the three normal equations (10.2).

Computer output

In computer output, many digits—often 10 or more—are shown for the constants in the multiple regression equation and for other measures. Usually more figures are shown than are justified by the usual rules of rounding according to numbers of significant digits. This is a standard procedure in multiple regression and correlation analysis, because many of the measures computed in such analyses are particularly sensitive to rounding. For simplicity of exposition here, we show only three decimal places for the constants in the regression equation.

Let us illustrate the use of the regression equation for estimation. Suppose we want to estimate food expenditures for a family from the same population as the sample studied. The family's income is \$6,000, and there are 4 persons in the family. Substituting $X_1 = 6$ and $X_2 = 4$ yields the following estimated expenditures on food:

$$\begin{aligned}\hat{Y} &= 7.918 + 2.363(6) - 1.024(4)\\ &= 18.00 \text{ (hundreds of dollars)}\\ &= 1{,}800 \text{ (dollars)}\end{aligned}$$

In two-variable analysis, we discussed the interpretation of the constants a and b in the regression equation. Let us consider the analogous interpretation of the constants a, b_1, and b_2 in the multiple regression equation. The constant a is again the Y intercept. However, now it is interpreted as the value of $\hat{Y}$ when X_1 and X_2 are both zero. The b values are referred to in multiple regression analysis as **net regression coefficients**. The b_1 coefficient measures the change in $\hat{Y}$ per unit change in X_1 when X_2 is held fixed, and b_2 measures the change in $\hat{Y}$ per unit change in X_2 when X_1 is held fixed.[2]

Net regression coefficients

Hence, in the present problem, the b_1 value of 2.363 indicates that if a family has an income \$1,000 greater than another's (a unit change in X_1) and *the families are the same size* (X_2 is held constant), then the estimated food expenditures of the higher-income family exceed those of the other by 2.363 hundreds of dollars, or by about \$236. Similarly, the b_2 value of -1.024 means that if a family has one person more than another (a unit change in X_2) and *the families have the same income* (X_1 is held constant), then the estimated food expenditures of the larger family are less than those of the smaller family by about \$102.

Two properties of these net regression coefficients are worth noting. The b_1 value of 2.363 hundreds of dollars implies that an increment of one unit in X_1, or a \$1,000 increment in income, occasions an increase of \$236.3 in $\hat{Y}$ estimated food expenditures, regardless of the size of the family (for families of the sizes studied). Hence, an increase of \$1,000 in income adds \$236.3 to estimated food expenditures, regardless of whether there are two or six people in the

[2] In the language of calculus, b_1 and b_2 are the partial derivatives of $\hat{Y}$ with respect to X_1 and X_2, respectively; that is,

$$\frac{\partial \hat{Y}}{\partial X_1} = b_1 \quad \text{and} \quad \frac{\partial \hat{Y}}{\partial X_2} = b_2$$

TABLE 10-2
Correlation coefficients for each pair of the three variables: expenditures on food (Y), family income (X_1), and family size (X_2)

	Y	X_1	X_2
Y	1.00		
X_1	0.933	1.00	
X_2	0.785	0.912	1.00

family. An analogous interpretation holds for b_2. These interpretations are embodied in the assumption of linearity and therefore follow from the fact that we used a *linear* multiple regression equation in this example.

A second property of regression coefficients is apparent from a comparison of the b value of 1.717—in the simple regression equation 9.8, $\hat{Y} = 7.698 + 1.717X$, previously obtained when family income X was the only independent variable—with the b_1 value of 2.363, the net regression coefficient of income in the multiple regression equation $\hat{Y} = 7.918 + 2.363X_1 - 1.024X_2$, when the family size variable is included in the regression equation. The coefficient $b = 1.717$ in the simple two-variable regression equation makes no explicit allowance for family size. The net regression coefficient $b_1 = 2.363$, on the other hand, "nets out" the effect of family size. A net regression coefficient may in general be greater or less than the corresponding regression coefficient in a two-variable analysis.

In this problem, the families with larger incomes were also the larger families. The positive correlation between income and family size is indicated by the correlation coefficient $r = 0.912$, shown in Table 10-2. The foregoing pattern exemplifies an important characteristic of regression coefficients, regardless of the number of independent variables included in the study.

> A regression coefficient for any specific independent variable (for example, income) measures not only the effect (of *income*) on the dependent variable, but also the effect attributable to *any other independent variables* that happen to be correlated with the independent variable but that have not been explicitly included in the analysis.

This is true for both two-variable and multiple regression analyses.

When independent variables are highly correlated, rather odd results may be obtained in a multiple regression analysis. For instance, a regression coefficient that is positive (or negative) in sign in a two-variable regression equation may change to a negative (or positive) sign for the same independent variable in a multiple regression equation containing other independent variables that are highly correlated with the one in question. For example, in this problem, the dependent variable, food expenditures (Y), is positively correlated with family size (X_2), as indicated by the correlation coefficient of +0.785 (Table 10-2). Hence, the regression coefficient for family size would also be positive in sign. However, the net regression coefficient for family size (b_2) in the three-variable regression equation is −1.024, and is thus negative in sign.

In the discussion of statistical inference in multiple regression, we shall see that the net regression coefficients for highly correlated independent variables tend to be unreliable. This is an important concept, because when independent variables are highly correlated, it is extremely difficult to separate the individual influences of each variable. Consider an extreme case. Suppose a two-variable regression and correlation analysis is carried out between a dependent variable, denoted Y, and an independent variable, denoted X_1. Furthermore, assume that we introduce another independent variable X_2, which has perfect positive correlation with X_1—that is, the correlation coefficient between X_1 and X_2 is +1. We now conduct a three-variable regression and correlation analysis. Clearly X_2 cannot account for or explain any additional variance in the dependent variable Y after X_1 has been taken into account. The same argument could be made if X_1 were introduced after X_2. As indicated in the ensuing discussion of statistical inference in multiple regression, the net regression coefficients, b_1 and b_2, in cases of high correlation between X_1 and X_2, will tend not to differ significantly from zero. Yet, if separate two-variable analyses had been run between Y and X_1 and Y and X_2, the individual regression coefficients might have differed significantly from zero. There is a great deal of concern in fields such as econometrics and applied statistics with this problem of correlation among independent variables, often referred to as **multicollinearity** or simply **collinearity**. One of the simplest solutions to the problem of two highly correlated independent variables is merely to discard one of them, but sometimes more sophisticated procedures are required.

Collinearity

The illustration in this section used only two independent variables. The general form of the linear multiple regression function for $k - 1$ independent variables $X_1, X_2, \ldots, X_{k-1}$ is

$$\hat{Y} = a + b_1X_1 + b_2X_2 + \cdots + b_{k-1}X_{k-1} \tag{10.4}$$

For convenience, the regression equation has been written with $k - 1$ independent variables. There are then k constants $a, b_1, b_2, \ldots, b_{k-1}$ to be determined in the regression equation. This simplifies somewhat the notation in later formulas, compared with the situation when k independent variables are included in the regression equation.

> A linear function that is fitted to data for two variables is referred to as a **straight line**; a linear function for three variables is a **plane**, and a linear function for four or more variables is a **hyperplane**.

Although we cannot visualize a hyperplane, its linear characteristics are analogous to those of the linear functions of two or three variables. With the use of electronic computers, it is possible to test and include large numbers of independent variables in a multiple regression analysis. However, good judgment and knowledge of the logical relationships involved must always be the main guides to deciding which variables to include in the construction of a regression equation.

10.3

STANDARD ERROR OF ESTIMATE

As in simple two-variable regression analysis, a measure of dispersion or scatter around the regression plane or hyperplane can be used as an indicator of the error of estimation. Probability assumptions similar in principle to those of the simple regression model must be introduced. The following are the usual assumptions made in a linear multiple regression analysis, illustrated for the case of two independent variables:

Assumptions

1. The Y values are assumed to be independent of one another.
2. The conditional distributions of Y given X_1 and X_2 are assumed to be normal.
3. These conditional distributions for each independent variable are assumed to have equal standard deviations.

The variance around the regression hyperplane defined in equation 10.4 is

Variance around the regression hyperplane

$$S^2_{Y.12\ldots(k-1)} = \frac{\Sigma(Y - \hat{Y})^2}{n - k} \tag{10.5}$$

where n is the number of observations and k is the number of constants in the regression equation. The divisor $n - k$ represents the number of degrees of freedom, and its use provides an unbiased estimator of the population variance.

Subscript notation

The subscript notation to S^2 lists the dependent variable to the left of the period and the $k - 1$ independent variables to the right. The subscripts $1, 2, \ldots, k - 1$ denote the variables $X_1, X_2, \ldots, X_{k-1}$, respectively. Hence, in our example involving the three variables Y, X_1, and X_2, the variance around the regression plane $\hat{Y} = 7.918 + 2.363X_1 - 1.024X_2$ is given by

Variance around the regression plane

$$S^2_{Y.12} = \frac{\Sigma(Y - \hat{Y})^2}{n - 3} \tag{10.6}$$

The standard error of estimate (often designated in computer printouts by the acronym SEE), which is the square root of this variance, is

Standard error of estimate

$$S_{Y.12} = \sqrt{\frac{\Sigma(Y - \hat{Y})^2}{n - 3}} \tag{10.7}$$

By means of calculations performed on a computer, the following result is obtained, rounded to three decimal places:

$$S_{Y.12} = 1.532 \text{ (hundreds of dollars)}$$

As in two-variable analysis, the standard error of estimate is the standard deviation of the observed values of Y around the fitted regression equation.

Standard errors of the conditional mean and standard errors of forecast do not generally appear in the computer output of standard packaged programs of multiple regression and correlation analysis.

10.4 COEFFICIENT OF MULTIPLE DETERMINATION

In two-variable correlation analysis, the degree of association between the two variables may be stated in terms of the adjusted coefficient of determination r_c^2, which was defined in equation 9.22 as

$$r_c^2 = 1 - \frac{s_{Y.X}^2}{s_Y^2}$$

In this form, r_c^2 measures the proportion of variance in the dependent variable explained by the regression equation relating Y to X; that is, r_c^2 measures the proportion of variance in the Y variable accounted for by, or associated with, the independent variable X.

An analogous measure, the **coefficient of multiple determination**, denoted R^2 with appropriate subscripts, quantifies the degree of association that exists when more than two variables are present. For the case of one dependent and two independent variables, the coefficient of multiple determination, corrected for degrees of freedom, is defined as

(10.8) $$R_{Y.12}^2 = 1 - \frac{S_{Y.12}^2}{s_Y^2}$$ **Coefficient of multiple determination**

where

(10.9) $$S_{Y.12}^2 = \frac{\Sigma(Y - \hat{Y})^2}{n - 3}$$ **Unexplained variance of Y**

and

(10.10) $$s_Y^2 = \frac{\Sigma(Y - \bar{Y})^2}{n - 1}$$ **Total variance of Y**

In keeping with convention, the subscript $R^2_{Y.12}$ lists the dependent variable to the left of the period and the independent variables to the right. We see from these definitions that $S^2_{Y.12}$ is the variance of Y values around the regression plane and s^2_Y is the variance of the Y values around their mean. Just as $\Sigma(Y - \hat{Y})^2$ in equation 10.9 was earlier referred to as "unexplained variation" and $\Sigma(Y - \bar{Y})^2$ in equation 10.10 as the "total variation" of the Y variable, $S^2_{Y.12}$ is referred to as the "unexplained variance" and s^2_Y as the "total variance." Hence, similar to the interpretation of r^2_c, we may interpret $R^2_{Y.12}$ as the proportion of variance in the dependent variable explained by the regression equation relating Y to X_1 and X_2. Alternatively, it measures the proportion of variance in the Y variable accounted for by all the independent variables combined.

We illustrate the calculation of $R^2_{Y.12}$ for the example with which we have been working. Calculated to six decimal places, $S_{Y.12}$ is equal to 1.532168 (hundreds of dollars). Hence,

$$S^2_{Y.12} = (1.532168)^2 = 2.347539$$

The standard deviation s_Y was computed to be 4.163 (hundreds of dollars). The total variance, or variance around $\bar{Y}$, is

$$s^2_Y = (4.163)^2 = 17.330569$$

Therefore, substituting into equation 10.8, the coefficient of multiple determination is

$$R^2_{Y.12} = 1 - \frac{2.347539}{17.330569} = 0.865$$

Thus, we have found that 86.5% of the variance in food expenditures has been explained by the linear regression equation relating that variable to family income and family size. The figure obtained in section 9.6 for the two-variable correlation coefficient, unadjusted for degrees of freedom, for food expenditures and family income was $r^2 = 0.870$. The corresponding figure adjusted for degrees of freedom is $r^2_c = 0.854$.[3] Comparing $R^2_{Y.12} = 0.865$ with the corresponding two-variable r^2_c value of 0.854, we find that the value of $R^2_{Y.12}$ is only 0.011, or about one percentage point, higher than the figure for r^2_c. This means that the addition of the second independent variable family size, has explained little of the variance in food expenditures, Y, beyond that which was already accounted for by family income alone. As we noted earlier, one reason for this is the high correlation between the independent variables. Once family income has been taken into account, since family size moves together with that variable, family size can do little to explain residual variation in food expenditures.

[3] The r^2_c figure is calculated as follows:

$$r^2_c = 1 - (1 - r^2)\left(\frac{n-1}{n-2}\right) = 1 - (1 - 0.870)\left(\frac{10-1}{10-2}\right) = 0.854$$

Two-Variable Correlation Coefficients

From the preceding discussion of the difficulties encountered in multiple correlation analysis when independent variables are intercorrelated, it is evident that it is good practice to compute coefficients of correlation or determination between each pair of independent variables that the analyst plans to enter into the regression equation. It is standard procedure in most multiple regression and correlation analysis computer programs to present a table of correlation coefficients for every pair of variables, including the dependent as well as all independent variables.

Printout

In the printout of computer programs, the correlation coefficients are often presented in the form of a triangular table, as shown in Table 10-2. The 1.00s along the diagonal of this table indicate that the correlation coefficient of each variable with itself is 1.00, that is, each variable is perfectly and directly correlated with itself.

10.5 INFERENCES ABOUT POPULATION NET REGRESSION COEFFICIENTS

In the preceding discussion of correlation and regression analysis, the various equations and measures were all stated in terms of sample values, rather than in terms of the corresponding population equations and characteristics. If the assumptions given at the beginning of the discussion of the standard error of estimate are met, then appropriate inferences and probability statements can be made concerning population parameters. In multiple regression analysis, a great deal of interest is centered on the reliability of the observed net regression coefficients. Just as in the two-variable case referred to in section 9.7, in which statistical inference about the population regression coefficient B was discussed, analogous hypothesis-testing and estimation techniques are available for regression coefficients, where three or more variables are involved.

In the two-variable problem, the regression coefficient b in the equation $\hat{Y} = a + bX$ is an estimate of the population parameter B in the population relationship $\mu_{Y.X} = A + BX$. Correspondingly, the regression coefficients in a three-variable problem, b_1 and b_2 in the equation $\hat{Y} = a + b_1X_1 + b_2X_2$, are estimates of the parameters B_1 and B_2 in a population relationship denoted $\mu_{Y.12} = A + B_1X_1 + B_2X_2$. The standard errors of the net regression coefficients, which represent the estimated standard deviations of the sampling distributions of b_1 and b_2 values, are given by

(10.11)
$$s_{b_1} = \frac{S_{Y.12}}{\sqrt{\Sigma(X_1 - \bar{X}_1)^2(1 - r_{12}^2)}}$$

and

(10.12)
$$s_{b_2} = \frac{S_{Y.12}}{\sqrt{\Sigma(X_2 - \bar{X}_2)^2(1 - r_{12}^2)}}$$

where all terms in equations 10.11 and 10.12 have the definitions stated above.

Substituting the required numerical values, we find

$$s_{b_1} = \frac{1.532168}{\sqrt{46[1 - (0.912)^2]}}$$
$$= 0.551$$

and

$$s_{b_2} = \frac{1.532168}{\sqrt{22[1 - (0.912)^2]}}$$
$$= 0.796$$

We can test hypotheses concerning B_1 and B_2 by computing t statistics in the usual way:

(10.13)
$$t_1 = \frac{b_1 - B_1}{s_{b_1}}$$

and

$$t_2 = \frac{b_2 - B_2}{s_{b_2}}$$

These t statistics approach normality as the sample size and number of degrees of freedom become large.

Hence, to test the hypotheses that the net regression coefficients are equal to zero, that is, that family income and family size have no effect on food expenditures, or

$$H_0: B_1 = 0 \qquad \text{and} \qquad H_0: B_2 = 0$$
$$H_1: B_1 \neq 0 \qquad \qquad \qquad H_1: B_2 \neq 0$$

we calculate

(10.14)
$$t_1 = \frac{b_1 - 0}{s_{b_1}} = \frac{b_1}{s_{b_1}} \quad \text{and} \quad t_2 = \frac{b_2 - 0}{s_{b_2}} = \frac{b_2}{s_{b_2}}$$

In the illustrative problem, we find

$$t_1 = \frac{2.363}{0.551} = 4.289 \quad \text{and} \quad t_2 = \frac{-1.024}{0.796} = -1.286$$

The number of degrees of freedom used to look up the critical t values for this test is $n - k$, which in this case is equal to $10 - 3 = 7$. This is the number of degrees of freedom used to estimate $S_{Y.12}$ in the calculation of s_{b_1} and s_{b_2}. The two-tailed critical t values at the 5% and 1% levels of significance are ± 2.365 and ± 3.499, respectively (Table A-6 of Appendix A). Since for b_1, the computed t_1 value of 4.289 exceeds the positive critical values, we conclude that b_1 differs significantly from zero at both the 5% and 1% levels of significance. Therefore, we reject the null hypothesis that $B_1 = 0$. The computed t_2 value of -1.286 for b_2 means that the b_2 value lies 1.286 estimated standard errors below zero. Comparing -1.286 with the critical values of -2.365 and -3.499, we conclude that b_2 does not differ significantly from zero at either the 5% or 1% level of significance. Hence, we accept the null hypothesis that $B_2 = 0$.

Summary

In summary, we conclude that family income X_1 has a statistically significant effect on food expenditures Y, but that after this income effect has been accounted for, family size X_2 does not have a statistically significant influence. This result is consistent with the previous discussion of the difficulty of measuring the separate effects of two intercorrelated independent variables.

An important point concerning the interpretation of the results of a multiple regression analysis follows from the above discussion. If the basic purpose of computing a regression equation is to make predictions of values of the dependent variable, then the reliability of the individual net regression coefficients is not of great consequence. On the other hand, if the purpose of the analysis is to measure accurately the separate effects of each of the independent variables on the dependent variable, then the reliability of the individual net regression coefficients is clearly important.

10.6 THE ANALYSIS OF VARIANCE

The **analysis of variance** in multiple regression analysis appraises the overall significance of the regression equation. It tests the null hypothesis that all of the true population regression (slope) coefficients equal zero. Hence, the null and alternative hypotheses are

H_0: All of the B_i values equal zero

H_1: Not all of the B_i values equal zero

As usual in an analysis of variance, the null hypothesis is accepted or rejected on the basis of an F test. In a simple two-variable regression analysis, the F test gives exactly the same result as the t test for the null hypothesis H_0: $B = 0$.

TABLE 10-3
General format of the analysis of variance in regression analysis

(1) Source of Variation	(2) Sum of Squares	(3) Degrees of Freedom	(4) Mean Square
Regression	$\Sigma(\hat{Y} - \bar{Y})^2$	$v_1 = k - 1$	$\Sigma(\hat{Y} - \bar{Y})^2/(k-1)$
Error	$\Sigma(Y - \hat{Y})^2$	$v_2 = n - k$	$\Sigma(Y - \hat{Y})^2/(n-k)$
Total	$\Sigma(Y - \bar{Y})^2$	$n - 1$	

$$F(v_1, v_2) = \frac{\Sigma(\hat{Y} - \bar{Y})^2/(k-1)}{\Sigma(Y - \hat{Y})^2/(n-k)}$$

In the interpretation of the coefficient of determination in section 9.6, we gave the following components of the total variation of the dependent variable:

(10.15) Total variation = Explained variation + Unexplained variation

$$\Sigma(Y - \bar{Y})^2 = \Sigma(\hat{Y} - \bar{Y})^2 + \Sigma(Y - \hat{Y})^2$$

On the right-hand side of equation 10.15, the $\Sigma(\hat{Y} - \bar{Y})^2$ term represents the variation in the dependent variable *explained* by the regression equation; the $\Sigma(Y - \hat{Y})^2$ term represents the variation in the dependent variable *not explained* by the regression equation. The *explained* variation is often referred to as the **regression sum of squares**, the *unexplained* variation is referred to as the **error sum of squares** or the **residual sum of squares**.

The analysis of variance is carried out as indicated by the standard format shown in Table 10-3.

- The first column shows the sources of variation. Note that we have used the terminology usually given in computer printouts, with **error** denoting the unexplained variation.
- The second column gives the sums of squares or variations.
- The third column shows the numbers of degrees of freedom that correspond to the sums of squares, that is, $k - 1$ for the regression sum of squares, where k is the number of constants in the regression equation, and $n - k$ for error.
- Column (4) gives the mean squares or variances, in this case, the regression (explained) variance and the error (unexplained) variance.
- The F value shown at the bottom of the table is the ratio of the regression variance to the error variance, or in other words, the ratio of the explained variance to the unexplained variance. If the null hypothesis is true that all of the B_i values equal zero, then this ratio is distributed according

FIGURE 10-2
Computer output for analysis of variance in the regression problem to predict family expenditures for food

SOURCE	SS	DF	MSS	F
REGRESSION	139.56725	2	69.78363	29.72628
ERROR	16.43275	7	2.34754	

to the F distribution given in equation 8.16 with $v_1 = k - 1$ and $v_2 = n - k$ degrees of freedom.

The decision rule for a critical F value denoted F_α is:

DECISION RULE

1. If $F(v_1, v_2) > F_\alpha$, reject the null hypothesis that all the $B_i = 0$.
2. If $F(v_1, v_2) \leqslant F_\alpha$, do not reject the null hypothesis.

Figure 10-2 shows a computer printout of the analysis of variance for the regression problem of predicting family food expenditures from family income and family size. Note that the number of degrees of freedom is $k - 1 = 3 - 1 = 2$ for the regression sum of squares because there are $k = 3$ constants to be estimated in the regression equation $\hat{Y} = a + b_1X_1 + b_2X_2$ relating food expenditures (Y) to income (X_1) and family size (X_2). There are $n - k = 10 - 3 = 7$ degrees of freedom for the error sum of squares because there are $n = 10$ observations in the problem and $k = 3$ constants in the regression equation. As indicated in Table 10-3, the F ratio is calculated by dividing the regression mean square (variance) by the error mean square (variance). In Figure 10-2, we have

$$F = \frac{139.56725/2}{16.43275/7} = \frac{69.78363}{2.34754} = 29.72628$$

We turn to Table A-8 of Appendix A, where degrees of freedom for the numerator are read across the top of the table, and degrees of freedom for the denominator are read down the side. For 2 and 7 degrees of freedom, respectively, we find critical F values of 4.74 at the 5% level of significance and 9.55 at the 1% level of significance. Since the computed F value shown for the computer output in Figure 10-2 is 29.72628, we reject the null hypothesis at both the 5% and 1% levels. These relationships are shown in Figure 10-3. Since we have rejected the null hypothesis that all of the slope coefficients are equal to zero, we have rejected a null hypothesis of no relationship between the dependent variable and all of the independent variables considered collectively.

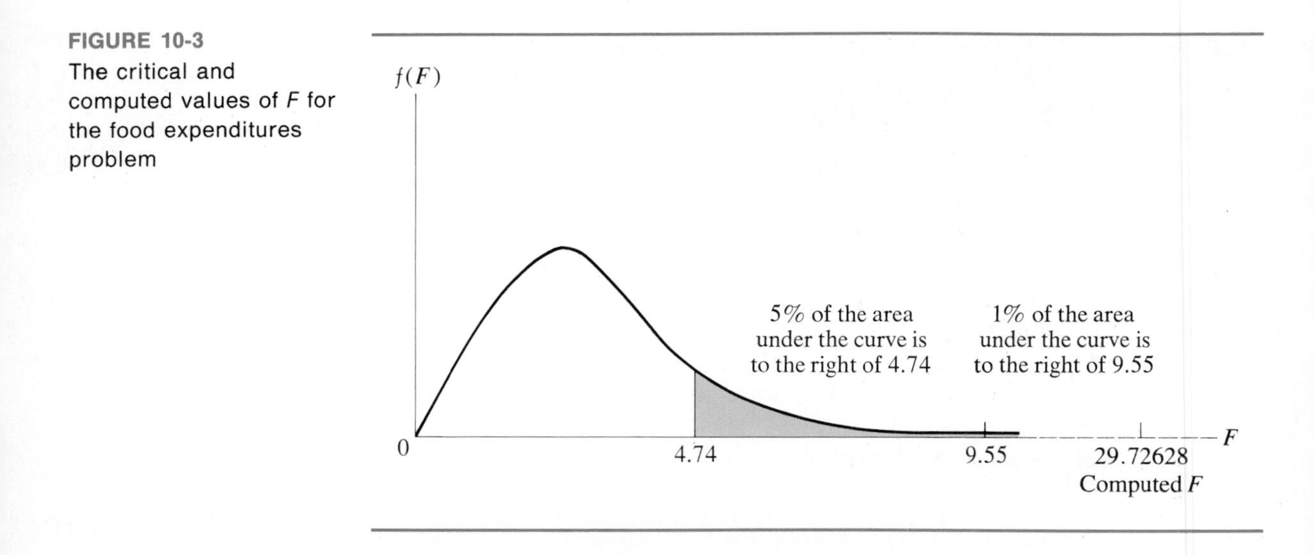

FIGURE 10-3 The critical and computed values of F for the food expenditures problem

Other Computer Output

Although we discuss the use of computers in multiple regression analysis in more detail later in this chapter, as an illustration we have reproduced in Figures 10-4 and 10-5 other selected computer output for the food expenditures versus income and family size problem. The problem was run using the Statistical Package for the Social Sciences (SPSS) written at Stanford University.

Figure 10-4 shows

1. The names of the variables; CONSTANT is the Y intercept or a value
2. The estimated b_1, b_2, and a values
3. The standard errors of the regression coefficients (s_{b_1} and s_{b_2})

The minor differences between the figures shown in the computer printout and those calculated earlier in this chapter are due to rounding.

FIGURE 10-4
Computer output for the variables in the regression equation to predict family expenditures for food

VARIABLES IN THE EQUATION

VAR.	B	STD. ERROR B
INCOME	2.36257	0.54957
FAMSIZE	−1.02339	0.79467
CONSTANT	7.91813	

FIGURE 10-5
Computer output for coefficients of multiple correlation, multiple determination, and the standard error of estimate

MULTIPLE R	0.94587
R SQUARE	0.89466
STANDARD ERROR	1.53217

Figure 10-5 shows

1. The multiple coefficient of correlation
2. The multiple coefficient of determination
3. The standard error of estimate

The R and R^2 values shown in the computer printout are unadjusted for degrees of freedom. Hence, the $R^2 = 0.89466$ value given in Figure 10-5 exceeds the $R^2_{Y.12} = 0.865$ figure computed earlier in this section. Denoting the R^2 value adjusted for degrees of freedom as R^2_c and the unadjusted figure as R^2, we have the following relationship between the two coefficients:

$$R_c^2 = 1 - (1 - R^2)\left(\frac{n-1}{n-k}\right) \tag{10.16}$$

The coefficient of multiple determination R^2 may be described as a measure of the effect on the dependent variable of all the independent variables combined. More specifically, R^2 measures the percentage of variance in the dependent variable that has been accounted for by all the independent variables combined. The square root of the coefficient of multiple determination $R = \sqrt{R^2}$ is referred to as the coefficient of multiple correlation. It is always positive. Since some of the individual independent variables may be positively correlated with the dependent variable and others negatively correlated, there would be no meaning in distinguishing between a positive and negative value for R. As in the case of r^2 and r in two-variable analysis, the R^2 figure is easier to interpret than R because R^2 is a percentage whereas R is not. We now consider some techniques and problems associated with multiple regression analysis.

10.7 DUMMY VARIABLE TECHNIQUES

Thus far in our discussion of regression analysis, the variables involved have been *quantitative*. However, variables of interest may be *qualitative* rather than quantitative. For example, in the family food expenditures example, suppose we wished to distinguish between urban and nonurban families, between home-owning and renting families, or between cases in which the head of the family is native born versus foreign born. Each of these examples represents a situation in which a classificatory or qualitative variable has two categories. If we wished to include one of these qualitative variables in the multiple regression analysis of family food expenditures versus family income and family size, we could set up a qualitative variable as a third independent variable as follows:

Variable X_3	Variable X_3	Variable X_3
$X_3 = 0$ if nonurban	$X_3 = 0$ if renter	$X_3 = 0$ if foreign born
$X_3 = 1$ if urban	$X_3 = 1$ if homeowner	$X_3 = 1$ if native born

Such *qualitative variables*, when included in regression analyses, are referred to as **dummy variables**. We illustrate the use of a dummy variable in the following example.

Assume that an airline wanted to relate its sales revenue (Y) to the incomes of its passengers (X_1) on the Boston to Miami route. Further assume that 6 months of the year are considered "off-peak months," while 6 months are considered "peak months." To account for this qualitative distinction, an analyst at the airline set up the following dummy variable: $X_2 = 0$ for off-peak months

TABLE 10-4

Data for regression analysis of airline sales revenue (Y), mean annual income of passengers (X_1), and a dummy variable (X_2) for off-peak (0) and peak months (1)

Month	Sales Revenue ($000)	Mean Annual Income of Passengers ($000)	Peak or Off-peak Month	Dummy Variable X_2
Jan.	4,480	45	Peak	1
Feb.	4,620	37	Peak	1
Mar.	4,550	49	Peak	1
Apr.	4,700	52	Peak	1
May	3,200	32	Off-peak	0
Jun.	3,360	35	Off-peak	0
Jul.	3,450	30	Off-peak	0
Aug.	3,400	28.5	Off-peak	0
Sep.	3,850	26	Off-peak	0
Oct.	3,940	34.5	Off-peak	0
Nov.	4,010	41.5	Peak	1
Dec.	3,700	33.5	Peak	1

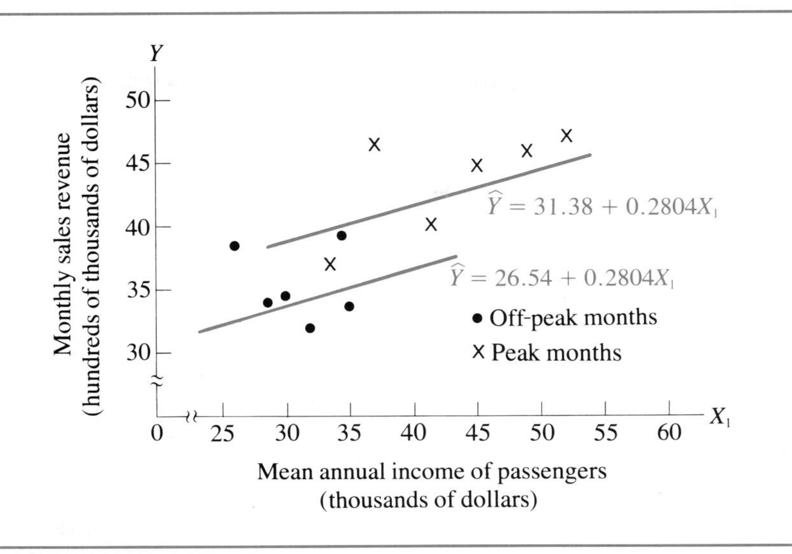

FIGURE 10-6
Scatter diagram of monthly sales revenue and mean annual income of passengers of an airline

and $X_2 = 1$ for peak months. The data for the year analyzed are shown in Table 10-4.

Figure 10-6 is a scatter diagram for the data on airline sales revenue and passenger income displayed in Table 10-4. Two distinct patterns are evident in the graph, one for the off-peak periods concentrated toward the lower left and one for the peak periods oriented toward the upper right. Therefore, it is generally evident that sales revenue varies directly with passenger income, but another factor also affects sales revenue on the Boston–Miami route, namely, whether the month of travel is a peak or off-peak period. To account for these relationships, we now set up the following linear regression equation:

(10.17)
$$\hat{Y} = a + b_1X_1 + b_2X_2$$

where $X_2 = 0$ for off-peak months
$= 1$ for peak months

Using ordinary least squares methods, we can solve for the constants a, b_1, and b_2 and analyze the implications of such a regression equation. If we substitute $X_2 = 0$ into equation 10.17 we obtain

(10.18)
$$\hat{Y} = a + b_1X_1$$

which we interpret as an equation that yields estimates of sales revenue for *off-peak* months. If we substitute $X_2 = 1$ into equation 10.17, we obtain

$$\hat{Y} = a + b_1X_1 + b_2$$

or

(10.19) $$\hat{Y} = (a + b_2) + b_1X_1$$

which we interpret as an equation that yields estimates of sales revenue for *peak* months.

The two regression equations 10.18 and 10.19 have the same slopes (b_1) but have different Y intercepts. Since the intercepts for the lines for off-peak periods and peak periods are, respectively, a and $a + b_2$, then b_2 can be interpreted as the extra amount of sales revenue, on the average, attributable to the peak months.

Using the method of least squares to estimate the values of the constants in equation 10.17, we obtain

(10.20) $$\hat{Y} = 26.54 + 0.2804X_1 + 4.84X_2$$

or, equivalently

(10.21) $$\hat{Y} = 26.54 + 0.2804X_1 \text{ for off-peak months}$$

(10.22) $$\hat{Y} = 31.38 + 0.2804X_1 \text{ for peak months}$$

Dummy variable

The regression lines corresponding to equations 10.21 and 10.22 are shown on the scatter diagram in Figure 10-6. In summary, by computing the one regression equation $\hat{Y} = a + b_1X_1 + b_2X_2$, in which X_2 was a **dummy variable**, we actually obtained two regression lines, one for off-peak periods and another for peak periods.

Note that in using the dummy variable X_2, we assume that it is reasonable to consider the slopes of the regression lines for off-peak and peak periods to be the same. If the slopes were markedly different, the dummy variable technique should not have been used and separate equations should have been computed for the off-peak and peak periods.

Let us now consider the implication of ignoring X_2, the dummy variable that accounted for whether a month was off-peak or peak. Referring again to the scatter diagram in Figure 10-6, we see that a single regression line fitted to the 12 points would have a much steeper slope than that of the two parallel lines. The slope of this regression line between sales revenue (Y) and passenger income (X_1) would be too great; that is, it would have an upward bias. Clearly, the effect that we would attribute to income alone should also be partially attributed to the factor of peak periods.

In the preceding example, we accounted for two qualitative categories, off-peak months and peak months. It is not appropriate to encode a dummy variable as 0, 1, 2, 3, and so forth, with more than two categories of a qualitative factor. By the use of such a code, we would imply no difference in the effect on the dependent variable of a change from zero to one, one to two, and two to three of the qualitative variable. This may be a totally unreasonable assumption. Hence, we may use a technique that establishes several dummy variables for one qualitative variable. We illustrate this technique in a simple example.

Suppose we wished to establish a linear regression model for the following variables relating to the production of a particular product in a company:

Y = average unit cost of production in dollars

X_1 = daily number of production workers

X_2 = daily total number of hours of work performed by the production workers

Assume that data on Y, X_1, and X_2 are available for three plants in the company.

To establish a coding for the three plants, we set up two dummy variables. that is, one fewer dummy variable than the number of categories in the qualitative variable. Labeling the two dummy variables X_3 and X_4, we denote the plants by the following code:

X_3	X_4	
1	0	when X_1 and X_2 are from plant 1
0	1	when X_1 and X_2 are from plant 2
0	0	when X_1 and X_2 are from plant 3

The linear equation would then take the following form:

(10.23) $$\hat{Y} = a + b_1X_1 + b_2X_2 + b_3X_3 + b_4X_4$$

Computation

The estimated average unit cost of production at plant 3, computed by substituting $X_3 = 0$ and $X_4 = 0$ in equation 10.23, would be equal to:

$$\hat{Y} = a + b_1X_1 + b_2X_2 + b_3(0) + b_4(0)$$

or

(10.24) $$\hat{Y} = a + b_1X_1 + b_2X_2$$

The estimated average unit cost of production at plant 1 is

$$\hat{Y} = a + b_1X_1 + b_2X_2 + b_3(1) + b_4(0)$$

or

(10.25) $$\hat{Y} = (a + b_3) + b_1X_1 + b_2X_2$$

Similarly, the estimated unit cost of production at plant 2 is

$$\hat{Y} = a + b_1X_1 + b_2X_2 + b_3(0) + b_4(1)$$

or

(10.26) $$\hat{Y} = (a + b_4) + b_1X_1 + b_2X_2$$

Summary

In summary, using this coding technique, we establish one multiple regression equation of the form in equation 10.23. However, that single equation yields three regression equations, one for each plant.

Interpretation

Let us now consider the interpretation of the slope coefficients for the dummy variables, that is, b_3 and b_4. Note that plant 3 was the category for which $X_3 = 0$ and $X_4 = 0$. Let us designate plant 3 as the "base plant." Comparing the Y intercept for plant 1 $(a + b_3)$ in equation 10.25 with that of the base plant, (a) in equation 10.24, we can state that, on the average, the unit cost of production at plant 1 is b_3 greater than at the base plant.

Analogously, referring to equation 10.26 for plant 2, and comparing the Y intercept of $(a + b_4)$ with that of the base plant, (a) in equation 10.24, we can state that, on the average, the unit cost of production at plant 2 is b_4 greater than at the base plant.

Let us assume, for example, that in this example of estimating average unit cost of production the multiple regression equation given in equation 10.27 had been derived.

$$\hat{Y} = 3.123 - 0.0051X_1 - 0.00036X_2 + 0.470X_3 + 0.237X_4 \tag{10.27}$$

The respective regression equations for plants 1, 2, and 3 are

Plant 1

$$\hat{Y} = 3.123 - 0.0051X_1 - 0.00036X_2 + 0.470(1) + 0.237(0)$$

or

$$\hat{Y} = 3.593 - 0.0051X_1 - 0.00036X_2$$

Plant 2

$$\hat{Y} = 3.123 - 0.0051X_1 - 0.00036X_2 + 0.470(0) + 0.237(1)$$

or

$$\hat{Y} = 3.360 - 0.0051X_1 - 0.00036X_2$$

Plant 3

$$\hat{Y} = 3.123 - 0.0051X_1 - 0.00036X_2 + 0.470(0) + 0.237(0)$$

or

$$\hat{Y} = 3.123 - 0.0051X_1 - 0.0036X_2$$

Hence, estimating the average unit cost of production for $X_1 = 100$ workers and $X_2 = 4{,}100$ total hours of work at each of the three plants, we have

Plant 1 $\hat{Y} = 3.593 - 0.0051(100) - 0.00036(4{,}100) = \$1.607$

Plant 2 $\hat{Y} = 3.360 - 0.0051(100) - 0.00036(4{,}100) = \$1.374$

Plant 3 $\hat{Y} = 3.123 - 0.0051(100) - 0.00036(4{,}100) = \$1.137$

Thus, we observe that the average unit cost at plant 3, the base plant, is \$1.137; the figure for plant 1 is \$1.607, or \$0.47 higher ($b_3 = 0.47$) than for the base plant; and the figure for plant 2 is \$1.374, or \$0.237 higher ($b_4 = 0.237$) than for the base plant. Of course, if either plant 1 or plant 2 had been selected as the base rather than plant 3, the same cost estimates would have resulted and the same relationships would have been observed among the cost figures for the three plants.

10.8 MULTICOLLINEARITY

Multicollinearity—or simply, **collinearity**—describes a problem that arises in multiple regression analysis when independent variables are highly correlated. In such situations, it is not possible to separate the individual effects of the independent variables on the dependent variable. To view the problem intuitively, we consider the extreme situation in which we correlate a dependent variable Y with an independent variable X_1 and then add another independent variable X_2 that is perfectly linearly correlated with X_1. Since X_1 and X_2 move together, it is impossible to disentangle the separate effects of these variables on Y. Furthermore, from a practical standpoint, once a regression equation has been established between Y and X_1, nothing is gained by adding X_2 to the equation. With X_1 already in the equation, X_2 does not account for any of the unexplained variance in the dependent variable.

When multicollinearity exists among the independent variables in a multiple regression equation, the net regression (slope) coefficients tend to be unreliable. Usually one or more of the independent variables will have a net regression coefficient that is not significantly different from zero. For example, in the problem in which we related family food expenditures (Y) to family income (X_1) and family size (X_2), we indicated that the net regression coefficient for income (b_1) differed significantly from zero, whereas the coefficient for family size (b_2) did not. A 0.912 correlation coefficient (Table 10-2) indicated the high correlation between income and family size. In this example, only b_2 did not differ significantly from zero. It would even have been possible for both net regression coefficients not to have differed significantly from zero, even though in separate two-variable regression equations—$\hat{Y} = a + b_1X_1$ and $\hat{Y} = a + b_2X_2$—the coefficients b_1 and b_2 were significantly different from zero.

Remedy 1: delete one or more variables

Although a number of steps can be taken to solve the problem of multicollinearity, the simplest is to delete one or more of the correlated variables. Judgment and significance tests for the net regression coefficients can determine which variables to drop. In general, if one of two highly correlated independent variables is dropped, the R^2 value will not change much. For example, when the food expenditures figure was regressed on income and family size, the R_c^2 value was 0.865. The r_c^2 value for food expenditures regressed on income alone was 0.854. Clearly, in this case, little predictive ability is lost by dropping the family size variable.

> A rule of thumb is often used to determine whether to delete a collinear variable: Drop the variable if R_c^2 increases upon its deletion.

It can be shown that this rule is equivalent to dropping the variable if the t statistic for its net regression coefficient is less than one.

Remedy 2: alter form of one or more variables

Another possible remedy for multicollinearity is to change the form of one or more of the independent variables. For example, if national income in current dollars is one of the independent variables and one or more other independent variables are highly correlated with it, then dividing income by population to yield a per capita national income variable or dividing income by a price index to yield national income in *real dollars* rather than *current dollars* may result in less correlated independent variables. Econometricians sometimes use more sophisticated measures. For example, when dealing with time-series data in estimating demand from income, prices, and other data, they may estimate one of the parameters in the regression equation independently from a cross-sectional study of family budget data. This combined use of time-series and cross-sectional data helps solve the multicollinearity problem, but gives rise to other methodological problems.

Summary

In summary, no single simplistic solution eliminates the multicollinearity problem. If we attempt to devise an explanatory model, and we use secondary data rather than the results of an original experiment, there may be no practical way to disentangle the separate effects of intercorrelated independent variables. On the other hand, if we construct a regression model for forecasting purposes, multicollinearity is of less concern, provided we can expect the correlations among the independent variables to persist in the future.

10.9 AUTOCORRELATION

Autocorrelation or **serial correlation** is a problem of regression analysis that is usually present when time-series data are used. Specifically, it arises because of

Autocorrelation

> the violation of the regression analysis assumption that successive observations of the dependent variable Y are independent.

As noted in section 9.1, this assumption means that a low (or high) value for a particular observation does not imply that the next Y value observed will also be low (or high). We may state the assumption in another way: A positive (or negative) value of a residual $(Y - \hat{Y})$ for a particular observation does not imply that the next observation will also have a positive (or negative) residual. The validity of this assumption for *cross-sectional data* is virtually assured because of random sampling. For example, in a regression analysis of the heights (X) and weights (Y) of a random sample of women, the fact that a particular woman's weight yields an observation that is above the regression line or curve should have no bearing on whether the next woman's weight is also a point above the regression line or curve. On the other hand, in the case of economic data, which are often cyclical in nature, if the observation for a particular period

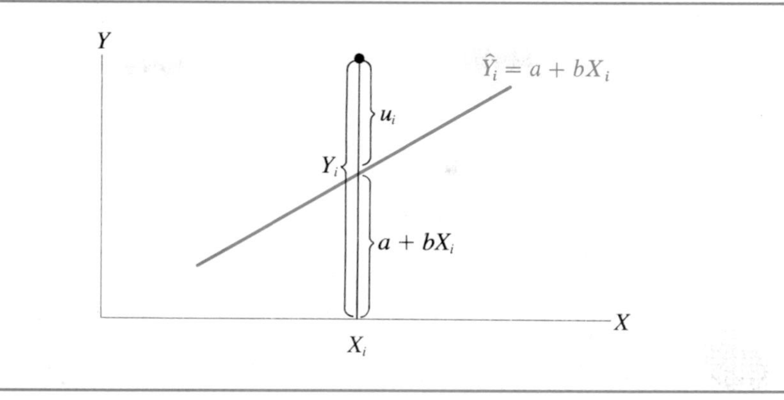

FIGURE 10-7
Alternative notation for the linear regression equation for the *i*th observation

lies above the regression line or curve—that is, it has a positive residual—it is likely that the next one also will. In fact, the observations will tend to be sequences or runs of positive residuals for points above the values predicted from the regression equation followed by runs of negative residuals. Stated differently, the residuals tend to be correlated among themselves (hence, the term, autocorrelation) rather than independent.

Durbin-Watson statistic

A widely used method of detecting the presence of autocorrelation involves the Durbin-Watson statistic. Before examining that statistic, let us consider a minor change in the notation that we have been using for the regression equation. We illustrate this change in terms of the two-variable linear regression equation that we have denoted $\hat{Y} = a + bX$. The following alternative symbology states the equation for the *i*th observation of the dependent variable:

$$Y_i = a + bX_i + u_i \quad (10.28)$$

As indicated in Figure 10-7, in this notation $a + bX_i$ is the height of the regression line for the *i*th observation of the independent variable, and u_i is the residual or deviation from the regression line. To see the meaning of u_i more clearly, note that if the regression equation is expressed as $\hat{Y}_i = a + bX_i$, then equation 10.28 becomes $Y_i = \hat{Y}_i + u_i$ and u_i is equal to $Y_i - \hat{Y}_i$, or the deviation of the observation Y_i from the predicted value $\hat{Y}_i$.

Using the aforementioned notation, the Durbin-Watson statistic, denoted d, is defined as

$$d = \frac{\sum_{i=2}^{n} (u_i - u_{i-1})^2}{\sum_{i=1}^{n} u_i^2} \quad (10.29)$$

where u_i = a residual from the regression equation in time period i
u_{i-1} = a residual from the regression equation in time period $i - 1$, that is, the period before i

> The Durbin-Watson statistic ordinarily tests the null hypothesis that *no positive autocorrelation* is present.

This is equivalent to the null hypothesis that the residuals are random.

J. Durbin and G. S. Watson have tabulated lower and upper critical values of the d statistic, d_L and d_U, such that

Critical values of d_L and d_U

1. If $d < d_L$, reject the null hypothesis of no positive autocorrelation and conclude that there is positive autocorrelation.
2. If $d > d_U$, accept the hypothesis of no positive autocorrelation and conclude that there is no positive autocorrelation.
3. If $d_L \leqslant d \leqslant d_U$, the test is inconclusive.

The critical values of d_L and d_U are given in Table A-11 of Appendix A for $\alpha = 0.05$ and $\alpha = 0.01$. These tabulated lower (d_L) and upper (d_U) bounds are given for n, the number of observations, and k, the number of independent variables.

For example, suppose we have a time series with $n = 25$ observations and $k = 3$ independent variables and we want to test the null hypothesis of no positive autocorrelation at the 5% significance level. In Appendix Table A-11, we find $d_L = 1.12$ and $d_U = 1.66$. Therefore, if the observed d value is less than 1.12, we reject the null hypothesis of no positive autocorrelation and conclude that the residuals exhibit positive autocorrelation. If the observed d value is greater than 1.66, we accept the null hypothesis of randomness, or no positive autocorrelation.

To see clearly why positive autocorrelation leads to low observed values of the Durbin-Watson statistic and why randomness leads to higher values, let us consider the two illustrative situations in Table 10-5. In Example A, the residuals (u_i) exhibit a trend. This might occur when the observations of Y are on the portion of a cycle rising above the regression line. The computed d value is $\frac{4}{55}$. On the other hand, in Example B, the same residuals are shown as in Example A, except now they are randomly distributed. The calculated value of d in Example B is $\frac{25}{55}$, more than 6 times the corresponding value given for Example A. Incidentally, note that there is *perfect direct autocorrelation* between u_i and u_{i-1} in Example A. That is, the correlation coefficient r equals one for the two following series:

Time Period	u_i	u_{i-1}
2	2	1
3	3	2
4	4	3
5	5	4

TABLE 10-5
Examples of a low and high value for the Durbin-Watson statistic (d)

Time Period	Example A: Low Value of d				Example B: High Value of d			
	u_i	$u_i - u_{i-1}$	$(u_i - u_{i-1})^2$	u_i^2	u_i	$u_i - u_{i-1}$	$(u_i - u_{i-1})^2$	u_i^2
1	1			1	1			1
2	2	1	1	4	5	4	16	25
3	3	1	1	9	3	−2	4	9
4	4	1	1	16	4	1	1	16
5	5	1	1	25	2	−2	4	4
			4	55			25	55

$$d = \frac{\sum_{i=2}^{5} (u_i - u_{i-1})^2}{\sum_{i=1}^{5} u_i^2} = \frac{4}{55} \qquad d = \frac{\sum_{i=2}^{5} (u_i - u_{i-1})^2}{\sum_{i=1}^{5} u_i^2} = \frac{25}{55}$$

The major difficulty in working with autocorrelated time-series data is that probability statements involving the use of standard errors assume that the $Y - \hat{Y}$ deviations are randomly distributed around the regression equation, and these statements can be seriously in error if the distribution is not random. For example, using the notation for a two-variable regression, the standard error of estimate $s_{Y.X}$ and the standard error of the regression coefficient s_b are both biased downwards. Therefore, if the residuals are serially correlated, the standard errors are deceptively small compared with what they should be if the residuals are randomly distributed. Therefore, invalid probability statements will be made about the regression equation and its slope coefficients, and F tests and t tests will not be strictly valid.

Remedy: change the variable

What techniques can remedy the problem of a considerable degree of autocorrelation? Probably the most widely used method is to work in terms of *changes* in the dependent and independent variables—referred to as **first differences**—rather than in terms of the original data themselves. For example, we might use the variable "year-to-year change in population" rather than population itself. The year-to-year changes are referred to as the *first differences in population.* Other remedies include transforming the variables, (for example, using logarithms), adding another variable, and using various modified versions of the first difference transformation. These latter techniques, which are not dealt with here, are referred to as **autoregressive schemes** and are treated in most texts in econometrics. Of course, combinations of these techniques are also used.

Other remedies

Our discussion has dealt with the problem of **positive autocorrelation**, a situation in which the successive terms in a time series are *directly* related. **Negative autocorrelation**, in which the successive terms are *inversely* related, is not a practical problem for economic data. In such a series, high and low values alternate in a systematic fashion in successive time periods. This phenomenon is not frequently encountered in economic activity.

10.10

ANALYSIS OF RESIDUALS

It is useful to study the residuals in a regression analysis for a variety of reasons. We can analyze these deviations of actual values from predicted values to determine the magnitudes of prediction errors, to check on the departures from regression model assumptions, and to make diagnoses concerning the steps to take in continuing the analysis.

Virtually all computer programs print out actual Y values, predicted $\hat{Y}$ values, and the residuals $Y - \hat{Y}$. Furthermore, most computer programs contain various options for graphing the residuals versus the order of observations and independent variables. We now consider a number of situations concerning residuals.

Residuals May Not Be Independent

Departures from the assumption of independence can be observed in a plot of the residuals versus the order of the original observations. Plotting this type of graph is always advisable, particularly for time-series data, in which the order is chronological time. In Figure 10-8, the residuals $Y - \hat{Y}$ are plotted against the order of observations for a time series. Usually, in computer

FIGURE 10-8
Positive autocorrelation in the residuals

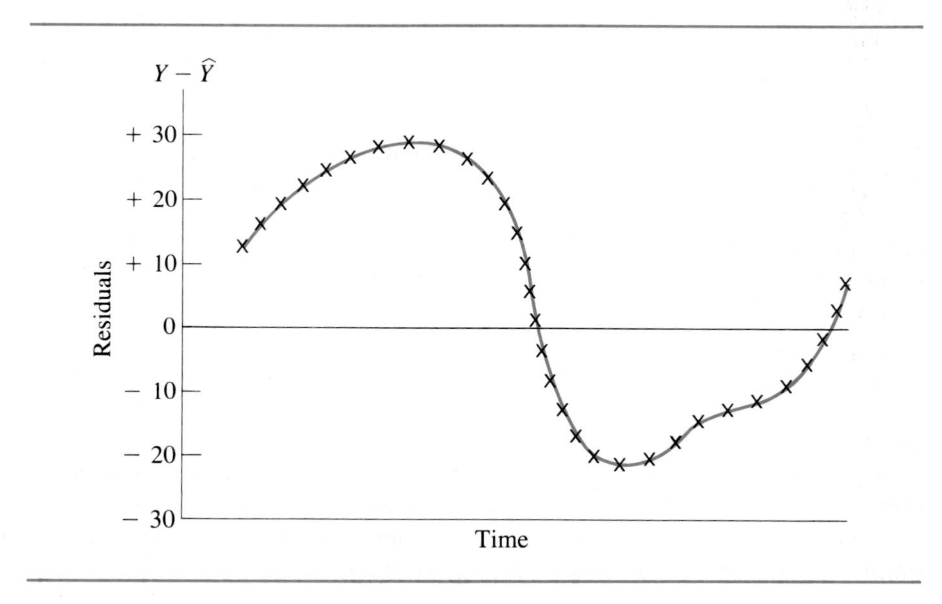

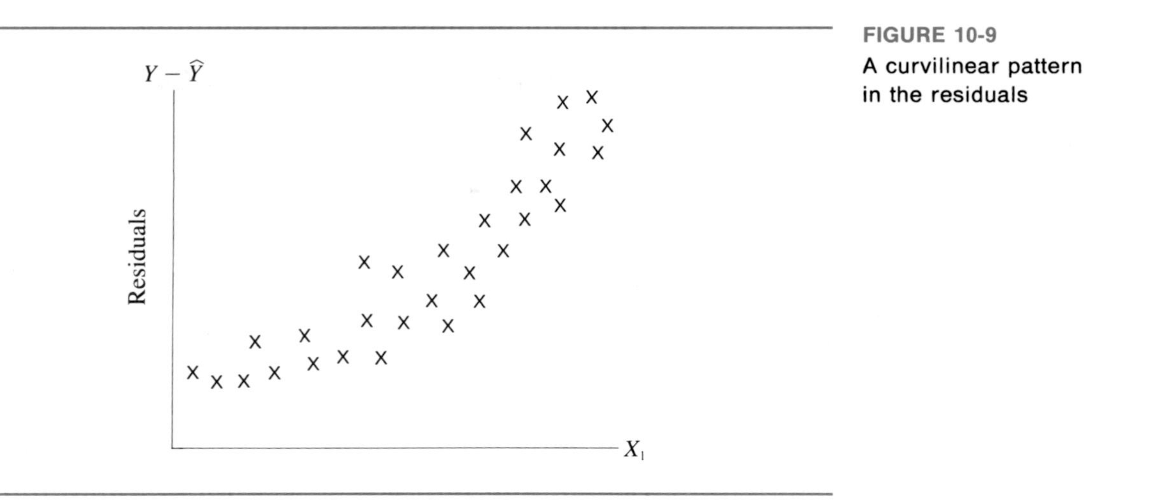

FIGURE 10-9
A curvilinear pattern in the residuals

printouts, the points are not joined by lines or curves, but they are so depicted here in order to clarify the pattern. The figure shows a run of positive residuals followed by a run of negative residuals and the beginning of another run of positive residuals. Such a configuration signals a problem of positive autocorrelation (direct serial correlation). In the section on autocorrelation, we considered some remedies for this problem.

A Functional Transformation May Be Required

Sometimes after a linear regression equation has been fitted to the data, the graph of the residuals indicates that some type of functional transformation may be appropriate. For example, Figure 10-9 depicts a curvilinear pattern when the residuals are plotted against the values of an independent variable X_1. In such a situation, if the fitted regression equation was of the form $\hat{Y} = a + b_1X_1 + \cdots$, on the next run, we might fit a second-degree parabolic function of the form $\hat{Y} = a + b_1X_1 + b_2X_1^2 + \cdots$. Most computer programs for regression analysis accomplish this by defining a new variable X_1^2 and including it as an additional independent variable. Another possibility for transforming a pattern such as that in Figure 10-9 is to use logarithms. That is, on the next run, we might fit a function of the form $\log \hat{Y} = a + b_1X_1 + \cdots$.

Outlier Observations Are Present

Sometimes, a regression equation may be a good fit to all but one or a few extreme observations, often referred to as *outliers*. For example, the plots in Figures 10-10(a) and (b) depict a single outlier observation in each graph.

As in many other problems of regression analysis, there is no single simplistic solution to the problem of extreme observations. We might be

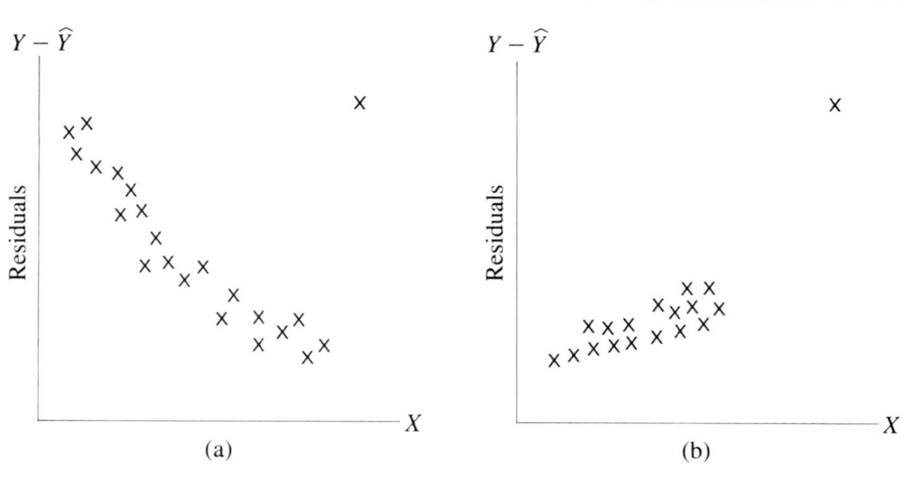

FIGURE 10-10
Two patterns of residuals that have single outlier observations

One solution: delete outlier

tempted merely to delete the outlier observations. However, it would clearly be an unscientific procedure to delete or ignore observations that in some way did not agree with a particular hypothesis or model. On the other hand, if we can demonstrate logically that the extreme observations were derived from a different conceptual universe than the other observations, then we may be justified in excluding such outliers. For example, if all of the points except the extreme observation pertained to women and the outlier was an observation for a man, or if all but one of the points pertained to towns with populations of 10,000 and under and the extreme observation related to New York City, then a logical justification might exist for excluding the outliers.

Compromise solution

A compromise position that is sometimes used is to run the analysis both with and without the extreme observations. Then, use of the analysis is a matter of *caveat emptor*, "let the buyer beware." That is, the analysis contains the results for both the inclusion and exclusion of outliers, and the investigator is free to draw his or her own conclusions.

Independent Variables May Have Been Omitted

One solution

Patterns in residuals or very large residuals may indicate that one or more independent variables have been omitted from the regression equation. To solve this problem on the next run, we should add one or more variables that might account for this variation. For example, suppose after running the regression analysis in Chapter 9 for family food expenditures as a function of family income, we discovered the most of the observations with negative deviations were from families headed by women. Correspondingly, many observations displaying positive deviations were from families headed by men. It would then make sense to introduce a dummy variable that might take on the value of zero for families headed by women and the value of one for families headed by men.

Another solution

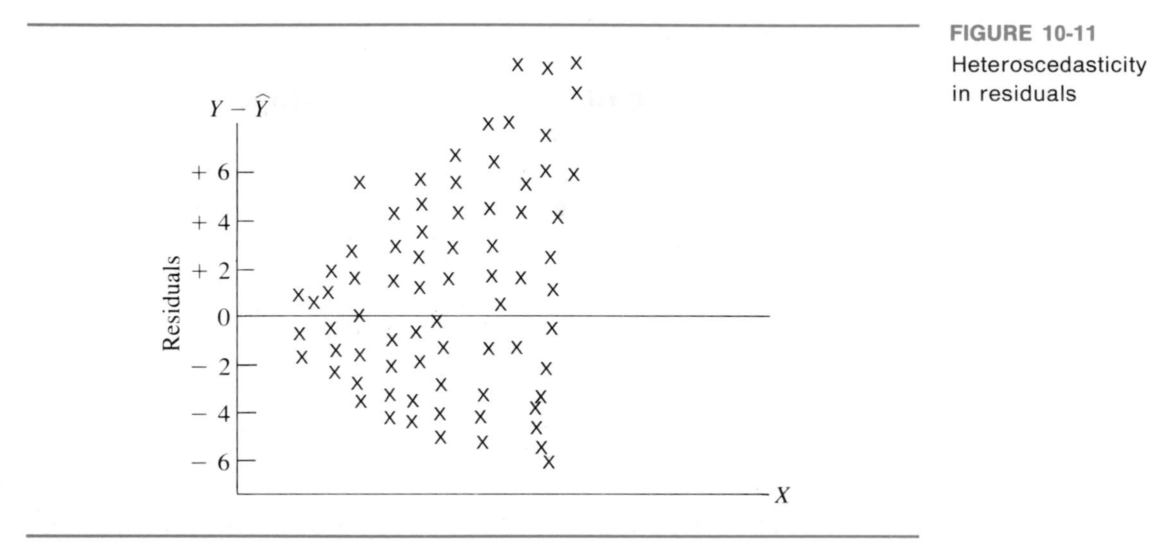

FIGURE 10-11
Heteroscedasticity in residuals

Similarly, the introduction of quantitative rather than dummy variables might assist in accounting for unexplained variation.

Other clues

Of course, a systematic pattern in residuals is only one of many possible indications that a variable has been omitted. Other clues would be a high standard error of estimate, a low R^2 value, and low t values for slope coefficients, which, on the basis of our knowledge of the situation, should have significant effects on the dependent variable.

The Error Terms May Not Have Constant Variance

Sometimes departures from linear regression model assumptions are clearly revealed by plots of the residuals. For example, let us assume that the equation $\hat{Y} = a + bX$ has been fitted to a set of data. The pattern of residuals displayed in Figure 10-11 indicates that the assumption of *homoscedasticity*—that is, constant variance of the residuals for all values of X—has been violated. The variance of residuals increases as the values of the independent variable increase. The corrections for **heteroscedasticity** (that is, departure from homoscedasticity) depend on whether the error variances for each set of independent variable observations are known from prior information. If not, then to estimate error variances, we need several observations of the dependent variable for each set of observations of the independent variables. One appropriate corrective technique, known as *weighted least squares*, is a special case of a general econometric technique called *generalized least squares.* These remedies for heteroscedasticity are outside the scope of our discussion, but may be found in most texts on econometrics.[4]

[4] For example, see R. S. Pindyck and D. L. Rubenfeld, *Econometric Models and Econometric Forecasts* (New York: McGraw-Hill, 1976), pages 97–102.

10.11

OTHER MEASURES IN MULTIPLE REGRESSION ANALYSIS

A number of other measures are sometimes calculated in a multiple regression and correlation analysis. Only brief reference will be made to them here.

Coefficients of partial correlation

It is possible to calculate measures that indicate the separate effect of each of the independent variables on the dependent variable, if the influence of all the other independent variables has been accounted for. For this purpose, it is conventional to compute **coefficients of partial correlation**. For example, the partial correlation coefficient for family income in our illustrative problem, designated $r_{Y1.2}$, would show the partial correlation between Y and X_1 after the effect of X_2 on Y had been removed. The square of this coefficient $r_{Y1.2}^2$ measures the reduction in variance brought about by introducing X_1 after X_2 has already been accounted for.

Beta coefficients

Sometimes, it is difficult to compare the differences in net regression coefficients because the independent variables are stated in different units. For example, in the illustrative example, b_1 indicates the average difference in food expenditures Y *per unit difference in family income* X_1, whereas b_2 indicates the average difference in food expenditures Y *per unit difference in family size.* (In both cases, the other independent variable is held constant.) Unit differences in X_1 and X_2 are in different units—$1,000 and one person, respectively. To improve comparability, we can state the regression equation in a different form, giving each of the variables in units of its own standard deviation. The transformed net regression coefficients are called **beta coefficients**. For example, in terms of beta coefficients, the linear regression equation for three variables would be

(10.30)
$$\frac{\hat{Y}}{s_Y} = \alpha + \beta_1 \frac{X_1}{s_{X_1}} + \beta_2 \frac{X_2}{s_{X_2}}$$

Thus, the beta coefficients are equal to

(10.31)
$$\beta_1 = b_1 \frac{s_{X_1}}{s_Y} \quad \text{and} \quad \beta_2 = b_2 \frac{s_{X_2}}{s_Y}$$

As an illustration of the meaning of the beta coefficients, β_1 measures the number of standard deviations that $\hat{Y}$ changes with each change of one standard deviation in X_1.[5]

Selected General Considerations

A great deal of care must be exercised in the use of multiple regression and correlation techniques. In the development of the model, theoretical analysis,

[5] The reader is alerted that the battle of notation must be continually fought. In Chapter 7, β denoted the probability of a Type II error. The specific meaning of this symbol must be determined from the context of the particular discussion.

Analysis, knowledge, and judgment

knowledge of the field of application, and logical judgment should aid in the selection of variables to be used in the study. Frequently, in business and economic applications, some of the relevant variables may not be easily quantifiable. Sometimes variables are not readily available and must be constructed from different sets of data.

Validity of assumptions

In the discussion of multiple regression and correlation analysis, we have confined ourselves to a linear model. We know that the underlying assumptions of this model should be checked for their validity. We should examine the graphs of the dependent variable against each of the independent variables at the outset of the analysis and check the plots of the $Y - \hat{Y}$ deviations against each of the independent variables after fitting the regression equation. Sometimes transformations—such as taking logarithms, reciprocals, or square roots of original observations—may provide better adherence to original assumptions and better fits of regression equations to the data. Of course, a linear regression equation simply represents a convenient approximation to the unknown "true" relationship. When linear relationships provide inadequate fits, curvilinear regression equations may be required.

Parsimony

The quest for a good fit of the regression equation to the data leads to adding more and more independent variables. However, cost considerations, difficulties of providing data in the implementation and monitoring of the model, and the search for a reasonably simple model ("parsimony") point toward the use of as few independent variables as possible. Since no mechanistic statistical procedure can resolve this dilemma and many other problems of multiple regression and correlation analysis, subjective judgment inevitably plays a large role. Statistical theory alone cannot determine which variables should be included in a regression analysis. Prior knowledge of the field of application is important in the initial selection of independent variables and in choices of variables to include or exclude based on the statistical analysis.

Dangers of extrapolation

We must guard against the dangers of extrapolation. There are subtle difficulties in multiple analysis compared with two-variable analysis. Even within the range of the data, certain combinations of values of the independent variables may not have been observed. Therefore, statistically valid estimates of the dependent variable cannot be made for these combinations of values.

10.12 THE USE OF COMPUTERS IN MULTIPLE REGRESSION ANALYSIS

The use of high-speed electronic computers has greatly simplified the testing and analysis of statistical relationships among variables. The libraries of most computer centers contain programs for various types of multivariate analysis including multiple regression analysis. In the past, the cost and tedious labor involved in multiple regression analyses involving more than two or three independent variables severely restricted the analyst's ability to test and experiment. Using modern computer programs, the analyst now has a much wider range of choice in selecting variables, in options for performing transformations, in adding and deleting variables at various stages of the analysis, and in testing curvilinear as well as linear relationships.

Stepwise regression analysis

Stepwise regression analysis is a versatile form of multiple regression analysis for which computers are particularly useful and for which a number of computer programs exist. In this type of analysis, at the first stage, the computer determines which of the independent variables is most highly correlated with the dependent variable. The computer printout then displays all the usual statistical measures for the two-variable relationship. At the next stage, the program selects the independent variable that accomplishes the greatest reduction in the unexplained variance remaining after the two-variable analysis. The computer printout then displays all the usual statistical measures for the three-variable relationship. The program continues in this stepwise fashion, at each stage entering the "best" independent variable in terms of ability to reduce the remaining unexplained variance. Analysis of variance tables and lists of residuals $(Y - \hat{Y})$ are provided at each stage. Obviously, without the use of computers and "canned" programs, the time needed to perform such an analysis by hand would be prohibitive.

Caution

One final comment is in order. The advent of electronic computers has opened up a greater choice than was formerly available in selection of variables, the inclusion of larger numbers of variables, more options in performing transformations of variables, and more testing and experimentation with different types of statistical relationships. However, because of these increased possibilities, it becomes even more important that care and good judgment be exercised to avoid misuse of methods and misinterpretation of findings.

10.13 A COMPUTER APPLICATION

As an example of a multiple regression analysis that deals with many of the problems of such analyses, we present a computer application that was carried out using an interactive program package.[6] Responses to queries posed in the program are the items following the questions. We state the problem as follows: The president of a carpet manufacturing firm is interested in predicting company sales. He has collected data on company sales and other variables by quarters of the year for 1978–1982. Hence, there are 20 observation periods

[6] This program was written by Professor James Pickands III at The Wharton School, University of Pennsylvania, and was run on the DEC System-10 computer. The problem was constructed by Professor Abba Krieger at The Wharton School, University of Pennsylvania.

FIGURE 10-12
Observations in 20 periods on 3 variables

QUARTER	Y	X1	X2
1 :	28.2	1.1	191
2 :	36.7	1.3	324.3
3 :	36.1	1.5	348.5
4 :	30.6	1.9	297.1
5 :	30.9	2	280.8
6 :	52.3	2.7	439.3
7 :	51.4	2.8	434.3
8 :	46.2	2.8	382.1
9 :	43.1	2.6	367.4
10 :	63.8	3.1	581.1
11 :	63.1	3.4	561.5
12 :	57.3	9.2	477.1
13 :	55	20.7	362
14 :	75.3	40.8	624.5
15 :	70.1	60.3	563.6
16 :	68.5	65.4	470.2
17 :	80.3	72.8	325.6
18 :	96.8	80.1	541.9
19 :	95.1	86.3	498.2
20 :	92.8	89.1	379.3

for the following variables:

Variable Number	
1	Y = sales (ten thousands of dollars)
2	X_1 = advertising budget (thousands of dollars)
3	X_2 = U.S. housing starts (thousands)

There was a supply shortage during the last two quarters of 1981. A printout of the data appears in Figure 10-12. (As a general matter, after you have typed the input data, you should print out and check the data before proceeding further.)

The first computer run involves fitting a linear regression equation, which in the standard notation that we have been using is expressed as $\hat{Y} = a + b_1X_1 + b_2X_2$. However, the computer program used here requires variables to be identified by numbers, so we have designated the dependent variable (Y) as 1, the first independent variable (X_1) as 2, and the second independent variable (X_2) as 3. This type of designation is often used in computer programs. Selected ouput of the first computer run is displayed in Figure 10-13.

FIGURE 10-13
Computer printout for the regression of carpet sales (variate 1) on advertising budget (variate 2) and U.S. housing starts (variate 3)

```
CONTINUE? YES
VARIATE STATISTICS:

              MEANS           VARIANCES         ST. DEVS.
   1:         58.68          471.04379         21.703543
   2:        27.495          1165.1331         34.134047
   3:        422.49          13418.566         115.83854

VARIATE CORRELATIONS:
   1:      1.000      0.884      0.634
   2:                 1.000      0.321
   3:                            1.000

CONTINUE? YES

WANT REGRESSION ANALYSIS? YES
IDENTIFY VARIATES BY NUMBER.

DEPENDENT VARIABLE?   1

ANY INDEPENDENT VARIABLE(S) OMITTED? NO
THE 0 COEFFICIENT IS THE INTERCEPT.
THE PROBABILITY OF OCCURRENCE ( α ) IS TWO-SIDED.
   VAR.            COEFF.            ST.DEV.            T        SIG ( α )
    0:        14.525849        5.7158829          2.54131     0.020205
    2:        0.48254797       0.046588056        10.3578     0.000000
    3:        0.07310586       0.013728065        5.32529     0.000071
S E OF RESIDUALS          6.56567
R SQUARED                 0.918117
  EFFECTS              SS        DF         MSS            F       SIG ( α )
REGRESSION     8216.9952         2      4.108E3       95.307     0.000000
ERROR           732.8368        17      4.311E1
TOTAL          8949.832         19
RESIDUALS:
SUM OF SQUARES              732.837
DURBIN-WATSON                0.973155
THERE ARE 12 POSITIVE RESIDUALS AND 8 NEGATIVE ONES.
```

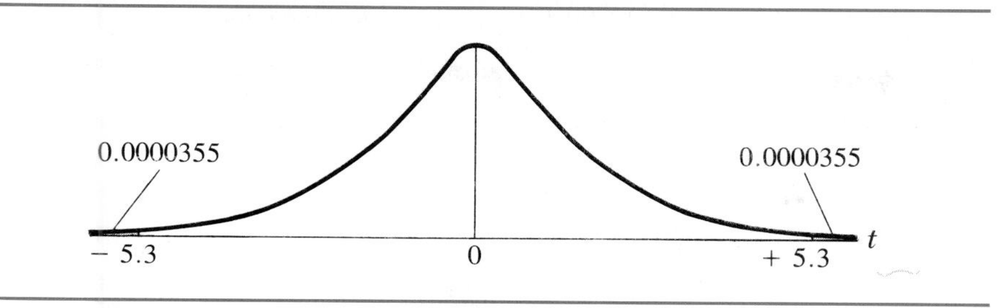

FIGURE 10-14
Significance test of the slope coefficient for the U.S. housing starts variable

The first batch of output in Figure 10-13, labeled VARIATE STATISTICS, gives the means, variances, and standard deviations of each of the variables. These self-explanatory items are followed by VARIATE CORRELATIONS, which are the correlation coefficients (r values) for each pair of variables. Reading from left to right, the implied column headings in the triangular matrix or table are variates 1, 2, and 3. Hence, the entries along the diagonal indicate that the correlation coefficient for each variable with itself is equal to 1.000. Note that on the basis of these gross correlations, carpet sales are more highly correlated with advertising budget ($r_{12} = 0.884$) than with U.S. housing starts ($r_{13} = 0.634$). Also, since the correlation coefficient between advertising budget and housing starts is low ($r_{23} = 0.321$), little multicollinearity exists between the independent variables.

Moving to the output headed VAR., COEFF., and so on, we can express the figures for the regression equation in the usual form as

$$\hat{Y} = 14.525849 + \underset{(0.046588056)}{0.48254797X_1} + \underset{(0.013728065)}{0.07310586X_2}$$

The figures in parentheses under the net regression coefficients (b values) are the standard errors of these slope coefficients (s_b values). As noted in the output, the zero coefficient is the intercept (a value). The entries under the column headed T are the $t = (b - 0)/s_b$ figures used to test the null hypothesis $H_0\colon B = 0$ for each of the slope coefficients. Referring to Figure 10-14 will be helpful in interpreting the entries under the column headed SIG(α). That figure depicts the t test for the slope coefficient of the U.S. housing variable. The t value has been abbreviated to one decimal place, that is, $t = 5.3$. The chart indicates the probability $P(|t| > 5.3) = 0.0000355 + 0.0000355 = 0.000071$. That is, the 0.000071 figure shown in the SIG(α) column of Table 10-14 is the probability of observing a t value greater than the 5.32529 figure in the preceding column. Therefore, if the SIG(α) figure is less than 0.05 or 0.01, the corresponding slope coefficient is significant at the 0.05 and 0.01 levels, respectively. The number of degrees of freedom for this test is $\nu = n - k = 20 - 3 = 17$.

The SE OF RESIDUALS is the standard error of estimate. In the notation that we have used in this chapter, $S_{Y.12} = 6.56567$ (tens of thousands of dollars). The next figure, uncorrected for degrees of freedom, is $R^2 = 0.918117$. The R^2 value indicates a strong relationship in which about 92% of the variation in carpet sales has been accounted for by the regression equation relating that variable to advertising budget and U.S. housing starts.

The analysis of variance table is in the standard form. In this problem, the addition of a SIG(α) figure indicates that the F ratio is significant at levels far below the conventional 0.05 and 0.01 critical values. One notational detail in the mean sum of squares (MSS) column, which displays the regression variance and error variance, is the use of $E3$ for 10^3 and $E1$ for 10^1. Thus, the F ratio is

$$F = \frac{4.108 \times 10^3}{4.311 \times 10^1} = \frac{4{,}108.00}{43.11} = 95.307$$

In view of the relatively high R^2 value, of course, we are not surprised that the F test is significant, indicating that the overall regression relationship is significant.

The computer then prints out the Durbin-Watson statistic $d = 0.973155$. Referring to Appendix Table A-11, we find $d_L = 1.10$ and $d_U = 1.54$ for $n = 20$ observations and $k = 2$ independent variables. Since the observed d value is less than $d_L = 1.10$, we reject the hypothesis of no positive autocorrelation and conclude that the residuals exhibit positive serial correlation. Referring to Figure 10-15, which shows the residuals $(Y - \hat{Y})$, we see clear evidence of positive autocorrelation—a sequence of negative residuals followed by a run of positive residuals followed again by negative residuals.

Summary

In summary, in this problem, company carpet sales were related to a company variable (advertising budget) and a macroeconomic variable (U.S. housing starts). A linear regression equation exhibited plausible slope coefficients, that is, carpet sales were directly related to both advertising budget and housing starts, and the slope coefficients were significant. The two independent variables accounted for about 92% of the variation in carpet sales and were not particularly correlated. The analysis of variance confirmed that the overall relationship between carpet sales and the independent variables was significant. However, a significant degree of positive serial correlation existed among the residuals.

In an attempt to use the observed autocorrelation to improve the prediction model, we add a new variable in a second computer run. This new variable is the value of the dependent variable lagged one quarter of the year. The rationale of this procedure is that if carpet sales were positively autocorrelated, then the sales in any quarter might represent a good predictor of the sales in the succeeding quarter. Hence, the next computer run involves the equation

$$\hat{Y}_T = a + b_1 X_{1T} + b_2 X_{2T} + b_3 Y_{T-1} \qquad (10.32)$$

where $\hat{Y}_T$ = the predicted value of Y for quarter T

Y_{T-1} = the value of Y in the preceding quarter

X_{1T} and X_{2T} = values of X_1 and X_2, respectively in period T

FIGURE 10-15
Computer printout of actual values of the dependent variable (Y), predicted values ($\hat{Y}$), and residuals ($Y - \hat{Y}$)

WANT A TABLE OF RESIDUALS AND PREDICTORS? YES

	DEP. VARS.	PREDICTORS	RESIDUALS
1:	28.2	29.019871	-0.8198708
2:	36.7	38.861392	-2.1613916
3:	36.1	40.727063	-4.627063
4:	30.6	37.162441	-6.562441
5:	30.9	36.01907	-5.1190702
6:	52.3	47.944133	4.3558673
7:	51.4	47.626858	3.7731418
8:	46.2	43.810732	2.3892677
9:	43.1	42.639567	0.46043346
10:	63.8	58.503563	5.2964371
11:	63.1	57.215452	5.8845476
12:	57.3	53.844096	3.455904
13:	55	50.978913	4.0210869
14:	75.3	79.868416	-4.5684156
15:	70.1	84.825954	-14.725954
16:	68.5	80.458861	-11.958861
17:	80.3	73.458609	6.8413912
18:	96.8	92.794007	4.0059935
19:	95.1	92.591078	2.5089222
20:	92.8	85.249925	7.5500747

It is interesting to note how the program accomplishes the lagged relationship. The question TRANSFORMATIONS, LAGS OR DIFFERENCES? in Figure 10-16 asks whether we wish to transform any of the variables, introduce any lags, or take differences, that is, use period-to-period differences—such as the change in carpet sales from one quarter to the next—as a variable rather than the original variable. The question WANT ANY FUNCTIONAL TRANSFORMATION(S)? asks whether we wish to change a variable, for example, change Y to log Y or to $\sqrt{Y}$. Finally, after we have indicated that we wish to introduce the dependent variable 1 as a lagged variable, note that we now have 19 observations and four variables, namely, the variables shown in equation 10.32. The first observation for the original three variables pertains to the second quarter of 1978, whereas the first observation for lagged carpet sales (Y_{T-1}) pertains to the first quarter of 1978.

The computer printout of the regression analysis after introduction of lagged carpet sales as variable 4 is shown in Figure 10-17. Examining the correlation matrix, we observe that carpet sales and lagged carpet sales are highly correlated ($r_{14} = 0.891$). Also, we note a high degree of collinearity between

FIGURE 10-16
Computer printout of the method of lagging the dependent variable

```
PLEASE NAME YOUR DATA SET. NAME? SALESDEM
DATA RETRIEVED FROM STORAGE.
THERE ARE 20 OBSERVATIONS AND 3 VARIATES.

WANT TO DISPLAY YOUR DATA? NO

WANT TO MAKE CORRECTIONS IN YOUR DATA? NO

TRANSFORMATIONS, LAGS OR DIFFERENCES? YES

WANT ANY FUNCTIONAL TRANSFORMATION(S)? NO

WANT LAGGED VARIABLES? YES
ONE OBSERVATION DELETED FOR EACH LAG.

WHICH VARIATE(S)?   1

MORE LAGGED VARIABLES? NO
THERE ARE 19 OBSERVATIONS AND 4 VARIATES.

WANT TO DISPLAY YOUR DATA? YES
```

1:	36.7	1.3	324.3	28.2
2:	36.1	1.5	348.5	36.7
3:	30.6	1.9	297.1	36.1
4:	30.9	2	280.8	30.6
5:	52.3	2.7	439.3	30.9
6:	51.4	2.8	434.3	52.3
7:	46.2	2.8	382.1	51.4
8:	43.1	2.6	367.4	46.2
9:	63.8	3.1	581.1	43.1
10:	63.1	3.4	561.5	63.8
11:	57.3	9.2	477.1	63.1
12:	55	20.7	362	57.3
13:	75.3	40.8	624.5	55
14:	70.1	60.3	563.6	75.3
15:	68.5	65.4	470.2	70.1
16:	80.3	72.8	325.6	68.5
17:	96.8	80.1	541.9	80.3
18:	95.1	86.3	498.2	96.8
19:	92.8	89.1	379.3	95.1

FIGURE 10-17
Computer printout for the regression of carpet sales (variate 1) on advertising budget (variate 2), U.S. housing starts (variate 3), and lagged carpet sales (variate 4).

```
VARIATE STATISTICS:

              MEANS           VARIANCES        ST. DEVS.
   1:      60.284211        442.88363        21.0448
   2:      28.884211       1189.1203         34.483624
   3:     434.67368       11030.263         105.02506
   4:      56.884211        429.13251        20.715514

VARIATE CORRELATIONS:
   1:    1.000   0.888   0.574   0.891
   2:            1.000   0.271   0.873
   3:                    1.000   0.407
   4:                            1.000

CONTINUE? YES

WANT REGRESSION ANALYSIS? YES
IDENTIFY VARIATES BY NUMBER.

DEPENDENT VARIABLE?   1

ANY INDEPENDENT VARIABLE(S) OMITTED? NO
THE 0 COEFFICIENT IS THE INTERCEPT.
THE PROBABILITY OF OCCURRENCE  ( α ) IS TWO-SIDED.
  VAR.          COEFF.            ST. DEV.          T          SIG ( α )
  0:        7.7770664          7.7500546        1.00349     0.333364
  2:        0.3490072          0.091631764      3.8088      0.001768
  3:        0.062346775        0.016043833      3.88603     0.001520
  4:        0.26942176         0.16074791       1.67605     0.111149
S  E OF RESIDUALS              6.4108
R  SQUARED                     0.922669
  EFFECTS              SS        DF          MSS           F      SIG ( α )
REGRESSION      7355.4297         3       2.452E3     59.6571    0.000000
ERROR            616.47553       15       4.110E1
TOTAL           7971.9053        18
RESIDUALS:
SUM OF SQUARES                 616.476
DURBIN-WATSON                    1.49385
THERE ARE 10 POSITIVE RESIDUALS AND 9 NEGATIVE ONES.
```

advertising budget and lagged carpet sales ($r_{24} = 0.873$). In the printout for the regression equation, the signs of the slope coefficients are all positive and therefore plausible. The slope coefficients for advertising budget and U.S. housing starts are significant at either a 0.05 or 0.01 level of significance, but lagged carpet sales is not significant at the 0.05 level. The standard error of estimate is slightly less than the previous run (6.4108 versus 6.56567), and the R^2 value is slightly higher (0.922669 versus 0.918117). If the R^2 values had been adjusted for degrees of freedom, they would have been even closer. Thus, perhaps surprisingly, the overall goodness of fit of the regression equation is virtually unchanged by the addition of the lagged carpet sales variable. Note that because we introduced a lagged value of the dependent variable as an independent variable, the Durbin-Watson test is no longer valid.[7]

Let us now assume that on examining the plot of carpet sales against advertising budget in Figure 10-18, we feel that the pattern appears somewhat curvilinear. Therefore, we decide to introduce a squared term for advertising budget, making the regression model read

$$\hat{Y}_T = a + b_1 X_{1T} + b_2 X_{1T}^2 + b_3 X_{2T} + b_4 Y_{T-1} \tag{10.33}$$

In this computer program, we introduce the squared term for advertising budget by adding that squared value to the equation as another variable, in this case, variable 5. The squared term for advertising budget is shown in equation 10.33 next to the term raised to the first power to emphasize the nature of the second-degree parabolic model that is fitted. Some selected computer printout information is shown in Figures 10-19 and 10-20 for the regression model given in equation 10.33.

We note in the correlation matrix shown in Figure 10-19 that the linear and squared terms of advertising budget are highly correlated ($r_{25} = 0.980$) and that the squared term is highly correlated to lagged carpet sales ($r_{45} = 0.869$). In Figure 10-20 we note that the only slope coefficient that differs significantly from zero at $\alpha = 0.05$ is that for U.S. housing starts (variable 3). The reason for these nonsignificant slope coefficients is undoubtedly the collinearity among the independent variables other than housing starts. The slope coefficient for advertising budget may be considered equal to zero because its t value is close to zero (-0.0529) and its SIG(α) value is high (0.912252). The standard error of estimate (6.01902) is slightly lower than the 6.4108 figure in the previous run, and the R^2 value of 0.936377 is slightly higher than the 0.922669 figure of the previous run.

At this point, we might consider dropping one or more of the collinear variables. However, in this illustration, the next step involves the addition of a dummy variable to take into consideration the supply shortage during the

[7] In J. Durbin and G. S. Watson, "Testing for Serial Correlation in Least Squares Regression," *Biometrica*, vol. 38, 1951, pp. 159–78, the authors state, "It should be emphasized that the tests described . . . apply only to regression models in which the independent variables can be regarded as 'fixed variables.' They do not, therefore, apply to autoregressive schemes and similar models in which lagged values of the dependent variable occur as independent variables."

FIGURE 10-18
Scatter diagram of carpet sales (vertical axis) and advertising budget (horizontal axis)

```
VARIATE NUMBER (S)?    1 2
```

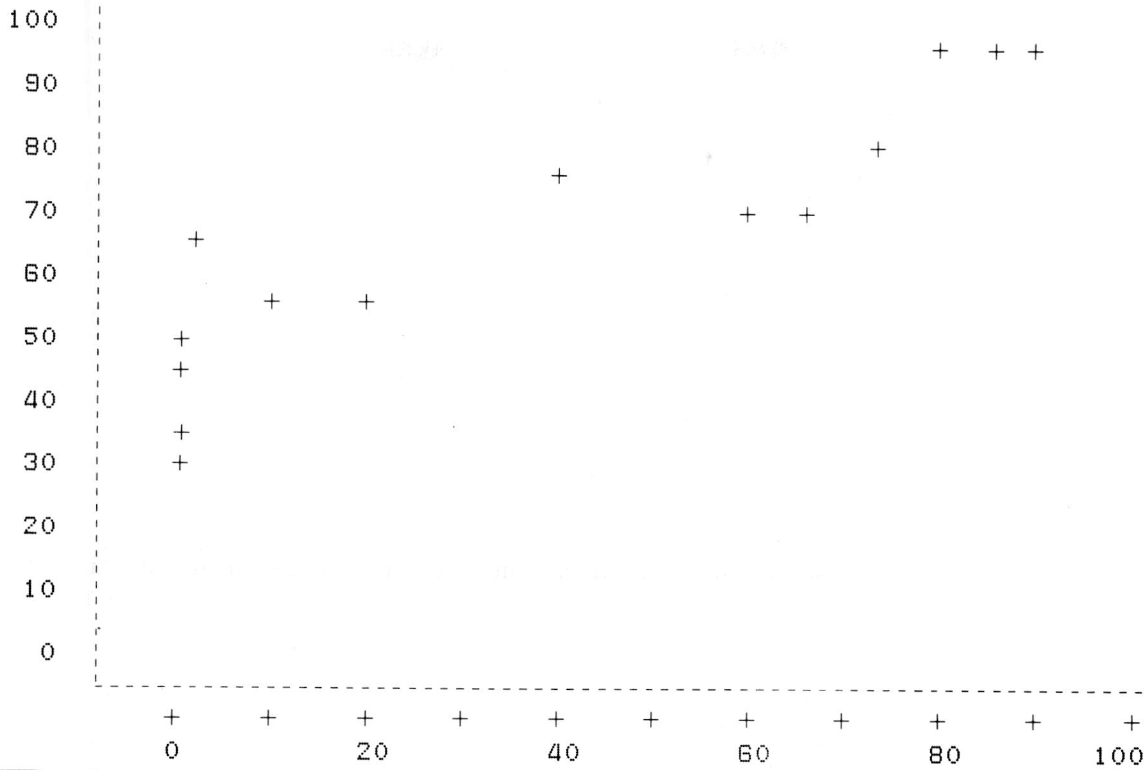

FIGURE 10-19
Selected computer printout of variate statistics and correlation matrix

CONTINUE? YES

VARIATE STATISTICS:

	MEANS	VARIANCES	ST. DEVS.
1:	60.284211	442.88363	21.0448
2:	28.884211	1189.1203	34.483624
3:	434.67368	11030.263	105.02506
4:	56.884211	429.13251	20.715514
5:	1960.8326	8287298.8	2878.7669

VARIATE CORRELATIONS:

1:	1.000	0.888	0.574	0.891	0.872
2:		1.000	0.271	0.873	0.980
3:			1.000	0.407	0.194
4:				1.000	0.869
5:					1.000

FIGURE 10-20
Selected computer printout for the regression of carpet sales (variate 1) on advertising budget (variate 2), U.S. housing starts (variate 3), lagged carpet sales (variate 4), and advertising budget squared (variate 5)

```
CONTINUE? YES

WANT REGRESSION ANALYSIS? YES
IDENTIFY VARIATES BY NUMBER.

DEPENDENT VARIABLE?    1

ANY INDEPENDENT VARIABLE(S) OMITTED? NO
THE 0 COEFFICIENT IS THE INTERCEPT.
THE PROBABILITY OF OCCURRENCE ( α ) IS TWO-SIDED.
  VAR.          COEFF.            ST. DEV.              T       SIG ( α )
  0:         7.8337969        7.2764977            1.0766    0.300557
  2:        -0.011898273      0.22491034          -0.0529    0.912252
  3:         0.075786894      0.01693492           4.4752    0.000571
  4:         0.18225733       0.15905017           1.1459    0.270748
  5:         0.004836697      0.0027849201         1.7367    0.101209
S E OF RESIDUALS              6.01902
R SQUARED                     0.936377
  EFFECTS              SS        DF          MSS          F      SIG ( α )
REGRESSION       7464.7054        4      1.866E3    51.5112     0.000000
ERROR             507.19988      14      3.623E1
TOTAL            7971.9053       18
RESIDUALS:
SUM OF SQUARES                507.2
DURBIN-WATSON                    1.71639
THERE ARE 11 POSITIVE RESIDUALS AND 8 NEGATIVE ONES.
```

last two quarters of 1981. Hence, we add a fifth independent variable that is equal to zero for all quarters but the last two quarters of 1981, when it is equal to one. Since the first observation was for the second quarter of 1978, this means that observations 14 and 15 are encoded one and all others are equal to zero. The coding is shown in Figure 10-21 and selected computer output for the resulting model is shown in Figures 10-22, 10-23, and 10-24.

The correlation matrix in Figure 10-22 does not contain any new information. The correlations of the dummy variable with the other variables yield little information, and the other correlations have not changed from the previous run. Moving to Figure 10-23 and the output pertaining to the regression equation, we see that all slope coefficients are significant at $\alpha = 0.05$ except those for advertising budget squared (variable 4) and lagged carpet sales (variable

FIGURE 10-21
Coding used for the dummy variable that pertains to a supply shortage in the third and fourth quarters of 1981

```
ENTER 1 NUMBER PER OBSERVATION.
OBSERVATION  1?  0
OBSERVATION  2?  0
OBSERVATION  3?  0
OBSERVATION  4?  0
OBSERVATION  5?  0
OBSERVATION  6?  0
OBSERVATION  7?  0
OBSERVATION  8?  0
OBSERVATION  9?  0
OBSERVATION 10?  0
OBSERVATION 11?  0
OBSERVATION 12?  0
OBSERVATION 13?  0
OBSERVATION 14?  1
OBSERVATION 15?  1
OBSERVATION 16?  0
OBSERVATION 17?  0
OBSERVATION 18?  0
OBSERVATION 19?  0
THERE ARE 19 OBSERVATIONS AND 6 VARIATES.
```

FIGURE 10-22
Selected computer printout of variate statistics and correlation matrix

```
THERE ARE 19 OBSERVATIONS AND 6 VARIATES.

CONTINUE? YES
VARIATE STATISTICS:

              MEANS           VARIANCES         ST. DEVS.
  1:       60.284211        4.4288363E2        21.0448
  2:       28.884211        1.1891203E3        34.483624
  3:      434.67368         1.1030263E4       105.02506
  4:       56.884211        4.2913251E2        20.715514
  5:     1960.8326          8.2872988E6      2878.7669
  6:        0.10526316      9.9415205E-2        0.31530177

VARIATE CORRELATIONS:
  1:    1.000   0.888   0.574   0.891   0.872   0.151
  2:            1.000   0.271   0.873   0.980   0.347
  3:                    1.000   0.407   0.194   0.276
  4:                            1.000   0.869   0.269
  5:                                    1.000   0.244
  6:                                            1.000
```

FIGURE 10-23
Selected computer printout for the regression of carpet sales (variate 1) on advertising budget (variate 2), U.S. housing starts (variate 3), lagged carpet sales (variate 4), advertising budget squared (variate 5), and dummy variable for supply shortage (variate 6)

```
CONTINUE? YES

WANT REGRESSION ANALYSIS? YES
IDENTIFY VARIATES BY NUMBER.

DEPENDENT VARIABLE?    1

ANY INDEPENDENT VARIABLE(S) OMITTED? NO
THE 0 COEFFICIENT IS THE INTERCEPT.
THE PROBABILITY OF OCCURRENCE ( α ) IS TWO-SIDED.
  VAR.          COEFF.          ST. DEV.             T       SIG ( α )
   0:          6.271E0        4.8457906          1.2941     0.216306
   2:          3.754E-1       0.17409938         2.1560     0.048393
   3:          7.678E-2       0.011248857        6.8260     0.000017
   4:          1.758E-1       0.10563578         1.6645     0.116749
   5:          7.541E-4       0.0020759933       0.3633     0.719922
   6:         -1.602E1        3.7002483         -4.3294     0.000872

S E OF RESIDUALS           3.99723
R SQUARED                  0.973944
  EFFECTS               SS         DF          MSS            F      SIG ( α )
REGRESSION        7764.1928         5       1.553E3      97.1867    0.000000
ERROR              207.71249       13       1.598E1
TOTAL             7971.9053        18
RESIDUALS:
SUM OF SQUARES             207.712
DURBIN-WATSON                1.70123
THERE ARE 10 POSITIVE RESIDUALS AND 9 NEGATIVE ONES.
```

5). If we were to use the rule of thumb referred to in the subsection on multicollinearity—that is, dropping a collinear variable if the t value for the slope coefficient is less than one—we would delete the squared advertising budget variable. This deletion would have little effect on the regression equation, the R^2 value, and the other statistics. Looking again at the scatter diagram in Figure 10-18, we see that the curvilinear effect of advertising budget was not very clearly delineated. The negative value for the slope coefficient of the supply shortage variable is reasonable because it indicates decreased sales for the quarters during which the supply shortage occurred.

FIGURE 10-24
Selected computer printout of a table of residuals

```
WANT A TABLE OF RESIDUALS AND PREDICTORS? YES
                    DEP. VARS.        PREDICTORS          RESIDUALS
     1:               36.7            36.620176           0.079824006
     2:               36.1            40.048459          -3.9484592
     3:               30.6            36.147395          -5.5473955
     4:               30.9            33.966543          -3.066543
     5:               52.3            46.454911           5.8450893
     6:               51.4            49.871806           1.5281935
     7:               46.2            45.705392           0.49460781
     8:               43.1            43.586427          -0.48642713
     9:               63.8            59.640068           4.1599321
    10:               63.1            61.888951           1.2110487
    11:               57.3            57.517475          -0.21747465
    12:               55              52.235735           2.7642653
    13:               75.3            80.464394          -5.1643941
    14:               70.1            72.144108          -2.0441076
    15:               68.5            66.455892           2.0441076
    16:               80.3            74.640358           5.6596423
    17:               96.8            96.905661          -0.1056607
    18:               95.1            99.556739          -4.4567388
    19:               92.8            91.54951            1.2504899

WANT A PLOT? YES

WANT INSTRUCTIONS? NO
```

The standard error of estimate has now been reduced to 3.99723 from 6.01902 in the preceding run, and the R^2 value has risen to 0.973944 from 0.936377.

The printout of a table of residuals is shown in Figure 10-24. The scatter plot of the residuals (variate number 0 in the program for plotting) shows that the deviations are randomly distributed (see Figure 10-25). No systematic pattern seems to be present.

Caveats

In this example of the development of a regression model for the prediction of carpet sales, the last computer run included 19 observations and 6 variables. In general, as more variables are added to the equation, the goodness of fit of

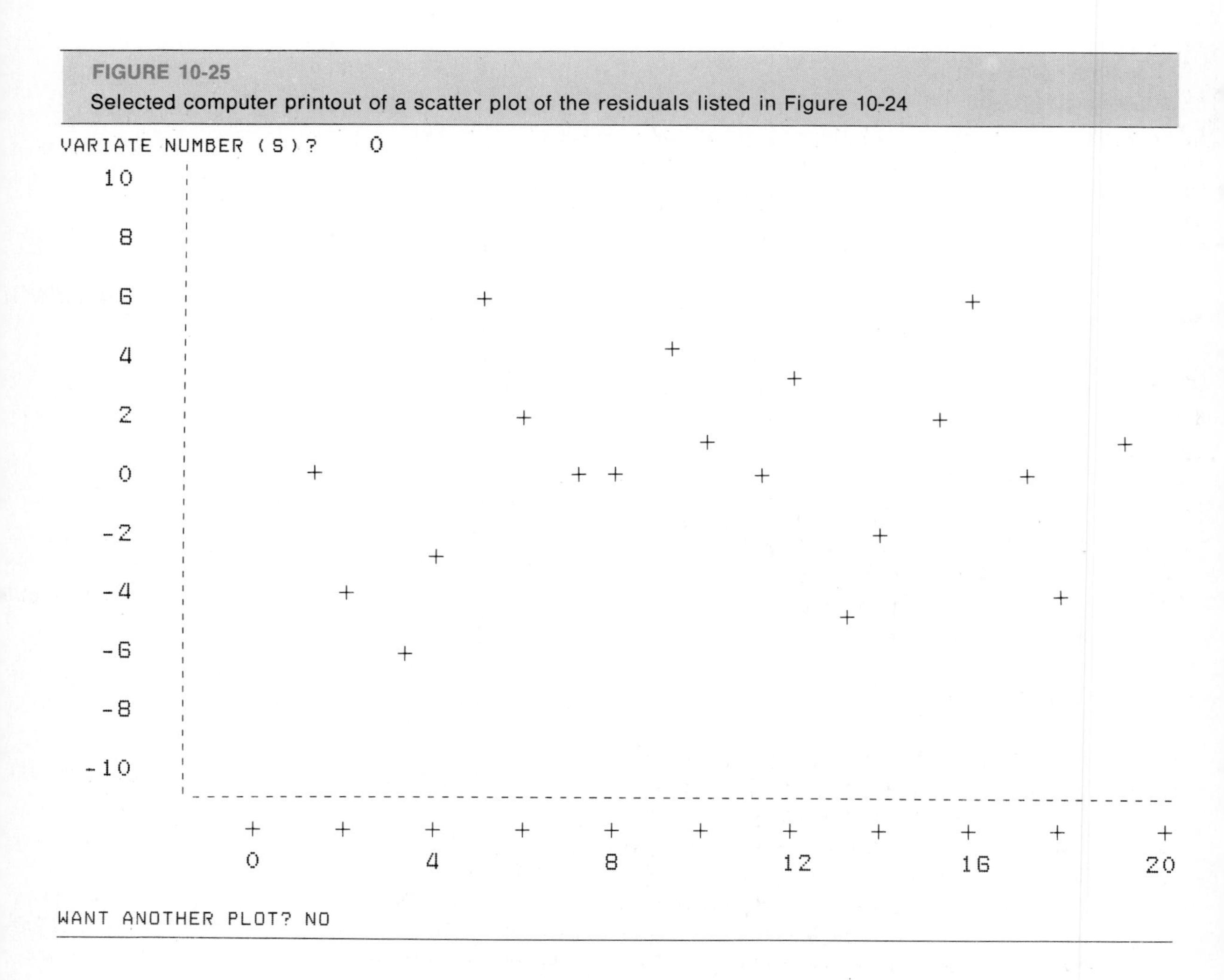

FIGURE 10-25
Selected computer printout of a scatter plot of the residuals listed in Figure 10-24

the model will tend to improve, as measured by R^2 and the standard error of estimate. After all, if we fit a straight line of the form $\hat{Y} = a + bX$ to two observations, we will have zero degrees of freedom ($n - k = 2 - 2 = 0$), and we will have a perfect fit. If we fit a plane of the form $\hat{Y} = a + b_1X_1 + b_2X_2$ to three observations, we will again have zero degrees of freedom ($n - k = 3 - 3 = 0$) and a perfect fit. Analogously, if we have 19 points and we fit a function with 19 variables, we will force a perfect fit. Such a model would clearly be nonsensical and would probably be a poor model for prediction purposes. In fact, with the same number of both observations and variables, we could include independent variables that are logically unrelated to the dependent variable, and we would still have a perfectly fitting model.

We must remember that our purpose in deriving a prediction model is not to obtain an equation that *at any cost* is the best fitting one to the past data, but rather to obtain a model that will predict well in the future.

Hence, we should strive for simplicity in the construction of the model. A regression equation with a small number of effective independent variables is preferable to another with a large number of less effective variables. In the development of the carpet sales regression model, we sequentially added variables in order to demonstrate a number of points. We did not delete any variables. In a real-world application, we undoubtedly would have tried to delete one or more variables to arrive at a more effective model for prediction purposes. Prior knowledge and reasoning concerning how independent variables affect the dependent variable should be major determinants for the inclusion of variables in the development of regression models.

Exercises 10.13

Note: A number of different forms of computer output for multiple regression analyses are illustrated in these exercises.

1. A study was carried out to inquire into the relationship between unemployment, the federal budget, and money supply. Annual time series data were used for a 21-year period. The following variables were defined, with the unemployment rate as the dependent variable:

Variable Number	
1	Unemployment rate (in percentages)
2	Ratio of federal expenditures to federal receipts (a number greater than one indicates a deficit)
3	Per capita money supply (in dollars)

Figure 10-26 shows the resulting selected computer output.

a. Interpret the regression coefficient for variable 2 (b value) specifically in terms of the variables and units in this problem.
b. Calculate and interpret the multiple coefficient of determination (R^2) specifically in terms of this problem.
c. Test the overall significance of the regression at the 1% significance level. State your conclusion.
d. If variable 2 were dropped, and a two-variable linear regression analysis were run for variables 1 and 3, what proportion of the variance in the dependent variable would be accounted for by the regression equation?
e. From the computer results given in Figure 10-26, an analyst concluded that fiscal and monetary policy have no effect on unemployment rates—since there seems to be little relationship between unemployment rates

FIGURE 10-26
Selected computer output for exercise 1

VAR.	COEFF.	STD. ERROR	T VALUE
2:	-0.28164	4.63681	-0.06074
3:	-0.03460	0.01821	-1.89985

INTERCEPT 1.89685

REGRESSION

DEGREES OF FREEDOM	2
SUM OF SQUARES	4.00000

ERROR

DEGREES OF FREEDOM	18
SUM OF SQUARES	18.90000

S E OF ESTIMATE 1.025

MEANS AND STANDARD DEVIATIONS

VAR.	MEAN	ST. DEV.
1:	4.68	1.130
2:	1.03	0.069
3:	196.95	26.292

SIMPLE CORRELATIONS

	1	2	3
1:	1.0000	0.1028	-0.3680
2:		1.0000	0.2061
3:			1.0000

and measures of fiscal policy (variable 2) and monetary policy (variable 3). Do you agree? Why or why not?

2. A study of effectiveness of various forms of advertising and promotion expenditure in increasing sales of a new refrigerator was conducted in 100 distinct market areas. The variables were:

1 = Number of sales per capita
2 = TV advertising expenditure per capita (in dollars)
3 = Newspaper advertising per capita (in dollars)
4 = Promotion per capita (in dollars)

A summary of the relevant output is presented in Figure 10-27.

FIGURE 10-27
Selected computer output for exercise 2

```
N = 100

VAR.      MEAN      VARIANCE
 1:       1.2         1.0
 2:       0           1.0
 3:       0           0.64
 4:       0           0.25

VAR.         COEFF.       STD. ERROR      T VALUE
 2:          5.575          0.037           151
 3:          9.8125         0.064           154
 4:        -22.25           0.147          -151

INTERCEPT    12
S E OF THE ESTIMATE     0.12
MULTIPLE R SQUARED    0.9866
F VALUE    2306.98

SIMPLE CORRELATIONS
               1          2          3          4
 1:          1.00       0.60       0.62       0.80
 2:                     1.00       0.50       0.80
 3:                                1.00       0.90
 4:                                           1.00
```

(**Note**: Variables 2, 3, 4 are given as deviations from the averages of the respective variables: For example, instead of using dollars spent on TV advertising in a market area, the number of dollars above or below the average of the 100 market areas spent on TV advertising was used.)

a. Does the model "fit the data"? Which independent variables are "important" in predicting sales per capita? Use $\alpha = 0.01$

b. What would you predict the sales per capita to be if, in another market area, the company spent:

   $1 more than the average per capita for TV advertising
   $1 more than the average per capita for newspaper advertising
   $2 less than the average per capita for promotion

   (all figures in comparison to the original 100 market areas)? Find a 90% confidence interval for the sales per capita in this market area.

c. To encourage sales, dealers in some market areas were given modest rebates for each refrigerator sold. In other market areas, competitive prizes were given to winning dealers. How would you incorporate this information into your analysis? How would you estimate the effect of the rebates?

d. In each market area, the levels of all 3 expenditure variables were set separately by the company. Criticize the choice of these levels, in terms of analyzing the separate effect of each of the 3 kinds of market efforts.

3. The Merrill Clinch Corporation ran a multiple regression and correlation analysis in which it related annual territory sales to 3 explanatory variables as follows:

$$Y = \text{territory sales (in millions of dollars)}$$

$$X_1 = \text{population (in millions)}$$

$$X_2 = \text{newspaper advertising (in tens of thousands of dollars)}$$

$$X_3 = \text{television advertising (in tens of thousands of dollars)}$$

FIGURE 10-28
Selected computer output for exercise 3

VAR.	MEAN	ST. DEV.
1:	30.0	12.2
2:	4.1	1.3
3:	5.2	1.1
4:	2.0	0.6

SIMPLE CORRELATIONS

	1	2	3	4
1:	1.00	0.73	0.65	0.14
2:		1.00	0.62	0.67
3:			1.00	0.92
4:				1.00

INTERCEPT 15.32

VAR.	COEFF.	STD. ERROR
2:	1.13	0.39
3:	2.50	1.03
4:	-0.02	0.52

SOURCE OF VARIATION
REGRESSION
SUM OF SQUARES 3982.9507
ERROR
SUM OF SQUARES 948.3216

The information in the selected computer output of Figure 10-28 refers to a least squares linear relationship established for these data. There were 34 observations (territories).

Variable designations: $Y = 1$, $X_1 = 2$, $X_2 = 3$, $X_3 = 4$

a. State a 95% confidence interval for territory sales for a territory that has a population of 3 million, newspaper advertising of \$50,000, and television advertising of \$20,000. **Hint**: Be mindful of the units in which the variables are stated.
b. Test the null hypothesis of no relationship in the multiple regression analysis (i.e., all B values are zero) by means of an F-test at the 1% significance level.
c. What percentage of variation in the dependent variable has been explained by the regression equation? Show your calculations.
d. Interpret $b_3 = -0.02$ specifically in terms of this problem. Give 90% confidence limits for B_3.
e. Do you feel that this would be a good model for predicting territory sales for this company? Explain why or why not. What specific suggestions would you make for improvement of the model?

4. A manufacturer of hand calculators is attempting to determine an optimal pricing strategy in this market. Many multiple regression analyses have been performed in order to determine factors influencing sales. One model used sales data from 27 cities for the following variables:

Y = the manufacturer's sales (in thousands of calculators)

X_1 = price of the manufacturer's calculator (in dollars)

X_2 = price of chief competitor's calculator (in dollars)

X_3 = monthly advertising expenditures (in thousands of dollars)

X_4 = college student population (in thousands)

Figure 10-29 shows the computer output that was obtained.

a. Test the overall significance of this regression at the 5% level. State your conclusion clearly.
b. Do the signs of the coefficients make sense? For each coefficient, explain in one short sentence.
c. What percentage of the variation in sales was attributable to the regression relationship?
d. Construct a 90% confidence interval for the coefficient of variable X_3.
e. Explain in words the exact meaning of the coefficient of variable X_2.
f. Construct a 90% prediction interval for monthly sales in a city where the calculator sells for \$20, the chief competitor's sells for \$17.50, advertising expenditures = \$15,000 and college population = 7,500. (You

FIGURE 10-29
Selected computer output for exercise 4

VAR.	MEAN	ST. DEV.
Y	2.50	0.35
X_1	21.40	2.00
X_2	18.50	1.50
X_3	7.65	2.30
X_4	3.10	1.20

VAR.	COEFF.	STD. ERROR
X_1	−0.47	0.08
X_2	+0.58	0.13
X_3	+1.25	0.29
X_4	+0.34	0.25

INTERCEPT -8.79
REGRESSION SUM OF SQUARES: 2.45
ERROR SUM OF SQUARES: 0.735

may use the standard error of estimate to approximate the standard error of forecast).

g. In some of the cities, a small gift item was given to every purchaser of the manufacturer's calculators. How could you incorporate this information into the analysis?

5. A new concept in public transportation, dial-a-ride, has been introduced on a test basis in some areas. A commuter phones in to dial-a-ride and indicates current location, destination, and time of travel. Vans are then routed through the area to meet these requests. Dial-a-ride has had mixed success. A study is conducted to predict the performance of dial-a-ride as measured by yearly number of rides per person (Y) from the following variables:

X_1 = per capita income (in thousands of dollars)

X_2 = train and bus route miles available in the area (in hundreds of miles)

X_3 = median age of residents in the area

Figure 10-30 shows the output from a linear multiple regression run using twenty areas.

a. What would be the predicted yearly number of rides per person in Wharton City where the per capita income is $10,000, the train and bus-route miles are 100, and the median age is 25?

FIGURE 10-30
Selected computer output for exercise 5

CORRELATIONS

Y	X_1	0.7206
Y	X_2	-0.7909
Y	X_3	0.2986
X_1	X_2	-0.5685
X_2	X_3	0.1956
X_3	X_3	-0.1487

VAR.	COEFF.	STD. ERROR
X_1	0.48956	0.19775
X_2	-3.56765	0.97030
X_3	0.02607	0.02323

INTERCEPT 4.70125

ANOVA TABLE

SOURCE	SS
REGRESSION	40.14471
ERROR	13.13729

b. What percentage of the variability in Y is explained by X_1, X_2, and X_3 combined?
c. Which of the variables are significant at $\alpha = 0.05$?
d. The researcher wants to see whether the attitudes toward dial-a-ride differ in the East, West, North, and South. Explain how you would incorporate this into your model. Be specific.

6. Explain in a few sentences why each of the following could occur.
   a. The regression of Y on X_1, X_2, and X_3 yields a significant F result, while the t results for X_2 and X_3 are insignificant. The regression eliminating X_2 and X_3, however, produces insignificant results.
   b. The F test is highly significant. However, the confidence interval for prediction in one predictor model is wide, even though the new value of X equals $\bar{X}$.
   c. A multiple regression analysis was run for 60 families in which $Y =$ food expenditures and in which 3 socioeconomic factors were the independent variables. The Durbin-Watson test indicated positive autocorrelation.

7. A microcomputer company wished to develop a regression model to assist it in manpower planning. Using a sample of 20 company records, a company

TABLE 10-6
Computer output for two regression models developed by a microcomputer company.

Descriptive statistics

Var.	Mean	Variance	St. Dev.
Y	8.19	28.20	5.31
X_1	5.70	12.25	3.50
X_2	6.40	13.91	3.73

Correlation matrix

	Y	X_1
X_1	0.935	1.0
X_2	−0.241	0.018

Two models were fit:

I. Dependent variable Y
   Independent variables X_1, X_2

II. Dependent variable Y
   Independent variables $X_1, X_2, (X_1 \times X_2)$
   That is, $X_1 \times X_2$ was included as a third independent variable.

Model I

Var.	Coeff.	Std. Error	t Value
Intercept	2.4052	0.8011	
X_1	1.4262	0.0902	
X_2	−0.3663	0.0845	

Analysis of Variance

Due to	SS	DF	MSS	F
Regression	503.203			
Error				
Total	535.338			

S E of estimate
Multiple R Squared
Durbin-Watson 1.86

Model II

Var.	Coeff.	Std. Error	t Value
Intercept	−0.3488	0.5397	
X_1	2.0654	0.0968	
X_2	0.02152	0.0666	
$(X_1 \times X_2)$	−0.09187	0.0124	

Analysis of Variance

Due to	SS	DF	MSS	F
Regression		3		
Error	7.2353			
Total				

S E of estimate
Multiple R Squared
Durbin-Watson 3.08

analyst obtained information on 3 variables:

Y = the time (in hours) spent on a service call to a customer

X_1 = the number of microcomputers to be serviced

X_2 = the number of months' experience of the person sent on the service call

Table 10-6 shows two models that were developed.

a. Discuss the difference between the two models with respect to:
   (1) which variables are significant in predicting service call time (use $\alpha = 0.05$).
   (2) the change in the proportion of the variance explained in Model II versus Model I.
   (3) the reason for the differences in results of Model I and Model II.

b. If a new company service record is chosen, based on "6 microcomputers serviced by a person with 4 months' experience," what is the prediction for the number of hours' duration of the service call for each model?

c. Test the overall significance of Model I and Model II at $\alpha = 0.05$.

8. A student carried out a multiple regression analysis to fulfill a requirement in a first-year MBA course. He used the following variables (observed for each country in a given year) to explain the total number of movie theaters on the basis of what he referred to as indicators of the economic levels in those countries:

Variable	
1	Y = number of movie theaters
2	X_1 = population (in millions)
3	X_2 = number of radio receivers per 1000 inhabitants
4	X_3 = number of TV receivers per 1000 inhabitants
5	X_4 = number of telephones per 100 inhabitants
6	X_5 = per capita income (in constant U.S. dollars)

Since islands constitute a large proportion of countries in the world and it was assumed that they would have characteristics different from other countries, they were removed from the population. The student then selected 30 countries at random from a resulting population of 100 countries and proceeded to run a two variable linear regression analysis between Y and X_5 "to test whether the data were compatible." On the basis of an examination of the residuals $(Y - \hat{Y})$, it was observed that 5 countries were outside of what appeared to be a good linear relationship. These outliers were removed, and the subsequent analysis was carried out on the basis of a sample of 25 countries.

FIGURE 10-31
Selected computer printout for exercise 8, showing the first regression analysis

MEANS AND VARIANCES

VAR.	MEAN	VARIANCE
1:	680.76	890670.5233
2:	14.236	352.2149
3:	122.16	19284.97333
4:	39.44	3796.91083
5:	4.104	62.3929
6:	457.08	325244.0767

SIMPLE CORRELATIONS (FOR EXAMPLE, r_{12} = 0.8003)

I	J	R(I,J)	I	J	R(I,J)
1	2	0.8003	2	6	0.5754
1	3	0.6633	3	4	0.8679
1	4	0.6798	3	5	0.7908
1	5	0.5382	3	6	0.8278
1	6	0.7279	4	5	0.7950
2	3	0.3074	4	6	0.8242
2	4	0.5126	5	6	0.9037
2	5	0.3291			

The first regression analysis yielded the selected results shown in Figure 10-31.

The student then deleted variables 4 and 5 and ran another linear regression with the selected results shown in Figure 10-32 (variable 1 was retained as the dependent factor).

FIGURE 10-32
Selected computer printout for exercise 8, with variables 4 and 5 deleted

VAR.	COEFF.	ST. ERROR
2:	36.0098	5.8
3:	3.9231	1.1
6:	-0.2682	0.3

INTERCEPT -188.53687

REGRESSION
SUM OF SQUARES 17913301.69

ERROR
SUM OF SQUARES 3462790.873

S E OF ESTIMATE 406.07241

a. Give your opinion, briefly but specifically, on the method of selection of the sample.
b. What proportion of variation in number of movie theaters was accounted for by the regression equation?
c. Explain specifically in terms of this problem the meaning of the slope coefficient (regression coefficient) for variable 6. Give a 90% confidence interval for the estimation of the population value of the slope coefficient for variable 6.
d. Carry out the F-test for this regression analysis, and state the test result specifically in terms of this problem.
e. The following calculation was carried out. Explain specifically what is measured by this computation.

$$1 - \frac{(406.07241)^2}{890670.5233} = 0.8149$$

9. Two different procedures are used for making rubber. Hardness of the rubber (Y) is known to be a linear function of how much compound is added (X) and the temperature at which the rubber was made (T) (i.e., for a given procedure we fit the equation $\hat{Y} = a + b_1X + b_2T$. We would like to compare the two procedures, so we fit the following model: $\hat{Y} = a + b_1X + b_2T + b_3Z$

$$\text{where} \quad Z = \begin{cases} 1 & \text{if the first procedure was used} \\ 0 & \text{otherwise} \end{cases}$$

Figure 10-33 shows part of the output of a multiple regression analysis.
a. Test the null hypothesis that the regression model is not significant. Use $\alpha = 0.05$.
b. Test to see whether the two procedures are equivalent. Use $\alpha = 0.05$.

FIGURE 10-33
Selected computer printout of a multiple regression analysis for exercise 9

VAR.	COEFF.	ST. DEV.
INTERCEPT.	33.840	2.441
X	-0.102	0.013
T	8.131	3.654
Z	-0.001	0.018

SOURCE	DF	SS
REGRESSION	3	1504.419
ERROR	16	176.381

NO. OF OBSERVATIONS 20

c. Is there sufficient evidence to show that there is a decrease in hardness when the amount of compound is increased for a given procedure and constant temperature? Use $\alpha = 0.05$.
d. Estimate the average hardness of the rubber when 3 ounces of compound are added at 1°C using the second procedure.
e. What proportion of the variation in hardness of rubber is explained by the regression model?

Computer Review Exercises for Chapters 9 and 10

Note: Access to a computer and a standard computer program for multiple regression analysis are needed to carry out these review exercises.

1. A multiple regression analysis was carried out to determine the extent to which academic performance of graduate management students can be predicted by variables such as GMAT scores, number of years of work experience, undergraduate grade point average (GPA), and average number of study hours per day in graduate school. The following model was fitted using the indicated variables:

$$\hat{Y} = a + b_1X_1 + b_2X_2 + b_3X_3 + b_4X_4$$

Y = graduate academic performance (grade point average):
3 = Distinguished, 2 = High Pass, 1 = Pass, 0 = No Credit

X_1 = GMAT scores

X_2 = number of years of work experience

X_3 = undergraduate GPA

X_4 = average number of study hours per day in graduate school

Data were obtained for a sample of 25 students. The data are listed in Figure 10-34.

a. Using a computer, fit the multiple linear regression model to the given data and obtain associated computer output.
b. Test the significance of the slope coefficients using $\alpha = 0.01$.
c. Test the overall significance of this regression model using $\alpha = 0.01$.
d. What percentage of the variation in graduate academic performance is accounted for by the regression relationship?
e. Revise the regression model that you have fitted, deleting years of work (X_2) and now using GMAT scores (X_1), undergraduate GPA (X_3), and graduate average study hours (X_4) in a linear regression equation.
f. What can you conclude in the comparison of the models fitted in parts (a) and (e)?

2. A multiple regression model was constructed to analyze employment in manufacturing in the Philippines, using annual data for a 20-year period beginning in 1960. The data are listed in Table 10-7.

FIGURE 10-34
Selected computer printout of data for exercise 1

STUDENT	GRADUATE ACADEMIC PERFORMANCE Y	GMAT SCORES X_1	YEARS OF WORK X_2	UNDERGRAD GPA X_3	AVERAGE NUMBER OF GRADUATE STUDY HOURS PER DAY X_4
1	2.00000	622	2.00000	3.00000	2.50000
2	2.80000	780	6.00000	3.70000	6.50000
3	2.80000	650	3.00000	3.50000	6.00000
4	2.00000	630	3.00000	3.50000	3.00000
5	2.80000	582	3.00000	3.90000	4.00000
6	1.60000	750	4.00000	3.70000	2.50000
7	2.00000	687	1.50000	3.74000	4.00000
8	2.00000	640	3.00000	3.70000	3.00000
9	1.80000	670	3.00000	3.50000	2.00000
10	2.20000	750	2.50000	3.65000	3.50000
11	2.00000	640	5.00000	3.10000	3.50000
12	2.40000	660	3.00000	3.70000	3.00000
13	1.60000	710	2.00000	3.10000	2.00000
14	1.80000	720	6.00000	3.30000	3.00000
15	2.40000	640	2.00000	3.50000	3.50000
16	1.80000	680	4.00000	3.30000	2.00000
17	2.00000	750	3.00000	3.56000	4.00000
18	2.00000	670	3.00000	2.80000	4.50000
19	2.20000	590	1.50000	3.50000	5.00000
20	1.80000	690	3.50000	2.80000	6.00000
21	2.20000	580	5.00000	3.60000	4.00000
22	2.40000	630	2.00000	3.40000	4.00000
23	1.80000	590	2.00000	3.10000	3.00000
24	2.80000	760	4.00000	3.60000	5.50000
25	2.00000	600	2.50000	3.20000	3.50000

a. Using a computer, fit a multiple linear regression function to the data given in Table 10-7; obtain associated computer output.
b. Interpret the slope coefficient for the implicit gross national product price deflator variable (b_2 value) specifically in terms of the variables and units in this problem.
c. Test the significance of the slope coefficients using $\alpha = 0.05$.
d. Do the signs of the coefficients make sense? Explain your answer for each coefficient in one short sentence.
e. Test the overall significance of this regression model using $\alpha = 0.05$.
f. Interpret the coefficient of multiple determination specifically in terms of this problem.
g. What do you learn from an examination of the correlation matrix?

TABLE 10-7
Employment in manufacturing and other economic variables for the Philippines, 1960–1979

Year	Employment in Manufacturing (1975 = 100.0) Y	Value Added in Manufacturing (in 1975 million pesos) X_1	Implicit Gross National Product Price Deflator (1975 = 1.0) X_2	Wage Index for Industrial Establishments (1975 = 100.0) X_3	Fixed Investment (in 1975 million pesos) X_4
1	67.300	10828.1	0.28314	42.200	1892.0
2	68.300	11326.0	0.29078	43.200	2282.0
3	69.400	12580.1	0.29402	44.500	2460.0
4	70.800	13807.9	0.31330	46.900	3131.0
5	72.000	13946.1	0.33750	47.400	3840.0
6	74.100	14697.9	0.34643	50.600	4134.0
7	73.300	15636.7	0.36706	54.500	4254.0
8	74.200	17053.9	0.38995	57.000	5255.0
9	76.200	17712.9	0.41921	63.400	5522.0
10	77.400	18500.9	0.43771	66.400	5732.0
11	77.300	20308.1	0.49765	73.600	6701.0
12	84.500	21528.8	0.56404	78.600	8154.0
13	81.800	22965.3	0.59984	83.300	8831.0
14	90.400	26231.4	0.70415	85.400	11049.0
15	97.300	27375.3	0.93277	92.300	18645.0
16	100.000	28543.9	1.00000	100.000	27800.0
17	103.100	30686.1	1.07254	105.100	32753.0
18	111.500	32599.1	1.17532	110.700	36415.0
19	116.300	34263.4	1.25823	115.200	41872.0
20	122.200	35912.3	1.29836	119.500	45130.0

3. A student ran a multiple regression analysis in an attempt to explain how gross profit of companies is affected by number of employees, property, and inventory. The student carried out a static analysis using data for the 23 firms listed in the Retail Stores-Specialty category of Moody's OTC Industrial Manual that supplied data on all 3 of these variables in a particular year. The following model was fitted using the indicated variables:

$$\hat{Y} = a + b_1X_1 + b_2X_2 + b_3X_3$$

Y = gross profit (in thousands of dollars)

X_1 = number of employees

X_2 = net property (in thousands of dollars)

X_3 = inventory at the end of the previous year (in thousands of dollars)

The data are listed in Table 10-8.

a. Using a computer, fit the multiple linear regression model to the given data and obtain associated computer output.

TABLE 10-8
Gross profit and other economic variables for a sample of 23 companies

Company	Gross Profit ($000) Y	Number of Employees X_1	Net Property ($000) X_2	Inventory at End of Previous Year ($000) X_3
1	2756	133	110	1967
2	1212	65	135	173
3	1132	120	420	711
4	1559	100	934	82
5	58730	420	53012	3623
6	48752	2830	21175	16324
7	7574	200	413	505
8	641	30	35	855
9	13767	512	11244	6673
10	80658	3700	34365	32100
11	10843	79	264	4463
12	71066	3640	41949	22697
13	13487	458	5737	3971
14	34013	312	2158	1965
15	14514	637	10755	2409
16	3007	150	799	2060
17	60892	4300	9284	32211
18	6357	150	881	12015
19	1485	51	700	739
20	13453	600	6197	10571
21	5857	18	2325	4508
22	5728	572	2423	1379
23	5812	310	1918	2049

b. Test the significance of the slope coefficients using $\alpha = 0.01$.
c. Test the overall significance of the regression model using $\alpha = 0.01$.
d. What percentage of the variation in gross profit is accounted for by the regression relationship?
e. Revise the regression model that you have fitted, eliminating the data for company 14. This company was determined to be an outlier because of its very large residual $(Y - \hat{Y})$ and a subsequent determination that this company was non-homogeneous in certain respects with the other included firms.
f. What can you conclude from a comparison of the models fitted in parts (a) and (e)? What would be a reasonable next step in your analysis?

4. Assume that the 75 families in Appendix E constitute a simple random sample of all families in a suburban area.

a. Display the output for a linear regression equation of the form

$$\hat{Y} = a + b_1x_1 + b_2x_2 + b_3x_3 + b_4x_4$$

where Y = annual food expenditures (in thousands of dollars)

b. Delete the variables whose slope coefficients do not differ significantly from zero at the 5% significance level. Rerun the regression analysis and display selected output for the linear regression equation.

c. Using the model obtained in part (b), calculate and interpret specifically in terms of your model:
   (1) An approximate 95% prediction interval for food expenditures for a family with the same values of independent variables as family number 20.
   (2) An approximate 95% confidence interval for one of your slope coefficients.

d. Again using the model derived in part (b), what do you learn from a comparison of R^2 and the two variable correlation coefficients? Explain.

11

Time Series

Decisions in private and public sector enterprises depend on perceptions of future outcomes that will affect the benefits and costs of possible alternative courses of action. Not only must managers forecast these future outcomes, but they must also plan and think through the nature of the activities that will permit them to accomplish their objectives. Clearly, managerial planning and decision making are inseparable from forecasting.

11.1 METHODS OF FORECASTING

Methods of forecasting vary considerably. For example, in forecasting customer demand, the prevailing methods include informal "seat of the pants" estimating, executive panels and composite opinions, consensus of sales force opinions, combined user responses, statistical techniques, and various combinations of these methods.

Statistical method of forecasting

Because of such factors as the increased complexity of business operations, the need for greater accuracy and timeliness, the dependence of outcomes on so many different variables, and the demonstrated utility of the techniques, management is increasingly turning to formal models—such as those provided by statistical methods—for assistance in the difficult task of peering into the future. A widely applied and extremely useful set of procedures is time-series analysis.

Time series for predicting

A **time series** is a set of statistical observations arranged in chronological order.

Examples include a weekly series of end-of-week stock prices, a monthly series of amounts of steel production, and an annual series of national income. Such time series are essentially historical series, whose values at any given time result from the interplay of large numbers of diverse economic, political, social, and other factors.

A first step in the prediction of any series involves an examination of past observations.

Time-series analysis deals with the methods of analyzing past data and then projecting the data to obtain estimates of future values.

The traditional, or "classical," methods of time-series analysis, which will be the primary subject of this chapter, are descriptive in nature and do not provide for probability statements concerning future events. However, these time-series models, although admittedly only approximate and not highly refined, have proven their worth when cautiously and sensibly applied. It is important to realize that these methods cannot simply be used mechanistically but must at all times be supplemented by sound subjective judgment.

Although this introduction has referred to the use of time-series analysis for the purposes of forecasting, planning, and control, these procedures also are often used for the simple purpose of historical description. For example, they may be usefully employed in an analysis in which interest centers on the differences in the nature of variations in different time series. The classical time-series model is described in the next section.

11.2 THE CLASSICAL TIME-SERIES MODEL

If we wished to construct an ideally satisfying mathematical model of an economic time series, we might seek to define and measure the many factors determining the variations in the time series and then proceed to state the mathematical relationships between these and the particular series in question.

Determinants of change

However, the determinants of change in an economic time series are multitudinous, including such factors as changes in population, consumer tastes, technology, investment or capital-goods formation, weather, customs, and numerous other economic and noneconomic variables. The enormity and impracticability of the task of measuring all these factors and then relating them mathematically to an economic time series militates against the use of this direct approach to time-series analysis. Hence, it is not surprising that a more indirect and practical approach has come into use. **Classical time-series analysis** is essentially a descriptive method that attempts to break down an economic time series into distinct components representing the effects of groups of explanatory factors such as those given earlier. These component variations are

Components of classical time series

1. Trend
2. Cyclical fluctuations
3. Seasonal variations
4. Irregular movements

Trend

Trend refers to a smooth upward or downward movement of a time series over a long period of time. Such movements are thought of as requiring a minimum of *about 15 or 20 years* to describe, and as being attributable to factors such as population change, technological progress, and large-scale shifts in consumer tastes.

Cyclical fluctuations

Cyclical fluctuations, or business cycle movements, are recurrent upward and downward movements around trend levels that have a duration of anywhere from *about 2–15 years*. There is no single simple explanation of business cycle activity, and there are different types of cycles of varying length and size. Not surprisingly, no satisfactory mathematical model has been constructed for either describing or forecasting these cycles, and perhaps none will ever be.

Seasonal variations

Seasonal variations are cycles that complete themselves within the period of *a calendar year* and then continue to repeat this basic pattern. The major factors producing these annually repetitive patterns of seasonal variations are weather and customs, the latter term broadly interpreted to include observance of various holidays such as Easter and Christmas. Series of monthly and quarterly data are ordinarily used to examine these seasonal variations. Hence, regardless of trend or cyclical levels, we can observe in the United States that each year more ice cream is sold during the summer months than during the winter, whereas more fuel oil for home heating is consumed in the winter than during the summer months. Both of these cases illustrate the effect of weather or climatic factors in determining seasonal patterns. Department store sales generally reveal a minor peak during the month in which Easter occurs and a larger peak in December, when Christmas occurs, reflecting the shopping customs of consumers. The techniques of measurement of seasonal variations that we will discuss are particularly well suited to the measurement of relatively stable patterns of seasonal variations, but they can be adapted to cases of changing seasonal movements as well.

Irregular movements

Irregular movements are fluctuations in time series that are short in duration, erratic in nature, and follow no regularly recurrent or other discernible pattern. These movements are sometimes referred to as **residual variations**, since, by definition, they represent what is left over in an economic time series after trend, cyclical, and seasonal elements have been accounted for. Irregular fluctuations result from sporadic, unsystematic occurrences such as erratic shifts in purchasing habits, accidents, strikes, and the like. Whereas in the classical

FIGURE 11-1
The components of a time series

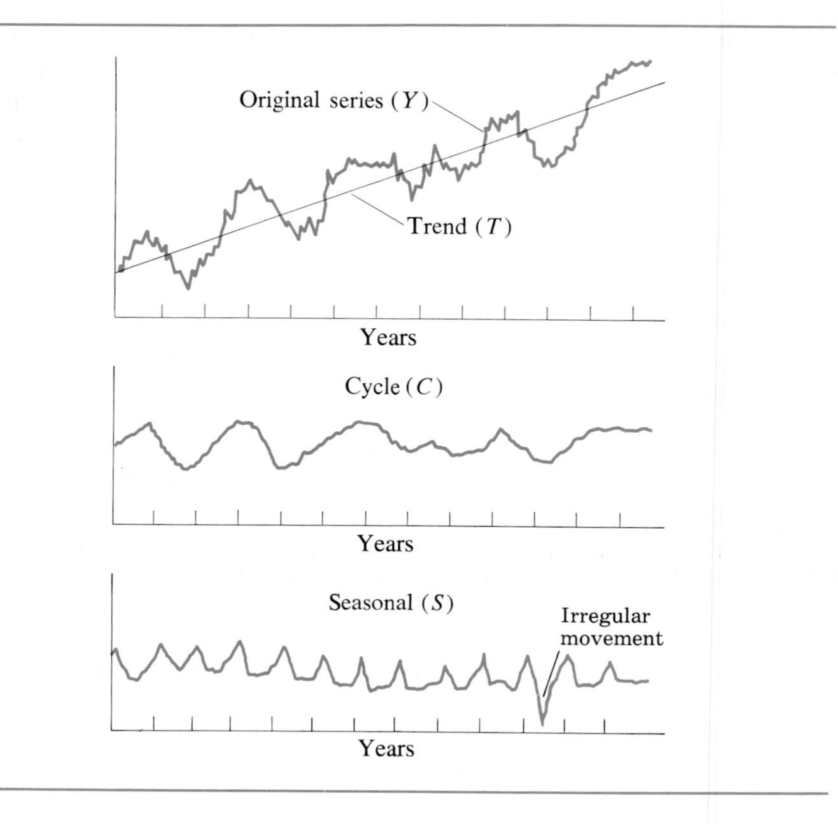

time-series model, the elements of trend, cyclical fluctuations, and seasonal variations are viewed as resulting from systematic influences leading to either gradual growth, decline, or recurrent movements, irregular movements are considered to be so erratic that it would be fruitless to attempt to describe them in terms of a formal model.

Figure 11-1 presents the typical patterns of the trend, cyclical, and seasonal components of a time series. Time is plotted on the horizontal axes of the graphs, and the values of the particular series (which might be retail sales, ice cream production, or airline revenues) are plotted on the vertical axes.

11.3 DESCRIPTION OF TREND

In the preceding section, we pointed out that the classical model involves the separate statistical treatment of the component elements of a time series. We begin our discussion by indicating how the description of the underlying trend is accomplished.

Before the trend of a particular time series can be determined, it is generally necessary to subject the data to some preliminary treatment. The amount of such adjustment depends to some extent on the period for which the data are stated. For example, if the time series is in monthly form, it may be necessary to revise the monthly data to account for the differing number of days per month. This may be accomplished by dividing the monthly figures by the number of days in the respective months, or by the number of working days per month, to state the data for each month on a per day basis.

Preliminary treatment of data

Even when the original data are in annual form, which is often the case when primary interest is centered on the long-term trend of the series, the data may require a considerable amount of preliminary treatment before a meaningful analysis can be carried out. Adjustments for changes in population size are often made by dividing the original series by population figures to state the series in per capita form. Frequently, comparisons of trends in per capita figures are far more meaningful than corresponding comparisons in unadjusted figures.

It is particularly important to scrutinize a time series and adjust it for differences in definitions of statistical units, the consistency and coverage of the reported data, and similar items. It is important to realize that we cannot simply proceed in a mechanical fashion to analyze a time series. Careful and critical preliminary treatment of such data is required to ensure meaningful results.

Matching a Measuring Technique to the Purpose of the Analysis for Fitting Trend Lines

The trend in a time series can be measured by the free-hand drawing of a line or curve that seems to fit the data, by fitting appropriate mathematical functions, or by the use of "moving average" methods. Moving averages are discussed later in this chapter in connection with seasonal variations.

A **free-hand curve** may be fitted to a time series by visual inspection. When this type of characterization of a trend is employed, the investigator is usually interested in a quick description of the underlying growth or decline in a series, without any careful further analysis. In many instances, this rapid graphic method may suffice. However, it clearly has certain disadvantages. Different investigators would surely obtain different results for the same time series. Indeed, even the same analyst would probably not sketch exactly the same trend line in two different attempts on the same series. This excessive amount of subjectivity in choice of a trend line is especially problematic if further quantitative analysis is planned. In the ensuing discussion, we will concentrate on mathematically fitted trend lines.

Quick graphic description

Even in the **mathematical measurement** of trend, the *purpose* of the analysis is of considerable importance in the selection of the appropriate trend line. Several different types of purposes can be specified.

Historical description

- Trend lines may be fitted for the purpose of historical description. If so, any line that fits well will suffice. The line need not have logical implications for forecasting purposes, nor should it be evaluated primarily by characteristics that might be desirable for other purposes.

Future projection

- If prediction or projection into the future is the purpose, particularly if long-term projection is desired, the selected line should have logical implications when it is extended into the future. The analyst must always carefully weigh the implications of the models being projected into the future as regards their reasonableness for the phenomena being described and predicted. For example, constant amounts of growth per unit time are implied if we project a *straight-line* trend into the future. This may not be a reasonable long-term projection for many series.

Study of nontrend elements

- Trend lines are fitted to economic data to describe and eliminate trend movements from the series in order to study nontrend elements. Thus, if the analyst's primary interest is to study cyclical fluctuations, freeing the original data of trend makes it possible to examine cyclical movements without the presence of the trend factor. For this purpose, any type of trend line that does a reasonably good job of bisecting the individual business cycles in the data would be appropriate.

Types of Trend Movements

There have been considerable variations in the trend movements of different economic and business series. Over long periods of time, some companies and industries have experienced periods of growth and then have gone into steep declines when more modern competitive processes and products have emerged in other companies and industries. Real GNP in the United States has exhibited a relatively constant rate of growth of about 3% per year since the early part of this century. Since real GNP is a measure of overall economic activity, it represents a type of average of all series for production of goods and services. Although many series have shown more or less similar trends to that of real GNP, sharp divergences have also occurred. One example of these differential movements is that the service sector of the economy has been growing relative to the agricultural sector. While employment in the service industries is increasing, the number of persons employed in agriculture is decreasing.

For a large number of American industries, numerous studies have revealed a trend that may be characterized as increasing at a decreasing percentage rate. Indeed, some investigators have adapted growth curves originally used to describe biological growth to depict the past change of many industrial series. These **growth curves** (of which the *Gompertz* and *logistic* are the best known) are S-shaped for increasing series plotted on graph paper with an arithmetic vertical scale, and are concave downward on a semilogarithmic chart. By convention, we graph time series on either arithmetic or semilogarithmic paper by plotting the variable of interest on the vertical axis and time on the

Growth curves

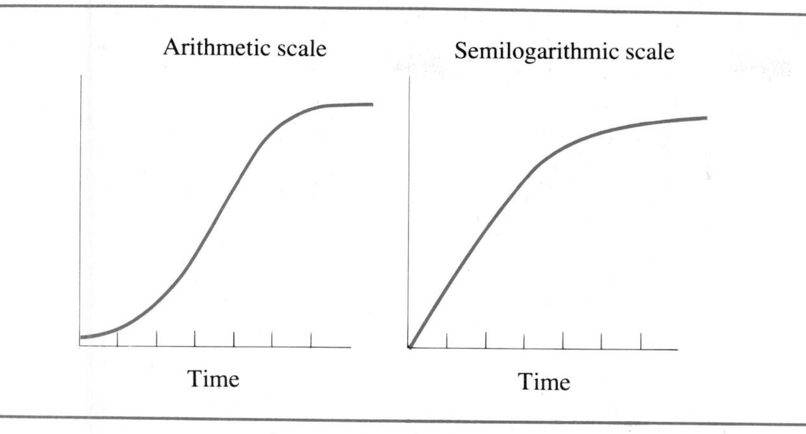

FIGURE 11-2
A growth curve plotted on arithmetic and on semilogarithmic paper

horizontal axis.[1] The general shape of such a growth curve, plotted on arithmetic graph paper and on semilogarithmic graph paper, is depicted in Figure 11-2. The so-called "law of growth" has been used to describe this type of change over time in an industry. On the arithmetic chart, in the early stages of the industry, the growth is slow at first and then becomes rapid, with the series increasing by increasing amounts. Then the industry moves through a point beyond which it increases by decreasing amounts, and finally it tapers off into a period of "maturity." Throughout all stages, as seen on the semilogarithmic chart, although the industry is growing, the increases are at a decreasing *percentage rate.* Various reasons for this type of industrial growth were propounded by the investigators who discerned analogous changes in biological and industrial time-series data. Since the equations of these growth curves are exponential in character, it is difficult to fit some of these curves by the method of least squares described later in this chapter.[2]

Law of growth

We indicated earlier that a great variety of types of trend movements exist in economic time series, and many of these trends cannot be adequately described or projected by means of growth curves. The growth curves have a number of desirable characteristics. For example, they have finite lower and upper limits, which are determined by the data to which the curves are fitted. However, no one family of curves is apt to be generally satisfactory for trend fitting purposes. In fact, the growth curves have been found to be quite inadequate for industrial

[1] The familiar arithmetic graph has an arithmetically ruled vertical scale on which equal vertical distances represent equal *amounts* of change. Semilogarithmic graphs have an arithmetic horizontal scale, but a logarithmically ruled vertical scale on which equal vertical distances represent equal *percentage rates* of change. For example, on an arithmetic graph, a straight line inclined upward depicts a series that is increasing with constant amounts of change. On a semilogarithmic graph, a straight line inclined upward depicts a series that is increasing at a constant percentage rate.

[2] A special technique known as "the method of selected points" is sometimes used for this purpose. Also, logarithmic transformations are often used.

growth prediction. The most commonly used polynomial-type trend lines, which are fitted by the method of least squares, are discussed in this chapter.

Selection of appropriate trend line

The purpose of fitting, the goodness of fit obtained, knowledge of the growth and decay processes involved, and trial-and-error experimentation are all essential ingredients in the selection of the appropriate trend line.

11.4 FITTING TREND LINES BY THE METHOD OF LEAST SQUARES

For situations in which it is desirable to have a mathematical equation to describe the trend of a time series, the most widely used method is to fit some type of polynomial function to the data. In this section, we illustrate the general method by means of simple examples, fitting a straight line and a second-degree parabola to time-series data by the method of least squares.

There are computer programs that allow for the fitting of trend lines of various types. One can use regression analysis programs to fit polynomial functions by the method of least squares to time series, treating time as the independent variable. The fitting of a couple of such functions by hand is illustrated in this chapter in order to give a somewhat better "feel" for the method and assumptions involved than if only computer outputs were shown. Also, the examples demonstrate the usefulness of encoding the time variable in terms of deviations from the mean (middle) time period.

The Method of Least Squares

The method of least squares, when used to fit trend lines to time-series data, is employed mainly because it is a simple, practical method that provides the best fits according to a reasonable criterion. However, we should recognize that the method of least squares does not have the same type of theoretical underpinning when applied to fitting trend lines as when used in regression and correlation analysis, described in Chapters 9 and 10. The major difficulty is that the usual probabilistic assumptions made in regression and correlation analysis are simply not met in the case of time-series data. For example, the illustrative problem in Chapter 9 involving the relationship between food expenditures and income for a sample of families included two possible theoretical models. In the first, both food expenditures and income were random variables; in the second, income was a controlled variable, that is, families of prespecified incomes were selected, and food expenditures was a random variable. In each model, the dependent variable was a random variable, and the model assumed conditional probability distributions of this random variable around the computed values of the dependent variable, which fell along the regression line. These computed

Major difficulty

Y values were the means of the conditional probability distributions. A number of assumptions are implicit in this type of model: Deviations from the regression line are considered to be random errors describable by a probability distribution. The successive observations of the dependent variable are assumed to be independent. For example, Family B's expenditures were assumed to be independent of Family A's, and so on.

Fitting trend lines to time-series data

In fitting trend lines to time-series data, the probabilistic assumptions of the method of least squares are not met.

If a trend line is fitted, for example, to an annual time series of department store sales, time is treated as the independent variable X and department store sales as the dependent variable Y. It is not reasonable to think of the deviation of actual sales in a given year from the computed trend value as a random error. Indeed, if the original data are annual, then deviations from trend would represent the operation of cyclical and irregular factors. (Seasonal factors would not be present in annual data, because by definition they complete themselves within a year.) Finally, the assumption of independence is not met in the case of time-series data. A department store's sales in a given year surely are not independent of what they were in the preceding year. In summary, we return to the point made at the outset of this discussion:

The method of least squares when used to fit trend lines is employed primarily because of its practicality, simplicity, and good fit characteristics rather than because of its justification from a theoretical viewpoint.

Fitting an Arithmetic Straight-Line Trend

As an example, we will fit a straight line by the method of least squares to an annual series on network compensation to television stations in the United States from 1972 to 1988. Although we wrote the equation of a straight line in the discussion of regression analysis in Chapter 9 as $\hat{Y} = a + bX$, in time-series analysis we will use the equation

$$Y_t = a + bx \tag{11.1}$$

The variables

The computed trend value is denoted Y_t, with the subscript "t" standing for trend. That is, Y_t is the computed trend figure for period x. In time-series analysis, the computations can be simplified by transforming the X variable, which is the independent variable "time," to a simpler variable with fewer digits. This is accomplished by stating the time variable in terms of deviations from the arithmetic mean time period, which is simply the middle period.

TABLE 11-1
Network compensation to television stations in the United States, 1972–1988

(1) Year	(2) x	(3) Network Compensation ($ millions) Y	(4) xY	(5) x^2	(6) Trend Y_t	(7) Percentage of Trend $\frac{Y}{Y_t} \cdot 100$
1972	−8	224	−1,792	64	217.0	103.2
1973	−7	233	−1,631	49	234.1	99.5
1974	−6	248	−1,488	36	251.2	98.7
1975	−5	258	−1,290	25	268.3	96.2
1976	−4	270	−1,080	16	285.4	94.6
1977	−3	288	−864	9	302.5	95.2
1978	−2	315	−630	4	319.6	98.6
1979	−1	344	−344	1	336.7	102.2
1980	0	369	0	0	353.8	104.3
1981	1	393	393	1	370.9	106.0
1982	2	406	812	4	388.0	104.6
1983	3	416	1,248	9	405.1	102.7
1984	4	425	1,700	16	422.2	100.7
1985	5	437	2,185	25	439.3	99.5
1986	6	450	2,700	36	456.4	98.6
1987	7	462	3,234	49	473.5	97.6
1988	8	476	3,808	64	490.6	97.0
	0	6,014	6,961	408		

$$a = \frac{\Sigma Y}{n} = \frac{6{,}014}{17} = 353.8$$

$$b = \frac{\Sigma xY}{\Sigma x^2} = \frac{6{,}961}{408} = 17.1$$

$$Y_t = 353.8 + 17.1x$$

$x = 0$ in 1980

x is in one-year intervals

Y is in millions of dollars

Sources: FCC TV financial data (1972–1980); *TV/Radio Age*, "Business Barometer" (1981–1983); estimates from Dick Gideon Enterprises (1984–1988)

The transformed time variable is denoted by lower-case x. In the illustrative example in Table 11-1, $x = 0$ in 1980, the middle year in the time series that runs from 1972 through 1988. The x values (or $X - \bar{X}$ figures) for years before and after 1980 are, respectively, −1, −2, −3, . . . , and 1, 2, 3, ... For example, the x value for 1981 is equal to one because $X - \bar{X} = 1981 - 1980 = 1$.

The constants

The constants in the trend equation are interpreted in a similar way to those in the straight line discussed in regression analysis; a is the computed trend figure for the period when $x = 0$, in this case, 1980; b is the slope of the trend line, or the amount of change in Y_t per unit change in x (per year in the present example). Because the sum of the deviations of a set of observations from their mean is equal to zero, $\Sigma x = 0$. This property makes the computation of the constants for the trend line simpler than in the corresponding case of the straight-line regression equation. In Chapter 9, the equations for fitting a straight line were given as

$$a = \bar{Y} - b\bar{X} \tag{11.2}$$

$$b = \frac{\Sigma XY - n\bar{X}\bar{Y}}{\Sigma X^2 - n\bar{X}^2} \tag{11.3}$$

In the least squares fitting of a straight-line trend equation, x is substituted for X. Since $\Sigma x = 0$, and therefore, $\bar{x} = 0$, the equations become

$$a = \bar{Y} = \frac{\Sigma Y}{n} \tag{11.4}$$

$$b = \frac{\Sigma xY}{\Sigma x^2} \tag{11.5}$$

Hence, the constant a is simply the mean of the Y values, and b is the quotient of two numbers easily determined from the original data. The calculations for fitting a straight-line trend to the time series on network compensation to television stations are given in Table 11-1. Columns (2) through (5) contain the basic computations for determining the values of a and b. In the calculations of these constants at the bottom of the table, $a = 353.8$ and $b = 17.1$. The trend equation is $Y_t = 353.8 + 17.1x$. The identification statements given below the trend equation—such as $x = 0$ in 1980 and Y is in millions of dollars—should always accompany the equation, since it is impossible to interpret the meaning of the trend line fully without them. The trend figures are determined by substituting the appropriate values of x into the trend equation. For example, the trend figure for 1972 is

$$Y_{t,1972} = 353.8 + 17.1(-8) = 217.0 \text{ millions of dollars}$$

Since the b value measures the change in Y_t per year, it can be added to each trend value to obtain the following year's figure. The trend figures are given in column (6) of Table 11-1.

The trend line is graphed in Figure 11-3. Any two points can be plotted to determine the line. Interpreting the values $a = 353.8$ and $b = 17.1$, we have a computed trend figure of \$353.8 million for network compensation to television stations in 1980 and an increase in trend of \$17.1 million per year. Since the line was fitted by the method of least squares, the sum of the squared deviations of the actual data from the trend line is less than from any other straight line

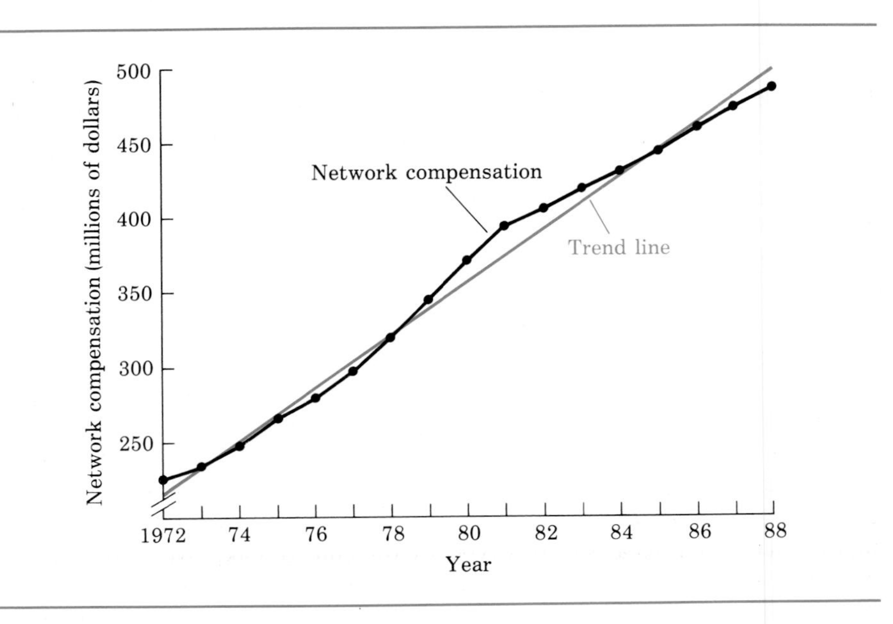

FIGURE 11-3 Straight-line trend fitted to network compensation to television stations in the United States, 1972–1988

that could have been fitted to the data, and the total of the deviations above the line is equal to the total below the line.

Odd numbers

A couple of technical points concerning the fitting procedure may be noted. Since the present illustration contained an odd number of years, the period at which $x = 0$, or the x origin, coincided with one of the years of data, and the x values were stated as 1, 2, 3, . . . for years after the x origin and -1, -2, -3, . . . for years before the origin.

Even numbers

If there had been an even number of years, then the mean time period, at which $x = 0$, would fall midway between the two central years. For example, suppose there had been one less year of data and the figures on network compensation to television stations were available only for 1972–1987. Then there would be 16 annual figures and $x = 0$ at $1979\frac{1}{2}$. The two central years, 1979 and 1980, would deviate from this origin by $-\frac{1}{2}$ and $+\frac{1}{2}$, respectively. To avoid the use of fractions, it is usual to state the deviations in terms of half-year intervals rather than a year. Hence, the x values for 1980, 1981, 1982, . . . would be 1, 3, 5, . . . and would be -1, -3, -5, . . . for 1979, 1978, 1977, ... The computation of the constants a and b would proceed in the usual way. However, a would be interpreted as the computed trend figure for a point midway between the two central years, and the b value would be the amount of change in trend per half year.

Time intervals

If the time intervals of the original data were not annual, the transformed time variable x would have to be appropriately interpreted. For example, if the data were stated in five-year averages and there were an odd number of such figures, then x would be in five-year intervals. If there were an even number of figures and the nonfractional method of stating x referred to above were used, then x would be in $2\frac{1}{2}$-year intervals.

Projection of the Trend Line

Projections of the computed trend line can be obtained by substituting the appropriate values of x into the trend equation. For example, the projected trend figure for 1995 for network compensation to television stations would be computed by substituting $x = 15$ in the previously determined trend equation. Hence,

$$Y_{t,1995} = 353.8 + 17.1(15) = \$610.3 \text{ (million)}$$

A rougher estimate of this trend figure would be obtained by extending the straight line graphically in Figure 11-3 to the year 1995. Remember that these projections are estimates of only the trend level in 1995 and not of the actual figure for network compensation to television stations in that year. If a prediction of the latter figure were desired, estimates of the nontrend factors would have to be combined with the trend estimate. This means that a prediction of cyclical fluctuations would have to be made and incorporated with the trend figure. Accurate forecasts of this type are difficult to make over extended time periods. However, insofar as managerial applications of trend analysis are concerned, for long-range planning purposes often all that is desired is a projection of the trend level of the economic variable of interest. For example, a good estimate of the trend of demand would be adequate for a business planning a plant expansion to anticipate demand many years into the future. Accompanying predictions of business cycle standings many years into the future would not be required; nor, for that matter, would they be realistically feasible.

Purpose

Cyclical Fluctuations

As previously indicated, when a time series consists of annual data, it contains trend, cyclical, and irregular elements. The seasonal variations are absent, since they occur within a year. Hence, deviations of the actual annual data from a computed trend line are attributable only to cyclical and irregular factors. Since the cyclical element is the dominant factor, a study of these deviations from trend essentially represents an examination of business cycle fluctuations. The deviations from trend are most easily observed by dividing the original data by the corresponding trend figures for the same period. By convention, the result of dividing an original figure by a trend value is multiplied by 100 to express the figure as a **percentage of trend**. Hence, if the original figure is exactly equal to the trend figure, the percentage of trend is 100; if the original figure exceeds the trend value, the percentage of trend is above 100; and if the original figure is less than the trend value, the percentage of trend is below 100.

Percentage of trend

The formula for percentage of trend figures is

$$\text{Percentage of trend} = \frac{Y}{Y_t} \cdot 100 \qquad (11.6)$$

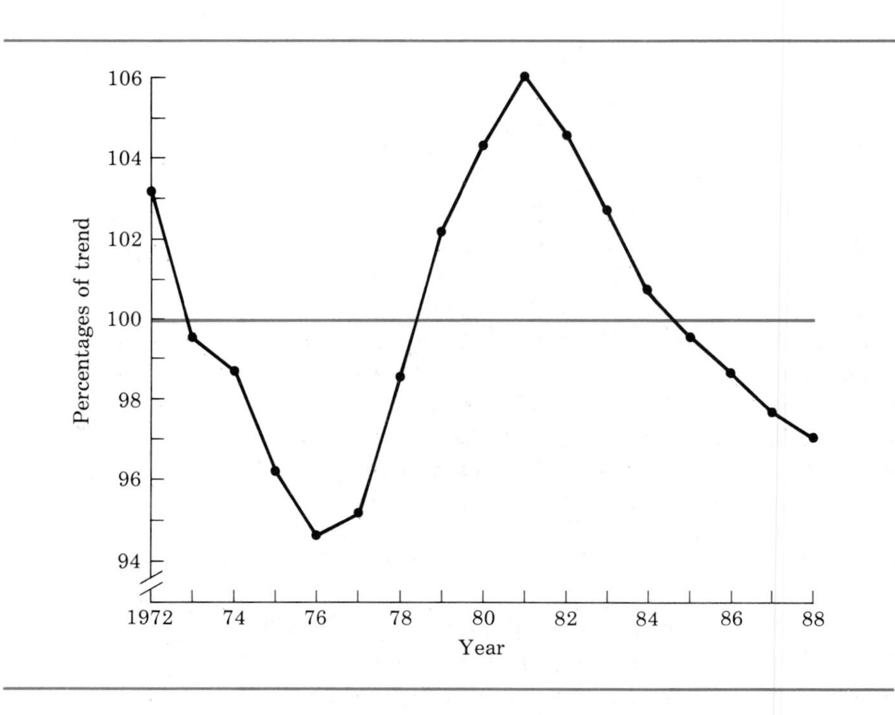

FIGURE 11-4
Percentages of trend for network compensation to television stations in the United States, 1972–1988

where Y = annual time-series data
Y_t = trend values

When converted to percentage of trend, the data contain only cyclical and irregular movements, since the division by trend eliminates that factor. A so-called multiplicative model for the analysis provides the rationale of this procedure. That is, the original annual figures are viewed as representing the combined effects of trend, cyclical, and irregular factors. In symbols, let T, C, and I represent trend, cyclical, and irregular factors, respectively, and let Y and Y_t mean the same as in equation 11.6. Dividing the original time series by the corresponding trend values yields

$$\frac{Y}{Y_t} = \frac{T \times C \times I}{T} = C \times I \tag{11.7}$$

The percentages of trend for the series on network compensation to television stations are given in column (7) of Table 11-1 and are plotted in Figure 11-4. On the figure, the upward trend movement is no longer present. Instead, the percentage of trend series fluctuates about the line labeled 100, which is the **trend level**. These percentages of trend are sometimes referred to as **cyclical relatives**; that is, the original data are stated relative to the trend figure. (Of course, strictly speaking, Y/Y_t is the cyclical relative, and the multiplication by 100 converts the relative to a percentage figure.) Another way of depicting cyclical fluctuations is in terms of relative cyclical residuals, which are percentage deviations from trend and are computed by the formula

Relative cyclical residuals

(11.8)
$$\text{Relative cyclical residual} = \frac{Y - Y_t}{Y_t} \cdot 100$$

For example, if we refer to the network compensation data in Table 11-1 for 1981, the actual figure is 393, the computed trend value is 370.9, and the percentage of trend is 106.0. The relative cyclical residual in this case is +6.0% indicating that actual network compensation is 6.0% above the trend figure because of cyclical and irregular factors. These residuals are positive or negative depending on whether the actual time-series figures fall above or below the computed trend values. The graph of relative cyclical residuals is visually identical to that of the percentage of trend values except that relative cyclical residuals are shown as fluctuations around a zero base line rather than around a base line of 100%.

Familiar charts

The familiar charts of business cycle fluctuations that often appear in publications such as the financial pages of newspapers and business periodicals are usually graphs of either percentages of trend or relative cyclical residuals. These charts may be studied for timing of peaks and troughs of cyclical activity, for amplitude of fluctuations, for duration of periods of expansion and contraction, and for other items of interest to the business cycle analyst.

Fitting a Second-Degree Trend Line

The preceding discussion on the fitting of a straight line pertains to the case in which the trend of a time series can be characterized as increasing or decreasing by constant amounts per time period. Actually very few economic time series exhibit this type of constant change over a long period of time (say, over a period of several business cycles). Therefore, it generally is necessary to fit other types of lines or curves to the given time series. Polynomial functions are particularly convenient to fit by the method of least squares. Frequently, a second-degree parabola provides a good description of the trend of a time series. In this type of curve, the amounts of change in the trend figures Y_t may increase or decrease per time period.

Second-degree parabola

A second-degree parabola may provide a good fit to a series whose trend is increasing by increasing amounts, increasing by decreasing amounts, and so on.

The procedure of fitting a parabola by the method of least squares involves the same general principles as fitting a straight line, but it entails somewhat more arithmetic. We illustrate the method of fitting a second-degree parabola to a time series in terms of the following illustration. It is important to note that

TABLE 11-2
Second-degree parabola fitted by the method of least squares to gross national product of the United States, 1971–1985

(1) Year	(2) x	(3) Y	(4) xY	(5) x^2Y	(6) x^2	(7) x^4	(8) Trend Y_t
		Gross National Product (in billions of dollars)					
1971	−7	1,078	−7,546	52,822	49	2,401	1,062.27
1972	−6	1,186	−7,116	42,696	36	1,296	1,174.31
1973	−5	1,326	−6,630	33,150	25	625	1,301.09
1974	−4	1,434	−5,736	22,944	16	256	1,442.61
1975	−3	1,549	−4,647	13,941	9	81	1,598.87
1976	−2	1,718	−3,436	6,872	4	16	1,769.87
1977	−1	1,918	−1,918	1,918	1	1	1,955.61
1978	0	2,164	0	0	0	0	2,156.09
1979	1	2,418	2,418	2,418	1	1	2,371.31
1980	2	2,632	5,264	10,528	4	16	2,601.27
1981	3	2,958	8,874	26,622	9	81	2,845.97
1982	4	3,069	12,276	49,104	16	256	3,105.41
1983	5	3,305	16,525	82,625	25	625	3,379.59
1984	6	3,680	22,080	132,480	36	1,296	3,668.51
1985	7	3,970	27,790	194,530	49	2,401	3,972.17
	0	34,405	58,198	672,650	280	9,352	

Sources: U.S. Department of Commerce, Bureau of Economic Analysis (1971–1983); estimates from Dick Gideon Enterprises (1984–1985)

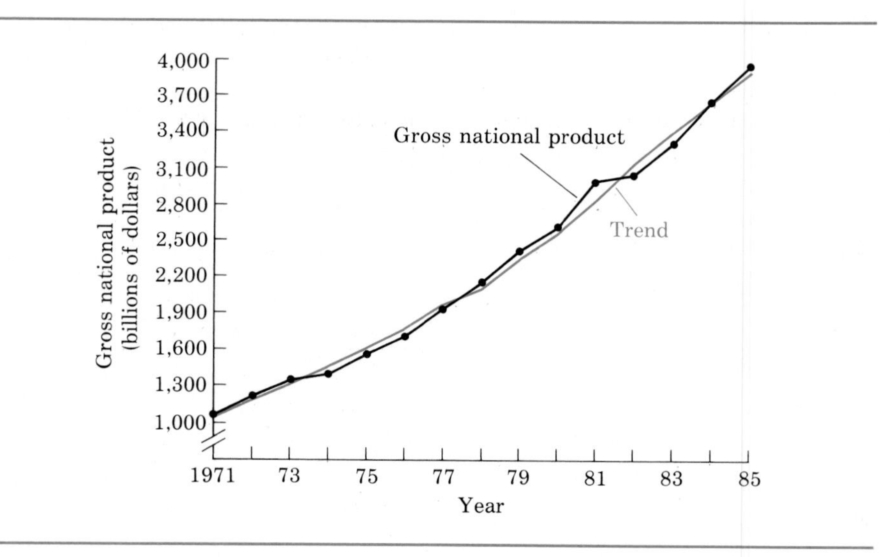

FIGURE 11-5
Second-degree parabola fitted to gross national product of the United States, 1971–1985

the time period in this example is too short to permit a valid description of long-term trend. However, the illustration is given for expository purposes to indicate the procedure involved.

EXAMPLE 11-1

A time series on the gross national product of the U.S. from 1971–1985 appears in Table 11-2. This time series is also graphed in Figure 11-5. The trend of these data may be described as increasing by increasing amounts. The general form of a second-degree parabola is $Y_t = a + bX + cX^2$. Analogous to the method of stating the equation for a straight-line trend, the trend line for a second-degree parabola may be written

$$Y_t = a + bx + cx^2 \tag{11.9}$$

where Y_t = trend values
a, b, c = constants to be determined
x = deviations from the middle time period

If we use the transformed variable x, which represents deviations from the mean time period, the equations for fitting a second-degree parabola are

$$\Sigma Y = na + c\Sigma x^2 \tag{11.10}$$

$$\Sigma x^2 Y = a\Sigma x^2 + c\Sigma x^4 \tag{11.11}$$

$$b = \frac{\Sigma xY}{\Sigma x^2} \tag{11.12}$$

Note that the constant b is determined by the same equation used for fitting a straight line. However, the constants a and c are found by solving equations 11.10 and 11.11 simultaneously.

In the present problem, because of the odd number of years in the data, $x = 0$ in the middle year, 1978. Solving for b by substituting the appropriate totals from Table 11-2, we have

$$b = \frac{58{,}198}{280} = 207.85$$

Substituting the appropriate figures into equations 11.10 and 11.11 gives

$$34{,}405 = 15a + 280c$$

$$672{,}650 = 280a + 9{,}352c$$

Multiplying the first equation by 56 and the second equation by 3 to equate the coefficients of a, we obtain

$$1{,}926{,}680 = 840a + 15{,}680c$$

$$2{,}017{,}950 = 840a + 28{,}056c$$

Subtracting the first equation from the second, we have

$$91{,}270 = 12{,}376c$$

and

$$c = \frac{91{,}270}{12{,}376} = 7.37$$

Substituting this value for c into the first equation yields

$$1{,}926{,}680 = 840a + 15{,}680(7.37)$$

$$a = 2{,}156.09$$

Therefore, the equation of the second-degree parabola fitted to the time series on gross national product of the United States is

$$Y_t = 2{,}156.09 + 207.85x + 7.37x^2 \tag{11.13}$$

where $x = 0$ in 1978
x is in one-year intervals
Y is in billions of dollars

The trend figures Y_t shown in column (8) of Table 11-2 are obtained by substituting the appropriate values of x in equation 11.13. The constants a, b, and c may be interpreted as follows: a is the computed trend figure at the time origin, that is, when $x = 0$; b is the slope of the parabola at the time origin; and c indicates the amount of acceleration or deceleration in the curve, or the amount by which the slope changes per time period.[3]

Dangers of mechanistic projections

Although the second-degree parabola appears from Figure 11-5 to provide a reasonably good fit to the data in this example, the dangers of a mechanistic projection of a trend line are clearly illustrated. The projected trend figures imply annual increases by increasing amounts. Therefore, we should not entertain the notion of extending the trend line into the future for forecasts unless an analysis of the underlying factors determining the trend of this series revealed reasons for a continuation of this type of trend.

Fitting Logarithmic Trend Lines

As discussed earlier, the equations of trend lines embody assumptions concerning the type of change that takes place over time. The arithmetic straight line assumes a trend that increases or decreases by constant amounts, whereas the second-degree parabola assumes that the change in these amounts of change is constant per unit time. It is often useful to describe the trend of an economic time series in terms of the percentage rates of change that are taking place, for example, by the use of logarithmic trend lines.

[3] The derivative of the second-degree parabola trend equation is

$$\frac{dY_t}{dx} = b + 2cx$$

Hence, the slope of the curve differs at each period x. When $x = 0$, $\frac{dY_t}{dx} = b$. Therefore, the slope at the time origin is b. The second derivative is $\frac{d^2Y_t}{dx^2} = 2c$. Thus, the acceleration, or rate of change in the slope, is $2c$ per time period.

If a time series increases or decreases at exactly a constant percentage rate, a straight line fitted to the logarithms of the data constitutes a perfect fit.

Constant percentage rate of increase or decrease

For example, suppose a time series has successive values of 10, 100, 1,000, and 10,000. This series may be characterized as increasing at a constant percentage rate of 900%. The logarithms of these values are log 10 = 1, log 100 = 2, log 1,000 = 3, and log 10,000 = 4. Hence, the logarithms are increasing by a constant amount, namely one unit, and a straight line could be drawn through the numbers 1, 2, 3, and 4. Some economic time series in the United States, although not changing exactly at a constant rate, have exhibited trends of approximately constant percentage increases over substantial periods of time. The equation of the logarithmic straight line that would describe the trend of such series is

$$\log Y_t = a + bx \tag{11.14}$$

The method of fitting this line is the same as for the arithmetic straight line, $Y_t = a + bx$, except that wherever Y appeared before, log Y now appears. The values of the constants a and b are computed as follows:

$$a = \frac{\Sigma \log Y}{n} \tag{11.15}$$

$$b = \frac{\Sigma x \log Y}{\Sigma x^2} \tag{11.16}$$

After a and b have been calculated, trend figures are determined by substituting values of x into the trend equation, computing log Y_t, and taking the antilogarithm to obtain Y_t. Although we will not present another example of the fitting process, since there really are no new principles involved, we will illustrate the calculation of a trend figure for the type of trend line under discussion.

Suppose the logarithmic trend line for a particular series had been determined as

$$\log Y_t = 2.3657 + 0.0170x$$

Then the logarithm of the trend figure for the year in which $x = 2$ would be given by substituting this value of x into the trend equation to obtain

$$\log Y_t = 2.3657 + 0.0170(2)$$

$$\log Y_t = 2.3997$$

Taking the antilog of this value yields the trend figure

$$Y_t = \text{antilog } 2.3997 = 251.0$$

The rate of change implied by this trend line can be obtained by calculating antilog $b - 1$. For example, in the above illustration the antilog of the slope coefficient b is

$$\text{antilog } 0.0170 = 1.040$$

This figure is the ratio of each trend figure to the preceding one. Subtracting 1.00 from this figure yields $1.04 - 1.00 = 0.04$. Hence, the trend figures increase by 4% per time period. If the series had been a declining one and the result of the above calculation was -0.04, this would mean that the trend figures decrease by 4% per time period.

Selecting constant-rate segments

A time series sometimes can be broken down into segments during which the rate of change has been approximately constant—even though the complete series does not exhibit a trend with constant rates of change throughout.

It is often useful in such cases to make comparisons of the rates of change similarly determined from different economic time series of interest.

Logarithmic second-degree parabolas

Logarithmic second-degree parabolas can also be fitted to time series in which the trend is increasing at an increasing percentage rate, increasing at a decreasing percentage rate, and so on. Ordinarily polynomials of third or higher degree are not fitted to time series in either arithmetic or logarithmic form because such curves permit too many changes in direction and tend to follow the cyclical fluctuations in the data as well as the trend. Therefore, these curves often do not have the required trend line characteristic of depicting the smooth, continuous movement underlying the cyclical swings in a time series.

Finding an appropriate trend line

In the attempt to find an appropriate trend line, the analyst should always plot the time series on both arithmetic and semilogarithmic graph paper. These two types of graphs may help in determining whether an arithmetic or logarithmic line would provide a better description of the trend.

Exercises 11.4

1. Discuss the nature and causes of component variations of an economic time series.
2. The sales of a consumer products company from 1982 to 1986 were:

Year	1982	1983	1984	1985	1986
Sales ($ millions)	35.0	40.0	44.0	49.9	55.0

a. For each year, compute the Y_t value for the equation

$$Y_t = a + bx$$

b. Are these Y_t values a good description of the trend of sales of this company? Why or why not?

c. Compute the relative cyclical residual for 1984 and explain what it means.

3. The following trend equation resulted from the fitting of a least squares parabola to the imports of a developing country.

$$Y_t = 49.17 + 4.23x - 0.19x^2$$

$x = 0$ in 1960

x is in $2\frac{1}{2}$-year intervals

Y is imports in millions of U.S. dollars

a. Assume the above trend line is "a good fit." What generalizations can you make concerning the way in which the imports of this country have grown in absolute amounts? Concerning the percentage rate at which imports have grown?

b. In part (a) you were instructed to assume the trend line was "a good fit." However, the actual amount of imports in 1985 was \$95,185,000. This is rather striking evidence that the equation is not "a good fit." Do you agree? Discuss.

4. Obtain trend figures for the 1986–1990 period by projecting a straight-line trend for U.S. population (age 16 and over), using the end-of-year data in the following table:

Year	U.S. Population Age 16 and Over (in thousands)	Year	U.S. Population Age 16 and Over (in thousands)
1961	121,343	1968	135,562
1962	122,981	1969	137,841
1963	125,154	1970	140,182
1964	127,224	1971	142,596
1965	129,236	1972	145,775
1966	131,180	1973	148,263
1967	133,319	1974	150,827

5. Assume that the following table presents the consumption (in billions of kilowatt hours) of electric power in a certain region of the United States:

Year	1935	1945	1955	1965	1975	1985
Consumption (billion kwh)	25	60	135	330	780	1855

a. What was the average amount of increase per decade in the series between 1935 and 1985? (Do not use the trend line.)
b. Fit a linear trend line by the method of least squares to (1) the original data and (2) the logarithms of the data.
c. In terms of this problem, interpret the meaning of the constants of the trend equation found for the original data in part (b).
d. Is the answer to part (a) consistent with the slope obtained for the original data in part (b)? Why or why not?
e. In 1957, consumption of electric power was 200 billion kilowatt hours. Compute and interpret the absolute cyclical residual and the relative cyclical residual for that year, using the straight-line trend fitted to the original data in part (b).

6. The asset values of ABD Pharmaceuticals over a 5-year period follow:

Year	1982	1983	1984	1985	1986
Assets ($ millions)	10	22	45	92	178

a. Fit a straight-line trend to these data by the method of least squares.
b. Interpret the meaning of the constants of the trend equation you computed in part (a).
c. With plotting these data, do you think that a straight-line trend is a good fit? Why or why not?

7. The following table presents the total population (in billions) of the United States taken during census years:

Year	1930	1940	1950	1960	1970	1980
Population (billions)	123	132	152	181	205	222

a. What was the average increase per decade of U.S. population between 1930 and 1980? (Do not use a trend line.)
b. Fit a linear trend by the method of least squares to the original data. Interpret the meaning of the constants of the trend equation in terms of problem.
c. Is the answer to part (a) consistent with the slope obtained for the original data in part (b)? Why or why not?
d. In 1975, total U.S. population was 214 billion. Compute and interpret the absolute cyclical residual and the relative cyclical residual for that year, using the straight line fitted to the original data in part (b).

11.5 MEASUREMENT OF SEASONAL VARIATIONS

For long-range planning and decision making, in terms of time-series components, executives of a business or government enterprise concentrate primarily on forecasts of trend movements. For intermediate planning—say, from about two to five years—business cycle fluctuations are of critical importance, too. For short-range planning, and for purposes of operational decisions and control, seasonal variations must also be taken into account.

Seasonal movements, as indicated in section 11.2, are periodic patterns of variation in a time series. Strictly speaking, the terms **seasonal movements** and **seasonal variations** can be applied to any regularly repetitive movements that occur in a time series when the interval of time for completion of a cycle is one year or less. Hence, this classification includes movements such as *daily* cycles in utilization of electrical energy and *weekly* cycles in the use of public transportation vehicles. However, these terms generally refer to the annual repetitive patterns of economic activity associated with climate and custom. As noted earlier, these movements are generally examined by using series of *monthly* or *quarterly* data.

Daily

Weekly

Monthly

Quarterly

Purpose of Analyzing Seasonal Variations

Just as was true in the study of trend movements, seasonal variations may be studied because *interest is primarily centered on these movements*, or they may be measured merely *in order to eliminate them*, so that business cycle fluctuations can be more clearly revealed. As an illustration of the first purpose, a company might analyze the seasonal variations in sales of a product it produces in order to iron out variations in production, scheduling, and personnel requirements. Another reason a company's interest may be primarily focused on seasonal variations is to budget a predicted annual sales figure by monthly or quarterly periods based on seasonal patterns observed in the past.

Focus of interest

On the other hand, as an illustration of the second purpose, an economist may wish to eliminate the usual month-to-month variations in series such as personal income, unemployment rates, and housing starts in order to study the underlying business cycle fluctuations present in these data.

Elimination to reveal other factors

Rationale of the Ratio to Moving Average Method

Seasonal variations can be measured by a number of techniques, but only the most widely used one, the so-called **ratio to moving average method**, will be discussed here. It is most frequently applied to monthly data, but we will illustrate its use for a series of quarterly figures, thus reducing substantially the required number of computations.

Obtaining seasonal indices

In understanding the rationale of the measurement of seasonal fluctuations, it is helpful to begin with the final product, the seasonal indices. When the raw data are for quarterly periods and a stable or regular seasonal pattern is present, the object of the calculations is to obtain four **seasonal indices**, each one indicating the seasonal importance of a quarter of the year. The arithmetic mean of these four indices is 100.0. Hence, a seasonal index of 105 for the first quarter means that the first quarter averages 5% higher than the average for the year as a whole. If the original data had been monthly, there would be 12 seasonal indices, which average 100.0, and each index would indicate the seasonal importance of a particular month. These indices are descriptive of the recurrent seasonal pattern in the original series.

As an example of how these seasonal indices might be used, we can refer to budgeting a predicted annual sales figure by quarterly periods. Suppose that $40 million of sales of particular products was budgeted for the next year, or an average of $10 million per quarter. If the quarterly seasonal indices based on an observed stable seasonal pattern were 97.0, 110.0, 85.0, and 108.0, then the amount of sales budgeted for each quarter would be

First quarter:	0.97 × $10 million	=	$ 9.7 million
Second quarter:	1.10 × 10 million	=	11.0 million
Third quarter:	0.85 × 10 million	=	8.5 million
Fourth quarter:	1.08 × 10 million	=	10.8 million

A problem

The essential problem in the measurement of seasonal variations is eliminating from the original data the nonseasonal elements in order to isolate the stable seasonal component. In trend analysis, when annual data were used and we wanted to arrive at cyclical fluctuations, a similar problem existed. It was solved by obtaining measures of trend and using them as base line or reference figures. Deviations from trend were then measures of cyclical (and irregular) movements. Analogously, when we have monthly or quarterly original data, which consist of all of the components of trend, cyclical, seasonal, and irregular movements, ideally we would like to obtain a series of base line figures that contain all the nonseasonal elements. Then deviations from the base line would represent the pattern of seasonal variations. Unsurprisingly, this ideal method of measurement is not feasible.

The solution

However, the practical method is to obtain a series of moving averages that roughly include the trend and cyclical components. Dividing the original data by these moving average figures eliminates the trend and cyclical elements and yields a series of figures that contain seasonal and irregular movements. These data are then averaged by months or by quarters to eliminate the irregular disturbances and to isolate the seasonal factor. This method of describing a pattern of stable seasonal movements is explained below.

Ratio to Moving Average Method

In order to derive a set of seasonal indices from a series characterized by a stable seasonal pattern, about five to nine years of monthly or quarterly data are required. A stable seasonal pattern means that the peaks and troughs generally occur in the same months or quarters of each year.

Computing seasonal indices for quarterly data

The ratio to moving average method of computing seasonal indices for quarterly data may be summarized as consisting of the following steps:

1. Derive a four-quarter moving average that contains the trend and cyclical components present in the original quarterly series.

> A **four-quarter moving average** is simply an annual average of the original quarterly data successively advanced one quarter at a time.

Deriving a four-quarter moving average

For example, the first moving average figure contains the first four quarters. Then the first quarter is dropped, and the second through fifth quarterly figures are averaged. The computation proceeds this way until the last moving average is calculated, containing the last four quarters of the original series. In the actual calculation, an adjustment is made to center the moving average figures so their timing corresponds to that of the original data.

Characteristics of moving averages

The reason these moving averages include the trend and cyclical components may perhaps be most easily understood by considering what these averages do *not* contain. Since they are annual averages, they do not contain seasonal movements (such fluctuations, by definition, average out over a one-year period). Furthermore, the irregular movements that raise the figures for certain months or quarters and lower them in others tend to cancel out when averaged over the year. Thus, only the trend and cyclical elements tend to be present in the moving averages.

2. Divide the original data for each quarter by the corresponding moving average figure.

These "ratio to moving average" numbers contain only the seasonal and irregular movements, since the trend and cyclical components are eliminated in the division by the moving average.

3. Arrange the ratio to moving average figures by quarters, that is, all the first quarters in one group, all the second quarters in another, and so forth.

Average these ratio to moving average figures for each quarter to eliminate the irregular movements, and thus to isolate the stable seasonal component. One type of average used for this purpose is referred to as a **modified mean**—an arithmetic mean of the ratio to moving average figures taken after dropping the highest and lowest extreme values.

4. Make an adjustment to force the four modified means to total 400, and thus average out to 100.0. The resulting four figures, one for each quarter of the year, constitute the seasonal indices for the series in question.

In symbols, this procedure may be summarized as follows. Let Y be the original quarterly observations; MA the moving average figures; and T, C, S, and I the trend, cyclical, seasonal, and irregular components, respectively. Dividing the original data by the moving average values gives

$$\frac{Y}{MA} = \frac{T \times C \times S \times I}{T \times C} = S \times I \tag{11.17}$$

Averaging these ratio to moving average figures (Y/MA) eliminates the irregular movements that tend to make the Y/MA values too high in certain years and too low in others. Hence, if the eliminations of the nonseasonal elements were perfect, the final seasonal indices would reflect only the seasonal variations. Of course, since the entire method is a rather rough and approximate procedure, the nonseasonal elements are generally not completely eliminated. The moving average usually contains the trend and *most* of the cyclical fluctuations. Therefore, the cyclical component may not be completely absent in the Y/MA values. Moreover, the modified means do not ordinarily remove all of the erratic disturbances attributed to the irregular component. Nevertheless, in the case of series with a stable seasonal pattern, the computed seasonal indices generally isolate the underlying seasonal pattern quite well.

Table 11-3 gives a quarterly series of national/regional spot TV time sales in the United States from 1975 to 1984. Examination of this series reveals that national/regional spot TV time sales tend to be highest during the second quarter and lowest during the first quarter. The calculation of quarterly seasonal indices will be illustrated in terms of this series.

The national/regional TV time sales figures have been listed in column (2) of Table 11-3 from the first quarter of 1975 through the second quarter of 1984. Our first task is the calculation of the four-quarter moving average. As indicated above, this moving average is calculated by averaging four quarters at a time, continually moving the average up by a quarter. However, because of a problem of centering of dates, a slightly different type of average, called a **two-of-a-four-quarter moving average**, is calculated. This problem is as follows. An average of four quarterly figures would be centered halfway between the dating of the second and third figures and would thus not correspond to the date of either of those figures. For example, the average of the four quarters of 1975, the first figures shown in column (2) of Table 11-3, would be centered midway between the second and third quarter dates, or at the center of the year, July 1, 1975. The original quarterly figures are centered at the middle of their respective time periods, for simplicity, say, February 15, May 15, August

A problem

15, and November 15. Hence, the dates of a simple four-quarter moving average would not correspond to those of the original data. This problem is easily solved by averaging the moving averages two at a time. For example, as we have seen, the first moving average obtainable from Table 11-3 is centered at July 1, 1975. The second moving average, which contains the last three quarters of 1975 and the first quarter of 1976, is centered at October 1, 1975. Averaging these two figures yields a figure centered at August 15, the same as the dating of the third quarter.

The solution

The easiest way to calculate this properly centered moving average is given in columns (3) through (5) of Table 11-3. Column (3) gives a four-quarter moving total. The first figure, 1,441.4, is the total of the first four quarterly figures—298.9, 383.7, 322.0, and 436.8. This figure is listed opposite the third quarter, 1975, although actually it is centered at July 1. The next four-quarter moving total is obtained by dropping the figure for the first quarter of 1975 and including the figure for the first quarter of 1976. Hence, 1,526.7 is the total of 383.7, 322.0, 436.8, and 384.2. The total of 1,441.4, and 1,526.7, or 2,968.1, is the first entry in column (4). This represents the total for the eight months that would be present in the averaging of the first two simple four-quarter moving averages. Dividing this total by eight yields the first two-of-a-four-quarter moving average figure of 371.0, properly centered at the middle of the third quarter, 1975.

Calculating

The moving averages given in column (5) of Table 11-3 are shown in Figure 11-6 along with the original data. It is useful to examine graphs in the calculation of seasonal indices, because we can observe visually what is accomplished in each major step of the procedure. We have noted earlier that the original data, if stated in monthly or quarterly form, contain all of the components of trend, cyclical, seasonal, and irregular movements. Although the period in this example is somewhat short for trend to be revealed, we can observe that national/regional spot TV time sales tend to rise throughout the entire period. The repetitive annual rhythm of the seasonal movements is clearly discernible. Irregular movements are also present. The moving average, which runs smoothly through the original data, can be observed to follow the trend movements rather closely, and if cyclical fluctuations were clearly indicated, we would be able to see how the moving average describes them as well. Another way to view this point is to note that the seasonal variations and (to a large degree) the irregular movements are absent from the smooth line that traces the path of the moving average. Note also that there are no moving average figures corresponding to the first two and the last two quarters of original data. Correspondingly, if the original data were in monthly form and a 12-month moving average were computed, there would be no moving averages to correspond to the first six months of data or to the last six months of data.

Using graphs

The ratio to moving average figures—that is, original data in column (2) divided by the moving averages in column (5) are given in column (6) of Table 11-3. As is customary, these figures have been multiplied by 100 to express them in percentage form. They are often referred to as **percentage of moving**

Percentage of moving average

TABLE 11-3

National/regional spot TV time sales in the United States by quarters, 1975–1984: computations for seasonal indices and deseasonalizing of original data

(1) Year/Quarter		(2) National/Regional Spot TV Time Sales ($ millions)	(3) Four-Quarter Moving Total	(4) Two-of-a-Four-Quarter Moving Total	(5) Moving Average (col. 4 × $\frac{1}{8}$)	(6) Original Data as Percentage of Moving Average $\left(\frac{\text{col. 2}}{\text{col. 5}} \times 100\right)$	(7) Seasonal Index	(8) Deseasonalized National/Regional Spot TV Time Sales $\left(\frac{\text{col. 2}}{\text{col. 7}} \times 100\right)$
1975	I	298.9					85.8	348.4
	II	383.7					115.2	333.1
	III	322.0	1,441.4	2,968.1	371.0	86.8	93.3	345.1
	IV	436.8	1.526.7	3,209.0	401.1	108.9	105.7	413.2
1976	I	384.2	1,682.3	3,490.7	436.3	88.1	85.8	447.8
	II	539.3	1,808.4	3,724.3	465.5	115.9	115.2	468.1
	III	448.1	1,915.9	3,869.6	483.7	92.6	93.3	480.3
	IV	544.3	1,953.7	3,904.3	488.0	111.5	105.7	514.9
1977	I	422.0	1,950.6	3,896.1	487.0	86.7	85.8	491.8
	II	536.2	1,945.5	3,905.8	488.2	109.8	115.2	465.5
	III	443.0	1,960.3	3,975.3	496.9	89.2	93.3	474.8
	IV	559.1	2,015.0	4,142.9	517.9	108.0	105.7	528.9
1978	I	476.7	2,127.9	4,338.3	542.3	87.9	85.8	555.6
	II	649.1	2,210.4	4,505.0	563.1	115.3	115.2	563.5
	III	525.5	2,294.6	4,621.0	577.6	91.0	93.3	563.2
	IV	643.3	2,326.4	4,744.3	593.0	108.5	105.7	608.6

1979	I	508.5	2,417.9	4,930.7	616.3	82.5	85.8	592.7
	II	740.6	2,512.8	5,077.1	634.6	116.7	115.2	642.9
	III	620.4	2,564.3	5,209.1	651.1	95.3	93.3	665.0
	IV	694.8	2,644.8	5,378.8	672.4	103.3	105.7	657.3
1980	I	589.0	2,734.0	5,557.7	694.7	84.8	85.8	686.5
	II	892.8	2,823.7	5,743.9	718.0	115.6	115.2	720.3
	III	710.1	2,920.2	5,919.7	740.0	96.0	93.3	761.1
	IV	791.3	2,999.5	6,114.8	764.4	103.5	105.7	748.6
1981	I	668.3	3,115.3	6,313.1	789.1	84.7	85.8	778.9
	II	945.6	3,197.8	6,500.1	812.5	116.4	115.2	820.8
	III	792.6	3,302.3	6,735.0	841.9	94.1	93.3	849.5
	IV	895.8	3,432.7	6,993.2	874.2	102.5	105.7	847.5
1982	I	798.7	3,560.5	7,246.1	905.8	88.2	85.8	930.9
	II	1,073.4	3,685.6	7,531.8	941.5	114.0	115.2	931.8
	III	917.7	3,846.2	7,760.5	970.1	94.6	93.3	983.6
	IV	1,056.4	3,914.3	7,954.7	994.3	106.2	105.7	999.4
1983	I	866.8	4,040.4	8,190.3	1,023.8	84.7	85.8	1,010.3
	II	1,199.5	4,149.9	8,360.2	1,045.0	114.8	115.2	1,041.2
	III	1,027.2	4,210.3	8,495.8	1,062.2	96.7	93.3	1,101.1
	IV	1,116.8	4,285.5	8,726.9	1,090.9	102.4	105.7	1,056.6
1984	I	942.0	4,441.4	9,024.4	1,128.1	83.5	85.8	1,097.9
	II	1,355.4	4,583.0	9,297.5	1,162.2	116.7	115.2	1,176.6
	III	1,168.8	4,714.5				93.3	1,252.7
	IV	1,248.3					105.7	1,181.0

Source: *Television/Radio Age*, page 62, New York, December 31, 1984.

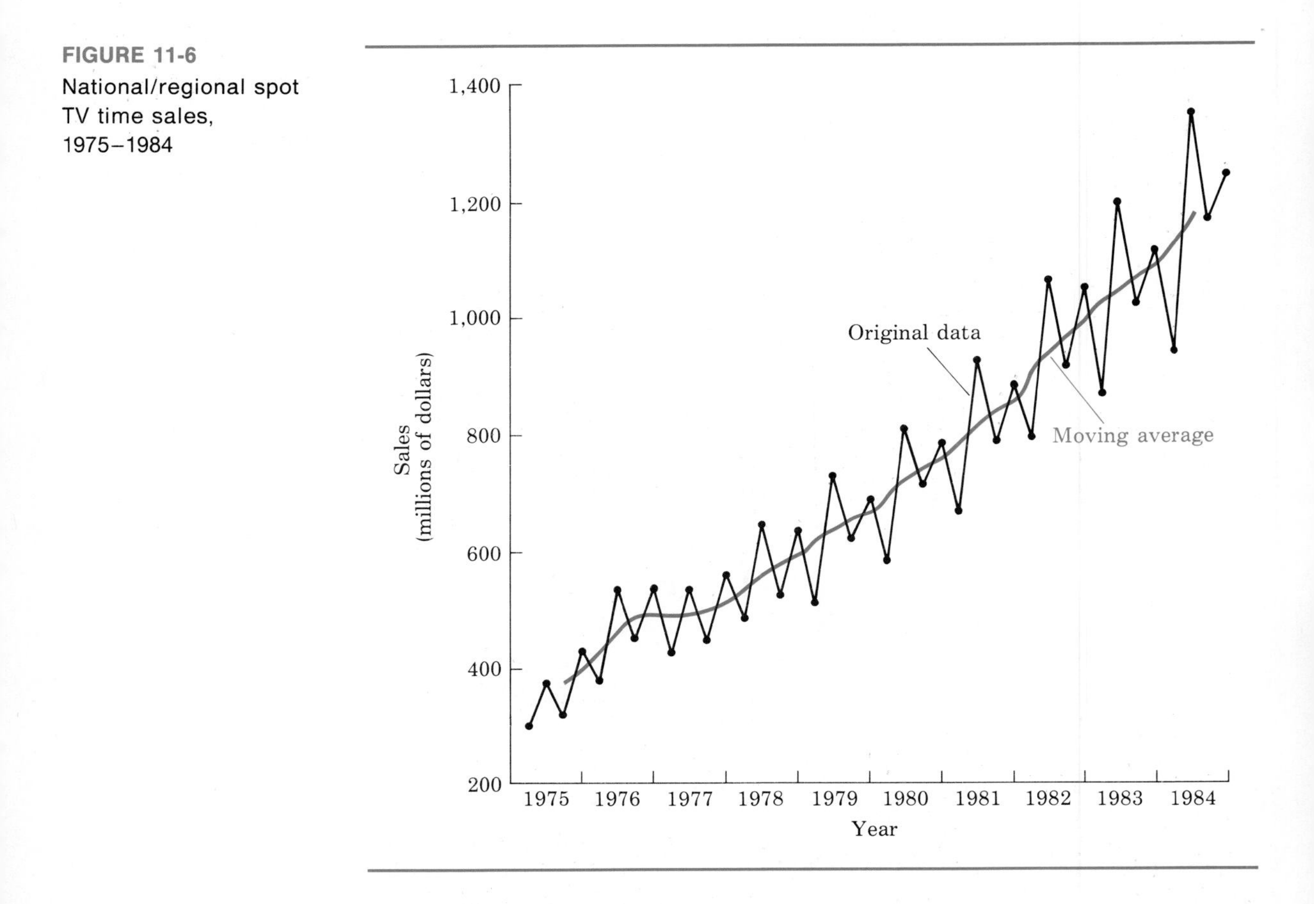

FIGURE 11-6
National/regional spot TV time sales, 1975–1984

average values and may be represented symbolically as $(Y/MA) \times 100$. These values are graphed in Figure 11-7. We can see in the graph that the trend and cyclical movements are no longer present in these figures. The 100-base line represents the level of the moving average, or the trend-cyclical base. The fluctuations above and below this base line clearly reveal the repetitive seasonal movement of the spot TV time sales. As noted earlier, the irregular component is also present in these figures.

The next step in the procedure is to remove the effect of irregular movements from the $(Y/MA) \times 100$ values. This is accomplished by averaging the percentages of moving average figures for the same quarter. That is, the first-quarter $(Y/MA) \times 100$ values are averaged, the second-quarter values are averaged, and so on. The average customarily used in this procedure is a modified mean, which is simply the arithmetic mean of the percentages of moving average figures for each quarter over the different years, after eliminating the lowest and highest figures. It is desirable to make these deletions particularly when the highest and lowest figures tend to be atypical because of erratic or irregular factors such as strikes, work stoppages, or other unusual occurrences.

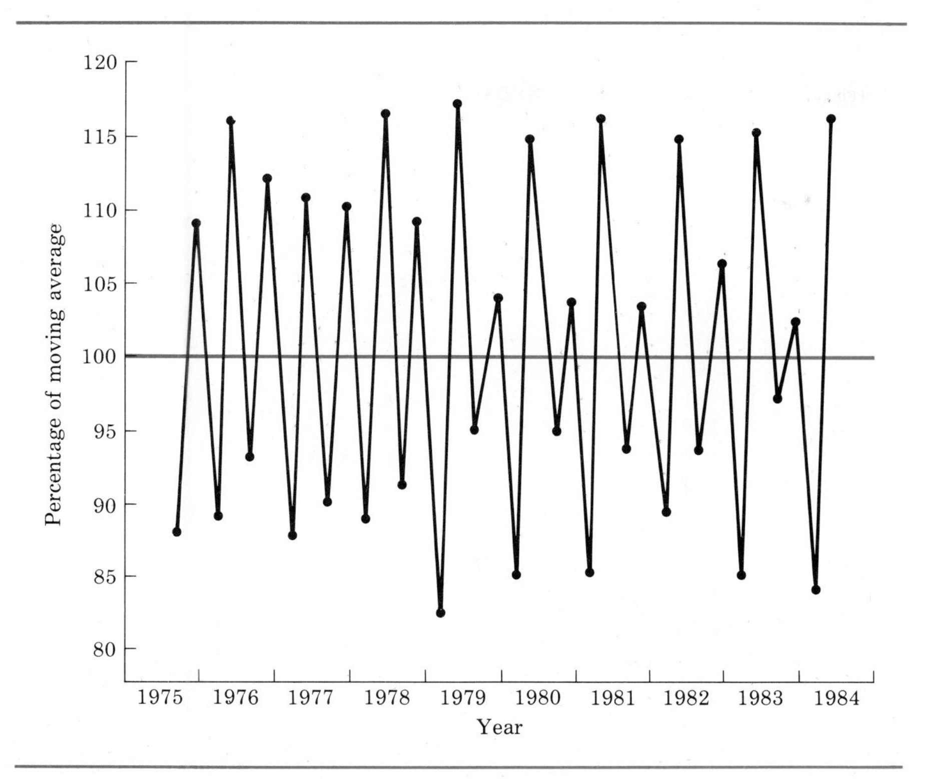

FIGURE 11-7
Percentage of moving average of national/regional spot TV time sales, 1975–1984

The percentages of moving average figures for each quarter are listed in Table 11-4. The highest and lowest figures have been deleted by a line drawn through them, and the modified means of the remaining values are shown for each quarter. These means are 85.8, 115.3, 93.3, and 105.8, respectively, for the first through fourth quarters. The total of these modified means is 400.2. Since it is desirable that the four indices total 400, in order that they average 100%, each of them is multiplied by an adjustment factor of $(\frac{400}{400.2})$. This adjustment forces a total of 400 by lowering each of the unadjusted figures by the same percentage. The final quarterly seasonal indices are shown on the bottom line of Table 11-4. In the case of monthly seasonal indices, a similar adjustment is made in order for the 12 monthly indices to total 1,200; thus, the average monthly index equals 100%.

As indicated earlier, if interest centers on the pattern of seasonal variations itself, the four quarterly indices represent the final product of the analysis. On the other hand, sometimes the purpose of measuring seasonal variations is to eliminate them from the original data in order to examine, for example, the cyclical movements. To **deseasonalize** the original data, or adjust these figures for seasonal movements, we simply divide them by the appropriate seasonal indices. This adjustment is shown in Table 11-3 for the spot TV time sales data by the division of the original figures in column (2) by the seasonal indices

Deseasonalized figures

TABLE 11-4
National/regional spot TV time sales, 1975–1984: calculation of quarterly seasonal indices from percentage of moving average figures

	Percentage of Moving Averages by Quarter			
Year	I	II	III	IV
1975			~~86.8~~	108.9
1976	88.1	115.9	92.6	~~111.5~~
1977	86.7	~~109.8~~	89.2	108.0
1978	87.9	115.3	91.0	108.5
1979	~~82.5~~	~~116.7~~	95.3	103.3
1980	84.8	115.6	96.0	103.5
1981	84.7	116.4	94.1	102.5
1982	~~88.2~~	114.0	94.6	106.2
1983	84.7	114.8	~~96.7~~	~~102.4~~
1984	83.5	~~116.7~~		
Modified means	85.8	115.3	93.3	105.8

Total of modified means = 400.2

$$\text{Adjustment factor} = \frac{400}{400.2} = 0.9995$$

Seasonal indices	I	II	III	IV
	85.8	115.2	93.3	105.7

in column (7). The result is multiplied by 100, since the seasonal index is stated as a percentage rather than as a relative.

Let us illustrate the meaning of a deseasonalized figure using the first line of figures in Table 11-3. The national/regional spot TV time sales in the first quarter of 1975 was $298.9 million. Dividing this figure by the seasonal index for the first quarter, 85.8, and multiplying by 100 yields $348.4 million. This figure is the national/regional spot TV time sales for the first quarter of 1975 adjusted for seasonal variations.

The figure, adjusted for seasonal variations, represents the level that such sales would have attained if the depressing effect of seasonality in the first quarter of the year had not been present.

All time-series components other than seasonal variations are present in these deseasonalized figures. This idea can be expressed symbolically as follows in terms of the multiplicative model of time-series analysis:

$$\frac{Y}{SI} = \frac{T \times C \times S \times I}{S} = T \times C \times I \qquad (11.18)$$

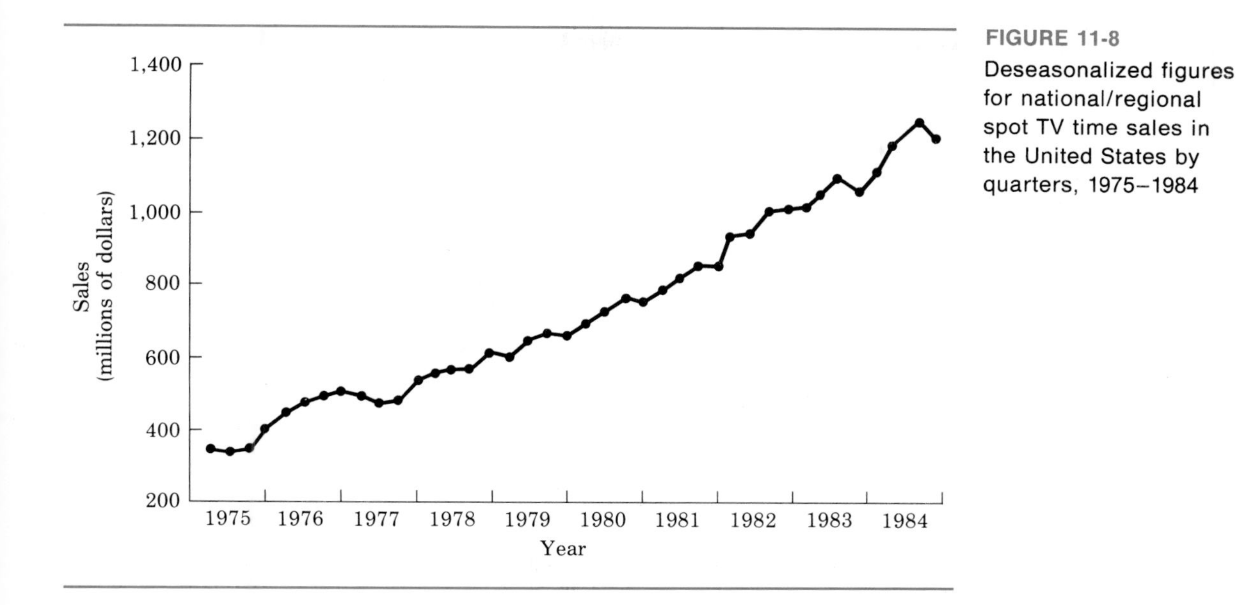

FIGURE 11-8
Deseasonalized figures for national/regional spot TV time sales in the United States by quarters, 1975–1984

The figures for national/regional spot TV time sales adjusted for seasonal movements are graphed in Figure 11-8. We can see the underlying trend and irregular movements in these data, and if cyclical fluctuations had been clearly present in the original data, we would see them as well. Note that compared with the plot of the original data in Figure 11-6, most of the repetitive seasonal movements are no longer present in the deseasonalized figures. However, ordinarily the adjustment for seasonality is not perfect, as is the case here. To the extent that seasonal indices do not completely portray the effect of seasonality, division of original data by seasonal indices will not entirely remove these influences.

Seasonal indices are often used for the purpose just discussed. Economic time series adjusted for seasonal variations are often charted in the *Federal Reserve Bulletin*, the *Survey of Current Business*, and other publications. Quarterly gross national product figures are often given as "seasonally adjusted at annual rates." These are simply deseasonalized quarterly figures multiplied by four to state the result in annual terms.

Exercises 11.5

1. Given the following data on unit sales for the Ultraponics Corporation

$$Y = \text{actual sales, February 1988} = 46{,}500$$

$$Y_t = \text{trend value, February 1988} = 50{,}000$$

$$SI = \text{February seasonal index} = 90$$

a. Express seasonally adjusted sales as a percentage of trend. What general factors account for the difference between your calculated value and 100%?
b. What is the meaning of the trend value?

2. The trend equation for sales of the Expo Corporation is

$$Y_t = 214 + 0.32x$$

where $x = 0$ in July 1983
x is in monthly intervals
Y is monthly sales in millions of dollars

Figures for certain months in 1987 are listed in a table.
a. What does the seasonal index of 101 for October mean?
b. For November 1987, isolate the effect of each component of a time series (trend, cyclical, irregular, and seasonal).

Month	Sales ($ millions)	Seasonal Index
September	190	86
October	236	101
November	268	110
December	392	165

3. The following data pertain to the number of automobiles sold by the Flakey Carl Dealership:

Number of autos sold in each quarter from 1982 to 1987

Quarter	Year 1982	1983	1984	1985	1986	1987
I	152	205	171	202	212	241
II	277	363	325	396	350	453
III	203	255	233	274	246	362
IV	174	182	180	238	208	355

a. Using the ratio to moving average method, determine constant seasonal indices for each of the four quarters.
b. Do you think constant seasonal indices should be employed in this problem? Why or why not?

c. Assuming the constant seasonal indices are appropriate, adjust the quarterly sales figures between 1982 and 1987 for seasonal variations.

4. Observe the following information for the production of Amareck, Inc., in 1988:

Month	Production (1,000 units)	Seasonal Index
January	40	110
February	38	100
March	36	90

Trend equation: $\text{Log } Y_t = 1.5500 + 0.0135x - 0.0007x^2$
$x = 0$ in December 1987
x is in one-month intervals
Y is production in thousands of units

a. Summarize in words the way in which the trend of the series is changing (based upon the equation given).
b. Would it be correct to conclude that Amareck, Inc., declined cyclically between January and March 1988? Show all relevant calculations.
c. Would you be willing to use the equation for forecasting? Discuss.

5. The sales manager of the Blue Giant Ice Cream Plant uses trend projection to estimate that annual sales of ice cream for 1988 will be 15% higher than 1987 levels. She expects that no sharp cyclical fluctuation will occur during the year, that the effect of trend within the year will be negligible, and that the past pattern of seasonal variation will continue. The 1987 quarterly sales volume of ice cream and quarterly seasonal indices are

	Spring	Summer	Fall	Winter
Ice Cream Sales (in 1,000 gallons)	25,500	31,875	25,725	20,160
Seasonal Index	90	125	105	80

a. Is the deseasonalized sales volume for the 1987 quarters consistent with the 1987 trend projection of 25 million gallons per quarter? Show calculations.
b. Prepare a forecast of quarterly ice cream sales for 1988.

6. a. Given the data tabulated for Consolidated Information Systems, compute seasonal indices for quarterly sales.
b. If 1988 sales in the fourth quarter are $10 million what would be the seasonally adjusted sales figure in that quarter?

Year	Percentage of Moving Averages by Quarter I	II	III	IV
1980	93.2	102.3	108.7	95.9
1981	94.8	101.0	109.5	94.8
1982	93.6	100.8	107.7	97.4
1983	90.1	103.4	107.8	96.5
1984	92.7	102.7	108.4	96.8
1985	94.0	101.1	105.2	93.1
1986	91.3	102.6	109.9	95.8
1987	92.2	103.1	107.9	97.7

7. The trend equation for the sales of Fleming, Inc., is

$$Y_t = 2{,}403 + 12.3x$$

where $x = 0$ in January 1984
x is in monthly intervals
Y is monthly sales in millions of dollars

Actual sales figures for certain months in 1987 follow:

Month	($ millions)	Seasonal Index
June	$2,530	86
July	3,045	101
August	3,424	110
September	4,242	165

a. In which month are sales the highest:
   (1) If they are not adjusted for seasonal variations?
   (2) If they are adjusted for seasonal variations?
b. For August 1987, isolate the effect of each component of the time series: (1) trend, (2) seasonal, (3) cyclical and irregular combined.

8. The following sentences refer to the ratio to moving average method of measuring seasonal variations when applied to a monthly series of data.
a. A 12-month moving total is computed to:
   (1) eliminate trend
   (2) make column totals equal to 1,200
   (3) cancel out seasonal variations over a period of 12 months
b. A two-item total of the 12-month total is taken to:
   (1) obtain moving average figures for the first and last six months
   (2) center the moving average properly
   (3) eliminate the rest of the random movements

c. This two-item total of a 12-month moving total is divided by:
 (1) 14 (2) 2 (3) 12 (4) 24

d. This moving average contains:
 (1) all of the trend, most of the cyclical, all of the seasonal, and some of the irregular (random) variations
 (2) all of the trend, most of the cyclical, and possibly some of the irregular (random) variations
 (3) most of the trend, most of the cyclical, and all of the irregular (random) variations

e. The original data are then divided by the moving average figures. These specific seasonal relatives (Y/MA values) contain:
 (1) all of the seasonal, possibly some of the cyclical, and almost all of the irregular (random) variations
 (2) only seasonal variations
 (3) all of the trend, most of the cyclical, and none of the irregular (random) variations

f. Modified means are taken of the ratio to moving average (Y/MA values) to:
 (1) eliminate the nonseasonal elements from the ratio to moving averages (Y/MA values)
 (2) eliminate seasonal elements
 (3) eliminate the trend in the ratio to moving averages (Y/MA values)
 (4) compensate for a changing seasonal pattern

g. To adjust the original data for seasonal variation, we compute:
 (1) original data × seasonal index
 (2) seasonal index ÷ original data
 (3) original data ÷ seasonal index

11.6 METHODS OF FORECASTING

Classical methods used in analyzing the separate components of an economic time series involve an implicit assumption that the various components act independently of one another.

For example, no specific procedures were established for taking into account cyclical influences on seasonal variations or long-run changes in the structure of business cycles. Special procedures can be established to gauge some of these interactions, but basically the model used in classical time-series analysis assumes independent sources of variation in economic time series and measures these sources separately. This *decomposition* or *separation* process, although often useful for descriptive or analytical purposes, is nevertheless artificial. Therefore, it is not surprising that for a complex problem such as economic forecasting, making mechanistic extrapolations based on classical time-series analysis alone will not suffice. However, time-series analysis frequently is a helpful starting point and an extremely useful supplement to other analytical and judgmental methods of forecasting.

Short-term forecasting

In short-term forecasting, a combined trend-seasonal projection often provides a convenient first step. For example, as a first approximation in a company's forecast of next year's sales by months, a trend projection for annual sales might be obtained. Then this figure might be allocated among months based on an appropriate set of seasonal indices. Of course, the underlying assumption in this procedure is the persistence of the historical pattern of trend and seasonal variations of this company's sales into the next year.

More complete forecasting

A more complete forecast might involve superimposing a cyclical prediction as well. For example, the first step may again involve a projection of trend to obtain an annual sales figure. Then an adjustment of this estimate may be made based on judgment of recent cyclical growth rates. Suppose that the past few years represented the expansion phase of a business cycle and the cyclical growth rate for the economy during the next year was predicted at about 4% by a group of economists. Assume further that the company in question had found in the past that these forecasts were quite accurate and applicable to the company's own cyclical growth rate—over and above its own forecast of trend levels. Then the company might increase its trend forecast by this 4% figure to obtain a trend-cyclical prediction. Again, if predictions by months are required, a monthly average could be obtained from the trend-cyclical forecast and seasonal indices could be applied to yield the monthly allocation. Ordinarily, no attempt would be made to predict the irregular movements. An alternative short-term forecasting technique that is similar to the one just described involves a projection of a 12-month moving average to obtain trend-cyclical forecasts, either monthly or as annual averages. Then seasonal indices are applied to obtain the monthly predictions.

Cyclical Forecasting and Business Indicators

Cyclical movements are more difficult to forecast than trend and seasonal elements. Cyclical fluctuations in a specific time series are strongly influenced by the general business cycle movements characteristic of large sectors of the overall economy. However, since there is considerable variability in the timing and amplitude with which many individual economic series trace out their cyclical swings, there is no simple mechanical method of projecting these movements.

Relatively "naive" methods such as the extension of the same percentage rate of increase or decrease in, say, sales for last year or during the past few years are often used. These may be quite accurate, particularly if the period for which the forecast is made occurs during the same phase of the business cycle as the periods from which the projections are made. However, the most difficult and most important items to forecast are the cyclical turning points at which reversals in direction occur. Obviously, managerial planning and implementation that presuppose a continuation of a cyclical expansion phase can give rise to serious problems if an unpredicted cyclical downturn occurs during the planning period.

Business indicators

Many statistical series produced by government and private sources have been extensively used as business indicators. Some of these series represent activity in specific areas of the economy such as employment in nonagricultural establishments or average hours worked per week in manufacturing. Others are broad measures of aggregate activity pertaining to the economy as a whole, as for example, gross national product and personal income. We noted earlier that economic series, while exhibiting a certain amount of resemblance in business cycle fluctuations, nevertheless display differences in timing and amplitude. The National Bureau of Economic Research and the U.S. Department of Commerce have studied these differences carefully and have specified a number of time series as statistical indicators of cyclical revivals and recessions.

Time series as statistical indicators

These time series have been classified into three groups.

- The first group consists of the so-called **leading series**. These series have usually reached their cyclical turning points prior to the analogous turns in general economic activity. The group includes series such as the layoff rate in manufacturing, the average workweek of production workers in manufacturing, the index of net business formation, the Standard & Poor's index of prices of 500 common stocks, and the index of homebuilding permits.
- The second group comprises **coinciding series** whose cyclical turns have roughly coincided with those of the general business cycle. Included are such series as the unemployment rate, the industrial production index, gross national product, and dollar sales of retail stores.
- Finally, the third group consists of the **lagging series**, those whose arrivals at cyclical peaks and troughs usually lag behind those of the general business cycle. This group includes series such as labor cost per unit of output, consumer installment debt, and bank interest on short-term business loans.

Note that rational explanations stemming from economic theory can be given for placing the various series into the respective groups, in addition to the empirical observations themselves. These statistical indicators are adjusted for seasonal movements. They are published monthly in *Business Conditions Digest* by the U.S. Department of Commerce. Another publication that carries the statistical indicators, as well as other time series with accompanying analyses, is *Economic Indicators*, published by the Council of Economic Advisors.

Probably the most widespread application of these cyclical indicators is as an aid in the prediction of the timing of *cyclical turning points*. If, for example, most of the leading indicators move in an opposite direction from the

Cyclical turning point

prevailing phase of the cyclical activity, this is taken to be a possible harbinger of a **cyclical turning point**. A subsequent similar movement by a majority of the roughly coincident indices would be considered a confirmation that a cyclical turn was in progress. These cyclical indicators, like all other statistical tools, have their limitations and must be used carefully. They are not completely consistent in their timing, and leading indicators sometimes give incorrect signals of forthcoming turning points because of erratic fluctuations in individual series. Furthermore, it is not possible to predict with any high degree of assurance the length of time between a signal given by the leading series group of an impending cyclical turning point and the turning point itself. There has been considerable variation in this lead time during past cycles of business activity.

Other Methods of Forecasting

Most individuals and companies engaged in forecasting do not depend on any single method, but rather utilize a variety of different approaches.

> We can place greater reliance on the consensus of a number of forecasts arrived at by relatively independent methods than on the results of any single technique.

Other methods of prediction range from informal judgmental techniques to highly sophisticated mathematical models. At the informal end of this scale, for example, sales forecasts are sometimes derived from the combined outlooks of the sales force of a company, from panels of executive opinion, or from a composite of both of these. At the other end of the scale are the formal mathematical models such as regression equations or complex econometric models. There is widespread use of various types of regression equations, by which firms attempt to predict the movements of their own company's or industry's activity on the basis of relationships to other economic and demographic factors. Often, for example, a company's sales are predicted on the basis of relationships with other series whose movements precede those of the sales series to be forecasted.

In recent years a wide variety of forecasting methods have been introduced, ranging from the relatively simple to the highly sophisticated. The availability of computers has made possible more extensive processing of data necessary in the derivation of forecasts. Although detailed discussion of forecasting techniques is outside the scope of this book, brief comments on several of these methods are given here.

Census II decomposition method

Perhaps the most detailed refinements of the classical time-series decomposition method can be found in **Census Method II**, developed by the U.S. Bureau of the Census. The elaborations of the classical decomposition method include better descriptions of trend-cyclical movements, permitting the analyst to make more accurate estimates of the combined effect of these components.

The procedure provides for the derivation of changing seasonal indices as well as for stable or constant indices. The Census method removes extremely large or small values (outliers) and smooths out irregular movements more effectively than classical decomposition. The Census method also includes tests for determining how well the decomposition has been carried out.

Box-Jenkins methods

The **Box-Jenkins methods** represent a three-stage set of procedures for modeling a time series. The first stage, known as *identification*, attempts to determine whether the time series can be described through a combination of moving average and autocorrelation terms. In the second stage, *estimation*, the time-series data are used to estimate the parameters of a tentative model. The third stage, *diagnostic testing*, consists of tests that examine the deviations from fitted models to determine the adequacy of the models. The forecasting model results from the modifications accomplished in stage three.

All time-series forecasting methods involve the processing of past data of the series to be predicted. If we project a 12-month moving average to obtain a forecast of trend-cyclical levels, past observations have been given equal weight. That is, the monthly figures are given equal weight in calculating the 12-month moving average figures. In the exponential smoothing method that is discussed at the end of this section, the most recent observations are weighted most heavily. **Generalized adaptive filtering** attempts to establish the optimal set of weights to be applied to past observations. No fixed weighting scheme is assumed. The techniques differ from those of Box-Jenkins in the methods of optimizing the parameters in the models and in the procedures used in the selection of models.

Generalized adaptive filtering

Among the most formal and mathematically sophisticated methods of forecasting in current use are econometric models. An **econometric model** is a set of two or more simultaneous mathematical equations that describe the interrelationships among the variables in the system. Some of the more complex models for prediction of movements in overall economic activity include dozens of individual equations. Special methods of solution for the parameters of these equation systems have been developed, since in many instances ordinary least squares techniques are not appropriate. Econometric models have in the past been used primarily at the first three levels in the hierarchy of forecasts, which includes the national economy, geographical regions, industries, companies, product lines, and products. However, in recent years econometric models have been used for prediction at the company level as well. Many companies, for example, have developed multiple regression equations for sales and other variables that are functions of industry and economy variables predicted by econometric models.

Econometric models

Management uses forecasting as an important ingredient of its planning, operational, and control functions. Invariably, judgment is applied to the results of various forecasting methods rather than a single method. Often, formal prediction techniques make a useful contribution by narrowing considerably the area within which intuitive judgment is applied.

Exponential smoothing

As stated earlier, all methods of forecasting involve processing past information in order to extract insights for prediction. Some methods, such as the classical time-series analysis discussed in this chapter, require extensive manipulation of past data. Other techniques, such as regression models or econometric models, require not only the use of considerable amounts of past data but also the fitting of formal mathematical models to these data. However, in some situations, we need a fast, simple, and rather mechanical way of forecasting for large numbers of items, without the computational burden of working with large quantities of data for each item for which forecasts are required. For example, in connection with inventory control, predictions of demand for thousands of items may be required for the determination of ordering points and quantities. **Exponential smoothing** is a forecasting method that meets the aforementioned requirements and is particularly appropriate for application on high-speed electronic computers. The technique simply weights together the current (most recent) actual and the current forecast figures to obtain the new forecast value. The basic formula for exponential smoothing is

$$F_t = wA_{t-1} + (1 - w)F_{t-1} \tag{11.19}$$

where F_t is the forecast for time period t
A_{t-1} is the actual figure for time period $t - 1$
F_{t-1} is the forecast for time period $t - 1$
w is a constant whose value is between 0 and 1

The time period $t - 1$ is the *current period*, and the prediction F_t is the *new forecast* for the next period. Therefore, equation 11.19 may be written in words as

$$\text{New forecast} = w(\text{Current actual}) + (1 - w)(\text{Current forecast}) \tag{11.20}$$

Note that the new forecast F_t is made in the current period $t - 1$; the current forecast F_{t-1} was made in the preceding period $t - 2$. The new forecast can be considered a weighted average of the current actual figure A_{t-1} and the forecast for the current period F_{t-1}. Another way of writing 11.19 is

$$F_t = A_{t-1} + (1 - w)(F_{t-1} - A_{t-1}) \tag{11.21}$$

In this form, the new forecast may be viewed as the current actual figure A_{t-1} plus some fraction of the most recent forecast error $(F_{t-1} - A_{t-1})$, that is, the difference between the current forecast and current actual figures.

An important advantage

An important advantage of this method of forecasting is that the only quantity that must be stored until the next period is the new forecast F_t. This modest storage requirement is an extremely useful feature for computer-based inventory management systems where, as mentioned earlier, there may be thousands of items for which forecasts are required. Despite the fact that exponential smoothing requires the storage of only one piece of data, that is, the new forecast, a bit of algebra demonstrates that each new forecast takes into account all past actual figures.

For example, we may write

$$F_2 = wA_1 + (1 - w)F_1$$

$$F_3 = wA_2 + (1 - w)F_2$$

$$F_4 = wA_3 + (1 - w)F_3$$

Substituting the first equation value for F_2 into the second equation and then the resulting value of F_3 into the third equation yields

$$F_4 = wA_3 + (1 - w)wA_2 + (1 - w)^2 wA_1 + (1 - w)^3 F_1 \quad (11.22)$$

Note that the last term of equation 11.22 is the only one that is not expressed in terms of actual figures (A_i). However, F_1 can be written to summarize the results of the actual figures A_0, A_{-1}, A_{-2}, and so on. Therefore, we can think of the forecast for any period as a weighted sum of all past actual figures, with the heaviest weight assigned to the most recent actual figure, the next highest to the second most recent one, and so on. We now see why the method is referred to as *exponential smoothing*, since the weighting factor $(1 - w)$ has successively higher exponents the farther the data recede into the past. From a management viewpoint, the fact that the most recent experience gets the greatest weight, with earlier periods getting respectively less weight, makes a great deal of sense.

Selecting the value of the weighting factor

One choice that the analyst must make in exponential smoothing is the value of the weighting factor, w. We can see from equations 11.19 and 11.22, if w is assigned a value close to zero, the most recent actual figure for, say, demand or sales will receive relatively little weight. Thus, forecasts made with a small w factor tend to have little variation from period to period. Hence, low values for w would tend to minimize the reactions of forecasts to relatively brief or erratic fluctuations in demand. This points up the basic trade-off involved in the choice of w.

If w is too low, the forecasts respond too slowly to basic long-lived changes in the demand pattern. On the other hand, if w is too high, the forecasts tend to react too sharply to random shifts in demand.

The values of w used in practice are relatively low, usually 0.3 or less, with a value of 0.1 often proving to be quite effective in a variety of forecasting situations. When policy decisions that are expected to change the demand pattern are made, appropriate changes can be made in w values.

As an example of the exponential smoothing procedure, let us examine Table 11-5. Columns (1) and (2) give the data for the weekly number of units of demand for a certain product. This series is characterized by a constant demand of 100 units per week except for a sudden sharp jump to 150 units in week 5. Columns (3) and (4) give the exponential smoothing forecasts for weight

TABLE 11-5
Exponential smoothing forecasts for weekly demand using weight factors of $w = 0.1$ and $w = 0.8$

(1) Week	(2) Actual Demand	(3) Next Forecast $w = 0.1$	(4) Next Forecast $w = 0.8$
1	$100 = A_1$	$100 = F_2$	$100 = F_2$
2	$100 = A_2$	$100 = F_3$	$100 = F_3$
3	$100 = A_3$	$100 = F_4$	$100 = F_4$
4	$100 = A_4$	$100 = F_5$	$100 = F_5$
5	$150 = A_5$	$105 = F_6$	$140 = F_6$
6	$100 = A_6$	$104.5 = F_7$	$108 = F_7$
7	$100 = A_7$	$104.05 = F_8$	$101.6 = F_8$
8	$100 = A_8$	$103.645 = F_9$	$100.32 = F_9$
9	$100 = A_9$	$103.2805 = F_{10}$	$100.064 = F_{10}$
10	$100 = A_{10}$	$102.95245 = F_{11}$	$100.0128 = F_{11}$

factors of $w = 0.1$ and $w = 0.8$, respectively. Note that the first actual demand figure A_1 is assumed to be the first forecast figure F_2. This is often the way an exponential smoothing set of forecasts is begun. Note too that the forecasts on each line of the table are the forecasts made in the given week for the following week.

As an illustration of the calculations, for example, the value of F_6 in column (3), which is 105, is derived as follows:

$$\begin{aligned} F_6 &= wA_5 + (1 - w)F_5 \\ &= (0.1)(150) + (0.9)(100) \\ &= 105 \end{aligned}$$

The effects of low and high values for the weighting factor w can be seen in Table 11-5. With the low value of $w = 0.1$, the forecasts show a sluggish response to the sudden spurt in demand to 150 units in week 5, with F_6 rising only to a value of 105. Furthermore, the effect of this spurt in actual demand lasts longer in the forecast series with $w = 0.1$ than with $w = 0.8$. With the higher value of $w = 0.8$, the forecast for F_6 jumps to 140 from the previous level of 100, and the effect dies out more rapidly than in the case of the lower weighting factor, as seen in columns (3) and (4). All the decimal places in the smoothing calculations have been shown in Table 11-5 for checking purposes. In practice, if demand can occur only in whole numbers of units, then only such integral values would be used as forecasts.

Summary

The method described above is the simplest and most basic form of exponential smoothing. It is particularly appropriate for economic time series that are relatively stable and do not have pronounced trend or seasonal movements. More complex forms of exponential smoothing are used, sometimes combined with spectral analysis (a method for determining which of various component cycles are most influential in a time series), to handle more complicated series.

Exponential smoothing techniques are often useful in situations in which a large number of economic time series must be forecast frequently.

It is a relatively simple method characterized by economy of storage and computational time.

Exercises 11.6

1. Assume that the following figures represent the daily number of units of demand for a product sold by a retail store on ten successive business days: 10, 10, 10, 20, 10, 10, 10, 10, 10, 10. Use the exponential smoothing procedure to prepare next-day forecasts using weighting factors of $w = 0.1$ and $w = 0.7$. Assume that the first actual demand figure is used as the first next-day forecast.
2. Assume that the daily number of units demanded of the product discussed in exercise 1 had been 10, 10, 10, 20, 20, 20, 20, 20, 20, 20. Again use the exponential smoothing procedure to prepare next-day forecasts using weighting factors of $w = 0.1$ and $w = 0.7$. As in exercise 1, assume that the first actual figure is used to "initialize" the set of forecasts.
3. Discuss the effects of using low versus high values of the weighting factor for the demand series given in exercises 1 and 2.
4. Exponential smoothing is used in security analysis as a mechanical means of forecasting company earnings. The accuracy of analysts' forecasts can then be compared with the accuracy of these mechanical forecasts in order to evaluate the analysts' performances. The actual earnings per share for the Independence Fabrics Corporation for 1975–1986 are listed. Use the exponential smoothing procedure to prepare next-year forecasts using a weighting factor of 0.4. Forecast earnings per share for 1987 and compare the forecasted value with the actual value of $6.28 per share. Use the first actual figure to begin the set of forecasts.

Year	Actual Earnings per Share	Year	Actual Earnings per Share
1975	$4.23	1981	$5.02
1976	4.67	1982	5.40
1977	4.64	1983	5.55
1978	4.81	1984	6.20
1979	4.72	1985	6.25
1980	4.95	1986	6.30

12

Index Numbers

In our daily lives, we often make judgments that involve summarizing how an economic variable changes with *time* or *place*. As an example of variation over time, a family's income may have increased by 40% over a five-year period. Suppose this was a period of inflation (generally rising prices). Has the family's "real income" increased? That is, can the family purchase more goods and services with its income than it could five years ago? If the general price level of items has increased more than the 40% figure, the family clearly cannot purchase as much. On the other hand, if prices have increased less than 40%, the family's real income has increased.

As an example of variation with changes in place, let us consider a company that wishes to transfer an executive from St. Louis to New York City. What should be the executive's minimum salary increase to allow for the higher cost of living in New York?

12.1 THE NEED FOR AND USE OF INDEX NUMBERS

Both cases mentioned above require measurements of general price levels. The prices of the numerous items purchased by the family have doubtless increased at different rates; a few may even have decreased. Similarly, although some prices in New York City are much higher than in St. Louis, some may be lower. How can we summarize in a single composite figure the average differences between the two time periods or the two cities? Index numbers answer questions of this type.

An **index number** is a summary measure that states a relative comparison between groups of related items. In its simplest form, an index number is nothing more than a **price relative**—a percentage figure that expresses the relationship between two numbers with one of the numbers used as the base.

Calculating price relatives

For example, in a time series of prices of a particular commodity, the prices may be expressed as percentages by dividing each figure by the price in the base period. In the calculation of economic indices, it is conventional to state these price relatives as numbers lying above and below 100, where the base period is 100%. For example, suppose the price of one pound of a certain brand of coffee was \$2.00 in 1978, \$3.22 in 1984, and \$4.04 in 1987. Then the price relatives of the three figures, with 1978 as the base (written, 1978 = 100), are

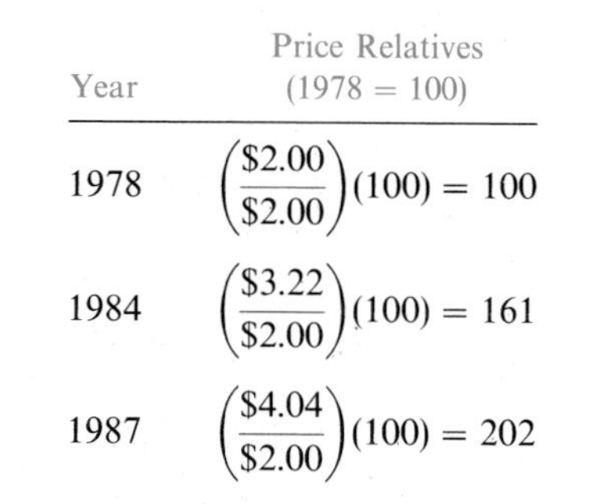

Year	Price Relatives (1978 = 100)
1978	$\left(\frac{\$2.00}{\$2.00}\right)(100) = 100$
1984	$\left(\frac{\$3.22}{\$2.00}\right)(100) = 161$
1987	$\left(\frac{\$4.04}{\$2.00}\right)(100) = 202$

As indicated, the price relative for any given year is obtained by dividing the price for that year by the base period figure. The resulting figure is multiplied by 100 to express the price relative in percentage form.

Interpreting price relatives

The price relatives may be interpreted as follows. In 1984, it would have cost 161% of the price in 1978 to purchase a pound of this brand of coffee; that is, the price increased 61% from 1978 to 1984. Similarly, in 1987, the price was 202% of the 1978 price, so the price had risen 102% from 1978 to 1987.

Composite index numbers

Of course, we are usually interested in price changes for more than one item. For example, in the cases of the family and the transferred executive cited earlier, we were interested in the prices of all commodities and services included in the cost of living; this means combining the price relatives for many different items into a single summary figure for each period. Similarly, we may wish to compute a food price index, a clothing price index, or an index of medical costs. Such summary figures constitute **composite index number** series. Since our discussion will pertain solely to composite indices, we will ordinarily use the term "index number" to mean "composite index number."

Uses for index numbers

Series of index numbers are extremely useful in the study and analysis of economic activity. Every economy, regardless of the political and social structure within which it operates, is engaged in the production, distribution, and consumption of goods and services. Convenient methods of aggregation, averaging, and approximation are required to summarize the myriad individual activities and transactions. Index numbers have proved to be useful tools in this connection. Thus, we find indices of industrial production, agricultural production, stock market prices, wholesale prices, consumer prices, prices of exports and imports, incomes of various types, and so on, in common use. Economic indices can be conveniently classified as indices of price, quantity, or value. The present discussion concentrates primarily on price indices, because most problems of construction, interpretation, and use of indices may be illustrated in terms of such measures. First, we deal with methods of index number construction, using the illustrative data of a simple example. Then we consider some general problems of index number construction.

12.2 AGGREGATIVE PRICE INDICES

In order to illustrate the construction and interpretation of price indices, we consider the artificial problem of constructing a price index for a list of four food commodities. A realistic counterpart of this problem is the Consumer Price Index (CPI) produced by the Bureau of Labor Statistics (BLS) of the U.S. Department of Labor. This index is used in many ways and provides the basis for many economic decisions. For example, fluctuations in the wages of millions of workers and the Social Security benefits of millions of people are partially based on changes that occur in the index figures. The series is also closely watched by monetary authorities as an indicator of inflationary price movements.

Unweighted Aggregates Index

In our simple illustration, we use a base period of 1980, and we are interested in the change in these prices from 1980 to 1987 for a typical family of four that purchased these products at retail prices in a certain city. The universe and other basic elements of the problem should be carefully defined. (For example, the price of a dozen eggs might be the price of a dozen Grade A, large white eggs.) However, in this problem we purposely leave these matters indefinite and concentrate on the construction and interpretation of the various indices. Table 12-1 shows the basic data of the problem and the calculation of the unweighted aggregates index. As indicated at the bottom of Table 12-1, the prices per unit are summed (or aggregated) for each year. Then one year (in this case, 1980) is selected as a base. The price index for any given year is obtained by dividing the sum of prices for that year by the sum for the base period. The resulting figure is multiplied by 100 to express the index in percentage form.

TABLE 12-1
Calculation of the unweighted aggregates index for food prices in 1980 and 1987

Food Commodity	Unit Price 1980 P_{80}	1987 P_{87}
Coffee (pound)	$2.82	$4.04
Bread (loaf)	0.89	1.19
Eggs (dozen)	0.90	1.15
Hamburger (pound)	1.90	2.39
	$6.51	$8.77

Unweighted Aggregates Index for 1987 on 1980 Base

$$\frac{\Sigma P_{87}}{\Sigma P_{80}} \cdot 100 = \frac{\$8.77}{\$6.51} \times 100 = 134.7$$

(Hence, the index takes the value 100 in the base period.) If the symbol P_0 is used to denote the price in a base period and P_n the price in a nonbase period, the general formula for the unweighted aggregates index may be expressed as follows:

Unweighted aggregates price index

$$\frac{\Sigma P_n}{\Sigma P_0} \cdot 100 \tag{12.1}$$

Let us interpret the index figure of 134.7 for 1987. In 1980, it would have cost $6.51 to purchase one pound of coffee, one loaf of bread, one dozen eggs, and one pound of hamburger. The corresponding cost in 1987 was $8.77. Expressing $8.77 as a percentage of $6.51, we find that in 1987 it would have cost 134.7% of the cost in 1980 to purchase one unit each of the specified commodities. In terms of percentage change, it would have cost 34.7% *more* in 1987 than in 1980 to purchase this "market basket" of goods.

The interpretation of the unweighted aggregates index is very straightforward. However, this type of index suffers from a serious limitation.

Limitation of unweighted aggregates index

The unweighted aggregates index is unduly influenced by the price variations of high-priced commodities.

The price totals in 1980 and 1987, respectively, were $6.51 and $8.77, an increase of $2.26. If we added to the list of commodities one that declined from $6.00 to $3.00 per unit from 1980 to 1987, the totals for 1980 and 1987 would then be $12.51 and $11.77. The price index figure for 1987 would be 94.1, indicating a 5.9% decline in prices. Although the prices of four commodities increased and only one decreased, the overall index shows a decline, because of the dominance of the price change in one high-priced commodity. Furthermore, this high-priced commodity may be relatively unimportant in the consumption pattern of the group to which the index pertains. We conclude that this so-called "unweighted index" has an inherent haphazard weighting scheme.

Another deficiency of this type of index is the arbitrary nature of the units for which the prices are stated. For example, if the price of eggs were stated per half-dozen rather than per dozen, the calculated price index figure would change. However, even if the prices of each commodity were stated in the same unit (say, per pound), the problem of the inherent haphazard weighting scheme would still remain; the index would be dominated by the commodities that happened to have high prices per pound. These may be the very commodities that are purchased least, because they are expensive. Clearly, explicit weights are needed to convert a simple aggregative index into an economically meaningful measure. We now turn to weighted aggregative price indices.

Weighted Aggregates Indices

To attribute the appropriate importance to each of the items included in an aggregative index, some reasonable weighting plan must be used. The weights to be used depend on the purposes of the index calculation, that is, on the economic question the index attempts to answer. In the case of a consumer food price index such as the one we have been discussing, reasonable weights would be the amounts of the individual food commodities purchased by the consumer units to whom the index pertains. These would constitute **quantity weights**, since they represent quantities of commodities purchased. The specific types of quantities used in an aggregative index would depend, of course, on the economic nature of the index computed. For example, an aggregative index of export prices would use quantities of commodities and services exported, whereas an index of import prices would use quantities imported.

Table 12-2 shows the prices of the same food commodities given in Table 12-1, but quantities consumed during the base period, 1980, are also shown. The figures in column Q_{80} (the symbol Q denotes quantity) represent average quantities consumed per week in 1980 by the consumer units to which the index pertains. Hence, they indicate an average consumption of one pound of coffee, three loaves of bread, and so on. The figures in the column labeled $P_{80}Q_{80}$ indicate the dollar expenditures for the quantities purchased in 1980. Correspondingly, the numbers in the column headed $P_{87}Q_{80}$ specify what it would cost to purchase these amounts of food in 1987. Therefore, the sums, $\Sigma P_{80}Q_{80} =$ \$9.19 and $\Sigma P_{87}Q_{80} =$ \$12.30, indicate what it would have cost to purchase the specified quantities of food commodities in 1980 and 1987. The index number for 1987 on a 1980 base is given by expressing the figure for $\Sigma P_{87}Q_{80}$ as a percentage of the $\Sigma P_{80}Q_{80}$ figure, yielding in this case a figure of 133.8, shown at the bottom of Table 12-2. Of course, the index number for the base period, 1980, would be 100.0.

TABLE 12-2
Calculation of the weighted aggregates index for food prices, using base period quantities consumed as weights (Laspeyres method)

Food Commodity	1980 P_{80}	1987 P_{87}	Quantity 1980 Q_{80}	$P_{80}Q_{80}$	$P_{87}Q_{80}$
Coffee (pound)	\$2.82	\$4.04	1	\$2.82	\$ 4.04
Bread (loaf)	0.89	1.19	3	2.67	3.57
Eggs (dozen)	0.90	1.15	2	1.80	2.30
Hamburger (pound)	1.90	2.39	1	1.90	2.39
		\$8.77		\$9.19	\$12.30

Weighted aggregates index, with base period weights, for 1987 on 1980 base

$$\frac{\Sigma P_{87}Q_{80}}{\Sigma P_{80}Q_{80}} \cdot 100 = \frac{\$12.30}{\$9.19} \times 100 = 133.8$$

This type of index measures the change in the total cost of a fixed market basket of goods. For example, the 133.8 figure indicates that in 1987 it would have cost 133.8% of what it cost in 1980 to purchase the weekly market basket of commodities representing a typical consumption pattern in 1980. Roughly speaking, this indicates an average price rise of 33.8% for this food market basket from 1980 to 1987. We can see that the corresponding index figure of 134.7% for the simple, or unweighted, index is quite close to the 133.8% figure for the weighted index. The reason for this is that in our example we have assumed that the prices of all four commodities moved in the same direction with the percentage changes all falling between about 25% and 45%. If there had been more dispersion in price movements (for example, if some prices increased while others decreased, as is often the case), the weighted index would have tended to differ more from the unweighted one.

The weighted aggregative index using base period weights is known as the **Laspeyres index**. The general formula for this type of index may be expressed as follows:

Weighted aggregates price index, base period weights (Laspeyres method)

$$\frac{\Sigma P_n Q_0}{\Sigma P_0 Q_0} \cdot 100 \tag{12.2}$$

where P_0 = price in a base period
P_n = price in a nonbase period
Q_0 = quantity in a base period

The Laspeyres index clearly illustrates the basic dilemma posed by the use of any weighting system. Since an aggregative price index attempts to measure price changes and contains data on both prices and quantities, it appears logical to hold the quantity factor constant in order to isolate change attributable to price movements. If both prices and quantities were permitted to vary, their changes would be entangled and it would be impossible to ascertain what part of the movement was due to price changes alone. However, by keeping quantities fixed as of the base period, the Laspeyres index assumes a frozen consumption pattern. As time goes on, this assumption becomes more unrealistic and untenable. The consumption pattern of the current period would seem to represent a more realistic set of weights from the economic viewpoint.

However, let us consider the implications of an aggregative index using current period (nonbase period) weights, which is known as the **Paasche index**. The general formula for a Paasche index is

Weighted aggregates price index, current period weights (Paasche method)

$$\frac{\Sigma P_n Q_n}{\Sigma P_0 Q_n} \cdot 100 \tag{12.3}$$

If such an index is prepared on an annual basis, the weights would change each year, since they would consist of current year quantity figures. The 1981 Paasche index would be computed by the formula $\Sigma P_{81}Q_{81}/\Sigma P_{80}Q_{81}$, the 1982 index would be $\Sigma P_{82}Q_{82}/\Sigma P_{80}Q_{82}$, and so on. The interpretation of any one of the resulting figures (as price change from the base period, assuming the consumption pattern of the current period) is clear. However, the use of changing current period weights destroys the possibility of obtaining unequivocal measures of year-to-year price change. For example, if we compare the Paasche formulas for the 1981 and 1982 indices (above), we can see that both prices and quantities have changed. Therefore, we can make no clear statement about price movements from 1981 to 1982. The use of current year weights makes year-to-year comparisons of price changes impossible.

Another practical disadvantage of using current period weights is the necessity of obtaining a new set of weights in each period. Let us consider the U.S. Bureau of Labor Statistics CPI to illustrate the difficulty of obtaining such weights. In order to obtain an appropriate set of weights for this index, the BLS conducts a massive sample survey of the expenditure patterns of families in a large number of cities. Such surveys have been carried out in 1917–1919, 1934–1936, 1950–1951, 1960–1961, and 1972–1973. They are large-scale, expensive undertakings, and it would be infeasible to conduct such surveys once a year or more frequently. Because of these disadvantages, the current period weighted aggregative method is not used in any well-known price index number series.

Because of these considerations and other factors, the most generally satisfactory type of price index is probably the weighted relative of aggregates index using a fixed set of weights. The term "fixed set of weights" rather than "base period weights" is used here, because the weights may pertain to a period different from the one that represents the base for measuring price changes. For example, one of the base periods for the CPI was 1957–1959, whereas the corresponding weights were derived from a 1960–1961 survey of consumer expenditures. The base year for the current CPI is 1967 = 100. The BLS revises its weighting system about every 10 years and also changes the reference base period for the measurement of price changes with about the same frequency. This procedure constitutes a workable solution to the need for both constant weights (in order to isolate price change) and up-to-date weights (in order to have a recent realistic description of consumption patterns).

The weighted [relative of] aggregates index using a fixed set of weights is referred to as the **fixed-weight aggregative index** and is defined by the formula

Weighted aggregates price index with fixed weights

$$\frac{\Sigma P_n Q_f}{\Sigma P_0 Q_f} \qquad (12.4)$$

where Q_f denotes a fixed set of quantity weights. The Laspeyres method may be viewed as a special case of this index, in which the period to which the weights refer is the same as the base period for prices. In order to clarify discussion of the two different periods, the term **weight base** is used for the period to which the quantity weights pertain, whereas the term **reference base** is used to designate the period from which the price changes are measured. Of course, a distinct advantage of a fixed-weight aggregative index is that the reference base period for measuring price changes may be changed without a corresponding change in the weight base. This is sometimes a useful and practical procedure, particularly in the case of some U.S. government indices that utilize data from censuses or large-scale sample surveys for changes in weights.

Exercises 12.2

1. Di Rullo's Restaurant keeps an index of the costs of major food commodities in order to ensure that menu prices are fair yet produce a profit. The average prices paid by Di Rullo's and the quantities consumed are listed for four food commodities in 1981 and 1986. Assume that suitable units were used for prices and quantities.
   a. Compute an appropriate weighted aggregates price index for 1986 on a 1981 base year (Laspeyres method).
   b. Compute an appropriate weighted aggregates price index for 1986 using 1986 weights (Paasche method).
   c. If the index of Di Rullo's menu prices was 122 in 1986 (1981 base year), what would you conclude concerning the restaurant's gross profit (revenue less cost of food) in 1986 compared to 1981? Assume that the items included in the index of food commodity prices represent a large

	1981		1986	
Food Commodity	Price	Quantity	Price	Quantity
Fish	$1.98	2,000	$2.15	2,500
Steak	5.80	1,500	6.35	1,600
Salad ingredients	0.75	4,000	0.70	4,000
Potatoes	3.00	8,000	2.65	7,500

proportion of the total food purchases, and therefore are representative of the price changes of all of the restaurant's food commodities.

2. Assume that the following is an index number series for sporting goods prices in the United States:

United States Sporting Goods Index

1975	1981	1982	1983	1984	1985
100	140	164	172	180	185

The owner of a small sporting goods shop in Longview asks you to construct an index for his store using data from a 1982 study which showed that the average store customer spent $150 on sports clothing, $100 on sports equipment, and $50 on accessories. You are also given the following data:

Commodity	1975	Average prices 1982	1985
Sports Clothes	$35.00	$50.00	$65.00
Equipment	40.00	50.00	65.00
Accessories	8.50	12.50	16.00

Using the weighted aggregates price index with base-period weights,

a. calculate the store index for 1975 and 1985, if 1982 is the base year.
b. compare the increase in price levels for the store from 1975 to 1985 with the corresponding increase determined from the U.S. index.

3. The O'Leary Metal Company uses three raw materials in its business. Listed are the average prices and quantities consumed of these three materials in 1979 and 1986.

a. Compute an appropriate weighted aggregates index for 1986 on a 1979 base.
b. A competitor reported a 1986 index for the same materials (1979 base year) of 120. Would you conclude that Jones paid more per unit in 1986 than this competitor? Why?

	1979		1986	
Materials	Price	Quantity	Price	Quantity
A	$20	$20	$25	30
B	1	100	2	120
C	5	50	6	70

12.3

AVERAGE OF RELATIVES INDICES

A second basic method of price index construction is the *average of relatives* procedure. In an **average of relatives** index, the first step involves the calculation of a price relative for each commodity by dividing its price in a nonbase period by the price in a base period. Then an average of these price relatives is calculated. As in the case of aggregative indices, an average of relatives index may be either unweighted or weighted. We consider first the unweighted indices, using the same data on prices as in the preceding section.

Unweighted Arithmetic Mean of Relatives Index

The price data previously shown in Tables 12-1 and 12-2 are repeated in Table 12-3. The first step in the calculation of any average of relatives price index is the calculation of price relatives, which express the price of each commodity as a percentage of the price in the base period. These price relatives for 1987 on a 1980 base, denoted $(P_{87}/P_{80}) \times 100$, are shown in column (4) of Table 12-3. Theoretically, once the price relatives are obtained, any average (including the arithmetic mean, median, and mode) could be used as a measure of their central tendency. The arithmetic mean has been most frequently used, doubtless because of its simplicity and familiarity. The calculation of the unweighted

TABLE 12-3
Calculation of the unweighted arithmetic mean of relatives index of food prices for 1987 on a 1980 base

(1) Food Commodity	(2) Unit Price 1980 P_{80}	(3) Unit Price 1987 P_{87}	(4) Price Relative $\frac{P_{87}}{P_{80}} \cdot 100$
Coffee (pound)	\$2.82	\$4.04	143.3
Bread (loaf)	0.89	1.19	133.7
Eggs (dozen)	0.90	1.15	127.8
Hamburger (pound)	1.90	2.39	125.8
			530.6

Unweighted arithmetic mean of relatives index for 1987, on a 1980 base

$$\frac{\Sigma\left(\frac{P_{87}}{P_{80}} \cdot 100\right)}{4} = \frac{530.6}{4} = 132.7$$

arithmetic mean of relatives for 1987 on a 1980 base appears at the bottom of Table 12-3. The formula for this unweighted arithmetic mean of relatives is equation 1.1, with the price relatives as the items to be averaged.

(12.5)

$$\frac{\Sigma\left(\frac{P_n}{P_0} \cdot 100\right)}{n}$$

Unweighted arithmetic mean of relatives index

where $\frac{P_n}{P_0} \cdot 100$ = the price relative for a commodity or service

n = the number of commodities and services

As in the case of the unweighted aggregative index, this "unweighted" index has an inherent weighting pattern. It is useful to consider the implications of this inherent weighting system. In the unweighted arithmetic mean of relatives, percentage increases are balanced against equal percentage decreases. For example, if we consider two commodities, one whose price increased by 50% and one whose price declined by 50% from 1980 to 1987, the respective price relatives for 1987 on a 1980 base would be 150 and 50. The unweighted arithmetic mean of these two figures is 100, indicating that, on the average, prices have remained unchanged. Thus, the unweighted arithmetic mean attaches the same weight to equal percentage changes in opposite directions. However, this method does not provide for explicit weighting in terms of the importance of the commodities whose prices have changed. Since it is widely recognized that explicit weighting is required to permit the individual items in an index to exert their proper influence, virtually none of the important government or private organization price indices are "unweighted." We now consider weighted average of relatives indices.

Weighted Arithmetic Mean of Relatives Indices

Although several averages can theoretically be used for calculating weighted averages of relatives, only the weighted arithmetic mean is ordinarily employed. The general formula for a weighted arithmetic mean of price relatives is

(12.6)

$$\frac{\Sigma\left(\frac{P_n}{P_0} \cdot 100\right)w}{\Sigma w}$$

Weighted arithmetic mean of relatives, general form

where w = the weight applied to the price relatives

Customarily, the weights used in this type of index are values, such as values consumed, produced, purchased, or sold. For example, in the food price index used as our illustrative problem, the weights are values consumed, that is, dollar expenditures on the individual food commodities by the typical family to whom the index pertains. It seems reasonable that the importance attached to the price change for each commodity be determined by the amounts spent on these commodities.

In index number construction, value = price × quantity.

For example, if a commodity has a price of $0.10 and the quantity consumed is three units, then the *value* of the commodity consumed is $0.10 × 3 = $0.30. Since prices and quantities can pertain to either a base period or a current period, the following systems of weights are all possibilities: P_0Q_0, P_0Q_n, P_nQ_0, and P_nQ_n. The weights P_0Q_0 and P_nQ_n are, respectively, base period values and current period values; the other two are mixtures of base and current period prices and quantities. Interestingly, the weighting systems P_0Q_0 and P_0Q_n, when used in the weighted arithmetic mean of relatives, result in indices that are algebraically identical to the Laspeyres and Paasche aggregative indices, respectively. This point is illustrated in equation 12.7, where base period weights P_0Q_0 are used.

Weighted arithmetic mean of relatives, with base period weights

$$\frac{\Sigma\left(\frac{P_n}{P_0}\right)P_0Q_0}{\Sigma P_0Q_0}\cdot 100 = \frac{\Sigma P_nQ_0}{\Sigma P_0Q_0}\cdot 100 \tag{12.7}$$

As is clear from equation 12.7, the P_0's in the numerator cancel, yielding the Laspeyres index. The calculation of the weighted arithmetic mean of relatives using 1980 base period value weights is given in Table 12-4 for the data of our illustrative problem. The numerical value of the index is, of course, exactly the same as that obtained previously for the weighted aggregative index with base period quantity weights (Laspeyres method) in Table 12-2.

Since the two indices in equation 12.7 are algebraically identical, it would seem immaterial which is used, but there are instances when it is more feasible to compute one than the other. For example, it is more convenient to use the weighted average of relatives than the Laspeyres index

- when value weights are easier to obtain than quantity weights
- when the basic price data are more easily obtainable in the form of relatives than absolute values
- when an overall index is broken down into a number of component indices and we wish to compare the individual components in the form of relatives.

As an illustration of the first situation, it is usually easier for manufacturing firms to furnish value of production weights in the form of "value added by manufacturing" (sales minus cost of raw materials) than to provide detailed data on quantities produced.

TABLE 12-4
Calculation of the weighted arithmetic mean of relatives index of food prices for 1987 on a 1980 base, using base period weights

(1) Food Commodity	(2) Prices 1980 P_{80}	(3) Prices 1987 P_{87}	(4) Price Relatives $\frac{P_{87}}{P_{80}} \cdot 100$	(5) Quantity 1980 Q_{80}	(6) $P_{80}Q_{80}$	(7) Weighted Price Relatives Col. 4 × Col. 6 $\left(\frac{P_{87}}{P_{80}} \cdot 100\right)(P_{80}Q_{80})$
Coffee (pound)	\$2.82	4.04	143.3	1	\$2.82	\$404.106
Bread (loaf)	0.89	1.19	133.7	3	2.67	356.979
Eggs (dozen)	0.90	1.15	127.8	2	1.80	230.040
Hamburger (pound)	1.90	2.39	125.8	1	1.90	239.020
					\$9.19	\$1,230.145

Weighted arithmetic mean of relatives for 1987, on 1980 base, using base period value weights

$$\frac{\Sigma\left(\frac{P_{87}}{P_{80}} \cdot 100\right)(P_{80}Q_{80})}{\Sigma P_{80}Q_{80}} = \frac{\$1,230.145}{\$9.19} = 133.8$$

As indicated earlier, the Paasche index and the weighted arithmetic mean of relatives with a P_0Q_n weighting system are algebraically identical. The reasons given for the wider use of the Laspeyres than the Paasche index apply to a similarly wider usage of weighted means of relatives with P_0Q_0 than with P_0Q_n weights. The other two possible value weighting systems, P_nQ_0 and P_nQ_n, create interpretational difficulties and therefore are not utilized in any important indices.

Exercises 12.3

1. A small electrical company produces three models of inexpensive household fans. Average unit selling prices and quantities sold in 1982 and 1986 are listed.
   a. Calculate the index of fan prices for 1982 on a base year of 1986. Use the weighted arithmetic mean of relatives method with base period weights.

Model	1982 Price	1982 Quantity (in thousands)	1986 Price	1986 Quantity (in thousands)
Economy	\$26	20	\$30	30
Model A	32	12	40	16
Model B	40	15	50	18

b. Explain, in words understandable to the lay person, precisely what the value of your index means.

2. a. Compute an index of Rondavi Wine prices for the data by the weighted arithmetic mean of relatives method with 1984 = 100, using base year weights.
   b. Compute a price index by the weighted aggregate method, with 1984 = 100, using base year weights. Compare your results in parts (a) and (b).

	Price (dollars per litre)		Production (millions of litres)	
	Green Nun	Blue Devil	Green Nun	Blue Devil
1984	$3.20	$3.05	2.45	2.60
1985	3.45	3.30	2.95	3.05
1986	3.55	3.50	2.80	3.30

3. A panel of economists studying the U.S. balance of payments with respect to a particular foreign country has assembled data on the average price and quantity imported by the United States of each of the five major export commodities of this foreign country for 1985 and 1986. Compute an index by the arithmetic mean of relatives method with 1985 as base year.

	1985		1986	
Commodity	Price	Quantity (millions of units)	Price	Quantity (millions of units)
1	$10	100	$16	200
2	6	200	14	150
3	25	50	30	100
4	20	175	15	150
5	10	125	15	250

4. A price index of two commodities is to be constructed from the listed data. A simple unweighted arithmetic mean of the two price relatives for 1986 on a 1984 base indicates that prices in 1986 were, on the average, 0.8% higher than in 1984. However, if a simple unweighted arithmetic mean of the two price relatives is calculated for 1984 on a 1986 base, the result indicates that prices in 1984 were, on the average, 0.8% higher than in 1986.
   a. How do you explain these paradoxical results?
   b. Compute what you consider to be the most generally satisfactory price index for 1986 using 1984 as a base year. You may use any of the given data you deem appropriate.

c. Explain precisely the meaning of the answer obtained in part (b).

Commodity	Unit Price		Quantities Consumed	
	1984	1986	1984	1986
A	$0.75	$0.85	6	8
B	1.70	1.50	4	5

12.4 GENERAL PROBLEMS OF INDEX NUMBER CONSTRUCTION

In a brief treatment, it is not feasible to discuss all the problems of index number construction. However, many of the important matters are included in the following two categories: (1) selection of items to be included and (2) choice of a base period.

Selection of items to be included

In the construction of price indices, as in other problems involving statistical methods, the definition of the problem and the statistical universe to be investigated are of paramount importance.

Most of the widely used price index number series are produced by government agencies or sizable private organizations and are used in many ways. Hence, it is not feasible to state a simple purpose for each price index that will clearly define the problem and the statistical universe. However, every index attempts to answer meaningful questions, and these general purposes of an index determine the specific items to be included. For example, the CPI attempts to answer a question concerning the average movement of certain prices over time. The specific nature of this question about price movements determines which items are included in the index. Similarly, many limitations of the use of the index stem from what the index does and does not attempt to measure.

Let us pursue the illustration of the CPI. Essentially, this index attempts to measure how much it would cost to purchase (at retail) a particular combination of goods and services compared with what it would have cost in a base period. More specifically, the combination of goods and services consists of items that represent a typical market basket of purchases by all urban consumers. Prior to 1978, the index covered only city wage earners and city clerical workers and their families, but in 1978 the index was expanded to include all urban consumers. The BLS continues to monitor the narrower index as well. By means of periodic consumer surveys, the BLS determines the goods and services purchased by the specified consumers and how they spread their spending among these items. The more broadly based CPI for All Urban Consumers (CPI-U) takes into account the buying patterns of professional and salaried workers, part-time workers, the self-employed, the unemployed, and retired people, in addition to wage earners and clerical workers. The most recent expenditures weights for both the CPI and the more broadly based index CPI-U were derived from the 1972–1973 Consumer Expenditures Survey.

Summary

> The general question the index purports to answer determines the items to be included.

Obviously, indices constructed for other purposes—such as indices of export prices or agricultural prices—would be based on different lists of items.

However, even when the general purpose of an index is clearly defined, many problems remain concerning the choice of items to be included. In the case of the CPI, the BLS has determined that urban consumers purchase thousands of items. However, the BLS includes only about 400 of these goods and services, having found that these few hundred accurately reflect the average change in the cost of the entire market basket. The choice of the commodities to be included in a price index is ordinarily not determined by usual sampling procedures. Each good or service cannot be considered a random sampling unit that is as representative as any other unit. Rather, an attempt is made to include practically all of the most important items and, by pricing these, to obtain a representative portrayal of the movement of the entire population of prices. If subgroup indices are required (for example, indices of food, housing, or medical care), as well as an overall consumers' price index, more items must be included than if only the overall index were desired. After the decisions have been made concerning the commodities to be included, sophisticated sampling procedures are often utilized to determine which specific prices will be included.

Choice of a base period

A second problem in the construction of a price index is the choice of a base period, that is, a period whose level of prices represents the base from which changes in prices are measured. As indicated earlier, the level of prices in the base period is taken as 100%. Price levels in nonbase periods are stated as percentages of the base period level.

Conventional calendar time with "normal" price levels

> The base period may be a conventional calendar time interval such as a month, a year, or even a period of years. Ideally, a period with "normal" price levels should be used.

Of course, it is virtually impossible to devise a meaningful definition of "normal" in almost any area of economic experience, but the time period selected should not be at or near the peaks or troughs of price fluctuations. Although there is nothing *mathematically* incorrect about using a base period with unusually low or high price levels, the use of such time intervals tends to produce distorted concepts, since comparisons are made with atypical periods.

The use of a period of years as a base produces an averaging effect on year-to-year variations. Any particular year may have unusual influences present, but if, say, a three- to five-year base period is used, these variations will tend to even out. Most of the U.S. government indices have used such time intervals (for example, 1935–1939, 1947–1949, and 1957–1959) as base periods.[1]

Another point to consider in choosing a base time interval is suggested by the three periods just mentioned:

The base period should not be too distant from the present.

Base period should not be too distant

The farther away we move from the base period, the less we know about economic conditions prevailing at that time. Consequently, comparisons with these remote periods lose significance and become rather tenuous in meaning. This is why producers of index number series, such as U.S. government agencies, shift their base periods every decade or so, in order to make comparisons with a base time interval in the recent past. Furthermore, it is desirable to shift the base from time to time because a period previously thought of as normal or average may no longer be so considered after a long lapse of time.

Other considerations

Other considerations may also be involved in choosing a base period for an index. If a number of important existing indices have a certain base period, then newly constructed indices may use the same time period for ease of comparison. Moreover, as new commodities are developed and indices are revised to include them, the base period may be shifted to a time interval that reflects the newer economic environment.

12.5 QUANTITY INDICES

The discussion in the preceding sections referred to price indices. Another important group of summary measures of economic change is represented by **quantity indices**, which measure changes in physical *quantities* such as the volume of industrial production, physical volume of imports and exports, quantities of goods and services consumed, and volume of stock transactions.

Virtually all currently used *quantity indices* measure the change in the *value* of a set of goods from the base period to the current period attributed only to changes in *quantities*, prices being held constant.

[1] A recent exception is the use of 1967 as a base. This was a year that was neither a peak nor a trough in business activity.

This corresponds to what is measured in weighted price indices as the change in the *value* of a set of goods attributed only to changes in *prices*, quantities being held constant. The same types of procedures used to calculate price indices are also employed to obtain quantity indices. Except for the case of the unweighted aggregates index, which would not be meaningful for a quantity index, corresponding quantity indices may be obtained by interchanging P's and Q's in the formulas given earlier in this chapter.

An unweighted average of relatives quantity index can be determined by establishing quantity relatives $(Q_n/Q_0) \times 100$ and calculating the arithmetic mean of these figures. An unweighted aggregative quantity index would not be meaningful, because it does not make sense to add up quantities stated in different units.

As was true for price indices, weighted quantity indices are preferable to unweighted ones. A weighted aggregative index of the Laspeyres type is given by the following formula:

Weighted relative of aggregates quantity index, base period weights (Laspeyres method)

$$\frac{\Sigma Q_n P_0}{\Sigma Q_0 P_0} \cdot 100 \tag{12.8}$$

Just as the corresponding Laspeyres price index measures the change in price levels from a base period assuming a fixed set of quantities produced or consumed in the base period, and so on, this quantity index measures the change in quantities produced or consumed, and so on, assuming a fixed set of prices that existed in the base period. As with price indices, a quantity index computed by the weighted average of relatives method, using the base period value weights given in equation 12.9, is algebraically identical to this Laspeyres index.

Weighted arithmetic mean of relatives quantity index, base period weights

$$\frac{\Sigma\left(\frac{Q_n}{Q_0} \cdot 100\right) Q_0 P_0}{\Sigma Q_0 P_0} \tag{12.9}$$

Let us interpret these two equivalent weighted indices by considering the Laspeyres version, given in equation 12.9. We continue the assumption that the raw data refer to quantities of food items consumed (during one week) and prices paid by a typical family in an urban area. The numerator of the index shows the value of the specified food items consumed in the nonbase year at base year prices. The denominator refers to the value of the food items consumed in the base year. Suppose a figure of 125 resulted from such an index. Since prices were kept constant, the increase would be solely attributable to an average increase of 25% in the quantity of these items consumed.

FRB Index of Industrial Production

Probably the most widely used and best-known quantity index in the United States is the Federal Reserve Board (FRB) Index of Industrial Production. This index measures changes in the physical volume of output of manufacturing, mining, and utilities. In addition to the overall index of industrial production, component indices are published by industry groupings such as Manufactures and Minerals, and by subcomponents such as Durable Manufactures and Nondurable Manufactures. Separate indices are reported for the output of consumer goods, output of equipment for business and government use, and output of materials. Based on the groupings used by the Standard Industrial Classification of the U.S. Bureau of the Budget, indices are also prepared for major industrial groups and subgroups. The indices are issued monthly, with 1967 as both reference base and weight base. The Index of Industrial Production is closely watched by business executives, economists, and financial analysts as a major indicator of the physical output of the economy.

The method of construction is the weighted arithmetic mean of relatives, using the base periods mentioned above. Numerous problems have had to be resolved concerning both the quantity relatives and value weights. Since many industries cannot easily provide physical output data for the quantity relatives, related data that tend to move more or less parallel to output (such as shipments and employee-hours worked) are sometimes used instead. The weights used are value-added data, which at the individual company level represent the sales of the firm minus all purchases of materials and services from other business firms. The index uses value-added rather than value of final product weights to avoid the problem of double counting. For example, if the value of the final product were used for a steel company that sells its steel to an automobile company, and the value of the final product of the automobile company were also used, the steel that went into making the automobile would be counted twice. Hence, the weights used follow the value-added approach, in which the values of so-called "intermediate products" produced at all stages prior to the final product are excluded. From the viewpoint of the economist, a firm's value added is conceptually equivalent to the total of its factor of production payments—wages, interest, rent, and profits.

Double counting

12.6 DEFLATION OF VALUE SERIES BY PRICE INDICES

One of the most useful applications of price indices is in adjusting series of dollar figures for changes in levels of prices. The result of this adjustment procedure, known as **deflation**, is a restatement of the original dollar figures in terms of so-called "constant dollars." The rationale of the procedure can be illustrated in terms of the simple example given in Table 12-5. Column (2) shows average (arithmetic mean) weekly wage figures for factory workers in a large city in 1980 and 1987. Such unadjusted dollar figures are said to be stated in "current dollars." Column (3) shows a CPI for the given city, with 1980 as reference base period. For simplicity of interpretation, let us assume that the

TABLE 12-5
Calculation of average weekly wages in 1980 constant dollars for factory workers in a large city, 1980 and 1987

(1) Year	(2) Average Weekly Wages	(3) Consumer Price Index (1980 = 100)	(4) "Real" Average Weekly Wages (1980 constant dollars)
1980	\$210.00	100	\$210.00
1987	\$352.10	151	233.18

CPI was computed by the Laspeyres method. As we note from column (2), average weekly wages of the given workers have increased from \$210.00 in 1980 to \$352.10 in 1987, a gain of 67.7%. But can these workers purchase 67.7% more goods and services with this increased income? If prices of the goods and services had remained unchanged between 1980 and 1987, then all other things being equal, the answer would be yes. But as we can see in column (3), prices rose 51% over this period. To determine what average weekly wages are in terms of 1980 constant dollars (dollars with 1980 purchasing power), we carry out the division \$352.10/151 = \$233.18. That is, we divide the 1987 weekly wage figure in current dollars by the 1987 CPI stated as a decimal figure (using a base of 1.00 rather than 100) to obtain the figure \$233.18 for average weekly wages in 1980 constant dollars. As indicated in the heading of column (4), the result of this adjustment for price change is referred to as **real wages** (in this case, "real average weekly wages"). The term *real* implies that a portion of the dollar increase in wages is absorbed by the increase in prices, so the adjustment attempts to isolate the "real change" in the volume of goods and services the weekly wages can purchase at base year prices. In summary, the dollar value figures in column (2) are divided by the price index figures in column (3) (stated on a base of 1.00) to obtain the real value figures in column (4). The same procedure would have been followed if there had been a series of figures in columns (2) and (3) (perhaps for several consecutive years) rather than just the current and base period figures.

Real wages

This division of a dollar figure by a price index is referred to as a **deflation of the current dollar value series**, whether a decrease or an increase occurs in going from figures in current dollars to constant dollars.

The rationale of the deflation procedure stems from the basic relationship value = price × quantity. The weekly wages in current dollars are value figures: they may be viewed as **value aggregates**, that is, as sums of prices of labor times quantities of such labor. By dividing such a figure by a price index, we attempt to isolate the change attributable to quantity or physical volume. Hence, we

Value aggregates

may think of the real average weekly wage figures as reflecting the changes in quantities of goods over which the wage figures have command.

This deflation procedure is widely used in business and economics. One interesting application is in measuring economic well-being and growth. For example, in comparing growth rates among countries, one of the most important indicators used is per capita growth in real GNP. Dividing GNP by population to yield per capita figures may be viewed as an adjustment for differences in population size. The division of the figures by a relevant price index to obtain real GNP is an adjustment for change in price levels. Per capita real GNP is an extremely useful measure of physical volume of production.

Of course, there are numerous limitations to the use of the deflation procedure. For example, in the weekly wages illustration, the market basket of commodities and services implicit in the consumer price index may not refer specifically to the factory workers to whom the weekly wages pertain. Even if the index had been constructed for this specific group of factory workers, it is still only an average, subject to all the interpretational problems of any measure of central tendency. Furthermore, inferences from such data must be made with care. For instance, an increase in real average weekly wages from one period to another does not imply an increase in economic welfare for the factory worker group in question; in the later period, there may be a less equitable distribution of this income, taxes may be higher (leading to lower disposable income), and so on. Nevertheless, despite such limitations and caveats, the deflation procedure is a useful, practical, and widely utilized tool of business and economic analysis.

We have seen how a price index may be used to remove from a value aggregate the change attributable to price movements. We may also view the deflation procedure as a method of adjusting a value figure for changes in the purchasing power of money. In this connection, we should note that

Purchasing power index

> a purchasing power index is conceptually the reciprocal of a price index.

For example suppose you have $40 to purchase shoes in a certain year when a pair of shoes costs $20. The $40 enables you to purchase two pairs of shoes. But if in a later year the price of shoes rises to $40, you can then purchase only one pair of shoes. Let us imagine a price index composed solely of the price of this pair of shoes. If the earlier year is the base period, the base period price index is 100 and the later period figure is 200. On the other hand, if the price of these shoes has doubled, the purchasing power of the dollar relative to shoes has halved. Hence, a purchasing power index that was 100 in the earlier base year should stand at 50 in the later period. If the indices are stated using 1.00 rather than 100 in the base period, the reciprocal relationship can be expressed as $2 \times \frac{1}{2} = 1$. That is, the doubling in the price index and the halving in the purchasing power index are reciprocals. This relationship between price and purchasing power indices is implied in such popular statements as that a dollar today is worth only 50 cents in terms of the dollar in some earlier period.

Exercises 12.6

1. a. Compute an index of the Penrite Company's fountain pen *production* (a quantity index) for the listed data by the weighted aggregates method, with 1984 = 100, using base year weights.
   b. Explain the meaning of the index you computed in part (a).

	Price		Production (in millions)	
	Elegante Model	Moderne Model	Elegante Model	Moderne Model
1984	$2.45	$2.15	1.28	1.40
1985	2.52	2.23	1.32	1.47
1986	2.60	2.41	1.31	1.52

2. a. Using the data of exercise 2 in section 12.3, compute an index of wine production by the weighted arithmetic mean of relatives method, with 1984 = 100, using base year weights.
   b. Compute a production index by the weighted aggregates method, on the same base, using base year weights.
3. Radco Electronics had gross sales of $2 million in 1982, $3.4 million in 1984, and $3.9 million in 1986. It uses the following price index of audio and video equipment and allied products as a price deflator.

	Price Index (1977 = 100)
1982	109.3
1984	115.6
1986	121.7

   By what percentage did real gross sales change from 1982 to 1986? From 1984 to 1986?
4. Assume the figures represent the Disposable Income and Consumer Price Index of a certain country. Adjust the series on disposable income so that it reflects changes in disposable income in 1980 constant dollars.

	Disposable Income ($ billions)	CPI (1975 = 100)
1970	$100	90
1975	160	100
1980	180	120
1985	220	125

5. DWS Electronics produces specialized types of electronic equipment supplies. The company asks you to assess its market share for a three-year period. Provide reasonable estimates of the company's market share using the data on industry sales of these types of electronic equipment and supplies, industry price indices, and the company's sales revenue. Note that the company's sales figures are given in 1967 dollars.

Year	Industry Sales ($000)	Industry Price Index (1967 = 100)	Company Sales in Constant 1967 Dollars ($000)
1984	19,440	257.8	3,304
1985	26,180	266.3	4,725
1986	32,333	275.5	5,901

12.7 SOME CONSIDERATIONS IN THE USE OF INDEX NUMBERS

Numerous problems arise in connection with the use of index numbers for analysis and decision purpose. A few of these are discussed below.

Shifting the base

For a variety of reasons, it is often necessary to change the reference base of an index number series from one period to another without returning to the original raw data and recomputing the entire series. This change of reference base period is usually referred to as **shifting the base.**

For example, we may want to compare several index number series computed on different base periods. Particularly if the several series are to be shown on the same graph, we may want them to have the same base period. In other situations, the shifting of a base period may simply reflect the desire to state the series in terms of a more recent period. The simple procedure for accomplishing the shift is illustrated in the following example, in which a food price index for a certain city with a reference base of 1980 is shifted to a new base period of 1986.

In Table 12-6, the original price index is shown in the first column stated on a base period of 1980. The shift to a 1986 base period is accomplished by dividing each figure in the original series by the index number for the desired new base period stated in decimal form. Hence, in this illustration, each index number on the old 1980 base is divided by 1.237, the 1986 figure stated as a

TABLE 12-6

	Food Price Index (1980 = 100)	Food Price Index (1986 = 100)
1979	97.3	78.7
1980	100.0	80.8
1981	103.7	83.8
1982	106.9	86.4
1983	109.4	88.4
1984	113.7	91.9
1985	118.3	95.6
1986	123.7	100.0
1987	129.8	104.9

decimal. Thus, the index number for 1979 shifted to the new base of 1986 is 78.7 $(\frac{97.3}{1.237})$; for 1980, the new figure is 80.8 $(\frac{100.0}{1.237})$.

Note that the relationships among the new index figures after the base is shifted are the same as in the old series. For example, the index number for 1980 exceeds that of 1979 by the same percentage in both series. That is, $(\frac{100.0}{97.3}) = (\frac{80.8}{78.7}) = 1.027$, and so on throughout the two series. However, a subtle problem arises with the weighting scheme. Let us suppose that the old series was computed using a Laspeyres type index. Hence, both the reference base and weight base periods are 1980. The procedure of dividing the series by the 1986 index number changes the reference period, but the weights still pertain to 1980. That is, the raw data originally collected for quantity weights pertained to 1980. The mere procedure of dividing the index numbers in the old series by one of its members does nothing to change these weights. Indeed, obtaining new weights for 1986 would involve a new data collection process. In summary, the new series has been shifted to a reference base period of 1986 for measuring price changes, but the weights are fixed at 1980. This point may be demonstrated algebraically as follows. Consider the Laspeyres price index for 1981 computed on the reference base of 1980. The formula for computing this figure may be written as $\Sigma P_{81}Q_{80}/\Sigma P_{80}Q_{80}$. Correspondingly, the formula for the 1986 price index on the 1980 base is $\Sigma P_{86}Q_{80}/\Sigma P_{80}Q_{80}$. (The multiplication by 100 has been dropped in these formulas to simplify the discussion.) To obtain the new price index figure for 1981 on a 1986 base, we divide the old 1981 figure by the old 1986 figure.

$$\frac{\Sigma P_{81}Q_{80}}{\Sigma P_{80}Q_{80}} \div \frac{\Sigma P_{86}Q_{80}}{\Sigma P_{80}Q_{80}} = \frac{\Sigma P_{81}Q_{80}}{\Sigma P_{86}Q_{80}}$$

Since the $\Sigma P_{80}Q_{80}$ values cancel, we see that the resulting index figure for 1981 is stated on a reference base of 1986, but the weights still pertain to 1980.

Despite the fact that weights are not changed by the procedure discussed here, this method of shifting reference bases is widely employed. It often represents the only practical way of shifting a base, and analysts ordinarily do not view as a matter of serious concern the fact that the weighting system remains unchanged.

Splicing

Sometimes an index number series is available for a period of time and then undergoes substantial revision, including a shift in the reference base period. In these cases, if we desire a continuous series going back through the period of the older series, the old and revised series must be *spliced* together. **Splicing** involves a similar arithmetic procedure to that for shifting a base. For example, suppose a price index number series was revised by inclusion of certain new products, exclusion of some old products, and change in definition of some other products. Table 12-7 shows such an old series on a reference base of 1981 and the revised series on a base of 1983. There must be an overlapping period of the old and revised series to provide for splicing them, and the period of overlap in this example is 1983. The splicing of the two series to obtain a continuous series on the new base of 1983 is accomplished by dividing each figure in the old series by the old index figure for 1983 stated in decimal form, that is, by 1.09. This restates the old series on the new base of 1983. The resulting spliced series is shown in the last column of Table 12-7. Had it been desired to state the continuous series on the old reference base of 1981, each figure in the revised index would be multiplied by 1.09.

Although the arithmetic procedure involved in splicing is simple, the interpretation of the resulting continuous series may be extremely difficult, particularly if long time periods are involved.

For example, it is difficult to specify precisely what is measured if a price index in the later period contains prices of frozen foods, clothing made from synthetic

TABLE 12-7

	Old Price Index (1981 = 100)	Revised Price Index (1983 = 100)	Spliced Price Index (1983 = 100)
1980	96.9		88.9
1981	100.0		91.7
1982	105.2		96.5
1983	109.0	100.0	100.0
1984		104.3	104.3
1985		106.8	106.8
1986		108.1	108.1
1987		110.1	110.1

fibers, television, and similar recently developed products, whereas the spliced indices for the earlier period (before these products were on the market) do not contain these products.

Despite such conceptual difficulties, splicing frequently represents the only practical method of comparing similar phenomena measured by indices over different time periods.

Quality changes

In the construction of an index such as the CPI, the basic data on prices are collected by trained investigators who price goods for which detailed specifications have been made. The same items are always priced in the same stores. However, as a result of technological and other improvements, a corresponding improvement in the quality of many commodities often occurs over time. It is difficult, and in many cases impossible, to make suitable adjustments in a price index for **quality changes**.

The artificial but practical procedure adopted by the Bureau of Labor Statistics considers a product's quality improved only if changes that increase the cost of producing the product have occurred.

For example, an automobile tire is not considered improved if it delivers increased mileage at the same cost of production. Because of such actual improvements in product quality, many analysts feel that over reasonably long periods of time, indices such as the CPI that have shown steady rises in price levels overstate the actual price increases of a fixed market basket of goods.

Uses of indices

Index number series are widely used in connection with decision making and analysis in business and government. One of the best known applications of a price index is the use of the CPI as an escalator in collective bargaining contracts. In this connection, millions of workers are covered by contracts that specify periodic changes in wage rates depending on the amount by which the CPI moves up or down. The Bureau of Labor Statistics Producers Price Index is similarly used for escalation clauses in contracts between business firms.

Much use is made of index numbers by individual companies as well as at the levels of entire industries and the overall economy. In certain industries, it is standard practice to key changes in selling prices to changes in indices of prices of raw materials and wage earnings. Assessments of past trends and current status and projection of future economic activity are made on the basis of appropriate indices. Economists follow many of the various indices in order to appraise the performance of the economy and to analyze its structure and behavior.

13

Nonparametric Statistics

Most of the methods discussed thus far have involved assumptions about the distributions of the populations sampled. For example, when certain hypothesis-testing techniques are used, it assumed that the observations are drawn from normally distributed populations. A number of useful techniques that do not make these restrictive assumptions have been developed. Such procedures are referred to as **nonparametric** or **distribution-free** tests. Many writers prefer the latter term, because it emphasizes the fact that the techniques are free of assumptions concerning the underlying population distribution. However, the two terms are generally used synonymously.

13.1 CHARACTERISTICS

Advantages

In addition to making less restrictive assumptions than the corresponding so-called "parametric" methods, nonparametric procedures are generally easy to carry out and understand. Furthermore, as is implied by the lack of underlying assumptions, they are applicable under a wide range of conditions. Many nonparametric tests are in terms of the *ranks* or *order* rather than the numerical values of the observations. Sometimes, even ordering is not required.

A "disadvantage"

However, when distribution-free procedures are applied where parametric techniques are possible, the nonparametric methods have one disadvantage. When these nonparametric procedures use ordering or ranking rather than the actual numerical values of the observations, they are ignoring a certain amount of information. As a result, nonparametric tests are somewhat less efficient than the corresponding standard tests. This means that in testing at a given level of significance, say $\alpha = 0.05$, the probability of a Type II error, β, would be greater for the nonparametric than for the parametric test.

Advocates of nonparametric tests argue that, despite the lessened efficiency of nonparametric tests, the analyst can have more confidence in these tests than in the standard ones that often require restrictive and somewhat unrealistic assumptions.

In this chapter, we consider a few simple and widely applied nonparametric techniques.

13.2 THE SIGN TEST

We saw in Chapter 7 that—in business and public administration and in social science research—the solution to many problems centers on a comparison between two different samples. In some of the hypothesis-testing techniques previously discussed, restrictive assumptions about the populations sampled were necessary. For example, the t test for the difference between two sample means assumes that the populations are normally distributed and have equal variances. Sometimes, one or both of these assumptions may be unwarranted. Furthermore, situations often exist in which quantitative measurements are impossible. In such cases, it may be possible to assign ranks or scores to the observations in each sample. In these situations, the **sign test** can be used. The name of the test indicates that the signs of observed differences (that is, positive or negative signs) are used rather than quantitative magnitudes.

We will illustrate the sign test in terms of data obtained from a panel of 60 beer-drinking consumers. Let us assume a blindfold test in which the tasters were asked to rate two brands of beer. (Wudbeiser and Diller) on a scale from 1 to 5, with 1 representing the best taste (excellent) and 5 representing the worst taste (poor) and the other scores denoting the appropriate intermediates. Table 13-1 shows a partial listing of the scores assigned by the panel members in this taste test.

Column (4) shows the signs of the difference between the scores assigned by each participant in columns (2) and (3). As indicated, a plus sign means a higher numerical score was assigned to Wudbeiser than to Diller beer, a minus sign means Diller beer was rated higher than Wudbeiser, and a zero denotes a tie score. Let us assume that the following results were obtained:

+ Scores	35
− Scores	15
0 Scores	10
Total	60

TABLE 13-1
Ranking scores assigned to taste of 2 brands of beer by a panel

(1) Panel Member	(2) Score for Wudbeiser	(3) Score for Diller	(4) Sign of Difference
A	3	2	+
B	4	1	+
C	2	4	−
D	3	3	0
E	1	2	−
⋮	⋮	⋮	⋮

Note: Best score = 1; worst score = 5. Hence, a plus sign means Diller is preferred; a minus sign means Wudbeiser is preferred.

Method

By means of the sign test, we can test the null hypothesis of no difference in rankings of the two brands of beer. More specifically, we can test the hypothesis that plus and minus signs are equally likely for the differences in rankings. If this null hypothesis were true, we would expect about equal numbers of plus and minus signs. We would reject the null hypothesis if too many of one type of sign occurred. If we use p to denote the probability of obtaining a plus sign, we can indicate the hypotheses as

$$H_0: p = 0.50$$

$$H_1: p \neq 0.50$$

Since tied cases are excluded in the sign test, the data used for the test consist of 35 pluses and 15 minuses. The problem is conceptually the same as one in which a coin has been tossed 50 times, yielding 35 heads and 15 tails, and we wish to test the hypothesis that the coin is fair. The binomial distribution is the theoretically correct one. However, we can use the large-sample method of section 7.2 consisting of the normal curve approximation to the binomial distribution. In terms of proportions, the mean and standard deviation of the sampling distribution are

$$\mu_{\bar{p}} = p = 0.50$$

$$\sigma_{\bar{p}} = \sqrt{\frac{pq}{n}} = \sqrt{\frac{(0.50)(0.50)}{50}} = 0.071$$

Assuming that the test is performed at the 5% level of significance ($\alpha = 0.05$), we would reject the null hypothesis if $z < -1.96$ or $z > 1.96$.

Since, in this problem, the observed proportion of plus signs is $\bar{p} = \frac{35}{50} = 0.70$, then

$$z = \frac{\bar{p} - p}{\sigma_{\bar{p}}} = \frac{0.70 - 0.50}{0.071} = 2.82$$

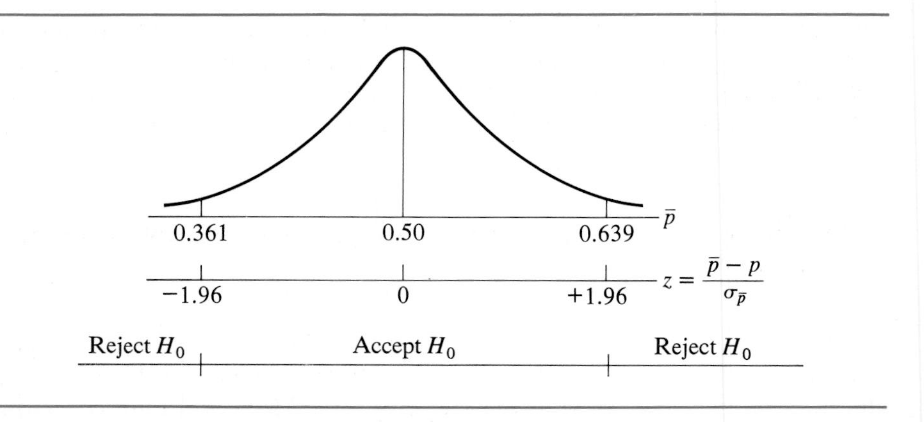

FIGURE 13-1
Sample distribution of a proportion for beer-tasting problem: $p = 0.50$, $n = 50$. This is a two-tailed test with $\alpha = 0.05$

Hence, we reject the null hypothesis that plus and minus signs are equally likely. Since the plus signs exceeded minus signs in the observed data, our interpretation of the experimental data is that Diller beer is preferred to Wudbeiser, according to the rank scores given by the consumer panel.

Note that the arithmetic could have been carried out in terms of critical limits of $\bar{p}$, rather than for z values. The critical limits for $\bar{p}$ are

$$p + 1.96\sigma_{\bar{p}} = 0.50 + (1.96)(0.071) = 0.639$$

$$p - 1.96\sigma_{\bar{p}} = 0.50 - (1.96)(0.071) = 0.361$$

Since the observed $\bar{p}$ of 0.70 exceeds 0.639, we reach the same conclusion and reject H_0. The testing procedure is shown in the usual way in Figure 13-1.

Two points may be made concerning the techniques used in this illustration of the sign test.

- Although a two-tailed test was appropriate for this problem, the sign test can also be used in one-tailed test situations.
- A normal curve approximation to the binomial distribution was used. For small samples, binomial probability calculations and tables should be used.

General Comments

As we have seen, the sign test is simple to apply. In the example, it was not feasible to obtain quantitative data for beer tasting, so rank scores were used.

> Because of its simplicity, the sign test is sometimes used instead of a standard test even when quantitative data are available.

For example, suppose we refer to the beer taste rankings for any particular individual as observations from a "matched pair." We might also have matched-pair observations when the data are, say, weights before and after a diet, grades on a preliminary scholastic aptitude test and on the regular aptitude test, or other pairs of quantitative values. In applying the sign test, only the signs of the differences in the matched-pair observations would be used, rather than the actual magnitudes of the differences. Of course, as noted earlier, some loss in the efficiency of the test would result.

In addition to being simple to apply, the sign test is applicable in a wide variety of situations.

As in the example given above, the two samples do not even have to be independent. Indeed, when matched-pair observations are used, the elements in the first sample are usually matched as closely as possible with the corresponding elements in the second sample. Furthermore, the sign test may be used in cases of qualitative classifications in which it may be difficult to use a ranking scheme such as that used in the beer-tasting problem. For example, after a treatment has been applied, an experimenter may classify subjects as improved (+), worse (−), or the same (0) and then use the sign test.

Exercises 13.2

1. The Campbell Soap company wishes to test consumer acceptance of a new, improved version of Twist bath soap. A panel of 200 consumers who were users of the old version of Twist tried the new version at home for a week. The results of the questionnaire filled out by the panel showed that 63 preferred improved Twist, 45 preferred old Twist, and 92 saw no difference between the soaps. Test the null hypothesis of no difference in consumer preference for the two versions of Twist. Use a two-tailed test at the 0.01 level of significance.
2. In a certain production process, the specified standard weight for a critical component was 13.7 pounds. In a production run of 240 of these components, 17 weighed 13.7 pounds exactly, 122 weighed more, and 101 weighed less. Assuming the sample values in excess of specifications are replaced by plus signs and those less than specifications are replaced by minus signs, test the null hypothesis that p (the proportion of pluses) equals 0.50. Use a two-tailed test at the 0.05 level of significance.
3. Thirty students at a college were asked to rank the quality of two alternative required courses on a scale of 0 to 10, with 10 as the highest score. One student assigned an 8 to the first course and a 3 to the second, while another student noted the first course 4 and the second course 8.

Course One: 8 4 3 3 9 9 9 8 7 4 6 7 7 9 7 10 4 8 9 7 8 5 5 2 8 3 5 4 7 7
Two: 3 8 4 6 6 4 5 10 7 7 6 8 7 5 9 4 5 8 2 5 9 5 6 9 9 7 8 5 2 9

Match the scores, assigning a plus sign when the rating for Course One exceeds that of Course Two and a minus sign when the rating for Course

Two is higher. Perform a two-tailed sign test at the 0.05 level of significance to test the hypothesis that the proportions of students favoring Course One and Course Two are equal to 0.50.

4. A research agency was interested in determining whether American citizens felt that there had been a decrease in the influence of the United States in world affairs over the past decade. In a simple random sample of 500 citizens, 296 felt that there had been a decrease, 150 felt that there had been an increase, and 54 thought that there had been no change. Use a minus sign to represent a perceived decrease, a plus sign for a perceived increase, and a 0 for "no change." Then apply a one-tailed test at the 0.02 significance level. Use the null hypothesis $H_0: p \leqslant 0.50$, where p is the probability of a perceived decrease in influence.

5. A marketing executive wanted to determine the opinions of the sales force on a proposed promotional program for a certain product as compared to a current program. The executive asked a randomly selected sample of 35 sales representatives to assign ratings from 0 to 10 for the effectiveness of each program, with 10 being the highest possible score. The first representative assigned a 9 to the new program and a 7 to the old, another representative rated the new program as a 7 and the old program as an 8.

Program		Program		Program	
New	Old	New	Old	New	Old
9	7	10	8	9	8
7	8	8	5	7	8
10	8	9	7	6	5
6	7	7	4	5	6
9	9	8	8	8	7
8	7	6	7	7	9
5	4	9	6	8	6
9	5	8	9	6	5
7	8	9	6	4	6
8	6	7	7	8	7
6	3	6	8	8	9
4	7	8	7		

Match the scores, assigning a plus sign when the rating for the new program exceeds that of the old, etc. Since the executive wished to place the burden of proof on the new program, it was decided to use a one-tailed test of the following type:

$$H_0: p \leqslant 0.50$$

$$H_1: p > 0.50$$

where p is the probability of obtaining a plus sign. Carry out the appropriate sign test using a 0.05 significance level.

6. A company made an improvement on a new product after experiencing a large number of customer returns of the product. The following data present the numbers of returns for six-month periods before and after the product improvement, by cities in which sales took place.
   a. Would you conclude that the product improvement resulted in a decrease in the numbers of product returns? Use a Wilcoxon matched-pairs signed rank test in a one-tailed test at the 5% significance level.
   b. Assuming that the normal distribution assumptions of the parametric t test for paired observations are satisfied, test the hypothesis of no difference between the average numbers of returns before and after the product improvement using this t test. Again, use a one-tailed test at the 5% significance level.

City	Returns before Improvement	Returns after Improvement
Atlanta	110	42
Boston	122	70
Chicago	467	301
Detroit	206	325
Los Angeles	340	283
Miami	76	38
New Orleans	134	75
New York	643	397
Philadelphia	389	227
San Francisco	291	183

13.3 THE WILCOXON MATCHED-PAIRS SIGNED RANK TEST

In section 13.2, we discussed the sign test, which is useful in testing for significant differences between paired observations. As we have seen, the sign test is carried out in terms of the signs of the differences between matched pairs of observations, without regard to the magnitudes of these differences. We turn now to another nonparametric test for significant differences between paired observations, the **Wilcoxon matched-pairs signed rank test**, which *does* take account of the magnitudes of the differences.

The Wilcoxon matched-pairs signed rank test is preferable to the sign test when the differences between paired observations can be *quantitatively* measured rather than merely assigned rankings.

Both the sign test and the Wilcoxon matched-pairs signed rank test may be considered substitutes for the analogous parametric t test for paired observations (see section 7.6). The nonparametric tests have the advantage of making no assumption about the population distribution, while the parametric t test for paired observations requires the assumption that the underlying population of differences is normally distributed.

We illustrate the Wilcoxon matched-pairs signed rank test by an example of the numerical grades on the midterm and final examinations in a certain course for 10 students selected at random from the course rolls. These data are given in Table 13-2. The Wilcoxon test is carried out in terms of the same paired differences as the parametric test, that is, $d = X_2 - X_1$, where the d values represent differences between two observations on the same individual or object. However, in the Wilcoxon test, the **absolute values** of the differences are obtained, pooled, and ranked from 1 to n, with the smallest difference being assigned the rank 1. These ranks are then given the sign (+ or −) of the corresponding value of d. If rankings are tied, the mean rank value is assigned to the tied items. (For example, in Table 13-2, because the sixth and seventh ranked items are tied, a rank of $\frac{6+7}{2} = 6.5$ is assigned to each item. The analogous procedure is used if more than two items are tied.) If the difference between the paired observations for an item is zero, as in the case of student

TABLE 13-2

Calculations for the Wilcoxon matched-pairs signed rank test of significance of differences in grades on 2 examinations for a simple random sample of 10 students

Student	Grade on Midterm Examination X_1	Grade on Final Examination X_2	Difference $d = X_2 - X_1$	Rank of $\lvert d \rvert$	Signed Rank Rank (+)	Signed Rank Rank (−)
1	75	72	−3	3		3
2	87	94	+7	6.5	6.5	
3	72	92	+20	9	9	
4	65	67	+2	2	2	
5	93	86	−7	6.5		6.5
6	85	85	0			
7	59	58	−1	1		1
8	73	79	+6	5	5	
9	64	69	+5	4	4	
10	71	82	+11	8	8	
Total (Σ)					34.5	10.5

6 in Table 13-2, that item is dropped, and the number of differences is correspondingly reduced. Since there is one such item in the present example, the effective sample size is $n = 10 - 1 = 9$.

As indicated in the last two columns of Table 13-2, the sums of the ranks are obtained separately for the positive and negative differences. These sums, denoted Σ rank $(+)$ and Σ rank $(-)$, form the basis for the null hypothesis H_0: Σ rank $(+) = \Sigma$ rank $(-)$. The null hypothesis is often referred to as one of "identical population distributions." More specifically, the hypothesis states that the population positive and negative differences are symmetrically distributed about a mean of zero. The smaller of the two ranked sums, conventionally known as **Wilcoxon's *T* statistic**, is the test statistic. Hence, in Table 13-2, the test statistic is $T = \Sigma$ rank $(-) = 10.5$.

Wilcoxon's *T* statistic

It can be shown that when n is large (at least 25), T is approximately normally distributed with the following mean and standard deviation:

$$\mu_T = \frac{n(n+1)}{4} \tag{13.1}$$

$$\sigma_T = \sqrt{\frac{n(n+1)(2n+1)}{24}} \tag{13.2}$$

Therefore, we can compute

$$z = \frac{T - \mu_T}{\sigma_T}$$

and carry out the test in the usual way.

The critical values of T are shown in Table A-9 of Appendix A, where, in keeping with the aforementioned convention, T is the smaller of the positive or negative rank sums. The table indicates that with $n = 9$ pairs, the null hypothesis of identical population distributions would be rejected at the 5% significance level using a two-tailed test at $T \leqslant 5$ ($T_{0.05} = 5$). Note that Table A-9 presents the *maximum* values that T can have and still be considered significant at the stated significance levels. In this example, since the calculated value of T (10.5) exceeds 5, the evidence on the midterm and final examinations does not permit us to reject the null hypothesis of identical population distributions. Hence, we conclude that the performances of this sample of students on the midterm and final examinations did not differ significantly.

Exercises 13.3

1. A large number of high school seniors take college Scholastic Aptitude Tests (SAT) more than once in an effort to improve their scores. The listed data represent the scores of a simple random sample of 12 students in a certain high school on the SAT mathematics test. Use the Wilcoxon matched-pairs signed rank test to determine whether the scores on the second test are significantly higher than the scores on the first test. Use $\alpha = 0.05$ in a one-tailed test.

Student	A	B	C	D	E	F	G	H	I	J	K	L
Score on first test	620	490	560	550	630	600	710	650	580	510	690	520
Score on second test	680	520	550	590	650	550	660	690	660	480	700	590

2. To determine the effectiveness of a large advertising campaign for Sudsy Soap laundry detergent, the manufacturer decided to examine the monthly sales figures before and after the promotion. The results are recorded for 10 U.S. cities. Use the Wilcoxon matched-pairs signed rank test to determine whether the advertising promotion was accompanied by an increase in number of sales. Use $\alpha = 0.05$ in a one-tailed test.

City	Sales before Promotion Campaign (thousands)	Sales after Promotion Campaign (thousands)
Los Angeles	22	30
San Francisco	16	19
Philadelphia	15	13
New York	32	28
Miami	18	17
St. Louis	10	10
Chicago	15	17
Dallas	25	28
Baltimore	17	16
Boston	9	14

3. After absorbing a large number of bad checks, a check cashing service instituted a new customer identification system. The number of bad checks for the six-month periods before and after the installation of the new system are listed by location of the service.
   a. Would you conclude that the new customer identification system resulted in a decrease in the number of bad checks? Use a Wilcoxon matched-pairs signed rank test in a one-tailed test at the 5% level.
   b. Assume that the normal distribution assumptions of the parametric t test for paired observations (discussed in section 7.6) are satisfied. Test the hypothesis of no difference between the average numbers of bad checks before and after the installation of the new system using this t test. Again, use a one-tailed test at the 5% significance level.

Location	1	2	3	4	5	6	7	8	9	10
Bad checks before new system	12	15	27	31	9	16	7	14	25	19
after new system	8	12	26	29	10	13	5	19	21	26

4. In a test involving 60 paired observations of portfolio performance before and after adoption of a new investment strategy, the lower of the two ranked sums was $T = \Sigma$ rank $(+) = 532$. Use the normal approximation to the Wilcoxon matched-pair signed rank test to determine whether the null hypothesis of identical population distributions should be rejected. Use $\alpha = 0.05$ in a two-tailed test. (You may assume that no paired observations are equal in performance.)

13.4 MANN–WHITNEY U TEST (RANK SUM TEST)

Another useful nonparametric technique involving a comparison of data from two samples is the **Mann–Whitney *U* test**.

> This procedure, often referred to as the **rank sum test**, is used to test whether two independent samples have been drawn from populations having the same *mean.*

Hence, the rank sum test may be viewed as a substitute for the parametric t test or the corresponding large-sample normal curve test for the difference between two means. The rank sum test uses more information than does the sign test in that it explicitly takes into account the rankings of measurements in each sample.

As an illustration of the use of the rank sum test, consider the data shown in Table 13-3. These data represent the grades obtained by management training program applicants on an aptitude test given by a large corporation. The samples consist of graduates of two different universities, referred to as H and W.

Method

The first step in the rank sum test is to merge the two samples, arraying the individual scores in rank order as shown in Table 13-4. The test is then carried out in terms of the sum of the ranks of the observations in either of the two

TABLE 13-3
Aptitude test grades obtained by graduates of 2 universities

H University	50	51	53	56	57	63	64	65	71	73	74	78	89	90	95
W University	70	76	77	80	81	82	83	86	87	88	92	93	96	98	99

TABLE 13-4
Array of aptitude test grades obtained by graduates of 2 universities

Rank	Grade	University	Rank	Grade	University	Rank	Grade	University
1	50	H	11	73	H	21	87	W
2	51	H	12	74	H	22	88	W
3	53	H	13	76	W	23	89	H
4	56	H	14	77	W	24	90	H
5	57	H	15	78	H	25	92	W
6	63	H	16	80	W	26	93	W
7	64	H	17	81	W	27	95	H
8	65	H	18	82	W	28	96	W
9	70	W	19	83	W	29	98	W
10	71	H	20	86	W	30	99	W

samples. The following symbolism is used:

n_1 = number of observations in sample number one

n_2 = number of observations in sample number two

R_1 = sum of the ranks of the items in sample number one

R_2 = sum of the ranks of the items in sample number two

Treating the data for H University as sample number one, we find that R_1 is the sum of ranks 1, 2, 3, 4, 5, 6, 7, 8, 10, 11, 12, 15, 23, 24, and 27, which is 158. Correspondingly, $R_2 = 307$.

If the null hypothesis that the two samples were drawn from the same population were true, we would expect the totals of the ranks (or equivalently, the mean ranks) of the two samples to be about the same. In order to carry out the test, a new statistic, U, is calculated. This test statistic, which depends only on the number of items in the samples and the total of the ranks in one of the samples, is defined as follows:

$$U = n_1 n_2 + \frac{n_1(n_1 + 1)}{2} - R_1 \tag{13.3}$$

The statistic U provides a measurement of the difference between the ranked observations of the two samples and yields evidence about the difference between the two population distributions. Very large or very small U values constitute evidence of the separation of the ordered observations of the two samples. Under the null hypothesis stated above, it can be shown that the sampling distribution of U has a mean equal to

$$\mu_U = \frac{n_1 n_2}{2} \tag{13.4}$$

and a standard deviation of

$$\sigma_U = \sqrt{\frac{n_1 n_2(n_1 + n_2 + 1)}{12}} \tag{13.5}$$

Furthermore, it can be shown the sampling distribution approaches normality rapidly and may be considered approximately normal when both n_1 and n_2 are in excess of about 10 items.

For the data in the present problem, substituting into equations 13.3, 13.4, and 13.5, we have

$$U = (15)(15) + \frac{(15)(15 + 1)}{2} - 158 = 187$$

$$\mu_U = \frac{(15)(15)}{2} = 112.5$$

$$\sigma_U = \sqrt{\frac{(15)(15)(15 + 15 + 1)}{12}} = 24.1$$

Proceeding in the usual manner, we calculate the standardized normal variate

$$z = \frac{U - \mu_U}{\sigma_U} = \frac{187 - 112.5}{24.1} = 3.09$$

Thus, if the test was originally set up as a two-tailed test at, say, the 1% significance level with a critical absolute value for z of 2.58, we would reject the null hypothesis that the samples were drawn from the same populations. On the other hand, if our alternative hypothesis predicted the direction of the difference, we would be dealing with a one-tailed test. For example, suppose our alternative hypothesis had stated that applicants from W University had a higher average aptitude test score ranking than did applicants from H University. The calculated z value would be the same as previously ($z = 3.09$), but the critical absolute value for z for a one-tailed test at the 1% significance level is 2.33. Since $3.09 > 2.33$, again we would reject the null hypothesis that the samples were drawn from the same population. Referring to the original sample data in Tables 13-3 and 13-4, we observe that W University has the higher average ranking ($R_2/n_2 = \frac{307}{15} = 20.5$ against $R_1/n_1 = \frac{158}{15} = 10.5$.). We now accept the alternative hypothesis of the one-tailed test and conclude that the population of applicants from W University has a higher average aptitude test score than does the corresponding population from H University.

General Comments

The above test was carried out in terms of the sum of the ranks for sample number one. That is, the U statistic was defined in terms of R_1. It could similarly have been defined in terms of R_2 as

$$U = n_1 n_2 + \frac{n_2(n_2 + 1)}{2} - R_2 \tag{13.6}$$

The subsequent test would have yielded the same z value as the one previously calculated, except that the sign would change. Of course, the conclusion would be exactly the same.

There were no ties in rankings in the example given. However, if such ties occur, the average rank value is assigned to the tied items. A correction is available for the calculation of σ_U when ties occur, but the effect is generally negligible for large samples.

As mentioned earlier, the rank sum test may be viewed as a substitute for the t test for the difference between two means. The rank sum test may be particularly useful in this connection, because it does not require the restrictive assumptions of the t test. For example, the t test assumption of population normality may not be valid in the preceding example.

Exercises 13.4

1. A college track star was interested in comparing his times for running the 100-yard dash during his senior and junior years. His times were recorded under essentially similar noncompetitive situations during the two years. When the times were merged for the two years and ranked, the following results were observed:

   Senior year: 1, 3, 4, 6, 8, 9, 12, 14, 16, 17, 18, 20, 23
   Junior year: 2, 5, 7, 10, 11, 13, 15, 19, 21, 22, 24, 25, 26, 27, 28

   Use the rank sum test for the null hypothesis that there is no difference between the true average times during the senior and junior years. Use $\alpha = 0.05$.

2. A market research director drew simple random samples of 15 salesmen from each of two sales regions of his company in order to compare sales figures. When last year's dollar values of sales made by these salesmen were arrayed for the two regions combined, the following rankings emerged:

   Region A: 1, 2, 4, 7, 8, 10, 12, 13, 14, 17, 21, 24, 26, 27, 28
   Region B: 3, 5, 6, 9, 11, 15, 16, 18, 19, 20, 22, 23, 25, 29, 30

   Use the rank sum test at the 0.01 level of significance to determine whether there is a significant difference in the average level of sales in the two samples.

3. The following data represent the weight losses of 28 different people during a one-week period. Half of the group used one diet and half used another diet. Use the rank sum procedure to test the hypothesis that the two samples were drawn from populations having the same average weight loss. Use $\alpha = 0.01$.

   Diet 1: 10.4 9.7 9.6 9.3 8.9 8.7 8.2 7.7 7.5 6.9 6.2 5.8 5.5 5.1
   Diet 2: 9.8 9.5 8.8 8.6 8.4 8.3 7.9 7.8 7.6 7.2 7.1 6.8 5.4 5.3

4. A simple random sample of 15 companies was drawn from each of two highly competitive industries. The capital expenditures ($ hundreds of thousands) made by the firms last year are recorded below.

Industry A: 33.3 18 38.7 48 52 30 38.4 42 25 44 36 51 35 26 40
Industry B: 46 17 24.6 24.3 37.8 39 14 23 33.8 37.1 45 13 27 21 31

Use the rank sum procedure to test the null hypothesis that there is no difference between the average levels of capital expenditures in the two industries. Use $\alpha = 0.05$.

13.5 ONE-SAMPLE TESTS OF RUNS

We have seen in Chapters 6, 7, and 8 that estimation procedures and parametric tests of hypotheses are predicated on the assumption that the observed data have been obtained from random samples. Indeed, in many instances, evidence of nonrandomness can represent an important phenomenon. As an example, in the frontier days of the Wild West, rather serious consequences were predictable if a card player questioned the randomness of the hands of cards dealt by another player. In many less exotic contexts as well, the randomness of selection of sampled items is of considerable import.

Let us consider a rather oversimplified situation. Suppose that in a certain city, the rolls of persons eligible for jury duty consisted of about 50% people 40 or older and 50% people under 40. Further, let us assume that the following sequence represents the order in which the first 48 persons were drawn from the rolls (B = below 40, F = 40 or older).

BBBBBBBBBBBB FFFFFFFFFFFF BBBBBBBBBBBB FFFFFFFFFFFF

On an intuitive basis, would you question the randomness of selection of these persons? Undoubtedly your answer is in the affirmative, but why? Note that there are 24 B's and 24 F's. Hence, the observed proportion of below-40's (and the proportion of 40-or-older's) is 50%, which does not differ from the known population proportion. Your suspicions concerning nonrandomness doubtless stem from the order of the items listed, rather than from their frequency of occurrence. Similarly, we would find a perfectly alternating sequence, BFBFBFBF . . . , suspect with respect to randomness of order of occurrence. The **theory of runs** has been developed to test samples of data for randomness, with emphasis on the *order* in which these events occur.

A **run** is defined as a sequence of identical occurrences (symbols) that are followed and preceded by different occurrences (symbols) or by none at all. Hence, in the listing of 48 symbols, there are four runs, the first run consisting of the first 12 B's, the second run consisting of 12 F's, and so on. Our intuitive feeling is that this represents too few runs. Analogously, in the perfectly alternating series, BFBFBFBF . . . , we would feel that there are too many runs to have occurred on the basis of chance alone.

Method

We illustrate the analytical procedure for the test of runs in terms of a somewhat less extreme illustration than the jury selection example. Let us assume that the following 42 symbols represent the successive occurrences of births of males (M) and females (F) in a certain hospital.

MM F M FFF MM FF M F MMM FF M FFF MM FF MM FF MMM FF MM FF MMM

Using the symbol r to denote the number of runs, we have $r = 21$. The runs (which, of course, may be of differing lengths) have been indicated by separation of sequences. As we have noted, if there are too few or too many runs, we have reason to doubt that their occurrences are random. The runs test is based on the idea that if there are n_1 symbols of one type and n_2 symbols of a second type, and r denotes the total number of runs, the sampling distribution of r has a mean of

$$\mu_r = \frac{2n_1n_2}{n_1 + n_2} + 1 \tag{13.7}$$

and a standard deviation of

$$\sigma_r = \sqrt{\frac{2n_1n_2(2n_1n_2 - n_1 - n_2)}{(n_1 + n_2)^2(n_1 + n_2 - 1)}} \tag{13.8}$$

If either n_1 or n_2 is larger than 20, the sampling distribution of r is closely approximated by the normal distribution. Hence, we can compute

$$z = \frac{r - \mu_r}{\sigma_r}$$

and proceed with the test in the usual manner.

In the present problem, where there are $n_1 = 22$ Ms, $n_2 = 20$ Fs, and $r = 21$, we have

$$\mu_r = \frac{(2)(22)(20)}{22 + 20} + 1 = 21.95$$

and

$$\sigma_r = \sqrt{\frac{(2)(22)(20)[(2)(22)(20) - 22 - 20]}{(22 + 20)^2(22 + 20 - 1)}} = 3.19$$

Therefore,

$$z = \frac{21 - 21.95}{3.19} \doteq -0.30$$

Testing at a significance level of, say, 5%, where a critical absolute value of 1.96 for z is required for rejection of the null hypothesis, we find that the randomness hypothesis cannot be rejected. In other words, the number of runs is neither small enough nor large enough for us to conclude that the sequence of male and female births is nonrandom.

General Comments

The runs test has many applications, including cases in which the sequential data are *numerical* in form rather than *symbolical* representations of attributes such as the letters used in the preceding illustrations.

- runs tests could be applied to sequences of random numbers, such as those in Table 4-1 on page 181. Such tests might be applied to sequences of random numbers generated on computers.
- One form of the test might be in terms of runs of numbers above the median and runs of numbers below the median. For the digits 0, 1, 2, 3, 4, 5, 6, 7, 8, and 9, the median is 4.5. Hence, runs could be determined for digits that fall above and below the median.
- The test could also be applied in terms of runs of odd-numbered digits and runs of even-numbered digits.
- Another alternative is to group the numbers into pairs of digits, so that the possible occurrences are 00, 01, . . . , 99. Here the median is 49.5, and runs tests similar to those suggested for the one-digit case could be applied.

Clearly, with a bit of imagination, an analyst can devise many useful and easily applied versions of the runs test.

Exercises 13.5

1. The following figures represent the monthly numbers of on-the-job accidents occurring in a certain factory over a 24-month period:

   2 2 1 3 4 4 1 3 1 2 4 4
   2 3 3 2 5 6 4 5 4 6 5 5

   Determine the median of this set of 24 figures. Label the numbers above and below the median as a and b, respectively. Perform a runs test at the 0.01 level of significance on the series of a's and b's. This type of test of runs above and below the median is particularly useful for determining the existence of trend patterns in data. If there is a trend, more a's will tend to appear in the early part of the series, and b's in the later part, or vice versa.

2. Toss a coin 40 times and record heads and tails as H and T, respectively. Test these data for randomness at the 0.05 level of significance.
3. In the table of random numbers (Table 4-1 on page 181), consider the 50 digits in the first line. Label the even digits *e* and the odd digits *o*. Carry out a runs test at both the 0.05 and 0.01 levels of significance.
4. Consider the same 50 digits as in Exercise 3. Now label the digits *a* or *b*, depending upon whether they are above or below the theoretical mean (4.5). Carry out a runs test at both the 0.05 and 0.01 levels of significance.
5. The one-sample runs test is sometimes used to test series of stock price changes for evidence of randomness to support a random walk theory. A certain stock has shown the following behavior during the past 25 business days: + + + − − + − − − + + − + − + − − + + − + + − + − (+ indicates a price increase, and − indicates a price decrease). Carry out a runs test at both the 0.05 and 0.01 levels of significance. Would you consider the stock price changes to be random events?

13.6 KRUSKAL-WALLIS TEST

In section 8.3, we noted that the one-factor analysis of variance represents an extension of the two-sample test for means and provides a test for whether several independent samples can be considered to have been drawn from populations having the same mean. Analogously, the Kruskal-Wallis one-factor analysis of variance by ranks is a nonparametric test that represents a generalization of the two-sample Mann-Whitney U rank sum test. The parametric analysis of variance discussed in section 8.3 assumes that the populations are normally distributed; otherwise, the F-test procedure is invalid. The Kruskal-Wallis test makes no assumptions about the population distribution.

The Kruskal-Wallis test is based on a test statistic calculated from ranks established by pooling the observations from c independent simple random samples, where $c > 2$. The null hypothesis is that the populations are identically distributed or alternatively, that the samples were drawn from c identical populations. We illustrate the procedure for the test by the following example.

Simple random samples of corporate treasurers in a certain industry were drawn from firms classified into three size categories (large, medium, and small). These executives, after being assured of the confidentiality of their replies, were asked to rate the overall quality of the Federal Reserve Board's performance in discount rate policy during the past six-month period on a scale from 0 to 100, with 0 denoting the lowest quality rating and 100 denoting the highest. The scores, classified by size of firm, and the rankings of the pooled sample scores are shown in Table 13-5. The result was the following pooled ranking, with the lowest score that was actually given represented by rank 1 and the

TABLE 13-5
Calculations for the Kruskal-Wallis one-factor analysis of variance: scores and ranks classified by size of firm

Large Firms (group 1)		Medium-sized Firms (group 2)		Small Firms (group 3)	
Score	Rank	Score	Rank	Score	Rank
78	12	68	6	82	14
95	20	77	11	65	5
85	16	84	15	50	1
87	17	61	3	93	19
75	10	62	4	70	7
90	18	72	8	60	2
80	13			73	9
$n_1 = 7$		$n_2 = 6$		$n_3 = 7$	
$R_1 = 106$		$R_2 = 47$		$R_3 = 57$	

highest by rank $n = 20$ (where n denotes the total number of pooled sample observations):

Score:	50	60	61	62	65	68	70	72	73	75	77	78	80	82	84	85	87	90	93	95
Rank:	1	2	3	4	5	6	7	8	9	10	11	12	13	14	15	16	17	18	19	20

The test statistic involves a comparison of the variation of the ranks of the sample groups. The Kruskal-Wallis test statistic is

$$K = \frac{12}{n(n+1)}\left(\Sigma \frac{R_j^2}{n_j}\right) - 3(n+1) \tag{13.9}$$

where n_j = the number of observations in the jth sample

$n = n_1 + n_2 + \cdots + n_c$

= the total number of observations in the c samples

R_j = the sum of the ranks for the jth sample

The sample sizes and rank sums are shown in Table 13-5 for each sample group. Substituting into equation 13.9, we compute the K statistic in the present example.

$$K = \frac{12}{20(20+1)}\left(\frac{106^2}{7} + \frac{47^2}{6} + \frac{57^2}{7}\right) - 3(20+1) = 6.641$$

It can be shown that the sampling distribution of K is approximately the same as the χ^2 distribution with $v = c - 1$ degrees of freedom (where c is the number of sample groups). In this example, where there are three sample groups, the number of degrees of freedom is $v = c - 1 = 3 - 1 = 2$. Testing the null hypothesis at the 5% level of significance ($\alpha = 0.05$) and using Table A-7 of Appendix A, we find the critical value of χ^2 to be $\chi^2_{0.05} = 5.991$. Hence, our decision rule is

If $K > 5.991$, reject the null hypothesis

If $K \leqslant 5.991$, accept the null hypothesis

Since $K = 6.641$ is greater than the critical value of 5.991, we reject the null hypothesis of identically distributed populations. Therefore, we conclude that there are significant differences by size of firm in the scores assigned by these three samples of corporate treasurers. Looking back at the scores given in Table 13-5, we find that treasurers of large firms tended to assign higher scores than did their counterparts in medium-sized or small firms.

Further Remarks

As in other nonparametric tests, when there are ties, observations are assigned the mean of the tied ranks. In the case of ties, a corrected K value K_c should be computed as follows:

$$K_c = \frac{K}{1 - \left[\dfrac{\Sigma(t_j^3 - t_j)}{(n^3 - n)}\right]} \tag{13.10}$$

where t_j is the number of tied scores in the jth sample.

Furthermore, for the χ^2 distribution to be applicable, the sample sizes (that is, the n_j values) should all be greater than five.

Exercises 13.6

1. The numbers of units produced by simple random samples of workers in three different plants are given in the table. The employees produced the same product in the same number of hours and under essentially the same conditions. Use the Kruskal–Wallis analysis of variance by ranks test to determine whether the three populations may be considered to be identically distributed. Use $\alpha = 0.05$.

Number of Units Produced		
Plant A	Plant B	Plant C
68	97	104
92	116	125
120	121	65
111	117	101
119	72	88
74	82	81
85	110	108
105		114

2. A manufacturer purchased large batches of a product from four subcontractors. There were the same number of articles in each batch. The table gives the numbers of defective articles in each batch. Would you conclude that there was no difference among subcontractors in numbers of defectives per batch? Use the Kruskal–Wallis analysis of variance by ranks test at a 1% significance level.

Number of Defectives per Batch			
Subcontractor A	Subcontractor B	Subcontractor C	Subcontractor D
12	30	15	18
6	28	17	27
10	7	20	13
0	25	3	22
2	24	19	16
4	29	21	23
	31	8	
	14		

3. In a test, the marketing research department of a firm sent three differently designed advertisements to equal-sized simple random samples of potential customers in six cities. The numbers of units of sales that resulted from business reply cards attached to these advertisements are as follows:

	Numbers of Units Sold		
City	Design A	Design B	Design C
1	38	64	55
2	59	75	82
3	30	36	80
4	52	77	66
5	61	69	73
6	43	67	47

a. Using the Kruskal–Wallis analysis of variance by ranks, test the hypothesis of no difference in effectiveness among the three advertisement designs in terms of numbers of sales that resulted. Use $\alpha = 0.01$.

b. Assuming that the normal distribution assumptions of analysis of variance are satisfied, use the parametric analysis of variance discussed in section 8.3 to test the hypothesis of no difference in average numbers of sales resulting from the three advertisements. Use $\alpha = 0.01$.

13.7

RANK CORRELATION

Nonparametric procedures can be useful in correlation analysis when the basic data are not available in the form of numerical magnitudes but rankings can be assigned. If two variables of interest can be ranked in separate ordered series, a **rank correlation coefficient** can be computed; this is a measure of the degree of correlation that exists between the two sets of ranks. We illustrate the method in terms of a simple random sample of individuals for whom rankings have been established for two variables concerning ability in two different sports activities.

Method

For illustrative purposes, we consider two extreme cases, the first case representing perfect *direct* correlation between two series, the second case representing perfect *inverse* correlation. Table 13-6 displays data on the rankings of a simple random sample of 10 individuals according to playing abilities in baseball and tennis. Clearly, this represents a case in which it would be extremely difficult, if not impossible, to obtain precise quantitative measures of these abilities, but in which rankings may be feasible. In rank correlation analysis, the rankings

TABLE 13-6

Rank correlation of baseball-playing ability with tennis-playing ability (perfect correlation case)

Individual	Rank in Baseball Ability X	Rank in Tennis Ability Y	Difference in Ranks $d = X - Y$	$d^2 = (X - Y)^2$
A	1	1	0	0
B	2	2	0	0
C	3	3	0	0
D	4	4	0	0
E	5	5	0	0
F	6	6	0	0
G	7	7	0	0
H	8	8	0	0
I	9	9	0	0
J	10	10	0	0
Total				0

$$r_r = 1 - \frac{6\Sigma d^2}{n(n^2 - 1)} = 1 - \frac{6(0)}{10(10^2 - 1)} = 1$$

may be assigned in order from high to low (with one representing the highest rating, two the next highest, and so on) or from low to high (with one representing the lowest rank, two the next lowest, and so on). The computed rank correlation coefficient will be the same, regardless of the rank ordering used. Let us assume in this case that one represents the highest or best rank, two the second highest, and so on.

The rank correlation coefficient (also referred to as the Spearman rank correlation coefficient) can be derived mathematically from one of the formulas for r, the sample correlation coefficient discussed in Chapter 9, where ranks are used for the observations of X and Y. We will use the symbol r_r to denote the rank correlation coefficient, computed by the following formula:

Computing the rank correlation coefficient

$$r_r = 1 - \frac{6\Sigma d^2}{n(n^2 - 1)} \tag{13.11}$$

where d = difference between the ranks for the paired observations
n = number of paired observations

The calculations of the rank correlation coefficients for the two extreme cases mentioned earlier are shown in Tables 13-6 and 13-7. In Table 13-6, there is perfect direct correlation in the rankings. That is, the individual who ranks highest in baseball playing ability is also best in tennis, and so on. On the

TABLE 13-7
Rank correlation of baseball-playing ability with tennis-playing ability (perfect inverse correlation case)

Individual	Rank in Baseball Ability X	Rank in Tennis Ability Y	Difference in Ranks $d = X - Y$	$d^2 = (X - Y)^2$
A	1	10	−9	81
B	2	9	−7	49
C	3	8	−5	25
D	4	7	−3	9
E	5	6	−1	1
F	6	5	1	1
G	7	4	3	9
H	8	3	5	25
I	9	2	7	49
J	10	1	9	81
Total				330

$$r_r = 1 - \frac{6\Sigma d^2}{n(n^2 - 1)} = 1 - \frac{6(330)}{10(10^2 - 1)} = -1$$

other hand, in Table 13-7, there is perfect inverse correlation. That is, the individual who ranks highest in baseball playing ability is worst in tennis, and so on. Note from the calculations shown in the two tables that in the case of perfect direct correlation between the ranks, $r_r = 1$; in perfect inverse correlation, $r_r = -1$. This is not surprising, because the rank correlation coefficient is derived mathematically from the sample correlation coefficient r. Hence, the range of possible values of these coefficients is the same. An r_r value of zero would analogously indicate no correlation between the rankings. Tied ranks are handled in the calculations by averaging in the usual way.

The significance of the rank correlation may be tested in the same way as for the sample correlation coefficient r. That is, we compute the statistic

$$t = \frac{r_r}{\sqrt{(1 - r_r^2)/(n - 2)}} \tag{13.12}$$

which has a t distribution for $n - 2$ degrees of freedom. For example, suppose in a situation such as the one above, r_r had been computed to be 0.90. Then substitution into (13.12) would yield

$$t = \frac{0.90}{\sqrt{(1 - 0.81)/(10 - 2)}} = 5.84$$

Let us assume that we are using a two-tailed test of the null hypothesis of zero correlation in the ranked data of the population. Then referring to Table A-6 of Appendix A, we find critical t values of 2.306 and 3.355 at the 5% and 1% levels of significance, respectively. We would reject the hypothesis of no rank correlation at both levels and conclude that a positive linear relationship exists between the rankings in baseball-playing ability and tennis-playing ability.

Exercises 13.7

1. Calculate the rank correlation coefficient for the following rankings of a league of 10 baseball teams in the pre-season and regular season competitions.

Team	A	B	C	D	E	F	G	H	I	J
Pre-season ranking	1	2	3	4	5	6	7	8	9	10
Regular season ranking	4	2	8	6	5	3	1	7	10	9

2. Calculate the rank correlation coefficient for the following performance rankings of 10 executives of a commercial bank during an economic boom and an economic recession.

Officer	A	B	C	D	E	F	G	H	I	J
Rank during boom	1	2	3	4	5	6	7	8	9	10
Rank during recession	5	6	9	3	8	4	10	1	7	2

3. Calculate the rank correlation coefficient for the following rankings of a simple random sample of companies with regard to total sales and return on equity.

Company	A	B	C	D	E	F	G	H	I	J	K	L
Sales rank	4	6	12	2	7	1	9	11	3	8	5	10
Return on equity rank	7	4	9	3	8	5	11	12	1	6	2	10

4. Calculate the rank correlation coefficient for the following results of a survey taken by an insurance company to examine the relationship between income and amount of life insurance held by heads of families. A simple random sample of 10 family heads was drawn. Assign average rank values to tied items.

Family	A	B	C	D	E	F	G	H	I	J
Insurance ($ thousands)	9	20	22	15	17	30	18	25	10	20
Income ($ thousands)	10	14	15	14	14	25	12	16	12	15

14

Decision Making Using Prior Information

In recent years, in addition to lively developments in classical, or traditional, statistical inference, there have been parallel developments in theory and methodology concerned with the problem of decision making under conditions of uncertainty. This modern formulation has come to be known as **statistical decision theory** or **Bayesian decision theory**. The latter term is often used to emphasize the role of Bayes' theorem in this type of decision analysis. The two ways of referring to modern decision analysis can be used interchangeably and will be so used in this book.

14.1 THE IMPORTANCE OF STATISTICAL DECISION THEORY

Statistical decision theory has become an important model for making rational selections among alternative courses of action when information is *incomplete* and *uncertain.* It is a prescriptive theory rather than a descriptive one. That is, it presents the principles and methods for making the best decisions under specified conditions, but it does not purport to describe how actual decisions are made in the real world.

14.2

STRUCTURE OF THE DECISION-MAKING PROBLEM

Managerial decision making has increased in complexity as the economy of the United States and the business units within it have grown larger and more intricate. However, Bayesian decision theory is based on the assumption that certain common characteristics of the decision problem can be discerned regardless of the type of decision (whether it involves long- or short-range consequences; whether it is in finance, production, marketing, or some other area; whether it is at a relatively high or low level of managerial responsibility). These characteristics constitute the formal description of the problem and provide the structure for a solution. The decision problem under study may be represented by a model comprising the following 5 elements:

The decision maker

- The agent charged with the responsibility for making the decision, the **decision maker** is viewed as an entity and may be a single individual, a corporation, a government agency, and so on.

Alternative courses of action

- The decision involves a selection among two or more alternative courses of action, referred to simply as **acts**. The problem is to choose the best of these alternative acts. Sometimes the decision maker must choose the best of alternative **strategies**, where each strategy is a decision rule indicating which act should be taken in response to a specific type of experimental or sample information.

Events

- Occurrences that affect the achievement of the objectives, **events** are viewed as lying outside the control of the decision maker, who does not know for certain which event will occur. The events constitute a mutually exclusive and complete set of outcomes; that is, one and only one of them can occur. Events are also referred to as **states of nature** or **states of the world** or, simply, **outcomes**.

Payoff

- A measure of net benefit to be received by the decision maker under particular circumstances, the payoffs are summarized in a **payoff table** or **payoff matrix**, which displays the consequences of each act selected and each event that occurs.

Uncertainty

- The indefiniteness concerning which events or states of nature will occur, **uncertainty** is indicated in terms of probabilities assigned to events. One of the distinguishing characteristics of Bayesian decision theory is the assignment of personalistic, or subjective, probabilities as well as other types of probabilities.

The payoff table, expressed symbolically in general terms, is given in Table 14-1. We assume that there are n alternative acts, denoted $A_1, A_2, \ldots, A_n$. These different possible courses of action are listed as column headings in the table. There are m possible events or states of nature, denoted $\theta_1, \theta_2, \ldots, \theta_m$. The payoffs resulting from each combination of an act and an event are desig-

TABLE 14-1
The payoff table

Event	Act A_1	A_2		A_n
θ_1	u_{11}	u_{12}	$\cdots$	u_{1n}
θ_2	u_{21}	u_{22}	$\cdots$	u_{2n}
$\vdots$	$\vdots$	$\vdots$	$\vdots$	$\vdots$
θ_m	u_{m1}	u_{m2}	$\cdots$	u_{mn}

nated by the symbol u with appropriate subscripts. The letter u has been used because it is the first letter of the word *utility*.

> The net benefit, or payoff, of the selection of an act and the occurrence of a state of nature can be treated most generally in terms of the *utility* of this consequence to the decision maker.

Summary

How these utili'ies are determined is a technical matter that is discussed later in this chapter. In summary, the utility of selecting act A_1 and having event θ_1 occur is denoted u_{11}; the utility of selecting act A_2 and having event θ_1 occur is u_{12}, and so on. Note that the first subscript in these utilities indicates the event that prevails and the second subscript denotes the act chosen. A convenient general notation is the symbol u_{ij}, which denotes the utility of selecting act A_j if subsequently event θ_i occurs. The rows of a table (or matrix) are commonly denoted by the letter i (where i can take on values $1, 2, \ldots, m$) and the columns are denoted by j (where j can take on values $1, 2, \ldots, n$).

If the event that will occur (for example, θ_3) were known with certainty beforehand, then the decision maker could simply look along row θ_3 in the payoff table and select the act that yields the greatest payoff. However, in the real world, since the states of nature lie beyond the control of the decision maker, he or she ordinarily does not know with certainty which specific event will occur. The choice of the best course of action in the face of this uncertainty is the crux of the decision maker's problem.

14.3 AN ILLUSTRATIVE EXAMPLE: THE INVENTOR'S PROBLEM

To illustrate the ideas discussed in the preceding section, we will use a simplified business decision problem.[1]

[1] This problem will be continued in later sections to exemplify other principles.

TABLE 14-2
Payoff table for the inventor's problem (units of $10,000 profit)

Event	A_1 Inventor Manufactures Device	A_2 Inventor Sells Patent Rights
θ_1: Strong sales	$80	$40
θ_2: Average sales	20	7
θ_3: Weak sales	−5	1

Inventor's problem

An inventor has patented a new device, and a bank is willing to lend the money to manufacture the device. Preliminary investigation establishes a suitable planning period of 5 years for the comparison of payoffs from this invention. According to the inventor's analysis, profits of $800,000 can be anticipated over the next 5 years if sales are strong; if sales are average, the inventor can expect to make $200,000; and if sales are weak, the inventor expects to lose $50,000. Nationwide Enterprises, Inc., has offered to purchase the patent rights. Based on the royalty arrangement, the inventor estimates that selling the patent rights may well bring a net profit of $400,000 if sales are strong, $70,000 if sales are average, and $10,000 if sales are weak.

The payoff table for the inventor's problem is given in Table 14-2.

The acts

The events

The payoffs

In this problem, the alternative acts, denoted A_1 and A_2, respectively, are for the inventor to manufacture the device or to sell the patent rights. The events or states of nature (denoted θ_1, θ_2, and θ_3, respectively) are strong sales, average sales, and weak sales for the 5-year planning period. The payoffs are in terms of the net profits that would accrue to the inventor under each act–event combination. To keep the numbers simple in this problem, the payoffs have been stated in units of $10,000; hence, a net profit of $800,000 has been recorded as $80, a net loss of $50,000 has been entered as −$5, and so on.[2]

The types of events used in the inventor's problem are, of course, simplified. Generally, an unlimited number of possible events could occur in the future relating to such matters as the customers, technological change, competitors, and so on, which lie beyond the decision maker's control. All of these states of nature may affect the potential payoffs of the alternative decisions to be made. However, in order to cut our way through the maze of complexities involved, and to construct a manageable framework of analysis for the problem, we can

[2] It is good practice in the comparison of economic alternatives to compare the present values of discounted cash flows or equivalent annual rates of return. Both of these methods take into account the time value of money; that is, the fact that a dollar received today is worth more than a dollar received in some future period. These are conceptually the types of monetary payoff values that should appear in the payoff table. This point is amply discussed in standard texts dealing with economy studies or investment analysis. To avoid a lengthy tangential discussion, we will not elaborate on the point here.

think of the variable "demand" as the resultant of all of these other underlying factors.

In the inventor's problem, 3 different levels of demand (strong, average, and weak) have been distinguished. It is helpful in this regard to think of demand as a variable.

Demand as a variable

In the inventor's problem, demand is a discrete random variable that can take on 3 possible values.

We could have considered demand as a discrete random variable taking on any finite or infinite number of values. (For example, it could have been stated in numbers of units demanded or in hundreds of thousands of units demanded.) Demand can also be treated as a continuous rather than a discrete variable. The conceptual framework of the solution to the decision problem remains the same, but the required mathematics differs somewhat from the case in which the events are stated in the form of a discrete variable.

14.4 CRITERIA OF CHOICE

Assuming that the inventor in our illustrative problem has carried out the thinking, experiments, data collection, and so on, required to construct the payoff matrix (Table 14-2), how should he or she now compare the alternative acts? Neither act is preferable to the other under all states of nature. For example, if event θ_1 occurs, that is, if sales are strong, the inventor would be better off to manufacture the device (act A_1), realizing a profit of \$800,000, compared with selling the patent rights (act A_2), which would yield a profit of only \$400,000. On the other hand, if event θ_3 occurs, and sales are weak, the preferable course of action would be to sell the patent rights, thereby earning a profit of \$10,000 compared with a loss of \$50,000. If the inventor knew with *certainty* which event was going to occur, the decision procedure would be simple: merely look along the row represented by that event and select the act that yields the highest payoff. However, the *uncertainty* with regard to which state of nature will prevail makes the decision problem an interesting one.

Maximin Criterion

Several different criteria for selecting the best act have been suggested. One of the earliest suggestions, made by mathematical statistician Abraham Wald,[3] is known as the *maximin criterion.*

[3] Abraham Wald, *Statistical Decision Functions* (New York: John Wiley & Sons, 1950).

Maximin criterion

Under the **maximin criterion** method, the decision maker assumes that once a course of action has been chosen, nature might be malevolent and might select the state of nature that minimizes the decision maker's payoff. The decision maker chooses the act that maximizes the payoff under the most pessimistic assumption concerning nature's activity.

In other words, Wald suggested that selecting the "best of the worst" is a reasonable form of protection. By this criterion, if the inventor chose act A_1, nature would cause event θ_3 to occur and the payoff would be a loss of \$50,000. If the decision maker chose A_2, nature would again cause θ_3 to occur, since that would yield the worst payoff—in this case, a profit of \$10,000. Comparing these worst, or minimum, payoffs, we have

Minimum payoffs (units of \$10,000)	−5	1

The decision maker now must—in the face of this sort of perverse nature—select the act that yields the greatest minimum payoff, namely, act A_2. That is, the inventor should sell the patent rights, for which the minimum payoff is \$10,000. Thus, the proposed decision procedure is to choose the act that yields the *maxi*mum of the *mini*mum payoffs—hence, the term **maximin**.

Obviously, the maximin is a pessimistic type of criterion. It is not reasonable to suppose that the executive would or should make decisions in this way. By following this decision procedure, the executive would always be concentrating on the worst things that could happen. In most situations, the maximin criterion would freeze the decision maker into complete inaction and would imply that it would be best to go out of business entirely. For example, let us consider an inventory stocking problem, in which the events are possible levels of demand, the acts are possible stocking levels (that is, the numbers of items to be stocked), and the payoffs are in terms of profits. If no items are stocked, the payoffs will be zero for every level of demand. For each of the other numbers of items stocked, we can assume that for some levels of demand, losses will occur. Since the worst that can happen if no items are stocked is that no profit will be made, and under all other courses of action the possibility of a loss exists, the maximin criterion would require the firm to carry no stock or, in effect, go out of business. Such a procedure is not necessarily irrational, and it might be consistent with some people's attitudes toward risk. However, the person who is willing to take some risks would regard such an arbitrary decision rule as completely unacceptable. A number of other decision criteria have been suggested by various writers, but, to avoid a lengthy digression, they will not be discussed here.[4]

[4] See, for example, Chapter 5 of D. W. Miller and M. K. Starr, *Executive Decisions and Operations Research* (Englewood Cliffs, N.J.: Prentice-Hall, 1960).

It seems reasonable to argue that a decision maker should take into account the probabilities of occurrence of the different possible states of nature. As an extreme example, if the state of nature that results in the minimum payoff for a given act has only one chance in a million of occurring, it would seem unwise to concentrate on the possibility of this occurrence. The decision procedures we will focus on include the probabilities of states of nature as an important part of the problem.

Expected Profit under Uncertainty

In a real decision-making situation, we may suppose that a decision maker would have some idea of the likelihood of occurrence of the various states of nature and that this knowledge would help in choosing a course of action. For example, in our illustrative problem, if the inventor felt confident that sales would be strong, he or she would move toward manufacturing the device, since the payoff under that act would exceed that of selling the patent rights. By the same reasoning, if the inventor were confident that sales would be weak, he or she would be influenced to sell the patent rights. If many possible events and many possible courses of action exist, the problem becomes complex, and the decision maker clearly needs some orderly method of processing all the relevant information. Such a systematic procedure is provided by the computation of the *expected* monetary value of each course of action and the selection of the act that yields the highest of these expected values. As we shall see, this procedure yields reasonable results in a wide class of decision problems. Furthermore, we will see how this method can be adjusted for the computation of expected utilities rather than expected monetary values in cases where the maximization of expected monetary values is not an appropriate criterion of choice.

Calculations for decision making

We now return to the inventor's problem to illustrate the calculations for decision making by maximization of the expected monetary value criterion. In this case, the maximization consists of selecting the act that yields the largest expected profit. Let us assume that the inventor carries out the following probability assignment procedure. On the basis of extensive investigation of past experience with similar devices, and on the basis of interviews with experts, the inventor concludes that the odds are 50:50 that sales will be average (that is, that the event we previously designated as θ_2 will occur). Furthermore, the inventor concludes that it is somewhat less likely that sales will be strong (event θ_1), than that they will be weak (event θ_3). On this basis, the inventor assigns the following subjective probability distribution to the events in question:

Event	Probability
θ_1: Strong sales	0.2
θ_2: Average sales	0.5
θ_3: Weak sales	0.3
	1.0

TABLE 14-3
Inventor's expected profits (units of $10,000 profit)

Event	Act A_1: Inventor Manufactures Device			Act A_2: Inventor Sells Patent Rights		
	Probability	Profit	Weighted Profit	Probability	Profit	Weighted Profit
θ_1: Strong sales	0.2	$80	$16.0	0.2	$40	$ 8.0
θ_2: Average sales	0.5	20	10.0	0.5	7	3.5
θ_3: Weak sales	0.3	−5	−1.5	0.3	1	0.3
	1.0		$24.5	1.0		$11.8
	Expected profit = 24.5 (ten thousands of dollars) = $245,000			Expected profit = 11.8 (ten thousands of dollars) = $118,000		

To determine the basis for choice between the inventor's manufacturing the device (act A_1) and selling the patent rights (act A_2), we compute the expected profit for each of these courses of action. These calculations are shown in Table 14-3. As indicated in that table, profit is treated as a variable that takes on different values depending on which event occurs. We compute its expected value in the usual way, according to equation 3.11. The "expected value of an act" is the weighted average of the payoffs under that act, where the weights are the probabilities of the various events that can occur.

We see from Table 14-3 that the inventor's expected profit in manufacturing the device is $245,000, whereas the expected profit in selling the patent rights is only $118,000. To maximize the expected profit, our inventor will select A_1 and will manufacture the device rather than sell the patent.

It is useful to have a brief term to refer to the expected benefit of choosing the optimal act under conditions of uncertainty. We shall refer to the expected value of the monetary payoff of the best act as the **expected profit under uncertainty**. Hence, in the foregoing problem, the expected profit under uncertainty is $245,000.

Summary

We can summarize the method of calculating the expected profit under uncertainty as follows:

1. Calculate the expected profit for each act as the weighted average of the profits under that act, where the weights are the probabilities of the various events that can occur.
2. The expected profit under uncertainty is the maximum of the expected profits calculated in step 1.

TABLE 14-4
Payoff table and opportunity loss table for the inventor's problem (units of $10,000)

Event	Payoff Table Acts		Opportunity Loss Table Acts	
	A_1	A_2	A_1	A_2
θ_1: Strong sales	$80*	$40	$0	$40
θ_2: Average sales	20*	7	0	13
θ_3: Weak sales	−5	1*	6	0

Expected Opportunity Loss

A useful concept in the analysis of decisions under uncertainty is that of opportunity loss.

An **opportunity loss** is the loss incurred because of failure to take the best possible action. Opportunity losses are calculated separately for each event that might occur.

Given the occurrence of a specific event, we can determine the best possible act. For a given event, the opportunity loss of an act is the difference between the payoff of that act and the payoff for the best act that could have been selected. For example, in the inventor's problem, if event θ_1 (strong sales) occurs, the best act is A_1, for which the payoff is $80 (in units of $10,000). The opportunity loss of that act is $80 − $80 = $0. The payoff for act A_2 is $40. The opportunity loss of act A_2 is the amount by which the payoff of the best act, $80, exceeds the $40 payoff of act A_2, which is $80 − $40 = $40.

It is convenient to asterisk the payoff of the best act for each event in the original payoff table in order to denote that opportunity losses are measured from these figures. Both the original payoff table and the opportunity loss table are given in Table 14-4 for the inventor's problem.

We can now proceed with the calculation of expected opportunity loss in a manner completely analogous to the calculation of expected profits. Again, we use the probabilities of events as weights and determine the weighted average opportunity loss for each act. Our goal is to select the act that yields the *minimum* expected opportunity loss. The calculation of the expected opportunity losses for the two acts in the inventor's problem is given in Table 14-5. The symbol EOL represents **expected opportunity loss**. Hence, EOL(A_1) and EOL(A_2) denote the expected opportunity losses of acts A_1 and A_2, respectively.

Expected opportunity loss (EOL)

TABLE 14-5
Expected opportunity losses for the inventor's problem (units of $10,000)

Event	Act A_1: Inventor Manufactures Device			Act A_2: Inventor Sells Patent Rights		
	Probability	Opportunity Loss	Weighted Opportunity Loss	Probability	Opportunity Loss	Weighted Opportunity Loss
θ_1: Strong sales	0.2	$0	$ 0	0.2	$40	$ 8.0
θ_2: Average sales	0.5	0	0	0.5	13	6.5
θ_3: Weak sales	0.3	6	1.8	0.3	0	0
	1.0		$1.8	1.0		$14.5
	EOL(A_1) = 1.8 (ten thousands of dollars) = $18,000			EOL(A_2) = 14.5 (ten thousands of dollars) = $145,000		

The inventor's EOL in manufacturing the device is $18,000, and in selling the patent rights, the EOL is $145,000. If the inventor selects the act that minimizes the EOL, he or she will choose A_1, that is, to manufacture the device. This is the same act selected under the criterion of maximizing expected profit.

It can be proved that the best act according to the criterion of maximizing expected profit is also best if the decision maker follows the criterion of minimizing expected opportunity loss.

The relationship between the maximum expected profit and the minimum expected opportunity loss will be examined later. Note that opportunity losses are not losses in the accountant's sense of profit and loss, because as we have seen, they occur even when only profits of different actions are compared for a given state. They represent opportunities foregone rather than monetary losses incurred.

Minimax Opportunity Loss

We noted earlier that various criteria of choice have been suggested for the decision problem. One that has been advanced in terms of opportunity losses is that of *minimax opportunity loss.*

Under the **minimax opportunity loss** method, the decision maker selects the act that minimizes the worst possible opportunity loss that can be incurred among the various acts.

As with the maximin criterion for payoffs, the minimax criterion for opportunity loss takes a pessimistic view toward which states of nature will occur. Once the opportunity loss table has been prepared, as in Table 14-4, the decision maker determines for each act the largest opportunity loss that can be incurred. For example, in the inventor's problem, for act A_2, it is \$400,000. The decision maker then chooses that act for which these worst possible losses are the least, that is, the act that *mini*mizes the *max*imum losses; hence, the term **minimax**. In the inventor's problem, the decision maker would choose act A_1, since the maximum possible opportunity loss (\$60,000) under this course of action is less than the corresponding worst loss under act A_2 (\$400,000). This criterion is also sometimes referred to as "minimax regret," where the opportunity losses are viewed as measures of regret for taking less than the best courses of action.[5]

In our illustrative problem, the minimax opportunity loss act, A_1, happens to be the same decision as would be made under the criterion of maximizing expected profit. However, this is not always the case. The minimax loss criterion, like the maximin payoff rule, singles out for each course of action the worst consequence that can befall the decision maker and then attempts to minimize this damage. As was true for the maximin payoff criterion, the minimax loss viewpoint yields results in many instances which imply that a business executive who faces risky ventures should simply go out of business.

Throughout the remainder of this book, we will use the Bayesian decision theory criterion of maximizing expected profit or its equivalent, minimizing expected opportunity loss.[6]

Exercises 14.4

1. If possible states of nature are:
   a. competitor will set his price higher
   b. competitor will set his price the same
   c. competitor will set his price lower

   what is wrong with assessing prior probabilities as 0.6, 0.3, and 0.2, respectively?
2. A new appliance store finds that in its first week of business it sold 5 major appliances, 10 home appliances, and 30 small appliances. Based solely on this past knowledge, what prior probability distribution would you formulate for the type of appliance to be sold?
3. As manager of a plant, you must decide to invest in either a cost reduction program or a new advertising campaign. Assume that you know the cost reduction program will increase the profit-to-sales ratio from the present 10% to 11%. The sales campaign, if successful, is expected to increase the

[5] See L. J. Savage, "The Theory of Statistical Decision," *Journal of the American Statistical Association*, vol. 46, 1951, pp. 55–57.

[6] We shall indicate in section 14.8 that the criterion is actually the maximization of expected utility. In the case of a decision maker who is indifferent to risk—that is, one who has a linear utility function—expected utility is maximized by maximizing expected profit.

present \$2 million of sales by 12%. The probability that the campaign will be successful is 0.8. What would be the better course of action?

4. R.B.A., Inc., is given the opportunity to submit a closed bid to the government to build certain electronic equipment. An examination of similar proposals made in the past revealed that the average profit per successful bid was \$175,000, and that R.B.A., Inc., received the contract (i.e., had the lowest bid) on 10% of its submitted bids.) The cost of preparing a bid is, on the average, \$10,000. Should R.B.A., Inc., prepare a bid?

5. In problem 4, suppose R.B.A., Inc., chose to prepare a bid. For this particular proposal, assume the company finds it can submit only the following four bids: \$1,600,000, \$1,700,000, \$1,800,000, or \$1,900,000. At \$1,600,000, expected profit is \$160,000. Each successive bid yields an increase in profit equal to the increase in the bid. From an examination of past accounting records, R.B.A., Inc., assesses the probabilities that the bids will be the lowest ones to be 0.4, 0.3, 0.2, 0.1, respectively. Which bid should be submitted?

6. A high-fashion store manager must decide how many sable coats to stock for the coming season. The coats cost \$3,000 and can be sold for \$8,000. Any coats left at the end of the season can be returned to the manufacturer for \$2000. Consult the profit table and expected probabilities to determine the optimal stocking level and the manager's expected profit.

		Number of Coats Stocked					
Probability	Number Demanded	0	1	2	3	4	5
0.15	0	0	−1*	−2	−3	−4	−5
0.15	1	0	5	4	3	2	1
0.20	2	0	5	10	9	8	7
0.20	3	0	5	10	15	14	13
0.15	4	0	5	10	15	20	19
0.15	5	0	5	10	15	20	25

* profits are in thousands of dollars

7. Fission, Inc., is given the opportunity to bid on a government project for a new nuclear energy plant. A review of previous bids on similar projects revealed that the average profit per successful bid was \$1.25 million and that Fission received the contract (that is, had the lowest bid) on 6% of its submitted bids. The cost of preparing a bid is, on the average, \$19,000. Should Fission, Inc., prepare a bid?

8. A large oil company is planning to open a new coal mine. This particular mine was not opened previously because market prices did not justify the expense of the special technology needed to extract marketable quantities.

With present forecasts of future energy needs, it is now a profitable venture. The total cost to open and run the mine will be \$17 million. By spending an extra \$10 million for the most refined equipment available, the profit margin per ton can be doubled. Based on estimated future demand and prices, total profits from spending \$17 million are estimated to be:

Probability	0.3	0.2	0.4	0.1
Profit (millions)	\$15	\$20	\$25	\$40

Should the oil company spend the extra \$10 million on the more refined equipment?

9. A homeowner must decide whether to purchase a new gas oven or a new electric oven. (Both gas and electric lines go into the home.) A gas oven will cost \$450 to purchase and install; an electric oven will cost \$350 to purchase and install. If both ovens have an expected life of 8 years and the cost per year is fixed after the first year, which oven should be purchased? Use the following table for the probabilities of yearly costs to operate a gas oven and an electric oven.

Gas Oven		Electric Oven	
Probability	Cost per Year	Probability	Cost per Year
0.2	\$72	0.1	\$84
0.3	96	0.2	96
0.3	120	0.2	108
0.2	144	0.2	120
		0.2	144
		0.1	168

14.5 EXPECTED VALUE OF PERFECT INFORMATION

Thus far in our discussion, we have considered situations in which the decision maker chooses among alternative courses of action on the basis of *prior information* without attempting to gather further information before making the decision. In other words, the probabilities used in computing the expected value of each act, as shown in Table 14-3, are **prior probabilities**, that is, probabilities established prior to obtaining additional information through sampling. The procedure of calculating expected value of each act based on these prior probabilities and selecting the optimal act is referred to in Bayesian decision theory as **prior analysis**. In Chapter 15 we consider how courses of action may be compared after these prior probabilities are revised on the basis of sample information, experimental data, or information resulting from tests of any sort. However, the analysis we have carried out so far provides a yardstick for measuring the value of perfect information concerning which events will occur. This yardstick will be referred to as the **expected value of perfect information**.

A yardstick for measuring

- To determine this value, we calculate the *expected profit with perfect information.*
- Then, we subtract the *expected profit under uncertainty*, (the calculation we previously examined) to find the *expected value of perfect information.*

These concepts will be explained in terms of the inventor's problem. We begin with the idea of expected profit with perfect information.

Expected profit with perfect information

The calculation of the *expected profit* of acting *with perfect information* is based on the expected payoff if the decision maker has access to a perfect predictor.

It is assumed that if this perfect predictor forecasts that a particular event will occur, then indeed that event will occur. The expected payoff under these conditions for the inventor's problem is given in Table 14-6.

To understand the meaning of this calculation, it is necessary to adopt a long-run relative frequency point of view. If the forecaster says the event "strong sales" will prevail, the decision maker can look along that row in the payoff table and select the act that yields the highest profit. In the case of strong sales, the best act is A_1, which yields a profit of $800,000. Hence, the figure $80 is

TABLE 14-6

Calculation of expected profit with perfect information for the inventor's problem (units of $10,000 profit)

Predicted Event	Profit	Probability	Weighted Profit
θ_1: Strong sales	$80	0.2	$16.0
θ_2: Average sales	20	0.5	10.0
θ_3: Weak sales	1	0.3	0.3
			$26.3

Expected profit with perfect information = $26.3 (ten thousands of dollars)
= $263,000

entered under the profit column in Table 14-6. The same procedure is used to obtain the payoffs for each of the other possible events. The probabilities shown in the next column are the original probability assignments to the three states of nature. From a relative frequency viewpoint, these probabilities are now interpreted as the proportion of times the perfect predictor would forecast the occurrences of the given states of nature if the present situation were faced repeatedly. Each time the predictor makes a forecast, the decision maker selects the optimal payoff.

The expected profit with perfect information is calculated (as in Table 14-6) by weighting the best payoffs by the probabilities and totaling the products.

The expected profit with perfect information in the inventor's problem is $263,000. This figure can be interpreted as the average profit the inventor could realize if he were faced with this decision problem repeatedly under identical conditions, and if he always took the best action after receiving the perfect indicator's forecast. Expected profit with perfect information has sometimes been called the "expected profit under certainty," but this term is somewhat misleading. The inventor is not *certain* to earn any one profit figure. The expected profit with perfect information should be interpreted as indicated in this discussion.

Expected value of perfect information (EVPI)

The expected value of perfect information, abbreviated EVPI, is defined as *the expected profit with perfect information* minus *the expected profit under uncertainty*.

The interpretation of the EVPI is clear from its calculation (shown in Table 14-7). In the inventor's problem, the expected payoff of selecting the optimal act under conditions of uncertainty is $245,000 (see Table 14-3). On the other hand, if the perfect predictor were available and the inventor acted according to those predictions, the expected payoff would be $263,000 (see Table 14-6). The difference of $18,000 represents the increase in expected profit

TABLE 14-7
Calculation of expected value of perfect information for the inventor's problem

Expected profit with perfect information	$263,000
Less: Expected profit under uncertainty	245,000
Expected value of perfect information (EVPI)	$ 18,000

EVPI = EOL of the optimal act under uncertainty = $18,000

attributable to the use of the perfect forecaster. Hence, the expected value of perfect information may be interpreted as the most the inventor should be willing to pay to get perfect information on the sales level for the device.

The expected opportunity loss of selecting the optimal act under conditions of uncertainty in the inventor's problem was shown earlier to be \$18,000 (see Table 14-5). That is, this figure represented the minimum value among the expected opportunity losses associated with each act. As shown in Table 14-7, this figure is equal to the expected value of perfect information. It can be mathematically proved that this equality holds in general. Another term used for the expected opportunity loss of the optimal act under uncertainty is the **cost of uncertainty**. This term highlights the "cost" attached to decision making under conditions of uncertainty. Expected profit would be larger if a perfect predictor were available and this uncertainty were removed. Hence, this cost of uncertainty is also equal to the expected value of perfect information. In summary, the following three quantities are equivalent:

Cost of uncertainty

- Expected value of perfect information (EVPI)
- Expected opportunity loss of the optimal act under uncertainty (EOL)
- Cost of uncertainty

14.6 REPRESENTATION BY A DECISION DIAGRAM

Decision tree diagram

It is useful to represent the structure of a decision problem under uncertainty by a **decision tree diagram**, also called a **decision diagram** or, briefly, a **tree**. The diagram depicts the problem in terms of a series of choices made in alternating order by the decision maker and by "chance." Forks at which the decision maker is in control of choice are referred to as **decision forks** (represented by a square); those at which chance is in control are called **chance forks** (represented by a circle). Forks may also be referred to as **branching points** or **junctures**.

A simplified decision diagram for the inventor's problem is given in Figure 14-1(a). After explaining this skeleton version, we will insert additional information to obtain a completed diagram. As we can see from Figure 14-1(a), the first choice is the decision maker's at branching point 1. He can follow either branch A_1 or branch A_2; that is, he can choose either act A_1 or A_2. Assuming that he follows path A_1, he comes to another juncture, branching point 2, which is a chance fork. Chance now determines whether the event that will occur is θ_1, θ_2, or θ_3. If chance takes him down the θ_1 path, the terminal payoff is \$800,000; the corresponding payoffs are indicated for the other paths. An analogous interpretation holds if he chooses to follow branch A_2. Thus, the decision diagram depicts the basic structure of the decision problem in schematic form. In Figure 14-1(b), additional information is superimposed on the diagram to represent the analysis and solution to the problem.

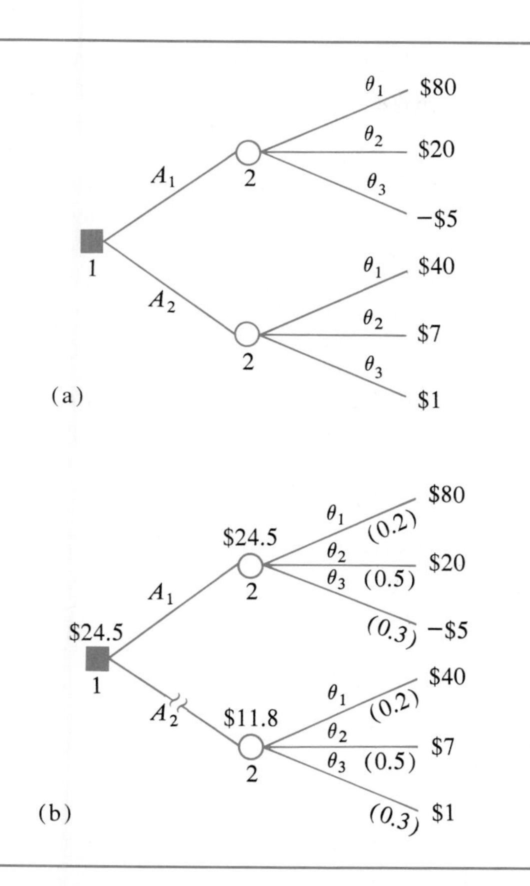

FIGURE 14-1
(a) Simplified decision diagram for inventor's problem
(b) Decision diagram for inventor's problem (payoffs are in units of $10,000 profit)

Backward induction

The decision analysis process represented by Figure 14-1(b) (and other decision diagrams to be considered later) is known as **backward induction**. We imagine ourselves as located at the right-hand side of the tree diagram, where the monetary payoffs are. Let us consider first the upper three paths denoted θ_1, θ_2, and θ_3. Below these symbols, in parentheses, we enter the respective probability assignments (0.2, 0.5, and 0.3) as given in Table 14-3. These represent the probabilities assigned by chance to following these three paths, after the decision maker has selected act A_1. Moving back to the chance fork from which these three paths emanate, we can calculate the expected monetary value of being located at that fork. This expected monetary value is $24.5 (in units of $10,000, as are the other obvious corresponding numbers) and is calculated in the usual way:

$$\$24.5 = (0.2)(\$80) + (0.5)(\$20) + (0.3)(-\$5)$$

This figure is entered at the upper chance fork. It represents the value of standing at that fork after choosing act A_1, as chance is about to select one of the three paths. The analogous figure entered at the lower chance fork is \$11.8. Therefore imagining ourselves as being transferred back to branching point 1, where the square represents a fork at which the decision maker can make a choice, we have the alternatives of selecting act A_1 or act A_2. Each of these acts leads us down a path at the end of which is a risky option whose expected profit has been indicated. Since following path A_1 yields a higher expected payoff than path A_2, we block off A_2 as a nonoptimal course of action. This is indicated on the diagram by the two wavy lines. Hence, A_1 is the optimal course of action, and it has the indicated expected payoff of \$24.5. This expected payoff of the best act, \$24.5, is shown above branching point 1 in Figure 14-1(b).

Thus, the decision tree diagram reproduces in compact schematic form the analysis given in Table 14-3. An analogous diagram would be constructed in terms of opportunity losses to reproduce the analysis of Table 14-5.

Exercises 14.6

1. Explain the meaning of *expected value of perfect information.*
2. Explain the difference between expected opportunity loss and expected value of perfect information.
3. Given an opportunity loss table, is it possible to compute the corresponding payoff table? Explain why or why not.
4. The following payoff matrix is in units of \$1000:

	Price of the Product			
Competitor's Price	A_1 \$1.25	A_2 \$1.35	A_3 \$1.45	A_4 \$1.55
S_1: \$1.40	20	32	28	15
S_2: \$1.30	22	18	25	20
S_3: \$1.20	28	20	22	30
S_4: \$1.10	35	25	26	32

The prior probabilities are:

State of Nature	Probability
S_1	0.2
S_2	0.4
S_3	0.3
S_4	0.1

Compute the EVPI by two different methods.

5. The following is a payoff table for profits in units of $1000:

Demand	Action A_1	A_2	A_3	A_4	A_5
S_1: High	112	142	93	85	108
S_2: Average	82	70	65	80	50
S_3: Low	55	30	60	42	50

where A_1 = open the store weekdays, evenings, and Saturdays
A_2 = open the store weekdays and Saturday
A_3 = open the store weekdays and Friday evening
A_4 = open the store weekdays and Wednesday and Friday evenings
A_5 = open the store only on weekdays

The prior probability distribution of demand is:

S_i	$P(S_i)$
S_1	0.4
S_2	0.4
S_3	0.2

a. Find the expected profit under certainty.
b. Find the expected profit under uncertainty.
c. How much would you pay for information that yields the true state of nature?

6. Trivia Press, Inc., has been offered an opportunity to publish a new novel. If the novel is a success, the firm can expect to earn $8 million over the next five years; if the novel is a failure, the firm can expect to lose $4 million over the next five years. After reading the novel, the publisher assesses the probability of success as $\frac{1}{3}$. Should Trivia Press publish the book? What is the expected value of perfect information?

7. An advertising firm submits for acceptance a campaign costing $55,000. The company's marketing manager estimates that if the campaign is received well by the public, profits will increase by $175,000; if it is received moderately well, profits will increase by $55,000; and if it is received poorly, profits will remain unchanged. Compute the appropriate opportunity loss table.

8. Assume that there are 10 urns, 7 of type *A* and 3 of type *B*. Type *A* urns contain 5 white balls and 5 black balls. Type *B* urns contain 8 white balls and 2 black balls. One of the 10 urns is to be selected at random. You are required to guess whether the urn selected is of type *A* or *B*. Assume you are willing to act on the basis of expected monetary value. You will receive payoffs and penalties according to the payoff table. Find and interpret the expected value of perfect information.

True State of Nature	Your Guess Type *A*	Type *B*
Type *A*	+$500	−$ 40
Type *B*	− 300	+ 800

9. As personnel manager of Lemon Motors you must decide whether to hire a new salesperson. Depending on sales performance, the payoff to the firm is

Sales	Payoff
High	$10,000
Average	3,000
Low	−13,000

a. If, judging from the sales application, you feel the probabilities attached to the salesperson's possible performance are high (0.3), average (0.4), and low (0.3), should you hire that person?
b. How much would you be willing to pay a perfect predictor to tell you what the salesperson's performance would be?

10. As marketing manager of a firm, you are trying to decide whether to open a new region for a product. Success of the product depends on demand in the new region. If demand is high, you expect to gain $100,000; if demand is average, you expect to gain $10,000; and if demand is low, you expect to lose $80,000. From your knowledge of the region and your product you feel the chances are 4 out of 10 that sales will be average, and equally likely that they will be high or low. Should you open the new region? How much would you be willing to pay to know the true state of nature?

11. In problem 7, if the president of the company, after examining the proposed campaign, feels that the probabilities that it would be received "well" and

"moderately well" are 0.4 and 0.2, respectively, what is the expected opportunity loss for each action? What is the optimal decision?

12. A brewer presently packages beer in old-style cans. He is debating whether to change the packaging of his beer for next year. He can adopt A_1, an easy-open aluminum can; A_2, a lift-top can; A_3, a new wide-mouth screw-top bottle; or retain A_4, the same old style cans. Profits resulting from each move depend on what the brewer's competitor does for the next year. The payoff matrix and prior probabilities, measured in $10,000 units, are tabulated.
    a. Find the expected opportunity loss for each act.
    b. Determine EVPI.
    c. Determine the optimal decision.

Prior Probability	Competitor Uses	Action			
		A_1	A_2	A_3	A_4
0.5	Old-style bottles	15	14	13	16
0.2	Easy-open cans	12	11	10	8
0.1	Lift-top cans	6	9	8	6
0.2	Screw-top bottles	5	6	8	5

14.7 THE ASSESSMENT OF PROBABILITY DISTRIBUTIONS

We indicated earlier that it is reasonable to assume that a decision maker would have some idea of the probabilities of the occurrences of the various states of nature that are relevant to the decision to be made. In this section, we discuss how such probabilities may be assessed.[7]

> If the random variable representing states of nature is discrete and there are only a small number of possible outcomes, then the decision maker will probably be able to assign probabilities directly to each possible outcome.

For example, if the decision maker must assess the probability that a particular manufactured article is defective, he or she may be able to assign a probability

[7] For a comprehensive discussion of the philosophy and practice of the assessment of subjective probabilities, see C. S. Spetzler and C. S. Staël Von Holstein, "Probability Encoding in Decision Analysis," *Management Science*, vol. 22, no. 3 (November 1975).

of, say, 0.1 that the article is defective; hence, the complementary probability would be 0.9 that the article is not defective. If there were three possible outcomes—such as seriously defective, moderately defective, and not defective—again the decision maker can assign a probability to each of these three possible outcomes. As an example in a quite different context, the decision maker might have to assign probabilities to the following events: (1) the U.S. Congress *will* pass the bill on education currently before it or (2) the U.S. Congress *will not* pass the bill. In all of these cases, the decision maker may use a combination of past empirical information and a subjective evaluation of the effects of additional knowledge in making the probability assignments.

On the other hand, the random variable of interest may take on a large number of possible values. For example, if the states of nature are values for a company's sales next year for a certain product, the relevant values to be considered by the decision maker may range from, say, 100,000 to 500,000 units. It would be rather meaningless to attempt to assess a probability for each of the 400,001 possible outcomes for sales, that is, for 100,000, 100,001, . . . , 500,000.

If the random variable has a large number of possible values, the decision maker should treat the random variable as continuous and should set up a cumulative distribution function by making probability assessments for a number of selected ranges of the random variable.

A distinction can be made between the following two situations:

1. The decision maker directly establishes a subjective cumulative probability distribution without formal processing of data.
2. There is a small quantity of past data, and the decision maker formally processes this information to set up a cumulative probability distribution.

We now discuss how these probability distributions may be constructed.

Direct Subjective Assessment

We first illustrate a situation in which a decision maker establishes a subjective cumulative probability distribution without formal processing of data. A decision maker, forecasting a company's sales of a certain product for next year, feels that sales may range from 100,000 to 500,000 units and wishes to establish a subjective cumulative probability distribution without explicitly using any data.

Finding the first three fractile values

The basis of the procedure is to focus attention at a few key points, or **fractiles**, in the distribution. For example, the 0.50 fractile is a value such that the decision maker believes the probability is $\frac{1}{2}$ that the random variable is equal to or less than that value. (Note that the 0.50 fractile is simply another name for the median.) It is useful in this context to think in terms of hypothetical gambles. For example, one gamble might pay $100 if the random variable in question is less than or equal to some value selected by the decision maker, and a second gamble might pay $100 if the random variable turns out to be more than that value. The 0.50 fractile is the point at which the decision maker is indifferent to the choice between the two gambles. Let us assume that after some serious reflection on these gambles, the decision maker selects 350,000 units as the median or 0.50 fractile value.

We continue with the subjective assessment process. The decision maker should now select the 0.25 and 0.75 fractiles. For example, the 0.25 fractile is a figure such that the probability is $\frac{1}{4}$ that the value of the random variable will lie below that figure and $\frac{3}{4}$ that the value will lie above it. The analogous interpretation applies for the 0.75 fractile. We return to choices among gambles to determine the 0.25 fractile value. The decision maker might be asked to assume that sales next year will be less than 350,000 units and should then be presented with a pair of gambles: The first wager will pay $100 if sales fall between a value that the decision maker chooses and the 0.50 fractile; the second will pay $100 if sales are equal to or less than the chosen figure. The dividing value of the random variable determined by this "indifference point" between the two gambles is the 0.25 fractile. Obviously, if the decision maker selects a tentative dividing value but then finds one gamble more attractive than the other, the 0.25 fractile value has not been determined. The assessment procedure should continue until the indifference point has been found. Let us assume that the decision maker chooses 250,000 as the 0.25 fractile. The 0.75 fractile should then be determined by a similar procedure; let us assume that the decision maker determines 400,000 as the 0.75 fractile value.

Three selected points (the 0.25, 0.50, and 0.75 fractiles) have now been found in the cumulative distribution function.

> The usual procedure at this juncture is to determine two more values at the extremes of the distribution and to use the resulting five points as a basis for sketching the function.

Selecting the extreme values

The extreme values often used are either the 0.01 and 0.99 or the 0.001 and 0.999 fractiles. In the present illustration, let us assume that the decision maker, when queried about the range of 100,000 to 500,000 units for next year's sales, stated that these were not absolute limits, that it was possible for next year's sales to fall outside them. However, after some further hard thinking, the decision maker selects these two values as the 0.01 and 0.99 fractiles. That is, the subjectively assigned odds are 99:1 that sales will exceed 100,000 (corresponding to a probability of 0.99). Similarly, the subjective odds are 99:1 that next year's sales will not exceed 500,000 units. Actually, these extreme fractiles are difficult to assess. Empirical experiments have demonstrated that people tend to make the distributions too tight; that is, the 0.01 (or 0.001) and 0.99 (or 0.999) fractiles are generally not low enough and not high enough, respectively. In formulating these assessments, we must make a conscientious effort to spread the extreme values sufficiently.

Testing for consistency of judgments

To monitor further the probability assessments, the decision maker should carry out some tests for consistency of judgments. For example, since the 0.25 and 0.75 fractiles in this example are 250,000 and 400,000, the decision maker is asserting a 0.50 chance that next year's sales will fall within the range of these two figures (known, incidentally, as the **interquartile range**). Hence these probability assessments imply that it is equally likely that next year's sales will fall inside this range or outside this range. Perhaps, on reflection, the decision maker will revise one or both of the two fractiles. After a few introspective checks of this sort, the five selected points should be plotted and a smooth curve drawn through them.

Sketching the function

Figure 14-2 shows a cumulative probability distribution that could be drawn through the 0.01, 0.25, 0.50, 0.75, and 0.99 fractiles referred to in our illustration. The curve is the typical S-shaped form obtained for cumulative probability distributions for continuous random variables.

Checking the probability assignments

Even after the graph has been drawn, the decision maker should continue to check or monitor the probability assignments. If the curve rises too slowly or too steeply in a certain portion of the random variable's range, the graph (and consequently the appropriate fractile values) should be adjusted. The reassessment procedure should continue until the decision maker feels that the curve appropriately describes the probability distribution.

Discarding past data

The direct subjective assessment of probability distributions is appropriate for the following situations.

- when there is little or no past historical data on which to base the construction of the function

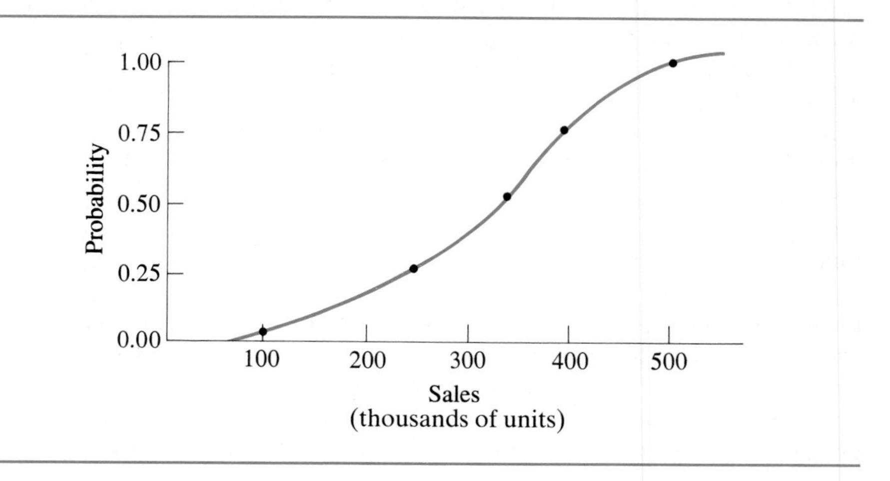

FIGURE 14-2
A cumulative probability distribution sketched through five selected fractiles

- when there *is* a substantial amount of past data but the decision maker is unwilling to use these data formally in the establishment of a probability distribution

The latter situation might arise if the decision maker decides that the factors that gave rise to the past historical data have changed so greatly that these data are not a reliable guide for assessing future probabilities. For example, suppose a company that purchases one of its raw materials from a supplier has a large quantity of past data on the percentages of defective materials in past shipments from the supplier. Assuming no important changes in the supplier's manufacturing process and in policies governing the production of the product, the purchasing company may feel justified in using the past relative frequencies of occurrence of percentage defectives per shipment to assess a probability distribution for the percentage of defectives that might occur on the *next* shipment. On the other hand, if the supplier has made an important change in its method of manufacture or in the work force that produces the product, the purchaser may feel that the past relative frequencies of occurrence are no longer relevant for assessing the probability distribution. The purchaser may then appropriately make a direct subjective assessment of the probability distribution.

Use of Past Data

We turn now to the first situation listed at the beginning of this section, namely that there is a *small* quantity of data and the decision maker formally processes these data to construct a cumulative probability distribution for the relevant states of nature. If there were a *large* quantity of relevant historical data in the form of relative frequencies of occurrence, the decision maker could simply use these figures to set up a discrete probability distribution. If only a *small* quantity of relevant data exists, how should these data be processed to construct the desired probability distribution? We will discuss here a systematic method for dealing with this type of situation.

Large quantity of data

Small quantity of data

Let us assume that a retail establishment asks you, a consultant, to prepare a probabilistic forecast of the number of telephone orders that would be received next week for a certain product. The product has been sold for the past 20 weeks, and Table 14-8(a) gives a record of the number of weeks in which stated numbers of telephone orders were received. If you set up a relative frequency distribution of this discrete random variable, you would obtain the function shown in Table 14-8(b).

Interpreting the relative frequency distribution of Table 14-8(b) as a probability distribution, you find yourself making some rather odd statements. For example, starting at the beginning of the distribution, you would state that receiving 21 orders has a probability of 0.05, receiving 22 and 23 orders has a probability of 0, receiving 24 orders has a probability of 0.10, and so on. However, you really do not believe that it is impossible for 22 or 23 telephone orders

TABLE 14-8

(a) Numbers of weeks in which specified numbers of telephone orders were received

Number of Telephone Orders	Number of Weeks That Number of Orders Was Received
21	1
22	0
23	0
24	2
25	2
26	3
27	3
28	1
29	0
30	0
31	1
32	1
33	1
34	1
35	0
36	0
37	1
38	1
39	1
40	0
41	0
42	1

(b) Relative frequency of occurrence of number of telephone orders

Number of Telephone Orders	Relative Frequency of Occurrence
21	0.05
24	0.10
25	0.10
26	0.15
27	0.15
28	0.05
31	0.05
32	0.05
33	0.05
34	0.05
37	0.05
38	0.05
39	0.05
42	0.05

to occur. You would recognize that these numbers of orders probably did not occur in the small sample of only 20 weeks of experience because of chance sampling fluctuations. If you work in terms of a cumulative probability distribution rather than the probability mass function of Table 14-8(b), you can develop a more reasonable interpretation of the data. The cumulative relative frequencies of occurrence for the telephone order data are shown in Table 14-9. These cumulative frequencies represent the proportion of weeks in which the indicated numbers of telephone orders or *fewer* were received. Interpreted as a probability, a cumulative frequency may be thought of as the probability that the random variable "number of telephone orders" is less than or equal to the specified value.

A graph of the cumulative frequency distribution of Table 14-9 is shown in Figure 14-3. In the present example, the raw data are ungrouped, and they are graphed as a step function in the figure. The upper point plotted on each vertical line in the graph is the cumulative frequency corresponding to the value

TABLE 14-9
Cumulative relative frequency of occurrence of numbers of telephone orders

Number of Telephone Orders	Proportion of Weeks in Which the Specified Number or Fewer Orders Were Received	Number of Telephone Orders	Proportion of Weeks in Which the Specified Number or Fewer Orders Were Received
21	0.05	32	0.70
22	0.05	33	0.75
23	0.05	34	0.80
24	0.15	35	0.80
25	0.25	36	0.80
26	0.40	37	0.85
27	0.55	38	0.90
28	0.60	39	0.95
29	0.60	40	0.95
30	0.60	41	0.95
31	0.65	42	1.00

of the random variable plotted on the horizontal axis. As with the original frequency distribution shown in Table 14-8(b), the irregularities of the step function graph of the cumulative relative frequency distribution in Figure 14-3 are doubtless the result of chance sampling variations. In an attempt to smooth out these sampling variations, you try to fit a smooth curve to the graph. Such a smooth curve is fitted to the step function data in Figure 14-3. In fitting this curve, you should draw through the *centers* of the flat portions ("flats") of the

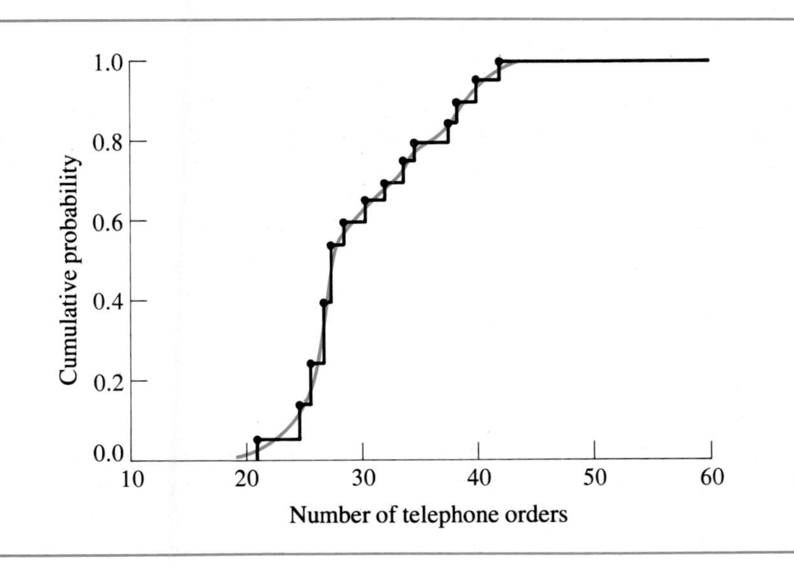

FIGURE 14-3
Continuous cumulative probability distribution fitted to the step function graph of cumulative relative frequency distribution in Table 14-8(b)

step function to show a representative cumulative probability for the *range* of values covered by these flats. The range of possible values shown on the graph should be wider than the analogous range observed in the small sample of data. Thus, for example, although the original data had a lowest value of 21 and a highest value of 42, the horizontal axis of Figure 14-3 ranges from 10 to 60 telephone orders. The lowest and highest values should reflect the decision maker's best judgment concerning the range of possible outcomes. In fact, the entire smooth cumulative probability distribution should represent the decision maker's judgment concerning the form of the curve (guided, of course, by the rough form suggested by the step function data).

Summary

In the preceding subsections, we outlined the following ways of constructing probability distributions for states of nature depending on the type and quantity of data available.

- If there are no historical data available, or if the decision maker prefers not to process the available data because of doubts concerning the constancy of the mechanisms producing these data, he or she may proceed to a direct subjective assessment of the cumulative probability distribution. As we have seen, by using selected fractiles, the decision maker can construct a continuous cumulative probability function.[8]

- When a relatively *small* quantity of available data exists, we have noted how a cumulative relative frequency step function could be graphed and how a continuous cumulative probability distribution could be fitted to that function.

If there are *large* quantities of relative frequency data available concerning states of nature, and if we are confident about the constancy of the causal systems producing the data, we can simply set up a discrete probability distribution from those data.

We discussed how to construct continuous cumulative probability distributions for events or states of nature in decision-making problems in which there are no data or relatively small amounts of data. Continuous probability distributions are difficult to deal with in the analysis of many practical decision problems because they assign probability densities to every possible value of the states of nature.[9]

[8] See R. S. Schlaifer, *Analysis of Decisions Under Uncertainty* (New York: McGraw-Hill, 1969) for a comprehensive discussion of the assessment of probability distributions based on the quantity and type of data available.

[9] More advanced texts in statistical decision theory deal with methods of analysis for continuous probability distributions whose formulas can be given algebraically (for example, the normal, gamma, and beta distributions).

Bracket Medians

In the earlier sections of this chapter, we discussed the structure of the decision making problem and decision diagram representation in terms of *discrete* probability distributions for states of nature.

The **bracket median** technique is a method of approximating a continuous cumulative probability distribution by a discrete distribution. This technique enables us to carry out decision analysis for states of nature represented by discrete random variables as outlined in this and subsequent chapters. As an illustration of the bracket median technique, we consider the continuous cumulative probability distribution for next year's sales of a certain product, shown in Figure 14-4. The basic rationale of the method is to break up the probability distribution into groups having equal probabilities. Thus, in Figure 14-4, the vertical scale of cumulative probabilities has five equal divisions, 0–0.2, 0.2–0.4, and so on. This procedure correspondingly divides the random variable sales into five equally likely groups, each having a probability of 0.2.

Our next step is to determine a representative value, or **bracket median**, for each of the five groups of the random variable and assign a 0.2 probability to each of these representative values. These bracket medians are determined by bisecting each of the probability groups along the vertical axis, drawing the broken lines shown at 0.1, 0.3, 0.5, 0.7, and 0.9 over to the curve. We then read down on the horizontal axis to determine the bracket medians. In Figure 14-4, these bracket medians are at 1.1, 1.8, 2.3, 2.8, and 3.4 millions of units of sales. The resulting discrete probability distribution is given in Table 14-10.

In this illustration, five bracket medians were used. Of course, we could obtain greater accuracy by using larger numbers of groups. For example, if we

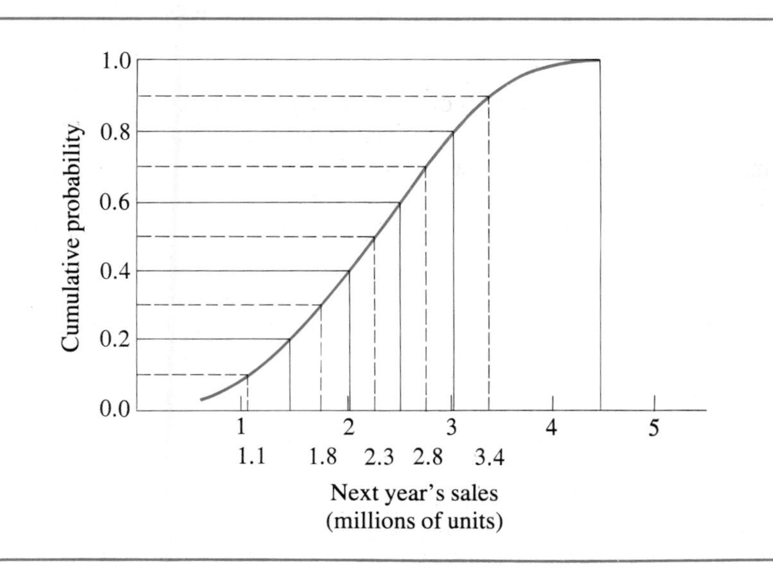

FIGURE 14-4
The representation of a continuous cumulative probability distribution by bracket medians

TABLE 14-10
Approximation of the continuous probability distribution of Figure 14-4 by a discrete distribution using bracket medians

Bracket Median Next Year's Sales (millions of units)	Probability
1.1	0.2
1.8	0.2
2.3	0.2
2.8	0.2
3.4	0.2
	1.0

used 10 groups, we would assign probabilities of 0.1 to the equally likely groups, and so on. Computer programs are available to carry out the bracket median procedure for large numbers of equally probable groupings. The resulting discrete probability distributions can be used in a decision analysis in the same way that the probability distribution of sales was used in the example given earlier in this chapter.

14.8 DECISION MAKING BASED ON EXPECTED UTILITY

In the decision analysis discussed up to this point, the criterion of choice was the maximization of expected monetary value. This criterion can be interpreted as a test of preference that selects the optimal act, that is, the act that yields the greatest long-run average profit. In a decision problem such as our example involving the inventor's choices, the optimal act is the one that would result in the largest long-run average profit if the same decision had to be made repeatedly under identical environmental conditions. In general, in such decision-making situations, the observed average payoff approaches the theoretical expected payoff as the number of repetitions increases. Gamblers, baseball managers, insurance companies, and others who engage in what is colloquially called "playing the percentages," are often characterized as using the aforementioned criterion. However, many of the most important personal and business decisions are made under unique sets of conditions, and on some of these occasions it may not be realistic to think in terms of many repetitions of the same decision situation. Indeed, in the business world, many of management's most important decisions are unique, high-risk, high-stake choice situations, whereas the less important, routine, repetitive decisions are ones customarily delegated to subordinates. Therefore, it is useful to have an apparatus for

One-time decision making

TABLE 14-11
Alternative courses of action with different expected monetary payoffs

Certainty Equivalent		Gamble
A_1: Receive \$0 for certain. (That is, you are certain to incur neither a gain nor a loss.)	or	A_2: Receive \$0.60 with probability $\frac{1}{2}$ and lose \$0.40 with probability $\frac{1}{2}$
B_1: Receive \$0 for certain. (That is, you are certain to incur neither a gain nor a loss.)	or	B_2: Receive \$60,000 with probability $\frac{1}{2}$ and lose \$40,000 with probability $\frac{1}{2}$
C_1: Receive a \$1 million gift for certain.	or	C_2: Receive \$2.1 million with probability $\frac{1}{2}$ and receive \$0 with probability $\frac{1}{2}$

dealing with one-time decision making. **Utility theory**, which we discuss in this section, provides such an apparatus, as well as a logical method for repetitive decision making.

Whether an individual, a corporation, or other entity would be willing to make decisions on the basis of the expected monetary value criterion depends on the decision maker's attitude toward risk situations. Several simple choice situations are presented in Table 14-11 to illustrate that in choosing between two alternative acts we might select the one with the lower expected value. We might make a choice for lower expected value because we feel that the risk of gain is too great—that the increase in expected gain of the act with greater expected monetary value does not sufficiently reward us for the additional risk involved. Table 14-11 gives three choice situations for alternative acts grouped in pairs. For each pair, a decision must be made between the two alternatives.

The illustrative choices are to be made once and only once. That is, the decision experiment is not to be repeated. We shall assume for simplicity that all monetary payoffs are tax free. Suppose you choose acts A_2, B_1, and C_1. In the choice between A_1 and A_2, you might argue as follows: "The expected value of act A_1 is \$0; the expected value of A_2 is $E(A_2) = (\frac{1}{2})(\$0.60) + (\frac{1}{2})(-\$0.40) = \$0.10$. A_2 has the higher expected value, and since I can sustain the loss of \$0.40 with equanimity, I am willing to accept the risk involved in the selection of this course of action."

A useful way of viewing the choice between A_1 and A_2 is to think of A_2 as an option in which a fair coin is tossed. If it lands "heads" you receive a payment of \$0.60. If it lands "tails" you must pay \$0.40. If you select act A_1, you are choosing to receive \$0 for certain rather than to play the game involved in flipping the coin. Hence, you neither lose nor gain anything. The \$0 figure is referred to as the **certainty equivalent** of the indicated gamble.

Certainty equivalent

On the other hand, you might very well choose act B_1 rather than B_2, even though the respective expected values are

$$E(B_1) = \$0$$

$$E(B_2) = \left(\frac{1}{2}\right)(\$60{,}000) + \left(\frac{1}{2}\right)(-\$40{,}000) = \$10{,}000$$

In this case, you might reason that even though act B_2 has the higher expected monetary value, a calamity of no mean proportions would occur if the coin landed tails, and you incurred a loss (say, a debt) of $40,000. Your present asset level might make such a loss intolerable. Hence, you would refuse to play the game. If you look at the difference between the two choices just discussed (A_1 versus A_2 and B_1 versus B_2) you will note that the only difference is in the amounts of the gains and losses. In A_2 we had the payoffs $0.60 and −$0.40. In B_2 the decimal point has been moved five places to the right for each of these numbers, making the monetary gains and losses much larger than in A_2. In all other respects, the wording of the choice between A_1 and A_2 and between B_1 and B_2 is the same. Nevertheless, as we shall note after the ensuing discussion of the choice between acts C_1 and C_2, it is not necessarily irrational to select act A_2 over A_1, where A_1 has the greater expected monetary value, and B_1 over B_2, where B_1 has the smaller expected monetary value.

In the choice between acts C_1 and C_2, you would probably select act C_1, which has the lower expected monetary value. That is, most people would doubtless prefer a certain gift of $1 million to a 50:50 chance at $2.1 million and $0, for which the expected payoff is

$$E(C_2) = \left(\frac{1}{2}\right)(\$2{,}100{,}000) + \left(\frac{1}{2}\right)(\$0) = \$1{,}050{,}000$$

In this case, you might argue that you would much prefer to have the $1 million for certain, and go home to contemplate your good fortune in peace, than to play a game where on the flip of a coin you might receive nothing at all. You might also feel that there are relatively few things that you could do with $2.1 million that you could not accomplish with $1 million. Hence, the increase in satisfaction to be derived even from winning on the toss of the coin in C_2 might not convince you to take the risk involved compared with the "sure thing" of $1 million in the selection of act C_1.

From the above discussion, we may conclude that it is reasonable to depart sometimes from the criterion of maximizing expected monetary values in making choices in risk situations. We cannot specify how a person *should* choose among alternative courses of action involving monetary payoffs, given only the type of information contained in Table 14-11. Our decisions will clearly depend upon our *attitudes toward risk*, which in turn will depend on a combination of factors such as level of assets, liking or distaste for gambling, and psychoemotional constitution.

Attitudes toward risk

Large and small corporations do (and should) have different attitudes toward risk.

If we single out the factor of level of assets, for example, a large corporation with a substantial asset level may undertake certain risky ventures that a smaller corporation with fewer assets would avoid. In the case of the large corporation, an outcome of a loss of a certain number of dollars might represent an unfortunate occurrence but as a practical matter would not materially change the nature of operation of the business. In the case of the small corporation, a loss of the same magnitude might constitute a catastrophe and might require the liquidation of the business. In comparing a venturesome management of a small company with a highly conservative management of a large company, however, the attitudes toward risky ventures might well be found to be the reverse of the norm just indicated.

We can summarize the problem concerning decision making in problems involving payoff that depend on risky outcomes as follows:

When monetary payoffs are inappropriate as a measuring device, it may be appropriate to substitute a set of values that reflects the decision maker's attitude toward risk.

A clever approach to this problem has been furnished by John Von Neumann and Oskar Morgenstern, who developed the so-called Von Neumann and Morgenstern utility measure.[10] In the next section, we consider how these utilities may be derived and the procedures for using them in decision analysis.

Construction of Utility Functions

We have seen that in certain risk situations we might prefer one course of action to another even though the act selected has a lower expected monetary value. In the language of decision theory, we prefer the selected act because it possesses greater expected **utility** than does the act that is not selected. The procedure used to establish the utility function of a decision maker requires a series of choices. In each choice, the decision maker opts for certainty or for uncertainty.

[10] The term *utility* as used by Von Neumann and Morgenstern and as used in this text differs from the economist's use of the same word. In traditional economics, utility refers to the inherent satisfaction delivered by a commodity and is measured in terms of psychic gains and losses. On the other hand, Von Neumann and Morgenstern conceived of utility as a measure of value used in the assessment of situations involving risk, which provides a basis for choice making. The two concepts can give rise to widely differing numerical measures of utility.

With certainty, one receives an amount of money denoted M (for "money"). With uncertainty, one gambles on receiving either an amount M_1 (with probability p) or an amount M_2 (with probability $1 - p$).

The question the decision maker must answer is, "What probability p for consequence M_1 would make me *indifferent* to receiving M for certain and to participating in the gamble involving the receipt of either M_1 with probability p or M_2 with probability $1 - p$?"

This probability assessment provides the assignment of a utility index to the monetary value M. The data obtained from the series of questions posed to the decision maker result in a set of **money-utility pairs** that can be plotted on a graph and that constitutes the decision maker's utility function for money.

Constructing a utility function

We will illustrate the procedure for constructing an individual's utility function by returning to one of the examples given in Table 14-12.

Suppose we ask the decision maker to choose between the following:

B_1: Receive \$0 for certain

B_2: Receive \$60,000 with probability $\frac{1}{2}$
and lose \$40,000 with probability $\frac{1}{2}$

Suppose the decision maker chooses to receive \$0 for certain. Our task then is to find out what probability for the receipt of the \$60,000 would make the decision maker indifferent to the choice between the gamble and the certain receipt of \$0. This will enable us to determine the utility assigned to \$0. The first step is the arbitrary assignment of "utilities" to the monetary consequences in the gamble, for example,

$$U(\$60{,}000) = 1$$

$$U(-\$40{,}000) = 0$$

where the symbol U denotes "utility" and $U(\$60{,}000) = 1$ is read "the utility of \$60,000 is equal to one."

The assignment of zero and one as the utilities of the lowest and highest outcomes of the gamble is entirely arbitrary. Any numbers can be assigned as long as the utility assigned to the higher monetary outcome is *greater than* that assigned to the lower outcome.

Thus, the utility scale has an arbitrary zero point, just like the 0° mark in temperature, which corresponds to different conditions depending on whether the Celsius or Fahrenheit scale is used.

The expected utility of the indicated gamble is

$$E[U(B_2)] = \frac{1}{2}[U(\$60{,}000)] + \frac{1}{2}[U(-\$40{,}000)] = \frac{1}{2}(1) + \frac{1}{2}(0) = \frac{1}{2}$$

Therefore, since the decision maker prefers \$0 for certain to this gamble, it follows that the utility assigned to \$0 is greater than $\frac{1}{2}$, or $U(\$0) > \frac{1}{2}$. In order to aid the decision maker in deciding how much greater than $\frac{1}{2}$ this utility is, we introduce the concept of a hypothetical lottery for calibrating the decision maker's utility assessment.

Let us assume that we have a box with 100 balls in it, 50 black and 50 white. The balls are identical in all other respects. Furthermore, we assume that if a ball is drawn at random from the box and its color is black, the decision maker receives a payoff of \$600,000. On the other hand, if the ball is white, the payoff is −\$40,000. We now have constructed a gamble denoted B_2.

The question is this: If we retain the total number of balls in the calibrating box at 100 but vary the composition in terms of the numbers of black and white balls, how many black balls would be required for a decision maker to be indifferent between receiving \$0 for certain and participating in the gamble? With 50 black balls in the box, the decision maker prefers \$0 for certain. With 100 black balls (and no white balls), the decision maker would obviously prefer the gamble, since it would result in a payoff of \$60,000 with certainty. For some number of black balls between 50 and 100, the decision maker should be indifferent, that is, at the threshold beyond which the gamble would become preferable to the certainty of the \$0 payoff. Suppose we begin replacing white balls with black balls, and for some time the decision maker is still unwilling to participate in the gamble. Finally, when there are 70 black balls and 30 white balls, the decision maker announces that the point of indifference has been reached. We now can calculate the utility assigned to \$0 as follows:

$$\begin{aligned} U(\$0) &= 0.70[U(\$60{,}000)] + 0.30[U(-\$40{,}000)] \\ &= 0.70(1) + 0.30(0) = 0.70 \end{aligned}$$

This utility calculation is a particular case of the general relationship

(14.1) $$U(M) = pU(M_1) + (1 - p)U(M_2)$$

where M is an amount of money received for certain and M_1 and M_2 are component prizes received in a gamble with respective probabilities p and $1 - p$.

We have now determined three money-utility ordered pairs: (−\$40,000, 0), (\$60,000, 1), and (\$0, 0.70). The first figure in each ordered pair represents a monetary payoff in dollars and the second figure represents the utility index assigned to this amount. The utility figures for other monetary payoffs between −\$40,000 and \$60,000 can be assessed in exactly the same way as for \$0, assuming that the patience of our long-suffering decision maker holds out. A relatively small number of points could be determined and the rest of the function interpolated. Suppose the utility function shown in Figure 14-5 results from the indifference probabilities assigned by the decision maker in the set of gambles

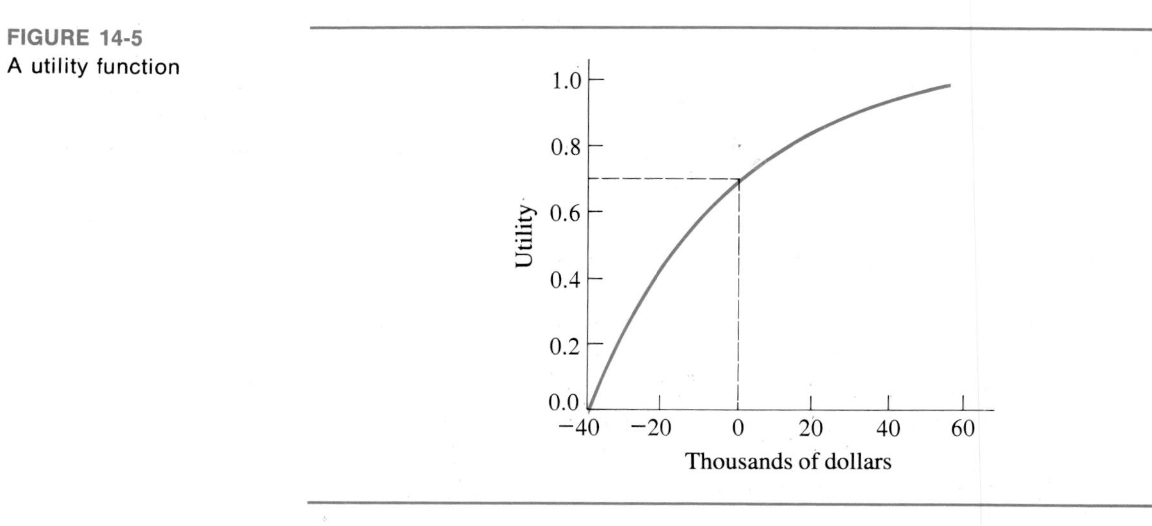

FIGURE 14-5
A utility function

proposed. The one point whose determination was illustrated ($0, 0.70) is depicted on the graph. This utility function can now be used to evaluate risk alternatives that might be presented to the decision maker. The expected utility of an alternative can be calculated by reading off the utility figure corresponding to each monetary outcome and then weighting these utilities by the probabilities that pertain to the outcomes. In other words, the utility figures can now be used by the decision maker in place of the original monetary values for calculation of expected utilities. For a person with this type of utility function, calculation of expected monetary values is clearly an inadequate guide for decision making.

The preceding discussion indicated one particular method of constructing a decision maker's utility function. That procedure was explained in terms of a basic gamble involving a maximum payoff of $60,000 and a minimum payoff of −$40,000, to which respective utility values of one and zero were assigned. Other points on the utility curve were obtained by determining the probabilities required for the receipt of the $60,000 to make the decision maker indifferent between the gamble and the certainty equivalent of specified dollar amounts ($0 was illustrated).

Five-point procedure of assessing utility functions

Another somewhat more practical method for assessing utility functions will now be discussed. This method may be referred to as the **five-point procedure,**[11] because in its simplest form, five points on the utility curve are determined, namely, the points pertaining to utility values 1.00, 0.75, 0.50, 0.25, and 0. Since it is helpful in understanding this method to have a particular example in mind, suppose you are considering two alternative investment proposals and wish to formalize your attitude toward risk by determining your utility function.

[11] See R. S. Schlaifer, *Analysis of Decisions Under Uncertainty* (New York: McGraw-Hill 1969) for a detailed discussion of this method of assessing points on a utility curve.

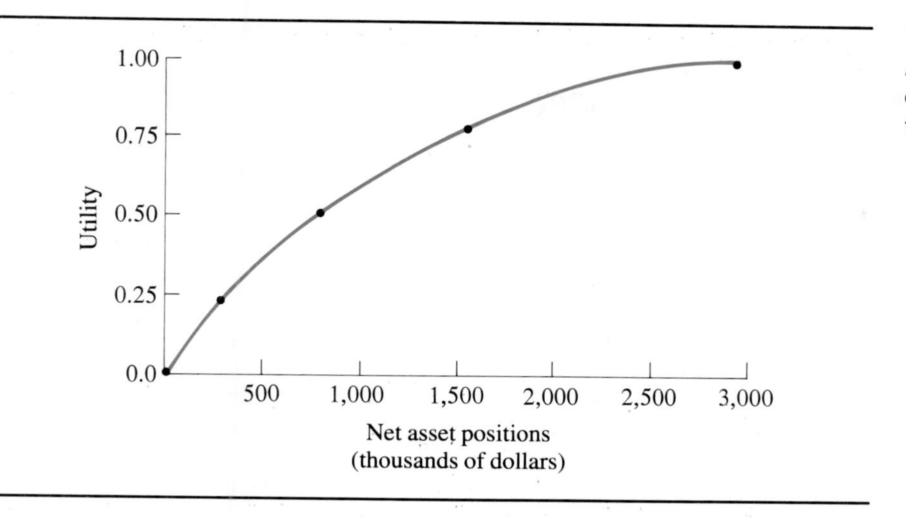

FIGURE 14-6
A utility function constructed by the five-point procedure

- You must first determine the **criterion** by which you should investigate your attitude toward risk.

A logical criterion would be your net asset position at the future time at which the consequences of the investment can be evaluated.[12]

- Then you should estimate the best and worst possible consequences in terms of your criterion.

Let us assume, for example, that under the best possible outcome of the two investment proposals, you believe your net asset position two years hence (including the results of the proposed investments) would be $2.5 million; the worst possible outcome would be a net asset position of $0.

- You should then choose a pair of **reference consequences** for your criterion of net assets.

The range of the reference consequences should be sufficiently wide to include the best and worst possible outcomes. Suppose you feel that reference consequence limits of $3 million and $0 are wide enough to include your possible net asset positions.

- We then arbitrarily assign utility values of one and zero to $3 million and to $0, respectively.

As in our previous example, we plot these as the two extreme points on our utility curve, shown in Figure 14-6.

[12] An alternative measure would be the *present value* of the net asset position at the future horizon data.

- We then obtain your certainty equivalents for three 50:50 gambles in the range between the two reference consequences.

Suppose that after considerable thought and discussion, you decide that you would take \$800,000 for certain in exchange for 50:50 chances at the extreme outcomes of \$3 million and \$0. Then the expected utility of the gamble would be

$$U(\$800{,}000) = \frac{1}{2}[U(\$3{,}000{,}000)] + \frac{1}{2}[U(\$0)]$$

$$= \frac{1}{2}(1) + \frac{1}{2}(0) = 0.50$$

- We could then plot the point (\$800,000, 0.50) on the graph.
- Now we can determine intermediate points between this 0.50 utility value and the two reference consequences.

Suppose that after suitable thought, you choose \$1.6 million as your certainty equivalent for a 50:50 gamble at consequences of \$3 million and \$800,000. We now have

$$U(\$1{,}600{,}000) = \frac{1}{2}[U(\$3{,}000{,}000)] + \frac{1}{2}[U(\$800{,}000)]$$

$$= \frac{1}{2}(1) + \frac{1}{2}(0.50) = 0.75$$

- This yields the point (\$1,600,000, 0.75) that can be plotted on the graph.
- Finally, let us assume that in a 50:50 gamble between \$800,000 and \$0, you choose a certainty equivalent of \$300,000.

Then we have

$$U(\$300{,}000) = \frac{1}{2}[U(\$800{,}000)] + \frac{1}{2}[U(\$0)]$$

$$= \frac{1}{2}(0.50) + \frac{1}{2}(0) = 0.25$$

We have now determined the following five points:

Utility	1.0	0.75	0.50	0.25	0
Net asset positions	\$3,000,000	\$1,600,000	\$800,000	\$300,000	\$0

The smooth curve drawn through these five points is shown in Figure 14-6.

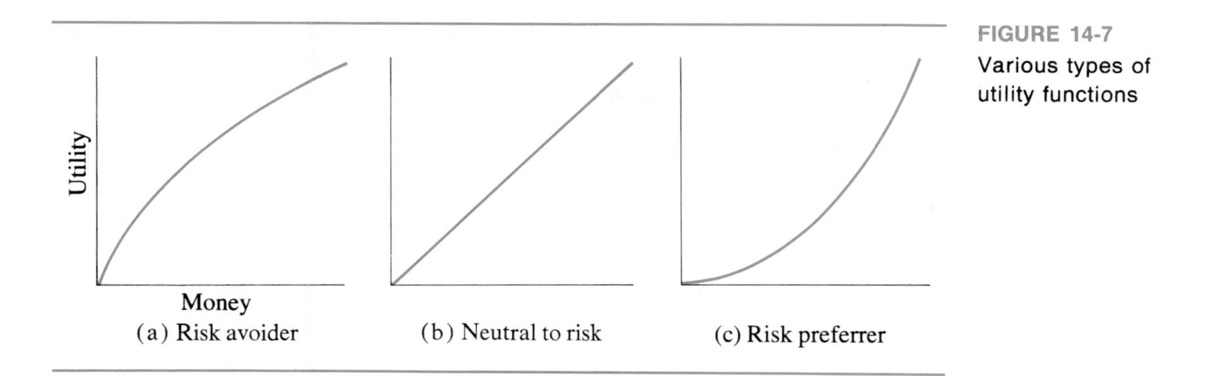

FIGURE 14-7
Various types of utility functions

In a real-world situation, when the five points are plotted they may not lie along a smooth curve. When presented with this evidence, the decision maker should resolve inconsistencies.

Furthermore, additional gambles should be posed for other intermediate points as consistency checks. The purpose of this checking procedure is to make sure that the utility curve appropriately represents the decision maker's attitude toward risk for all outcomes in the range of the two reference consequences.

Characteristics and Types of Utility Functions

The utility functions depicted in Figures 14-5 and 14-6 rise consistently from the lower left to the upper right side of the chart. That is, the utility curves have positive slopes. This general characteristic of utility functions simply implies that people ordinarily attach greater utility to a large sum of money than to a small sum.[13] Economists have noted this psychological trait in traditional demand theory and have referred to it as a *positive marginal utility for money*. The concave downward shape shown in Figure 14-6 illustrates the utility curve of an individual who has a diminishing marginal utility for money, although the marginal utility is always positive. This type of utility curve is characteristic of a risk avoider, as indicated in Figure 14-7(a). A risk avoider would prefer a small but *certain* monetary gain to a gamble that involves either a large but unlikely gain or a large and likely (not unlikely) loss. The linear function in Figure 14-7(b) depicts the behavior of a person who is neutral or indifferent to risk. For such a person every increase of, say, $1,000 has an associated constant increase in utility. This type of individual would use the

Risk avoider

Indifferent to risk

[13] The infinite variety of human behavior is demonstrated by the fact that conduct that runs counter to a generalization is occasionally observed. A newspaper reported that an heir to a fortune of $30 million committed suicide at the age of 23, after indicating in a letter that his wealth prevented him from living a normal life.

criterion of **maximizing expected monetary value** in decision making, because this would also maximize expected utility.

Risk preferrer

Figure 14-7(c) shows the utility curve for a risk preferrer. This type of person willingly accepts gambles that have a smaller expected monetary value than an alternative payoff received with certainty. For such an individual, the attractiveness of a possible large payoff in the gamble tends to outweigh the fact that the probability of such a payoff may indeed be very small.

Empirical research suggests that most individuals have utility functions in which for small changes in money amounts the slope does not change much. Over these ranges of money outcomes, the utility function may be considered approximately linear and constant in slope. However, in considering courses of action in which one of the consequences is adverse or in which one of the payoffs is large, individuals can be expected to depart from the maximization of expected monetary values as a guide to decision making.

For many business decisions in which the monetary consequences represent only a small fraction of the total assets of the business unit, the use of maximization of expected monetary payoff constitutes a reasonable approximation to the decision-making criterion of maximization of expected utility.

In other words, in such cases, the utility function may often be treated as approximately linear over the range of monetary payoffs considered.

Assumptions Underlying Utility Theory

The utility measure we have discussed was derived by evoking the decision maker's preferences between sums of money obtainable with certainty and lotteries or gambles involving a set of basic alternative monetary outcomes. This procedure entails a number of assumptions.

- It is assumed that an individual, when faced with the types of choices discussed, can determine three things: whether act A_1 is preferable to act A_2, whether A_2 is preferred to A_1, or whether these acts A_1 and A_2 are regarded indifferently. If A_1 is preferred to A_2, then the utility assigned to A_1 should exceed the utility assigned to A_2; if A_2 is preferred, the utility assigned to A_2 should be greater.

The principle of transitivity

- Another behavioral assumption is that an individual who prefers A_1 to A_2 and A_2 to A_3, will also prefer A_1 to A_3. This is referred to as the **principle of transitivity**. The assumption extends also to indifference relationships. That is, the decision maker who is indifferent to the choice between A_1 and A_2 and between A_2 and A_3 should also be indifferent to the choice between A_1 and A_3.

The principle of substitution

- Furthermore, it is assumed that if one is indifferent to the choice between a payoff or consequence of an act that replaces another consequence, one should also be indifferent to the choice between the old and new acts. This is often referred to as the **principle of substitution**.
- Finally, it is assumed that the utility function is **bounded**, which means that utility cannot increase or decrease without limit. As a practical matter, this simply means that the range of possible monetary values is limited. For example, at the lower end the range may be limited by a bankruptcy condition.

We may argue that human beings do not always exhibit the type of consistency implied by these assumptions. However, the point is that in the construction of our utility function if it is observed that we are behaving inconsistently and these incongruities are indicated to us and if we are "reasonable" or "rational," we should adjust our choices accordingly. If we insist on being irrational and refuse to adjust the choices that violate the underlying assumptions of utility theory, then a utility function cannot be constructed for us and we cannot use maximization of expected utility as a criterion of rationality in choice making. It is important to keep in mind that the theory discussed here does not purport to describe the way people actually *do* behave in the real world, but rather specifies how they *should* behave if their decisions are to be consistent with their own expressed judgments as to preferences among consequences. Indeed, we may argue that since human beings are fallible and do make mistakes, it is useful to have prescriptive procedures that police their behavior and provide ways in which the behavior can be improved.

A Brief Note on Scales

The Von Neumann–Morgenstern utility scales are examples of *interval scales.*

Interval scales have a constant unit of measurement but an arbitrary zero point. Differences between scale values can be expressed as multiples of one another, but individual values cannot.

In decision making using utility measures, if a different zero point and a different scale are selected, the same choices will be made. A constant can be added to each utility value, and each utility value can be multiplied by a constant, without changing the properties of the utility function. Thus, if a is any constant, b is a positive constant, and x is an amount of money,

$$U_2(x) = a + bU_1(x)$$

and $U_2(x)$ is as legitimate a measure of utility as $U_1(x)$.

EXAMPLE 14-1

The familiar scales for temperature are examples of interval scales. We cannot say that 100°C is twice as hot as 50°C. The corresponding Fahrenheit measures would not exhibit a ratio of 2:1. On the other hand, we can say that the intervals or differences between 100°C and 50°C and between 75°C and 50°C are in a 2:1 ratio. Using the relationship $F = \frac{9}{5}C + 32°$, we have

$$C = 100°; \quad F = (\tfrac{9}{5})(100°) + 32° = 212°$$

$$C = 75°; \quad F = (\tfrac{9}{5})(75°) + 32° = 167°$$

$$C = 50°; \quad F = (\tfrac{9}{5})(50°) + 32° = 122°$$

The difference between 100° and 50° = 50°; between 75° and 50° = 25°. The ratio of 50° to 25° is 2:1.

The difference between 212° and 122° = 90°; between 167° and 122° = 45°. The ratio of 90° to 45° is 2:1.

Exercises 14.8

1. a. The expected monetary return of the decision to buy life insurance is negative. Thus, it is irrational to buy life insurance. Do you agree or disagree? Explain.
   b. If a large company does not carry automobile insurance, why do you think this is so?
2. If the following investment outcomes have the given utilities

Prospect	A	B	C	D	E
Units of Utility	10	8	5	3	2

   would you prefer C for certain to
   a. a chance of getting A with 0.4 probability and E with 0.6 probability?
   b. a chance of getting B with 0.5 probability and E with 0.5 probability?
   c. a chance of getting A with 0.3 probability and D with 0.7 probability?
   d. a chance of getting B with 0.4 probability and E with 0.6 probability?
3. You have a choice of placing your money in the bank and receiving interest equal to 10 units of utility or investing in Rerox stock. With a probability of 0.4, Rerox will yield gains equal to 45 units of utility and with a probability of 0.6, it will cause a loss of 15 units of utility.
   a. What is the expected utility of the prospect "buy the stock"?
   b. Should you buy Rerox or put the money in the bank?
4. Drillwell Oil Company is debating what it should do with an option on a parcel of land. If it takes the option, the firm can drill with 100% interest or with 50% interest (i.e., all costs and profits are split with another firm). It costs $50,000 to drill a well and $20,000 to operate a producing well

until it is dry. The oil is worth $1 a barrel. Assume that the well is either dry or produces 200,000 or 500,000 barrels of oil. The firm assesses the probability of each outcome as 0.8, 0.1, and 0.1, respectively.

a. Based on expected monetary return, what is the best action?
b. Supposing Drillwell's management has the following utility function, what is the best decision based on expected utility?

Dollars	Units of Utility
−50,000	−30
−25,000	−10
65,000	25
130,000	60
215,000	120
430,000	200

5. The IRS has audited your last year's income tax and has sent you a bill for $225 for back taxes. You now have the choice of paying the bill or disputing the audit. If you dispute it, it will cost you $20 for an accountant's fee to prepare your case. After preliminary talks with your accountant, you feel the chances of your winning the dispute are 0.05.
   a. Should you dispute the case based on monetary expectations?
   b. Assume that large losses of money are disastrous to you as a struggling student. This is reflected in your utility function, which indicates that $U(-\$20)$ is -4 units of utility, $U(-\$225)$ is -425 units of utility, and $U(-\$245)$ is -440 units of utility. Based on expected utility, what is your best course of action?
6. A drug manufacturer has developed a new drug named Curitol. Tests have shown it to be extremely effective with almost no side effects. However, it has been tested for only three years, and long-range side effects are really unknown. The research department feels the probability the drug will have any serious long-range effects is 0.01. The Food and Drug Administration (FDA) must first clear the drug for sale. Assume that the FDA evaluates the loss to society because of serious long-range side effects as −900,000 units of utility, the gain to society because of the use of the drug as 8,000 units of utility, and the gain attributable to the economic advantages of production of a new drug as 1,000 units of utility. If the FDA accepts the firm's appraisal of the probability of long-range side effects, should it "accept" the drug?
7. An investor who is considering buying a franchised furniture business estimates that the business will yield either a loss of $50,000 or a profit of $100,000, $200,000, or $500,000 every five years, with probabilities 0.5, 0.2,

0.1, 0.2, respectively. The investor's utility function is found to be

Dollars	Units of Utility	Dollars	Units of Utility	Dollars	Units of Utility
−50,000	−40	25,000	1.2	150,000	7.5
−37,500	−10	50,000	2.5	200,000	10.0
−25,000	−4.2	75,000	5.8	250,000	12.5
−12,500	−2	100,000	6.0	375,000	26.0
0	0	125,000	6.2	500,000	40.0

a. Graph the utility function and interpret the shape of the curve.
b. Based on expected utility value, should the investor buy or not?
c. Suppose the investor can buy the whole franchise or $\frac{3}{4}$ or $\frac{1}{2}$ or $\frac{1}{4}$ of it (i.e., a $\frac{1}{4}$ interest means that the investor receives $\frac{1}{4}$ of all profits and pays $\frac{1}{4}$ of all losses). What is the best investment decision?

15

Decision Making with Posterior Probabilities

The discussion in Chapter 14 may be referred to as **prior analysis**, that is, decision making in which expected payoffs of acts are computed on the basis of *prior probabilities.* In this chapter, we discuss **posterior analysis**, in which expected payoffs are calculated on the basis of *revisions of prior probabilities.*

15.1 POSTERIOR PROBABILITIES

Bayes' theorem is utilized to accomplish revision of prior probabilities, which then become **posterior probabilities**. In this context, *prior* and *posterior* are relative terms. For example, subjective prior probabilities may be revised to incorporate the additional evidence of a particular sample. The revised probabilities then constitute posterior probabilities. If these probabilities are in turn revised on the basis of another sample or some experimental evidence, they represent prior probabilities relative to the new sample information, and the revised probabilities are posteriors.

Selecting terms of analyses

As in Chapters 14 through 17, for simplicity of presentation, we have carried out decision analyses in terms of **monetary payoffs** rather than *utility figures.* As noted in section 14.8, this use of monetary values constitutes an assumption that the utility function is approximately linear over the range of monetary payoffs considered. Of course, when this assumption is not valid, the analyses should be performed with appropriate utility values substituted for monetary payoffs.

Reducing the cost of uncertainty

The basic purpose of attempting to incorporate more evidence through sampling is to reduce the expected cost of uncertainty. If the expected cost of uncertainty (or the expected opportunity loss of the optimal act) is high, then it will ordinarily be wise to engage in sampling. Sampling in this context includes statistical sampling, experimentation, testing, and any other methods used to acquire additional information.

The general method of incorporating sample evidence into the decision-making process can be illustrated in terms of two types of sample information, where

- The reliability of the sample information is specified.
- The sample size is specified.

We will illustrate the first type in terms of the inventor's problem in Chapter 14. Then we will consider an acceptance sampling problem in which the additional evidence is based on a sample of a given size.

15.2 POSTERIOR ANALYSIS: A SPECIFIED RELIABILITY ILLUSTRATION

Reconsidering the inventor's problem

In the problem discussed in Chapter 14, suppose the inventor decided not to rely solely on prior probabilities concerning the demand for the new device, but to have a market research organization conduct a sample survey of potential consumers to gather additional evidence for the probable level of sales for the product. Let us assume that the survey can result in three sample outcomes, denoted x_1, x_2, and x_3, corresponding to the three states of nature, sales levels θ_1, θ_2, and θ_3. Specifically, the possible results may be

x_1: Sample indicates strong sales

x_2: Sample indicates average sales

x_3: Sample indicates weak sales

The survey is conducted, and the sample indicates an average level of sales, that is, x_2 is observed. Assume that on the basis of previous surveys of this type, the market research organization can assess the reliability of the sample evidence in the following terms: In the past, when the actual sales level after a new device was placed on the market was average, sample surveys properly indicated an average level of demand 80% of the time. However, when the actual level was strong sales, about 10% of the sample surveys incorrectly indicated demand as average; and when the actual level was weak sales, about 20% of the sample surveys gave an indication of average sales. These relative frequencies represent conditional probabilities that the sample evidence indicates "average sales," given the three possible actual events concerning sales level. They can be sym-

TABLE 15-1
Computation of posterior probabilities in the inventor's problem for the sample indication of an average level of sales

Event θ_i	$P(\theta_i)$	Conditional Probability $P(x_2\|\theta_i)$	Joint Probability $P(\theta_i)P(x_2\|\theta_i)$	Posterior Probability $P(\theta_i\|x_2)$
θ_1: Strong sales	0.2	0.1	0.02	0.042
θ_2: Average sales	0.5	0.8	0.40	0.833
θ_3: Weak sales	0.3	0.2	0.06	0.125
	1.0		0.48	1.000

bolized as follows:

$$P(x_2|\theta_1) = 0.1$$
$$P(x_2|\theta_2) = 0.8$$
$$P(x_2|\theta_3) = 0.2$$

The revision by means of Bayes' theorem of the prior probabilities assigned to the three sales levels on the basis of the observed sample evidence x_2 (average sales) is given in Table 15-1. In terms of equation 2.17 for Bayes' theorem, x_2 plays the role of B, the sample observation; θ_i replaces A_i, the possible events, or states of nature. After the joint probabilities are calculated, they are divided by their total (in this case, 0.48) to yield posterior, or revised, probabilities for the possible events. We note the effect of the weighting given to the sample evidence by Bayes' theorem in the revision of the prior probabilities by comparing the posterior probabilities with the corresponding "priors" in Table 15-1. With a sample indication of average sales, the 0.5 prior probability of the event "average sales" was revised upward to 0.833. Correspondingly, the probabilities of events "strong sales" and "weak sales" declined from 0.2 to 0.042 and from 0.3 to 0.125, respectively.

Decision Making after the Observation of Sample Evidence

The revised probabilities calculated in Table 15-1 can now be used to compute the **posterior expected profits** of the inventor's alternative courses of action. In Table 14-3, expected payoffs were computed based on the subjective prior probabilities assigned to the possible events. We can now refer to these as **prior expected profits**. The calculation of the posterior expected profits (using the revised, or posterior, probabilities as weights) is displayed in Table 15-2. By convention, we denote prior probabilities as $P_0(\theta_i)$ and posterior probabilities as $P_1(\theta_i)$. That is, the subscript zero is used to denote prior probabilities and the subscript one to signify posterior probabilities. A decision diagram of this

TABLE 15-2
Calculation of posterior or expected profits in the inventor's problem using revised probabilities of events (units of $10,000)

	Act A_1: Inventor Manufactures Device			Act A_2: Inventor Sells Patent Rights		
Event	Probability $P_1(\theta_i)$	Profit	Weighted Profit	Probability $P_1(\theta_i)$	Profit	Weighted Profit
θ_1: Strong sales	0.042	$80	3.360	0.042	$40	1.680
θ_2: Average sales	0.833	20	16.660	0.833	7	5.831
θ_3: Weak sales	0.125	−5	−0.625	0.125	1	0.125
	1.000		19.395	1.000		7.636
	Posterior expected profit A_1 = $19.395 (ten thousands of dollars) = $193,950			Posterior expected profit A_2 = $7.636 (ten thousands of dollars) = $76,360		

posterior analysis is given in Figure 15-1. Note that the decision tree is essentially the same as the one for the prior analysis given in Figure 14-1(b), except that posterior probabilities have been substituted for prior probabilities.

Since the posterior expected profit of act A_1 exceeds that of A_2, the better of the two courses of action remains that of the inventor manufacturing the device himself. However, after the sample indication of "average sales," the expected profit of act A_1 has decreased from $245,000 based on the prior prob-

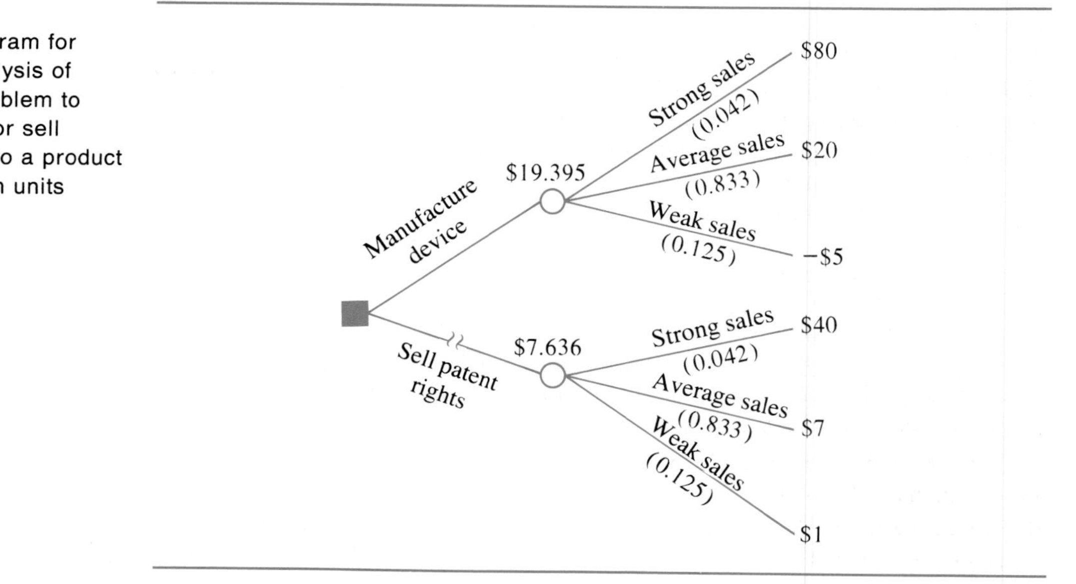

FIGURE 15-1
Decision diagram for posterior analysis of inventor's problem to manufacture or sell patent rights to a product (payoffs are in units of $10,000)

abilities to $193,950 based on the revised probabilities. Moreover, the difference in the expected profits of the two acts has narrowed somewhat. The $245,000 and $193,950 figures are, respectively, the **prior expected profit under uncertainty** and the **posterior expected profit under uncertainty**. The optimal course of action under a posterior analysis may be different from that of the prior analysis. In the present example, if the sample indication had been "weak sales," with appropriate conditional probabilities, the posterior expected profit of A_2 may have exceeded that of A_1. (Assume some figures and demonstrate this point.)

Insight can be gained into the cost of uncertainty and the value of obtaining additional information by calculating the **posterior expected value of perfect information**, which is simply the expected payoff, using posterior probabilities, of decision making in conjunction with a perfect predictor. We can now refer to the EVPI calculated in Chapter 14 (Table 14-7) as the **prior EVPI**. The prior EVPI of $18,000 means that the decision maker should have been willing to pay up to $18,000 for perfect information to eliminate his uncertainty concerning states of nature. Since no sample could be expected to yield perfect information, the decision maker cannot yet determine the worth of obtaining additional information through sampling. Expected value of sample information is discussed in Chapter 16. However, the prior EVPI of $18,000 sets an upper limit to the worth of obtaining perfect information and eliminating uncertainty concerning events. After the decision maker obtains additional information through sampling, he can calculate the **posterior EVPI**. The change in EVPI may be used to evaluate the decision to be made and the worth of attempting to obtain further information.

The posterior EVPI is computed to be $7,500 in Table 15-3(a). Analogously to prior analysis, the posterior EVPI is calculated by subtracting the posterior expected profit under uncertainty from the posterior expected profit with perfect information. The alternative determination of the posterior EVPI as the expected opportunity loss of the optimal act using posterior probabilities is given in Table 15-3(b). The only difference between this calculation and the similar calculation in Table 14-5 for the prior expected opportunity loss for act A_1 is the substitution of posterior probabilities for the prior probabilities.

In summary, the EVPI has been reduced from $18,000 to $7,500 by the information obtained from the sample. Whereas the decision maker should have been willing to pay up to $18,000 for a perfect predictor prior to having the sample information, the expected value of perfect information is only $7,500 after the sample indication of "average sales." In other words, the decision maker has reduced his cost of uncertainty; the availability of a perfect forecaster is not as valuable as it was prior to the sample survey. Although this problem resulted in a decrease from the prior EVPI to the posterior EVPI, there might very well have been an increase. This could occur if there were a marked difference between the posterior and prior probability distributions, and a reversal took place in the optimal act after the incorporation of sample

TABLE 15-3

(a) Calculation of the posterior expected value of perfect information for the inventor's problem (units of $10,000)

Event	Profit	Posterior Probability	Weighted Profit
θ_1: Strong sales	$80	0.042	3.360
θ_2: Average sales	20	0.833	16.660
θ_3: Weak sales	1	0.125	0.125
		1.000	20.145

Posterior expected profit with perfect information

= $20.145 (ten thousands of dollars)

= $201,450

Posterior expected profit with perfect information	$201,450
Less: Posterior expected profit under uncertainty	193,950
Posterior EVPI	$ 7,500

(b) Posterior expected opportunity loss of the optimal act for the inventor's problem (units of $10,000)

Event	Act A_1: Inventor Manufactures Device Probability	Opportunity Loss	Weighted Opportunity Loss
θ_1: Strong sales	0.042	0	0
θ_2: Average sales	0.833	0	0
θ_3: Weak sales	0.125	6	0.75
			0.75

Posterior EOL of the optimal act

= Posterior EVPI

= $0.75 (ten thousands of dollars)

= $7,500

information. An increase in the EVPI after inclusion of knowledge gained from sampling can be interpreted to mean that the additional evidence has increased the doubt concerning the decision. Of course, the determination of the prior and posterior EVPIs by alternative methods is not necessary in practice, but the computations have been shown here to indicate the relationships involved.

Exercises 15.2

1. Given

State of Nature	$P(\theta)$	$P(X\|\theta)$	$P(\theta)P(X\|\theta)$	$P(\theta\|X)$
θ_1: Housing starts will increase next year	0.5	0.6		
θ_2: Housing starts will remain at the same level or will decline		0.4		

If X is the result of a survey of 100 construction companies, fill in the blanks and interpret the data.

2. A foreign exchange broker has ascertained the following probabilities for the spot rate of the Dutch guilder one month from now:

S	$P(S)$
$S_1 = \$.35$	0.15
$S_2 = .36$	0.25
$S_3 = .37$	0.30
$S_4 = .38$	0.20
$S_5 = .39$	0.10

A sample observation X with the following properties occurs:

$$P(X|S_1) = 0.90 \qquad P(X|S_2) = 0.80 \qquad P(X|S_3) = 0.45$$

$$P(X|S_4) = 0.20 \qquad P(X|S_5) = 0.10$$

What are the revised probabilities?

3. A firm is trying to decide whether to embark on a new advertising campaign. Management assigns the following prior probability distribution:

State of Nature	Probability
S_1: Successful	0.5
S_2: Unsuccessful	0.5

A sample result with the following probability of occurrence is observed:

0.3 if S_1 is true
0.6 if S_2 is true

Revise the prior probability distribution in the light of this new information.

4. In a certain situation, before sampling, the best act is A_1 and the EVPI is \$100. After a sample was drawn, the best act was still A_1 and the revised EVPI was \$50. The cost of sampling was \$20. Can you conclude that the actual value of the sample information to the decision maker was \$30? Explain your answer.

5. Given

State of Nature	$P(X\|\theta)$	$P(\theta\|X)$
θ_1	0.75	0.50
θ_2	0.25	0.50

where θ_1 is "product is superior to competitor's" and θ_2 is "product is as good as or inferior to competitor's." An experiment was run with the results given above. What was the prior probability distribution before sampling?

6. There are two possible actions to take, A_1 and A_2, and there are two states of nature, S_1 and S_2. A_1 is preferred if S_1 is true, and A_2 is preferred if S_2 is true. If the prior probabilities are $P(S_1) = 0.7$ and $P(S_2) = 0.3$ and you observe a sample observation S such that the $P(S|S_1) = 0.9$ and $P(S|S_2) = 0.2$, can you conclude that A_1 is the better act? Explain your answer.
7. Management's prior probability assessment of demand for a newly developed product is 0.55 for high demand, 0.25 for average demand, and 0.20 for low demand. A survey taken to help determine the true demand for the product indicates that demand is high. The reliability of the survey is such that it will indicate high demand 80% of the time when demand is actually high, 60% of the time when demand is actually average, and 15% of the time when demand is actually low. In the light of this information, what would be the reassessed probabilities of the three states of nature?
8. A bond trader is considering the sale of inventory holdings of New York City obligations. The profit on such a transaction would be $5,000 if undertaken immediately. If she waits six months before selling, the profit will depend on the direction of change in interest rates during that period. The potential profits and the trader's subjective probabilities for each possible event are shown below:

Event	Probability $P(\theta_i)$	Profit
θ_1: Higher interest rates	0.45	−$15,000
θ_2: No change	0.35	$ 5,000
θ_3: Lower interest rates	0.20	$18,000

What action should she take?

Now suppose the latest business statistics show that the number of companies planning to raise capital through debt during the next six months is very large. The trader knows that the probability of this situation given expectations of lower interest rates is 0.90, the probability given no expected change in interest rates is 0.25, and the probability given expectations of higher rates is 0.10. Compute the revised probabilities of the three events and the posterior expected profits of the two possible actions.

9. There are two states of nature and two alternative actions open to a firm. The payoff matrix in units of $ thousands and the initial probabilities are

Action	State of Nature S_1	S_2	$P(S_i)$
A_1	100	10	$P(S_1) = 0.3$
A_2	20	30	$P(S_2) = 0.7$

The firm is planning to construct an information system, that is, an organized system of data collection, storage, and analysis. It can construct two possible information structures, A and B. Both systems yield information either of type M_1 or type M_2. The reliability (that is, $P(M_i|S_j)$) of the information from each system is as follows:

	System A M_1	System A M_2
S_1	0.9	0.1
S_2	0.2	0.8

	System B M_1	System B M_2
S_1	0.6	0.4
S_2	0.4	0.6

a. What is the maximum amount the firm should pay for an information system?
b. Given that you receive information of type M_1 from system A, what are the revised prior probabilities, the best action, and the revised EVPI?
c. Given that you receive information of type M_1 from system B, what are the revised prior probabilities, the best action, and the revised EVPI?

15.3 POSTERIOR ANALYSIS: AN ACCEPTANCE SAMPLING ILLUSTRATION

As the second illustration of posterior analysis, we consider a problem in acceptance sampling of a manufactured product. Let us assume that the Renny Corporation inspects incoming lots of articles produced by a supplier in order to determine whether to accept or reject these lots. In the past, incoming lots from this supplier have contained 10%, 20%, or 30% defective articles. On a relative frequency basis, lots with these percentages of defectives have occurred 50%, 30%, and 20% of the time, respectively. The Renny Corporation feels justified in using these past percentages as prior probabilities for a lot just delivered by the supplier. Renny Corporation draws a simple random sample of 10 units with replacement from the incoming lot, and two defectives are found.

The Renny Corporation has found from past experience that it should accept lots that have 10% defectives and should reject those that have 20% or 30% defectives. That is, the costs of rework make it uneconomical to accept lots with more than 10% defectives. On the basis of a careful analysis of past costs, the Renny Corporation constructed the payoff matrix in terms of opportunity losses shown in Table 15-4. The two possible courses of action are act

TABLE 15-4
Payoff matrix showing opportunity losses for accepting and rejecting lots with specified proportions of defectives

Event p (lot proportion of defectives)	Act	
	A_1 Reject	A_2 Accept
0.10	$200	$ 0
0.20	0	100
0.30	0	200

A_1, to reject the incoming lot, and act A_2, to accept the incoming lot. The three states of nature are the possible proportions of defectives in the lot—namely, 0.10, 0.20, and 0.30. (For convenience, we consider the defectiveness of the lot in terms of decimals rather than percentages.) The proportion of defectives in the lot is denoted p and will be treated as a discrete random variable that can take on only the three given values. Of course, it is rather unrealistic to assume that an incoming lot can be characterized only by a 0.10, 0.20, or 0.30 proportion of defectives. However, for convenience of exposition, we make that assumption here. The same general principles would hold if more realistic assumptions were made, as, for example, that the proportion of defectives could take on values at intervals of one percentage point—namely, 0.00, 0.01, . . . , 1.00—or that a continuous probability distribution for proportion of defectives was approximated by the bracket median method described in section 14.7. We now proceed to apply some of the principles of decision analysis we have learned.

Suppose the Renny Corporation had to take action concerning acceptance of the present lot before drawing the sample of ten units from this lot. Assuming that the firm is willing to make its decision on the basis of prior information, what action should be taken? In the absence of additional information, the company could reasonably use the past relative frequencies of lot proportion of defectives as prior probability assignments. The prior analysis of the company's two courses of action based on expected opportunity losses is given in Table 15-5 and is depicted in Figure 15-2.[1] As shown in the table, when the past relative frequencies are used as prior probabilities, the expected opportunity loss of rejecting the lot is $100 and that of accepting the lot is $70. Hence, the optimal act is A_2, to accept the lot. The $70 figure is also the prior EVPI, since it represents the expected opportunity loss of the optimal act using the prior probability distribution.

[1] The notation $P_0(p)$ is used in Table 15-5 for the prior probability distribution of the random variable p, and $P_1(p)$ is used in Table 15-6 for the corresponding posterior probability distribution. Moreover, the random variable is referred to by lower-case p in this and subsequent illustrations. This is a departure from the convention of using capital letters to denote random variables and lower-case letters to represent the values taken on by the random variable. This is done to avoid confusion because of the common use of the capital P to mean "probability."

TABLE 15-5
Prior expected opportunity losses for the Renny Corporation problem

	Act A_1: Reject the Lot			Act A_2: Accept the Lot		
Event p	Prior Probability $P_0(p)$	Opportunity Loss	Weighted Opportunity Loss	Prior Probability $P_0(p)$	Opportunity Loss	Weighted Opportunity Loss
0.10	0.50	$200	$100	0.50	$ 0	$ 0
0.20	0.30	0	0	0.30	100	30
0.30	0.20	0	0	0.20	200	40
	1.00		$100	1.00		$70

$\text{EOL}(A_1) = \$100$ $\qquad$ $\text{EOL}(A_2) = \$70$

Prior EVPI = $70

We turn now to the posterior analysis. Assume that the Renny Corporation draws the simple random sample of 10 units with replacement from the incoming lot and observes two defectives. If we take this sample evidence into account, what is the optimal course of action?

Following the same general procedure as in the inventor's problem, we use the sample evidence to revise the prior probabilities assigned to the possible lot proportions of defectives. The application of Bayes' theorem for this purpose is shown in Table 15-6. The conditional probabilities shown in the third column of Table 15-6 are often referred to as "likelihoods." That is, they represent the likelihoods of obtaining two defectives in 10 units in a simple random sample

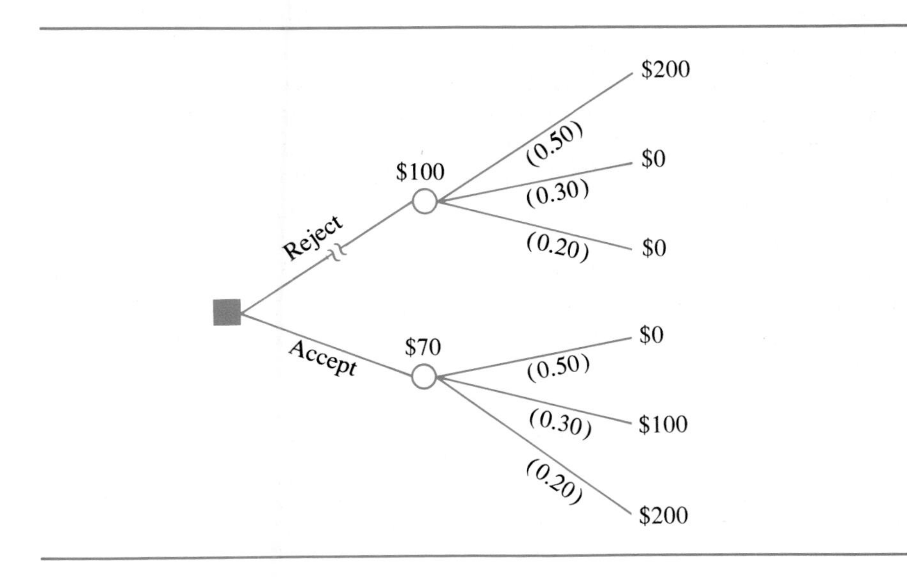

FIGURE 15-2
Decision diagram for prior analysis of the Renny Corporation problem of whether to accept or reject a lot produced by a supplier

TABLE 15-6
Computation of posterior probabilities for the Renny Corporation problem incorporating evidence based on a sample of size 10

Event p	Prior Probability $P_0(p)$	Conditional Probability $P(X = 2 \mid n = 10, p)$	Joint Probability $P_0(p)P(X = 2 \mid n = 10, p)$	Posterior Probability $P_1(p)$
0.10	0.50	0.1937	0.09685	0.4136
0.20	0.30	0.3020	0.09060	0.3869
0.30	0.20	0.2335	0.04670	0.1995
	1.00		0.23415	1.0000

drawn with replacement from the assumed incoming lots. When the basic random variable is the parameter p of a Bernoulli process, as in this problem, the likelihoods of the observed "number of successes" in the sample are computed by the binomial distribution. The likelihood figures in Table 15-6 were obtained from Table A-1 of Appendix A. The notation $P(X = 2 \mid n = 10, p)$ means "the probability that the random variable 'number of successes' is equal to 2 in 10 trials of a Bernoulli process whose parameter is p." We will use this type of symbolism in this and other problems for likelihood calculations. With the evidence of two defectives in a sample of 10 units, or 20% defectives in the sample, the prior probability that the lot contains 20% defectives is revised upward from 0.30 to 0.3869, as indicated in Table 15-6. Correspondingly, the probabilities that the lots contain 10% or 30% defectives are revised downward.

Expected payoffs of the two acts can now be recomputed using the posterior probabilities. These computations are shown in Table 15-7 and are

TABLE 15-7
Posterior expected opportunity losses for the Renny Corporation problem

	Act A_1: Reject the Lot			Act A_2: Accept the Lot		
Event p	Posterior Probability $P_1(p)$	Opportunity Loss	Weighted Opportunity Loss	Posterior Probability $P_1(p)$	Opportunity Loss	Weighted Opportunity Loss
0.10	0.4136	$200	$82.72	0.4136	$ 0	$ 0
0.20	0.3869	0	0	0.3869	100	38.69
0.30	0.1995	0	0	0.1995	200	39.90
	1.0000		$82.72	1.0000		$78.59

$\text{EOL}(A_1) = \$82.72$

$\text{EOL}(A_2) = \$78.59$

Posterior EVPI = $78.59

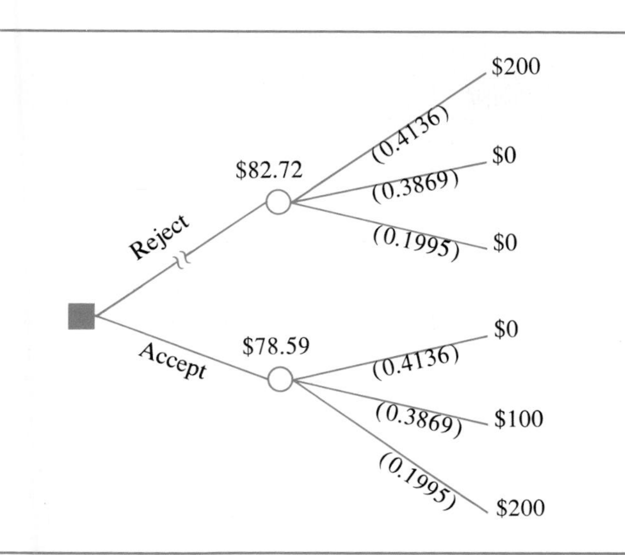

FIGURE 15-3
Decision diagram for posterior analysis of the Renny Corporation problem of whether to accept on reject a lot produced by a supplier

displayed in Figure 15-3. The optimal act is still A_2, accept the lot. However, the posterior expected opportunity losses of the two acts are much closer together than were the prior ones. Furthermore, the posterior expected opportunity loss of the optimal act, A_2, is $78.59, which represents an increase from the prior expected opportunity loss of the optimal act, $70. In other words, the posterior EVPI now exceeds the prior EVPI, which indicates that the value of having a perfect predictor available has increased.

Effect of Sample Size

We can use this acceptance sampling problem to illustrate an important point in Bayesian decision analysis, namely, the effect of sample size on the posterior probability distribution. Suppose that instead of a simple random sample size of 10 units being drawn from the incoming lot, a similar sample of 100 units was drawn. Furthermore, let us assume that 20 defectives were found in the sample; in other words, the fraction of defectives in this larger sample is 0.20, just as it was in the smaller sample. The computation of posterior probabilities using Bayes' theorem and the information from the sample of size 100 is given in Table 15-8. The conditional probabilities in Table 15-8 were obtained from a table of binomial probabilities that includes $n = 100$.

Although 0.30 was the prior probability assigned to the state of nature that the incoming lot proportion of defectives was 0.20, in the last column of Table 15-8, the revised probability is 0.9336. Hence, because of the implicit weight given to the sample evidence by Bayes' theorem, it is much more probable that the defective lot fraction is 0.20 according to the posterior distribution

TABLE 15-8
Computation of posterior probabilities for the Renny Corporation problem incorporating evidence based on a sample of size 100

Event p	Prior Probability $P_0(p)$	Conditional Probability $P(X = 20 \mid n = 100, p)$	Joint Probability $P_0(p)P(X = 20 \mid n = 100, p)$	Posterior Probability $P_1(p)$
0.10	0.50	0.0012	0.00060	0.0188
0.20	0.30	0.0993	0.02979	0.9336
0.30	0.20	0.0076	0.00152	0.0476
	1.00		0.03191	1.0000

than according to the prior distribution. Furthermore, comparing the posterior probability distributions in Tables 15-6 and 15-8, a much higher probability (0.9336) is assigned to the event $p = 0.20$ after 20% defectives have been observed in a sample of size 100 than when that percentage of defectives is found in a sample of 10 units (0.3869). A generalization of this result is that as sample size increases, the posterior distribution of the random variable "proportion of defectives" is influenced more by the sample evidence and less by the prior distribution.

Prior and Posterior Means

Let us consider a somewhat more formal explanation of the relationship between the prior distribution, the sample evidence, and the posterior distribution. This explanation can be given in terms of the change that takes place

TABLE 15-9
Calculation of the prior mean for the defective proportion in the Renny Corporation problem

Event p	Prior Probability $P_0(p)$	$pP_0(p)$
0.10	0.50	0.05
0.20	0.30	0.06
0.30	0.20	0.06
	1.00	0.17

Prior mean $= E_0(p) = 0.17$ defectives

TABLE 15-10
Calculation of posterior means for the proportion of defectives in the Renny Corporation problem

	Posterior Distribution Incorporating Sample Evidence $X = 2, n = 10$		Posterior Distribution Incorporating Sample Evidence $X = 20, n = 100$	
Event p	Posterior Probability $P_1(p)$	$pP_1(p)$	Posterior Probability $P_1(p)$	$pP_1(p)$
0.10	0.4136	0.04136	0.0188	0.00188
0.20	0.3869	0.07738	0.9336	0.18672
0.30	0.1995	0.05985	0.0476	0.01428
	1.0000	0.17859	1.0000	0.20288
	Posterior mean $= E_1(p) = 0.17859$		Posterior mean $= E_1(p) = 0.20288$	

between the mean of the prior distribution and the mean of the corresponding posterior distribution. For brevity, we will refer to the mean of a prior distribution of a random variable representing states of nature as a *prior mean* or *prior expected value* and the corresponding mean of a posterior distribution as a *posterior mean* or *posterior expected value.* The prior mean is obtained by the usual method for computing the mean of any probability distribution. The calculation of the prior mean for the acceptance sampling problem is given in Table 15-9. Thus, the prior mean is 0.17 defective articles. The notation $E_0(p)$ is used for the prior mean, the letter E denoting "expected value" and the subscript 0 denoting "prior distribution." Analogously, the mean of the posterior distribution is denoted $E_1(p)$. The corresponding computations of the posterior means for the cases in which the posterior distributions reflect sample evidence of two defectives in a sample of 10 units and 20 defectives in a sample of 100 units are given in Table 15-10. The prior mean is the expected proportion of defectives in the supplier's lot based on the Renny Corporation's prior assessment, that is, before the use of sample information. The posterior mean is the expected proportion of defectives in the supplier's lot based on the Renny Corporation's posterior assessment, that is, after incorporation of the sample information.

Rounding off the results obtained in Table 15-10, we observe posterior means of 0.179 defectives based on the sample evidence of two defectives in a sample of 10 units, and 0.203 based on sample evidence of 20 defectives in a sample of 100. Hence, in the case of the smaller sample size, the posterior mean lies closer to the prior mean of 0.17 defectives than to the sample evidence of 0.20 defectives. On the other hand, when the larger sample is employed, the

TABLE 15-11
Posterior expected opportunity losses for the Renny Corporation problem: the posterior probabilities incorporate sample evidence $X = 20$, $n = 100$

	Act A_1: Reject the Lot			Act A_2: Accept the Lot		
Event p	Posterior Probability $P_1(p)$	Opportunity Loss	Weighted Opportunity Loss	Posterior Probability $P_1(p)$	Opportunity Loss	Weighted Opportunity Loss
0.10	0.0188	$200	$3.76	0.0188	$ 0	$ 0
0.20	0.9336	0	0	0.9336	100	93.36
0.30	0.0476	0	0	0.0476	200	9.52
	1.0000		$3.76	1.0000		$102.88

$\text{EOL}(A_1) = \$3.76$ $\qquad$ $\text{EOL}(A_2) = \$102.88$

Posterior EVPI = $3.76

posterior mean falls closer to the sample evidence of 0.20 defectives than to the mean of the prior distribution.[2] This empirical finding agrees with our previous statement that as the sample size increases, the posterior distribution is progressively more influenced by the sample evidence and less by the prior distribution.

It is instructive to determine the optimal act using posterior probabilities that incorporate the evidence of 20 defectives in a sample of size 100. Table 15-11 gives the posterior expected opportunity losses based on posterior probabilities. The posterior expected opportunity losses of act A_1 and act A_2 are further apart than in any of the preceding cases. The optimal act is A_1, reject the lot, with a posterior expected opportunity loss of only $3.76 compared with the corresponding figure of $102.88 for A_2. Comparing this decision in favor of act A_1 with the analogous choice of act A_2 based on prior expected opportunity losses illustrates a reversal of decision that takes place because of sample evidence. The low $3.76 figure, which can be interpreted as the posterior EVPI, indicates that after observing 20 defectives in a sample of 100 units, the Renny Corporation would not be wise to spend much money accumulating additional evidence before making its decision concerning acceptance or rejection of the lot.

[2] The posterior mean of 0.203 exceeds both the value of the prior mean, 0.17, and the sample value, 0.20 defectives. An examination of the results for $p = 0.10$ and $p = 0.30$ in Table 15-8 gives the reason. The likelihood of 20 defectives in a sample of 100 for $p = 0.30$ is more than six times the similar likelihood for $p = 0.10$. Despite a lower prior probability for $p = 0.30$ than for $p = 0.10$, the joint probability and, therefore, the posterior probability in the case of 0.30 far exceed those for $p = 0.10$. The net effect is to pull $E_1(p)$ somewhat closer to 0.30 than to 0.10.

Exercises 15.3

Note: For ease of computations in these exercises, assume sampling with replacement (binomial sampling distributions) where appropriate.

1. Let p be the true percentage of customers who will purchase a new product. Assume that p can take on the values given below with the respective prior probabilities.

p	$P_0(p)$
0.05	0.65
0.15	0.35

A simple random sample of 15 customers is drawn. The customers are asked whether they would purchase the product. What are the revised probabilities if

a. one customer would purchase?
b. two customers would purchase?
c. three customers would purchase?

2. Let μ equal the average number of pedestrians injured per month in a process for which the Poisson distribution is a suitable model. The state of nature, μ, can assume the values given below with the respective prior probabilities.

μ	$P_0(\mu)$
5	0.6
6	0.4

If during a particular month the number of pedestrians injured was 6, what are the revised probabilities?

3. A small retail company is considering putting in a credit system. Let p be the proportion of new accounts that will be uncollectable if a credit system is installed. The states of nature and their respective prior probabilities are

p	$P_0(p)$
0.01	0.1
0.05	0.4
0.10	0.4
0.15	0.1

Assume that a credit system is installed. What are the revised probabilities if credit is extended to 100 customers and
a. 4 are uncollectable accounts?
b. 12 are uncollectable accounts?
Use a normal curve approximation.

4. Let p be the proportion of defective transistors in a lot offered to a radio manufacturer. Given:

p	$P_0(p)$	Opportunity Loss of Accepting Lot
0.10	0.40	0
0.20	0.30	100
0.25	0.20	300
0.30	0.10	600

A sample of 20 is taken, and 4 are found defective. What is the expected opportunity loss of the action "accept,"
a. if action is taken before sampling?
b. if action is taken after sampling?

5. In a previous problem describing an opportunity for Trivia Press, Inc., to publish a new novel, the probability of success for the novel was assessed as $\frac{1}{3}$. The gain from a successful novel would be $8 million; the loss if the novel is unsuccessful would be $4 million. The publisher decides to send a copy of the book to 10 critics for their opinions.

   The results are slightly unfavorable; 6 out of 10 dislike the book. If the book should be a failure, the probability that a critic would dislike it is 0.8, and if the book should be a success, there is a 50:50 chance that a critic would like it. In view of this added information, would you recommend publication of the book?

6. The government is trying to decide whether to build a reservoir in Dodd County or in Todd County. Dodd County has a population of 800,000, and Todd has a population of 1,200,000. The cost of building the reservoir is the same in each location. The immediate expected benefit of the reservoir is computed as $10 per person in the county immediately *after* the reservoir is built. Building the reservoir will displace people, and the loss due to displacement is estimated at $50 per person. The reservoir in Dodd would displace 15,000 people, and in Todd it would displace 92,000 people. Of those displaced, the percentage of people who would move out of Todd County is between 10% and 20%, and out of Dodd County, between 5% and 15%. From economic studies and comparisons with previous reservoir sites, the prior probabilities of the proportion that would move out are

estimated for each county as follows:

θ	Dodd $P_0(\theta)$	Todd $P_0(\theta)$
0.05	0.3	0.0
0.10	0.4	0.3
0.15	0.3	0.4
0.20	0.0	0.3

a. Write a mathematical expression for the benefit derived from each reservoir, letting x represent the proportion of displaced people who would move out of the county.
b. Based on prior beliefs, where should the reservoir be built?
c. In each county, a random sample of 20 persons was drawn. These people were asked whether they would move out of the county if displaced. Three people in Dodd and five in Todd said that they would move. Based on this new information, where should the reservoir be built?

7. A survey was taken by Noody's Investor Service to determine the percentage of large corporations that use debt only as a last resort for financing, with the aim of keeping their financial ratings as high as possible. The Investor Service's subjective probabilities associated with the possible percentage values, before considering the survey's result, were as follows:

p	$P_0(p)$
0.10	0.25
0.20	0.30
0.30	0.45

a. If the survey included 20 companies and 20% (four companies) maintained the conservative viewpoint on debt described above, what would be the revised probabilities after incorporating sample information?
b. If the survey included 100 companies and indicated the same percentage, what would be the revised probabilities? (See Table 15-8 for the conditional probabilities.)

8. Calculate the prior mean for the percentage of companies following the conservative financing practice described in exercise 7. Calculate the posterior mean both for parts (a) and (b).

16

Devising Optimal Strategies prior to Sampling

In Chapter 14, we discussed **prior analysis**, a method for decision making prior to the incorporation of additional information obtained through sampling or experimentation. In Chapter 15, we studied **posterior analysis**, a corresponding method for decision making after additional information has been obtained through sampling or experimentation. In posterior analysis, we draw a sample and choose the best act, taking into account both the sample data and prior probabilities of the events that affect payoffs.

16.1 PREPOSTERIOR ANALYSIS

In this chapter, we will consider **preposterior analysis**, which determines whether it is worthwhile to collect sample or experimental data at all. If data collection is worthwhile, this method further specifies the best courses of action for each possible type of sample or experimental outcome, and how large a sample to take. This type of analysis leads to a more complex but more interesting view of decision making than those we have considered so far.

Terminal decision

In both prior and posterior analysis, a final decision is made among the alternative courses of action based on the information at hand. A decision that makes a final disposition of the choice of a best act is referred to in Bayesian analysis as a **terminal decision**. The act itself is referred to as a **terminal act**.

In many business decision-making situations, the wise course of action is not to choose a terminal act, but rather to delay making a terminal decision in order to obtain further information.

Preposterior analysis, in addition to establishing whether additional information should be obtained (and, if so, how much), also delineates the optimal decision rule to employ based on the possible types of evidence that can be produced by the additional information if it is gathered. An obvious difficulty in this type of decision procedure is that the specific outcome of a sample (or additional information) is unknown prior to taking the sample. Yet, the decision whether sample information will be worthwhile must be made prior to drawing the sample. Because of the cost associated with acquiring additional knowledge through a sampling process, it pays to take a sample only if the anticipated worth of this sample information exceeds its cost. The methods discussed in this chapter provide a procedure for determining the **expected value of sample information**. We will begin by considering a simple problem involving a single-stage sample of fixed size. In section 16.2, we discuss two examples of preposterior analysis.

16.2 EXAMPLES OF PREPOSTERIOR ANALYSIS

A preposterior analysis, as the name implies, is an investigation that must be carried out *before* sample information is obtained and therefore *prior to* the availability of posterior probabilities based on a particular sample outcome. However, this type of analysis takes into account all possible sample results and computes the expected worth (or expected opportunity loss) of a strategy that assumes that the best acts are selected depending on the type of sample information observed. We will illustrate preposterior analysis by an oversimplified example in order to convey the basic principles of the procedure without getting bogged down in computational detail.

A Marketing Example

The A. B. Westerhoff Company problem

The A. B. Westerhoff Company, a firm that manufactures consumer products, considered marketing a new product it had developed. However, the company wanted to appraise the advisability of engaging a market research firm to help determine whether sufficient consumer demand existed to warrant placing the product on the market. The market research firm offered to conduct a nationwide survey of consumers to obtain an appropriate indication of the market for the product. The fee for the survey was $15,000. The A. B. Westerhoff

Company also wished to carry out an analysis that would specify whether it was better to act now (on the basis of prior betting odds as to the success or failure of the product and estimated payoffs) or to engage the market research firm and then act on the basis of the survey indication.

The company analyzed the situation as follows. The two available actions were

a_1: Market the product

a_2: Do not market the product

The states of nature, or possible events, were defined as

θ_1: Successful product

θ_2: Unsuccessful product

Although only two states of nature for success of the product are used in this example, many states could be employed to indicate degrees of success. Similarly more than two courses of action might be considered.

The company decided to view the problem in terms of opportunity losses of incorrect action. Based on appropriate estimates, the opportunity loss matrix (in thousands of dollars) shown in Table 16-1 was constructed.

The company, on the basis of its past experience with products of this type, assessed the odds that the product would be a failure at 3:1. That is, the company assigned prior probabilities to the success and failure of the product as follows:

$$P(\theta_1) = P(\text{successful product}) = \frac{1}{4} = 0.25$$

$$P(\theta_2) = P(\text{unsuccessful product}) = \frac{3}{4} = 0.75$$

The prior analysis in Table 16-2 was conducted to determine the optimal course of action if no additional information was obtained. As shown in Table 16-2, the optimal course of action was a_2 (do not market). The expected opportunity loss for this course of action was $50,000. A decision diagram of this prior analysis is given in Figure 16-1.

TABLE 16-1

Payoff table showing opportunity losses for the A. B. Westerhoff Company problem (thousands of dollars)

Event	Market a_1	Do Not Market a_2
θ_1: Successful product	$ 0	$200
θ_2: Unsuccessful product	160	0

TABLE 16-2
Calculation of prior expected opportunity losses for the A. B. Westerhoff Company problem (thousands of dollars)

Event	a_1: Market the Product			a_2: Do Not Market the Product		
	Probability $P(\theta_i)$	Opportunity Loss	Weighted Opportunity Loss	Probability $P(\theta_i)$	Opportunity Loss	Weighted Opportunity Loss
θ_1: Successful product	0.25	$ 0	$ 0	0.25	$200	$50
θ_2: Unsuccessful product	0.75	160	120	0.75	0	0
	1.00		$120	1.00		$50
	EOL(a_1) = $120 (thousands of dollars)			EOL(a_2) = $50 (thousands of dollars)		
				EOL of the optimal act = EOL(a_2) = $50 (thousands of dollars)		

The A. B. Westerhoff Company then turned its attention to the problem of analyzing its expected opportunity losses if its action was based on the survey results obtained by the market research firm. It requested the market research firm to indicate the nature and reliability of the consumer survey's results. The market research firm replied that the survey would yield one of the following three types of indications:

X_1: Favorable
X_2: Intermediate
X_3: Unfavorable

That is, an X_1 indication meant an observed level of consumer demand in the survey that was favorable to the success of the product, an X_3 result was unfavorable, and an X_2 indication meant a situation falling between levels X_1 and X_3 and classified as "intermediate." Although we are using the verbal classifications "favorable," "intermediate," and "unfavorable" here for simplicity of reference, each of these indications will represent a specific numerical range

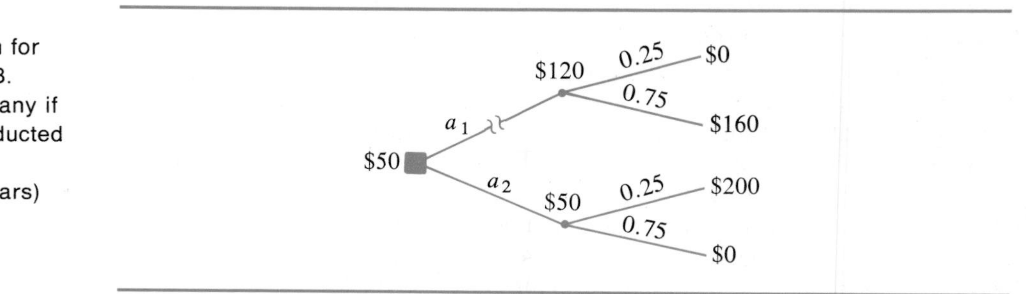

FIGURE 16-1
Decision diagram for action by the A. B. Westerhoff Company if no survey is conducted (all losses are in thousands of dollars)

TABLE 16-3
Conditional probabilities of 3 types of survey evidence for the A. B. Westerhoff Company problem of deciding whether to conduct a survey

Event	Conditional Probability $P(X_j \mid \theta_i)$			
	X_1	X_2	X_3	Total
θ_1: Successful product	0.72	0.16	0.12	1.00
θ_2: Unsuccessful product	0.08	0.12	0.80	1.00

of demand. We will explain later how this type of sample is used in decision making.

Anticipated reliability of survey indications

As a description of the anticipated *reliability* of the survey indications, the market research firm supplied the array of conditional probabilities shown in Table 16-3. The entries in this table are values of $P(X_j \mid \theta_i)$ based on past relative frequencies in similar types of surveys. That is, they represent the conditional probabilities of each type of sample evidence given that the product actually was successful or unsuccessful. For example, the $P(X_1 \mid \theta_1) = 0.72$ entry in the upper left-hand corner of the table means the probability is 72% that the survey will yield a favorable indication, given a successful product. The subscript j is used for the sample indication, denoted X_j, analogous to the use of the subscript i for the states of nature, denoted θ_i. It may not always be feasible to use past relative frequencies as the conditional probability assignments as we are doing here. The use of past relative frequencies as probabilities for future events always involves the assumption of a continuation of the same conditions as existed when the relative frequencies were established. As we have seen in other examples, probability functions such as the binomial distribution are often appropriate for computing these "likelihoods" or conditional probabilities of sample results.

The A. B. Westerhoff Company decided to carry out a preposterior analysis to determine whether to engage the market research firm to conduct the survey. For this purpose, it was necessary to compare (1) the expected opportunity loss of purchasing the survey and then selecting a terminal act to (2) the expected opportunity loss of terminal action without the survey. The latter expected loss figure was obtained from the prior analysis. As an intermediate step, the company computed the joint probability distribution of the sample evidence X_j and the events θ_i, as shown in Table 16-4. These joint probabilities were obtained by multiplying the prior (marginal) probabilities by the appropriate conditional probabilities. For example, 0.18, the upper left entry in the joint probability distribution in Table 16-4, was obtained by multiplying 0.25 by 0.72. In symbols, $P(X_1 \text{ and } \theta_1) = P(\theta_1)P(X_1 \mid \theta_1)$, and so on.

TABLE 16-4
Calculation of the joint probability distribution of survey evidence and events for the A. B. Westerhoff Company problem of deciding whether to conduct a survey

Event θ_i	Prior Probability $P(\theta_i)$	Conditional Probability $P(X_j\|\theta_i)$			Joint Probability $P(X_j \text{ and } \theta_i)$			Total
		X_1	X_2	X_3	X_1	X_2	X_3	
θ_1	0.25	0.72	0.16	0.12	0.18	0.04	0.03	0.25
θ_2	0.75	0.08	0.12	0.80	0.06	0.09	0.60	0.75
Total	1.00				0.24	0.13	0.63	1.00

Interesting points about joint probability distribution

It is useful to consider the following interesting points about the joint probability distribution to understand the subsequent analysis.

- As in any joint bivariate probability distribution, the totals in the margins of the table are marginal probabilities.
- The row totals are the prior probabilities $P(\theta_1) = 0.25$ and $P(\theta_2) = 0.75$.
- The column totals are the marginal probabilities of the survey evidence; that is, $P(X_1) = 0.24$, $P(X_2) = 0.13$, and $P(X_3) = 0.63$.
- Conditional probabilities of events, given the survey evidence—that is, probabilities of the form $P(\theta_i|X_j)$—can be computed by dividing the joint probabilities by the appropriate column totals, $P(X_j)$.

For example, the probability of a successful product, given a favorable survey indication, is

$$P(\theta_1|X_1) = \frac{P(\theta_1 \text{ and } X_1)}{P(X_1)} = \frac{0.18}{0.24}$$

and so on. These probabilities can be viewed as revised or posterior probability assignments to the events θ_1 and θ_2, given the survey evidence X_1, X_2, and X_3. The calculation of these posterior probabilities represents an application of Bayes' theorem (equations 2.12 and 2.16), as shown in the following method of symbolizing the probability $P(\theta_1|X_1)$ just calculated:

$$P(\theta_1|X_1) = \frac{P(\theta_1 \text{ and } X_1)}{P(X_1)} = \frac{P(\theta_1)P(X_1|\theta_1)}{P(\theta_1)P(X_1|\theta_1) + P(\theta_2)P(X_1|\theta_2)}$$

That is, the joint probability $P(\theta_1 \text{ and } X_1) = 0.18$ was calculated in Table 16-4 by multiplying the marginal probability $P(\theta_1) = 0.25$ by the conditional probability $P(X_1|\theta_1) = 0.72$; the marginal probability $P(X_1) = 0.24$ was obtained by adding the joint probability $P(\theta_1)P(X_1|\theta_1) = 0.18$ and $P(\theta_2)P(X_1|\theta_2) = 0.06$.

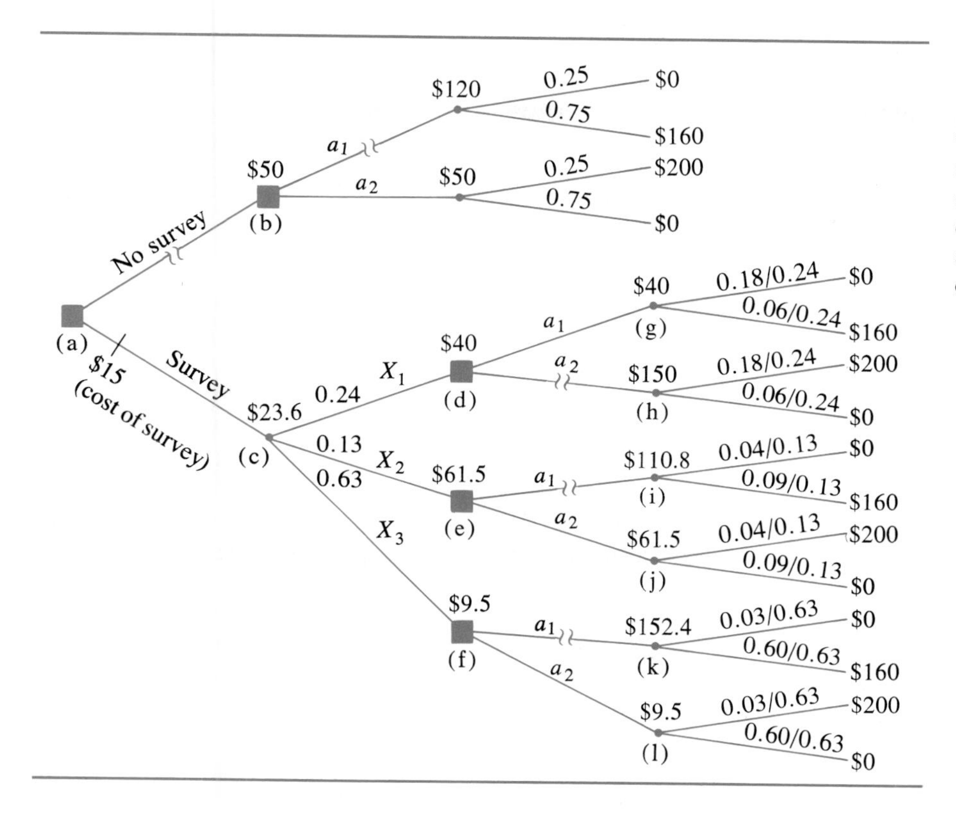

FIGURE 16-2
Decision diagram for preposterior analysis for the A. B. Westerhoff Company problem of deciding whether to conduct a survey (payoffs are in thousands of dollars)

The A. B. Westerhoff Company proceeded with its analysis and constructed the decision diagram shown in Figure 16-2. All monetary figures are in thousands of dollars. Returning to the beginning of the problem, the first choice is to purchase the survey or not to purchase it. Therefore, starting at node (a) at the left and following the "no survey" branch of the tree, we move along to node (b). From node (b) to the right is the decision tree depicted in Figure 16-1 for the prior analysis with no survey. As indicated in that figure, the $50 entry at node (b) is the expected opportunity loss of choosing the optimal terminal act without conducting the survey.

Expected Value of Sample Information

On the other hand, suppose that at the outset [at node (a)], the decision is to conduct the survey, and we move down to point (c). The results of the survey then determine which branch to follow. The three branches emanating from node (c), which represent X_1, X_2, and X_3 types of survey information, are marked with their respective marginal probabilities, 0.24, 0.13, and 0.63. Suppose type X_1 information (that is, a "favorable" indication) were observed.

Moving ahead to point (d), we can choose either act a_1 or act a_2, that is, market or not market the product. If act a_1 is selected, we move to node (g); if act a_2 is selected, we move to node (h). At these points, posterior or revised probability questions must be answered. For example, if act a_1 is chosen, the probabilities shown on the two branches stemming from node (g), $\frac{0.18}{0.24}$ and $\frac{0.06}{0.24}$, are the conditional (posterior) probabilities, given type X_1 information, that the product is successful or unsuccessful, or in symbols, $P(\theta_1|X_1)$ and $P(\theta_2|X_1)$. Thus, they represent revised probabilities of these two events after observation of a particular type of sample evidence. These conditional probabilities are calculated from Table 16-4, as indicated earlier, by dividing joint probabilities by the appropriate marginal probabilities. Looking forward from node (g) and applying the posterior probabilities ($\frac{0.18}{0.24}$ and $\frac{0.06}{0.24}$) as weights to the losses attached to the events "successful product" and "unsuccessful product," we obtain an expected opportunity loss payoff of \$40 for act a_1. This figure is entered at node (g). Comparing it with the corresponding figure of \$150 for a_2, we block off act a_2 as being nonoptimal. Therefore, \$40 is carried down to node (d), representing the payoff for the optimal act upon observation of type X_1 information. Similar calculations yield \$61.5 and \$9.5 at nodes (e) and (f) for X_2 and X_3 types of information. Weighting these three payoffs by the marginal probabilities of X_1, X_2, and X_3 indications (0.24, 0.13, and 0.63, respectively), we obtain a loss of \$23.6 as the expected payoff of conducting the survey and taking optimal action after the observation of the sample evidence.

Comparing the \$23.6 figure with the \$50 obtained under the "no survey" option, we see that it would be worthwhile to pay up to \$50 − \$23.6 = 26.4 (thousands of dollars) for the survey.

Expected value of sample information (EVSI)

The difference represented by the 26.4 (thousands of dollars) is referred to as the **expected value of sample information**, denoted EVSI. Hence, if we are considering a choice between immediate terminal action without obtaining sample information and a decision to sample and then select a terminal act, EVSI is the *expected amount by which the terminal opportunity loss is reduced by the information to be derived from the sample.* This EVSI is a gross figure, since it has not taken into account the cost of obtaining the survey information.

Expected net gain of sample information (ENGS)

To calculate the **expected net gain of sample information**, which we denote as ENGS, we subtract the cost of obtaining the sample information from the expected value of this sample information. In general,

(16.1) $$\text{ENGS} = \text{EVSI} - \text{Cost of sample information}$$

and in this problem,

$$\text{ENGS} = \$26.4 - \$15 = 11.4 \text{ (thousands of dollars)}$$

In conclusion, since the expected value of sample information was \$26,400 and the cost of the survey was \$15,000, the ENGS was \$11,400. Therefore, the A. B. Westerhoff Company decided that it was worthwhile to engage the market research firm to conduct the survey.

It is important to recognize that the EVSI and ENGS computations have been made with respect to the particular prior probability distribution used in the analysis. If different prior probabilities were used, the survey would have had correspondingly different EVSI and ENGS values. Sensitivity analysis, discussed in section 16.5, can test how sensitive the alternative actions are to the size of the prior probabilities.

Some Considerations in Preposterior Analysis

A few points concerning preposterior analysis arise from consideration of the foregoing example. First of all, we note that in this problem, the optimal action if no survey were conducted was a_2, not to market the product. On the other hand, as seen from Figure 16-2, if the survey were carried out and a favorable indication of demand, X_1, were obtained, the best course of action would be a_1, to market the product. The fact that for at least one of the survey outcomes the decision maker may possibly change the selected act gives the survey some value. Clearly, if a decision maker's course of action cannot be modified regardless of the experimental outcome, then the experiment is without value.

Second, the calculations in the preceding problem were carried out in terms of opportunity losses. If the payoffs had been expressed in terms of profits, the obvious equivalent analysis would have been required. For example, instead of the EVSI being the expected amount by which the terminal opportunity loss is reduced by the information to be derived from the sample, it would be the *expected amount by which the terminal profit is increased* by this information, Hence, expected profit of terminal action without sampling and expected profit of choosing a terminal action after sampling would be calculated. The first quantity would be subtracted from the second to yield the same EVSI figure obtained in the opportunity loss analysis.

Third, in this problem, the terms "sample" and "survey" were used interchangeably. Actually, even if the survey had represented a complete enumeration, the same analysis could have been carried out. Thus, the term "sample" in this context is used in a general sense to include any sample from size one up to a complete census. Indeed, the sample outcomes X_1, X_2, and X_3 need not have arisen from a statistical sampling procedure but could represent any set of experimental outcomes. Therefore, a preposterior analysis may be viewed as yielding the "expected value of experimental information" rather than the "expected value of sample information." However, since the latter term is conventionally used in Bayesian decision analysis, we have resisted the temptation to adopt a new term.

Finally, another generalization suggests itself. Only one survey of a fixed size was considered in this problem. Many surveys of different types and different sizes could have been considered. The preposterior expected opportunity loss would then be calculated for each of these different surveys. The minimum of these figures would be subtracted from the prior expected opportunity loss to yield the EVSI. In order to obtain the ENGS, the total expected opportunity loss for each survey is calculated by adding the cost of the survey to the corresponding preposterior expected opportunity loss. Then these figures are subtracted from the prior expected loss. The maximum of these differences is the ENGS, since it represents the expected net gain of the survey with the lowest total expected opportunity loss. Theoretically, the number of possible surveys or experiments might be considered infinite, but obviously there would be a delimitation at the outset based on factors of practicability or feasibility. On the other hand, the use of computers increases considerably the number of possible alternatives that can practically be compared.

Exercises 16.2

1. Define and compare EVSI and ENGS.
2. Let θ_1 be the state of nature "the unemployment rate will decrease over the next six months" and θ_2 "the unemployment rate will stay the same or increase over the next six months." The chief economist for the Bank of Boston will predict either X, "the rate will decrease," or Y, "the rate will stay the same or increase." Fill in the missing entries of the following table:

θ	$P(\theta)$	$P(X\|\theta)$	$P(Y\|\theta)$	$P(\theta)P(X\|\theta)$	$P(\theta)P(Y\|\theta)$	$P(\theta\|X)$	$P(\theta\|Y)$
θ_1	0.2	0.7					
θ_2	0.8	0.2					

3. Let the two possible states of nature be θ_1: the average level of stock market prices will advance, and θ_2: the average level of stock market prices will stay the same or decrease. Your prior belief is that there is a 60% chance that market prices will advance. In a week the forecast of a well-known econometric model will be published. Either an advance or no advance will be indicated. The model is correct 80% of the time. Find the posterior probabilities for θ for each of these indications.
4. Given the illustration of extensive analysis with payoffs in terms of opportunity losses that is shown in Figure 16-3,
   a. fill in all the missing entries.
   b. interpret each part of the tree.
   c. find the EVPI before and after sampling.
   d. find the EVSI.
   e. find the expected net gain from sampling.
5. A motel located near the site of a soon-to-be-opened world's fair is contemplating the construction of some temporary extra rooms. The rooms are of no value after the fair is closed, since the present size of the motel is more than adequate for normal demand. Let θ_1 be "demand for the

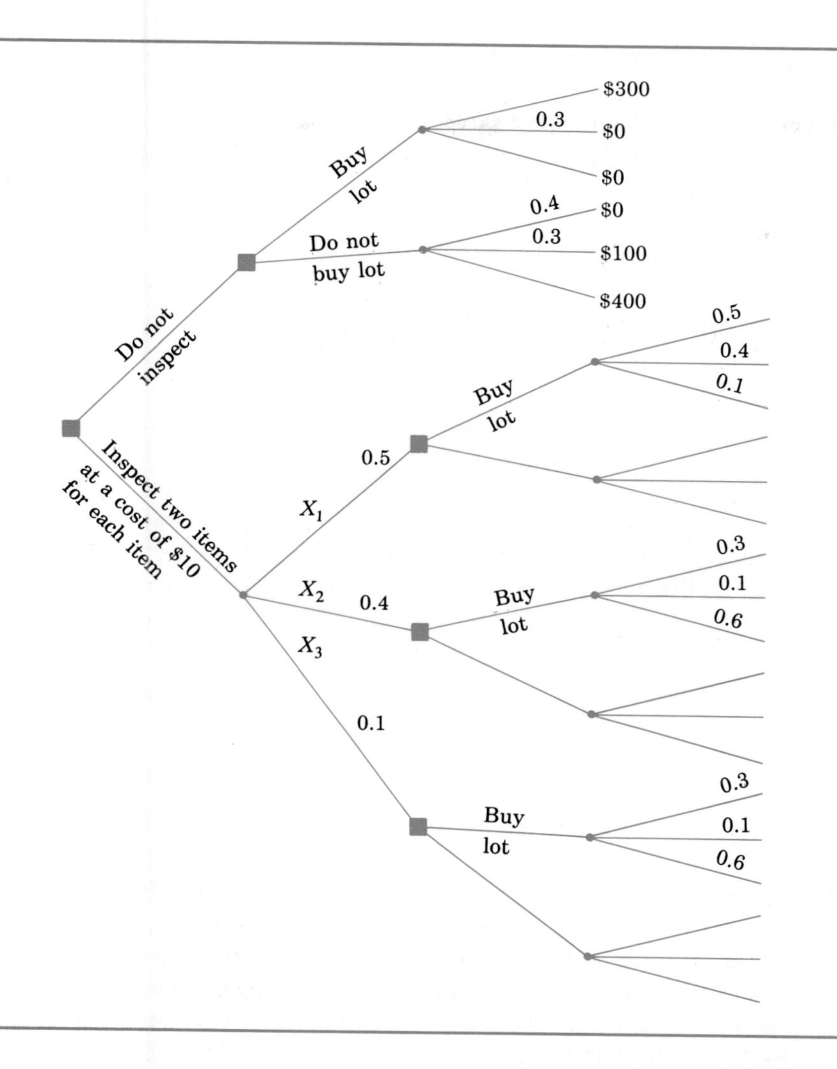

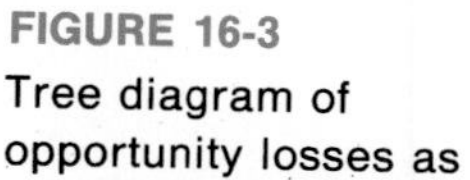

FIGURE 16-3
Tree diagram of opportunity losses as payoffs (see exercise 4)

temporary rooms is at least enough to cover the cost of building them," and θ_2 be, "demand for the temporary rooms is not sufficient to warrant construction." A survey can be taken that would yield two possible indications, X_1 or X_2. The following is known:

State of Nature	$P(\theta)$	$P(X_1 \mid \theta)$	Opportunity Loss Table: Build	Opportunity Loss Table: Do Not Build
θ_1	0.5	0.55	0	6
θ_2	0.5	0.50	5	0

a. Find
   1. the optimal act before sampling
   2. the optimal act if X_1 occurs
   3. the optimal act if X_2 occurs
   4. $P(X_1)$ and $P(X_2)$
   5. the EVSI

b. Should the survey be taken?

6. A mutual fund is contemplating the sale of a certain common stock. A study can be made that will yield two possible results, X_1 or X_2. Given the following information:

				Opportunity Loss (units of $10,000)	
State of Nature	$P(\theta)$	$P(X_1\|\theta)$	$P(X_2\|\theta)$	Sell	Do Not Sell
θ_1: Stock price up	0.70	0.85	0.15	8	0
θ_2: Stock price down or same	0.30	0.60	0.40	0	8

a. Find
   1. the optimal act before sampling
   2. the optimal act if X_1 results
   3. the optimal act if X_2 results
   4. $P(X_1)$ and $P(X_2)$
   5. the EVSI

b. If the cost of the study is $1000, what is the expected net gain from the study?

7. Assume that there are three possible states, S_1 (business will be a success), S_2 (business will have limited success), and S_3 (business will not be a success); with prior probabilities of occurrence 0.58, 0.25, and 0.17, respectively. An investigation costing $10,000 is run for which there are three outcomes, X, Y, and Z. The probabilities of these outcomes are 0.6, 0.3, and 0.1, respectively. Draw a decision tree diagram and make all the necessary entries on the branches dealing with outcome X.

	Opportunity Loss Table (in $000)		Posterior Probabilities of S_i Given the Respective Outcomes		
	Invest	Do Not Invest	X	Y	Z
S_1	0	400	0.8	0.3	0.1
S_2	0	100	0.1	0.5	0.4
S_3	500	0	0.1	0.2	0.5
			1.0	1.0	1.0

8. A pharmaceutical firm classifies its drugs as low-volume, medium-volume, and high-volume products. The firm introduces many new drugs each year. When a new drug is released, an initial marketing plan (that is, stocking of the drug, advertising, etc.) is put into effect. It is important that the firm initiate the proper marketing plan. For example, if a drug has a high-volume potential but a low-volume marketing plan is initiated, the drug will not realize its potential. The opportunity loss table for different marketing plans is

	State of Nature (in $000)		
Marketing Plan Used	Low Volume	Medium Volume	High Volume
Low	0	200	700
Medium	200	0	500
High	400	200	0

A new drug, Novatol, is to be released. The marketing vice-president feels the probability that it will be high-volume is 0.2 and the probability that it will be low-volume is 0.1. A forecasting model has been developed that has the following reliability [that is, P(indication/state of nature)]

	State of Nature		
Model Indicates	Low Volume	Medium Volume	High Volume
Low	0.85	0.10	0.05
Medium	0.10	0.80	0.10
High	0.05	0.10	0.85

What is the expected value of information from the forecasting model for the drug Novatol?

9. A large manufacturing firm is planning to build a new plant in either Georgia or Pennsylvania. The cost of the required land would be $750,000 less in Georgia due to the lower cost of the land and special tax incentives. However, in Pennsylvania a skilled labor force would be available, whereas that might not be the case in Georgia. It costs the company $5,000 to train and pay an employee until the worker is productive. It would be cheaper to move a skilled labor force into the area, but it is against company policy to relocate labor. The firm needs 500 skilled workers. Assume that either 300, 350, or 400 skilled employees can be hired in Georgia and the rest would have to be trained. Management feels that

the probabilities of hiring these numbers of employees are 0.3, 0.4, and 0.3, respectively. A survey of the labor situation can be undertaken. The reliability of the survey is

Survey Indicates	State of Nature		
	300	350	400
300	0.8	0.3	0.1
350	0.1	0.6	0.2
400	0.1	0.1	0.7

What is the maximum amount that the firm should pay for the survey?

10. In exercise 6, suppose the mutual fund felt that the prior probability the common stock price would increase was 0.95 instead of 0.70. What would be the expected net gain from the study? Compare it to the expected net gain if the prior probability is 0.70.
11. A mail order house with a fixed "market" of 100,000 people is deciding whether to sell a new line of goods. If more than 30% of its customers will purchase, the company should market the line, and if fewer than 30% purchase, it should not. For simplicity, assume that either 20% or 40% will purchase the new line. The manager of the firm believes the probability that only 20% of the market will buy is 0.6. The payoff table (in units of $10,000) is

	20% Will Purchase	40% Will Purchase
Market new line	−3	5
Do not market new line	0	0

a. One person is selected at random from the 100,000 and asked whether he will purchase. What is the expected value of this information?
b. Suppose two persons are selected at random from the 100,000 and asked whether they will purchase. What is the expected value of this information?

16.3 EXTENSIVE-FORM AND NORMAL-FORM ANALYSES

The type of preposterior investigation carried out in the A. B. Westerhoff Company problem is known as **extensive-form analysis**. It is perhaps easiest to characterize this type of analysis in terms of a decision tree diagram, such as

Figure 16-2. In that diagram, a prior analysis is given in the upper part of the tree with the resulting prior expected opportunity loss figure entered at node (b). An extensive-form analysis is given in the lower part of the tree with the resulting preposterior expected opportunity loss entered at node (c).

For purposes of comparison, we summarize the procedures for prior, posterior, and extensive-form preposterior analysis. This comparison assumes that experimental (sample) information is collected in a single-stage procedure as in the A. B. Westerhoff Company problem. Each type of analysis may be thought of as starting at the right side of a decision tree diagram and then proceeding inward by the process referred to in Chapter 14 as backward induction. See Table 16-5 for a comparison of the three types of analysis.

TABLE 16-5
The process of backward induction

<table>
<tr><th>Prior Analysis</th><th>Posterior Analysis</th><th>Preposterior Analysis—Extensive-Form</th></tr>
<tr><td colspan="3">1. Sketch a tree and depict states of nature at the right tips.</td></tr>
<tr><td colspan="3">2. Assign payoffs to each of the states for each possible action.</td></tr>
<tr><td>3. Assign prior probabilities to each state of nature.</td><td>Assign posterior probabilities to each state of nature based on a specific outcome of the experimental information.</td><td>a. Assign marginal probabilities to each possible experimental (sample) outcome.
b. Assign posterior probabilities to each state of nature given specific experimental outcomes.
c. For each experimental outcome, carry out steps 4 and 5.</td></tr>
<tr><td colspan="3">4. Calculate expected terminal payoffs for each act.</td></tr>
<tr><td colspan="3">5. Select the act with the highest expected terminal payoff.</td></tr>
<tr><td colspan="2">6.</td><td>For each type of experimental information, weight the expected terminal payoffs of the best act by the marginal probability of occurrence of that type of information, and add these products to yield an overall preposterior expected payoff.
The difference between the preposterior expected payoff and a prior expected payoff is the EVSI.</td></tr>
</table>

Concept of a Strategy

If many experiments are possible, rather than one as assumed in Table 16-5, the extensive-form analysis also specifies which of the possible experiments should be carried out. Moreover, the extensive-form analysis supplies a decision rule that selects an optimal act for each possible outcome of the chosen experiment. For example, in Figure 16-2, this decision rule was as follows:

$$X_1 \to a_1$$

$$X_2 \to a_2$$

$$X_3 \to a_2$$

where $X_1 \to a_1$ means that if experimental outcome X_1 is observed, select act a_1, and so on. Such a decision rule is referred to as a **strategy**.

Mathematically, a strategy can be defined as a function in which an act is assigned to each possible experimental outcome.

We now turn to an alternative procedure to extensive-form analysis, known as **normal-form analysis**, in which the problem begins with a listing of all possible strategies that might be employed. Normal-form analysis then makes an explicit comparison of the worth of all of these strategies to arrive at the same optimal strategy as was derived in extensive-form analysis.

Normal-Form Analysis

In the preceding section, extensive-form analysis was applied to the A. B. Westerhoff Company problem. In order to indicate the nature of the normal-form procedure and the relationship between extensive and normal forms of analysis, we will now solve the same problem by the latter method. Then some factors relating to the choice between the two procedures will be given.

All possible strategies or decision rules for the A. B. Westerhoff Company problem are listed in Table 16-6. The strategies (denoted $s_1, s_2, \ldots, s_8$) indi-

TABLE 16-6
A listing of all possible strategies in the A. B. Westerhoff Company problem of deciding whether to conduct a survey

Sample Outcome	Strategy s_1	s_2	s_3	s_4	s_5	s_6	s_7	s_8
X_1	a_1	a_1	a_1	a_1	a_2	a_2	a_2	a_2
X_2	a_1	a_1	a_2	a_2	a_1	a_1	a_2	a_2
X_3	a_1	a_2	a_1	a_2	a_1	a_2	a_1	a_2

cate the acts taken in response to each sample outcome. Hence, for example, strategy s_3 is

$$X_1 \rightarrow a_1$$
$$X_2 \rightarrow a_2$$
$$X_3 \rightarrow a_1$$

and s_4 is the one previously described. In this problem, there are 8 possible strategies, 3 sample outcomes (X_1, X_2, and X_3), and 2 possible acts (a_1 and a_2). In general, the number of possible strategies is given by n^r, where n denotes the number of acts and r is the number of sample outcomes. Hence, in this case, there are $2^3 = 8$ possible strategies.

Clearly, some of the strategies in Table 16-6 are not very sensible. For example, strategies s_1 and s_8 select the same act regardless of the experimental outcome (survey indication). Strategy s_5 selects act a_2, not to market the product, if the "favorable" survey indication X_1 is obtained, but perversely it would market the product if the intermediate or "unfavorable" indications X_2 or X_3 are obtained. On the other hand, strategies s_2 and s_4 seem to be quite logical and would probably be the only ones a reasonable person would seriously consider on an intuitive basis if he or she contemplated using the experimental information at all.

Continuing with the normal-form analysis, we now compute the expected payoff of each possible strategy. The method consists of calculating for each strategy the expected opportunity loss conditional on the occurrence of each state of nature.

These conditional expected losses are referred to in Bayesian decision analysis as **risk**.

We will use that term or its equivalent, **conditional expected losses**. The weighted average, or expected value, of these risks, using prior probabilities of states as weights, yields the **expected opportunity loss**, or **expected risk**, of the strategy. Table 16-7 shows the calculation of the risks associated with strategies s_2 and s_4. For example, let us consider the calculations for strategy s_2. The left-hand portion of the table is the payoff table in terms of opportunity losses. In the next section are **probabilities of action**, which are probabilities of taking actions specified by the given strategy based on the observations of the possible types of sample evidence. A convenient notation for designating a strategy is given in the column heading of this section of the table. The notation $s_2(a_1, a_1, a_2)$ means strategy s_2 consists in taking act a_1 if the sample indication X_1 is observed; again selecting a_1 if X_2 is observed; and choosing a_2 if X_3 is observed. That is, the first element within the parentheses denotes the action to be taken upon observing the first sample indication, the second element specifies the action to be taken upon observing the second sample indication, and so on.

TABLE 16-7
Calculation of risks, or conditional expected opportunity losses, for strategies s_2 and s_4 ($000)

State of Nature	Strategy s_2					
	Opportunity Loss		Probability of Action $s_2(a_1, a_1, a_2)$			Risk (conditional expected loss) $R(s_2 \mid \theta_i)$
	a_1	a_2	a_1	a_1	a_2	
θ_1	$ 0	$200	0.72	0.16	0.12	$24.00
θ_2	160	0	0.08	0.12	0.80	32.00

$$R(s_2 \mid \theta_1) = 0.72(\$0) + 0.16(\$0) + 0.12(\$200) = \$24.00$$
$$R(s_2 \mid \theta_2) = 0.08(\$160) + 0.12(\$160) + 0.80(\$0) = \$32.00$$

State of Nature	Strategy s_4					
	Opportunity Loss		Probability of Action $s_4(a_1, a_2, a_2)$			Risk (conditional expected loss) $R(s_4 \mid \theta_i)$
	a_1	a_2	a_1	a_2	a_2	
θ_1	$ 0	$200	0.72	0.16	0.12	$56.00
θ_2	160	0	0.08	0.12	0.80	12.80

$$R(s_4 \mid \theta_1) = 0.72(\$0) + 0.16(\$200) + 0.12(\$200) = \$56.00$$
$$R(s_4 \mid \theta_2) = 0.08(\$160) + 0.12(\$0) + 0.80(\$0) = \$12.80$$

The risk, or expected opportunity loss, given the occurrence of a specific state of nature, is calculated by multiplying the probabilities of action by the respective losses incurred if these actions are taken and summing the products. Thus, the risk, or expected opportunity loss, associated with the use of strategy s_2, given that state of nature θ_1 occurs—denoted $R(s_2 \mid \theta_1)$—is calculated to be $24.00, as shown in the top half of Table 16-7. Correspondingly, the expected opportunity loss of strategy s_2, conditional on the occurrence of θ_2—denoted $R(s_2 \mid \theta_2)$—is calculated to be $32.00. The analogous calculation of risks for strategy s_4 is given in the bottom half of Table 16-7.

We may also think of these risks as simply *calculations for each state of nature of the loss due to taking the wrong act times the probability that the wrong act will be taken.* For example, let us consider $R(s_2 \mid \theta_1)$, the conditional expected

TABLE 16-8

Alternative calculation of risks, or conditional expected opportunity losses, for strategies s_2 and s_4 ($000)

	Strategy $s_2(a_1, a_1, a_2)$		
State of Nature	Opportunity Loss of Wrong Act	Probability of Wrong Act Given θ_i	Conditional Expected Opportunity Loss $R(s_2\|\theta_i)$
θ_1	$200	0.12	$24.00
θ_2	160	0.20	32.00

	Strategy $s_4(a_1, a_2, a_2)$		
State of Nature	Opportunity Loss of Wrong Act	Probability of Wrong Act Given θ_i	Conditional Expected Opportunity Loss $R(s_4\|\theta_i)$
θ_1	$200	0.28	$56.00
θ_2	160	0.08	12.80

loss of strategy s_2, given that θ_1 occurs. Now, if θ_1 occurs, the correct (best) course of action is a_1; the incorrect act is a_2. The $24.00 figure for $R(s_2|\theta_1)$ is merely $200, the loss of taking act a_2, times 0.12, the total probability of selecting a_2 under the strategy s_2, given the occurrence of θ_1. Similarly, the $32.00 figure for $R(s_2|\theta_2)$ is equal to $160, the loss of taking act a_1, times 0.20, the total probability of selecting a_1 under strategy s_2, conditional on the occurrence of θ_2. The risk calculations of Table 16-7 are shown in Table 16-8, utilizing this alternative conception.

A summary of the risks, or conditional expected opportunity losses, for all eight strategies is given in Table 16-9. Let us review the interpretation of

TABLE 16-9

Risks, or conditional expected opportunity losses, for the 8 strategies in the A. B. Westerhoff Company problem ($000)

State of Nature	Strategy							
	s_1	s_2	s_3	s_4	s_5	s_6	s_7	s_8
θ_1	$ 0	$24.00	$ 32.00	$56.00	$144.00	$168.00	$176.00	$200.00
θ_2	160.00	32.00	140.80	12.80	147.20	19.20	128.00	0

TABLE 16-10
Expected opportunity losses of the 8 strategies in the A. B. Westerhoff Company problem

Strategy	s_1	s_2	s_3	s_4	s_5	s_6	s_7	s_8
EOL ($)	120.00	30.00	113.60	23.60*	146.40	56.40	140.00	50.00

* The minimum loss figure

these figures, using strategy s_4 as an example. In a relative frequency sense (that is, in a large number of identical situations of successful new products, θ_1), if survey evidence X_1, X_2, and X_3 occur with the specified probabilities and if strategy s_4 is employed, the average opportunity loss per product would be $56.00. A similar interpretation holds for unsuccessful products, θ_2, with an average loss of $12.80. Now, we can weight these average losses by the prior probabilities of successful and unsuccessful products to obtain an overall expected opportunity loss for strategy s_4. The prior probabilities were given as $P(\theta_1) = 0.25$ and $P(\theta_2) = 0.75$. Since θ_1 occurs with probability 0.25 and the conditional expected loss if θ_1 occurs is $56.00, and since θ_2 occurs with probability 0.75 and the conditional expected loss if θ_2 occurs is $12.80, the expected opportunity loss of using strategy s_4 is

$$\text{EOL}(s_4) = (0.25)(\$56.00) + (0.75)(\$12.80) = \$23.60$$

In symbols, we have

$$\text{EOL}(s_4) = P(\theta_1)R(s_4|\theta_1) + P(\theta_2)R(s_4|\theta_2)$$

In general form, the expected opportunity loss of the kth strategy s_k is

$$\text{(16.2)} \quad \text{EOL}(s_k) = P(\theta_1)R(s_k|\theta_1) + P(\theta_2)R(s_k|\theta_2) + \cdots + P(\theta_m)R(s_k|\theta_m) = \sum_{i=1}^{m} P(\theta_i)R(s_k|\theta_i)$$

The decision rule for which this expected opportunity loss is a minimum is known as the **Bayes' strategy**. The expected opportunity losses of the 8 strategies in the A. B. Westerhoff Company problem as calculated from equation 16.2 are given in Table 16-10. The optimal strategy, or the one with the lowest expected opportunity loss, is s_4. An asterisk has been placed beside the $23.60 expected opportunity loss associated with decision rule s_4 to indicate that it is the minimum loss figure.

Summary

In summary, using normal-form analysis, the optimal strategy in this problem is $s_4(a_1, a_2, a_2)$; this strategy has an expected opportunity loss of $23.60 (thousands of dollars). Referring to Figure 16-2, we see that this is exactly the same solution arrived at by extensive-form analysis. In that figure, the $23.6 is shown at node (c) and the optimal strategy (a_1, a_2, a_2) is determined by noting for the survey outcomes X_1, X_2, and X_3 the forks that have not been blocked off.

We can now summarize the steps involved in a normal form-analysis, in which sample evidence is obtained from a single sample (experiment) as in the foregoing example.

Normal-form analysis

1. List all possible strategies in terms of actions to be taken upon observation of sample outcomes.
2. Calculate the conditional expected opportunity loss (risk) for each state of nature. The probabilities used in this calculation are conditional probabilities of sample outcomes, given states of nature.
3. Compute the (unconditional) expected opportunity loss of each strategy by weighting the conditional expected opportunity losses by the prior probabilities of states of nature.
4. Select the strategy that has the minimum expected opportunity loss.

This summary of normal-form analysis has been given in terms of a single experiment. If more than one experiment is conducted, the decision maker should carry out steps (1) through (4) and then select the experiment that yields the lowest expected opportunity loss. Furthermore, the summary has been expressed in terms of opportunity losses. If payoffs of utility or profits were used, the same procedures would be followed except that the decision maker would maximize expected utility or profit rather than minimize expected opportunity loss.

16.4 COMPARISON OF EXTENSIVE-FORM AND NORMAL-FORM ANALYSES

As we have seen, the extensive and normal forms of analysis are equivalent approaches. In both procedures, an expected opportunity loss (or expected profit) is calculated before experimental results are observed. This expected payoff anticipates the selection of optimal acts after the observation of experimental outcomes. However, the two types of analysis differ in the way in which the various components of the procedures are performed. These differences give rise to advantages and disadvantages in the two forms of analysis.

We saw in the A. B. Westerhoff Company example that the extensive-form solution can be calculated more rapidly. This follows from the fact that *in the extensive-form approach, it is not necessary to carry out expected loss calculations for every possible decision rule.* Because nonoptimal courses of action posterior to the observation of sample evidence are blocked off, only the expected loss for the optimal strategy need be determined. In some problems, the number of decision rules or strategies that must be evaluated in a normal-form analysis may be large. For example, in the problem discussed in this chapter, the number of possible strategies was $2^3 = 8$, because there were two acts and three experimental outcomes. If there had been 3 acts ($n = 3$) and 4 experimental outcomes ($r = 4$), normal-form calculations would be required for $n^r = 3^4 = 81$ strategies.

On the other hand, the normal form of analysis may appeal to a decision maker who feels uneasy about making the subjective probability assessments involved in preposterior analysis. In many problems, the conditional probabilities of sample outcomes, given states of nature, may be based on relative frequencies, as in the A. B. Westerhoff Company case, or on an appropriate probability distribution, as in a problem discussed later in this chapter. However, the prior probabilities $P(\theta_i)$, will most likely represent subjective or judgmental assignments in most problems. In normal-form analysis, these prior probabilities are applied as the last step in the calculations. Therefore, it is possible to proceed all the way to this point without introducing subjective probability assessments. We can then judge how sensitive the choice of the optimal strategy is to the prior probability assignments. That is, we can determine by how much the magnitudes of the prior probabilities may be permitted to change without a shift occurring in the optimal decision rule to be employed. This procedure is discussed in the next section.

Exercises 16.4

1. Define and contrast the extensive-form and normal-form approaches.
2. For exercise 4 of section 16.2, construct a table showing all possible strategies and select the optimal strategy if the two items are inspected.
3. For exercise 7 of section 16.2, the following strategy table has been constructed. The two alternative courses of action are a_1: invest in business and a_2: do not invest. Fill in those strategies that are missing and select the optimal strategy.

Sample outcome	s_1	s_2	s_3	s_4	s_5	s_6	s_7	s_8
X	a_1	a_1	a_1	a_1	a_2	a_2	a_2	a_2
Y		a_2	a_1	a_1		a_2	a_1	a_1
Z		a_1	a_2	a_1		a_1	a_2	a_1

4. In exercise 8 of section 16.2, there are 27 strategies. Assume that the optimal strategy is one of the following 3:

Model Indication	s_5	s_{15}	s_{23}
Low	L	M	L
Medium	M	M	M
High	M	H	H

L, M, and H mean low, medium, and high marketing plans used, respectively. Compute conditional expected losses and expected opportunity losses to determine which of these 3 strategies is best.

5. The following table pertains to exercise 9 of section 16.2. Fill in the missing strategies. The two courses of action are a_1: build in Pennsylvania and a_2: build in Georgia.

Sample Outcome	s_1	s_2	s_3	s_4	s_5	s_6	s_7	s_8
300	a_1	a_1	a_1	a_1	a_2	a_2	a_2	a_2
350	a_1	a_1	a_2		a_1	a_1		a_2
400		a_1	a_2			a_1		a_1

Compute the conditional expected losses as well as the expected opportunity losses for strategies s_1, s_3, and s_4. Which strategy is optimal?

6. Suppose there are 3 states of nature, θ_1, θ_2, and θ_3; 2 actions, a_1 and a_2; and 3 outcomes from a forecast, X, Y, and Z. The outcome X indicates that θ_1 will occur, outcome Y indicates θ_2 will occur, and outcome Z indicates θ_3 will occur. The opportunity loss table and a forecasting model with the following reliability have been developed.

State of Nature	a_1	a_2	Model Indicates		
			X	Y	Z
θ_1	0	1300	0.75	0.15	0.10
θ_2	300	0	0.09	0.50	0.41
θ_3	800	0	0.07	0.09	0.84

Of the 8 strategies only 2 are plausible: $s_3(a_1, a_2, a_2)$ and $s_5(a_1, a_1, a_2)$. Use normal-form analysis to determine the optimal strategy. In computing the conditional expected losses for the 2 strategies, use the *loss* of taking the wrong act times the probability of taking the wrong act. The prior probabilities for θ_1, θ_2, and θ_3 are 0.35, 0.30, and 0.35, respectively.

16.5 SENSITIVITY ANALYSIS

In the new product decision problem discussed in this chapter, only two decision rules appeared reasonable on an intuitive basis, namely, s_2 and s_4. We saw in Table 16-10 that these strategies yielded the two lowest expected opportunity losses, with figures of \$30.00 and \$23.60, respectively, for s_2 and s_4. However, we can observe in Table 16-7 that although the conditional expected loss for s_2 is lower than s_4 if θ_1 occurs (\$24.00 versus \$56.00, respectively), the opposite is true if θ_2 occurs (\$32.00 versus \$12.80, respectively). Because these conditional expected losses were weighted by prior probabilities of states of nature, $P(\theta_1)$ and $P(\theta_2)$, clearly the choice between strategies s_2 and s_4 is dependent on the magnitudes of these prior probabilities. We can test how *sensitive* the choice between decision rules s_2 and s_4 is to the size of these prior probabilities.

The **sensitivity analysis** tests how sensitive the solution to a decision problem is to changes in the data for the variables of the problem.

In the present problem, a sensitivity test can be accomplished by solving for a *break-even value for the prior probability of one of the two states of nature* such that the expected opportunity losses of the two strategies will be equal. If the prior probability rises above this break-even value, s_2 is the optimal act rather than s_4. We illustrate this procedure in the new product decision problem. As indicated earlier, the (unconditional) expected opportunity losses of the two strategies in question were computed as follows:

$$\text{EOL}(s_2) = (0.25)(\$24.00) + (0.75)(\$32.00) = \$30.00$$

$$\text{EOL}(s_4) = (0.25)(\$56.00) + (0.75)(\$12.80) = \$23.60$$

If we now substitute p for the 0.25 value of $P(\theta_1)$, and $1 - p$ for the 0.75 figure of $P(\theta_2)$, we can solve for the value of p for which $\text{EOL}(s_2) = \text{EOL}(s_4)$. Making this substitution and equation the resulting expressions for expected opportunity loss, we have

$$p(\$24.00) + (1 - p)(\$32.00) = p(\$56.00) + (1 - p)(\$12.80)$$

Carrying out the multiplications, we obtain

$$\$24.00p + \$32.00 - \$32.00p = \$56.00p + \$12.80 - \$12.80p$$

Collecting terms, we find the break-even value of p.

$$\$51.20p = \$19.20$$

$$p = \frac{19.20}{51.20} = 0.375$$

Summary

In summary, we conclude that if $p = P(\theta_1) = 0.375$, then the expected opportunity losses of strategies s_2 and s_4 would be equal, and we would have no preference between them. In the A. B. Westerhoff Company new product decision problem, $p = P(\theta_1) = 0.25$. Hence, we observe that this subjective prior probability assignment could have varied up to a value of 0.375 and s_4 would still have been a better strategy than s_2. However, for p values in excess of 0.375, s_2 has a lower expected opportunity loss and is therefore the better rule.

Of course, break-even values of p could be determined between s_4 and other strategies as well. In the case of a strategy such as s_5, whose conditional expected values given θ_1 and θ_2 are each higher than the corresponding figures for s_4 (\$144.00 versus \$56.00 and \$147.20 versus \$12.80, respectively), it is impossible for the weighted average or expected opportunity loss to be less than that of s_4, regardless of the weights p and $1 - p$. In this situation, the strategy s_4 is said to *dominate* s_5. As a general definition, a strategy, say s_1, is a **dominating strategy** with respect to s_2 if

Dominating strategy

$$R(\theta|s_1) \leqslant R(\theta|s_2) \qquad \text{for all values of } \theta$$

and

$$R(\theta|s_1) < R(\theta|s_2) \qquad \text{for at least one value of } \theta$$

In words, s_1 is said to dominate s_2 if the conditional expected losses (risks) for s_1 are *equal to or less than* the corresponding figures for s_2 for every state of nature *and* if the conditional expected loss for s_1 is *less than* the corresponding figure for s_2 for at least one value of θ.

The preceding discussion illustrated the use of sensitivity analysis to determine the effects of variations in prior probabilities on the selection of a best act. We could also determine the effects of changes in payoff matrix entries on our choice of a best act.

In general, in any decision analysis problem, it is useful to test how sensitive the solution is to changes in all of the important variables of the problem.

In problems involving numerous states of nature, experimental outcomes, and courses of action, the analyses may require many calculations. The use of computers in such situations is usually a practical necessity.

16.6 AN ACCEPTANCE SAMPLING EXAMPLE

As another illustration of preposterior analysis, we will return to the problem of acceptance sampling of a manufactured product discussed in section 15.2. Since the problem was posed in posterior analysis, we must make some changes to convert it to a problem in preposterior analysis. We assume, as previously, that the Renny Corporation inspects incoming lots of articles produced by a supplier in order to determine whether to accept or reject these lots. We also retain the assumptions that in the past, incoming lots from this supplier have

TABLE 16-11
Basic data for acceptance sampling problem for the Renny Corporation

Event p (lot proportion of defectives)	Prior Probability $P_0(p)$	Opportunity Loss a_1 Reject	Opportunity Loss a_2 Accept
0.10	0.50	\$200	\$ 0
0.20	0.30	0	100
0.30	0.20	0	200
	1.00		

Prior $\text{EOL}(a_1) = (0.50)(\$200) + (0.30)(\$0) + (0.20)(\$0) = \$100$

Prior $\text{EOL}(a_2) = (0.50)(\$0) + (0.30)(\$100) + (0.20)(\$200) = \$70$

contained 10%, 20%, or 30% defectives and that lots with these percentages of defectives have occurred 50%, 30%, and 20% of the time, respectively. The same payoff matrix as shown in Table 15-4 is assumed. Table 15-5 showed the calculation of prior expected opportunity losses for the two acts,

a_1: Reject the incoming lot[1]

a_2: Accept the incoming lot

if action were taken without sampling. For convenience, this information on events, prior probabilities, payoffs, and prior expected opportunity losses is summarized in Table 16-11. The preferable action without sampling is act a_2, "accept the incoming lot," which has a prior expected opportunity loss of \$70 as opposed to act a_1, "reject the incoming lot," which has a prior expected opportunity loss of \$100.

Deciding whether or not to sample

We would like to know whether it is better to make the decision concerning acceptance or rejection of the incoming lot without sampling or to draw and inspect a random sample of items from the incoming shipment and then make the decision. We recognize this problem as one of preposterior analysis, specifically one that requires an evaluation of the *expected value of sample information* (EVSI) and the *expected net gain of sampling* (ENGS). We consider drawing a sample of two articles without replacement and inspecting these two articles at a cost of \$5. Hence, our problem is to choose between a terminal decision without sampling and a decision to examine a fixed-size sample of two articles and then select a terminal act.

[1] The alternative actions have been denoted here by lower-case rather than capital letters to maintain consistency with the notation we have used in specifying strategies.

Of course, a sample of only two articles may be entirely too small to be realistic in many cases. On the other hand, in the case of certain manufactured complex assemblies of large unit cost where the testing procedure is destructive (for example, in missile testing or in space probing where the test vehicle is not retrievable), it may only be feasible to test a small number of items. However, the principles illustrated in this problem are perfectly general. The assumption of a larger sample size would merely increase the computational burden. In section 16.7, we will consider appropriate decision theory procedures for determining an optimal sample size.

As in the preceding example in this section, we will solve the problem in two different ways, first through the use of extensive-form analysis, and then by normal-form analysis.

Extensive-Form Analysis

The decision diagram for the preposterior analysis of the Renny Corporation problem is given in Figure 16-4. As indicated in the legend, the alternative actions in this problem have been denoted

a_1: Reject the lot

a_2: Accept the lot

The possible sample outcomes based on a random sample of two articles drawn from the incoming lot are denoted

$X = 0$ defectives

$X = 1$ defective

$X = 2$ defectives

Two choices

At node (a), the two basic choices are shown: Not to engage in sampling inspection, or to sample and inspect two articles from the incoming lot. The prior analysis for the "no sampling" choice is shown in the upper portion of the tree. As previously observed in Table 16-11, the prior expected opportunity loss of rejection, a_1, is \$100 and that of acceptance, a_2, is \$70. Hence, the non-optimal act a_1 has been blocked off on the branch emanating from node (b) and the \$70 figure for a_2 has been entered at node (b). This \$70 figure is the expected opportunity loss of choosing the optimal act without sampling.

The alternative choice

The alternative choice is to sample and inspect two items prior to making the terminal decision of acceptance or rejection of the lot. To carry out the calculations for this part of the analysis, we begin at the right tips of the tree and proceed inward by backward induction. The expected values of \$116.60 and \$55.80 shown at nodes (g) and (h), for example, are the expected opportunity losses of acts a_1 and a_2, respectively, after observing no defectives ($X = 0$) in the sample of two articles. The probabilities used to calculate these expected values are the conditional (posterior)probabilities, given $X = 0$, that the lot proportion

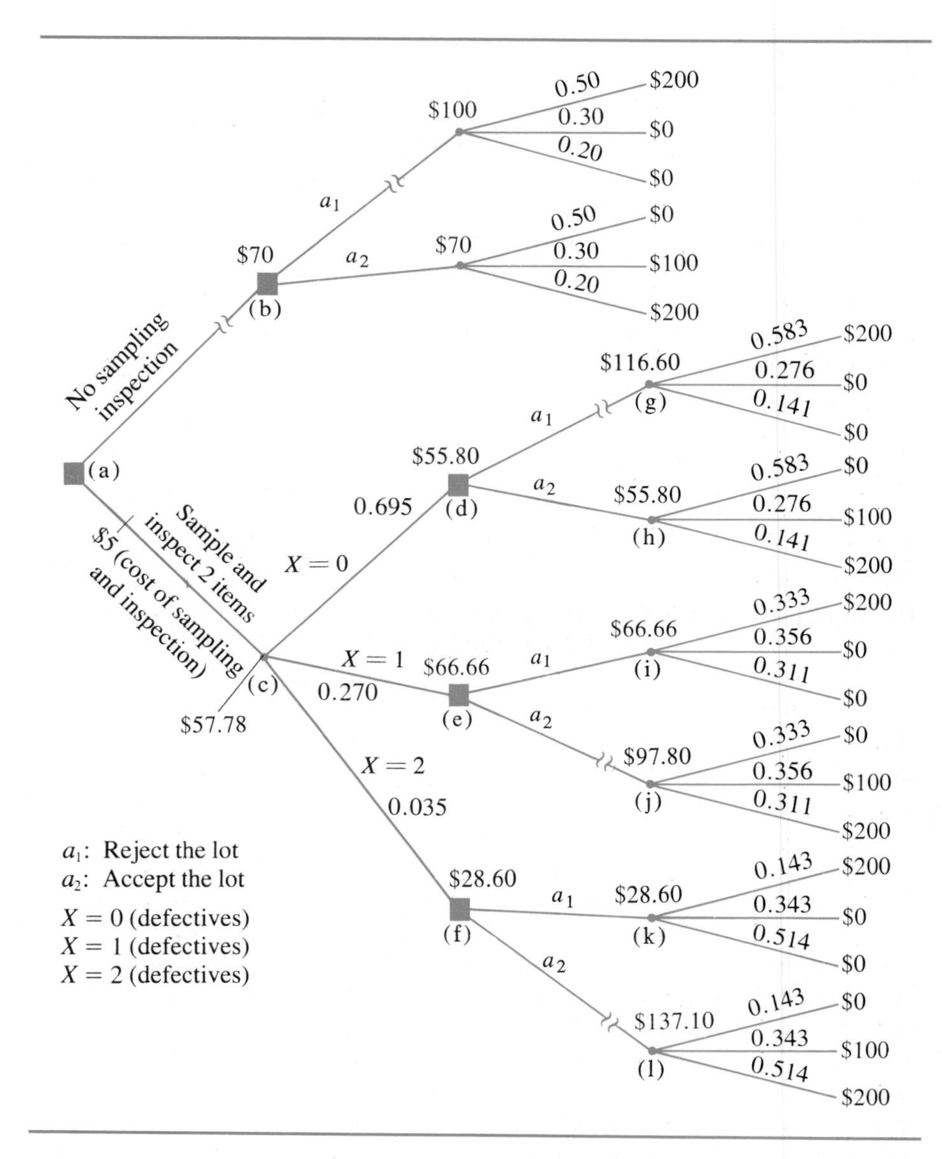

FIGURE 16-4
Decision diagram for preposterior analysis of the lot acceptance decision for the Renny Corporation problem

of defectives, p, is 0.10, 0.20, and 0.30, respectively. In symbols,

$$P(p = 0.10 | X = 0) = 0.583$$

$$P(p = 0.20 | X = 0) = 0.276$$

$$P(p = 0.30 | X = 0) = 0.141$$

The calculation of these posterior probabilities, shown in Table 16-13, will be explained shortly. Since acceptance of the lot (act a_2) has a lower expected loss than does rejection (act a_1), given that no defectives have been observed in the sample, the lower figure, \$55.80, is entered at node (d). Furthermore, act a_1 is blocked off as nonoptimal. Corresponding figures of \$66.66 and \$28.60 for expected losses of optimal acts after observing one and two defectives are shown at nodes (e) and (f), respectively. Rejection, act a_1, is shown to be optimal after observation of either one or two defectives. Weighting the three payoffs shown at nodes (d), (e), and (f) by the marginal probabilities of observing zero, one, and two defectives (0.695, 0.270, and 0.035, respectively), we find a figure of \$57.78 as the expected opportunity loss of sampling and inspecting two articles and taking terminal action after observing the sample evidence. Before proceeding with the calculation of EVSI and ENGS, let us consider the method of calculating the marginal and posterior probabilities mentioned in this paragraph.

Obtaining the joint probability distribution of the sample evidence and events

In order to calculate the marginal probabilities of the sample evidence, $P(X = 0)$, $P(X = 1)$, and $P(X = 2)$, and the posterior probabilities of the form $P(p|X)$, we must obtain the joint probability distribution of the sample evidence and events, that is, $P(X \text{ and } p)$, for $X = 0$, 1, and 2, and $p = 0.10$, 0.20, and 0.30. As in the A. B. Westerhoff Company problem, these joint probabilities are obtained by multiplying prior probabilities of events—in this case, $P_0(p)$—by the appropriate conditional probabilities of sample evidence, given events—in this case, $P(X|p)$. In the A. B. Westerhoff Company problem, we assumed that these P(sample outcomes|events) values were based on past relative frequencies of occurrence in surveys previously conducted by a market research firm. In the present problem, we must calculate these $P(X|p)$ values by using an appropriate probability distribution. Let us assume that the incoming lot from which the sample of two articles is drawn is large relative to the sample size. Then, as indicated in section 3.4, we can assume that even if the sample is selected without replacement, the drawings of the articles may be considered trials of a Bernoulli process, and the binomial distribution may be used to calculate probabilities of sample outcomes. That is, we may assume that since there is so little change in the population (lot) because of the drawing of the first article, the probability of obtaining a defective item on the second draw is the same as on the first draw. If the incoming lot is not large relative to the sample size—that is, if it is not at least 10 times the sample size—then the hypergeometric distribution is appropriate. The $P(X|p)$ values for the present problem (that is, the conditional probabilities of observing zero, one, or two defectives, given lot proportion defectives of 0.10, 0.20, and 0.30 calculated by the use of the binomial distribution) are shown in Table 16-12. As indicated in the table, the probabilities of zero, one, and two defectives in a sample of size 2 from a lot that contains 0.10 defectives are given by the respective terms of the binomial distribution whose parameters are $n = 2$, $p = 0.10$, and corresponding calculations provide the probabilities for the cases in which $p = 0.20$ and 0.30.

Computation of the joint probability distribution $P(X \text{ and } p)$

Table 16-13 shows the computation of the joint probability distribution $P(X \text{ and } p)$. This joint distribution is obtained by multiplying the prior probabilities, $P_0(p)$, by the conditional probabilities, $P(X|p)$, derived in Table 16-12. For example, 0.405, the upper left entry in the joint probability distribution in

TABLE 16-12

Conditional probabilities of specified numbers of defectives in a sample of 2 articles from an incoming lot in the Renny Corporation problem

$$P(X = 0|p = 0.10) = (0.90)^2 = 0.81$$
$$P(X = 1|p = 0.10) = 2(0.90)(0.10) = 0.18$$
$$P(X = 2|p = 0.10) = (0.10)^2 = 0.01$$
$$1.00$$

$$P(X = 0|p = 0.20) = (0.80)^2 = 0.64$$
$$P(X = 1|p = 0.20) = 2(0.80)(0.20) = 0.32$$
$$P(X = 2|p = 0.20) = (0.20)^2 = 0.04$$
$$1.00$$

$$P(X = 0|p = 0.30) = (0.70)^2 = 0.49$$
$$P(X = 1|p = 0.30) = 2(0.70)(0.30) = 0.42$$
$$P(X = 2|p = 0.30) = (0.30)^2 = 0.09$$
$$1.00$$

Table 16-13, was obtained by multiplying 0.50 by 0.81. In symbols, $P(X = 0$ and $p = 0.10) = P_0(p = 0.10)P(X = 0|p = 0.10)$. We can now obtain the marginal probabilities of the sample outcomes of zero, one, and two defectives. The column totals in the joint probability distribution are $P(X = 0) = 0.695$, $P(X = 1) = 0.270$, and $P(X = 2) = 0.035$, respectively. These marginal probabilities are entered in the decision diagram in Figure 16-4 on the branches emanating from node (c).

The calculations of posterior, or revised, probabilities of lot proportion of defectives, given the sample outcomes $X = 0$, 1, and 2, are shown in Table 16-14. These posterior probabilities were obtained by dividing the joint probabilities by the appropriate column totals, $P(X)$. For example, the probability that the incoming lot contains 0.10 defectives, given that no defectives were observed in the sample of two articles, is

$$P(p = 0.10|X = 0) = \frac{P(X = 0 \text{ and } p = 0.10)}{P(X = 0)} = \frac{0.405}{0.695} = 0.583$$

TABLE 16-13

Calculation of the joint probability distribution of sample outcomes and events in the Renny Corporation problem

Event p (lot proportion of defectives)	Prior Probability $P_0(p)$	Conditional Probability $P(X\|p)$			Joint Probability $P(X$ and $p)$			
		$X = 0$	$X = 1$	$X = 2$	$X = 0$	$X = 1$	$X = 2$	Total
0.10	0.50	0.81	0.18	0.01	0.405	0.090	0.005	0.50
0.20	0.30	0.64	0.32	0.04	0.192	0.096	0.012	0.30
0.30	0.20	0.49	0.42	0.09	0.098	0.084	0.018	0.20
					0.695	0.270	0.035	1.00

TABLE 16-14
Calculation of the posterior probabilities of lot proportion of defectives in the Renny Corporation problem

$P(p = 0.10 \mid X = 0) = \frac{0.405}{0.695} = 0.583$	$P(p = 0.10 \mid X = 1) = \frac{0.090}{0.270} = 0.333$	$P(p = 0.10 \mid X = 2) = \frac{0.005}{0.035} = 0.143$
$P(p = 0.20 \mid X = 0) = \frac{0.192}{0.695} = 0.276$	$P(p = 0.20 \mid X = 1) = \frac{0.096}{0.270} = 0.356$	$P(p = 0.20 \mid X = 2) = \frac{0.012}{0.035} = 0.343$
$P(p = 0.30 \mid X = 0) = \frac{0.098}{0.695} = \frac{0.141}{1.000}$	$P(p = 0.30 \mid X = 1) = \frac{0.084}{0.270} = \frac{0.311}{1.000}$	$P(p = 0.30 \mid X = 2) = \frac{0.018}{0.035} = \frac{0.514}{1.000}$

As previously indicated in connection with Table 16-4, the calculation of these posterior probabilities represents an application of Bayes' theorem. The posterior probabilities are shown in Figure 16-4 on the branches stemming from nodes (g), (h), (i), (j), (k), and (l).

We can now complete the extensive-form analysis. We noted earlier that the \$70 at node (b) in Figure 16-4 represents the expected loss of terminal action without sampling. Correspondingly, the \$57.78 at node (c) is the expected loss of sampling and inspecting two articles and then taking terminal action. Hence, the *expected value of sample information* is

$$\text{EVSI} = \$70 - \$57.78 = \$12.22$$

The expected amount by which the terminal opportunity loss of action without sampling is reduced by sampling two articles and taking action after inspection of the sample is \$12.22. Since the sampling and inspection of two articles costs \$5.00, the *expected net gain of sample information* is

$$\text{ENGS} = \$12.22 - \$5.00 = \$7.22$$

Therefore, since the expected net gain of sample information is \$7.22 for terminal action after sampling and inspecting two articles compared with terminal action without sampling, the Renny Corporation should follow the former course of action.

Normal-Form Analysis

We now turn to the normal-form analysis of the Renny Corporation problem, which we have just solved by extensive-form procedures. As in the A. B. Westerhoff problem, we begin the normal-form analysis by listing all possible strategies, or decision rules. These are shown in Table 16-15. We can see on an intuitive basis that some of the strategies are illogical and would therefore tend to have relatively large expected opportunity losses associated with them. For example, strategy $s_2(a_1, a_1, a_2)$ rejects the incoming lot if the sample of two

List all strategies

TABLE 16-15
A listing of all possible strategies in the Renny Corporation problem

Sample Outcome (number of defectives)	Strategy s_1	s_2	s_3	s_4	s_5	s_6	s_7	s_8
$X = 0$	a_1	a_1	a_1	a_1	a_2	a_2	a_2	a_2
$X = 1$	a_1	a_1	a_2	a_2	a_1	a_1	a_2	a_2
$X = 2$	a_1	a_2	a_1	a_2	a_1	a_2	a_1	a_2

a_1: Reject the lot
a_2: Accept the lot

articles yields zero or one defective, but accepts the lot if two defectives are observed. Strategy $s_4(a_1, a_2, a_2)$ also employs perverse logic, since it rejects the lot on an observation of zero defectives but accepts it if one or two defectives are observed. Strategy $s_1(a_1, a_1, a_1)$ rejects the lot, and $s_8(a_2, a_2, a_2)$ accepts the lot, regardless of the sample observation. The two most logical decision rules appear to be strategy $s_5(a_2, a_1, a_1)$, which accepts the lot if zero defectives are observed in the sample and rejects the lot if one or two defectives are observed, and strategy $s_7(a_2, a_2, a_1)$, which accepts the lot if zero or one defective is observed and rejects the lot if two defectives are found.

Select most-logical strategies

Calculate risks for each strategy

The next step is to calculate for each strategy the risks or conditional expected losses associated with the occurrence of each event (lot proportion of defectives). We will illustrate these calculations for strategies s_5 and s_7 by computing, for each proportion of defectives, the *loss* of taking the wrong act times the *probability* of taking the wrong act. These risks, denoted $R(s_5|p)$ and $R(s_7|p)$, are shown in Table 16-16. For example, let us consider strategy $s_5(a_2, a_1, a_1)$.

TABLE 16-16
Calculation of risks, or conditional expected opportunity losses, for strategies s_5 and s_7

	Strategy $s_5(a_2, a_1, a_1)$			Strategy $s_7(a_2, a_2, a_1)$		
Event p (lot proportion of defectives)	Opportunity Loss of Wrong Act	Probability of Wrong Act Given p	Risk (conditional expected loss) $R(s_5|p)$	Opportunity Loss of Wrong Act	Probability of Wrong Act Given p	Risk (conditional expected loss) $R(s_7|p)$
0.10	$200	0.19	$38	$200	0.01	$ 2
0.20	100	0.64	64	100	0.96	96
0.30	200	0.49	98	200	0.91	182

TABLE 16-17
Risks, or conditional expected opportunity losses, for the 8 strategies in the Renny Corporation problem

Event p (lot proportion of defectives)	Strategy							
	s_1	s_2	s_3	s_4	s_5	s_6	s_7	s_8
0.10	$200	$198	$164	$162	$38	$ 36	$ 2	$ 0
0.20	0	4	32	36	64	68	96	100
0.30	0	18	84	102	98	116	182	200

Referring back to the original payoff matrix in Table 16-11, we recall that the correct course of action if $p = 0.10$ is to accept, and if $p = 0.20$ or 0.30, to reject the lot. Strategy $s_5(a_2, a_1, a_1)$ accepts the lot on the observation of no defectives and rejects it otherwise. From Table 16-13 we find a 0.81 conditional probability of observing no defectives, given that the lot proportion of defectives is $p = 0.10$. Hence, with strategy s_5, the probability of making the wrong decision, given a lot proportion of defectives $p = 0.10$, is $1 - 0.81 = 0.19$. The loss associated with rejection if $p = 0.10$ is $200 (Table 16-11). Therefore, as shown in Table 16-16, the risk of strategy s_5, given $p = 0.10$, is $R(s_5|p = 0.10) = \$200 \times 0.19 = \$38$. Analogous calculations produce the other risks shown in Table 16-16. A summary of the risks for all eight strategies is presented in Table 16-17.

We can now calculate the expected loss of each strategy using equation 16.2. Adapting the notation of that equation to that used in the present problem, we obtain as the expected opportunity loss of the kth strategy

$$\text{EOL}(s_k) = \sum_p P_0(p)R(s_k|p) \tag{16.3}$$

That is, for each strategy, we weight the conditional expected losses associated with each lot proportion of defectives (Table 16-17) by the prior probabilities of such incoming lots (Table 16-13). Hence, the expected opportunity losses (expected risks) of strategies s_5 and s_7 are

$$\text{EOL}(s_5) = (0.50)(\$38) + (0.30)(\$64) + (0.20)(\$98) = \$57.80$$

$$\text{EOL}(s_7) = (0.50)(\$2) + (0.30)(\$96) + (0.20)(\$182) = \$66.20$$

The expected opportunity losses of all eight strategies are given in Table 16-18. The optimal strategy is seen to be $s_5(a_2, a_1, a_1)$, since it has the minimum expected opportunity loss of $57.80. The strategy, which accepts the lot if no defectives are observed in the sample and rejects the lot if one or two defectives are found, is the same as the one found in the extensive-form analysis depicted

TABLE 16-18
Expected opportunity losses of the 8 strategies in the Renny Corporation problem

Strategy	s_1	s_2	s_3	s_4	s_5	s_6	s_7	s_8
EOL ($)	100.00	103.80	108.40	112.20	57.80*	61.60	66.20	70.00

* The minimum loss figure

in the decision diagram in Figure 16-4. The minor difference between the expected opportunity losses of the optimal strategy—$57.78 in the extensive-form analysis and $57.80 in the normal-form analysis—is attributable to rounding of decimal places.

Exercises 16.6

1. Trivia Press, Inc., estimates that the gain from a successful novel would be $8 million; the loss from an unsuccessful novel would be $4 million. The publisher sends a copy of the book to 10 critics for their opinions. The results are slightly unfavorable: Six out of 10 critics dislike the book. If the book should be a failure, the probability that a critic would dislike it is 0.8, and if the book should be a success, there is a 50:50 chance the critic would like it. At what value of prior probability of success would the decision concerning publication change?
2. A certain manufacturing process produces lots of 500 units each. In each lot, either 10% or 30% of the 500 items are defective. The quality control department, which inspects each lot before shipment, knows that 80% of the lots produced in the past contained 10% defectives. If accepted for shipment, the lot produces a profit of $500, but if rejected, the lot is sold for scrap at cost. If a 30% defective lot is sent out, the firm estimates its loss in goodwill, and hence in future orders, at $1,500. It costs $3 to test an item, but the test is not destructive (that is, the item is still good after the test and can be sold). What is the expected net gain from testing 3 items drawn at random and what is the best decision rule? Use a binomial probability distribution to compute conditional probabilities.
3. You are given the following information:

State of Nature	s_1	s_2
θ_1	50	70
θ_2	100	40

At what value of the prior probability of θ_1 would you be indifferent between the 2 strategies?

16.7 OPTIMAL SAMPLE SIZE

In the preceding acceptance sampling problem, we indicated how the expected value of sample information can be derived prior to the actual drawing of a sample. This EVSI figure was obtained by subtracting the expected opportunity loss of the best terminal act without sampling from the expected loss of the optimal strategy with sampling. More precisely, the latter value is the expected opportunity loss of a decision to sample and then take optimal terminal action after observation of the sample outcome. In that problem, we assumed a fixed sample size of two articles. However, as previously indicated (page 671), the method of analysis presented is a general one, and the only practical effect of an assumption of a larger sample size would have been an increase in the computational burden. Nevertheless, can an *optimal sample size* be derived in a problem such as the one presented? The answer is yes, and the general method for obtaining such an optimal value follows.

Obtaining an optimal sample size

As might be suspected intuitively, an increase in sample size brings about an increase in the EVSI. However, an increase in sample size also results in an increase in the cost of sampling. (The term "cost of sampling" here means the total cost of sampling and inspection.) Therefore, we would like to find the sample size for which the difference between the EVSI and the cost of sampling is the largest—this is the **optimal sample size**. An equivalent and more convenient approach is to minimize *total loss*, where the total loss associated with any sample size n is defined as

Minimizing total loss

(16.4)
$$\text{Total loss} = \begin{matrix}\text{Cost of}\\ \text{sampling}\end{matrix} + \begin{matrix}\text{Expected opportunity loss}\\ \text{of the optimal strategy}\end{matrix}$$

Let us consider how the quantities in equation 16.4 might be calculated. The cost of sampling would ordinarily not be difficult to calculate. In many instances, this cost may be entirely variable, that is, proportional to the number of articles sampled. In that case, the cost of sampling would be equal to

Cost of sampling when cost is proportional to size of sample

(16.5)
$$C = vn$$

where C = cost of sampling
v = cost of sampling each unit
n = number of units in the sample

In the Renny Corporation example, where the cost of sampling was \$5.00 for two articles, if costs were entirely variable, the cost of sampling each unit would be \$2.50. Hence, $v = \$2.50$ and $n = 2$. The cost of sampling for a sample of size 10 would be $C = \$2.50 \times 10 = \$25.00$, and so on.

Cost of sampling with fixed and variable costs

In certain situations, a portion of the total cost of sampling might be fixed and the remaining part variable. In that case, the cost of sampling would be

$$C = f + vn \tag{16.6}$$

where C = cost of sampling
f = fixed cost
v = cost of sampling each unit
n = number of units in the sample

Thus, for example, in a case in which the fixed cost of sampling is \$10 and the variable cost per unit is \$2, the cost of sampling 20 units would be

$$C = \$10 + (\$2)(20) = \$50$$

Other formulas can be derived for more complex situations.

More convenient notation

Turning now to the other term in equation 16.4 for **total loss**, namely, the **expected opportunity loss of the optimal strategy**, we first introduce more convenient notation. Let us assume we are dealing with an acceptance sampling problem in which the sample size is 10. Although in the normal-form analysis of the Renny Corporation problem we considered every possible strategy for the acceptance of the incoming lot, it can be shown that the only strategies worth considering as potential optimal decision rules are those that accept the incoming lot if a certain number of defectives or less are observed and reject the lot otherwise. It is conventional to refer to this critical number as "the acceptance number," denoted c. Hence, in the present problem, in which the sample size is 10, the possible values for the acceptance number c are 0, 1, 2, . . . , 10. For example, if $c = 0$, the lot is accepted if no defectives are observed in a sample of size $n = 10$ and rejected otherwise. If $c = 1$, the lot is accepted if *one or fewer* defectives are observed and rejected otherwise. If $c = 2$, the lot is accepted if *two or fewer* defectives are observed and rejected otherwise, and so on. Therefore, we can characterize the optimal strategy for each sample size by two figures: c, the acceptance number, and n, the sample size. For example, the c, n pair, denoted (c, n), for an optimal strategy with an acceptance number $c = 2$ and sample size $n = 10$ would be (2, 10). In the Renny Corporation problem, the c, n pair was (0, 2) for the optimal strategy s_5.

Theoretically, we could calculate for every possible sample size the expected opportunity loss of the optimal strategy (c, n). Adding this figure to the cost of sampling, we would obtain the total loss associated with each sample size. We would then select as the optimal sample size the one that yielded the *minimum total loss*.

At first, this might seem to be an impossible procedure, because if the population is infinite, the sample size could conceivably take on any positive integral value. However, n must be a finite value because of the cost of sampling involved. The expected value of sample information (EVSI) cannot exceed the expected value of perfect information (EVPI). As we saw in section 14.5, EVPI is the expected opportunity loss of the optimal act prior to sampling. That is, it is the expected loss of the best terminal action without sampling. It would never be worthwhile to take a sample so large that the cost of sampling exceeded the EVPI. Hence, n for the optimal sample size will be a finite number.

Binomial sampling

The acceptance sampling problem we have been discussing can be described as involving **binomial sampling**. That is, the conditional probabilities of sample outcomes of the form $P(X|p)$ were calculated in the Renny Corporation problem by the binomial probability distribution. In this type of problem, the optimal strategies for the various sample sizes must be calculated by the methods we have indicated, which may be characterized as trial-and-error procedures. That is, no simple general formula enables us to derive the optimal (c, n) pairs for every problem. Therefore, a considerable amount of calculation may be involved to determine the optimal sample size. For sufficiently large and important problems, the use of computers may represent the only practical method of carrying out the computations.

16.8 GENERAL COMMENTS

Preposterior analysis

In this chapter, we have discussed decision-making procedures for the selection of optimal strategies prior to obtaining experimental data or sample information. The method used, preposterior analysis, anticipates the adoption of best actions after the observation of the experimental or sample results. The general principles of this type of analysis have been discussed using two examples.

- The first illustration involved the introduction of a new product. Experimental evidence in the form of sample survey results was obtainable for revising prior probabilities of occurrence of the states of nature "successful product" and "unsuccessful product."
- The second example involved a decision concerning acceptance or rejection of an incoming lot. Here, information in the form of results of a random sample drawn from a lot was obtainable for revising prior probabilities of the states of nature "lot proportion of defectives."

In the first example, an empirical joint frequency distribution on the success or failure of past new products and survey results provided information for the calculation of probabilities required for the problem solution. In the second example, the decision depended on the proportion of defectives, denoted p. The observable sample data were represented by the random variable "number of defectives," denoted X, in a sample drawn from the incoming lot. This random variable was taken to be binomially distributed.

States of nature

In both of these problems, the states of nature and the experimental or sample evidence were discrete random variables. Another class of problems comprises situations in which the states of nature and sample evidence are represented by continuous random variables. For example, the decision may depend on a parameter μ, the mean of a population, and the observed sample evidence may be a sample mean $\bar{X}$. The mathematics required for the solution of this class of problems is outside the scope of this book. However, the same basic principles discussed in this chapter for extensive-form and normal-form analysis are applicable, regardless of whether the random variables representing states of nature and experimental outcomes are discrete or continuous.

17

Comparison of Classical and Bayesian Statistics

Topics in classical statistical inference and Bayesian decision theory were discussed in earlier chapters. In this chapter, we compare some aspects of classical statistics and Bayesian statistics.

17.1 COMPARING CLASSICAL AND BAYESIAN METHODS

Classical statistics is a broad term that includes the two main topics of classical statistical inference (hypothesis testing and confidence interval estimation) as well as other topics, such as classical regression analysis discussed in Chapters 9 and 10. **Bayesian statistics** is also a broad term that analogously may be thought of as including Bayesian decision theory, Bayesian estimation, and other topics such as Bayesian regression analysis.

Although the terminologies of classical and Bayesian statistics differ, there are many similarities in the structure of the problems they address and in their methods of analysis. However, important differences, particularly in their methods of analysis, are a matter of continuing discussion and study.

To compare these two important types of statistical analysis, we consider in section 17.2 an illustrative problem that presents a comparison of classical hypothesis-testing methods and Bayesian decision theory; in section 17.3, we compare classical and Bayesian estimation procedures. In conclusion, section 17.4 presents some general comments on both the common ground and the differences between these two schools of thought.

Testing the Null Hypothesis against the Alternative Hypothesis

Classical approach

To introduce the comparison, let us consider a standard hypothesis-testing problem. Suppose we wish to test the null hypothesis, H_0: $p \leqslant p_0$ (where p_0 is a known or hypothesized population proportion), against the alternative hypothesis, H_1: $p > p_0$. For example, we might test the hypothesis that p (the proportion of defectives in a shipment of a manufactured product) is less than or equal to 0.03 against the alternative hypothesis that $p > 0.03$. Using classical hypothesis-testing methods, we could design a decision rule, which would tell us whether to accept or to reject the null hypothesis on the basis of a random sample drawn from the shipment. We would fix α, the desired maximum probability of making a Type I error, and (using the power curve) we could determine the risks of making Type II errors for values of p for which the alternative hypothesis H_1 is true. Table 17-1 summarizes the relationship between actions concerning these hypotheses and the truth or falsity of the hypotheses. For convenience, the table is given in terms of the null hypothesis H_0. However, it is understood that when H_0 is true, H_1 is false and when H_0 is false, H_1 is true. We refer to the truth or falsity of H_0 as the prevailing "state of nature." As indicated in the column headings of the table, the symbols a_1 and a_2 denote the actions "accept H_0" and "reject H_0," respectively.

We see that this hypothesis-testing problem includes three components.

1. states of nature representing the truth or falsity of the null hypothesis
2. actions a_1 and a_2, which accept or reject the null hypothesis
3. sample or experimental data, which—when examined in the light of a decision rule—lead to one of the actions a_1 or a_2

Bayesian approach

Let us rephrase the example in terms of Bayesian decision theory. We are dealing with a two-action problem involving acts a_1 and a_2, where the states of nature are the possible values of the proportion of defectives p. Although p varies along a continuum, and may be considered a continuous random variable, we assume for comparative purposes that only two states of nature are distinguished, namely, θ_1: $p \leqslant 0.03$ and θ_2: $p > 0.03$. Hence, θ_1 and θ_2 correspond to truth and falsity of H_0, respectively, in the classical hypothesis-testing

TABLE 17-1
Relationships between actions concerning a null hypothesis and the truth or falsity of the hypothesis

State of Nature	Action Concerning the Null Hypothesis a_1: Accept H_0	a_2: Reject H_0
H_0 is true	No error	Type I error
H_0 is false	Type II error	No error

TABLE 17-2
Payoff table in terms of opportunity losses
for the two-action problem to accept or reject a shipment

	Act	
State of Nature	a_1	a_2
θ_1	0	$L(a_1 \mid \theta_1)$
θ_2	$L(a_1 \mid \theta_2)$	0

problem. Finally, a random sample can be drawn from the shipment, and the observed sample or experimental data can be used to help choose the better action of a_1 and a_2. Therefore, the same three components of the decision-theory problem are present as were discussed in the hypothesis-testing problem.

1. states of nature
2. alternative actions
3. experimental data that aid in the choice of actions

Furthermore, Table 17-2, a payoff table for this problem in terms of opportunity losses, is similar to Table 17-1. The symbols $L(a_2|\theta_1)$ and $L(a_1|\theta_2)$ denote the opportunity loss of action a_2 given that θ_1 is the true state of nature and a_1 given that θ_2 is the true state of nature. The zeros in the other two cells of the table indicate no opportunity loss when the correct action is taken for the specified states of nature. Actually, payoffs would ordinarily be treated as a function of p and would vary with p. However, we have assumed in this discussion that only two states of nature are distinguished.

Difference in approach

Now, let us consider the difference in the two approaches. In hypothesis testing, the choice of α, the significance level, establishes the decision rule and is the overriding feature in the choice between alternative actions. In symbols, $\alpha = P(a_2|H_0 \text{ is true})$. That is, α is the conditional probability of rejecting the null hypothesis given that it is true. Hence, a major criterion of choice among actions in hypothesis testing is the relative frequency of occurrence of this type of error. But how is α chosen? In many applications, conventional significance levels such as 0.05 and 0.01 are used uncritically with little or no thought given to underlying considerations. However, it would be unfair to criticize a methodological approach simply because there are misuses of it.

Classical statistics

In classical statistics, the investigator must consider the relative seriousness of both Type I and Type II errors in establishing alternative hypotheses and significance levels at which these hypotheses are to be tested.

Also, the investigator is aided by prior knowledge concerning the likelihood that H_0 and H_1 are true. For example, in the problem just discussed, why did the investigator not set up the null hypothesis as, say, H_0: $p \leqslant 0.001$ or H_0: $p \leqslant 0.60$? In this particular acceptance sampling problem, we may know that two hypotheses such as these would be utterly ridiculous because of the extremely low and extremely high proportion of defectives implied. We may be virtually certain that the first of these hypotheses is false and that the second is true. Consequently, it would not be useful to set up the hypotheses in these forms.

Prior knowledge concerning the likelihood of truth of the competing hypotheses also helps the investigator in establishing the significance level. Hence, if it is considered likely that the null hypothesis is true, we will tend to set α at a low figure in order to maintain a low probability of erroneously rejecting that hypothesis.

Bayesian decision theory

Advocates of Bayesian decision theory criticize the classical hypothesis-testing procedures for informality and for excessive reliance on unaided intuition and judgment.

The Bayesians argue that the structure of their decision theory represents a logical extension of classical hypothesis testing, since it explicitly provides for the assignment of prior probability distributions to states of nature and incorporates losses into the formal structure of the problem. These decision theorists contend that losses should be considered in classical hypothesis testing in evaluating the relative seriousness of Type I and Type II errors. But how can losses be considered if no explicit loss function is formulated?

The following acceptance sampling problem compares the Bayesian approach with classical hypothesis testing. The problem demonstrates that nonoptimal decisions may be made if tests of hypotheses are conducted in the usual manner (establishing decision rules of rejecting or failing to reject hypotheses at preset levels of significance).

17.2 A COMPARATIVE PROBLEM

Let us assume a situation in which a company inspects incoming lots of articles produced by a particular supplier. Acceptance sampling inspection is carried out to decide whether to accept or reject these incoming lots by selecting a random sample of n articles from each lot. As in previous problems of this type, we make the simplifying assumption that only a few levels of proportion of de-

TABLE 17-3

Payoff matrix showing opportunity losses for actions of acceptance and rejection

State of Nature (p = lot proportion of defectives)	Prior Probability	Act: a_1 Reject	Act: a_2 Accept
0.02	0.50	$200	$ 0
0.05	0.25	0	300
0.08	0.25	0	500
	1.00		

fectives are possible, in this case, 0.02, 0.05, and 0.08. On the basis of an analysis of past costs, the company constructed the payoff table in terms of opportunity losses depicted in Table 17-3. From long experience, the firm has determined that lots containing 0.02 defectives are "good" and should be accepted. Hence, as indicated in Table 17-3, "accept" is the best act in the case of a 0.02 defective lot, and the opportunity loss is $0. On the other hand, "reject" is the optimal act for lots containing 0.05 and 0.08 defectives, and correspondingly the opportunity loss is $0 for such correct action.

On the basis of past performance, it has been determined that half of this supplier's lots are 2% defective, a fourth are 5% defective, and a fourth are 8% defective. In the absence of any further information, these past relative frequencies are adopted as the prior probabilities that such lots will be submitted by the supplier for acceptance or rejection.

Comparison procedure

To compare the approaches of Bayesian decision theory and traditional hypothesis testing, we will first carry out a study of possible single sampling plans (see page 645) by extensive- and normal-form preposterior analyses. The results of these analyses will determine the optimal sampling plan or strategy. Then a hypothesis-testing solution will be given, and a comparison will be made of the two approaches.

Sampling plan

A decision tree diagram is given in Figure 17-1, beginning with the decision to sample and inspect n items. We move to branch point (b), where the results of the sampling inspection determine which branch to follow. The possible results of sampling have been classified into three categories. The number of defectives, denoted X, may have been less than or equal to some number c_1, where $c_1 < n$. It may have been greater than c_1 but less than or equal to c_2, where $c_1 < c_2 < n$. Finally, the number of defectives may have been greater than c_2. These three types of results, for purposes of brevity, are referred to as type L (low), type M (middle), and type H (high) information, respectively. In Table 17-4, a joint frequency distribution is given for sample results and states of nature. We will assume that these frequencies were derived from a large

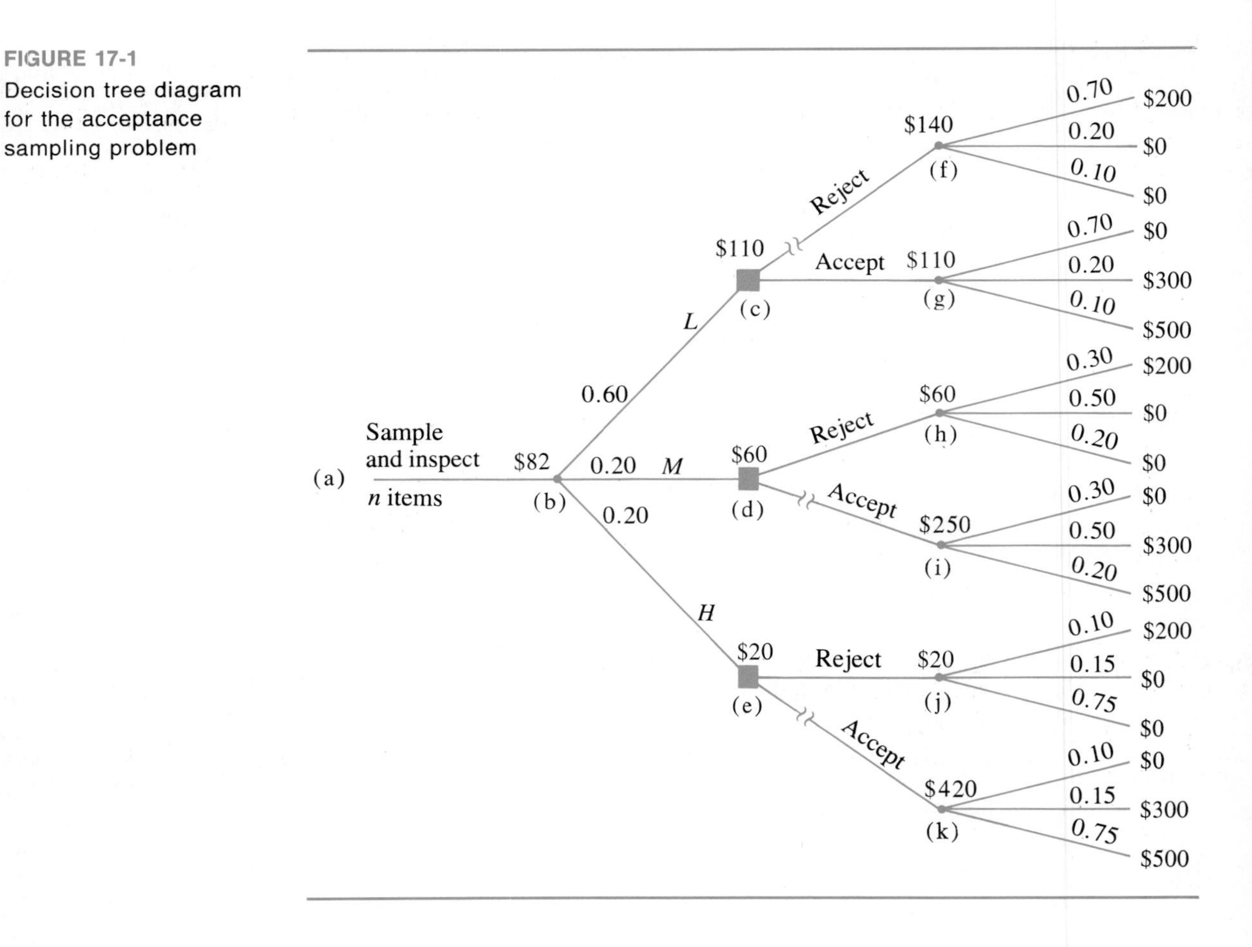

FIGURE 17-1
Decision tree diagram for the acceptance sampling problem

TABLE 17-4
Joint frequency distribution of sample results and states of nature

State of Nature (p = lot proportion of defectives)	Sample Results: Type L $X \leqslant c_1$	Type M $c_1 < X \leqslant c_2$	Type H $X > c_2$	Total
0.02	0.42	0.06	0.02	0.50
0.05	0.12	0.10	0.03	0.25
0.08	0.06	0.04	0.15	0.25
	0.60	0.20	0.20	1.00

TABLE 17-5
Calculation of posterior probabilities of states of nature

State of Nature (p = lot proportion of defectives)	Type L Information ($X \leqslant c_1$)			
	Prior Probability $P_0(p)$	Conditional Probability $P(L\|p)$	Joint Probability $P_0(p)P(L\|p)$	Posterior Probability $P_1(p)$
0.02	0.50	0.84	0.42	0.70
0.50	0.25	0.58	0.12	0.20
0.08	0.25	0.24	0.06	0.10
			0.60	1.00

number of past observations and therefore may represent probability in the relative frequency sense.[1] For example, in 0.42 of the lots inspected in the past, the number of defectives observed was c_1 or less, and the lots contained 0.02 defectives. In terms of marginal frequencies, type L information ($X \leqslant c_1$) was observed in 0.60 of the lots, type M ($c_1 < X \leqslant c_2$) was observed in 0.20 of the lots, and type H ($X > c_2$) was observed in 0.20 of the lots.

Extensive-form analysis

Returning to the decision tree, we find the three branches representing L, M, and H types of information emanating from node (b) marked with their respective probabilities, 0.60, 0.20, and 0.20. We will give a brief explanation of the extensive-form analysis, using type L information as an example. If type L information is observed, we move to branch point (c), where we can either accept or reject the lot. If we reject, we move to node (f); if we accept, to node (g). The probabilities shown on the three branches stemming from (f)—that is, 0.70, 0.20, and 0.10—are the posterior probabilities, given type L information, that the lots contain 0.02, 0.05, and 0.08 defectives, respectively. The calculation of these posterior probabilities by Bayes' theorem is given in Table 17-5. These probabilities can also be derived from Table 17-4 by dividing joint probabilities by the appropriate marginal probabilities, for example, $0.70 = (\frac{0.42}{0.60})$.

Backward induction technique

We now use the standard backward induction technique (see section 14.6) to obtain the expected opportunity loss of the optimal strategy. Looking forward from node (f) and using the posterior probabilities 0.70, 0.20, and 0.10 as weights attached to the three states of nature (0.20, 0.05, and 0.08 defective lots), we obtain an expected opportunity loss of \$140 for the act "reject." Comparing this figure with the corresponding \$110 figure for "accept," we block off the action "reject" as nonoptimal. Therefore, \$110 is entered at node (c), representing the payoff for the optimal act upon observing type L ($X \leqslant c_1$) information. Similar calculations yield \$60 and \$20 at (d) and (e) for types M and H

[1] Alternatively, the conditional probabilities of sample results (likelihoods) derivable from this table may be thought of as having been calculated from an appropriate probability distribution such as the binomial or hypergeometric distribution. The basic methodological discussion remains unchanged.

TABLE 17-6
Possible decision rules based on information derived from single samples of size *n*

Sample Information	Strategy s_1	s_2	s_3	s_4	s_5	s_6	s_7	s_8
Type L ($X \leqslant c_1$)	R	A	A	A	A	R	R	R
Type M ($c_1 < X \leqslant c_2$)	R	A	A	R	R	A	R	A
Type H ($X > c_2$)	R	A	R	R	A	A	A	R

information. Weighting these three payoffs by the marginal probabilities of obtaining types L, M, and H information, we obtain a loss of \$82 as the expected payoff of sampling and inspecting n items. The cost of sampling and inspection would then have to be added if, for example, we wished to make a comparison with the expected loss of terminal action without sampling. However, for our comparison of the Bayesian decision theory approach with hypothesis testing, we will focus attention on the \$82 figure, which has been entered at node (b). We note, in summary, that \$82 is the expected loss of the optimal strategy, which accepts the lot if type L information is observed and rejects it otherwise.

Normal-form analysis

We turn now to normal-form analysis, in which all possible decision rules or strategies will be considered as a means of commenting on traditional hypothesis-testing procedures. The eight possible strategies implicit in the decision tree diagram in Figure 17-1 are enumerated in Table 17-6. (An R denotes reject; an A denotes accept.) For example, strategy s_3 means accept the lot if type L or type M information is observed, that is, if c_2 or fewer defectives are found. Strategy s_4 signifies acceptance if c_1 or fewer defectives are observed. Thus, a choice between strategies s_3 and s_4 means (in acceptance sampling terms) a selection between a single sampling plan with an acceptance number of c_2 and a plan with an acceptance number of c_1. (The conclusion of the extensive-form analysis was that s_4 is the optimal strategy, that is, s_4 has the minimum expected opportunity loss.)

Decision rules may not appear to be logical

As in previous problems, certain decision rules do not make much sense. For example, strategy s_1 would reject the lot and strategy s_2 would accept the lot, regardless of the information revealed by the sample. Strategy s_6 would reject the lot if a small number of defectives ($X \leqslant c_1$) were observed in the sample, but would accept the lot for larger numbers of defectives ($X > c_1$). The only strategies that appear to be at all logical are s_3 and s_4.

Hypothesis-testing procedure

Now, let us suppose this problem had been approached from the standpoint of a hypothesis-testing procedure. The two alternative hypotheses would be

$$H_0: p = 0.02$$

$$H_1: p = 0.05 \quad \text{or} \quad 0.08$$

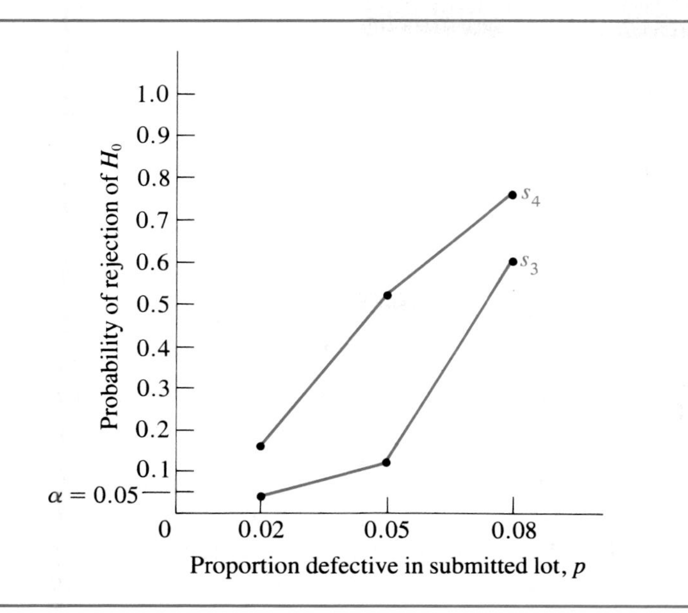

FIGURE 17-2
Power curves for strategies s_3 and s_4

Acceptance or rejection of the null hypothesis H_0 would mean acceptance or rejection of the lot, respectively. As indicated earlier, the company conducting the acceptance sampling wishes to accept lots that are 0.02 defective and to reject otherwise. Hence, the rejection of a good lot (one that contains 0.02 defectives) constitutes a Type I error. Let us assume that the company decides to test the null hypothesis at a 0.05 significance level. That is, the company specifies that it wants to reject lots containing 0.02 defectives no more than 5% of the time. We will examine what this selection of $\alpha = 0.05$ implies concerning the choice of a decision rule.

Power curves

Power curves may be plotted for each of the strategies or decision rules given in Table 17-6. However, we will show only the ones for s_3 and s_4, since none of the other strategies are worthy of further consideration. The power curves are depicted in Figure 17-2.

Let us consider how the power curves are plotted by taking the points for $p = 0.02$ as an example. Strategy s_3 accepts H_0 (accepts the lot) if type L or type M information is observed. From Table 17-4, we find that the conditional probability of observing type L or type M information, given $p = 0.02$, is $(\frac{0.42}{0.50}) + (\frac{0.06}{0.50}) = 0.96$. Therefore, the probability that H_0 will be accepted, given $p = 0.02$, is 0.96. The probability that H_0 will be rejected, given $p = 0.02$, is $1 - 0.96 = 0.04$. Symbolically, for strategy s_3, $P(\text{Rejection of } H_0|p = 0.02 = 0.04$. Analogously, we find that for strategy s_4, $P(\text{Rejection of } H_0|p = 0.02) = 1 - (\frac{0.42}{0.50}) = 0.16$.

Now, if we impose the condition of a 0.05 significance level, that is, lots containing 0.02 defectives should be rejected no more than 5% of the time, we find that strategy s_3 meets this criterion but strategy s_4 does not! Therefore, traditional hypothesis-testing procedures would require the use of strategy s_3, which has been shown to be nonoptimal. Looking at Figure 17-1, we can see why this is so. Under strategy s_3, if type M information is observed, the lot must be accepted, incurring an expected loss of $250, whereas under strategy s_4, if type M information is observed, the lot is rejected, with a loss of only $60. In summary, the expected opportunity losses of the two strategies are

$$\text{EOL}(s_3) = (0.60)(\$110) + (0.20)(\$250) + (0.20)(\$20) = \$120$$

$$\text{EOL}(s_4) = (0.60)(\$110) + (0.20)(\$60) + (0.20)(\$20) = \$82$$

A major criticism

The major criticism of traditional hypothesis-testing procedures implied by this example is that too much burden is placed on significance levels as a means of deciding between alternative acts.

> The inclusion of economic costs, or more generally, opportunity losses is not a standard procedure in the decision-making process.

Another illustrative problem is given in the next section, which more thoroughly contrasts the two sets of procedures. This is followed by a general comparative discussion.

Exercises 17.2

1. On what major grounds do advocates of Bayesian decision theory criticize classical hypothesis-testing procedures?
2. Suppose the following table is a joint frequency distribution for a particular problem:

State of Nature	Sample Result			Total
	Type L	Type M	Type H	
θ_1	0.36	0.07	0.01	0.44
θ_2	0.10	0.20	0.02	0.32
θ_3	0.04	0.03	0.17	0.24
Total	0.50	0.30	0.20	1.00

The only strategies that make sense are $s_2(A, R, R)$ and $s_7(A, A, R)$ where A means accept H_0 and R means reject H_0. As an example, $s_2(A, R, R)$ means accept H_0 if type L information is observed, but reject H_0 if type M or H information is observed. Construct power curves for both of these strategies. If the probability of a Type I error has been set at 0.05, which strategy satisfies the requirement?

The alternative hypotheses are

$$H_0: \theta = \theta_1$$

$$H_1: \theta = \theta_2 \quad \text{or} \quad \theta_3$$

3. Complete the following joint frequency distribution table. Determine which of the following strategies satisfy the requirement that the Type I error be no larger than 0.10: $s_4(R, R, A)$, $s_6(R, A, A)$. Draw power curves for both strategies. The hypotheses and the symbols have the same meaning as in exercise 2.

	Sample Result			
State of Nature	Type L	Type M	Type H	Total
θ_1		0.12		
θ_2	0.07		0.08	0.38
θ_3			0.17	0.29
Total	0.26	0.43		

4. The Stalwart Appliance Center regularly inspects incoming shipments of small appliances from its supplier to regulate the acceptance of defective items. From an analysis of past costs and past performance, the company has developed a payoff table showing prior probabilities and opportunity losses:

State of Nature (p = lot proportion of defectives)	Prior Probability	Act	
		A_1 Reject	A_2 Accept
0.01	0.45	\$400	\$ 0
0.02	0.35	0	700
0.05	0.20	0	900

The company has set two limits, x_1 and x_2, for classification of the results of an inspection into one of three categories:

Type L: $X \leqslant x_1$

Type M: $x_1 < X \leqslant x_2$

Type H: $X > x_2$

where X is the number of defectives in a sample of n items.

From a large number of past samples, the following results are obtained:

State of Nature (p = lot proportion of defectives)	Type L $X \leqslant x_1$	Type M $x_1 < X \leqslant x_2$	Type H $X > x_2$	Total
0.01	0.30	0.11	0.04	0.45
0.02	0.15	0.13	0.07	0.35
0.05	0.05	0.06	0.09	0.20
	0.50	0.30	0.20	1.00

a. Draw a decision tree for this problem and make all necessary entries. What is the optimal decision (reject or accept shipment) if Type M information is obtained from a sample?

b. Construct power curves for the two strategies $s_3(A_2, A_2, A_1)$ and $s_4(A_2, A_1, A_1)$. If the probability of a Type I error has been set at 0.10, which strategy satisfies the requirement? The alternative hypotheses are

$$H_0: p = 0.01$$
$$H_1: p = 0.02 \quad \text{or} \quad 0.05$$

17.3 CLASSICAL AND BAYESIAN ESTIMATION

In the preceding section, a comparison was made between hypothesis-testing procedures in classical statistical inference and the corresponding approaches in Bayesian decision theory. In this section, a comparison is made between the estimation techniques in the two approaches.

In Chapter 6, a brief description was given of classical point estimation techniques, that is, methods in which a population parameter value is estimated by a single statistic computed from the observations in a sample. For example, the mean of a sample may be used as the best single estimate of a population mean. In most practical problems, it is not sufficient to have merely a point estimate. If we were given two different point estimates of a population parameter and no further information, we could not distinguish the degree of reliability of these estimates. Yet one estimate might be based on a sample of size 10,000 and the other on a sample of size 10. Clearly, these estimates differ greatly in reliability. As we have seen, traditional statistics handles the problem of indicating reliability by the use of the confidence interval procedure. In this section, we will compare this classical technique to the corresponding Bayesian approach. However, before making this comparison, we comment briefly on point estimation techniques in the two approaches.

Point Estimation

In section 6.2, criteria of goodness of estimation were discussed. We have become familiar with point estimators such as the observed sample proportion of successes $\bar{p}$ in a Bernoulli process, which is used as the estimator of the population parameter p, and the observed sample mean $\bar{x}$ in a process described by the normal distribution, which is used as the estimator of the population mean μ.

Bayesian decision theory

Bayesian decision theory takes a different approach to the problem of point estimation. It views estimation as a straightforward problem of decision making.

The estimator is the decision rule, the estimate is the action, and the possible values that the population parameter can assume are the states of nature.

For example, the sample mean $\bar{x}$ might be the estimator (decision rule), 10.6 might be the estimate (action), and the possible values that the population mean μ can assume are the parameter values (states of nature). In this formulation, the unknown population parameter is treated as a random variable.

To clarify the method, we will introduce some notation. Let θ be the true value of the parameter we want to estimate and $\hat{\theta}$ the estimate or action. A loss is involved if the value of $\hat{\theta}$ differs from θ, and the amount of the loss is some function of the difference between $\hat{\theta}$ and θ. Two possible loss functions might be

Linear loss function

$$L(\hat{\theta}|\theta) = |\hat{\theta} - \theta| \tag{17.1}$$

and

Quadratic loss function

$$L(\hat{\theta}|\theta) = (\hat{\theta} - \theta)^2 \tag{17.2}$$

where $L(\hat{\theta}|\theta)$ is the loss involved in estimating (taking action) $\hat{\theta}$ when the parameter value (state of nature) is θ.

Somewhat more generally, the loss functions 17.1 and 17.2 may be written as

$$L(\hat{\theta}|\theta) = k(\theta)|\hat{\theta} - \theta| \tag{17.3}$$

and

$$L(\hat{\theta}|\theta) = k(\theta)(\hat{\theta} - \theta)^2 \tag{17.4}$$

respectively, where $k(\theta)$ is a constant for a particular value of θ. This constant may be in money units, utility units, and so on. For simplicity, in the ensuing discussion, we will assume $k(\theta) = 1$ unit of utility, sometimes referred to as a **utile**. Therefore, we are dealing with functions of the form of equations 17.1 and 17.2, and the losses are given in units of utility.

Expression 17.1 is referred to as a **linear loss function**; expression 17.2 is a **quadratic loss function** (or **squared error loss function**). The nature of these functions can be illustrated by simple examples. Assume that the true value of the parameter θ is 10. Consider the losses involved if we estimate this parameter incorrectly as $\hat{\theta} = 11$ and $\hat{\theta} = 12$. For these two estimates, the respective linear loss functions 17.1 are

$$L(11|10) = |11 - 10| = 1$$

and

$$L(12|10) = |12 - 10| = 2$$

On the other hand, the quadratic loss function, 17.2 is equal to

$$L(11|10) = (11 - 10)^2 = 1$$

and

$$L(12|10) = (12 - 10)^2 = 4$$

In other words, in the linear case the loss in overestimating by two units is *twice* as much as in overestimating by one unit. In the quadratic case, the loss in overestimating by two units is *four* times as much as in overestimating by one unit. Note that in both functions, an underestimate of a given size, say two units, is as serious as an overestimate of the same size. Such loss functions are said to be **symmetrical**.

The ideas of these two loss functions were referred to earlier in section 1.17 in somewhat different forms. There, we were concerned with guessing the value of an observation selected at random from a frequency distribution. The penalty of an incorrect guess or estimate was referred to as the "cost of error." That "cost" corresponds to "loss" in the present discussion. We pointed out that if the cost of error varies directly with the size of error regardless of sign (the linear loss function 17.1), the median is the "best guess," since it minimizes average absolute deviations. On the other hand, if the cost of error varies according to the square of the error (the quadratic loss function 17.2), the mean should be the estimated value, since the average of the squared deviations around it is less than around any other figure. Note that least squares methods of estimation in classical statistics assume a quadratic loss function, since they obtain estimates for which the average squared error is minimized. Whether or not this is an appropriate loss function for the particular problem involved is rarely investigated.

Bayesian method of point estimation

The Bayesian method of point estimation begins with setting up whatever loss function appears to be appropriate. Then these losses are used in the standard decision procedure. Risks (conditional expected losses) are computed for each decision rule, or estimator. Prior probabilities are assigned to states of nature, or parameter values. Expected risks are computed for each decision rule. Then the estimator for which the expected risk is the least is the one chosen.

No Bayesian point estimators will be derived here, but one result is of particular interest. If the parameter p of a Bernoulli process is estimated using a squared error loss function, and a uniform or rectangular (continuous) prior probability distribution for p is assumed (that is, all values between zero and one are assumed to be equally likely), then the Bayesian estimator of p, denoted $\hat{p}$, is

$$\hat{p} = \frac{X + 1}{n + 2} \tag{17.5}$$

where X = the number of successes
n = the number of trials

This value of $\hat{p}$ is also the mean or expected value of the posterior distribution of p if the prior distribution of p is assumed to be rectangular (and continuous) and the sample evidence is an observation of X successes in n trials. The estimate of p that we used earlier was simply the observed proportion of successes X/n. If the sample size n is large, these two estimates are approximately equal. Furthermore, there are other prior probability distributions besides the rectangular (uniform) distribution for which the mean of the posterior distribution will have a difference of this magnitude when compared with an estimator derived by classical methods. This brings out a very interesting point. From the Bayesian point of view, the standard use of the X/n estimate in such situations carries with it assumptions concerning the nature of the prior distribution of p. The Bayesian decision analyst would argue that some of these prior distributions are unreasonable in the context of particular problems. For example, a rectangular prior distribution implies that all values of the parameter (in its admissible range) are equally likely. Such an implication may be quite unrealistic based on the analyst's prior knowledge.

Interval Estimation

We turn now to a consideration of interval estimation in classical and Bayesian statistics. We have seen that in estimating a population parameter in classical confidence interval estimation, an interval is set up on the basis of a sample of n observations and a so-called confidence coefficient is associated with this interval. Suppose, for example, we wanted to make a confidence interval estimate of p, the proportion of all customers on an importer's mailing list who would purchase special jars of cocktail onions if advertisements were sent to them, and let us assume that we want to base this estimate on the proportion $\bar{p}$ who purchased the jars in a simple random sample of 100 drawn from the list. We could establish a 95% confidence interval around $\bar{p}$ for the estimation of p in the usual way. Let us review the interpretation of this confidence interval.

Classical approach

According to the classical school, it is definitely *incorrect* to say that the probability is 95% that the parameter is included in the interval. The population parameter is a particular value and therefore cannot be considered a random variable. Indeed, in all of classical statistics, it is forbidden to make conditional probability statements about a *population parameter* given the value of a sample statistic, such as $P(p|\bar{p})$. Permissible statements concern conditional probabilities of sample statistics given the value of a population parameter. For example, in a problem involving a Bernoulli process, we could compute probabilities of the type $P(\bar{p}|p)$.

Returning to the importer's problem, from the classical viewpoint, the confidence interval estimate of p cannot be interpreted as a probability statement about the proportion of all customers on the mailing list who would purchase the product. Since the interval is considered to be the random variable, the confidence coefficient refers to the concept that 95% of the intervals so constructed would bracket or include the true value of the population parameter. Thus, on a relative frequency basis, 95% of the statements made on the basis of such intervals would be correct. Furthermore, in keeping with the classical viewpoint, only the evidence of this particular sample can be used in establishing the confidence interval. Prior knowledge of any sort is not a part of the estimation procedure. Finally, just as in hypothesis testing, the use of the sample observations must be determined prior to the examination of these observations.

Bayesian approach

The Bayesian approach to this general problem contrasts sharply with the classical procedure. The Bayesian argues that if the value of the population parameter is unknown, then it can and should be treated as a random variable. In a setting such as the importer's problem, we would view the population parameter p as a random variable affecting a decision that must be made. Hence, we would be willing to compute conditional probabilities of the type $P(p|\bar{p})$. Furthermore, we would state that these conditional probabilities are relevant to the decision maker, rather than those of the form $P(\bar{p}|p)$. For example, in problems similar to that of the importer's, we might be interested in the probability that at least a certain proportion of the population would purchase the product based on the sample evidence. We would not be interested in the reverse conditional probability concerning a proportion in the sample given some postulated value for the population. The Bayesian decision analyst would argue that the confidence interval information is not particularly relevant. The decision maker is not interested in the proportion of correct statements that would be made in the long run, but rather in making a correct decision in this particular case.

The Bayesian approach also maintains that the analyst should not be restricted to the evidence of the particular sample that has been drawn but that the sample evidence should be incorporated with prior information through the use of Bayes' theorem to produce a posterior probability distribution. This leads to the Bayesian approach to the problem that classical inference solves by confidence interval estimation. The Bayesian procedure begins with the assignment of a prior probability distribution to the parameter being estimated. Then a sample is drawn, and the sample evidence is used to revise the prior probability distribution. The revision generates a posterior probability distribution. Then statements such as the following can be made: The probability is 0.90 that p lies between 0.04 and 0.06. The probability is 0.95 that the value of p is 0.07 or less, and so on. We may have a large number of possible values for p if that random variable is discrete, or we may have a probability density function over p if the random variable is continuous. The principle remains the same. If the prior distribution was a subjective probability distribution, then the posterior probabilities similarly represent revised degrees of belief or betting odds.

An interesting result occurs that is analogous to a relationship indicated earlier between classical and Bayesian point estimation when a rectangular prior distribution was assumed.

If a rectangular distribution is assumed for the random variable p, and if the sample size is large, then there is a close coincidence between the statements made under the classical and Bayesian schools of thought.

Specifically, for example, the posterior probability that p lies in a 0.95 confidence interval is approximately 0.95. A roughly rectangular or uniform prior distribution is often referred to by Bayesian decision theorists as a "diffuse" or "gentle" prior distribution. Such a distribution implies roughly equal likelihood of occurrence of all values of the random variable in its admissible range. This type of distribution is thought of as an appropriate subjective prior distribution when the decision maker has virtually no knowledge of the value of the parameter being estimated. Doubtless, such states of almost complete lack of knowledge about parameter values are rare. Hence, Bayesian decision theorists argue that the uncritical use of confidence interval estimates may imply unreasonable assumptions about the investigator's prior knowledge concerning the parameter being estimated.

17.4 SOME REMARKS ON CLASSICAL AND BAYESIAN STATISTICS

As might be surmised from the material in this chapter, some controversy has arisen between adherents of the classical, or orthodox, school of statistics and advocates of the Bayesian viewpoint. In this section, we will comment on some of the areas of common ground and some of the points of difference between the two schools of thought.

Common ground

- Despite differences in terminology, both schools conceptualize a problem of decision making consisting of states of nature and actions that must be taken in the light of sample or experimental evidence about these states of nature.

- Both schools use conditional probabilities of sample outcomes, given states of nature (population parameters) for the decision process. These conditional probabilities provide the error characteristic curve from which the classicist chooses the decision rule. Informally, we should take into account the relative seriousness of Type I and Type II errors by considering the entire error characteristic curve, but since losses virtually always vary with population parameter values, it is not clear how we can actually do this.

Points of difference

- The Bayesian approach supplements or completes the classical analysis by formally providing a loss function that specifies the seriousness of errors in selecting acts and by assigning prior probabilities to states of nature on either an objective or a subjective basis. However, serious measurement problems are clearly present both in the establishment of loss functions and prior probability distributions.

- Some classicists have affirmed that hypothesis testing is not a decision problem, but rather one of drawing conclusions or inferences. However, other classical adherents have specifically formulated hypothesis testing as an action problem. In any event, it is not always clear whether a problem is one of inference or decision making.

- An important area of disagreement between non-Bayesian and Bayesian analysts is the matter of subjective prior probability distributions. The non-Bayesians argue that the only legitimate types of probabilities are "objective" or relative frequency of occurrence probabilities. They find it difficult to accept the idea that subjective or personalistic probabilities should be processed together with the relative frequencies, as in the Bayesian's use of Bayes' theorem, to arrive at posterior probabilities. The Bayesian argues that in actual decision making we do exactly that type of analysis. We have prior betting odds on events that influence the payoffs of our actions. On the observation of sample or experimental information, we revise these prior betting odds. This argument centers on "descriptive" behavior, that is, a purported description of how people actually behave. However, the Bayesian goes further, saying that Bayesian procedures are "normative" or "prescriptive," that is, they specify how a reasonable person *should* choose among alternatives to be consistent with one's own evaluations of payoffs and degrees of belief attached to uncertain events. The Bayesian also argues that if we rigidly maintain that only objective probabilities have meaning, we prevent ourselves from handling some of the most important uncertainties involved in problems of decision making. This latter point is surely a cogent one, particularly in areas such as business and economic decision making.

- The problem of how to assign prior probabilities is troublesome, even to convinced Bayesians, and is a subject of ongoing research. There are unresolved problems involved in determining whether all events should be considered equally likely under ignorance, how to pose questions to a decision maker to derive that individual's distribution of betting odds, or more generally, how best to quantify judgments about uncertainty.

- The Bayesian turns the tables on the orthodox school, which makes an accusation of excessive subjectivity, and directs a similar charge against classical statistics. The choices of hypotheses to test, probability distributions to use, significance and confidence levels, and what data to collect in order to obtain a relative frequency distribution are all inextricably interwoven with subjective judgments.

The preceding indication of some of the disagreements between the classical and Bayesian schools tends to emphasize a polarization of points of view. However, the fact is that even within each of these schools there are philosophical and methodological disagreements as well. These diversities of viewpoint between and within schools of thought make statistical analysis for decision making a lively and growing field.

Appendix A

Statistical Tables

TABLE A-1

Selected values of the binomial cumulative distribution function

$$F(c) = P(X \leqslant c) = \sum_{x=0}^{c} \binom{n}{x} (1-p)^{n-x} p^x$$

Example If $p = 0.20$, $n = 7$, $c = 2$, then $F(2) = P(X \leqslant 2) = 0.8520$.

		p									
n	c	0.05	0.10	0.15	0.20	0.25	0.30	0.35	0.40	0.45	0.50
2	0	0.9025	0.8100	0.7225	0.6400	0.5625	0.4900	0.4225	0.3600	0.3025	0.2500
	1	0.9975	0.9900	0.9775	0.9600	0.9375	0.9100	0.8775	0.8400	0.7975	0.7500
3	0	0.8574	0.7290	0.6141	0.5120	0.4219	0.3430	0.2746	0.2160	0.1664	0.1250
	1	0.9928	0.9720	0.9392	0.8960	0.8438	0.7840	0.7182	0.6480	0.5748	0.5000
	2	0.9999	0.9990	0.9966	0.9920	0.9844	0.9730	0.9571	0.9360	0.9089	0.8750
4	0	0.8145	0.6561	0.5220	0.4096	0.3164	0.2401	0.1785	0.1296	0.0915	0.0625
	1	0.9860	0.9477	0.8905	0.8192	0.7383	0.6517	0.5630	0.4752	0.3910	0.3125
	2	0.9995	0.9963	0.9880	0.9728	0.9492	0.9163	0.8735	0.8208	0.7585	0.6875
	3	1.0000	0.9999	0.9995	0.9984	0.9961	0.9919	0.9850	0.9744	0.9590	0.9375
5	0	0.7738	0.5905	0.4437	0.3277	0.2373	0.1681	0.1160	0.0778	0.0503	0.0312
	1	0.9774	0.9185	0.8352	0.7373	0.6328	0.5282	0.4284	0.3370	0.2562	0.1875
	2	0.9988	0.9914	0.9734	0.9421	0.8965	0.8369	0.7648	0.6826	0.5931	0.5000
	3	1.0000	0.9995	0.9978	0.9933	0.9844	0.9692	0.9460	0.9130	0.8688	0.8125
	4	1.0000	1.0000	0.9999	0.9997	0.9990	0.9976	0.9947	0.9898	0.9815	0.9688

Source: From Irwin Miller and John E. Freund, *Probability and Statistics for Engineers*, Second Edition, © 1977, pp. 477–481. Reprinted by permission of Prentice-Hall, Inc., Englewood Cliffs, NJ.

TABLE A-1 (continued)

		p									
n	c	0.05	0.10	0.15	0.20	0.25	0.30	0.35	0.40	0.45	0.50
6	0	0.7351	0.5314	0.3771	0.2621	0.1780	0.1176	0.0754	0.0467	0.0277	0.0156
	1	0.9672	0.8857	0.7765	0.6554	0.5339	0.4202	0.3191	0.2333	0.1636	0.1094
	2	0.9978	0.9842	0.9527	0.9011	0.8306	0.7443	0.6471	0.5443	0.4415	0.3438
	3	0.9999	0.9987	0.9941	0.9830	0.9624	0.9295	0.8826	0.8208	0.7447	0.6562
	4	1.0000	0.9999	0.9996	0.9984	0.9954	0.9891	0.9777	0.9590	0.9308	0.8906
	5	1.0000	1.0000	1.0000	0.9999	0.9998	0.9993	0.9982	0.9959	0.9917	0.9844
7	0	0.6983	0.4783	0.3206	0.2097	0.1335	0.0824	0.0490	0.0280	0.0152	0.0078
	1	0.9556	0.8503	0.7166	0.5767	0.4449	0.3294	0.2338	0.1586	0.1024	0.0625
	2	0.9962	0.9743	0.9262	0.8520	0.7564	0.6471	0.5323	0.4199	0.3164	0.2266
	3	0.9998	0.9973	0.9879	0.9667	0.9294	0.8740	0.8002	0.7102	0.6083	0.5000
	4	1.0000	0.9998	0.9988	0.9953	0.9871	0.9712	0.9444	0.9037	0.8471	0.7734
	5	1.0000	1.0000	0.9999	0.9996	0.9987	0.9962	0.9910	0.9812	0.9643	0.9375
	6	1.0000	1.0000	1.0000	1.0000	0.9999	0.9998	0.9994	0.9984	0.9963	0.9922
8	0	0.6634	0.4305	0.2725	0.1678	0.1001	0.0576	0.0319	0.0168	0.0084	0.0039
	1	0.9428	0.8131	0.6572	0.5033	0.3671	0.2553	0.1691	0.1064	0.0632	0.0352
	2	0.9942	0.9619	0.8948	0.7969	0.6785	0.5518	0.4278	0.3154	0.2201	0.1445
	3	0.9996	0.9950	0.9786	0.9437	0.8862	0.8059	0.7064	0.5941	0.4770	0.3633
	4	1.0000	0.9996	0.9971	0.9896	0.9727	0.9420	0.8939	0.8263	0.7396	0.6367
	5	1.0000	1.0000	0.9998	0.9988	0.9958	0.9887	0.9747	0.9502	0.9115	0.8555
	6	1.0000	1.0000	1.0000	0.9999	0.9996	0.9987	0.9964	0.9915	0.9819	0.9648
	7	1.0000	1.0000	1.0000	1.0000	1.0000	0.9999	0.9998	0.9993	0.9983	0.9961
9	0	0.6302	0.3874	0.2316	0.1342	0.0751	0.0404	0.0207	0.0101	0.0046	0.0020
	1	0.9288	0.7748	0.5995	0.4362	0.3003	0.1960	0.1211	0.0705	0.0385	0.0195
	2	0.9916	0.9470	0.8591	0.7382	0.6007	0.4628	0.3373	0.2318	0.1495	0.0898
	3	0.9994	0.9917	0.9661	0.9144	0.8343	0.7297	0.6089	0.4826	0.3614	0.2539
	4	1.0000	0.9991	0.9944	0.9804	0.9511	0.9012	0.8283	0.7334	0.6214	0.5000
	5	1.0000	0.9999	0.9994	0.9969	0.9900	0.9747	0.9464	0.9006	0.8342	0.7461
	6	1.0000	1.0000	1.0000	0.9997	0.9987	0.9957	0.9888	0.9750	0.9502	0.9102
	7	1.0000	1.0000	1.0000	1.0000	0.9999	0.9996	0.9986	0.9962	0.9909	0.9805
	8	1.0000	1.0000	1.0000	1.0000	1.0000	1.0000	0.9999	0.9997	0.9992	0.9980
10	0	0.5987	0.3487	0.1969	0.1074	0.0563	0.0282	0.0135	0.0060	0.0025	0.0010
	1	0.9139	0.7361	0.5443	0.3758	0.2440	0.1493	0.0860	0.0464	0.0232	0.0107
	2	0.9885	0.9298	0.8202	0.6778	0.5256	0.3828	0.2616	0.1673	0.0996	0.0547
	3	0.9990	0.9872	0.9500	0.8791	0.7759	0.6496	0.5138	0.3823	0.2660	0.1719
	4	0.9999	0.9984	0.9901	0.9672	0.9219	0.8497	0.7515	0.6331	0.5044	0.3770
	5	1.0000	0.9999	0.9986	0.9936	0.9803	0.9527	0.9051	0.8338	0.7384	0.6230
	6	1.0000	1.0000	0.9999	0.9991	0.9965	0.9894	0.9740	0.9452	0.8980	0.8281
	7	1.0000	1.0000	1.0000	0.9999	0.9996	0.9984	0.9952	0.9877	0.9726	0.9453
	8	1.0000	1.0000	1.0000	1.0000	1.0000	0.9999	0.9995	0.9983	0.9955	0.9893
	9	1.0000	1.0000	1.0000	1.0000	1.0000	1.0000	1.0000	0.9999	0.9997	0.9990

TABLE A-1 (continued)

		p									
n	*c*	0.05	0.10	0.15	0.20	0.25	0.30	0.35	0.40	0.45	0.50
11	0	0.5688	0.3138	0.1673	0.0859	0.0422	0.0198	0.0088	0.0036	0.0014	0.0005
	1	0.8981	0.6974	0.4922	0.3221	0.1971	0.1130	0.0606	0.0302	0.0139	0.0059
	2	0.9848	0.9104	0.7788	0.6174	0.4552	0.3127	0.2001	0.1189	0.0652	0.0327
	3	0.9984	0.9815	0.9306	0.8389	0.7133	0.5696	0.4256	0.2963	0.1911	0.1133
	4	0.9999	0.9972	0.9841	0.9496	0.8854	0.7897	0.6683	0.5328	0.3971	0.2744
	5	1.0000	0.9997	0.9973	0.9883	0.9657	0.9218	0.8513	0.7535	0.6331	0.5000
	6	1.0000	1.0000	0.9997	0.9980	0.9924	0.9784	0.9499	0.9006	0.8262	0.7256
	7	1.0000	1.0000	1.0000	0.9998	0.9988	0.9957	0.9878	0.9707	0.9390	0.8867
	8	1.0000	1.0000	1.0000	1.0000	0.9999	0.9994	0.9980	0.9941	0.9852	0.9673
	9	1.0000	1.0000	1.0000	1.0000	1.0000	1.0000	0.9998	0.9993	0.9978	0.9941
	10	1.0000	1.0000	1.0000	1.0000	1.0000	1.0000	1.0000	1.0000	0.9998	0.9995
12	0	0.5404	0.2824	0.1422	0.0687	0.0317	0.0138	0.0057	0.0022	0.0008	0.0002
	1	0.8816	0.6590	0.4435	0.2749	0.1584	0.0850	0.0424	0.0196	0.0083	0.0032
	2	0.9804	0.8891	0.7358	0.5583	0.3907	0.2528	0.1513	0.0834	0.0421	0.0193
	3	0.9978	0.9744	0.9078	0.7946	0.6488	0.4925	0.3467	0.2253	0.1345	0.0730
	4	0.9998	0.9957	0.9761	0.9274	0.8424	0.7237	0.5833	0.4382	0.3044	0.1938
	5	1.0000	0.9995	0.9954	0.9806	0.9456	0.8822	0.7873	0.6652	0.5269	0.3872
	6	1.0000	0.9999	0.9993	0.9961	0.9857	0.9614	0.9154	0.8418	0.7393	0.6128
	7	1.0000	1.0000	0.9999	0.9994	0.9972	0.9905	0.9745	0.9427	0.8883	0.8062
	8	1.0000	1.0000	1.0000	0.9999	0.9996	0.9983	0.9944	0.9847	0.9644	0.9270
	9	1.0000	1.0000	1.0000	1.0000	1.0000	0.9998	0.9992	0.9972	0.9921	0.9807
	10	1.0000	1.0000	1.0000	1.0000	1.0000	1.0000	0.9999	0.9997	0.9989	0.9968
	11	1.0000	1.0000	1.0000	1.0000	1.0000	1.0000	1.0000	1.0000	0.9999	0.9998
13	0	0.5133	0.2542	0.1209	0.0550	0.0238	0.0097	0.0037	0.0013	0.0004	0.0001
	1	0.8646	0.6213	0.3983	0.2336	0.1267	0.0637	0.0296	0.0126	0.0049	0.0017
	2	0.9755	0.8661	0.6920	0.5017	0.3326	0.2025	0.1132	0.0579	0.0269	0.0112
	3	0.9969	0.9658	0.8820	0.7473	0.5843	0.4206	0.2783	0.1686	0.0929	0.0461
	4	0.9997	0.9935	0.9658	0.9009	0.7940	0.6543	0.5005	0.3530	0.2279	0.1334
	5	1.0000	0.9991	0.9925	0.9700	0.9198	0.8346	0.7159	0.5744	0.4268	0.2905
	6	1.0000	0.9999	0.9987	0.9930	0.9757	0.9376	0.8705	0.7712	0.6437	0.5000
	7	1.0000	1.0000	0.9998	0.9988	0.9944	0.9818	0.9538	0.9023	0.8212	0.7095
	8	1.0000	1.0000	1.0000	0.9998	0.9990	0.9960	0.9874	0.9679	0.9302	0.8666
	9	1.0000	1.0000	1.0000	1.0000	0.9999	0.9993	0.9975	0.9922	0.9797	0.9539
	10	1.0000	1.0000	1.0000	1.0000	1.0000	0.9999	0.9997	0.9987	0.9959	0.9888
	11	1.0000	1.0000	1.0000	1.0000	1.0000	1.0000	1.0000	0.9999	0.9995	0.9983
	12	1.0000	1.0000	1.0000	1.0000	1.0000	1.0000	1.0000	1.0000	1.0000	0.9999
14	0	0.4877	0.2288	0.1028	0.0440	0.0178	0.0068	0.0024	0.0008	0.0002	0.0001
	1	0.8470	0.5846	0.3567	0.1979	0.1010	0.0475	0.0205	0.0081	0.0029	0.0009
	2	0.9699	0.8416	0.6479	0.4481	0.2811	0.1608	0.0839	0.0398	0.0170	0.0065
	3	0.9958	0.9559	0.8535	0.6982	0.5213	0.3552	0.2205	0.1243	0.0632	0.0287
	4	0.9996	0.9908	0.9533	0.8702	0.7415	0.5842	0.4227	0.2793	0.1672	0.0898

TABLE A-1 (continued)

n	c	0.05	0.10	0.15	0.20	0.25	0.30	0.35	0.40	0.45	0.50
						p					
	5	1.0000	0.9985	0.9885	0.9561	0.8883	0.7805	0.6405	0.4859	0.3373	0.2120
	6	1.0000	0.9998	0.9978	0.9884	0.9617	0.9067	0.8164	0.6925	0.5461	0.3953
	7	1.0000	1.0000	0.9997	0.9976	0.9897	0.9685	0.9247	0.8499	0.7414	0.6047
	8	1.0000	1.0000	1.0000	0.9996	0.9978	0.9917	0.9757	0.9417	0.8811	0.7880
	9	1.0000	1.0000	1.0000	1.0000	0.9997	0.9983	0.9940	0.9825	0.9574	0.9102
	10	1.0000	1.0000	1.0000	1.0000	1.0000	0.9998	0.9989	0.9961	0.9886	0.9713
	11	1.0000	1.0000	1.0000	1.0000	1.0000	1.0000	0.9999	0.9994	0.9978	0.9935
	12	1.0000	1.0000	1.0000	1.0000	1.0000	1.0000	1.0000	0.9999	0.9997	0.9991
	13	1.0000	1.0000	1.0000	1.0000	1.0000	1.0000	1.0000	1.0000	1.0000	0.9999
15	0	0.4633	0.2059	0.0874	0.0352	0.0134	0.0047	0.0016	0.0005	0.0001	0.0000
	1	0.8290	0.5490	0.3186	0.1671	0.0802	0.0353	0.0142	0.0052	0.0017	0.0005
	2	0.9638	0.8159	0.6042	0.3980	0.2361	0.1268	0.0617	0.0271	0.0107	0.0037
	3	0.9945	0.9444	0.8227	0.6482	0.4613	0.2969	0.1727	0.0905	0.0424	0.0176
	4	0.9994	0.9873	0.9383	0.8358	0.6865	0.5155	0.3519	0.2173	0.1204	0.0592
	5	0.9999	0.9978	0.9832	0.9389	0.8516	0.7216	0.5643	0.4032	0.2608	0.1509
	6	1.0000	0.9997	0.9964	0.9819	0.9434	0.8689	0.7548	0.6098	0.4522	0.3036
	7	1.0000	1.0000	0.9996	0.9958	0.9827	0.9500	0.8868	0.7869	0.6535	0.5000
	8	1.0000	1.0000	0.9999	0.9992	0.9958	0.9848	0.9578	0.9050	0.8182	0.6964
	9	1.0000	1.0000	1.0000	0.9999	0.9992	0.9963	0.9876	0.9662	0.9231	0.8491
	10	1.0000	1.0000	1.0000	1.0000	0.9999	0.9993	0.9972	0.9907	0.9745	0.9408
	11	1.0000	1.0000	1.0000	1.0000	1.0000	0.9999	0.9995	0.9981	0.9937	0.9824
	12	1.0000	1.0000	1.0000	1.0000	1.0000	1.0000	0.9999	0.9997	0.9989	0.9963
	13	1.0000	1.0000	1.0000	1.0000	1.0000	1.0000	1.0000	1.0000	0.9999	0.9995
	14	1.0000	1.0000	1.0000	1.0000	1.0000	1.0000	1.0000	1.0000	1.0000	1.0000
16	0	0.4401	0.1853	0.0743	0.0281	0.0100	0.0033	0.0010	0.0003	0.0001	0.0000
	1	0.8108	0.5147	0.2839	0.1407	0.0635	0.0261	0.0098	0.0033	0.0010	0.0003
	2	0.9571	0.7892	0.5614	0.3518	0.1971	0.0994	0.0451	0.0183	0.0066	0.0021
	3	0.9930	0.9316	0.7899	0.5981	0.4050	0.2459	0.1339	0.0651	0.0281	0.0106
	4	0.9991	0.9830	0.9209	0.7982	0.6302	0.4499	0.2892	0.1666	0.0853	0.0384
	5	0.9999	0.9967	0.9765	0.9183	0.8103	0.6598	0.4900	0.3288	0.1976	0.1051
	6	1.0000	0.9995	0.9944	0.9733	0.9204	0.8247	0.6881	0.5272	0.3660	0.2272
	7	1.0000	0.9999	0.9989	0.9930	0.9729	0.9256	0.8406	0.7161	0.5629	0.4018
	8	1.0000	1.0000	0.9998	0.9985	0.9925	0.9743	0.9329	0.8577	0.7441	0.5982
	9	1.0000	1.0000	1.0000	0.9998	0.9984	0.9929	0.9771	0.9417	0.8759	0.7728
	10	1.0000	1.0000	1.0000	1.0000	0.9997	0.9984	0.9938	0.9809	0.9514	0.8949
	11	1.0000	1.0000	1.0000	1.0000	1.0000	0.9997	0.9987	0.9951	0.9851	0.9616
	12	1.0000	1.0000	1.0000	1.0000	1.0000	1.0000	0.9998	0.9991	0.9965	0.9894
	13	1.0000	1.0000	1.0000	1.0000	1.0000	1.0000	1.0000	0.9999	0.9994	0.9979
	14	1.0000	1.0000	1.0000	1.0000	1.0000	1.0000	1.0000	1.0000	1.0000	0.9997
	15	1.0000	1.0000	1.0000	1.0000	1.0000	1.0000	1.0000	1.0000	1.0000	1.0000

TABLE A-1 (continued)

n	c	p = 0.05	0.10	0.15	0.20	0.25	0.30	0.35	0.40	0.45	0.50
17	0	0.4181	0.1668	0.0631	0.0225	0.0075	0.0023	0.0007	0.0002	0.0000	0.0000
	1	0.7922	0.4818	0.2525	0.1182	0.0501	0.0193	0.0067	0.0021	0.0006	0.0001
	2	0.9497	0.7618	0.5198	0.3096	0.1637	0.0774	0.0327	0.0123	0.0041	0.0012
	3	0.9912	0.9174	0.7556	0.5489	0.3530	0.2019	0.1028	0.0464	0.0184	0.0064
	4	0.9988	0.9779	0.9013	0.7582	0.5739	0.3887	0.2348	0.1260	0.0596	0.0245
	5	0.9999	0.9953	0.9681	0.8943	0.7653	0.5968	0.4197	0.2639	0.1471	0.0717
	6	1.0000	0.9992	0.9917	0.9623	0.8929	0.7752	0.6188	0.4478	0.2902	0.1662
	7	1.0000	0.9999	0.9983	0.9891	0.9598	0.8954	0.7872	0.6405	0.4743	0.3145
	8	1.0000	1.0000	0.9997	0.9974	0.9876	0.9597	0.9006	0.8011	0.6626	0.5000
	9	1.0000	1.0000	1.0000	0.9995	0.9969	0.9873	0.9617	0.9081	0.8166	0.6855
	10	1.0000	1.0000	1.0000	0.9999	0.9994	0.9968	0.9880	0.9652	0.9174	0.8338
	11	1.0000	1.0000	1.0000	1.0000	0.9999	0.9993	0.9970	0.9894	0.9699	0.9283
	12	1.0000	1.0000	1.0000	1.0000	1.0000	0.9999	0.9994	0.9975	0.9914	0.9755
	13	1.0000	1.0000	1.0000	1.0000	1.0000	1.0000	0.9999	0.9995	0.9981	0.9936
	14	1.0000	1.0000	1.0000	1.0000	1.0000	1.0000	1.0000	0.9999	0.9997	0.9988
	15	1.0000	1.0000	1.0000	1.0000	1.0000	1.0000	1.0000	1.0000	1.0000	0.9999
	16	1.0000	1.0000	1.0000	1.0000	1.0000	1.0000	1.0000	1.0000	1.0000	1.0000
18	0	0.3972	0.1501	0.0536	0.0180	0.0056	0.0016	0.0004	0.0001	0.0000	0.0000
	1	0.7735	0.4503	0.2241	0.0991	0.0395	0.0142	0.0046	0.0013	0.0003	0.0001
	2	0.9419	0.7338	0.4797	0.2713	0.1353	0.0600	0.0236	0.0082	0.0025	0.0007
	3	0.9891	0.9018	0.7202	0.5010	0.3057	0.1646	0.0783	0.0328	0.0120	0.0038
	4	0.9985	0.9718	0.8794	0.7164	0.5187	0.3327	0.1886	0.0942	0.0411	0.0154
	5	0.9998	0.9936	0.9581	0.8671	0.7175	0.5344	0.3550	0.2088	0.1077	0.0481
	6	1.0000	0.9988	0.9882	0.9487	0.8610	0.7217	0.5491	0.3743	0.2258	0.1189
	7	1.0000	0.9998	0.9973	0.9837	0.9431	0.8593	0.7283	0.5634	0.3915	0.2403
	8	1.0000	1.0000	0.9995	0.9957	0.9807	0.9404	0.8609	0.7368	0.5778	0.4073
	9	1.0000	1.0000	0.9999	0.9991	0.9946	0.9790	0.9403	0.8653	0.7473	0.5927
	10	1.0000	1.0000	1.0000	0.9998	0.9988	0.9939	0.9788	0.9424	0.8720	0.7597
	11	1.0000	1.0000	1.0000	1.0000	0.9998	0.9986	0.9938	0.9797	0.9463	0.8811
	12	1.0000	1.0000	1.0000	1.0000	1.0000	0.9997	0.9986	0.9942	0.9817	0.9519
	13	1.0000	1.0000	1.0000	1.0000	1.0000	1.0000	0.9997	0.9987	0.9951	0.9846
	14	1.0000	1.0000	1.0000	1.0000	1.0000	1.0000	1.0000	0.9998	0.9990	0.9962
	15	1.0000	1.0000	1.0000	1.0000	1.0000	1.0000	1.0000	1.0000	0.9999	0.9993
	16	1.0000	1.0000	1.0000	1.0000	1.0000	1.0000	1.0000	1.0000	1.0000	0.9999
19	0	0.3774	0.1351	0.0456	0.0144	0.0042	0.0011	0.0003	0.0001	0.0000	0.0000
	1	0.7547	0.4203	0.1985	0.0829	0.0310	0.0104	0.0031	0.0008	0.0002	0.0000
	2	0.9335	0.7054	0.4413	0.2369	0.1113	0.0462	0.0170	0.0055	0.0015	0.0004
	3	0.9868	0.8850	0.6841	0.4551	0.2630	0.1332	0.0591	0.0230	0.0077	0.0022
	4	0.9980	0.9648	0.8556	0.6733	0.4654	0.2822	0.1500	0.0696	0.0280	0.0096

TABLE A-1 (continued)

		p									
n	*c*	0.05	0.10	0.15	0.20	0.25	0.30	0.35	0.40	0.45	0.50
	5	0.9998	0.9914	0.9463	0.8369	0.6678	0.4739	0.2968	0.1629	0.0777	0.0318
	6	1.0000	0.9983	0.9837	0.9324	0.8251	0.6655	0.4812	0.3081	0.1727	0.0835
	7	1.0000	0.9997	0.9959	0.9767	0.9225	0.8180	0.6656	0.4878	0.3169	0.1796
	8	1.0000	1.0000	0.9992	0.9933	0.9713	0.9161	0.8145	0.6675	0.4940	0.3238
	9	1.0000	1.0000	0.9999	0.9984	0.9911	0.9674	0.9125	0.8139	0.6710	0.5000
	10	1.0000	1.0000	1.0000	0.9997	0.9977	0.9895	0.9653	0.9115	0.8159	0.6762
	11	1.0000	1.0000	1.0000	1.0000	0.9995	0.9972	0.9886	0.9648	0.9129	0.8204
	12	1.0000	1.0000	1.0000	1.0000	0.9999	0.9994	0.9969	0.9884	0.9658	0.9165
	13	1.0000	1.0000	1.0000	1.0000	1.0000	0.9999	0.9993	0.9969	0.9891	0.9682
	14	1.0000	1.0000	1.0000	1.0000	1.0000	1.0000	0.9999	0.9994	0.9972	0.9904
	15	1.0000	1.0000	1.0000	1.0000	1.0000	1.0000	1.0000	0.9999	0.9995	0.9978
	16	1.0000	1.0000	1.0000	1.0000	1.0000	1.0000	1.0000	1.0000	0.9999	0.9996
	17	1.0000	1.0000	1.0000	1.0000	1.0000	1.0000	1.0000	1.0000	1.0000	1.0000
20	0	0.3585	0.1216	0.0388	0.0115	0.0032	0.0008	0.0002	0.0000	0.0000	0.0000
	1	0.7358	0.3917	0.1756	0.0692	0.0243	0.0076	0.0021	0.0005	0.0001	0.0000
	2	0.9245	0.6769	0.4049	0.2061	0.0913	0.0355	0.0121	0.0036	0.0009	0.0002
	3	0.9841	0.8670	0.6477	0.4114	0.2252	0.1071	0.0444	0.0160	0.0049	0.0013
	4	0.9974	0.9568	0.8298	0.6296	0.4148	0.2375	0.1182	0.0510	0.0189	0.0059
	5	0.9997	0.9887	0.9327	0.8042	0.6172	0.4164	0.2454	0.1256	0.0553	0.0207
	6	1.0000	0.9976	0.9781	0.9133	0.7858	0.6080	0.4166	0.2500	0.1299	0.0577
	7	1.0000	0.9996	0.9941	0.9679	0.8982	0.7723	0.6010	0.4159	0.2520	0.1316
	8	1.0000	0.9999	0.9987	0.9900	0.9591	0.8867	0.7624	0.5956	0.4143	0.2517
	9	1.0000	1.0000	0.9998	0.9974	0.9861	0.9520	0.8782	0.7553	0.5914	0.4119
	10	1.0000	1.0000	1.0000	0.9994	0.9961	0.9829	0.9468	0.8725	0.7507	0.5881
	11	1.0000	1.0000	1.0000	0.9999	0.9991	0.9949	0.9804	0.9435	0.8692	0.7483
	12	1.0000	1.0000	1.0000	1.0000	0.9998	0.9987	0.9940	0.9790	0.9420	0.8684
	13	1.0000	1.0000	1.0000	1.0000	1.0000	0.9997	0.9985	0.9935	0.9786	0.9423
	14	1.0000	1.0000	1.0000	1.0000	1.0000	1.0000	0.9997	0.9984	0.9936	0.9793
	15	1.0000	1.0000	1.0000	1.0000	1.0000	1.0000	1.0000	0.9997	0.9985	0.9941
	16	1.0000	1.0000	1.0000	1.0000	1.0000	1.0000	1.0000	1.0000	0.9997	0.9987
	17	1.0000	1.0000	1.0000	1.0000	1.0000	1.0000	1.0000	1.0000	1.0000	0.9998
	18	1.0000	1.0000	1.0000	1.0000	1.0000	1.0000	1.0000	1.0000	1.0000	1.0000

TABLE A-2
Selected values of the binomial probability distribution

$$P(x) = \binom{n}{x}(1-p)^{n-x}p^x$$

Example If $p = 0.15$, $n = 4$, and $x = 3$, then $P(3) = 0.0115$. When $p > 0.5$, the value of $P(x)$ for a given n, x, and p is obtained by finding the tabular entry for the given n, with $n - x$ in place of the given x and $1 - p$ in place of the given p.

		p									
n	*x*	0.05	0.10	0.15	0.20	0.25	0.30	0.35	0.40	0.45	0.50
1	0	0.9500	0.9000	0.8500	0.8000	0.7500	0.7000	0.6500	0.6000	0.5500	0.5000
	1	0.0500	0.1000	0.1500	0.2000	0.2500	0.3000	0.3500	0.4000	0.4500	0.5000
2	0	0.9025	0.8100	0.7225	0.6400	0.5625	0.4900	0.4225	0.3600	0.3025	0.2500
	1	0.0950	0.1800	0.2550	0.3200	0.3750	0.4200	0.4550	0.4800	0.4950	0.5000
	2	0.0025	0.0100	0.0225	0.0400	0.0625	0.0900	0.1225	0.1600	0.2025	0.2500
3	0	0.8574	0.7290	0.6141	0.5120	0.4219	0.3430	0.2746	0.2160	0.1664	0.1250
	1	0.1354	0.2430	0.3251	0.3840	0.4219	0.4410	0.4436	0.4320	0.4084	0.3750
	2	0.0071	0.0270	0.0574	0.0960	0.1406	0.1890	0.2389	0.2880	0.3341	0.3750
	3	0.0001	0.0010	0.0034	0.0080	0.0156	0.0270	0.0429	0.0640	0.0911	0.1250
4	0	0.8145	0.6561	0.5220	0.4096	0.3164	0.2401	0.1785	0.1296	0.0915	0.0625
	1	0.1715	0.2916	0.3685	0.4096	0.4219	0.4116	0.3845	0.3456	0.2995	0.2500
	2	0.0135	0.0486	0.0975	0.1536	0.2109	0.2646	0.3105	0.3456	0.3675	0.3750
	3	0.0005	0.0036	0.0115	0.0256	0.0469	0.0756	0.1115	0.1536	0.2005	0.2500
	4	0.0000	0.0001	0.0005	0.0016	0.0039	0.0081	0.0150	0.0256	0.0410	0.0625
5	0	0.7738	0.5905	0.4437	0.3277	0.2373	0.1681	0.1160	0.0778	0.0503	0.0312
	1	0.2036	0.3280	0.3915	0.4096	0.3955	0.3602	0.3124	0.2592	0.2059	0.1562
	2	0.0214	0.0729	0.1382	0.2048	0.2637	0.3087	0.3364	0.3456	0.3369	0.3125
	3	0.0011	0.0081	0.0244	0.0512	0.0879	0.1323	0.1811	0.2304	0.2757	0.3125
	4	0.0000	0.0004	0.0022	0.0064	0.0146	0.0284	0.0488	0.0768	0.1128	0.1562
	5	0.0000	0.0000	0.0001	0.0003	0.0010	0.0024	0.0053	0.0102	0.0185	0.0312
6	0	0.7351	0.5314	0.3771	0.2621	0.1780	0.1176	0.0754	0.0467	0.0277	0.0156
	1	0.2321	0.3543	0.3993	0.3932	0.3560	0.3025	0.2437	0.1866	0.1359	0.0938
	2	0.0305	0.0984	0.1762	0.2458	0.2966	0.3241	0.3280	0.3110	0.2780	0.2344
	3	0.0021	0.0146	0.0415	0.0819	0.1318	0.1852	0.2355	0.2765	0.3032	0.3125
	4	0.0001	0.0012	0.0055	0.0154	0.0330	0.0595	0.0951	0.1382	0.1861	0.2344
	5	0.0000	0.0001	0.0004	0.0015	0.0044	0.0102	0.0205	0.0369	0.0609	0.0938
	6	0.0000	0.0000	0.0000	0.0001	0.0002	0.0007	0.0018	0.0041	0.0083	0.0156
7	0	0.6983	0.4783	0.3206	0.2097	0.1335	0.0824	0.0490	0.0280	0.0152	0.0078
	1	0.2573	0.3720	0.3960	0.3670	0.3115	0.2471	0.1848	0.1306	0.0872	0.0547
	2	0.0406	0.1240	0.2097	0.2753	0.3115	0.3177	0.2985	0.2613	0.2140	0.1641
	3	0.0036	0.0230	0.0617	0.1147	0.1730	0.2269	0.2679	0.2903	0.2918	0.2734
	4	0.0002	0.0026	0.0109	0.0287	0.0577	0.0972	0.1442	0.1935	0.2388	0.2734

TABLE A-2 (continued)

		p									
n	x	0.05	0.10	0.15	0.20	0.25	0.30	0.35	0.40	0.45	0.50
	5	0.0000	0.0002	0.0012	0.0043	0.0115	0.0250	0.0466	0.0774	0.1172	0.1641
	6	0.0000	0.0000	0.0001	0.0004	0.0013	0.0036	0.0084	0.0172	0.0320	0.0547
	7	0.0000	0.0000	0.0000	0.0000	0.0001	0.0002	0.0006	0.0016	0.0037	0.0078
8	0	0.6634	0.4305	0.2725	0.1678	0.1001	0.0576	0.0319	0.0168	0.0084	0.0039
	1	0.2793	0.3826	0.3847	0.3355	0.2670	0.1977	0.1373	0.0896	0.0548	0.0312
	2	0.0515	0.1488	0.2376	0.2936	0.3115	0.2965	0.2587	0.2090	0.1569	0.1094
	3	0.0054	0.0331	0.0839	0.1468	0.2076	0.2541	0.2786	0.2787	0.2568	0.2188
	4	0.0004	0.0046	0.0185	0.0459	0.0865	0.1361	0.1875	0.2322	0.2627	0.2734
	5	0.0000	0.0004	0.0026	0.0092	0.0231	0.0467	0.0808	0.1239	0.1719	0.2188
	6	0.0000	0.0000	0.0002	0.0011	0.0038	0.0100	0.0217	0.0413	0.0703	0.1094
	7	0.0000	0.0000	0.0000	0.0001	0.0004	0.0012	0.0033	0.0079	0.0164	0.0312
	8	0.0000	0.0000	0.0000	0.0000	0.0000	0.0001	0.0002	0.0007	0.0017	0.0039
9	0	0.6302	0.3874	0.2316	0.1342	0.0751	0.0404	0.0207	0.0101	0.0046	0.0020
	1	0.2985	0.3874	0.3679	0.3020	0.2253	0.1556	0.1004	0.0605	0.0339	0.0176
	2	0.0629	0.1722	0.2597	0.3020	0.3003	0.2668	0.2162	0.1612	0.1110	0.0703
	3	0.0077	0.0446	0.1069	0.1762	0.2336	0.2668	0.2716	0.2508	0.2119	0.1641
	4	0.0006	0.0074	0.0283	0.0661	0.1168	0.1715	0.2194	0.2508	0.2600	0.2461
	5	0.0000	0.0008	0.0050	0.0165	0.0389	0.0735	0.1181	0.1672	0.2128	0.2461
	6	0.0000	0.0001	0.0006	0.0028	0.0087	0.0210	0.0424	0.0743	0.1160	0.1641
	7	0.0000	0.0000	0.0000	0.0003	0.0012	0.0039	0.0098	0.0212	0.0407	0.0703
	8	0.0000	0.0000	0.0000	0.0000	0.0001	0.0004	0.0013	0.0035	0.0083	0.0176
	9	0.0000	0.0000	0.0000	0.0000	0.0000	0.0000	0.0001	0.0003	0.0008	0.0020
10	0	0.5987	0.3487	0.1969	0.1074	0.0563	0.0282	0.0135	0.0060	0.0025	0.0010
	1	0.3151	0.3874	0.3474	0.2684	0.1877	0.1211	0.0725	0.0403	0.0207	0.0098
	2	0.0746	0.1937	0.2759	0.3020	0.2816	0.2335	0.1757	0.1209	0.0763	0.0439
	3	0.0105	0.0574	0.1298	0.2013	0.2503	0.2668	0.2522	0.2150	0.1665	0.1172
	4	0.0010	0.0112	0.0401	0.0881	0.1460	0.2001	0.2377	0.2508	0.2384	0.2051
	5	0.0001	0.0015	0.0085	0.0264	0.0584	0.1029	0.1536	0.2007	0.2340	0.2461
	6	0.0000	0.0001	0.0012	0.0055	0.0162	0.0368	0.0689	0.1115	0.1596	0.2051
	7	0.0000	0.0000	0.0001	0.0008	0.0031	0.0090	0.0212	0.0425	0.0746	0.1172
	8	0.0000	0.0000	0.0000	0.0001	0.0004	0.0014	0.0043	0.0106	0.0229	0.0439
	9	0.0000	0.0000	0.0000	0.0000	0.0000	0.0001	0.0005	0.0016	0.0042	0.0098
	10	0.0000	0.0000	0.0000	0.0000	0.0000	0.0000	0.0000	0.0001	0.0003	0.0010
11	0	0.5688	0.3138	0.1673	0.0859	0.0422	0.0198	0.0088	0.0036	0.0014	0.0005
	1	0.3293	0.3835	0.3248	0.2362	0.1549	0.0932	0.0518	0.0266	0.0125	0.0054
	2	0.0867	0.2131	0.2866	0.2953	0.2581	0.1998	0.1395	0.0887	0.0513	0.0269
	3	0.0137	0.0710	0.1517	0.2215	0.2581	0.2568	0.2254	0.1774	0.1259	0.0806
	4	0.0014	0.0158	0.0536	0.1107	0.1721	0.2201	0.2428	0.2365	0.2060	0.1611

TABLE A-2 (continued)

n	x	p = 0.05	0.10	0.15	0.20	0.25	0.30	0.35	0.40	0.45	0.50
	5	0.0001	0.0025	0.0132	0.0388	0.0803	0.1321	0.1830	0.2207	0.2360	0.2256
	6	0.0000	0.0003	0.0023	0.0097	0.0268	0.0566	0.0985	0.1471	0.1931	0.2256
	7	0.0000	0.0000	0.0003	0.0017	0.0064	0.0173	0.0379	0.0701	0.1128	0.1611
	8	0.0000	0.0000	0.0000	0.0002	0.0011	0.0037	0.0102	0.0234	0.0462	0.0806
	9	0.0000	0.0000	0.0000	0.0000	0.0001	0.0005	0.0018	0.0052	0.0126	0.0269
	10	0.0000	0.0000	0.0000	0.0000	0.0000	0.0000	0.0002	0.0007	0.0021	0.0054
	11	0.0000	0.0000	0.0000	0.0000	0.0000	0.0000	0.0000	0.0000	0.0002	0.0005
12	0	0.5404	0.2824	0.1422	0.0687	0.0317	0.0138	0.0057	0.0022	0.0008	0.0002
	1	0.3413	0.3766	0.3012	0.2062	0.1267	0.0712	0.0368	0.0174	0.0075	0.0029
	2	0.0988	0.2301	0.2924	0.2835	0.2323	0.1678	0.1088	0.0639	0.0339	0.0161
	3	0.0173	0.0852	0.1720	0.2362	0.2581	0.2397	0.1954	0.1419	0.0923	0.0537
	4	0.0021	0.0213	0.0683	0.1329	0.1936	0.2311	0.2367	0.2128	0.1700	0.1208
	5	0.0002	0.0038	0.0193	0.0532	0.1032	0.1585	0.2039	0.2270	0.2225	0.1934
	6	0.0000	0.0005	0.0040	0.0155	0.0401	0.0792	0.1281	0.1766	0.2124	0.2256
	7	0.0000	0.0000	0.0006	0.0033	0.0115	0.0291	0.0591	0.1009	0.1489	0.1934
	8	0.0000	0.0000	0.0001	0.0005	0.0024	0.0078	0.0199	0.0420	0.0762	0.1208
	9	0.0000	0.0000	0.0000	0.0001	0.0004	0.0015	0.0048	0.0125	0.0277	0.0537
	10	0.0000	0.0000	0.0000	0.0000	0.0000	0.0002	0.0008	0.0025	0.0068	0.0161
	11	0.0000	0.0000	0.0000	0.0000	0.0000	0.0000	0.0001	0.0003	0.0010	0.0029
	12	0.0000	0.0000	0.0000	0.0000	0.0000	0.0000	0.0000	0.0000	0.0001	0.0002
13	0	0.5133	0.2542	0.1209	0.0550	0.0238	0.0097	0.0037	0.0013	0.0004	0.0001
	1	0.3512	0.3672	0.2774	0.1787	0.1029	0.0540	0.0259	0.0113	0.0045	0.0016
	2	0.1109	0.2448	0.2937	0.2680	0.2059	0.1388	0.0836	0.0453	0.0220	0.0095
	3	0.0214	0.0997	0.1900	0.2457	0.2517	0.2181	0.1651	0.1107	0.0660	0.0349
	4	0.0028	0.0277	0.0838	0.1535	0.2097	0.2337	0.2222	0.1845	0.1350	0.0873
	5	0.0003	0.0055	0.0266	0.0691	0.1258	0.1803	0.2154	0.2214	0.1989	0.1571
	6	0.0000	0.0008	0.0063	0.0230	0.0559	0.1030	0.1546	0.1968	0.2169	0.2095
	7	0.0000	0.0001	0.0011	0.0058	0.0186	0.0442	0.0833	0.1312	0.1775	0.2095
	8	0.0000	0.0000	0.0001	0.0011	0.0047	0.0142	0.0336	0.0656	0.1089	0.1571
	9	0.0000	0.0000	0.0000	0.0001	0.0009	0.0034	0.0101	0.0243	0.0495	0.0873
	10	0.0000	0.0000	0.0000	0.0000	0.0001	0.0006	0.0022	0.0065	0.0162	0.0349
	11	0.0000	0.0000	0.0000	0.0000	0.0000	0.0001	0.0003	0.0012	0.0036	0.0095
	12	0.0000	0.0000	0.0000	0.0000	0.0000	0.0000	0.0000	0.0001	0.0005	0.0016
	13	0.0000	0.0000	0.0000	0.0000	0.0000	0.0000	0.0000	0.0000	0.0000	0.0001
14	0	0.4877	0.2288	0.1028	0.0440	0.0178	0.0068	0.0024	0.0008	0.0002	0.0001
	1	0.3593	0.3559	0.2539	0.1539	0.0832	0.0407	0.0181	0.0073	0.0027	0.0009
	2	0.1229	0.2570	0.2912	0.2501	0.1802	0.1134	0.0634	0.0317	0.0141	0.0056
	3	0.0259	0.1142	0.2056	0.2501	0.2402	0.1943	0.1366	0.0845	0.0462	0.0222
	4	0.0037	0.0349	0.0998	0.1720	0.2202	0.2290	0.2022	0.1549	0.1040	0.0611

TABLE A-2 (continued)

n	x	p = 0.05	0.10	0.15	0.20	0.25	0.30	0.35	0.40	0.45	0.50
	5	0.0004	0.0078	0.0352	0.0860	0.1468	0.1963	0.2178	0.2066	0.1701	0.1222
	6	0.0000	0.0013	0.0093	0.0322	0.0734	0.1262	0.1759	0.2066	0.2088	0.1833
	7	0.0000	0.0002	0.0019	0.0092	0.0280	0.0618	0.1082	0.1574	0.1952	0.2095
	8	0.0000	0.0000	0.0003	0.0020	0.0082	0.0232	0.0510	0.0918	0.1398	0.1833
	9	0.0000	0.0000	0.0000	0.0003	0.0018	0.0066	0.0183	0.0408	0.0762	0.1222
	10	0.0000	0.0000	0.0000	0.0000	0.0003	0.0014	0.0049	0.0136	0.0312	0.0611
	11	0.0000	0.0000	0.0000	0.0000	0.0000	0.0002	0.0010	0.0033	0.0093	0.0222
	12	0.0000	0.0000	0.0000	0.0000	0.0000	0.0000	0.0001	0.0005	0.0019	0.0056
	13	0.0000	0.0000	0.0000	0.0000	0.0000	0.0000	0.0000	0.0001	0.0002	0.0009
	14	0.0000	0.0000	0.0000	0.0000	0.0000	0.0000	0.0000	0.0000	0.0000	0.0001
15	0	0.4633	0.2059	0.0874	0.0352	0.0134	0.0047	0.0016	0.0005	0.0001	0.0000
	1	0.3658	0.3432	0.2312	0.1319	0.0668	0.0305	0.0126	0.0047	0.0016	0.0005
	2	0.1348	0.2669	0.2856	0.2309	0.1559	0.0916	0.0476	0.0219	0.0090	0.0032
	3	0.0307	0.1285	0.2184	0.2501	0.2252	0.1700	0.1110	0.0634	0.0318	0.0139
	4	0.0049	0.0428	0.1156	0.1876	0.2252	0.2186	0.1792	0.1268	0.0780	0.0417
	5	0.0006	0.0105	0.0449	0.1032	0.1651	0.2061	0.2123	0.1859	0.1404	0.0916
	6	0.0000	0.0019	0.0132	0.0430	0.0917	0.1472	0.1906	0.2066	0.1914	0.1527
	7	0.0000	0.0003	0.0030	0.0138	0.0393	0.0811	0.1319	0.1771	0.2013	0.1964
	8	0.0000	0.0000	0.0005	0.0035	0.0131	0.0348	0.0710	0.1181	0.1647	0.1964
	9	0.0000	0.0000	0.0001	0.0007	0.0034	0.0116	0.0298	0.0612	0.1048	0.1527
	10	0.0000	0.0000	0.0000	0.0001	0.0007	0.0030	0.0096	0.0245	0.0515	0.0916
	11	0.0000	0.0000	0.0000	0.0000	0.0001	0.0006	0.0024	0.0074	0.0191	0.0417
	12	0.0000	0.0000	0.0000	0.0000	0.0000	0.0001	0.0004	0.0016	0.0052	0.0139
	13	0.0000	0.0000	0.0000	0.0000	0.0000	0.0000	0.0001	0.0003	0.0010	0.0032
	14	0.0000	0.0000	0.0000	0.0000	0.0000	0.0000	0.0000	0.0000	0.0001	0.0005
	15	0.0000	0.0000	0.0000	0.0000	0.0000	0.0000	0.0000	0.0000	0.0000	0.0000
16	0	0.4401	0.1853	0.0743	0.0281	0.0100	0.0033	0.0010	0.0003	0.0001	0.0000
	1	0.3706	0.3294	0.2097	0.1126	0.0535	0.0228	0.0087	0.0030	0.0009	0.0002
	2	0.1463	0.2745	0.2775	0.2111	0.1336	0.0732	0.0353	0.0150	0.0056	0.0018
	3	0.0359	0.1423	0.2285	0.2463	0.2079	0.1465	0.0888	0.0468	0.0215	0.0085
	4	0.0061	0.0514	0.1311	0.2001	0.2252	0.2040	0.1553	0.1014	0.0572	0.0278
	5	0.0008	0.0137	0.0555	0.1201	0.1802	0.2099	0.2008	0.1623	0.1123	0.0667
	6	0.0001	0.0028	0.0180	0.0550	0.1101	0.1649	0.1982	0.1983	0.1684	0.1222
	7	0.0000	0.0004	0.0045	0.0197	0.0524	0.1010	0.1524	0.1889	0.1969	0.1746
	8	0.0000	0.0001	0.0009	0.0055	0.0197	0.0487	0.0923	0.1417	0.1812	0.1964
	9	0.0000	0.0000	0.0001	0.0012	0.0058	0.0185	0.0442	0.0840	0.1318	0.1746
	10	0.0000	0.0000	0.0000	0.0002	0.0014	0.0056	0.0167	0.0392	0.0755	0.1222
	11	0.0000	0.0000	0.0000	0.0000	0.0002	0.0013	0.0049	0.0142	0.0337	0.0667
	12	0.0000	0.0000	0.0000	0.0000	0.0000	0.0002	0.0011	0.0040	0.0115	0.0278
	13	0.0000	0.0000	0.0000	0.0000	0.0000	0.0000	0.0002	0.0008	0.0029	0.0085
	14	0.0000	0.0000	0.0000	0.0000	0.0000	0.0000	0.0000	0.0001	0.0005	0.0018

TABLE A-2 (continued)

n	x	0.05	0.10	0.15	0.20	0.25	0.30	0.35	0.40	0.45	0.50
						p					
	15	0.0000	0.0000	0.0000	0.0000	0.0000	0.0000	0.0000	0.0000	0.0001	0.0002
	16	0.0000	0.0000	0.0000	0.0000	0.0000	0.0000	0.0000	0.0000	0.0000	0.0000
17	0	0.4181	0.1668	0.0631	0.0225	0.0075	0.0023	0.0007	0.0002	0.0000	0.0000
	1	0.3741	0.3150	0.1893	0.0957	0.0426	0.0169	0.0060	0.0019	0.0005	0.0001
	2	0.1575	0.2800	0.2673	0.1914	0.1136	0.0581	0.0260	0.0102	0.0035	0.0010
	3	0.0415	0.1556	0.2359	0.2393	0.1893	0.1245	0.0701	0.0341	0.0144	0.0052
	4	0.0076	0.0605	0.1457	0.2093	0.2209	0.1868	0.1320	0.0796	0.0411	0.0182
	5	0.0010	0.0175	0.0668	0.1361	0.1914	0.2081	0.1849	0.1379	0.0875	0.0472
	6	0.0001	0.0039	0.0236	0.0680	0.1276	0.1784	0.1991	0.1839	0.1432	0.0944
	7	0.0000	0.0007	0.0065	0.0267	0.0668	0.1201	0.1685	0.1927	0.1841	0.1484
	8	0.0000	0.0001	0.0014	0.0084	0.0279	0.0644	0.1134	0.1606	0.1883	0.1855
	9	0.0000	0.0000	0.0003	0.0021	0.0093	0.0276	0.0611	0.1070	0.1540	0.1855
	10	0.0000	0.0000	0.0000	0.0004	0.0025	0.0095	0.0263	0.0571	0.1008	0.1484
	11	0.0000	0.0000	0.0000	0.0001	0.0005	0.0026	0.0090	0.0242	0.0525	0.0944
	12	0.0000	0.0000	0.0000	0.0000	0.0001	0.0006	0.0024	0.0081	0.0215	0.0472
	13	0.0000	0.0000	0.0000	0.0000	0.0000	0.0001	0.0005	0.0021	0.0068	0.0182
	14	0.0000	0.0000	0.0000	0.0000	0.0000	0.0000	0.0001	0.0004	0.0016	0.0052
	15	0.0000	0.0000	0.0000	0.0000	0.0000	0.0000	0.0000	0.0001	0.0003	0.0010
	16	0.0000	0.0000	0.0000	0.0000	0.0000	0.0000	0.0000	0.0000	0.0000	0.0001
	17	0.0000	0.0000	0.0000	0.0000	0.0000	0.0000	0.0000	0.0000	0.0000	0.0000
18	0	0.3972	0.1501	0.0536	0.0180	0.0056	0.0016	0.0004	0.0001	0.0000	0.0000
	1	0.3763	0.3002	0.1704	0.0811	0.0338	0.0126	0.0042	0.0012	0.0003	0.0001
	2	0.1683	0.2835	0.2556	0.1723	0.0958	0.0458	0.0190	0.0069	0.0022	0.0006
	3	0.0473	0.1680	0.2406	0.2297	0.1704	0.1046	0.0547	0.0246	0.0095	0.0031
	4	0.0093	0.0700	0.1592	0.2153	0.2130	0.1681	0.1104	0.0614	0.0291	0.0117
	5	0.0014	0.0218	0.0787	0.1507	0.1988	0.2017	0.1664	0.1146	0.0666	0.0327
	6	0.0002	0.0052	0.0301	0.0816	0.1436	0.1873	0.1941	0.1655	0.1181	0.0708
	7	0.0000	0.0010	0.0091	0.0350	0.0820	0.1376	0.1792	0.1892	0.1657	0.1214
	8	0.0000	0.0002	0.0022	0.0120	0.0376	0.0811	0.1327	0.1734	0.1864	0.1669
	9	0.0000	0.0000	0.0004	0.0033	0.0139	0.0386	0.0794	0.1284	0.1694	0.1855
	10	0.0000	0.0000	0.0001	0.0008	0.0042	0.0149	0.0385	0.0771	0.1248	0.1669
	11	0.0000	0.0000	0.0000	0.0001	0.0010	0.0046	0.0151	0.0374	0.0742	0.1214
	12	0.0000	0.0000	0.0000	0.0000	0.0002	0.0012	0.0047	0.0145	0.0354	0.0708
	13	0.0000	0.0000	0.0000	0.0000	0.0000	0.0002	0.0012	0.0045	0.0134	0.0327
	14	0.0000	0.0000	0.0000	0.0000	0.0000	0.0000	0.0002	0.0011	0.0039	0.0117
	15	0.0000	0.0000	0.0000	0.0000	0.0000	0.0000	0.0000	0.0002	0.0009	0.0031
	16	0.0000	0.0000	0.0000	0.0000	0.0000	0.0000	0.0000	0.0000	0.0001	0.0006
	17	0.0000	0.0000	0.0000	0.0000	0.0000	0.0000	0.0000	0.0000	0.0000	0.0001
	18	0.0000	0.0000	0.0000	0.0000	0.0000	0.0000	0.0000	0.0000	0.0000	0.0000

TABLE A-2 (continued)

		p									
n	x	0.05	0.10	0.15	0.20	0.25	0.30	0.35	0.40	0.45	0.50
19	0	0.3774	0.1351	0.0456	0.0144	0.0042	0.0011	0.0003	0.0001	0.0000	0.0000
	1	0.3774	0.2852	0.1529	0.0685	0.0268	0.0093	0.0029	0.0008	0.0002	0.0000
	2	0.1787	0.2852	0.2428	0.1540	0.0803	0.0358	0.0138	0.0046	0.0013	0.0003
	3	0.0533	0.1796	0.2428	0.2182	0.1517	0.0869	0.0422	0.0175	0.0062	0.0018
	4	0.0112	0.0798	0.1714	0.2182	0.2023	0.1491	0.0909	0.0467	0.0203	0.0074
	5	0.0018	0.0266	0.0907	0.1636	0.2023	0.1916	0.1468	0.0933	0.0497	0.0222
	6	0.0002	0.0069	0.0374	0.0955	0.1574	0.1916	0.1844	0.1451	0.0949	0.0518
	7	0.0000	0.0014	0.0122	0.0443	0.0974	0.1525	0.1844	0.1797	0.1443	0.0961
	8	0.0000	0.0002	0.0032	0.0166	0.0487	0.0981	0.1489	0.1797	0.1771	0.1442
	9	0.0000	0.0000	0.0007	0.0051	0.0198	0.0514	0.0980	0.1464	0.1771	0.1762
	10	0.0000	0.0000	0.0001	0.0013	0.0066	0.0220	0.0528	0.0976	0.1449	0.1762
	11	0.0000	0.0000	0.0000	0.0003	0.0018	0.0077	0.0233	0.0532	0.0970	0.1442
	12	0.0000	0.0000	0.0000	0.0000	0.0004	0.0022	0.0083	0.0237	0.0529	0.0961
	13	0.0000	0.0000	0.0000	0.0000	0.0001	0.0005	0.0024	0.0085	0.0233	0.0518
	14	0.0000	0.0000	0.0000	0.0000	0.0000	0.0001	0.0006	0.0024	0.0082	0.0222
	15	0.0000	0.0000	0.0000	0.0000	0.0000	0.0000	0.0001	0.0005	0.0022	0.0074
	16	0.0000	0.0000	0.0000	0.0000	0.0000	0.0000	0.0000	0.0001	0.0005	0.0018
	17	0.0000	0.0000	0.0000	0.0000	0.0000	0.0000	0.0000	0.0000	0.0001	0.0003
	18	0.0000	0.0000	0.0000	0.0000	0.0000	0.0000	0.0000	0.0000	0.0000	0.0000
	19	0.0000	0.0000	0.0000	0.0000	0.0000	0.0000	0.0000	0.0000	0.0000	0.0000
20	0	0.3585	0.1216	0.0388	0.0115	0.0032	0.0008	0.0002	0.0000	0.0000	0.0000
	1	0.3774	0.2702	0.1368	0.0576	0.0211	0.0068	0.0020	0.0005	0.0001	0.0000
	2	0.1887	0.2852	0.2293	0.1369	0.0669	0.0278	0.0100	0.0031	0.0008	0.0002
	3	0.0596	0.1901	0.2428	0.2054	0.1339	0.0716	0.0323	0.0123	0.0040	0.0011
	4	0.0133	0.0898	0.1821	0.2182	0.1897	0.1304	0.0738	0.0350	0.0139	0.0046
	5	0.0022	0.0319	0.1028	0.1746	0.2023	0.1789	0.1272	0.0746	0.0365	0.0148
	6	0.0003	0.0089	0.0454	0.1091	0.1686	0.1916	0.1712	0.1244	0.0746	0.0370
	7	0.0000	0.0020	0.0160	0.0545	0.1124	0.1643	0.1844	0.1659	0.1221	0.0739
	8	0.0000	0.0004	0.0046	0.0222	0.0609	0.1144	0.1614	0.1797	0.1623	0.1201
	9	0.0000	0.0001	0.0011	0.0074	0.0271	0.0654	0.1158	0.1597	0.1771	0.1602
	10	0.0000	0.0000	0.0002	0.0020	0.0099	0.0308	0.0686	0.1171	0.1593	0.1762
	11	0.0000	0.0000	0.0000	0.0005	0.0030	0.0120	0.0336	0.0710	0.1185	0.1602
	12	0.0000	0.0000	0.0000	0.0001	0.0008	0.0039	0.0136	0.0355	0.0727	0.1201
	13	0.0000	0.0000	0.0000	0.0000	0.0002	0.0010	0.0045	0.0146	0.0366	0.0739
	14	0.0000	0.0000	0.0000	0.0000	0.0000	0.0002	0.0012	0.0049	0.0150	0.0370
	15	0.0000	0.0000	0.0000	0.0000	0.0000	0.0000	0.0003	0.0013	0.0049	0.0148
	16	0.0000	0.0000	0.0000	0.0000	0.0000	0.0000	0.0000	0.0003	0.0013	0.0046
	17	0.0000	0.0000	0.0000	0.0000	0.0000	0.0000	0.0000	0.0000	0.0002	0.0011
	18	0.0000	0.0000	0.0000	0.0000	0.0000	0.0000	0.0000	0.0000	0.0000	0.0002
	19	0.0000	0.0000	0.0000	0.0000	0.0000	0.0000	0.0000	0.0000	0.0000	0.0000
	20	0.0000	0.0000	0.0000	0.0000	0.0000	0.0000	0.0000	0.0000	0.0000	0.0000

TABLE A-3

Selected values of the Poisson cumulative distribution

$$F(c) = P(X \leqslant c) = \sum_{x=0}^{c} \frac{\mu^x e^{-\mu}}{x!}$$

Example If $\mu = 1.00$, then $F(2) = P(X \leqslant 2) = 0.920$.

μ \ c	0	1	2	3	4	5	6	7	8	9
0.02	0.980	1.000								
0.04	0.961	0.999	1.000							
0.06	0.942	0.998	1.000							
0.08	0.923	0.997	1.000							
0.10	0.905	0.995	1.000							
0.15	0.861	0.990	0.999	1.000						
0.20	0.819	0.982	0.999	1.000						
0.25	0.779	0.974	0.998	1.000						
0.30	0.741	0.963	0.996	1.000						
0.35	0.705	0.951	0.994	1.000						
0.40	0.670	0.938	0.992	0.999	1.000					
0.45	0.638	0.925	0.989	0.999	1.000					
0.50	0.607	0.910	0.986	0.998	1.000					
0.55	0.577	0.894	0.982	0.998	1.000					
0.60	0.549	0.878	0.977	0.997	1.000					
0.65	0.522	0.861	0.972	0.996	0.999	1.000				
0.70	0.497	0.844	0.966	0.994	0.999	1.000				
0.75	0.472	0.827	0.959	0.993	0.999	1.000				
0.80	0.449	0.809	0.953	0.991	0.999	1.000				
0.85	0.427	0.791	0.945	0.989	0.998	1.000				
0.90	0.407	0.772	0.937	0.987	0.998	1.000				
0.95	0.387	0.754	0.929	0.984	0.997	1.000				
1.00	0.368	0.736	0.920	0.981	0.996	0.999	1.000			
1.10	0.333	0.699	0.900	0.974	0.995	0.999	1.000			
1.20	0.301	0.663	0.879	0.966	0.992	0.998	1.000			
1.30	0.273	0.627	0.857	0.957	0.989	0.998	1.000			
1.40	0.247	0.592	0.833	0.946	0.986	0.997	0.999	1.000		
1.50	0.223	0.558	0.809	0.934	0.981	0.996	0.999	1.000		
1.60	0.202	0.525	0.783	0.921	0.976	0.994	0.999	1.000		
1.70	0.183	0.493	0.757	0.907	0.970	0.992	0.998	1.000		
1.80	0.165	0.463	0.731	0.891	0.964	0.990	0.997	0.999	1.000	
1.90	0.150	0.434	0.704	0.875	0.956	0.987	0.997	0.999	1.000	
2.00	0.135	0.406	0.677	0.857	0.947	0.983	0.995	0.999	1.000	
2.20	0.111	0.355	0.623	0.819	0.928	0.975	0.993	0.998	1.000	
2.40	0.091	0.308	0.570	0.779	0.904	0.964	0.988	0.997	0.999	1.000
2.60	0.074	0.267	0.518	0.736	0.877	0.951	0.983	0.995	0.999	1.000
2.80	0.061	0.231	0.469	0.692	0.848	0.935	0.976	0.992	0.998	0.999
3.00	0.050	0.199	0.423	0.647	0.815	0.916	0.966	0.988	0.996	0.999

Source: From Eugene L. Grant, *Statistical Quality Control*, Copyright 1964 by McGraw-Hill Book Company. Used with permission of McGraw-Hill Book Company.

TABLE A-3 (continued)

μ \ c	0	1	2	3	4	5	6	7	8	9
3.20	0.041	0.171	0.380	0.603	0.781	0.895	0.955	0.983	0.994	0.998
3.40	0.033	0.147	0.340	0.558	0.744	0.871	0.942	0.977	0.992	0.997
3.60	0.027	0.126	0.303	0.515	0.706	0.844	0.927	0.969	0.988	0.996
3.80	0.022	0.107	0.269	0.473	0.668	0.816	0.909	0.960	0.984	0.994
4.00	0.018	0.092	0.238	0.433	0.629	0.785	0.889	0.949	0.979	0.992
4.20	0.015	0.078	0.210	0.395	0.590	0.753	0.867	0.936	0.972	0.989
4.40	0.012	0.066	0.185	0.359	0.551	0.720	0.844	0.921	0.964	0.985
4.60	0.010	0.056	0.163	0.326	0.513	0.686	0.818	0.905	0.955	0.980
4.80	0.008	0.048	0.143	0.294	0.476	0.651	0.791	0.887	0.944	0.975
5.00	0.007	0.040	0.125	0.265	0.440	0.616	0.762	0.867	0.932	0.968
5.20	0.006	0.034	0.109	0.238	0.406	0.581	0.732	0.845	0.918	0.960
5.40	0.005	0.029	0.095	0.213	0.373	0.546	0.702	0.822	0.903	0.951
5.60	0.004	0.024	0.082	0.191	0.342	0.512	0.670	0.797	0.886	0.941
5.80	0.003	0.021	0.072	0.170	0.313	0.478	0.638	0.771	0.867	0.929
6.00	0.002	0.017	0.062	0.151	0.285	0.446	0.606	0.744	0.847	0.916

μ \ c	10	11	12	13	14	15	16
2.80	1.000						
3.00	1.000						
3.20	1.000						
3.40	0.999	1.000					
3.60	0.999	1.000					
3.80	0.998	0.999	1.000				
4.00	0.997	0.999	1.000				
4.20	0.996	0.999	1.000				
4.40	0.994	0.998	0.999	1.000			
4.60	0.992	0.997	0.999	1.000			
4.80	0.990	0.996	0.999	1.000			
5.00	0.986	0.995	0.998	0.999	1.000		
5.20	9.982	0.993	0.997	0.999	1.000		
5.40	0.977	0.990	0.996	0.999	1.000		
5.60	0.972	0.988	0.995	0.998	0.999	1.000	
5.80	0.965	0.984	0.993	0.997	0.999	1.000	
6.00	0.957	0.980	0.991	0.996	0.999	0.999	1.000

μ \ c	0	1	2	3	4	5	6	7	8	9
6.20	0.002	0.015	0.054	0.134	0.259	0.414	0.574	0.716	0.826	0.902
6.40	0.002	0.012	0.046	0.119	0.235	0.384	0.542	0.687	0.803	0.886
6.60	0.001	0.010	0.040	0.105	0.213	0.355	0.511	0.658	0.780	0.869
6.80	0.001	0.009	0.034	0.093	0.192	0.327	0.480	0.628	0.755	0.850
7.00	0.001	0.007	0.030	0.082	0.173	0.301	0.450	0.599	0.729	0.830

TABLE A-3 (continued)

μ \ c	0	1	2	3	4	5	6	7	8	9
7.20	0.001	0.006	0.025	0.072	0.156	0.276	0.420	0.569	0.703	0.810
7.40	0.001	0.005	0.022	0.063	0.140	0.253	0.392	0.539	0.676	0.788
7.60	0.001	0.004	0.019	0.055	0.125	0.231	0.365	0.510	0.648	0.765
7.80	0.000	0.004	0.016	0.048	0.112	0.210	0.338	0.481	0.620	0.741
8.00	0.000	0.003	0.014	0.042	0.100	0.191	0.313	0.453	0.593	0.717
8.50	0.000	0.002	0.009	0.030	0.074	0.150	0.256	0.386	0.523	0.653
9.00	0.000	0.001	0.006	0.021	0.055	0.116	0.207	0.324	0.456	0.587
9.50	0.000	0.001	0.004	0.015	0.040	0.089	0.165	0.269	0.392	0.522
10.00	0.000	0.000	0.003	0.010	0.029	0.067	0.130	0.220	0.333	0.458

μ \ c	10	11	12	13	14	15	16	17	18	19
6.20	0.949	0.975	0.989	0.995	0.998	0.999	1.000			
6.40	0.939	0.969	0.986	0.994	0.997	0.999	1.000			
6.60	0.927	0.963	0.982	0.992	0.997	0.999	0.999	1.000		
6.80	0.915	0.955	0.978	0.990	0.996	0.998	0.999	1.000		
7.00	0.901	0.947	0.973	0.987	0.994	0.998	0.999	1.000		
7.20	0.887	0.937	0.967	0.984	0.993	0.997	0.999	0.999	1.000	
7.40	0.871	0.926	0.961	0.980	0.991	0.996	0.998	0.999	1.000	
7.60	0.854	0.915	0.954	0.976	0.989	0.995	0.998	0.999	1.000	
7.80	0.835	0.902	0.945	0.971	0.986	0.993	0.997	0.999	1.000	
8.00	0.816	0.888	0.936	0.966	0.983	0.992	0.996	0.998	0.999	1.000
8.50	0.763	0.849	0.909	0.949	0.973	0.986	0.993	0.997	0.999	0.999
9.00	0.706	0.803	0.876	0.926	0.959	0.978	0.989	0.995	0.998	0.999
9.50	0.645	0.752	0.836	0.898	0.940	0.967	0.982	0.991	0.996	0.998
10.00	0.583	0.697	0.792	0.864	0.917	0.951	0.973	0.986	0.993	0.997

μ \ c	20	21	22
8.50	1.000		
9.00	1.000		
9.50	0.999	1.000	
10.00	0.998	0.999	1.000

μ \ c	0	1	2	3	4	5	6	7	8	9
10.50	0.000	0.000	0.002	0.007	0.021	0.050	0.102	0.179	0.279	0.397
11.00	0.000	0.000	0.001	0.005	0.015	0.038	0.079	0.143	0.232	0.341
11.50	0.000	0.000	0.001	0.003	0.011	0.028	0.060	0.114	0.191	0.289
12.00	0.000	0.000	0.001	0.002	0.008	0.020	0.046	0.090	0.155	0.242
12.50	0.000	0.000	0.000	0.002	0.005	0.015	0.035	0.070	0.125	0.201
13.00	0.000	0.000	0.000	0.001	0.004	0.011	0.026	0.054	0.100	0.166
13.50	0.000	0.000	0.000	0.001	0.003	0.008	0.019	0.041	0.079	0.135
14.00	0.000	0.000	0.000	0.000	0.002	0.006	0.014	0.032	0.062	0.109
14.50	0.000	0.000	0.000	0.000	0.001	0.004	0.010	0.024	0.048	0.088
15.00	0.000	0.000	0.000	0.000	0.001	0.003	0.008	0.018	0.037	0.070

TABLE A-3 (continued)

	10	11	12	13	14	15	16	17	18	19
10.50	0.521	0.639	0.742	0.825	0.888	0.932	0.960	0.978	0.988	0.994
11.00	0.460	0.579	0.689	0.781	0.854	0.907	0.944	0.968	0.982	0.991
11.50	0.402	0.520	0.633	0.733	0.815	0.878	0.924	0.954	0.974	0.986
12.00	0.347	0.462	0.576	0.682	0.772	0.844	0.899	0.937	0.963	0.979
12.50	0.297	0.406	0.519	0.628	0.725	0.806	0.869	0.916	0.948	0.969
13.00	0.252	0.353	0.463	0.573	0.675	0.764	0.835	0.890	0.930	0.957
13.50	0.211	0.304	0.409	0.518	0.623	0.718	0.798	0.861	0.908	0.942
14.00	0.176	0.260	0.358	0.464	0.570	0.669	0.756	0.827	0.883	0.923
14.50	0.145	0.220	0.311	0.413	0.518	0.619	0.711	0.790	0.853	0.901
15.00	0.118	0.185	0.268	0.363	0.466	0.568	0.664	0.749	0.819	0.875

	20	21	22	23	24	25	26	27	28	29
10.50	0.997	0.999	0.999	1.000						
11.00	0.995	0.998	0.999	1.000						
11.50	0.992	0.996	0.998	0.999	1.000					
12.00	0.988	0.994	0.997	0.999	0.999	1.000				
12.50	0.983	0.991	0.995	0.998	0.999	0.999	1.000			
13.00	0.975	0.986	0.992	0.996	0.998	0.999	1.000			
13.50	0.965	0.980	0.989	0.994	0.997	0.998	0.999	1.000		
14.00	0.952	0.971	0.983	0.991	0.995	0.997	0.999	0.999	1.000	
14.50	0.936	0.960	0.976	0.986	0.992	0.996	0.998	0.999	0.999	1.000
15.00	0.917	0.947	0.967	0.981	0.989	0.994	0.997	0.998	0.999	1.000

μ \ c	4	5	6	7	8	9	10	11	12	13
16.00	0.000	0.001	0.004	0.010	0.022	0.043	0.077	0.127	0.193	0.275
17.00	0.000	0.001	0.002	0.005	0.013	0.026	0.049	0.085	0.135	0.201
18.00	0.000	0.000	0.001	0.003	0.007	0.015	0.030	0.055	0.092	0.143
19.00	0.000	0.000	0.001	0.002	0.004	0.009	0.018	0.035	0.061	0.098
20.00	0.000	0.000	0.000	0.001	0.002	0.005	0.011	0.021	0.039	0.066
21.00	0.000	0.000	0.000	0.000	0.001	0.003	0.006	0.013	0.025	0.043
22.00	0.000	0.000	0.000	0.000	0.001	0.002	0.004	0.008	0.015	0.028
23.00	0.000	0.000	0.000	0.000	0.000	0.001	0.002	0.004	0.009	0.017
24.00	0.000	0.000	0.000	0.000	0.000	0.000	0.001	0.003	0.005	0.011
25.00	0.000	0.000	0.000	0.000	0.000	0.000	0.001	0.001	0.003	0.006

	14	15	16	17	18	19	20	21	22	23
16.00	0.368	0.467	0.566	0.659	0.742	0.812	0.868	0.911	0.942	0.963
17.00	0.281	0.371	0.468	0.564	0.655	0.736	0.805	0.861	0.905	0.937
18.00	0.208	0.287	0.375	0.469	0.562	0.651	0.731	0.799	0.855	0.899
19.00	0.150	0.215	0.292	0.378	0.469	0.561	0.647	0.725	0.793	0.849
20.00	0.105	0.157	0.221	0.297	0.381	0.470	0.559	0.644	0.721	0.787
21.00	0.072	0.111	0.163	0.227	0.302	0.384	0.471	0.558	0.640	0.716

TABLE A-3 (continued)

	14	15	16	17	18	19	20	21	22	23
22.00	0.048	0.077	0.117	0.169	0.232	0.306	0.387	0.472	0.556	0.637
23.00	0.031	0.052	0.082	0.123	0.175	0.238	0.310	0.389	0.472	0.555
24.00	0.020	0.034	0.056	0.087	0.128	0.180	0.243	0.314	0.392	0.473
25.00	0.012	0.022	0.038	0.060	0.092	0.134	0.185	0.247	0.318	0.394

	24	25	26	27	28	29	30	31	32	33
16.00	0.978	0.987	0.993	0.996	0.998	0.999	0.999	1.000		
17.00	0.959	0.975	0.985	0.991	0.995	0.997	0.999	0.999	1.000	
18.00	0.932	0.955	0.972	0.983	0.990	0.994	0.997	0.998	0.999	1.000
19.00	0.893	0.927	0.951	0.969	0.980	0.988	0.993	0.996	0.998	0.999
20.00	0.843	0.888	0.922	0.948	0.966	0.978	0.987	0.992	0.995	0.997
21.00	0.782	0.838	0.883	0.917	0.944	0.963	0.976	0.985	0.991	0.994
22.00	0.712	0.777	0.832	0.877	0.913	0.940	0.959	0.973	0.983	0.989
23.00	0.635	0.708	0.772	0.827	0.873	0.908	0.936	0.956	0.971	0.981
24.00	0.554	0.632	0.704	0.768	0.823	0.868	0.904	0.932	0.953	0.969
25.00	0.473	0.553	0.629	0.700	0.763	0.818	0.863	0.900	0.929	0.950

	34	35	36	37	38	39	40	41	42	43
19.00	0.999	1.000								
20.00	0.999	0.999	1.000							
21.00	0.997	0.998	0.999	0.999	1.000					
22.00	0.994	0.996	0.998	0.999	0.999	1.000				
23.00	0.988	0.993	0.996	0.997	0.999	0.999	1.000			
24.00	0.979	0.987	0.992	0.995	0.997	0.998	0.999	0.999	1.000	
25.00	0.966	0.978	0.985	0.991	0.994	0.997	0.998	0.999	0.999	1.000

TABLE A-4
Four-place common logarithms

N	0	1	2	3	4	5	6	7	8	9	Proportional Parts 1	2	3	4	5	6	7	8	9
10	0000	0043	0086	0128	0170	0212	0253	0294	0334	0374	4	8	12	17	21	25	29	33	37
11	0414	0453	0492	0531	0569	0607	0645	0682	0719	0755	4	8	11	15	19	23	26	30	34
12	0792	0828	0864	0899	0934	0969	1004	1038	1072	1106	3	7	10	14	17	21	24	28	31
13	1139	1173	1206	1239	1271	1303	1335	1367	1399	1430	3	6	10	13	16	19	23	26	29
14	1461	1492	1523	1553	1584	1614	1644	1673	1703	1732	3	6	9	12	15	18	21	24	27
15	1761	1790	1818	1847	1875	1903	1931	1959	1987	2014	3	6	8	11	14	17	20	22	25
16	2041	2068	2095	2122	2148	2175	2201	2227	2253	2279	3	5	8	11	13	16	18	21	24
17	2304	2330	2355	2380	2405	2430	2455	2480	2504	2529	2	5	7	10	12	15	17	20	22
18	2553	2577	2601	2625	2648	2672	2695	2718	2742	2765	2	5	7	9	12	14	16	19	21
19	2788	2810	2833	2856	2878	2900	2923	2945	2967	2989	2	4	7	9	11	13	16	18	20
20	3010	3032	3054	3075	3096	3118	3139	3160	3181	3201	2	4	6	8	11	13	15	17	19
21	3222	3243	3263	3284	3304	3324	3345	3365	3385	3404	2	4	6	8	10	12	14	16	18
22	3424	3444	3464	3483	3502	3522	3541	3560	3579	3598	2	4	6	8	10	12	14	15	17
23	3617	3636	3655	3674	3692	3711	3729	3747	3766	3784	2	4	6	7	9	11	13	15	17
24	3802	3820	3838	3856	3874	3892	3909	3927	3945	3962	2	4	5	7	9	11	12	14	16
25	3979	3997	4014	4031	4048	4065	4082	4099	4116	4133	2	3	5	7	9	10	12	14	15
26	4150	4166	4183	4200	4216	4232	4249	4265	4281	4298	2	3	5	7	8	10	11	13	15
27	4314	4330	4346	4362	4378	4393	4409	4425	4440	4456	2	3	5	6	8	9	11	13	14
28	4472	4487	4502	4518	4533	4548	4564	4579	4594	4609	2	3	5	6	8	9	11	12	14
29	4624	4639	4654	4669	4683	4698	4713	4728	4742	4757	1	3	4	6	7	9	10	12	13
30	4771	4786	4800	4814	4829	4843	4857	4871	4886	4900	1	3	4	6	7	9	10	11	13
31	4914	4928	4942	4955	4969	4983	4997	5011	5024	5038	1	3	4	6	7	8	10	11	12
32	5051	5065	5079	5092	5105	5119	5132	5145	5159	5172	1	3	4	5	7	8	9	11	12
33	5185	5198	5211	5224	5237	5250	5263	5276	5289	5302	1	3	4	5	6	8	9	10	12
34	5315	5328	5340	5353	5366	5378	5391	5403	5416	5428	1	3	4	5	6	8	9	10	11
35	5441	5453	5465	5478	5490	5502	5514	5527	5539	5551	1	2	4	5	6	7	9	10	11
36	5563	5575	5587	5599	5611	5623	5635	5647	5658	5670	1	2	4	5	6	7	8	10	11
37	5682	5694	5705	5717	5729	5740	5752	5763	5775	5786	1	2	3	5	6	7	8	9	10
38	5798	5809	5821	5832	5843	5855	5866	5877	5888	5899	1	2	3	5	6	7	8	9	10
39	5911	5922	5933	5944	5955	5966	5977	5988	5999	6010	1	2	3	4	5	7	8	9	10
40	6021	6031	6042	6053	6064	6075	6085	6096	6107	6117	1	2	3	4	5	6	8	9	10
41	6128	6138	6149	6160	6170	6180	6191	6201	6212	6222	1	2	3	4	5	6	7	8	9
42	6232	6243	253	6263	6274	6284	6294	6304	6314	6325	1	2	3	4	5	6	7	8	9
43	6335	6345	6355	6365	6375	6385	6395	6405	6415	6425	1	2	3	4	5	6	7	8	9
44	6435	6444	6454	6464	6474	6484	6493	6503	6513	6522	1	2	3	4	5	6	7	8	9
45	6532	6542	6551	6561	6571	6580	6590	6599	6609	6618	1	2	3	4	5	6	7	8	9
46	6628	6637	6646	6656	6665	6675	6684	6693	6702	6712	1	2	3	4	5	6	7	7	8
47	6721	6730	6739	6749	6758	6767	6776	6785	6794	6803	1	2	3	4	5	5	6	7	8
48	6812	6821	6830	6839	6848	6857	6866	6875	6884	6893	1	2	3	4	4	5	6	7	8
49	6902	6911	6920	6928	6937	6946	6955	6964	6972	6981	1	2	3	4	4	5	6	7	8
50	6990	6998	7007	7016	7024	7033	7042	7050	7059	7067	1	2	3	3	4	5	6	7	8
51	7076	7084	7093	7101	7110	7118	7126	7135	7143	7152	1	2	3	3	4	5	6	7	8
52	7160	7168	7177	7185	7193	7202	7210	7218	7226	7235	1	2	2	3	4	5	6	7	7
53	7243	7251	7259	7267	7275	7284	7292	7300	7308	7316	1	2	2	3	4	5	6	6	7
54	7324	7332	7340	7348	7356	7364	7372	7380	7388	7396	1	2	2	3	4	5	6	6	7
N	0	1	2	3	4	5	6	7	8	9	1	2	3	4	5	6	7	8	9

TABLE A-4 (continued)

N	0	1	2	3	4	5	6	7	8	9	Proportional Parts 1	2	3	4	5	6	7	8	9
55	7404	7412	7419	7427	7435	7443	7451	7459	7466	7474	1	2	2	3	4	5	5	6	7
56	7482	7490	7497	7505	7513	7520	7528	7536	7543	7551	1	2	2	3	4	5	5	6	7
57	7559	7566	7574	7582	7589	7597	7604	7612	7619	7627	1	2	2	3	4	5	5	6	7
58	7634	7642	7649	7657	7664	7672	7679	7686	7694	7701	1	1	2	3	4	4	5	6	7
59	7709	7716	7723	7731	7738	7745	7752	7760	7767	7774	1	1	2	3	4	4	5	6	7
60	7782	7789	7796	7803	7810	7818	7825	7823	7839	7846	1	1	2	3	4	4	5	6	6
61	7853	7860	7868	7875	7882	7889	7896	7903	7910	7917	1	1	2	3	4	4	5	6	6
62	7924	7931	7938	7945	7952	7959	7966	7973	7980	7987	1	1	2	3	3	4	5	6	6
63	7993	8000	8007	8014	8021	8028	8035	8041	8048	8055	1	1	2	3	3	4	5	5	6
64	8062	8069	8075	8082	8089	8096	8102	8109	8116	8122	1	1	2	3	3	4	5	5	6
65	8129	8136	8142	8149	8156	8162	8169	8176	8182	8189	1	1	2	3	3	4	5	5	6
66	8195	8202	8209	8215	8222	8228	8235	8241	8248	8254	1	1	2	3	3	4	5	5	6
67	8261	8267	8274	8280	8287	8293	8299	8306	8312	8319	1	1	2	3	3	4	5	5	6
68	8325	8331	8338	8344	8351	8357	8363	8370	8376	8382	1	1	2	3	3	4	4	5	6
69	8388	8395	8401	8407	8414	8420	8426	8432	8439	8445	1	1	2	2	3	4	4	5	6
70	8451	8457	8463	8470	8476	8482	8488	8494	8500	8506	1	1	2	2	3	4	4	5	6
71	8513	8519	8525	8531	8537	8543	8549	8555	8561	8567	1	1	2	2	3	4	4	5	5
72	8573	8579	8585	8591	8597	8603	8609	8615	8621	8627	1	1	2	2	3	4	4	5	5
73	8633	8639	8645	8651	8657	8663	8669	8675	8681	8686	1	1	2	2	3	4	4	5	5
74	8692	8698	8704	8710	8716	8722	8727	8733	8739	8745	1	1	2	2	3	4	4	5	5
75	8751	8756	8762	8768	8774	8779	8785	8791	8797	8802	1	1	2	2	3	3	4	5	5
76	8808	8814	8820	8825	8831	8837	8842	8848	8854	8859	1	1	2	2	3	3	4	5	5
77	8865	8871	8876	8882	8887	8893	8899	8904	8910	8915	1	1	2	2	3	3	4	4	5
78	8921	8927	8932	8938	8943	8949	8954	8960	8965	8971	1	1	2	2	3	3	4	4	5
79	8976	8982	8987	8993	8998	9004	9009	9015	9020	9025	1	1	2	2	3	3	4	4	5
80	9031	9036	9042	9047	9053	9058	9063	9069	9074	9079	1	1	2	2	3	3	4	4	5
81	9085	9090	9096	9101	9106	9112	9117	9122	9128	9133	1	1	2	2	3	3	4	4	5
82	9138	9143	9149	9154	9159	9165	9170	9175	9180	9186	1	1	2	2	3	3	4	4	5
83	9191	9196	9201	9206	9212	9217	9222	9227	9232	9238	1	1	2	2	3	3	4	4	5
84	9243	9248	9253	9258	9263	9269	9274	9279	9284	9289	1	1	2	2	3	3	4	4	5
85	9294	9299	9304	9309	9315	9320	9325	9330	9335	9340	1	1	2	2	3	3	4	4	5
86	9345	9350	9355	9360	9365	9370	9375	9380	9385	9390	1	1	2	2	3	3	4	4	5
87	9395	9400	9405	9410	9415	9420	9425	9430	9435	9440	0	1	1	2	2	3	3	4	4
88	9445	9450	9455	9460	9465	9469	9474	9479	9484	9489	0	1	1	2	2	3	3	4	4
89	9494	9499	9504	9509	9513	9518	9523	9528	9533	9538	0	1	1	2	2	3	3	4	4
90	9542	9547	9552	9557	9562	9566	9571	9576	9581	9586	0	1	1	2	2	3	3	4	4
91	9590	9595	9600	9605	9609	9614	9619	9624	9628	9633	0	1	1	2	2	3	3	4	4
92	9638	9643	9647	9652	9657	9661	9666	9671	9675	9680	0	1	1	2	2	3	3	4	4
93	9685	9689	9694	9699	9703	9708	9713	9717	9722	9727	0	1	1	2	2	3	3	4	4
94	9731	9736	9741	9745	9750	9754	9759	9763	9768	9773	0	1	1	2	2	3	3	4	4
95	9777	9782	9786	9791	9795	9800	9805	9809	9814	9818	0	1	1	2	2	3	3	4	4
96	9823	9827	9832	9836	9841	9845	9850	9854	9859	9863	0	1	1	2	2	3	3	4	4
97	9868	9872	9877	9881	9886	9890	9894	9899	9903	9908	0	1	1	2	2	3	3	4	4
98	9912	9917	9921	9926	9930	9934	9939	9943	9948	9952	0	1	1	2	2	3	3	4	4
99	9956	9961	9965	9969	9974	9978	9983	9987	9991	9996	0	1	1	2	2	3	3	3	4
N	0	1	2	3	4	5	6	7	8	9	1	2	3	4	5	6	7	8	9

TABLE A-5

Areas under the standard normal probability distribution between the mean and successive values of z

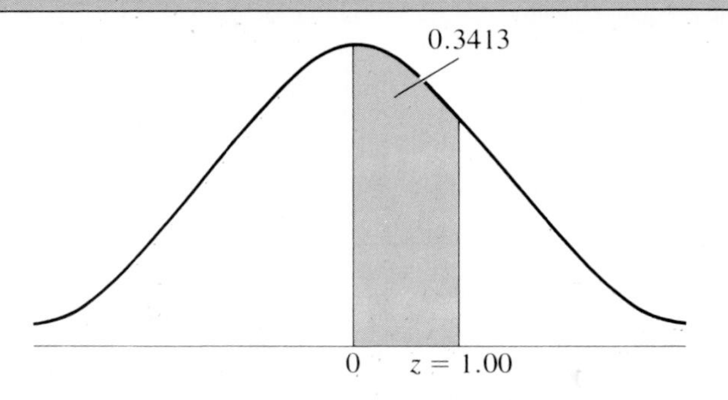

Example If $z = 1.00$, then the area between the mean and this value of z is 0.3413.

z	0.00	0.01	0.02	0.03	0.04	0.05	0.06	0.07	0.08	0.09
0.0	0.0000	0.0040	0.0080	0.0120	0.0160	0.0199	0.0239	0.0279	0.0319	0.0359
0.1	0.0398	0.0438	0.0478	0.0517	0.0557	0.0596	0.0636	0.0675	0.0714	0.0753
0.2	0.0793	0.0832	0.0871	0.0910	0.0948	0.0987	0.1026	0.1064	0.1103	0.1141
0.3	0.1179	0.1217	0.1255	0.1293	0.1331	0.1368	0.1406	0.1443	0.1480	0.1517
0.4	0.1554	0.1591	0.1628	0.1664	0.1700	0.1736	0.1772	0.1808	0.1844	0.1879
0.5	0.1915	0.1950	0.1985	0.2019	0.2054	0.2088	0.2123	0.2157	0.2190	0.2224
0.6	0.2257	0.2291	0.2324	0.2357	0.2389	0.2422	0.2454	0.2486	0.2518	0.2549
0.7	0.2580	0.2612	0.2642	0.2673	0.2704	0.2734	0.2764	0.2794	0.2823	0.2852
0.8	0.2881	0.2910	0.2939	0.2967	0.2995	0.3023	0.3051	0.3078	0.3106	0.3133
0.9	0.3159	0.3186	0.3212	0.3238	0.3264	0.3289	0.3315	0.3340	0.3365	0.3389
1.0	0.3413	0.3438	0.3461	0.3485	0.3508	0.3531	0.3554	0.3577	0.3599	0.3621
1.1	0.3643	0.3665	0.3686	0.3708	0.3729	0.3749	0.3770	0.3790	0.3810	0.3830
1.2	0.3849	0.3869	0.3888	0.3907	0.3925	0.3944	0.3962	0.3980	0.3997	0.4015
1.3	0.4032	0.4049	0.4066	0.4082	0.4099	0.4115	0.4131	0.4147	0.4162	0.4177
1.4	0.4192	0.4207	0.4222	0.4236	0.4251	0.4265	0.4279	0.4292	0.4306	0.4319
1.5	0.4332	0.4345	0.4357	0.4370	0.4382	0.4394	0.4406	0.4418	0.4429	0.4441
1.6	0.4452	0.4463	0.4474	0.4484	0.4495	0.4505	0.4515	0.4525	0.4535	0.4545
1.7	0.4554	0.4564	0.4573	0.4582	0.4591	0.4599	0.4608	0.4616	0.4625	0.4633
1.8	0.4641	0.4649	0.4656	0.4664	0.4671	0.4678	0.4686	0.4693	0.4699	0.4706
1.9	0.4713	0.4719	0.4726	0.4732	0.4738	0.4744	0.4750	0.4756	0.4761	0.4767
2.0	0.4772	0.4778	0.4783	0.4788	0.4793	0.4798	0.4803	0.4808	0.4812	0.4817
2.1	0.4821	0.4826	0.4830	0.4834	0.4838	0.4842	0.4846	0.4850	0.4854	0.4857
2.2	0.4861	0.4864	0.4868	0.4871	0.4875	0.4878	0.4881	0.4884	0.4887	0.4890
2.3	0.4893	0.4896	0.4898	0.4901	0.4904	0.4906	0.4909	0.4911	0.4913	0.4916
2.4	0.4918	0.4920	0.4922	0.4925	0.4927	0.4929	0.4931	0.4932	0.4934	0.4936
2.5	0.4938	0.4940	0.4941	0.4943	0.4945	0.4946	0.4948	0.4949	0.4951	0.4952
2.6	0.4953	0.4955	0.4956	0.4957	0.4959	0.4960	0.4961	0.4962	0.4963	0.4964
2.7	0.4965	0.4966	0.4967	0.4968	0.4969	0.4970	0.4971	0.4972	0.4973	0.4974
2.8	0.4974	0.4975	0.4976	0.4977	0.4977	0.4978	0.4979	0.4979	0.4980	0.4981
2.9	0.4981	0.4982	0.4982	0.4983	0.4984	0.4984	0.4985	0.4985	0.4986	0.4986
3.0	0.49865	0.4987	0.4987	0.4988	0.4988	0.4989	0.4989	0.4989	0.4990	0.4990
4.0	0.49997									

TABLE A-6
Student's *t* distribution

Example For 15 degrees of freedom, the *t* value that corresponds to an area of 0.05 in both tails combined is 2.131.

Degrees of Freedom	Area in Both Tails Combined			
	0.10	0.05	0.02	0.01
1	6.314	12.706	31.821	63.657
2	2.920	4.303	6.965	9.925
3	2.353	3.182	4.541	5.841
4	2.132	2.776	3.747	4.604
5	2.015	2.571	3.365	4.032
6	1.943	2.447	3.143	3.707
7	1.895	2.365	2.998	3.499
8	1.860	2.306	2.896	3.355
9	1.833	2.262	2.821	3.250
10	1.812	2.228	2.764	3.169
11	1.796	2.201	2.718	3.106
12	1.782	2.179	2.681	3.055
13	1.771	2.160	2.650	3.012
14	1.761	2.145	2.624	2.977
15	1.753	2.131	2.602	2.947
16	1.746	2.120	2.583	2.921
17	1.740	2.110	2.567	2.898
18	1.734	2.101	2.552	2.878
19	1.729	2.093	2.539	2.861
20	1.725	2.086	2.528	2.845
21	1.721	2.080	2.518	2.831
22	1.717	2.074	2.508	2.819
23	1.714	2.069	2.500	2.807
24	1.711	2.064	2.492	2.797
25	1.708	2.060	2.485	2.787
26	1.706	2.056	2.479	2.779
27	1.703	2.052	2.473	2.771
28	1.701	2.048	2.467	2.763
29	1.699	2.045	2.462	2.756
30	1.697	2.042	2.457	2.750
40	1.684	2.021	2.423	2.704
60	1.671	2.000	2.390	2.660
120	1.658	1.980	2.358	2.617
Normal Distribution	1.645	1.960	2.326	2.576

Source: From Table III of Fisher and Yates: *Statistical Tables for Biological, Agricultural and Medical Research*, published by Longman Group, Ltd., London (1974) 6th edition (previously published by Oliver and Boyd, Ltd., Edinburgh), and by permission of the authors and publishers.

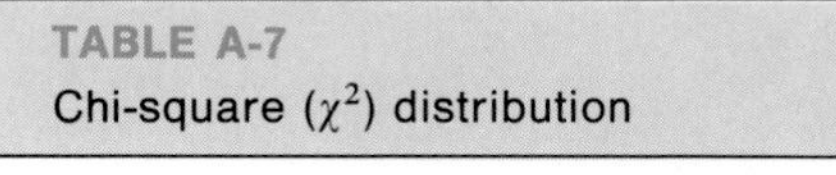

TABLE A-7
Chi-square (χ^2) distribution

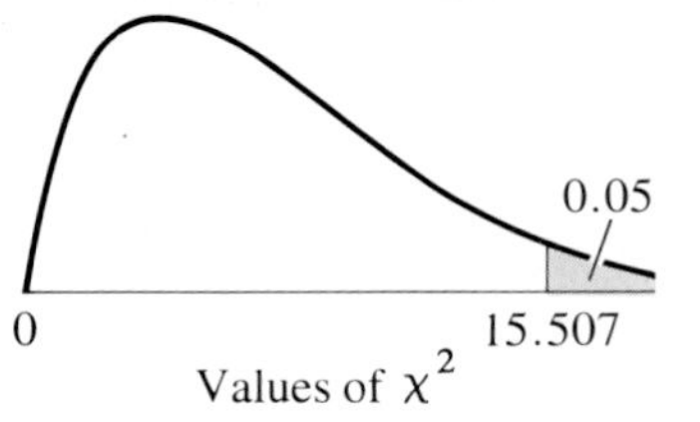

Example In a chi-square distribution with $v = 8$ degrees of freedom, the area to the right of a chi-square value of 15.507 is 0.05.

Degrees of Freedom	Area in Right Tail				
v	0.20	0.10	0.05	0.02	0.01
1	1.642	2.706	3.841	5.412	6.635
2	3.219	4.605	5.991	7.824	9.210
3	4.642	6.251	7.815	9.837	11.345
4	5.989	7.779	9.488	11.668	13.277
5	7.289	9.236	11.070	13.388	15.086
6	8.558	10.645	12.592	15.033	16.812
7	9.803	12.017	14.067	16.622	18.475
8	11.030	13.362	15.507	18.168	20.090
9	12.242	14.684	16.919	19.679	21.666
10	13.442	15.987	18.307	21.161	23.209
11	14.631	17.275	19.675	22.618	24.725
12	15.812	18.549	21.026	24.054	26.217
13	16.985	19.812	22.362	25.472	27.688
14	18.151	21.064	23.685	26.873	29.141
15	19.311	22.307	24.996	28.259	30.578
16	20.465	23.542	26.296	29.633	32.000
17	21.615	24.769	27.587	30.995	33.409
18	22.760	25.989	28.869	32.346	34.805
19	23.900	27.204	30.144	33.687	36.191
20	25.038	28.412	31.410	35.020	37.566
21	26.171	29.615	32.671	36.343	38.932
22	27.301	30.813	33.924	37.659	40.289
23	28.429	32.007	35.172	38.968	41.638
24	29.553	33.196	36.415	40.270	42.980
25	30.675	34.382	37.652	41.566	44.314
26	31.795	35.563	38.885	42.856	45.642
27	32.912	36.741	40.113	44.140	46.963
28	34.027	37.916	41.337	45.419	48.278
29	35.139	39.087	42.557	46.693	49.588
30	36.250	40.256	43.773	47.962	50.892

Source: From Table IV of Fisher and Yates: *Statistical Tables for Biological, Agricultural and Medical Research*, published by Longman Group, Ltd., London (1974) 6th edition (previously published by Oliver and Boyd, Ltd., Edinburgh), and by permission of the authors and publishers.

TABLE A-8
F distribution

Example In an F distribution with $v_1 = 5$ and $v_2 = 6$ degrees of freedom, the area to the right of an F value of 4.39 is 0.05. The value on the F scale to the right of which lies 0.05 of the area is in lightface type. The value on the F scale to the right of which lies 0.01 of the area is in boldface type. For the numerator, v_1 = number of degrees of freedom; v_2 = number of degrees of freedom for the denominator.

0.05
0
4.39
Values of F

v_2 \ v_1	1	2	3	4	5	6	7	8	9	10	11	12	14	16	20	24	30	40	50	75	100	200	500	∞	v_1 / v_2
1	161	200	216	225	230	234	237	239	241	242	243	244	245	246	248	249	250	251	252	253	253	254	254	254	1
	4,052	**4,999**	**5,403**	**5,625**	**5,764**	**5,859**	**5,928**	**5,981**	**6,022**	**6,056**	**6,082**	**6,106**	**6,142**	**6,169**	**6,208**	**6,234**	**6,261**	**6,286**	**6,302**	**6,323**	**6,334**	**6,352**	**6,361**	**6,366**	
2	18.51	19.00	19.16	19.25	19.30	19.33	19.36	19.37	19.38	19.39	19.40	19.41	19.42	19.43	19.44	19.45	19.46	19.47	19.47	19.48	19.49	19.49	19.50	19.50	2
	98.49	**99.00**	**99.17**	**99.25**	**99.30**	**99.33**	**99.36**	**99.37**	**99.39**	**99.40**	**99.41**	**99.42**	**99.43**	**99.44**	**99.45**	**99.46**	**99.47**	**99.48**	**98.48**	**99.49**	**99.49**	**99.49**	**99.50**	**99.50**	
3	10.13	9.55	9.28	9.12	9.01	8.94	8.88	8.84	8.81	8.78	8.76	8.74	8.71	8.69	8.66	8.64	8.62	8.60	8.58	8.57	8.56	8.54	8.54	8.53	3
	34.12	**30.82**	**29.46**	**28.71**	**28.24**	**27.91**	**27.67**	**27.49**	**27.34**	**27.23**	**27.13**	**27.05**	**26.92**	**26.83**	**26.69**	**26.60**	**26.50**	**26.41**	**26.35**	**26.27**	**26.23**	**26.18**	**26.14**	**26.12**	
4	7.71	6.94	6.59	6.39	6.26	6.16	6.09	6.04	6.00	5.96	5.93	5.91	5.87	5.84	5.80	5.77	5.74	5.71	5.70	5.68	5.66	5.65	5.64	5.63	4
	21.20	**18.00**	**16.69**	**15.98**	**15.52**	**15.21**	**14.98**	**14.80**	**14.66**	**14.54**	**14.45**	**14.37**	**14.24**	**14.15**	**14.02**	**13.93**	**13.83**	**13.74**	**13.69**	**13.61**	**13.57**	**13.52**	**13.48**	**13.46**	
5	6.61	5.79	5.41	5.19	5.05	4.95	4.88	4.82	4.78	4.74	4.70	4.68	4.64	4.60	4.56	4.53	4.50	4.46	4.44	4.42	4.40	4.38	4.37	4.36	5
	16.26	**13.27**	**12.06**	**11.39**	**10.97**	**10.67**	**10.45**	**10.29**	**10.15**	**10.05**	**9.96**	**9.89**	**9.77**	**9.68**	**9.55**	**9.47**	**9.38**	**9.29**	**9.24**	**9.17**	**9.13**	**9.07**	**9.04**	**9.02**	
6	5.99	5.14	4.76	4.53	4.39	4.28	4.21	4.15	4.10	4.06	4.03	4.00	3.96	3.92	3.87	3.84	3.81	3.77	3.75	3.72	3.71	3.69	3.68	3.67	6
	13.74	**10.92**	**9.78**	**9.15**	**8.75**	**8.47**	**8.26**	**8.10**	**7.98**	**7.87**	**7.79**	**7.72**	**7.60**	**7.52**	**7.39**	**7.31**	**7.23**	**7.14**	**7.09**	**7.02**	**6.99**	**6.94**	**6.90**	**6.88**	
7	5.59	4.74	4.35	4.12	3.97	3.87	3.79	3.73	3.68	3.63	3.60	3.57	3.52	3.49	3.44	3.41	3.38	3.34	3.32	3.29	3.28	3.25	3.24	3.23	7
	12.25	**9.55**	**8.45**	**7.85**	**7.46**	**7.19**	**7.00**	**6.84**	**6.71**	**6.62**	**6.54**	**6.47**	**6.35**	**6.27**	**6.15**	**6.07**	**5.98**	**5.90**	**5.85**	**5.78**	**5.75**	**5.70**	**5.67**	**5.65**	
8	5.32	4.46	4.07	3.84	3.69	3.58	3.50	3.44	3.39	3.34	3.31	3.28	3.23	3.20	3.15	3.12	3.08	3.05	3.03	3.00	2.98	2.96	2.94	2.93	8
	11.26	**8.65**	**7.59**	**7.01**	**6.63**	**6.37**	**6.19**	**6.03**	**5.91**	**5.82**	**5.74**	**5.67**	**5.56**	**5.48**	**5.36**	**5.28**	**5.20**	**5.11**	**5.06**	**5.00**	**4.96**	**4.91**	**4.88**	**4.86**	
9	5.12	4.26	3.86	3.63	3.48	3.37	3.29	3.23	3.18	3.13	3.10	3.07	3.02	2.98	2.93	2.90	2.86	2.82	2.80	2.77	2.76	2.73	2.72	2.71	9
	10.56	**8.02**	**6.99**	**6.42**	**6.06**	**5.80**	**5.62**	**5.47**	**5.35**	**5.26**	**5.18**	**5.11**	**5.00**	**4.92**	**4.80**	**4.73**	**4.64**	**4.56**	**4.51**	**4.45**	**4.41**	**4.36**	**4.33**	**4.31**	
10	4.96	4.10	3.71	3.48	3.33	3.22	3.14	3.07	3.02	2.97	2.94	2.91	2.86	2.82	2.77	2.74	2.70	2.67	2.64	2.61	2.59	2.56	2.55	2.54	10
	10.04	**7.56**	**6.55**	**5.99**	**5.64**	**5.39**	**5.21**	**5.06**	**4.95**	**4.85**	**4.78**	**4.71**	**4.60**	**4.52**	**4.41**	**4.33**	**4.25**	**4.17**	**4.12**	**4.05**	**4.01**	**3.96**	**3.93**	**3.91**	
11	4.84	3.98	3.59	3.36	3.20	3.09	3.01	2.95	2.90	2.86	2.82	2.79	2.74	2.70	2.65	2.61	2.57	2.53	2.50	2.47	2.45	2.42	2.41	2.40	11
	9.65	**7.20**	**6.22**	**5.67**	**5.32**	**5.07**	**4.88**	**4.74**	**4.63**	**4.54**	**4.46**	**4.40**	**4.29**	**4.21**	**4.10**	**4.02**	**3.94**	**3.86**	**3.80**	**3.74**	**3.70**	**3.66**	**3.62**	**3.60**	
12	4.75	3.88	3.49	3.26	3.11	3.00	2.92	2.85	2.80	2.76	2.72	2.69	2.64	2.60	2.54	2.50	2.46	2.42	2.40	2.36	2.35	2.32	2.31	2.30	12
	9.33	**6.93**	**5.95**	**5.41**	**5.06**	**4.82**	**4.65**	**4.50**	**4.39**	**4.30**	**4.22**	**4.16**	**4.05**	**3.98**	**3.86**	**3.78**	**3.70**	**3.61**	**3.56**	**3.49**	**3.46**	**3.41**	**3.38**	**3.36**	

Source: From George W. Snedecor and William G. Cochran, *Statistical Methods*, Seventh Edition, © 1980 by the Iowa State University Press, Ames, 1A, 50010. Reprinted by permission.

TABLE A-8 (continued)

v_2 \ v_1	1	2	3	4	5	6	7	8	9	10	11	12	14	16	20	24	30	40	50	75	100	200	500	∞	v_1 / v_2
13	4.67	3.80	3.41	3.18	3.02	2.92	2.84	2.77	2.72	2.67	2.63	2.60	2.55	2.51	2.46	2.42	2.38	2.34	2.32	2.28	2.26	2.24	2.22	2.21	13
	9.07	**6.70**	**5.74**	**5.20**	**4.86**	**4.62**	**4.44**	**4.30**	**4.19**	**4.10**	**4.02**	**3.96**	**3.85**	**3.78**	**3.67**	**3.59**	**3.51**	**3.42**	**3.37**	**3.30**	**3.27**	**3.21**	**3.18**	**3.16**	
14	4.60	3.74	3.34	3.11	2.96	2.85	2.77	2.70	2.65	2.60	2.56	2.53	2.48	2.44	2.39	2.35	2.31	2.27	2.24	2.21	2.19	2.16	2.14	2.13	14
	8.86	**6.51**	**5.56**	**5.03**	**4.69**	**4.46**	**4.28**	**4.14**	**4.03**	**3.94**	**3.86**	**3.80**	**3.70**	**3.62**	**3.51**	**3.43**	**3.34**	**3.26**	**3.21**	**3.14**	**3.11**	**3.06**	**3.02**	**3.00**	
15	4.54	3.68	3.29	3.06	2.90	2.79	2.70	2.64	2.59	2.55	2.51	2.48	2.43	2.39	2.33	2.29	2.25	2.21	2.18	2.15	2.12	2.10	2.08	2.07	15
	8.68	**6.36**	**5.42**	**4.89**	**4.56**	**4.32**	**4.14**	**4.00**	**3.89**	**3.80**	**3.73**	**3.67**	**3.56**	**3.48**	**3.36**	**3.29**	**3.20**	**3.12**	**3.07**	**3.00**	**2.97**	**2.92**	**2.89**	**2.87**	
16	4.49	3.63	3.24	3.01	2.85	2.74	2.66	2.59	2.54	2.49	2.45	2.42	2.37	2.33	2.28	2.24	2.20	2.16	2.13	2.09	2.07	2.04	2.02	2.01	16
	8.53	**6.23**	**5.29**	**4.77**	**4.44**	**4.20**	**4.03**	**3.89**	**3.78**	**3.69**	**3.61**	**3.55**	**3.45**	**3.37**	**3.25**	**3.18**	**3.10**	**3.01**	**2.96**	**2.98**	**2.86**	**2.80**	**2.77**	**2.75**	
17	4.45	3.59	3.20	2.96	2.81	2.70	2.62	2.55	2.50	2.45	2.41	2.38	2.33	2.29	2.23	2.19	2.15	2.11	2.08	2.04	2.02	1.99	1.97	1.96	17
	8.40	**6.11**	**5.18**	**4.67**	**4.34**	**4.10**	**3.93**	**3.79**	**3.68**	**3.59**	**3.52**	**3.45**	**3.35**	**3.27**	**3.16**	**3.08**	**3.00**	**2.92**	**2.86**	**2.79**	**2.76**	**2.70**	**2.67**	**2.65**	
18	4.41	3.55	3.16	2.93	2.77	3.66	2.58	2.51	2.46	2.41	2.37	2.34	2.29	2.25	2.19	2.15	2.11	2.07	2.04	2.00	1.98	1.95	1.93	1.92	18
	8.28	**6.01**	**5.09**	**4.58**	**4.25**	**4.01**	**3.85**	**3.71**	**3.60**	**3.51**	**3.44**	**3.37**	**3.27**	**3.19**	**3.07**	**3.00**	**2.91**	**2.83**	**2.78**	**2.71**	**2.68**	**2.62**	**2.59**	**2.57**	
19	4.38	3.52	3.13	2.90	2.74	2.63	2.55	2.48	2.43	2.38	2.34	2.31	2.26	2.21	2.15	2.11	2.07	2.02	2.00	1.96	1.94	1.91	1.90	1.88	19
	8.18	**5.93**	**5.01**	**4.50**	**4.17**	**3.94**	**3.77**	**3.63**	**3.52**	**3.43**	**3.36**	**3.30**	**3.19**	**3.12**	**3.00**	**2.92**	**2.84**	**2.76**	**2.70**	**2.63**	**2.60**	**2.54**	**2.51**	**2.49**	
20	4.35	3.49	3.10	2.87	2.71	2.60	2.52	2.45	2.40	2.35	2.31	2.28	2.23	2.18	2.12	2.08	2.04	1.99	1.96	1.92	1.90	1.87	1.85	1.84	20
	8.10	**5.85**	**4.94**	**4.43**	**4.10**	**3.87**	**3.71**	**3.56**	**3.45**	**3.37**	**3.30**	**3.23**	**3.13**	**3.05**	**2.94**	**2.86**	**2.77**	**2.69**	**2.63**	**2.56**	**2.53**	**2.47**	**2.44**	**2.42**	
21	4.32	3.47	3.07	2.84	2.68	2.57	2.49	2.42	2.37	2.32	2.28	2.25	2.20	2.15	2.09	2.05	2.00	1.96	1.93	1.89	1.87	1.84	1.82	1.81	21
	8.02	**5.78**	**4.87**	**4.37**	**4.04**	**3.81**	**3.65**	**3.51**	**3.40**	**3.31**	**3.24**	**3.17**	**3.07**	**2.99**	**2.88**	**2.80**	**2.72**	**2.63**	**2.58**	**2.51**	**2.47**	**2.42**	**2.38**	**2.36**	
22	4.30	3.44	3.05	2.82	2.66	2.55	2.47	2.40	2.35	2.30	2.26	2.23	2.18	2.13	2.07	2.03	1.98	1.93	1.91	1.87	1.84	1.81	1.80	1.78	22
	7.94	**5.72**	**4.82**	**4.31**	**3.99**	**3.76**	**3.59**	**3.45**	**3.35**	**3.26**	**3.18**	**3.12**	**3.02**	**2.94**	**2.83**	**2.75**	**2.67**	**2.58**	**2.53**	**2.46**	**2.42**	**2.37**	**2.33**	**2.31**	
23	4.28	3.42	3.03	2.80	2.64	2.53	2.45	2.38	2.32	2.28	2.24	2.20	2.14	2.10	2.04	2.00	1.96	1.91	1.88	1.84	1.82	1.79	1.77	1.76	23
	7.88	**5.66**	**4.76**	**4.26**	**3.94**	**3.71**	**3.54**	**3.41**	**3.30**	**3.21**	**3.14**	**3.07**	**2.97**	**2.89**	**2.78**	**2.70**	**2.62**	**2.53**	**2.48**	**2.41**	**2.37**	**2.32**	**2.28**	**2.26**	
24	4.26	3.40	3.01	2.78	2.62	2.51	2.43	2.36	2.30	2.26	2.22	2.18	2.13	2.09	2.02	1.98	1.94	1.89	1.86	1.82	1.80	1.76	1.74	1.73	24
	7.82	**5.61**	**4.72**	**4.22**	**3.90**	**3.67**	**3.50**	**3.36**	**3.25**	**3.17**	**3.09**	**3.03**	**2.93**	**2.85**	**2.74**	**2.66**	**2.58**	**2.49**	**2.44**	**2.36**	**2.33**	**2.27**	**2.23**	**2.21**	
25	4.24	3.38	2.99	2.76	2.60	2.49	2.41	2.34	2.28	2.24	2.20	2.16	2.11	2.06	2.00	1.96	1.92	1.87	1.84	1.80	1.77	1.74	1.72	1.71	25
	7.77	**5.57**	**4.68**	**4.18**	**3.86**	**3.63**	**3.46**	**3.32**	**3.21**	**3.13**	**3.05**	**2.99**	**2.89**	**2.81**	**2.70**	**2.62**	**2.54**	**2.45**	**2.40**	**2.32**	**2.29**	**2.23**	**2.19**	**2.17**	
26	4.22	3.37	2.98	2.74	2.59	2.47	2.39	2.32	2.27	2.22	2.18	2.15	2.10	2.05	1.99	1.95	1.90	1.85	1.82	1.78	1.76	1.72	1.70	1.69	26
	7.72	**5.53**	**4.64**	**4.14**	**3.82**	**3.59**	**3.42**	**3.29**	**3.17**	**3.09**	**3.02**	**2.96**	**2.86**	**2.77**	**2.66**	**2.58**	**2.50**	**2.41**	**2.36**	**2.28**	**2.25**	**2.19**	**2.15**	**2.13**	

TABLE A-8 (continued)

v_2 \ v_1	1	2	3	4	5	6	7	8	9	10	11	12	14	16	20	24	30	40	50	75	100	200	500	∞	v_1 / v_2
27	4.21	3.35	2.96	2.73	2.57	2.46	2.37	2.30	2.25	2.20	2.16	2.13	2.08	2.03	1.97	1.93	1.88	1.84	1.80	1.76	1.74	1.71	1.68	1.67	27
	7.68	**5.49**	**4.60**	**4.11**	**3.79**	**3.56**	**3.39**	**3.26**	**3.14**	**3.06**	**2.98**	**2.93**	**2.83**	**2.74**	**2.63**	**2.55**	**2.47**	**2.38**	**2.33**	**2.25**	**2.21**	**2.16**	**2.12**	**2.10**	
28	4.20	3.34	2.95	2.71	2.56	2.44	2.36	2.29	2.24	2.19	2.15	2.12	2.06	2.02	1.96	1.91	1.87	1.81	1.78	1.75	1.72	1.69	1.67	1.65	28
	7.64	**5.45**	**4.57**	**4.07**	**3.76**	**3.53**	**3.36**	**3.23**	**3.11**	**3.03**	**2.95**	**2.90**	**2.80**	**2.71**	**2.60**	**2.52**	**2.44**	**2.35**	**2.30**	**2.22**	**2.18**	**2.13**	**2.09**	**2.06**	
29	4.18	3.33	2.93	2.70	2.54	2.43	2.35	2.28	2.22	2.18	2.14	2.10	2.05	2.00	1.94	1.90	1.85	1.80	1.77	1.73	1.71	1.68	1.65	1.64	29
	7.60	**5.42**	**4.54**	**4.04**	**3.73**	**3.50**	**3.33**	**3.20**	**3.08**	**3.00**	**2.92**	**2.87**	**2.77**	**2.68**	**2.57**	**2.49**	**2.41**	**2.32**	**2.27**	**2.19**	**2.15**	**2.10**	**2.06**	**2.03**	
30	4.17	3.32	2.92	2.69	2.53	2.42	2.34	2.27	2.21	2.16	2.12	2.09	2.04	1.99	1.93	1.89	1.84	1.79	1.76	1.72	1.69	1.66	1.64	1.62	30
	7.56	**5.39**	**4.51**	**4.02**	**3.70**	**3.47**	**3.30**	**3.17**	**3.06**	**2.98**	**2.90**	**2.84**	**2.74**	**2.66**	**2.55**	**2.47**	**2.38**	**2.29**	**2.24**	**2.16**	**2.13**	**2.07**	**2.03**	**2.01**	
32	4.15	3.30	2.90	2.67	2.51	2.40	2.32	2.25	2.19	2.14	2.10	2.07	2.02	1.97	1.91	1.86	1.82	1.76	1.74	1.69	1.67	1.64	1.61	1.59	32
	7.50	**5.34**	**4.46**	**3.97**	**3.66**	**3.42**	**3.25**	**3.12**	**3.01**	**2.94**	**2.86**	**2.80**	**2.70**	**2.62**	**2.51**	**2.42**	**2.34**	**2.25**	**2.20**	**2.12**	**2.08**	**2.02**	**1.98**	**1.96**	
34	4.13	3.28	2.88	2.65	2.49	2.38	2.30	2.23	2.17	2.12	2.08	2.05	2.00	1.95	1.89	1.84	1.80	1.74	1.71	1.67	1.64	1.61	1.59	1.57	34
	7.44	**5.29**	**4.42**	**3.93**	**3.61**	**3.38**	**3.21**	**3.08**	**2.97**	**2.89**	**2.82**	**2.76**	**2.66**	**2.58**	**2.47**	**2.38**	**2.30**	**2.21**	**2.15**	**2.08**	**2.04**	**1.98**	**1.94**	**1.91**	
36	4.11	3.26	2.86	2.63	2.48	2.36	2.28	2.21	2.15	2.10	2.06	2.03	1.98	1.93	1.87	1.82	1.78	1.72	1.69	1.65	1.62	1.59	1.56	1.55	36
	7.39	**5.25**	**4.38**	**3.89**	**3.58**	**3.35**	**3.18**	**3.04**	**2.94**	**2.86**	**2.78**	**2.72**	**2.62**	**2.54**	**2.43**	**2.35**	**2.26**	**2.17**	**2.12**	**2.04**	**2.00**	**1.94**	**1.90**	**1.87**	
38	4.10	3.25	2.85	2.62	2.46	2.35	2.26	2.19	2.14	2.09	2.05	2.02	1.96	1.92	1.85	1.80	1.76	1.71	1.67	1.63	1.60	1.57	1.54	1.53	38
	7.35	**5.21**	**4.34**	**3.86**	**3.54**	**3.32**	**3.15**	**3.02**	**2.91**	**2.82**	**2.75**	**2.69**	**2.59**	**2.51**	**2.40**	**2.32**	**2.22**	**2.14**	**2.08**	**2.00**	**1.97**	**1.90**	**1.86**	**1.84**	
40	4.08	3.23	2.84	2.61	2.45	2.34	2.25	2.18	2.12	2.07	2.04	2.00	1.95	1.90	1.84	1.79	1.74	1.69	1.66	1.61	1.59	1.55	1.53	1.51	40
	7.31	**5.18**	**4.31**	**3.83**	**3.51**	**3.29**	**3.12**	**2.99**	**2.88**	**2.80**	**2.73**	**2.66**	**2.56**	**2.49**	**2.37**	**2.29**	**2.20**	**2.11**	**2.05**	**1.97**	**1.94**	**1.88**	**1.84**	**1.81**	
42	4.07	3.22	2.83	2.59	2.44	2.32	2.24	2.17	2.11	2.06	2.02	1.99	1.94	1.89	1.82	1.78	1.73	1.68	1.64	1.60	1.57	1.54	1.51	1.49	42
	7.27	**5.15**	**4.29**	**3.80**	**3.49**	**3.26**	**3.10**	**2.96**	**2.86**	**2.77**	**2.70**	**2.64**	**2.54**	**2.46**	**2.35**	**2.26**	**2.17**	**2.08**	**2.02**	**1.94**	**1.91**	**1.85**	**1.80**	**1.78**	
44	4.06	3.21	2.82	2.58	2.43	2.31	2.23	2.16	2.10	2.05	2.01	1.98	1.92	1.88	1.81	1.76	1.72	1.66	1.63	1.58	1.56	1.52	1.50	1.48	44
	7.24	**5.12**	**4.26**	**3.78**	**3.46**	**3.24**	**3.07**	**2.94**	**2.84**	**2.75**	**2.68**	**2.62**	**2.52**	**2.44**	**2.32**	**2.24**	**2.15**	**2.06**	**2.00**	**1.92**	**1.88**	**1.82**	**1.78**	**1.75**	
46	4.05	3.20	2.81	2.57	2.42	2.30	2.22	2.14	2.09	2.04	2.00	1.97	1.91	1.87	1.80	1.75	1.71	1.65	1.62	1.57	1.54	1.51	1.48	1.46	46
	7.21	**5.10**	**4.24**	**3.76**	**3.44**	**3.22**	**3.05**	**2.92**	**2.82**	**2.73**	**2.66**	**2.60**	**2.50**	**2.42**	**2.30**	**2.22**	**2.13**	**2.04**	**1.98**	**1.90**	**1.86**	**1.80**	**1.76**	**1.72**	
48	4.04	3.19	2.80	2.56	2.41	2.30	2.21	2.14	2.08	2.03	1.99	1.96	1.90	1.86	1.79	1.74	1.70	1.64	1.61	1.56	1.53	1.50	1.47	1.45	48
	7.19	**5.08**	**4.22**	**3.74**	**3.42**	**3.20**	**3.04**	**2.90**	**2.80**	**2.71**	**2.64**	**2.58**	**2.48**	**2.40**	**2.28**	**2.20**	**2.11**	**2.02**	**1.96**	**1.88**	**1.84**	**1.78**	**1.73**	**1.70**	

TABLE A-8 (continued)

ν_2 \ ν_1	1	2	3	4	5	6	7	8	9	10	11	12	14	16	20	24	30	40	50	75	100	200	500	∞	ν_1 / ν_2
50	4.03	3.18	2.79	2.56	2.40	2.29	2.20	2.13	2.07	2.02	1.98	1.95	1.90	1.85	1.78	1.74	1.69	1.63	1.60	1.55	1.52	1.48	1.46	1.44	50
	7.17	**5.06**	**4.20**	**3.72**	**3.41**	**3.18**	**3.02**	**2.88**	**2.78**	**2.70**	**2.62**	**2.56**	**2.46**	**2.39**	**2.26**	**2.18**	**2.10**	**2.00**	**1.94**	**1.86**	**1.82**	**1.76**	**1.71**	**1.68**	
55	4.02	3.17	2.78	2.54	2.38	2.27	2.18	2.11	2.05	2.00	1.97	1.93	1.88	1.83	1.76	1.72	1.67	1.61	1.58	1.52	1.50	1.46	1.43	1.41	55
	7.12	**5.01**	**4.16**	**3.68**	**3.37**	**3.15**	**2.98**	**2.85**	**2.75**	**2.66**	**2.59**	**2.53**	**2.43**	**2.35**	**2.23**	**2.15**	**2.06**	**1.96**	**1.90**	**1.82**	**1.78**	**1.71**	**1.66**	**1.64**	
60	4.00	3.15	2.76	2.52	2.37	2.25	2.17	2.10	2.04	1.99	1.95	1.92	1.86	1.81	1.75	1.70	1.65	1.59	1.56	1.50	1.48	1.44	1.41	1.39	60
	7.08	**4.98**	**4.13**	**3.65**	**3.34**	**3.12**	**2.95**	**2.82**	**2.72**	**2.63**	**2.56**	**2.50**	**2.40**	**2.32**	**2.20**	**2.12**	**2.03**	**1.93**	**1.87**	**1.79**	**1.74**	**1.68**	**1.63**	**1.60**	
65	3.99	3.14	2.75	2.51	2.36	2.24	2.15	2.08	2.02	1.98	1.94	1.90	1.85	1.80	1.73	1.68	1.63	1.57	1.54	1.49	1.46	1.42	1.39	1.37	65
	7.04	**4.95**	**4.10**	**3.62**	**3.31**	**3.09**	**2.93**	**2.79**	**2.70**	**2.61**	**2.54**	**2.47**	**2.37**	**2.30**	**2.18**	**2.09**	**2.00**	**1.90**	**1.84**	**1.76**	**1.71**	**1.64**	**1.60**	**1.56**	
70	3.98	3.13	2.74	2.50	2.35	2.23	2.14	2.07	2.01	1.97	1.93	1.89	1.84	1.79	1.72	1.67	1.62	1.56	1.53	1.47	1.45	1.40	1.37	1.35	70
	7.01	**4.92**	**4.08**	**3.60**	**3.29**	**3.07**	**2.91**	**2.77**	**2.67**	**2.59**	**2.51**	**2.45**	**2.35**	**2.28**	**2.15**	**2.07**	**1.98**	**1.88**	**1.82**	**1.74**	**1.69**	**1.62**	**1.56**	**1.53**	
80	3.96	3.11	2.72	2.48	2.33	2.21	2.12	2.05	1.99	1.95	1.91	1.88	1.82	1.77	1.70	1.65	1.60	1.54	1.51	1.45	1.42	1.38	1.35	1.32	80
	6.96	**4.88**	**4.04**	**3.56**	**3.25**	**3.04**	**2.87**	**2.74**	**2.64**	**2.55**	**2.48**	**2.41**	**2.32**	**2.24**	**2.11**	**2.03**	**1.94**	**1.84**	**1.78**	**1.70**	**1.65**	**1.57**	**1.52**	**1.49**	
100	3.94	3.09	2.70	2.46	2.30	2.19	2.10	2.03	1.97	1.92	1.88	1.85	1.79	1.75	1.68	1.63	1.57	1.51	1.48	1.42	1.39	1.34	1.30	1.28	100
	6.90	**4.82**	**3.98**	**3.51**	**3.20**	**2.99**	**2.82**	**2.69**	**2.59**	**2.51**	**2.43**	**2.36**	**2.26**	**2.19**	**2.06**	**1.98**	**1.89**	**1.79**	**1.73**	**1.64**	**1.59**	**1.51**	**1.46**	**1.43**	
125	3.92	3.07	2.68	2.44	2.29	2.17	2.08	2.01	1.95	1.90	1.86	1.83	1.77	1.72	1.65	1.60	1.55	1.49	1.45	1.39	1.36	1.31	1.27	1.25	125
	6.84	**4.78**	**3.94**	**3.47**	**3.17**	**2.95**	**2.79**	**2.65**	**2.56**	**2.47**	**2.40**	**2.33**	**2.23**	**2.15**	**2.03**	**1.94**	**1.85**	**1.75**	**1.68**	**1.59**	**1.54**	**1.46**	**1.40**	**1.37**	
150	3.91	3.06	2.67	2.43	2.27	2.16	2.07	2.00	1.94	1.89	1.85	1.82	1.76	1.71	1.64	1.59	1.54	1.47	1.44	1.37	1.34	1.29	1.25	1.22	150
	6.81	**4.75**	**3.91**	**3.44**	**3.14**	**2.92**	**2.76**	**2.62**	**2.53**	**2.44**	**2.37**	**2.30**	**2.20**	**2.12**	**2.00**	**1.91**	**1.83**	**1.72**	**1.66**	**1.56**	**1.51**	**1.43**	**1.37**	**1.33**	
200	3.89	3.04	2.65	2.41	2.26	2.14	2.05	1.98	1.92	1.87	1.83	1.80	1.74	1.69	1.62	1.57	1.52	1.45	1.42	1.35	1.32	1.26	1.22	1.19	200
	6.76	**4.71**	**3.88**	**3.41**	**3.11**	**2.90**	**2.73**	**2.60**	**2.50**	**2.41**	**2.34**	**2.28**	**2.17**	**2.09**	**1.97**	**1.88**	**1.79**	**1.69**	**1.62**	**1.53**	**1.48**	**1.39**	**1.33**	**1.28**	
400	3.86	3.02	2.62	2.39	2.23	2.12	2.03	1.96	1.90	1.85	1.81	1.78	1.72	1.67	1.60	1.54	1.49	1.42	1.38	1.32	1.28	1.22	1.16	1.13	400
	6.70	**4.66**	**3.83**	**3.36**	**3.06**	**2.85**	**2.69**	**2.55**	**2.46**	**2.37**	**2.29**	**2.23**	**2.12**	**2.04**	**1.92**	**1.84**	**1.74**	**1.64**	**1.57**	**1.47**	**1.42**	**1.32**	**1.24**	**1.19**	
1,000	3.85	3.00	2.61	2.38	2.22	2.10	2.02	1.95	1.89	1.84	1.80	1.76	1.70	1.65	1.58	1.53	1.47	1.41	1.36	1.30	1.26	1.19	1.13	1.08	1,000
	6.66	**4.62**	**3.80**	**3.34**	**3.04**	**2.82**	**2.66**	**2.53**	**2.43**	**2.34**	**2.26**	**2.20**	**2.09**	**2.01**	**1.89**	**1.81**	**1.71**	**1.61**	**1.54**	**1.44**	**1.38**	**1.28**	**1.19**	**1.11**	
∞	3.84	2.99	2.60	2.37	2.21	2.09	2.01	1.94	1.88	1.83	1.79	1.75	1.69	1.64	1.57	1.52	1.46	1.40	1.35	1.28	1.24	1.17	1.11	1.00	∞
	6.63	**4.60**	**3.78**	**3.32**	**3.02**	**2.80**	**2.64**	**2.51**	**2.41**	**2.32**	**2.24**	**2.18**	**2.07**	**1.99**	**1.87**	**1.79**	**1.69**	**1.59**	**1.52**	**1.41**	**1.36**	**1.25**	**1.15**	**1.00**	

TABLE A-9

Critical values of T in the Wilcoxon matched-pairs signed-ranks test

Critical values of T at various levels of probability

The symbol T denotes the smaller sum of ranks associated with differences that are all of the same sign. For any given N (number of ranked differences), the obtained T is significant at a given level if it is equal to or *less than* the value shown in the table.

	Level of Significance for One-tailed Test					Level of Significance for One-tailed Test			
	0.05	0.025	0.01	0.005		0.05	0.025	0.01	0.005
	Level of Significance for Two-tailed Test					Level of Significance for Two-tailed Test			
N	0.10	0.05	0.02	0.01	N	0.10	0.05	0.02	0.01
5	0	—	—	—	28	130	116	101	91
6	2	0	—	—	29	140	126	110	100
7	3	2	0	—	30	151	137	120	109
8	5	3	1	0	31	163	147	130	118
9	8	5	3	1	32	175	159	140	128
10	10	8	5	3	33	187	170	151	138
11	13	10	7	5	34	200	182	162	148
12	17	13	9	7	35	213	195	173	159
13	21	17	12	9	36	227	208	185	171
14	25	21	15	12	37	241	221	198	182
15	30	25	19	15	38	256	235	211	194
16	35	29	23	19	39	271	249	224	207
17	41	34	27	23	40	286	264	238	220
18	47	40	32	27	41	302	279	252	233
19	53	46	37	32	42	319	294	266	247
20	60	52	43	37	43	336	310	281	261
21	67	58	49	42	44	353	327	296	276
22	75	65	55	48	45	371	343	312	291
23	83	73	62	54	46	389	361	328	307
24	91	81	69	61	47	407	378	345	322
25	100	89	76	68	48	426	396	362	339
26	110	98	84	75	49	446	415	379	355
27	119	107	92	83	50	466	434	397	373

(Slight discrepancies will be found between the critical values appearing in the table above and in Table 2 of the 1964 revision of F. Wilcoxon and R.A. Wilcox, *Some Rapid Approximate Statistical Procedures*, New York, Lederle Laboratories, 1964. The disparity reflects the latter's policy of selecting the critical value nearest a given significance level, occasionally overstepping that level. For example, for $N = 8$, the probability of a T of three equals 0.0390 (two-tail), and the probability of a T of four equals 0.0546 (two-tail). Wilcoxon and Wilcox select a T of four as the critical value at the 0.05 level of significance (two-tail), whereas Table A-9 reflects a more conservative policy by setting a T of three as the critical value at this level.)

Source: From Frank Wilcoxon and Roberta A. Wilcox, *Some Rapid Approximate Statistical Procedures*. Revised 1964 by Lederle Laboratories, Pearl River, NY. Reproduced with the permission of the American Cyanamid Company.

TABLE A-10
Table of exponential functions

x	e^x	e^{-x}	x	e^x	e^{-x}
0.00	1.000	1.000	3.00	20.086	0.050
0.10	1.105	0.905	3.10	22.198	0.045
0.20	1.221	0.819	3.20	24.533	0.041
0.30	1.350	0.741	3.30	27.113	0.037
0.40	1.492	0.670	3.40	29.964	0.033
0.50	1.649	0.607	3.50	33.115	0.030
0.60	1.822	0.549	3.60	36.598	0.027
0.70	2.014	0.497	3.70	40.447	0.025
0.80	2.226	0.449	3.80	44.701	0.022
0.90	2.460	0.407	3.90	49.402	0.020
1.00	2.718	0.368	4.00	54.598	0.018
1.10	3.004	0.333	4.10	60.340	0.017
1.20	3.320	0.301	4.20	66.686	0.015
1.30	3.669	0.273	4.30	73.700	0.014
1.40	4.055	0.247	4.40	81.451	0.012
1.50	4.482	0.223	4.50	90.017	0.011
1.60	4.953	0.202	4.60	99.484	0.010
1.70	5.474	0.183	4.70	109.95	0.009
1.80	6.050	0.165	4.80	121.51	0.008
1.90	6.686	0.150	4.90	134.29	0.007
2.00	7.389	0.135	5.00	148.41	0.007
2.10	8.166	0.122	5.10	164.02	0.006
2.20	9.025	0.111	5.20	181.27	0.006
2.30	9.974	0.100	5.30	200.34	0.005
2.40	11.023	0.091	5.40	221.41	0.005
2.50	12.182	0.082	5.50	244.69	0.004
2.60	13.464	0.074	5.60	270.43	0.004
2.70	14.880	0.067	5.70	298.87	0.003
2.80	16.445	0.061	5.80	330.30	0.003
2.90	18.174	0.055	5.90	365.04	0.003
3.00	20.086	0.050	6.00	403.43	0.002

TABLE A-11

Values of d_L and d_U for the Durbin–Watson Test for $\alpha = 0.05$

n = number of observations
k = number of independent variables

	$k = 1$		$k = 2$		$k = 3$		$k = 4$		$k = 5$	
n	d_L	d_U	d_L	d_U	d_L	d_U	d_L	d_U	d_L	d_U
15	1.08	1.36	0.95	1.54	0.82	1.75	0.69	1.97	0.56	2.21
16	1.10	1.37	0.98	1.54	0.86	1.73	0.74	1.93	0.62	2.15
17	1.13	1.38	1.02	1.54	0.90	1.71	0.78	1.90	0.67	2.10
18	1.16	1.39	1.05	1.53	0.93	1.69	0.82	1.87	0.71	2.06
19	1.18	1.40	1.08	1.53	0.97	1.68	0.86	1.85	0.75	2.02
20	1.20	1.41	1.10	1.54	1.00	1.68	0.90	1.83	0.79	1.99
21	1.22	1.42	1.13	1.54	1.03	1.67	0.93	1.81	0.83	1.96
22	1.24	1.43	1.15	1.54	1.05	1.66	0.96	1.80	0.86	1.94
23	1.26	1.44	1.17	1.54	1.08	1.66	0.99	1.79	0.90	1.92
24	1.27	1.45	1.19	1.55	1.10	1.66	1.01	1.78	0.93	1.90
25	1.29	1.45	1.21	1.55	1.12	1.66	1.04	1.77	0.95	1.89
26	1.30	1.46	1.22	1.55	1.14	1.65	1.06	1.76	0.98	1.88
27	1.32	1.47	1.24	1.56	1.16	1.65	1.08	1.76	1.01	1.86
28	1.33	1.48	1.26	1.56	1.18	1.65	1.10	1.75	1.03	1.85
29	1.34	1.48	1.27	1.56	1.20	1.65	1.12	1.74	1.05	1.84
30	1.35	1.49	1.28	1.57	1.21	1.65	1.14	1.74	1.07	1.83
31	1.36	1.50	1.30	1.57	1.23	1.65	1.16	1.74	1.09	1.83
32	1.37	1.50	1.31	1.57	1.24	1.65	1.18	1.73	1.11	1.82
33	1.38	1.51	1.32	1.58	1.26	1.65	1.19	1.73	1.13	1.81
34	1.39	1.51	1.33	1.58	1.27	1.65	1.21	1.73	1.15	1.81
35	1.40	1.52	1.34	1.58	1.28	1.65	1.22	1.73	1.16	1.80
36	1.41	1.52	1.35	1.59	1.29	1.65	1.24	1.73	1.18	1.80
37	1.42	1.53	1.36	1.59	1.31	1.66	1.25	1.72	1.19	1.80
38	1.43	1.54	1.37	1.59	1.32	1.66	1.26	1.72	1.21	1.79
39	1.43	1.54	1.38	1.60	1.33	1.66	1.27	1.72	1.22	1.79
40	1.44	1.54	1.39	1.60	1.34	1.66	1.29	1.72	1.23	1.79
45	1.48	1.57	1.43	1.62	1.38	1.67	1.34	1.72	1.29	1.78
50	1.50	1.59	1.46	1.63	1.42	1.67	1.38	1.72	1.34	1.77
55	1.53	1.60	1.49	1.64	1.45	1.68	1.41	1.72	1.38	1.77
60	1.55	1.62	1.51	1.65	1.48	1.69	1.44	1.73	1.41	1.77
65	1.57	1.63	1.54	1.66	1.50	1.70	1.47	1.73	1.44	1.77
70	1.58	1.64	1.55	1.67	1.52	1.70	1.49	1.74	1.46	1.77
75	1.60	1.65	1.57	1.68	1.54	1.71	1.51	1.74	1.49	1.77
80	1.61	1.66	1.59	1.69	1.56	1.72	1.53	1.74	1.51	1.77
85	1.62	1.67	1.60	1.70	1.57	1.72	1.55	1.75	1.52	1.77
90	1.63	1.68	1.61	1.70	1.59	1.73	1.57	1.75	1.54	1.78
95	1.64	1.69	1.62	1.71	1.60	1.73	1.58	1.75	1.56	1.78
100	1.65	1.69	1.63	1.72	1.61	1.74	1.59	1.76	1.57	1.78

Source: From J. Durbin and G.S. Watson, "Testing for Serial Correlation in Least Squares Regression," *Biometrika*, 38 June 1951. Reproduced by permission of the Biometrika Trustees.

TABLE A-11 (continued)
Values of d_L and d_U for the Durbin–Watson Test for $\alpha = 0.01$

	$k = 1$		$k = 2$		$k = 3$		$k = 4$		$k = 5$	
n	d_L	d_U	d_L	d_U	d_L	d_U	d_L	d_U	d_L	d_U
15	0.81	1.07	0.70	1.25	0.59	1.46	0.49	1.70	0.39	1.96
16	0.84	1.09	0.74	1.25	0.63	1.44	0.53	1.66	0.44	1.90
17	0.87	1.10	0.77	1.25	0.67	1.43	0.57	1.63	0.48	1.85
18	0.90	1.12	0.80	1.26	0.71	1.42	0.61	1.60	0.52	1.80
19	0.93	1.13	0.83	1.26	0.74	1.41	0.65	1.58	0.56	1.77
20	0.95	1.15	0.86	1.27	0.77	1.41	0.68	1.57	0.60	1.74
21	0.97	1.16	0.89	1.27	0.80	1.41	0.72	1.55	0.63	1.71
22	1.00	1.17	0.91	1.28	0.83	1.40	0.75	1.54	0.66	1.69
23	1.02	1.19	0.94	1.29	0.86	1.40	0.77	1.53	0.70	1.67
24	1.04	1.20	0.96	1.30	0.88	1.41	0.80	1.53	0.72	1.66
25	1.05	1.21	0.98	1.30	0.90	1.41	0.83	1.52	0.75	1.65
26	1.07	1.22	1.00	1.31	0.93	1.41	0.85	1.52	0.78	1.64
27	1.09	1.23	1.02	1.32	0.95	1.41	0.88	1.51	0.81	1.63
28	1.10	1.24	1.04	1.32	0.97	1.41	0.90	1.51	0.83	1.62
29	1.12	1.25	1.05	1.33	0.99	1.42	0.92	1.51	0.85	1.61
30	1.13	1.26	1.07	1.34	1.01	1.42	0.94	1.51	0.88	1.61
31	1.15	1.27	1.08	1.34	1.02	1.42	0.96	1.51	0.90	1.60
32	1.16	1.28	1.10	1.35	1.04	1.43	0.98	1.51	0.92	1.60
33	1.17	1.29	1.11	1.36	1.05	1.43	1.00	1.51	0.94	1.59
34	1.18	1.30	1.13	1.36	1.07	1.43	1.01	1.51	0.95	1.59
35	1.19	1.31	1.14	1.37	1.08	1.44	1.03	1.51	0.97	1.59
36	1.21	1.32	1.15	1.38	1.10	1.44	1.04	1.51	0.99	1.59
37	1.22	1.32	1.16	1.38	1.11	1.45	1.06	1.51	1.00	1.59
38	1.23	1.33	1.18	1.39	1.12	1.45	1.07	1.52	1.02	1.58
39	1.24	1.34	1.19	1.39	1.14	1.45	1.09	1.52	1.03	1.58
40	1.25	1.34	1.20	1.40	1.15	1.46	1.10	1.52	1.05	1.58
45	1.29	1.38	1.24	1.42	1.20	1.48	1.16	1.53	1.11	1.58
50	1.32	1.40	1.28	1.45	1.24	1.49	1.20	1.54	1.16	1.59
55	1.36	1.43	1.32	1.47	1.28	1.51	1.25	1.55	1.21	1.59
60	1.38	1.45	1.35	1.48	1.32	1.52	1.28	1.56	1.25	1.60
65	1.41	1.47	1.38	1.50	1.35	1.53	1.31	1.57	1.28	1.61
70	1.43	1.49	1.40	1.52	1.37	1.55	1.34	1.58	1.31	1.61
75	1.45	1.50	1.42	1.53	1.39	1.56	1.37	1.59	1.34	1.62
80	1.47	1.52	1.44	1.54	1.42	1.57	1.39	1.60	1.36	1.62
85	1.48	1.53	1.46	1.55	1.43	1.58	1.41	1.60	1.39	1.63
90	1.50	1.54	1.47	1.56	1.45	1.59	1.43	1.61	1.41	1.64
95	1.51	1.55	1.49	1.57	1.47	1.60	1.45	1.62	1.42	1.64
100	1.52	1.56	1.50	1.58	1.48	1.60	1.46	1.63	1.44	1.65

Appendix B

Symbols, Subscripts, and Summations

Symbols

In statistics, **symbols** such as X, Y, and Z are used to represent different sets of data. Hence, if we have data for five families, we might let

$$X = \text{family income}$$

$$Y = \text{family clothing expenditures}$$

$$Z = \text{family savings}$$

Subscripts

Subscripts are used to represent individual observations within these sets of data. Thus, X_i represents the income of the ith family, where i takes on the values 1, 2, 3, 4, and 5. In this notation X_1, X_2, X_3, X_4, and X_5 stand for the incomes of the first family, the second family, and so on. The data are arranged in some order, such as by size of income, the order in which the data were gathered, or any other way suitable to the purposes or convenience of the investigator. The subscript i is a variable used to index the individual data observations. Therefore, X_i, Y_i, and Z_i represent the income, clothing expenditures, and savings of the ith family. For example, X_2 represents the income of the second family, Y_2 clothing expenditures of the second (same) family, and Z_5 the savings of the fifth family.

Now, let us suppose that we have data for two different samples, say the net worths of 100 corporations and the test scores of 20 students. To refer to individual observations in these samples, we can let X_i denote the net worth of the ith corporation, where i assumes values from 1 to 100. (This latter idea is indicated by the notation $i = 1, 2, 3, \ldots, 100$.) We can also let Y_j denote the test score of the jth student, where $j = 1, 2, 3, \ldots, 20$. The different subscript letters make it clear that different samples are involved. Letters such as X, Y, and Z generally represent the different variables or types of measurements involved, whereas subscripts such as i, j, k, and l designate individual observations.

Summations

We now turn to the method of expressing **summations** of sets of data. Suppose we want to add a set of four observations, denoted X_1, X_2, X_3, and X_4. A convenient way of designating this addition is

$$\sum_{i=1}^{4} X_i = X_1 + X_2 + X_3 + X_4$$

where the symbol Σ (Greek capital "sigma") means "the sum of." Hence, the symbol

$$\sum_{i=1}^{4} X_i$$

is read "the sum of the X_i's, i going from 1 to 4." For example, if $X_1 = 3$, $X_2 = 1$, $X_3 = 10$, and $X_4 = 5$,

$$\sum_{i=1}^{4} X_i = 3 + 1 + 10 + 5 = 19$$

In general, if there are n observations, we write

$$\sum_{i=1}^{n} X_i = X_1 + X_2 + \cdots + X_n$$

EXAMPLE B-1

Let $X_1 = -2$, $X_2 = 3$, and $X_3 = 5$. Find

a. $\sum_{i=1}^{3} X_i$ b. $\sum_{j=1}^{3} X_j^2$ c. $\sum_{j=1}^{3} (2X_j + 3)$

Solutions

a. $\sum_{i=1}^{3} X_i = X_1 + X_2 + X_3 = -2 + 3 + 5 = 6$

b. $\sum_{j=1}^{3} X_j^2 = X_1^2 + X_2^2 + X_3^2 = (-2)^2 + (3)^2 + (5)^2 = 38$

c. $\sum_{j=1}^{3} (2X_j + 3) = (2X_1 + 3) + (2X_2 + 3) + (2X_3 + 3)$

$= (-4 + 3) + (6 + 3) + (10 + 3) = -1 + 9 + 13 = 21$

EXAMPLE B-2

Prove

a. $\sum_{i=1}^{n} aX_i = a \sum_{i=1}^{n} X_i$ b. $\sum_{i=1}^{n} a = na$ c. $\sum_{i=1}^{n} (X_i + Y_i) = \sum_{i=1}^{n} X_i + \sum_{i=1}^{n} Y_i$

where a is a constant.

Solutions

a. $\sum_{i=1}^{n} aX_i = aX_1 + aX_2 + \cdots + aX_n = a(X_1 + X_2 + \cdots + X_n) = a \sum_{i=1}^{n} X_i$

b. $\sum_{i=1}^{n} a = a \sum_{i=1}^{n} 1 = a\underbrace{(1 + 1 + \cdots + 1)}_{n \text{ terms}} = na$

c. $\sum_{i=1}^{n} (X_i + Y_i) = X_1 + Y_1 + X_2 + Y_2 + \cdots + X_n + Y_n$

$$= (X_1 + X_2 + \cdots + X_n) + (Y_1 + Y_2 + \cdots + Y_n) = \sum_{i=1}^{n} X_i + \sum_{i=1}^{n} Y_i$$

These three summation properties are listed as rules 1, 2, and 3 at the end of this appendix.

Double summations

Double summations are used to indicate summations of more than one variable, where different subscript indexes are involved. For example, the symbol

$$\sum_{j=1}^{3} \sum_{i=1}^{2} X_i Y_j$$

means "the sum of the products of X_i and Y_j where $i = 1, 2$ and $j = 1, 2, 3$." Thus, we can write

$$\sum_{j=1}^{3} \sum_{i=1}^{2} X_i Y_j = X_1 Y_1 + X_2 Y_1 + X_1 Y_2 + X_2 Y_2 + X_1 Y_3 + X_2 Y_3$$

Simplified Summation Notations

In this text, simplified summation notations are often used in which subscripts are eliminated. For example, ΣX, ΣX^2, and ΣY^2 are used instead of

$$\sum_{i=1}^{n} X_i \qquad \sum_{i=1}^{n} X_i^2 \qquad \text{and} \qquad \sum_{i=1}^{n} Y_i^2$$

Also in this text, subscripts have ordinarily been dropped in the case of probability distributions. The statement that the sum of the probabilities is equal to one is given by

$$\sum_{i=1}^{3} f(x_i) = 1$$

In this textbook, we use the customary simplified notation, as shown in these two examples of a discrete probability distribution.

Standard Notation		Simplified Notation	
x_i	$f(x_i)$	x	$f(x)$
$x_1 = 0$	0.2	0	0.2
$x_2 = 1$	0.3	1	0.3
$x_3 = 2$	0.5	2	0.5
	1.0		1.0

The corresponding summation statement is $\sum_x f(x) = 1$ where $\sum_x$ means "sum over all values of x." The notation is also further simplified by writing

$$\sum f(x) = 1$$

Summation Properties

Rule 1

$$\sum_{i=1}^{n} aX_i = a \sum_{i=1}^{n} X_i$$

Rule 2

$$\sum_{i=1}^{n} a = \underbrace{a + a + \cdots + a}_{n \text{ terms}} = na$$

Rule 3

$$\sum_{i=1}^{n} (X_i + Y_i) = \sum_{i=1}^{n} X_i + \sum_{i=1}^{n} Y_i$$

Appendix C

Properties of Expected Values and Variances

In keeping with notational conventions used in this text, a and b represent constants, whereas X represents a random variable. The symbols $E(X)$ and $\sigma^2(X)$ denote the expected value and variance of the random variable X, and the symbols $E(X_1 + X_2 + \cdots + X_n)$ and $\sigma^2(X_1 + X_2 + \cdots + X_n)$ denote the expected value and variance of the sum of the random variables $X_1, X_2, \ldots, X_n$, and so forth.

Rule 1 declares that the expected value of a constant is equal to that constant.

$$E(a) = a$$

Rule 1

Rule 2 states that the expected value of a constant times a random variable is equal to the constant times the expected value of the random variable.

$$E(bX) = bE(X)$$

Rule 2

Rule 3 combines rules 1 and 2.

$$E(a + bX) = a + bE(X)$$

Rule 3

A brief proof for rule 3 illustrates a general method of proofs for expected values.

Let X denote a discrete random variable that takes on values $x_1, x_2, \ldots, x_i, \ldots, x_n$ with probabilities $f(x_1), f(x_2), \ldots, f(x_i), \ldots, f(x_n)$. Then, using the definition of an expected value given in equation 3.11 in Chapter 3, we have

$$E(a + bX) = \sum_{i=1}^{n} (a + bx_i)f(x_i) = \sum_{i=1}^{n} af(x_i) + \sum_{i=1}^{n} bx_i f(x_i)$$

$$= a \sum_{i=1}^{n} f(x_i) + b \sum_{i=1}^{n} x_i f(x_i)$$

$$= a(1) + bE(X) = a + bE(X)$$

Rule 4 says the expected value of a sum equals the sum of the expected values.

Rule 4

$$E(X_1 + X_2 + \cdots + X_n) = E(X_1) + E(X_2) + \cdots + E(X_n)$$

where $X_1, X_2, \ldots, X_n$ are random variables. The X_i's are not restricted in any way; that is, they may be either independent or dependent.

Expressing this rule in somewhat different symbols, we have

$$E\left[\sum_{i=1}^{n} (X_i)\right] = \sum_{i=1}^{n} [E(X_i)]$$

Treating the expected value and summation symbols as operators (that is, as defining specific operations on the X_i's), we have the result that the summation sign and expected value symbol are interchangeable operators.

Rule 5 states that the variance of a constant is equal to zero.

Rule 5

$$\sigma^2(a) = 0$$

Rule 6 says the variance of a constant times a random variable is equal to the constant squared times the variance of the random variable.

Rule 6

$$\sigma^2(bX) = b^2\sigma^2(X)$$

Rule 7 combines rules 5 and 6.

Rule 7

$$\sigma^2(a + bX) = b^2\sigma^2(X)$$

As in the case of rule 3 for the expected value, a simple application of the definition of a variance yields the desired result. The proof is left to the reader as an exercise.

Rule 8 states that for n independent random variables, the variance of a sum is equal to the sum of the variances.

Rule 8

$$\sigma^2(X_1 + X_2 + \cdots + X_n) = \sigma^2(X_1) + \sigma^2(X_2) + \cdots + \sigma^2(X_n)$$

where $X_1, X_2, \ldots, X_n$ are independent random variables; that is, every pair of X_i's is independent.

Thus, if the X_i's are independent, the variance of a sum is equal to the sum of the variances.

Expressing rule 8 in summation terminology, we obtain

$$\sigma^2\left[\sum_{i=1}^{n}(X_i)\right] = \sum_{i=1}^{n}[\sigma^2(X_i)]$$

The variance and summation symbols are interchangeable operators if the X_i's are independent.

Rule 9 is derived by applying rules 7 and 8.

Rule 9

$$\sigma^2(a_1X_1 + a_2X_2) = a_1^2\sigma^2(X_1) + a_2^2\sigma^2(X_2)$$

if X_1 and X_2 are independent.

Special cases of rule 9 are given as rules 10 and 11. In **rule 10**, $a_1 = +1$ and $a_2 = +1$.

Rule 10

$$\sigma^2(X_1 + X_2) = \sigma^2(X_1) + \sigma^2(X_2)$$

if X_1 and X_2 are independent. In **rule 11**, $a_1 = +1$ but $a_2 = -1$.

Rule 11

$$\sigma^2(X_1 - X_2) = \sigma^2(X_1) + \sigma^2(X_2)$$

if X_1 and X_2 are independent.

Rule 12 says that the variance of a sample mean is equal to the population variance divided by the sample size.

Rule 12

$$\sigma^2(\bar{X}) = \frac{\sigma^2}{n}$$

where X is a random variable, μ and σ are its mean and standard deviation, respectively, and $\bar{X}$ is the arithmetic mean in a sample of n independent observations of X. If $X_1, X_2, \ldots, X_n$ denote the n observations, then

$$\bar{X} = \frac{1}{n}\sum_{i=1}^{n} X_i$$

Proof: Rule 12 may be proved in a few steps.

1. $$\sigma^2(\bar{X}) = \sigma^2\left(\frac{1}{n}\sum_{i=1}^{n} X_i\right) = \frac{1}{n^2}\sigma^2\left(\sum_{i=1}^{n} X_i\right) \qquad \text{(by rule 6)}$$

2. $$= \frac{1}{n^2}\sum_{i=1}^{n}[\sigma^2(X_i)] \qquad \text{(by rule 8)}$$

But since every X_i has the same probability distribution as X, then $\sigma^2(X_i) = \sigma^2(X)$ for each i. Hence,

3. $$\sum_{i=1}^{n}[\sigma^2(X_i)] = n\sigma^2$$

Substituting step 3 into step 2 gives

4. $$\sigma^2(\bar{X}) = \left(\frac{1}{n^2}\right)(n\sigma^2) = \frac{\sigma^2}{n}$$

which completes the proof.

Let us express rule 12 in the language and symbolism of sampling theory. If a simple random sample of size n is drawn from an infinite population (or a finite population with replacement) with standard deviation σ, the variance of the sample mean is given by

5. $$\sigma_{\bar{X}}^2 = \frac{\sigma^2}{n}$$

Rule 13 states that the expected value of the sample mean is equal to the population mean where the same conditions prevail as in rule 12; X is a random variable and μ and σ are its mean and standard deviation.

Rule 13

$$E(\bar{X}) = \mu$$

Proof: Rule 13 is easily proved as follows:

1. $$E(\bar{X}) = E\left(\frac{1}{n}\sum_{i=1}^{n} X_i\right)$$

$$= \frac{1}{n}\left[E\left(\sum_{i=1}^{n} X_i\right)\right] \qquad \text{(by rule 2)}$$

2. $$= \frac{1}{n}\left[\sum_{i=1}^{n} E(X_i)\right] \qquad \text{(by rule 4)}$$

But since every X_i has the same probability distribution as X, then $E(X_i) = E(X)$ for each i. Hence,

3. $$\sum_{i=1}^{n} E(X_i) = nE(X) = n\mu$$

Substituting step 3 into step 2 gives

4. $$E(\bar{X}) = \left(\frac{1}{n}\right)(n\mu) = \mu$$

As in rule 12, **let us express this result in terms of sampling theory.** If a simple random sample of size n is drawn from an infinite population (or a finite population with replacement) with mean μ, the expected value (arithmetic mean) of the sample mean is given by

5. $$E(\bar{X}) = \mu_{\bar{x}} = \mu$$

The expected value of the sample variance defined with divisor $n - 1$ is equal to the population variance.

This result is expressed in symbols in **rule 14.**

Rule 14

$$E(s^2) = E\left[\frac{\Sigma(X_i - \bar{X})^2}{n-1}\right] = \sigma^2$$

where the same conditions prevail as in rules 12 and 13.

Appendix D

Shortcut Formulas to Use When Class Intervals Are Equal in Size

Shortcut Calculation of the Mean

A shortcut calculation known as the step-deviation method is useful when class intervals in a frequency distribution are of equal size. The method results in simpler arithmetic than the direct definitional formula, particularly if the class intervals and frequencies involve a large number of digits.

The **step-deviation method** of computing the mean involves three basic steps:

Step-deviation method

1. Selection of an assumed (or arbitrary) mean.
2. Calculation of an average deviation from this assumed mean.
3. Addition of this average deviation as a correction factor to the assumed mean to obtain the true mean. This correction factor is positive if the assumed mean lies below the true mean, negative if above.

To accomplish step 2, a midpoint of a class (preferably near the center of the distribution) is taken as the assumed mean. Then deviations of the midpoints of the other classes are taken from the assumed mean in class interval units. These deviations are denoted d. After the d values are averaged, the result is multiplied by the size of the class interval to return to the units of the original data.

TABLE D-1
Calculation of the arithmetic mean for grouped data by the step-deviation method for bottle weights data

Weight (in ounces)	Number of bottles f	d	fd
14.0 and under 14.5	8	−4	−32
14.5 and under 15.0	10	−3	−30
15.0 and under 15.5	15	−2	−30
15.5 and under 16.0	18	−1	−18
16.0 and under 16.5	22	0	0
16.5 and under 17.0	14	1	14
17.0 and under 17.5	8	2	16
17.5 and under 18.0	5	3	15
	100		−65

$\bar{X}_a = 16.25$

$$\bar{X} = \bar{X}_a + \left(\frac{\Sigma fd}{n}\right)(i) = 16.25 + \left(\frac{-65}{100}\right)(0.5) = 15.93 \text{ ounces}$$

The formula for the step-deviation method is given in equation D.1.

Step-deviation method for the arithmetic mean (grouped data)

$$\bar{X} = \bar{X}_a + \left(\frac{\Sigma fd}{n}\right)(i) \tag{D.1}$$

where $\bar{X}$ = the arithmetic mean
$\bar{X}_a$ = the assumed arithmetic mean
f = frequencies
d = deviations of midpoints from the assumed mean in class interval units
n = the number of observations
i = the size of a class interval

The step-deviation method is illustrated in Table D-1 for the same frequency distribution of bottle weights given in Table 1-6 on page 24, where the arithmetic mean was computed by the direct definitional method. In Table D-1, the assumed arithmetic mean is 16.25. The values in the d column indicate the number of class intervals below or above the one in which the assumed mean is taken. Note that the mean calculated in Table D-1 is the same as that calculated in Table 1-6.

Shortcut Calculation of the Standard Deviation

Just as in the case of the arithmetic mean, the step-deviation method of calculating the standard deviation is useful when class intervals in a frequency distri-

TABLE D-2
Calculation of the standard deviation for grouped data by the step-deviation method for bottle weights data

Weight (in ounces)	Number of Bottles f	d	fd	fd^2
14.0 and under 14.5	8	−4	−32	128
14.5 and under 15.0	10	−3	−30	90
15.0 and under 15.5	15	−2	−30	60
15.5 and under 16.0	18	−1	−18	18
16.0 and under 16.5	22	0	0	0
16.5 and under 17.0	14	1	14	14
17.0 and under 17.5	8	3	15	45
17.5 and under 18.0	5	3	15	45
	100		−65	387

$\bar{X} = 16.25$

$$s = (i)\sqrt{\frac{\Sigma fd^2 - \dfrac{(\Sigma fd)^2}{n}}{n-1}} = (0.5)\sqrt{\frac{387 - \dfrac{(-65)^2}{100}}{100-1}} = (0.5)(1.866)$$

$$= 0.933 \text{ ounce}$$

bution are the same size. The saving in computational effort accomplished by the use of the step-deviation method is illustrated for the bottle weights data given in Table D-1.

As with the calculation for the arithmetic mean, the procedure involves taking deviations of midpoints of classes from an assumed mean and stating them in class interval units. Only one additional column of values, fd^2, is required to compute the standard deviation by the step-deviation method compared with the corresponding arithmetic mean computation given in Table D-1. The formula for the step-deviation method is given in equation D.2. All of the symbols have the same meaning as in equation D.1 for the arithmetic mean and the computation in Table D-1. The standard deviation is equal to 0.933 ounce, the same value that was obtained by the direct definitional method in Table 1-14 on page 44.

$$s = (i)\sqrt{\frac{\Sigma fd^2 - \dfrac{(\Sigma fd)^2}{n}}{n-1}} \qquad \text{(D.2)}$$

Step-deviation method for the sample standard deviation (grouped data)

The use of equation D.2 is illustrated in Table D-2. The same assumed mean, $\bar{X}_a = 16.25$ ounces, was used in Table D-2 as in the calculation of the arithmetic mean given in Table D-1.

Appendix E

Data Bank for Computer Exercises

These data are the base for the computer exercises in sections 1.18, 6.5, 7.2, and 10.13, dealing with descriptive statistics, confidence interval estimation, hypothesis testing, and multiple regression analysis.

Family	Annual Food Expenditures ($000)	Annual Income ($000)	Family Size	Age of the Highest Income Earner (in years)	Home Owner (0) or Renter (1)
1	4.7	24	3	32	1
2	5.2	29	3	28	1
3	6.1	30	2	25	0
4	4.8	23	1	43	1
5	10.1	52	4	50	0
6	9.2	61	2	55	0
7	6.5	33	3	32	0
8	5.4	28	2	28	1
9	7.8	41	1	37	0
10	9.8	53	6	54	0
11	4.9	42	3	30	1
12	7.3	44	4	31	0
13	5.2	26	1	28	1
14	3.2	12	5	48	0
15	3.4	18	3	42	0
16	7.2	47	1	32	1
17	15.6	112	6	60	0
18	13.7	85	5	47	0
19	5.1	27	2	33	0
20	2.9	13	2	29	1
21	3.8	19	1	26	1

Family	Annual Food Expenditures ($000)	Annual Income ($000)	Family Size	Age of the Highes Income Earner (in years)	Home Owner (0) or Renter (1)
22	7.2	38	1	45	1
23	4.9	25	4	43	1
24	10.2	62	3	30	0
25	10.0	54	4	55	0
26	4.8	28	3	33	1
27	4.7	29	2	29	1
28	5.3	34	1	26	0
29	4.4	30	1	25	1
30	4.3	25	1	31	1
31	10.3	57	6	48	0
32	7.6	45	4	55	0
33	7.3	47	3	31	1
34	5.1	36	1	32	1
35	3.3	19	4	29	1
36	4.6	28	4	49	0
37	2.8	14	2	43	1
38	3.0	20	5	33	1
39	8.0	49	3	35	0
40	13.8	87	3	63	0
41	12.4	72	2	34	0
42	2.5	12	1	23	1
43	4.3	28	2	27	1
44	3.1	14	1	25	1
45	3.1	19	1	28	1
46	7.7	39	4	30	0
47	4.2	27	2	51	0
48	10.1	64	5	45	1
49	9.6	53	5	47	0
50	4.7	27	3	28	1
51	5.5	28	3	29	1
52	6.1	33	4	32	0
53	5.4	29	1	25	1
54	4.8	24	1	27	1
55	9.8	55	7	46	0
56	6.9	43	5	48	0
57	8.0	45	4	52	0
58	5.8	34	3	36	1
59	2.9	17	1	29	1
60	5.1	26	2	32	0
61	3.2	15	1	24	1
62	4.1	21	1	28	1
63	7.5	50	2	42	0
64	13.1	78	3	58	0

Family	Annual Food Expenditures ($000)	Annual Income ($000)	Family Size	Age of the Highest Income Earner (in years)	Home Owner (0) or Renter (1)
65	5.5	27	1	68	0
66	5.1	31	2	33	1
67	12.5	73	2	43	0
68	4.5	29	3	38	1
69	3.2	20	1	31	1
70	7.5	38	4	35	0
71	9.7	51	5	51	0
72	5.3	33	3	29	1
73	10.2	53	4	52	0
74	4.8	43	3	30	1
75	7.1	49	1	33	1

Glossary of Symbols

Numbers in parentheses refer to the section in which the symbol is introduced.

General Mathematical Symbols	Meaning
$\|d\|$	Absolute value of d (13.3)
e	A constant equal to 2.71828 . . . ; the base of the Naperian, or natural, logarithm system (3.7)
$\log X$	Logarithm of X to the base 10 (9.8)
$n!$	n factorial, or $(n)(n-1)\cdots(2)(1)$ (2.4)
π	A constant equal to 3.1416 . . . (5.4)
Σ	Sum of (see Appendix B) (1.9)
$\sum_{i=1}^{n}$	Sum of the terms that follow, from $i = 1$ to $i = n$ (see Appendix B) (1.9)
$\sum_{x}$ or $\sum_x$	Sum of the terms that follow, for all values x takes on (3.1)
$\sum_{j}\sum_{i}$	Summation of terms that follow, first over all values of i, then over all values of j (8.3)
$x \leqslant a$	x is less than or equal to a (3.1)
$x \geqslant a$	x is greater than or equal to a (3.1)

Probability and Statistical Symbols	Meaning
A	Y intercept in a two-variable linear regression model. The value of a in a sample regression equation is an estimator of A (9.1)
$\bar{A}$	Complement of event A (9.1)
$A_1, A_2, \ldots, A_n$	Alternative acts (14.2)
A_t	Actual figure for time period t in exponential smoothing (11.6)
a	Y intercept calculated from a sample of observations; computed value of Y when $X = 0$ in a two-variable regression equation; computed value of Y when values of all independent variables are 0 in a multiple regression equation (9.4)
a, b	Constants in a straight-line trend equation (11.4)
a, b, c	Constants in a parabolic trend equation (11.4)
α	Probability of a Type I error, or of rejecting H_0 when it is true; the significance level (7.2)

Symbol	Meaning
B	Regression coefficient in a two-variable linear regression model. The value of b in a sample regression equation is an estimator of B (9.1)
b	Regression coefficient calculated from a sample of observations; slope of the regression line in a sample two-variable regression equation (9.4)
b_1, b_2	Sample net regression coefficients; the coefficients of independent variables X_1 and X_2 (10.1)
β	Probability of a Type II error, or of accepting H_0 when it is false (7.2)
β_1, β_2	Beta coefficients in a multiple regression equation; β_1 measures the number of standard deviations of change in $\hat{Y}$ for each change of one standard deviation in X_1, when X_2 is held constant (10.11)
C	Correction term in shortcut computation of an analysis of variance (8.3)
C	Effect of the cyclical factor in time-series analysis (11.4)
$\chi^2 = \sum \frac{(f_o - f_t)^2}{f_t}$	χ^2 statistic in a test of independence or goodness of fit (8.1)
CV	Coefficient of variation (1.16)
c	Number of columns in an arrangement of data to which analysis of variance is applied (8.2)
D	Tolerated sampling error in a determination of sample size (6.5)
d	Durbin–Watson statistic; a statistic used in tests for autocorrelation in time series (10.9)
$\bar{d}$	Mean difference of pairs of observations made on the same individuals or objects (7.6)
$d = X - Y$	Difference between ranks of paired observations in rank correlation (13.7)
ENGS	Expected net gain of sample information (16.2)
EOL(A_j)	Expected opportunity loss of act A_j (14.4)
EVPI	Expected value of perfect information (14.5)
EVSI	Expected value of sample information (16.2)
$E(X)$	Expected value of the random variable X, that is, the expected value of the probability distribution of X (3.9)
F	F ratio; the ratio of the between-treatment variance to the within-treatment variance (8.3)
F	F ratio; the ratio of explained variance to unexplained variance (10.6)
$F(\nu_1, \nu_2)$	F ratio with ν_1 and ν_2 degrees of freedom for the numerator and denominator, respectively (8.3)
f	Number of observations (frequency) in a class interval of a frequency distribution (1.9)
$f = \frac{n}{N}$	Sampling fraction (5.5)
f_o	Observed frequency in a χ^2 test (8.1)
f_t	Theoretical, or expected, frequency in χ^2 test (8.1)
F_t	Forecast for time period t in exponential smoothing (11.6)
$F(x) = P(X \leqslant x)$	Cumulative probability that random variable X is less than or equal to x (3.1)
f_{Md}	Frequency of the class containing the median (1.11)
f_p	Frequencies in classes preceding the one containing the median (1.11)
$f(x) = P(X = x)$	Probability that random variable X is equal to the value x (3.1)
$f(Y\|X)$	Conditional probability distribution of Y given X in the linear regression model (9.1)
H	Value of the highest observation (1.2)
H_0	Null hypothesis; basic hypothesis being tested (7.1)
H_1	Alternative hypothesis; rejection of H_0 implies tentative acceptance of H_1 (7.1)
I	Effect of the irregular factor in time-series analysis (11.4)
i	Size of the class interval (1.2)
K	Kruskal–Wallis test statistic in a one-factor analysis of variance by ranks (13.6)
k	Number of classes in a frequency distribution (1.2)
L	Value of the lowest observation (1.2)
$L(a_1\|\theta_1)$	Opportunity loss of act a_1 given state of nature θ_1 (17.1)

Symbol	Meaning
L_{Md}	Lower limit of class containing the median (1.11)
$L(\hat{\theta}\|\theta)$	Loss involved in estimating $\hat{\theta}$ when the parameter value is θ (17.3)
MA	Moving average figures in seasonal variation analysis (11.5)
Md	Median (1.11)
m	Midpoint of a class interval of a frequency distribution (1.9)
μ	Arithmetic mean of a population (1.9)
μ	Arithmetic mean of a probability distribution (3.7)
$\mu_{n\bar{p}}$	Mean of sampling distribution of number of occurrences (5.2)
$\mu_{\bar{p}}$	Mean of sampling distribution of a proportion (5.2)
$\mu_{\bar{p}_1 - \bar{p}_2}$	Mean of sampling distribution of the difference between two proportions (6.3)
μ_r	Mean of sampling distribution of the number of runs r (13.5)
μ_T	Mean of sampling distribution of T in the Wilcoxon matched-pairs signed rank test (13.3)
μ_U	Mean of sampling distribution of U in the Mann–Whitney U test (rank sum test) (13.4)
$\mu_{\bar{x}_1 - \bar{x}_2}$	Mean of sampling distribution of the difference between two sample means (6.3)
$\mu_{Y.X}$	Mean of conditional probability distribution of Y given X in the linear regression model; the population value that corresponds to the $\hat{Y}$ value computed from sample observations (9.1)
N	Number of observations in a population (1.9)
n	Number of observations in a sample (1.9)
$\binom{n}{x}$	Number of combinations of n objects taken x at a time (2.4)
${}_nP_x$	Number of permutations of n objects taken x at a time (2.4)
ν	*nu*: Number of degrees of freedom (6.4)
$P(A)$	Probability of event A (2.1)
$P(A_1 \text{ or } A_2)$	Probability of the occurrence of at least one of events A_1 and A_2 (2.2)

Symbol	Meaning
$P(A_1 \text{ and } A_2)$	Joint probability of events A_1 and A_2 (2.2)
$P(B_1\|A_1)$	Conditional probability of event B_1 given A_1 (2.2)
P_{n}	Price in a nonbase period in an index number formula (12.2)
P_0	Price in a base period in an index number formula (12.2)
$P_0(p)$	Prior probability distribution of random variable p (15.2)
$P_1(p)$	Posterior probability distribution of random variable p (15.2)
$P(X = x\|Y = y)$	Conditional probability that random variable X is equal to the value x given that random variable Y is equal to y (3.13)
$P(X = x \text{ and } Y = y)$	Joint probability that X takes on the value x and Y takes on the value y (3.13)
p	Probability of a success on a given trial (binomial distribution); also used as population proportion of successes (3.4)
$\bar{p} = \dfrac{x}{n}$	Proportion of successes in a sample of size n (5.2)
$\hat{p}$	Weighted mean of two sample proportions (7.3)
Q_{f}	A fixed set of weights in an index number formula (12.2)
Q_{n}	Quantity in a nonbase period in an index number formula (12.2)
Q_0	Quantity in a base period in an index number formula (12.2)
$q = 1 - p$	Probability of failure on a given trial (binomial distribution); also used as population proportion of failures (3.4)
R_1, R_2	Sum of ranks of the items in samples 1 and 2, respectively, in the Mann–Whitney U test (rank sum test) (13.4)
$R(s_1\|\theta_1)$	Risk or expected opportunity loss associated with the use of strategy s_1, given that state of nature θ_1 occurs (16.3)
$R^2_{Y.12\ldots(k-1)}$	Sample coefficient of multiple determination for a regression equation involving $k - 1$ independent variables $X_1, X_2, \ldots, X_{k-1}$ and dependent variable Y (10.4)
r	Number of rows in an arrangement of data to which analysis of variance is applied (8.2)

Symbol	Meaning
r	Sample correlation coefficient (9.6)
r^2	Sample coefficient of determination (9.6)
r_c^2	Corrected or adjusted sample coefficient of determination (9.6)
r_j	Number of observations in the jth column (8.3)
r_r	Rank correlation coefficient (13.7)
r_{12}	Correlation coefficient for variables X_1 and X_2 (10.5)
ρ	*rho*: Population correlation coefficient (9.6)
ρ^2	Population coefficient of determination (9.6)
S	Sample space (2.1)
SI	Seasonal index (11.5)
SSA	Between-treatment sum of squares (8.3)
SSE	Within-treatment (error) sum of squares (8.3)
SST	Total sum of squares (8.3)
$S^2_{Y.12\ldots(k-1)}$	Sample variance around a regression equation involving $k - 1$ independent variables $X_1, X_2, \ldots, X_{k-1}$ and dependent variable Y (10.3)
s	Standard deviation of a sample (1.15)
s^2	Variance of a sample (1.15)
$s_1, s_2, \ldots$	Strategies (16.3)
s_b	Standard error of regression coefficient b (9.7)
s_d	Standard deviation of differences between pairs of observations made on the same individuals or objects (7.6)
$s_{\bar{d}}$	Standard error of $\bar{d}$, the mean difference of pairs of observations made on the same individuals or objects (7.6)
$s_{\bar{p}}$	Estimated standard error of a proportion (6.3)
s_{IND}	Standard error of forecast in a two-variable regression analysis. Used to establish prediction intervals for individual Y values
$s_{\bar{p}_1 - \bar{p}_2}$	Estimated or approximate standard error of the difference between two proportions (6.3)
$s_{\bar{x}_1 - \bar{x}_2}$	Estimated or approximate standard error of the difference between two means (6.3)
s_Y	Standard deviation of a sample of Y values (9.5)
$s_{\hat{Y}}$	Estimated standard error of the conditional mean in a two-variable regression analysis. Used to establish confidence intervals for a conditional mean (9.5)
$s_{Y.X}$	Standard error of estimate. Measures scatter of observed values of Y around the corresponding $\hat{Y}$ values on a two-variable regression line. An estimator of $\sigma_{Y.X}$ (9.5)
σ	Standard deviation of a population (1.15)
σ	Standard deviation of a probability distribution (3.10)
σ^2	Variance of a population (1.15)
σ^2	Variance of a probability distribution (3.10)
$\sigma_{n\bar{p}}$	Standard deviation of sampling distribution of number of occurrences (5.2)
$\sigma_{\bar{p}}$	Standard error of a proportion, that is, standard deviation of sampling distribution of a proportion (5.2)
$\sigma_{\bar{p}_1 - \bar{p}_2}$	Standard error of the difference between two proportions (7.3)
σ_r	Standard error of sampling distribution of the number of runs r (13.5)
σ_T	Standard deviation of sampling distribution of T in the Wilcoxon matched-pairs signed rank test (13.3)
σ_U	Standard error of U statistic in the Mann–Whitney U test (rank sum test) (13.4)
$\sigma_{\bar{x}}$	Standard error of the mean, that is, standard deviation of sampling distribution of the mean (5.5)
$\sigma_{\bar{x}_1 - \bar{x}_2}$	Standard error of the difference between two means (6.3)
$\sigma_{Y.X}$	Standard deviation of conditional probability distribution of Y given X in the linear regression model; the population value that corresponds to the $s_{Y.X}$ value computed from sample observations (9.1)
T	Grand total of all observations in a shortcut computation of an analysis of variance (8.3)
T	Effect of the trend factor in time-series analysis (11.4)

Symbol	Meaning
T	Wilcoxon's T statistic; the smaller of two ranked sums (13.3)
T_j	Total of the observations in the jth column (8.3)
$t = \frac{\bar{x} - \mu}{s/\sqrt{n}}$	The t statistic, distributed according to the Student t distribution with v degrees of freedom (6.4)
θ	Population parameter (6.2)
$\hat{\theta}$	Estimator of θ (6.2)
$\theta_1, \theta_2, \ldots, \theta_m$	States of nature (14.2)
U	A measure of the difference between the ranked observations of two samples in the Mann–Whitney U test (rank sum test) (13.4)
$U(x)$	Utility of monetary payoff x (14.8)
u_{ij}	Utility of selecting act A_j when state of nature θ_i occurs (14.2)
w	Weight applied to an observation (1.10)
X	Value of an observation (1.9)
$\bar{\bar{X}}$	Grand (arithmetic) mean of all observations (8.3)
X_{ij}	Value of an observation in the ith row and jth column (8.3)
$\bar{X}_j$	Arithmetic mean of the jth column of observations (8.3)
$\bar{X}_w$	Weighted arithmetic mean (1.10)
$\bar{X}, \bar{x}$	Arithmetic mean of a sample (1.9)
x	Number of successes in a sample of size n (5.2)
$\hat{Y}$	Computed value of Y in a sample regression equation (9.4)
Y_t	Computed trend value for the time-series variable Y (11.4)
$z = \frac{x - \mu}{\sigma}$	Standard score; deviation of value of an observation from the arithmetic mean of a distribution expressed in multiples of the standard deviation (5.4)

Bibliography

Statistics for the General Reader

Bailey, M. *Reducing Risks to Life.* Washington, DC: American Enterprise Institute, 1980.

Bickel, P.J., and O'Connell, J.W. "Sex Bias in Graduate Admissions: Data from Berkeley," *Science*, Vol. 187 (February 1975), pp. 398–404.

Campbell, S.K. *Flaws and Fallacies in Statistical Thinking.* Englewood Cliffs, NJ: Prentice-Hall, 1974.

Fairley, W.B., and Mosteller, F., eds. *Statistics and Public Policy.* Reading, MA: Addison-Wesley, 1977.

Federer, W.T. *Statistics and Society.* New York: Marcel Dekker, 1973.

Gilbert, J.P., McPeek, B., and Mosteller, F. "Statistics and Ethics in Surgery and Anesthesia," *Science*, Vol. 198 (November 1977), pp. 684–89.

Hamburg, M. *Basic Statistics: A Modern Approach*, 3rd ed. San Diego, CA: Harcourt Brace Jovanovich, 1985.

Huff, D. *How to Lie with Statistics.* New York: W.W. Norton, 1954.

"Is Vitamin C Really Good for Colds?" *Consumer Reports*, Vol. 41, No. 2 (February 1976), pp. 68–70.

Kemp, K.W. *Dice, Data and Decisions: Introductory Statistics.* New York: Halstead, 1984.

Larson, R.J., and Stroup, D.F. *Statistics in the Real World.* New York: Macmillan, 1976.

Levinson, H.C. *Chance, Luck, and Statistics.* New York: Dover Publications, 1963.

McNeil, B.J., Weichselbaum, R., and Pareker, S.G. "Fallacy of the Five-Year Survival in Lung Cancer," *New England Journal of Medicine*, Vol. 229 (December 1978), pp. 1397–401.

Moore, D.S. *Statistics: Concepts and Controversies.* San Francisco: Freeman, 1979.

Moroney, M.J. *Facts from Figures.* New York: Penguin Books, 1956.

Mosteller, F., Pieters, R.S., Kruskal, W.H., Rising, G.R., Link, R.F., Carlson, R., and Zelinka, M. *Statistics by Example.* Reading, MA: Addison-Wesley, 1973.

Odell, J.W. *Basic Statistics: An Introduction to Problem Solving with Your Personal Computer.* Blue Ridge Summit, PA: TAB Books, Inc., 1984.

Reichman, W.J. *Use and Abuse of Statistics.* New York: Oxford University Press, 1962.

Tanur, J.M., et al., eds. *Statistics: A Guide to the Unknown.* San Francisco: Holden-Day, 1977. Also revised as three paperbacks: (1) *Statistics: A Guide to Business and Economics;* (2) *Statistics: A Guide to Biological and Health Sciences;* (3) *Statistics: A Guide to Political and Social Issues.*

Tufte, E.R. *Data Analysis for Politics and Policy.* Englewood Cliffs, NJ: Prentice-Hall, 1974.

U.S. Surgeon-General, *Smoking and Health: A Report of the Surgeon-General.* Washington, DC: U.S. Department of Health, Education and Welfare, 1979.

Wallis, W.A., and Roberts, H.V. *The Nature of Statistics.* New York: The Free Press, 1965.

Aldrich, J.H., and Nelson, F.D. *Linear Probability, Logit and Probit Models.* Beverly Hills, CA: Sage Publicatons, Inc. 1984.

Clarke, B.A., and Disney, Ralph L. *Probability and Random Processes: A First Course with Applications.* New York: John Wiley & Sons, 1985.

Probability

Feller, W. *An Introduction to Probability Theory and Its Applications*, 3rd ed. Vol. I. New York: John Wiley & Sons, 1968.

Freund, J.E. *Introduction to Probability.* Encino, CA: Dickenson Publishing, 1973.

Galambos, J. *Introductory Probability Theory.* New York: Dekker, 1984.

Goldberg, S. *Probability: An Introduction.* Englewood Cliffs, NJ: Prentice-Hall, 1960.

Hacking, I. *The Emergence of Probability: A Philosophical Study of Early Ideas about Probability, Induction and Statistical Inference.* New York: Cambridge University Press, 1984.

Hodges, J.L., and Lehman, E.L. *Basic Concepts of Probability and Statistics.* San Francisco: Holden-Day, 1964.

Hodges, J.L., and Lehman, E.L. *Elements of Finite Probability.* San Francisco: Holden-Day, 1965.

Hoel, P.G., Port, S.C., and Stone, C.J. *Introduction to Probability Theory.* Boston: Houghton Mifflin, 1972.

Hogg, R.V., and Tanis, E.A. *Probability and Statistical Inference.* New York: Macmillan, 1983.

Jeffreys, H. *Theory of Probability*, 3rd ed. New York: Oxford University Press, 1983.

Malcolm, J.G. "Practical Application of Bayes' Formulas." Annual Reliability and Maintainability Symposium. IEEE: 1983, pp. 180–86.

Mosteller, F., Rourke, R., and Thomas, G., Jr. *Probability and Statistics.* Reading, MA: Addison-Wesley, 1961.

Newman, R.W., and White, R.M. *Reference Anthropometry of Army Men.* Report No. 180. Lawrence, MA: Environmental Climatic Research Laboratory.

Parzen, E. *Modern Probability Theory and Its Applications.* New York: John Wiley & Sons, 1960.

Regression and Correlation Analysis

Berry, W., and Feldman, S. *Multiple Regression in Practice.* Beverly Hills, CA: Sage, 1985.

Breiman, L., and Freedman, D. "How Many Variables Should Be Entered in a Regression Equation?" *Journal of the American Statistical Association* Vol. 78 (March 1983), pp. 131–36.

Chatterjee, S., and Price, B. *Regression Analysis by Example.* New York: John Wiley & Sons, 1978.

Cohen, J., and Cohen, P. *Applied Multiple Regression: Correlation Analysis for the Behavioral Sciences*, 2nd ed. Hillsdale, NJ: L. Erlbaum Associates, 1983.

Dixon, W.J. *BMDP: Biomedical Computer Programs.* Berkeley: University of California Press, 1975.

Draper, N.R., and Smith, H. *Applied Regression Analysis*, 2nd ed. New York: John Wiley & Sons, 1981.

Edwards, A.L. *Multiple Regression Analysis and the Analysis of Variance and Covariance*, 2nd ed. New York: W.H. Freeman, 1985.

Ezekiel, M., and Fox, K.A. *Methods of Correlation and Regression Analysis*, 3rd ed. New York: John Wiley & Sons, 1959.

Johnston, J. *Econometric Methods*, 2nd ed. New York: McGraw-Hill, 1971.

Mosteller, F., and Tukey, J.W. *Data Analysis and Regression: A Second Course in Statistics.* Reading, MA: Addison-Wesley, 1977.

Neter, J., and Kutner, M.H. *Applied Linear Regression Analysis.* Homewood, IL: Richard D. Irwin, 1983.

Neter, J., and Wasserman, W. *Applied Linear Statistical Models.* Homewood, IL: Richard D. Irwin, 1974.

Nie, N., Hull, C.H., Jenkins, J.G., Steinbrenner, K., and Bent, D.H. *Statistical Package for the Social Sciences*, 2nd ed. New York: McGraw-Hill, 1975.

Ryan, T.A., Joiner, B.L., and Ryan, B.F. *Minitab Student Handbook.* N. Scituate, MA: Duxbury Press, 1976.

SAS Institute, Inc. *SAS User's Guide, 1979 Edition.* Cary, NC: 1979.

Schroeder, L.D. *Understanding Regression Analysis.* Beverly Hills, CA: Sage, 1986.

Weisberg, S. *Applied Linear Regression*, 2nd ed. New York: John Wiley & Sons, 1985.

Williams, E.J. *Regression Analysis.* New York: John Wiley & Sons, 1959.

Wonnacott, R.J., and Wonnacott, T.H. *Econometrics*, 2nd ed. New York: John Wiley & Sons, 1979.

Wonnacott, T.H., and Wonnacott, R.J. *Regression: A Second Course in Statistics.* New York: John Wiley & Sons, 1981.

Younger, M. *First Course in Linear Regression*, 2nd ed. Boston: PWS Publishers, 1985.

Time-Series Analysis and Forecasting

Bowerman, B.L., and O'Connell, R.T. *Forecasting and Time Series.* Boston: Duxbury Press, 1979.

Box, G.E.P., and Jenkins, G.M. *Time-Series Analysis, Forecasting, and Control.* San Francisco: Holden-Day, 1969.

Brown, Robert G. *Smoothing, Forecasting, and Prediction of Discrete Time Series.* Englewood Cliffs, NJ: Prentice-Hall, 1963.

Chambers, J.C., Mullick, S.K., and Smith, D.D. *An Executive's Guide to Forecasting.* New York: John Wiley & Sons, 1974.

Chatfield, C. *Analysis of Time Series: An Introduction*, 3rd ed. New York: Methuen, Inc., 1984.

Cryer. *Time-Series Analysis with Minitab.* Boston: PWS Publishers, 1986.

Daniels, L.M. *Business Forecasting for the 1980s—and Beyond.* Baker Library, Reference List No. 31. Boston: Harvard Business School, 1980.

Gilchrist, W. *Statistical Forecasting.* New York: John Wiley & Sons, 1976.

Granger, C.W.J. *Forecasting in Business and Economics.* New York: Academic Press, 1980.

Makridakis, S. *The Forecasting Accuracy of Major Time Series Methods.* New York: John Wiley & Sons, 1984.

Makridakis, S., Wheelwright, S.C., and McGee, V.E. *Forecasting: Methods and Applications*, 2nd ed. New York: John Wiley & Sons, 1983.

Milne, T.E. *Business Forecasting—A Managerial Approach.* London: Longman Group, Ltd., 1975.

Nelson, C.R. *Applied Time-Series Analysis for Managerial Forecasting.* San Francisco: Holden-Day, 1973.

1980 Supplement to Economic Indicators. Washington, DC: U.S. Government Printing Office, 1980.

Roberts, H.V. *Time Series and Forecasting with IDA.* New York: McGraw-Hill, 1984.

Sullivan, W.G., and Claycombe, W.W. *Fundamentals of Forecasting.* Reston, VA: Reston Publishing, 1977.

Wheelwright, S.C., and Makridakis, S. *Forecasting Methods for Management,* 3rd ed. New York: John Wiley & Sons, 1980.

Woods, D., and Fildes, R. *Forecasting for Business: Methods and Applications.* New York: Longman Group, Ltd., 1976.

Index Numbers

Allen, R.E.D. *Index Numbers in Theory and Practice.* Chicago: Aldine Publishing, 1975.

Fisher, I. *Making of Index Numbers,* 3rd reprint. New York: Kelly, 1927.

Banerjee, K.S. *Cost of Living Index Numbers: Practice, Precision, and Theory.* New York: Dekker, 1975.

Mudgett, Bruce D. *Index Numbers.* New York: John Wiley & Sons, 1951.

U.S. Department of Labor, Bureau of Labor Statistics. *Monthly Labor Review,* Vol. 104, No. 9 (September 1981), pp. 20–25.

Analysis of Variance and Design of Experiments

Anderson, V.L., and McLean, R.A. *Design of Experiments: A Realistic Approach.* New York: Dekker, 1974.

Cochran, G., and Cox, M. *Experimental Designs,* 2nd ed. New York: John Wiley & Sons, 1957.

Cox, R. *Planning of Experiments.* New York: John Wiley & Sons, 1958.

Finney, D.J. *Experimental Design and Its Statistical Basis.* Chicago: University of Chicago Press, 1955.

Guenther, W.C. *Analysis of Variance.* Englewood Cliffs, NJ: Prentice-Hall, 1964.

Fisher, R.A. *The Design of Experiments.* New York: Hafner, 1974.

Hicks, C.R. *Fundamental Concepts in the Design of Experiments,* 3rd ed. New York: Holt, Rinehart & Winston, 1982.

Mendenhall, W. *An Introduction to Linear Models and the Design and Analysis of Experiments.* Belmont, CA: Wadsworth, 1968.

Morrison, Donald F. *Applied Linear Statistical Methods.* Englewood Cliffs, NJ: Prentice-Hall, 1983.

Neter, J., and Wasserman, W. *Applied Linear Statistical Models.* Homewood, IL: Richard D. Irwin, 1974.

Wolach, A.H. *Basic Analysis of Variance Programs for Microcomputers.* Belmont, CA: Wadsworth Publishers, 1983.

Wolter, K.M. *Introduction to Variance Estimation.* New York: Springer-Verlag, 1985.

Nonparametric Statistics

Bradley, J.V. *Distribution-Free Statistical Tests.* Englewood Cliffs, NJ: Prentice-Hall, 1968.

Conover, W.J. *Practical Nonparametric Statistics*, 2nd ed. New York: John Wiley & Sons, 1980.

Gibbons, J.D. *Nonparametric Statistical Inference.* New York: McGraw-Hill, 1970.

Hajek, J. *Nonparametric Statistics.* San Francisco: Holden-Day, 1969.

Krishnaiah, P.R. and Sen, P.K., ed. *Nonparametric Methods.* New York: Elsevier, 1985.

Noether, G.E. *Introduction to Statistics: A Nonparametric Approach*, 2nd ed. Boston: Houghton Mifflin, 1976.

Puri, M.L. and Sen, P.K. *Nonparametric Methods in General Linear Models.* New York: John Wiley & Sons, 1985.

Siegel, S. *Nonparametric Statistics for the Behavioral Sciences.* New York: McGraw-Hill, 1956.

Decision Analysis

Aitchison, J. *Choice Against Chance.* Reading, MA: Addison-Wesley, 1970.

Baird, B.F. *Introduction to Decision Analysis.* Boston: Duxbury Press, 1978.

Brown, R.V., Kahr, A.S., and Peterson, C. *Decision Analysis for the Manager.* New York: Holt, Rinehart & Winston, 1974.

Bunn, D.W. *Applied Decision and Analysis.* New York: McGraw-Hill, 1984.

Chernoff, H., and Moses, L. *Elementary Decision Theory.* New York: John Wiley & Sons, 1959.

Edwards, W., and Tversky, A., eds. *Decision Making: Selected Readings.* Harmondsworth, England: Penguin Books, 1967.

Forester, J. *Statistical Selection of Business Strategies.* Homewood, IL: Richard D. Irwin, 1968.

Gupta, M.M., and Sanchez, E., eds. *Approximate Reasoning in Decision Analysis.* New York: Elsevier, 1983.

Harvard Business Review. Statistical Decision Series, Parts I–IV. Boston: Harvard University Press, 1951–1970.

Howard, R.A. "The Decision to Seed Hurricanes," *Science,* Vol. 176 (June 1972), pp. 1191–1201.

Keeney, R.L., and Raiffa, H. *Decisions with Multiple Objectives: Preferences and Value Tradeoffs.* New York: John Wiley & Sons, 1976.

Kostolansky, J., and Bailey, A.D. *Decision Analysis, Including Modeling and Information Systems.* Baton Rouge, LA: Malibu Publications, 1984.

Lindley, D.V. *Introduction to Probability and Statistics from a Bayesian Viewpoint.* Part 2, "Inference." New York: Cambridge University Press, 1965.

Lindley, D.V. *Making Decisions.* London: Wiley-Interscience, 1971.

Luce, R.D., and Raiffa, H. *Games and Decisions.* New York: John Wiley & Sons, 1957.

Morris, W.T. *Management Science: A Bayesian Introduction.* Englewood Cliffs, NJ: Prentice-Hall, 1968.

Pessemier, E.A. *New Product Decisions—An Analytic Approach.* New York: McGraw-Hill, 1966.

Pratt, J., Raiffa, H., and Schlaifer, R. *Introduction to Statistical Decision Theory.* New York: McGraw-Hill, 1965.

Raiffa, H. *Decision Analysis: Introductory Lectures on Choices Under Uncertainty,* New York: Random, 1986.

Raiffa, H., and Schlaifer, R. *Applied Statistical Decision Theory.* Cambridge, MA: Division of Research, Graduate School of Business Administration, Harvard University, 1961.

Risk Analysis—Proceedings of the United States Army Operations Research Symposium, May 15–18, 1972. Washington, DC: Office of the Chief of Research and Development, Department of the Army, 1972.

Schlaifer, R. *Probability and Statistics for Business Decisions.* New York McGraw-Hill, 1959.

Vatter, R.A., Bradley, S.P., Frey, S.C., Jr., and Jackson, B.B. *Quantitative Methods in Management: Text and Cases.* Homewood, IL: Richard D. Irwin, 1978.

Winkler, R.L. *An Introduction to Bayesian Inference and Decision.* New York: Holt, Rinehart & Winston, 1972.

Statistical Tables

Beyer, W.H. *Handbook of Tables for Probability and Statistics,* 2nd ed. Cleveland: The Chemical Rubber Company, 1968.

Burington, R.S., and May, D.C. *Handbook of Probability and Statistics with Tables,* 2nd ed. New York: McGraw-Hill, 1970.

Fisher, R.A., and Yates, F. *Statistical Tables for Biological, Agricultural, and Medical Research*, 6th ed. London: Longman Group, Ltd., 1978.

Hald, A. *Statistical Tables and Formulas.* New York: John Wiley & Sons, 1952.

Kres, H., and Wadsack, P., trans. *Statistical Tables for MultivariateAnalysis: A Handbook with References to Applications.* New York: Springer-Verlag, 1983.

Lindley, D.V., and Scott, W.F. *New Cambridge Elementary Statistical Tables.* New York: Cambridge University Press, 1984.

National Bureau of Standards. *Tables of the Binomial Probability Distribution.* Applied Mathematical Series 6. Washington, DC: U.S. Department of Commerce, 1950.

Odeh, O. *Tables for Tests, Confidence Limits and Plans Based on Proportions.* New York: Dekker, 1983.

Owen, D. *Handbook of Statistical Tables.* Reading, MA: Addison-Wesley, 1962.

Pearson, E.S., and Hartley, H.O. *Biometrika Tables for Statisticians*, 2nd ed. Cambridge, England: Cambridge University Press, 1962.

RAND Corporation. *A Million Random Digits with 100,000 Normal Deviates.* New York: The Free Press of Glencoe, 1955.

Romig, H.G. *50–100 Binomial Tables.* New York: John Wiley & Sons, 1953.

White, J., and Yeats, A. *Tables for Statisticians.* New York: University of Queensland Press, 1984.

Zehna, P.W., and Barr, D.R. *Tables of the Common Probability Distributions.* Monterey, CA: U.S. Naval Postgraduate School, 1970.

Dictionary of Statistical Terms

Freund, J., and Williams, F. *Dictionary/Outline of Basic Statistics.* New York: McGraw-Hill, 1966.

Kendall, M.G., and Buckland, W.R. *A Dictionary of Statistical Terms*, 4th ed. New York: Longman, Inc., 1983.

Omuircheartaigh, C., and Francis, D.P. *Statistics: A Dictionary of Terms and Ideas.* New York: Hippocrene Books, Inc., 1983.

Tietjen, G.L. *A Topical Dictionary of Statistics.* New York: Methuen, Inc., 1985.

Solutions to Even-Numbered Exercises

CHAPTER 1

Exercises 1.6

2. a.

Size of Holding (in dollars)	Percentage of Number of Stockholders	Percentage of Dollar Value of Holdings
Less than 200	45.2	2.2
Less than 400	64.0	4.9
Less than 600	73.8	7.2
Less than 1,000	86.3	12.0
Less than 5,000	98.4	23.1
Less than 20,000	99.3	27.4
Less than 50,000	99.5	30.8
Less than 100,000	99.7	37.6

b. Because the distributions are open ended.

4. a.

Employment (in thousands)	Number of States
Under 250	5
250 and under 500	10
500 and under 1,000	9
1,000 and under 1,500	7
1,500 and under 2,000	8
2,000 and under 3,000	4
3,000 and under 4,000	3
4,000 and under 5,000	3
over 5,000	3
Total	52

b. The class limits are stated in round numbers for convenience and ease of interpretation. The use of unequal size classes permits the display of detail at the lower end of the distribution where there is a clustering of values. With this type of data, it is difficult to obtain a smooth progression of frequencies.

6. a. Number of cases of beer produced per 8-hour shift:

Number of Cases	Total Frequency Combined	Frequency Line 1	Frequency Line 2
5,500 and under 6,000	2	1	1
6,000 and under 6,500	3	2	1
6,500 and under 7,000	9	7	2
7,000 and under 7,500	4	2	2
7,500 and under 8,000	9	2	7
8,000 and under 8,500	3	1	2
	30	15	15

Distribution for Line 1 and Line 2 combined

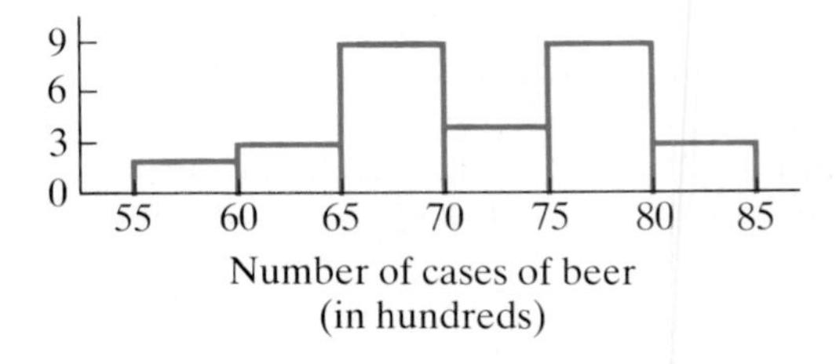

b. Distributions for Line 1 and for Line 2

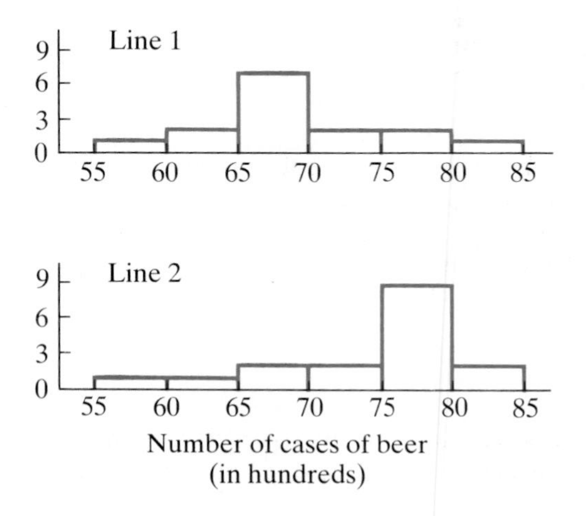

c. The answer in part (b) is preferable because a bimodal distribution results if the data for the two production lines are merged.

Exercises 1.10

2. No. For example, suppose 99% of West End's total dollar credit extended consists of personal loans and 99% of East End's business is industrial loans. Then the overall debt ratios are:

West End: $(0.99)(0.05) + (0.01)(0.03) = 0.0498$
East End: $(0.01)(0.06) + (0.99)(0.04) = 0.0402$

4. a. For the first strategy, average cost of United Aerodynamics is $\$\frac{5,275}{125} = \$42.20$; average cost of Mitton Industries is $\$\frac{4,675}{125} = \$37.40$.
For the second strategy, average cost of United Aerodynamics is $\$\frac{5,039}{125} = \$40.31$; average cost of Mitton Industries is $\$\frac{4,959}{141} = \$35.17$.
b. The second strategy achieved the lower average cost for both stocks.
c. For the second strategy, since the number of shares bought each time varied inversely with the stock price, higher weights were placed on the lower stock prices in the calculation of the average cost per share. This resulted in a lower overall average for each stock than in the first strategy, in which the average cost per share was the unweighted mean of the 5 share prices. That is, if equal weights are used in computing a weighted arithmetic mean, the resulting figure is the same as the unweighted mean.

Exercises 1.13

2.

	Mean	Median	Mode
a.	$4.70	$4.50	$8.00
b.	$5.00	$5.00	$5.00
c.	$17.50	$4.00	$4.00

4. a. Because the large account balances at the upper end of the distribution would make it impractical to have enough closed classes to encompass all account balances.
b. Modal class: $2,000–$2,499 c. Median class: $1,500–$1,999

Exercises 1.18

2. a. All three stocks have mean and median prices of $47. Thus, there is no difference simply on the basis of measures of central tendency.

b. s (Conservocorp) $= \sqrt{\frac{68}{9}} = \$2.75$
s (Mesocorp) $= \sqrt{\frac{158}{9}} = \$4.19$
s (Specucorp) $= \sqrt{\frac{400}{9}} = \$6.67$
The standard deviation is a measure of price fluctuation. Hence, it can be considered as a measure of the risk associated with the stock.

4. a. $\bar{X} = \dfrac{(0)(73) + (1)(74) + (2)(37) + (3)(12) + (4)(4)}{200} = 1$ unit

b. $s^2 = \dfrac{(0-1)^2(73) + (1-1)^2(74) + (2-1)^2(37) + (3-1)^2(12) + (4-1)^2(4)}{199}$

$s^2 = 0.9749$ units squared
$s = 0.99$ units

6. a. $\bar{X} = 15$; Median $= 15$

b. $\bar{X} = \dfrac{(9)(12) + (9)(17) + (3)(22) + (3)(27)}{24} = 17$

Mean production on the other days is greater than mean production on Mondays.

c. CV Monday $= \frac{3.85}{15} = 25.7\%$
CV Other $= \frac{5.11}{17} = 30.1\%$
No. Other days have greater relative dispersion in production than do Mondays.

Chapter 1 Computer Exercises

2. $\bar{X} = 37.947$ (thousand dollars)
$s = 19.4$ (thousand dollars)

4. *Food expenditures* Standard score $= \frac{3.02}{6.5173} = 0.463$
Income Standard score $= \frac{19.4}{37.947} = 0.511$
The income series is relatively more variable than food expenditures.

CHAPTER 2

Exercises 2.1

2. a. This is a subjective probability.
b. Subjective probabilities are useful in business decision making. If more information is gathered, subjective probabilities can be revised and will become more reliable.

54 | | 1

55

56 | 1

$36 + 25 + 9 + 4 + 1 + 1 + 1 + 9 + 36 + 36$

$s = 3.97$

$100 + 81 + 25 + 9 + 1 + 9 + 9 + 36 + 49 + 81$

$s = 6.325$

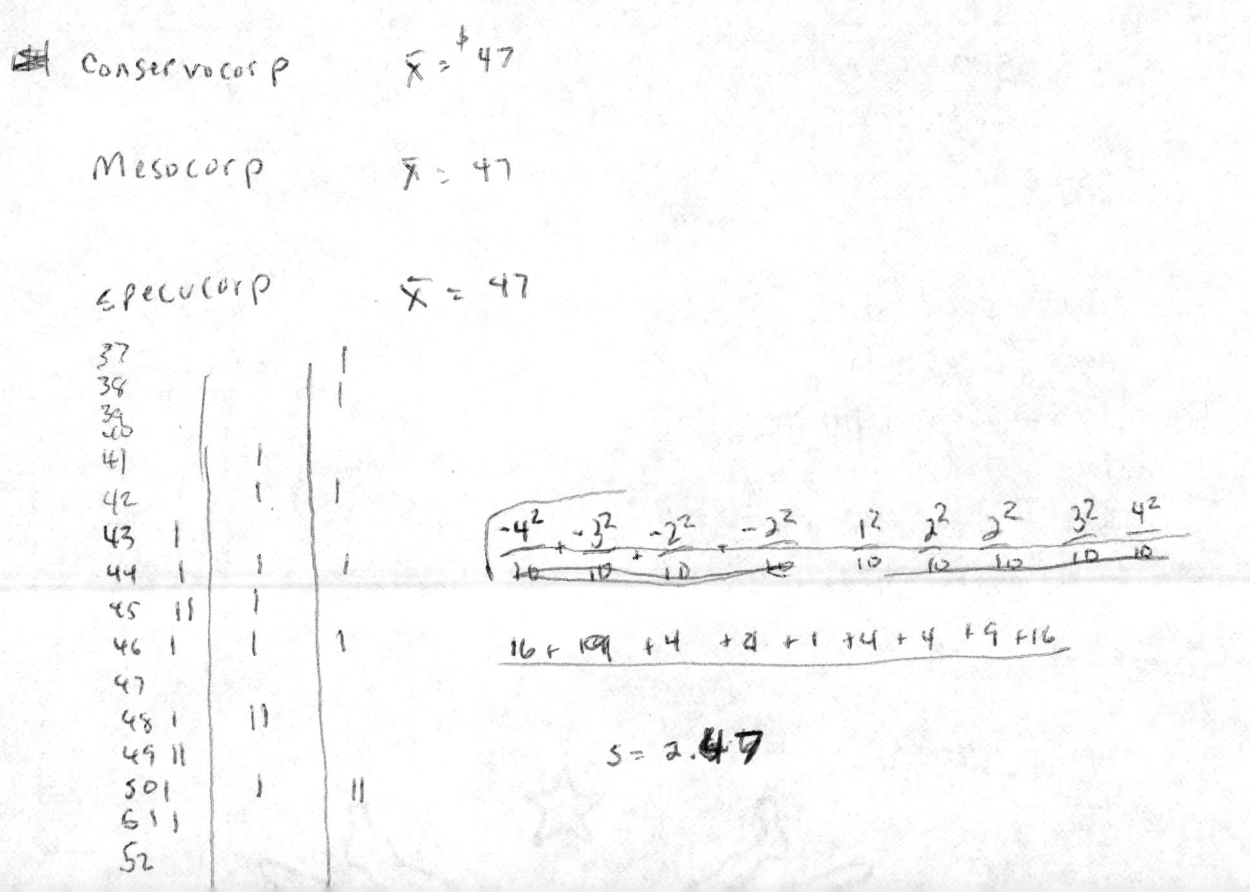

page 48 # 2
Conservocorp $\bar{x}$ = $47
Mesocorp $\bar{x}$ = 47
Specucorp $\bar{x}$ = 47
37
38
39
40
41
42
43
44
45
46
47
48
49
50
51
52
16 + 9 + 4 + 4 + 1 + 4 + 4 + 9 + 16
s = 2.47

PAGE 19 # 6

(a)

$$i = \frac{8500 - 5500}{6}$$

500

5500 - 6000	III	3
6001 - 6500	II	2
6501 - 7000	~~IIII~~ ~~IIII~~	10
7001 - 7500	III	3
7501 - 8000	~~IIII~~ IIII	9
8001 - 8500	III	3
		30

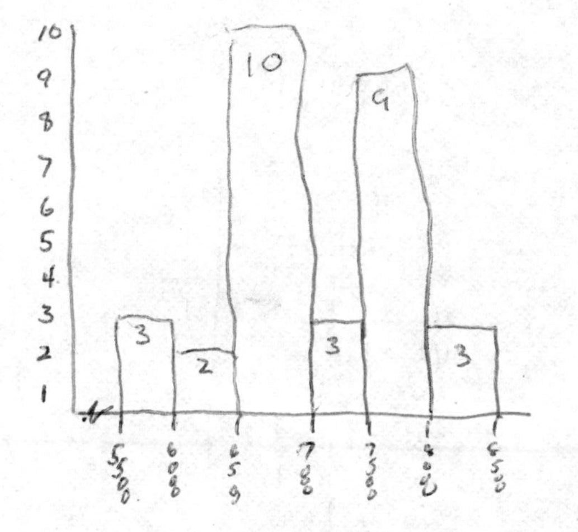

(b)

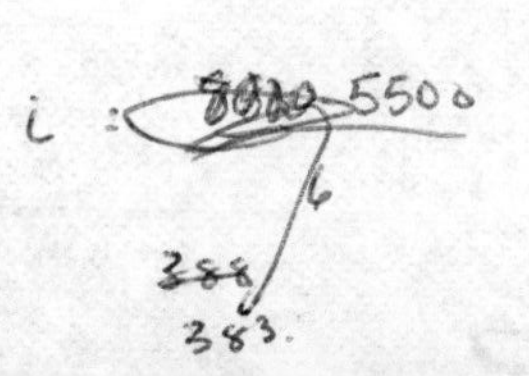

4. $\{I, EI, EEI, EEEI, \ldots\}$ $\quad$ (E = effective; I = ineffective)

6. Code used for change in national income

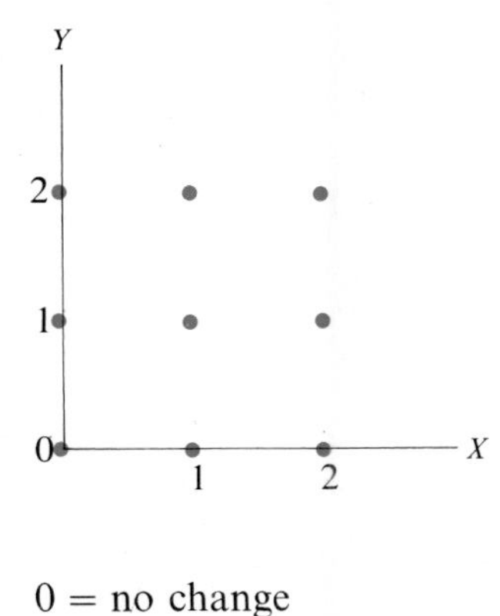

0 = no change
1 = increase
2 = decrease

Exercises 2.2

2. Event A = a defective item from the first machine
Event B = a defective item from the second machine
Event C = a defective item from the third machine
Because these events are independent,

$$\begin{aligned} P(A \text{ and } B \text{ and } C) &= P(A)P(B)P(C) \\ &= (0.05)(0.02)(0.03) \\ &= 0.00003 \end{aligned}$$

4. a. $P(\bar{A} \text{ and } \bar{B}) = 1 - \frac{1}{6} - \frac{2}{9} - \frac{1}{3} = \frac{5}{18}$

b. $P(A) = \frac{1}{6} + \frac{2}{9} = \frac{7}{18}$

$P(B) = \frac{1}{6} + \frac{1}{3} = \frac{1}{2}$

$P(A \text{ and } B) = \frac{1}{6} \neq P(A)P(B)$
Therefore A and B are *not* independent.

c. $P(A \text{ and } B) = \frac{1}{6} \neq 0.$
Therefore, A and B are not mutually exclusive events.

6. $P(X) = P(A \text{ will buy } B) = 0.7$
$P(\bar{X}) = P(A \text{ will not buy } B) = 0.3$
Therefore, the odds that Company A will not buy Company B are 3:7.

8. a. $\frac{18}{38}$ $\quad$ b. $\frac{1}{38}$ $\quad$ c. $\frac{18}{38}$ $\quad$ d. 18:20 = 9:10 $\quad$ e. 1:37 $\quad$ f. Win: $\frac{18}{38}$
Lose: $\frac{20}{38}$

10. $(0.8)^5 = 0.328$

12. a. $\frac{220}{1,000} = \frac{11}{50}$ b. $\frac{400}{1,000} = \frac{2}{5}$ c. $\frac{400}{1,000} = \frac{2}{5}$ d. $\frac{255}{400} = \frac{51}{80}$ e. $\frac{125}{400} = \frac{5}{16}$
f. No.

Opinion	Region			Total
	East	Midwest	West	
Opposed	240	210	150	600
Not opposed	160	140	100	400
Total	400	350	250	1,000

Exercises 2.3

2. $\dfrac{(0.5)(0.7)}{(0.5)(0.7) + (0.5)(0.3)} = 0.70$

4. $\dfrac{(0.8)(0.75)}{(0.8)(0.75) + (0.2)(0.25)} = 0.92$

Exercises 2.4

2. a. $4! = 24$ b. ${}_4P_2 = \dfrac{4!}{(4-2)!} = 12$

4. a. $\dbinom{5}{3} = \dfrac{5!}{3!2!} = 10$ b. ${}_5P_2 = \dfrac{5!}{(5-2)!} = 20$ c. $\dfrac{\dbinom{3}{1}\dbinom{2}{1}}{\dbinom{5}{2}} = \dfrac{6}{10}$

6. a. $10^4 = 10{,}000$ b. $26(10^3) = 26{,}000$ c. $36^4 = 1{,}679{,}616$

8. a. $\dbinom{5}{3} = 10$ b. $(5)(4)(3) = 60$ c. $\dfrac{(3)(2)(1)}{(5)(4)(3)} = \dfrac{1}{10}$ or $\dfrac{1}{\dbinom{5}{3}} = \dfrac{1}{10}$

10. $4! = 4 \times 3 \times 2 \times 1 = 24$

12. $\dbinom{10}{3}(3!) = 720$ or $(10)(9)(8) = 720$

CHAPTER 3

Exercises 3.1

2.

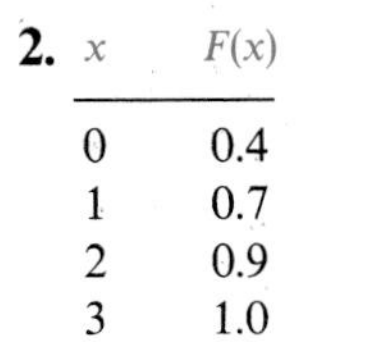

x	$F(x)$
0	0.4
1	0.7
2	0.9
3	1.0

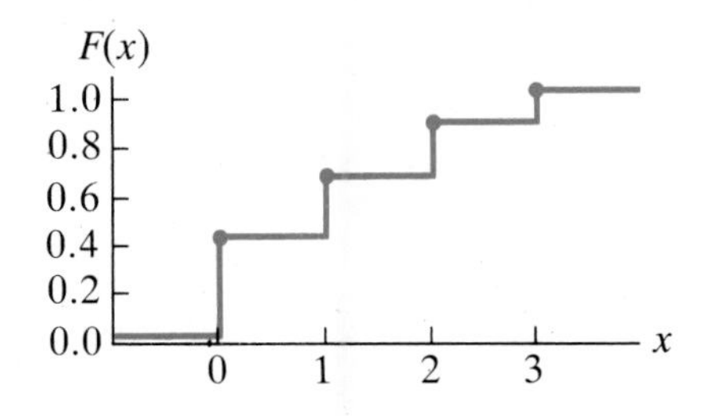

4. a. No. $\sum_x f(x)$ must equal 1 b. Yes
c. No; $f(x)$ cannot be a negative number. d. Yes

6.

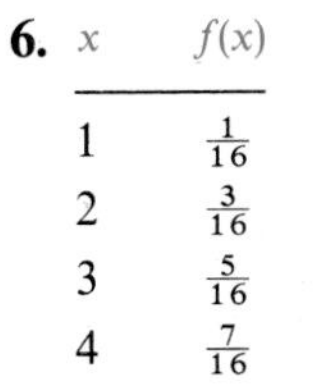

x	$f(x)$
1	$\frac{1}{16}$
2	$\frac{3}{16}$
3	$\frac{5}{16}$
4	$\frac{7}{16}$

8. a. $f(x) = \dfrac{\binom{10}{x}\binom{40}{2-x}}{\binom{50}{2}}$ where $x = 0, 1, 2$

b. $P(X = 2) = \dfrac{\binom{10}{2}\binom{40}{0}}{\binom{50}{2}} = 0.0367$

Exercises 3.3

2. a. $f(x) = \frac{1}{96}, x = 1, \ldots, 96$ b. $\frac{70}{96}$ c. $\frac{40}{96}$ d. $\frac{41}{96}$ e. $\frac{25}{96}$

Exercises 3.4

2. $P(X \geq 2) = 1 - F(1) = 1 - \sum_{x=0}^{1} \binom{10}{x}(0.95)^{10-x}(0.05)^x$

$= 1 - 0.9139 = 0.0861$

4. a. $\binom{10}{1}(0.9)^9(0.1)^1 = 0.3874$ b. $\sum_{x=1}^{10} \binom{10}{x}(0.9)^{10-x}(0.1)^x = 0.6513$

c. $\binom{10}{0}(0.9)^{10}(0.1)^0 = 0.3487$

6. a. $\binom{8}{4}(0.65)^4(0.35)^4 = 0.1875$

b. $P(X \geq 4) = 1 - F(3) = 1 - 0.7064 = 0.2936$

c. No, stock price movements cannot ordinarily be considered independent. Hence, the binomial is not the appropriate probability distribution.

8. The distribution is symmetrical when $p = 0.5$, skewed to the right when $p = 0.2$ and skewed to the left when $p = 0.8$. The graphs for $p = 0.2$ and $p = 0.8$ are equally skewed in opposite directions around p.

10. a. $P(X \leq 2) = 0.3980$ b. $P(X \geq 1) = 1 - F(0) = 1 - 0.0352 = 0.9648$

c. $P(X = 1) = 0.1319$ d. $P(2 \leq X \leq 4) = F(4) - F(1)$

$= 0.8358 - 0.1671$

$= 0.6687$

12. a. $\binom{15}{3}(0.75)^{12}(0.25)^3 = 0.2252$

b. $\sum_{x=0}^{4} \binom{15}{x}(0.75)^{15-x}(0.25)^x = 0.6865$

c. $\sum_{x=11}^{15} \binom{15}{x}(0.75)^{15-x}(0.25)^x = 0.0001$

Exercises 3.5

2. a. $\frac{5!}{2!3!0!} = (0.1)^2(0.8)^3(0.1)^0 = 0.0512$

b. $\frac{5!}{0!5!0!}(0.1)^0(0.8)^5(0.1)^0 = (0.8)^5 = 0.32768$

Exercises 3.6

2. $\binom{5}{1}\binom{4}{1}\binom{3}{1}\Big/\binom{12}{3} = \frac{3}{11}$

4. a. $\binom{4}{1}\binom{4}{1}\binom{4}{1}\binom{4}{1}\binom{4}{1}\binom{4}{1}\Big/\binom{24}{6} = \frac{4{,}096}{134{,}596} \doteq 0.0304$

b. $\binom{4}{4}\binom{20}{2}\Big/\binom{24}{6} = \frac{190}{134{,}596} = 0.0014$

c. $\binom{4}{0}\binom{20}{6}\Big/\binom{24}{6} = \frac{38{,}760}{134{,}596} = 0.2880$

6. a. $P(X = 1) = \dfrac{\binom{5}{1}\binom{45}{3}}{\binom{50}{4}} = 0.308$

b. $P(X \leqslant 2) = P(X = 0) + P(X = 1) + P(X = 2)$

$= \left[\binom{5}{0}\binom{45}{4} + \binom{5}{1}\binom{45}{3} + \binom{5}{2}\binom{45}{2}\right]\Big/\binom{50}{4}$

$= 0.998$

Exercises 3.7

2. $\mu = \frac{5}{60} \times 24 = 2$

a. $P(X = 3) = \dfrac{2^3e^{-2}}{3!} = 0.1804$ b. $P(X = 0) = \dfrac{2^0e^{-2}}{0!} = 0.1353$

4. a. $1 - P(X = 0) = 1 - \dfrac{2^0e^{-2}}{0!} = 1 - 0.1353 = 0.8647$

b. $\mu = (10)(\frac{2}{5}) = 4$
$P(X > 4) = 1 - P(X \leqslant 4)$
$= 1 - 0.629 = 0.371$

6. μ = one customer per minute

a. $P(X = 2) = \dfrac{1^2e^{-1}}{2!} = 0.1839$

b. $P(X \leqslant 2) = \dfrac{1^0e^{-1}}{0!} + \dfrac{1^1e^{-1}}{1!} + \dfrac{1^2e^{-1}}{2!} = 0.9197$

8. a. The probability that all computers will be working is

$$\binom{20}{20}(0.05)^0(0.95)^{20} = 0.3585$$

b. The probability that at least two computers will not be working is

$$1 - \sum_{x=0}^{1} \binom{20}{x}(0.95)^{20-x}(0.05)^x = 1 - 0.7358 = 0.2642$$

c. The corresponding Poisson probabilities are

(1) $$f(0) = \frac{1^0 e^{-1}}{0!} = 0.368$$

(2) $$1 - \sum_{x=0}^{1} \frac{1^x e^{-1}}{x!} = 1 - 0.736 = 0.264$$

The Poisson approximation in (2) which involves a sum of terms is much closer than in (1) which involves only a single term.

Exercises 3.12

2. $f(x) = \frac{1}{11}x$; 10, 11, . . . , 20
$E(X) = \Sigma x f(x) = 15$ items

4. Let X = annual profits in millions of dollars

x	$f(x)$
-1	$\frac{3}{10}$
$+1$	$\frac{1}{2}$
$+5$	$\frac{1}{5}$

$E(X) = \frac{3}{10}(-1) + \frac{1}{2}(1) + \frac{1}{5}(5) = 1.2$
$E(X^2) = \frac{3}{10}(-1)^2 + \frac{1}{2}(1)^2 + \frac{1}{5}(5)^2 = 5.8$
$\sigma_x^2 = 5.8 - (1.2)^2 = 4.36$
$\sigma_x = 2.0881$

The expected profit is \$1.2 million
The standard deviation is \$2.0881 million

6. a. $\mu = E(X) = 5$ magazines
$\sigma^2 = 30.92 - (5)^2 = 5.92$
$\sigma = 2.43$ magazines

b. The probability is at least 75% that the number of magazines in inventory is $5 \pm 2(2.43) = 0.14$ to 9.86. The probability is at least 89% that the number of magazines in inventory is $5 \pm 3(2.43) = 0$ to 12.29.

8. a. $E(\text{gain}) = (\$36)(0.998) + (-\$14{,}964)(0.002) = \$6$

b. No. On any single policy, the company will either earn the premium of \$36 or pay out \$14,964 (\$15,000 − \$36). The \$6 represents the average return per policy if an infinite number of policies were issued.

10. a. Let $X_1, X_2, \ldots, X_{20}$ represent the respective commissions per week for the 20 sales representatives. Then $X_1 + X_2 + \ldots + X_{20}$ = total commissions per week for the 20 sales representatives combined.

$$E(X_1 + X_2 + \cdots + X_{20}) = E(X_1) + E(X_2) + \cdots + E(X_{20}) = 20(\$260) = \$5{,}200$$

$$\sigma^2(X_1 + X_2 \cdots + X_{20}) = \sigma^2(X_1) + \sigma^2(X_2) + \cdots + \sigma^2(X_{20}) = 20(10{,}000) = 200{,}000$$

$$\sigma(X_1 + X_2 + \cdots + X_{20}) = \sqrt{200{,}000} = \$447.21$$

b. $E(X + \$50) = E(X) + \$50 = \$260 + \$50 = \$310$
$\sigma^2(X + 50) = \sigma^2(X) = 10{,}000$
$\sigma(X + 50) = \sqrt{10{,}000} = \$100$
Note that this question pertains to the mean and standard deviation per sales representative and not to the total commissions for the 20 sales representatives combined.

c. $E(1.10X) = 1.10E(X) = 1.10(\$260) = \$286$
$\sigma^2(1.10X) = (1.10)^2\sigma^2(X) = (1.10)^2(10{,}000) = 12{,}100$
$\sigma(1.10X) = \sqrt{12{,}100} = \$110$

Exercises 3.13

2. a. $\frac{46}{90}$ b. $\frac{60}{90}$ c. $\frac{4}{30}$

4. a. $E(X) = (0)(0.2) + (1)(0.6) + (2)(0.2) = 1.0$
$E(Y) = (0)(0.2) + (1)(0.6) + (2)(0.2) = 1.0$
$E(X + Y) = E(X) + E(Y) = 1.0 + 1.0 = 2.0$
The expected number of sales for Exotic Motors and Phlegmatic Motors combined is 2.0 per day.

b. $E(Y|X = 1) = (0)\frac{1}{6} + (1)\frac{5}{6} + (2)(0) = \frac{5}{6}$ sales per day
$\sigma^2(Y|X = 1) = (0)^2\frac{1}{6} + (1)^2\frac{5}{6} + (2)^2(0) - (\frac{5}{6})^2 = \frac{5}{36}$
$\sigma(Y|X = 1) = \sqrt{\frac{5}{36}} \approx 0.37$ sales per day

c. Profit $= \$1{,}500X + \$1{,}200Y$. The only cases in which profit would exceed \$4,000 are when $X = 2$ and $Y = 1$ and $X = 2$ and $Y = 2$. The probability $= 0 + 0.2 = 0.2$.

CHAPTER 4

Exercises 4.2

2. See the discussion in section 4.2 for the answer to problem 2.

Exercises 4.4

2. See the discussion in section 4.4

4. a. The control group would consist of a group of weight lifters who eat identical diets minus the liquid protein and perform the same daily exercises.

b. The control group would consist of sixth grade students in this city who would be taught arithmetic by the old method. The treatment group would be the sixth grade students in the city taught by the new method.

c. To test the assertion "marathon runners do not suffer heart attacks," no control group is really needed. Rather, a large and ongoing sample to find any marathon runners who have suffered a heart attack (which would disprove the hypothesis) is needed. However, to test the hypothesis that marathon runners have a reduced risk of heart attacks would require an extensive cross-matched sample and control (by age, sex, lifestyle, family history, and so on).

6. a. Conceptually, all areas other than low-income urban areas would constitute a control group. The specific control group(s) used would depend on the types of comparison desired. For example, urban areas other than low-income ones might be used.

b. Perhaps the most direct control group would be a part of the plant in which safety lights are not installed and in which similar types of work and working conditions prevail.

8. a. Conceptually, the population of interest is all male consumers in the New York City area. We are interested in the proportion of this population who possess the characteristic "would buy a certain type of man's suit." As a practical matter, the population might be defined as those male consumers who shop in the stores and departments of stores that sell this type of man's suit.

b. This is a dynamic population, since men enter and leave this population primarily because of shifting shopping patterns and changing tastes.

Exercises 4.5

2. a. False. No objective measure of random error is possible with judgment sampling, because of the nonrandom method of sampling.

b. False. Sampling can reduce the time spent and the likelihood of error.

c. True

CHAPTER 5

Exercises 5.2

2.

$\mu_{n\bar{p}}$	$\sigma_{n\bar{p}}$
8 persons	$\sqrt{(20)(0.4)(0.6)} = 2.19$ persons
16 persons	$\sqrt{(40)(0.4)(0.6)} = 3.10$ persons
32 persons	$\sqrt{(80)(0.4)(0.6)} = 4.38$ persons

4. $\mu_{\bar{p}} = 0.80$
$\sigma_{\bar{p}} = \sqrt{pq/n} = \sqrt{(0.80)(0.20)/100} = 0.04$

Exercises 5.4

2. a. $z_1 = \frac{45-50}{10} = -0.50$
$z_2 = \frac{62-50}{10} = 1.20$
$P(45 \leqslant x \leqslant 62) = P(-0.50 \leqslant z \leqslant 1.20)$
$= 0.1915 + 0.3849$
$= 0.5764$

b. $P(x < 50) = P(z < 0)$
$= 0.50$

c. $z_0 = \frac{67-50}{10} = 1.70$
$P(x > 67) = P(z > 1.70)$
$= 0.5000 - 0.4554 = 0.0446$

d. $z_0 = \frac{38-50}{10} = -1.20$
$P(x < 38) = P(z < -1.20)$
$= 0.5000 - 0.3849 = 0.1151$

4. $z_0 = \frac{10-15}{2.5} = -2.0$
$P(x < 10) = P(z < 2.0) = 0.0228$

6. a. $z_0 = \frac{1,750-1,520}{125} = 1.84$
$P(x > 1,750) = P(z > 1.84) = 0.5000 - 0.4671 = 0.0329$

b. $z_0 = \frac{1,480-1,520}{125} = -0.32$
$P(x \leqslant 1,480) = P(z \leqslant -0.32) = 0.5000 = 0.1255 = 0.3745$

c. $z_1 = \frac{1,420-1,520}{125} = -0.80$
$z_2 = \frac{1,480-1,520}{125} = -0.32$
$P(1,420 \leqslant x \leqslant 1,480) = P(-0.80 \leqslant z \leqslant -0.32)$
$= 0.2881 - 0.1255 = 0.1626$

d. $z_1 = \frac{1,420-1,520}{125} = -0.80$
$z_2 = \frac{1,750-1,520}{125} = 1.84$
$P(1420 \leqslant x \leqslant 1750) = P(-0.80 \leqslant z \leqslant 1.84)$
$= 0.2881 + 0.4671 = 0.7552$

8. $\mu = np = (15)(0.4) = 6$ tires

$\sigma = \sqrt{npq} = \sqrt{(15)(0.4)(0.6)} = 1.9$ tires

a. $z_1 = \frac{4.5-6}{1.9} = -0.79$

$z_2 = \frac{5.5-6}{1.9} = -0.26$

$P(X = 5) = P(-0.79 \leqslant z \leqslant -0.26)$

$= 0.2852 - 0.1026 = 0.1826$

b. $z_1 = \frac{3.5-6}{1.9} = -1.32$

$z_2 = \frac{7.5-6}{1.9} = 0.79$

$P(4 \leqslant x \leqslant 7) = P(-1.32 \leqslant z \leqslant 0.79)$

$= 0.4066 + 0.2852$

$= 0.6918$

10. a. With $n = 20$ and $p = 0.05$, from Appendix Table A-1, we have

$P(x \geqslant 2) = 1 - F(1) = 1 - 0.7358 = 0.2642$

b. $P\left(z \geqslant \dfrac{1.5 - 1}{\sqrt{(0.05)(0.95)(20)}}\right) = 0.3015$

Exercises 5.5

2. a. $z_0 = \dfrac{2{,}100 - 1{,}900}{200} = 1$

$P(x > 2{,}100) = P(z > 1) = 0.1587$

b. $z_0 = \dfrac{1{,}850 - 1{,}900}{200/\sqrt{30}} = -1.37$

$P(x < 1{,}850) = P(z < -1.37) = 0.0853$

4. a. $z_0 = \dfrac{\$3{,}000 - \$3{,}160}{\$800/\sqrt{50}} = -1.41$

$P(\bar{x} > \$3{,}000) = P(z > -1.41) = 0.9207$

b. $P(\bar{x} < 3{,}000) = P(z < -1.41)$

$= 0.0793$

c. $z_1 = \dfrac{\$3{,}200 - \$3{,}160}{\$800/\sqrt{50}} = 0.35$

$z_2 = \dfrac{\$3{,}300 - \$3{,}160}{\$800/\sqrt{50}} = 1.24$

$P(\$3{,}200 < \bar{x} < \$3{,}300) = P(0.35 < z < 1.24)$

$= 0.3925 - 0.1368 = 0.2557$

6. $z_0 = \dfrac{48 - 44}{16/\sqrt{64}} = 2$

$P(\bar{x} > 48) = P(z > 2) = 0.0228$

8. a. Agree
b. Disagree. The probability of obtaining heads on *exactly* half the tosses approaches 0 as the number of tosses approaches infinity.
c. Disagree. The Central Limit Theorem applies to the distribution of the means of samples, not to the population distribution.
d. Agree
e. Disagree. The Central Limit Theorem applies to the approach of the sampling distribution of the mean to normality as n becomes large.

10. a. $\mu_{\bar{x}} = \$68{,}900,\ \sigma_{\bar{x}} = \dfrac{\$4210}{\sqrt{25}}\sqrt{\dfrac{100-25}{100-1}} = \$732.87$

b. $\mu_{\bar{x}} = \$68{,}900;\ \sigma_{\bar{x}} = \dfrac{\$4210}{\sqrt{50}}\sqrt{\dfrac{100-50}{100-1}} = \$423.12$

c. $\mu_{\bar{x}} = \$68{,}900;\ \sigma_{\bar{x}} = 0$, since only one sample is possible.

Review Exercises for Chapters 1 through 5

2. a. *Poisson distribution*
If $\mu = 2$ per month, then $\mu = 1$ per 2 weeks
$\mu = 1;\ P(X \geqslant 1) = 1 - P(X = 0) = 1 - 0.368 = 0.632$
b. If $\mu = 2$ per month, $P(X \geqslant 1) = 1 - P(X = 0) = 1 - 0.135 = 0.865$
Binomial distribution
Let $p = 0.865 = P(\text{at least one accident in a given month})$
$q = 0.135$ and $n = 5$

$$f(4) = \binom{5}{4}(0.135)^1(0.865)^4 = 0.378$$

4. a. Treating the top 4 as a category, and using the hypergeometric distribution, we obtain:

$$P(\text{All men in top 4}) = \frac{\binom{4}{4}\binom{2}{0}}{\binom{6}{4}} = \frac{1}{15}$$

Alternatively
Using the multiplication rule for dependent events, we obtain:

$$P(\text{All men in top 4}) = \left(\frac{4}{6}\right)\left(\frac{3}{5}\right)\left(\frac{2}{4}\right)\left(\frac{1}{3}\right) = \frac{1}{15}$$

b. $P(0 \text{ or } 1 \text{ woman}) = \left[\binom{4}{3}\binom{2}{0} + \binom{4}{2}\binom{2}{1}\right] \Big/ \binom{6}{3} = \dfrac{4+12}{20} = \dfrac{4}{5}$

6. a. Binomial distribution:
$n = 10$, $p = 0.1$ $P(X \geqslant 1) = 1 - F(0) = 1 - 0.3487 \doteq 0.6513$
b. $P(X \leqslant 2) = 0.9298$ c. $\mu = np = 100(0.1) = 10$
$P(X \geqslant 1) = 1 - F(0) \simeq 1$

8. a. Poisson distribution: $\mu = 5 + \frac{20{,}000}{10{,}000} = 7$ sales per month
$P(X \leqslant 6) = F(6) = 0.45$
b. $\mu = 14$ sales per 2 months
$P(X > 12) = 1 - F(12) = 1 - 0.358 = 0.642$
c. Binomial distribution: $n = 12$, $p = 0.45$
$P(X = 6) = F(6) - F(5) = 0.7393 - 0.5269 = 0.2124$

10. a. $z_0 = \dfrac{35 - 30}{3} = 1.67$ $P(z > 1.67) = 0.5000 - 0.4525 = 0.0475$

b. $z_0 = \dfrac{31 - 30}{3/\sqrt{36}} = 2$ $P(z < 2) = 0.5000 + 0.4772 = 0.9772$

c. The answer to part (a) requires the assumption of a normally distributed population. The answer to part (b) does not require that assumption, because the Central Limit Theorem permits us to treat the sampling distribution of means as normal for large samples ($n > 30$).

12. a. *Normal curve approximation*
$\mu = (20)(0.20) = 4.0$
$\sigma = \sqrt{(20)(0.20)(0.80)} = 1.79$

$$z_0 = \frac{0.5 - 4.0}{1.79} = -1.96$$

$$z_1 = \frac{1.5 - 4.0}{1.79} = -1.40$$

$P(0.5 < X < 1.5) = P(-1.96 < z < -1.40) = 0.4750 - 0.4192 = 0.0558$
Binomial probability
$n = 20$, $p = 0.20$, $q = 0.80$
$P(X = 1) = 0.0576$
The normal curve approximation, 0.0558, is about 3.1% less than the binomial probability of 0.0576.

b. $\sigma_{\bar{p}} = \sqrt{\dfrac{(0.20)(0.80)}{100}} = 0.04$

$$z_0 = \frac{0.08 - 0.20}{0.04} = -3.00$$

$$z_1 = \frac{0.13 - 0.20}{0.04} = -1.75$$

$$P(0.08 < X < 0.20) = P(-3.00 < z < -1.75) = 0.49865 - 0.4599 = 0.03875$$

c. $2.05 = \dfrac{X_0 - 0.20}{0.04}$

$X_0 = 0.20 + 2.05(0.04) = 0.282 = 28.2\%$

14. a. $P(Y \geqslant 1 | X = 1) = \dfrac{0.10}{0.30} + \dfrac{0.05}{0.30} = 0.5$

b. $P(Y = 1 \text{ or } Y = 0 | X = 1 \text{ or } X = 2) = \dfrac{0.15 + 0.10 + 0.10 + 0.15}{0.30 + 0.25} = 0.91$

c. No, X and Y are not independent random variables.
For example, $P(X = 0 \text{ and } Y = 1) \neq P(X = 0)\ P(Y = 1)$
Numerically, $0.15 \neq (0.40)(0.40)$

d. .$E(X) = (0)(0.40) + (1)(0.30) + (2)(0.25) + (3)(0.05) = 0.95$
$E(X^2) = (0)^2(0.40) + (1)^2(0.30) + (2)^2(0.25) + (3)^2(0.05) = 1.75$
$\sigma^2 = 1.75 - (0.95)^2 = 0.85$

CHAPTER 6

Exercises 6.2

2. The larger the sample size, the greater is the cost of sampling. Therefore, most practical estimation situations involve trade-offs between precision and cost.

4. a. False. If a consistent, biased estimator is more efficient (i.e., has smaller sampling variability) than an unbiased estimator, the consistent, biased estimator may be preferable.

b. True

c. False. If two competing estimators are both *unbiased*, then the one with the smaller variance (for a given sample size) is said to be relatively more efficient.

Exercises 6.3

2. $110 \pm 1.96\left(\dfrac{20}{\sqrt{100}}\right)$ or 106.08 to 113.92 pounds

4. $\$40{,}100 \pm 1.65\left(\dfrac{\$4{,}100}{\sqrt{100}}\right)$ or \$39,424 to \$40,777

6. $(\$36{,}000 - \$32{,}000) \pm 1.96\sqrt{\dfrac{(2{,}800)^2}{35} + \dfrac{(2{,}500)^2}{40}}$ or \$2,791 to \$5,209

8. $(12.8 - 10.1) \pm 2\sqrt{\frac{(2.6)^2}{40} + \frac{(2.2)^2}{50}}$ or 1.67 to 3.73 years

10. a. $\$23{,}500 \pm 1.96\left(\frac{5{,}700}{\sqrt{900}}\right)$ or \$23,128 to \$23,872

b. $23{,}000 \pm 1.96\left(\frac{4{,}200}{\sqrt{400}}\right)$ or \$22,588 to \$23,412

c. $(\$23{,}500 - \$23{,}000) \pm 1.96\sqrt{\frac{(5{,}700)^2}{900} + \frac{(4200)^2}{400}}$ or $-\$55$ to \$1,055

Exercises 6.4

2. a. $51 \pm 3.355\ (16.70/\sqrt{9}) = 51 \pm 18.02$, or from 32.98 minutes to 69.02 minutes.

b. It is assumed in the derivation of the t distribution that the underlying population is normally distributed, and this assumption would be violated by a highly skewed population distribution. Hence, the range set up in part (a) would not have exactly a 99% confidence coefficient associated with it.

4. $\$750{,}000 \pm 3.106\left(\frac{\$100{,}000}{\sqrt{12}}\right)$ or \$660,338 to \$839,662

$\$750{,}000 \pm 2.807\left(\frac{\$100{,}000}{\sqrt{24}}\right)$ or \$692,702 to \$807,298

Exercises 6.5

2. $n = \frac{(2.58)^2(4)^2}{(2)^2} = 26.6 \simeq 27$ customers

4. $n = \frac{(2.00)^2(0.2)(0.8)}{(0.05)^2} = 256$ business persons

6. a. 95.5% of all the possible samples he could draw will yield a correct interval estimate, whereas 4.5% of all possible samples will lead to an incorrect interval estimate.

b. $n = \frac{(2)^2(\$0.50)^2}{(\$0.05)^2} = 400$ employees

8. The cost of taking a larger sample might be uneconomical when compared with the return on the sample information.

10. a. $n = \dfrac{(1.65)^2(\$500)^2}{(\$50)^2} = 272.25 \simeq 273$ letters and telegrams

b. $n = \dfrac{(2.58)^2(\$500)^2}{(50)^2} = 665.64 \simeq 666$ letters and telegrams

Computer Exercises 6.5

2. *98% Confidence Intervals*

Annual Income: $37.947 \pm 2.33\left(\dfrac{19.4}{\sqrt{75}}\right) = 32.728$ to 43.166 (thousand dollars)

Family Size: $2.7733 \pm 2.33\left(\dfrac{1.55}{\sqrt{75}}\right) = 2.36$ to 3.19 persons

Age of Highest Income Earner: $37.520 \pm 2.33\left(\dfrac{10.9}{\sqrt{75}}\right) = 34.6$ to 40.5 years

CHAPTER 7

Exercises 7.2

2. The statement is incorrect, since there are two types of errors that can occur, Type I and Type II. If α (the probability of a Type I error) is made extremely small, the Type II error probability (with a fixed sample size) will tend to become quite large. The two errors, for a fixed sample size, move inversely to one another. As an extreme case, for example, if we always accept H_0 regardless of test results, the α error is 0, but the β error is equal to 1.

4. Power is defined as the probability of rejecting H_0 when H_1 is the true state of nature. Power $= 1 - \beta$.

6. a. Critical value $= 37 - 1.28\left(\dfrac{11}{\sqrt{100}}\right) = 35.6$ hours

Decision Rule

1. Reject H_0 if $\bar{x} < 35.6$ hours
2. Accept H_0 otherwise

b. H_0: $\mu \geqslant 37$
H_1: $\mu < 37$

c. Decision: Since $\bar{x} = 35$ hours < 35.6, we reject H_0 and the shipment.

8. H_0: $p = 0.8$
H_1: $p \neq 0.8$
$\alpha = 0.01$

Critical values $= 0.8 \pm (2.57)\sqrt{\dfrac{(0.8)(0.2)}{625}} = (0.76, 0.84)$

Decision Rule
1. Reject H_0 if $p < 0.76$ or $\bar{p} > 0.84$
2. Accept H_0 otherwise $(0.76 \leqslant \bar{p} \leqslant 0.84)$

10. a. H_0: $\mu = 120$
H_1: $\mu \neq 120$
$\alpha = 0.05$

Critical values $= 120 \pm 1.96(10/\sqrt{50})\sqrt{\dfrac{130 - 50}{130 - 1}} = 120 \pm 2.18$

Decision Rule
1. Reject H_0 if $\bar{x} < 117.82$ or $\bar{x} > 122.18$
2. Accept H_0 if $117.82 \leqslant \bar{x} \leqslant 122.18$

Since $\bar{x} = 118$, we accept H_0 and conclude that the number of ski customers at the Snow Mountain resort has not changed.

b. $P(|\bar{x} - 120| > 3) = P(|z| > 2.69) = 0.0072$
Therefore, the level of significance or p-value is 0.0072.

12. H_0: $\mu \leqslant 0.7$ grains
H_1: $\mu > 0.7$ grains
$\alpha = 0.01$
Critical value $= 0.7 + 2.33(0.05/\sqrt{225}) = 0.7078$ grains

Decision Rule
1. Reject H_0 (and lot) if $\bar{x} > 0.7078$ grains
2. Accept H_0 (and lot) otherwise

14. a. $P(\text{accept } H_0 | \mu = \$61)$

$$= P\left(\frac{\$58.72 - \$61}{\$1} < z < \frac{\$61.28 - \$61}{\$1}\right) = 0.5990$$

b. The power of the test when $\mu = \$61$ is the probability of rejecting the null hypothesis, H_0: $\mu = \$60$, for $\mu = \$61$. The power for $\mu = \$61$ is $1 - \beta = 1 - 0.5990 = 0.4010$.

c. $\$1.28 = z(\$20/\sqrt{400})$
$z = 1.28$
$\alpha = 0.20$ level of significance

16. a. The Type I error here is an incorrect decision to stop marketing the product.

b. Probability of Type I error $= P(\bar{p} \leqslant 0.15 | p = 0.20)$

$$= P\left(z \leqslant \frac{0.15 - 0.20}{\sqrt{(0.2)(0.8)/200}}\right)$$

$$= P(z \leqslant -1.77) = 0.0384$$

18. No. Since the assistant accepted the null hypothesis H_0: $\mu \geqslant \$10{,}000$, the possible error is Type II. Therefore, if β is large, the evidence is not "strong" in favor of the second proposal. Furthermore, because the standard deviation of the second proposal is greater than that of the first proposal, the risk is correspondingly greater.

Computer Exercises 7.2

2. H_0: $\mu \leqslant 34.0$ (thousand dollars)
H_1: $\mu > 34.0$
$\alpha = 0.01$
Critical value $= 34.0 + 2.33(19.4/\sqrt{75}) = 39.2220$

Decision Rule
1. Reject H_0 if $\bar{x} > 39.220$
2. Accept H_0 if $\bar{x} \leqslant 39.220$

Since $\bar{x} = 37.947 < 39.220$, accept H_0. We cannot consider the claim valid at the 0.01 level of significance.

Exercises 7.3

2. H_0: $p_1 = p_2$ $\qquad \hat{p} = \dfrac{22 + 18}{50 + 50} = 0.4$
H_1: $p_1 \neq p_2$
$\alpha = 0.05$
Critical value $= 0 \pm 1.96\sqrt{(0.4)(0.6)[\frac{1}{50} + \frac{1}{50}]}$

Decision Rule:
1. Reject H_0 if $|\bar{p}_1 - \bar{p}_2| > 0.192$
2. Accept H_0 if $|\bar{p}_1 - \bar{p}_2| \leqslant 0.192$

Decision: Accept H_0 because $|0.44 - 0.36| < 0.192$ and conclude "no real difference."

4. a. $H_0: \mu_1 = \mu_2$
$\mu_1 \neq \mu_2$
$\alpha = 0.02$
Critical values $= 0 \pm (2.33)\sqrt{(140)^2/100 + (120)^2/150} = 0 \pm 39.82$

Decision Rule
1. Reject H_0 if $\bar{x}_1 - \bar{x}_2 < -39.82$ or $\bar{x}_1 - \bar{x}_2 > 39.82$
2. Accept H_0 if $-39.82 \leqslant \bar{x}_1 - \bar{x}_2 \leqslant 39.82$

Decision: Reject H_0, since $1500 - 1450 = 50 > 39.82$

b – d.

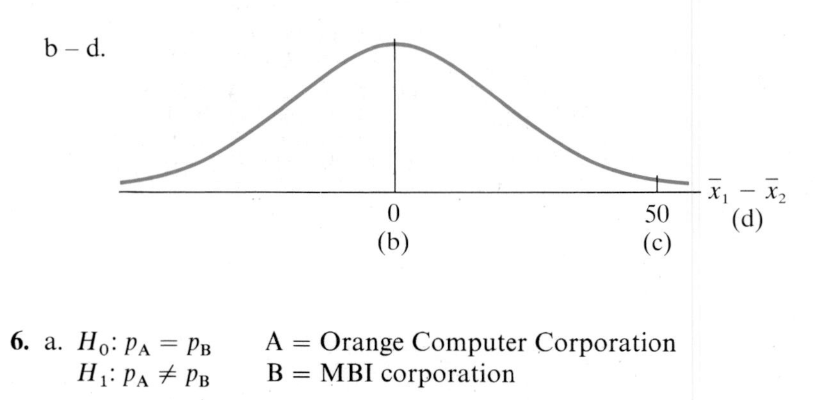

6. a. $H_0: p_A = p_B$ A = Orange Computer Corporation
$H_1: p_A \neq p_B$ B = MBI corporation

$$\hat{p} = \frac{120 + 231}{200 + 300} = 0.702$$

Critical values $= 0 \pm 1.96\sqrt{(0.702)(0.298)(\frac{1}{200} + \frac{1}{300})}$
$= \pm 0.082$

Decision Rule
1. Reject H_0 if $|\bar{p}_A - \bar{p}_B| > 0.082$
2. Accept H_0 if $|\bar{p}_A - \bar{p}_B| \leqslant 0.082$

Reject H_0 since $|\bar{p}_A - \bar{p}_B| = |0.60 - 0.77| = 0.17 > 0.082$

b. Type I error means concluding that there is a difference in the proportions of Orange Computer Corporation employees and MBI Corporation employees who favor the merger, when in fact, the proportions are the same.

Exercises 7.4

2. The test statistic should be a t statistic with 19 degrees of freedom.

4. $H_0: \mu_1 = \mu_2$
$H_1: \mu_1 \neq \mu_2$
$\alpha = 0.01$

$$\hat{s}^2 = \frac{6(54) + 3(60)}{6 + 3} = 56$$

$$s_{\bar{x}_1 - \bar{x}_2} = \sqrt{\frac{56}{7} + \frac{56}{4}} = 4.69$$

$$t = \frac{35 - 44}{4.69} = -1.92$$

Critical values for $t = \pm 3.25$

a. Since $t = -1.92 > -3.25$, we agree that there was not a statistically significant difference between the sample averages.
b. No, because if there was a small difference in the population means, it would still be very likely that the difference of the sample means would fall between -15.24 and 15.24. That is, the power of the test is very low when $\mu_1 - \mu_2$ differs from zero by a small amount, and the sample sizes are so small.

Exercises 7.5

2. $n = \left[\frac{(1.65 + 2.33)\$9}{(\$30 - \$28)}\right]^2 \approx 321$

Decision Rule

1. If $\bar{x} < \$29.17$, reject H_0: $\mu = \$30$
2. If $\bar{x} \geqslant \$29.17$, accept H_0: $\mu = \$30$

Exercises 7.7

2. Let μ_a and μ_b equal the mean population weights after and before diet, respectively.
H_0: $\mu_a = \mu_b$
H_1: $\mu_a > \mu_b$

$$s_{\bar{d}} = \frac{-2}{\sqrt{10}} = -0.63 \text{ pounds}$$

Decision Rule

1. Reject H_0 if $t < -1.833$
2. Accept H_0, otherwise

Decision: Since $t = \frac{-2}{0.63} < -1.833$, reject H_0 ($t = -1.833$ for 9 degrees of freedom in a one-tailed test). Conclude that the diet is effective.

CHAPTER 8

Exercises 8.1

2. H_0: The distribution is binomial with $p = \frac{1}{2}$
H_1: The distribution is not binomial with $p = \frac{1}{2}$
$\chi^2 = 10.934$
$\nu = 6$
$\chi^2_{0.05} = 12.592$
Hence, accept H_0 and conclude that the binomial distribution with $p = \frac{1}{2}$ is a good fit.

4. H_0: The distribution is Poisson with $\mu = 2.8$
H_1: The distribution is not Poisson with $\mu = 2.8$

Number of Calls (1)	Observed Number of Days f_o (2)		(Col. 1 × Col. 2) (3)	$f(x)$ (4)	Expected Number of Days f_t (5)	
0	5		0	0.061	6.1	
1	7		7	0.170	17.0	
2	30		60	0.238	23.8	
3	40		120	0.223	22.3	
4	7		28	0.156	15.6	
5	4		20	0.087	8.7	
6	5	} 7	30	0.041	4.1	} 6.3
7	1		7	0.016	1.6	
8	1		8	0.006	0.6	
Total	100		280	0.998	99.8	

Mean $= \frac{280}{100} = 2.8$ calls
$\chi^2 = 29.102$
$\nu = 5$
$\chi^2_{0.01} = 15.086$
Hence, reject H_0, and conclude that the Poisson distribution with $\mu = 2.8$ is not a "good fit."

6. H_0: The probability distribution is uniform
H_1: The probability distribution is not uniform

Brand Preferred	Observed Frequency f_o	Expected Frequency f_t
A	27	20
B	16	20
C	22	20
D	18	20
E	17	20
Total	100	100

$\chi^2 = 4.100$
$\nu = 4$
$\chi^2_{0.05} = 9.488$
$\chi^2_{0.01} = 13.277$

Hence, accept H_0 at both $\alpha = 0.05$ and $\alpha = 0.01$. The data are consistent with the hypothesis of no real difference in taste preference among the five brands.

Exercises 8.3

2. H_0: The proportions hired were the same for all three categories of applicants.
H_1: The proportions hired were not the same for all three categories of applicants.

f_o	f_t
175	135
100	120
25	45
275	315
300	280
125	105
1,000	1,000

$\chi^2 = 34.392$
$\nu = 2$
$\chi^2_{0.05} = 5.991$
Therefore, reject H_0 at $\alpha = 0.05$ and conclude that the proportions hired were not the same for the three categories of applicants.

4. H_0: Starting salary and grade point average are independent.
H_1: Starting salary and grade point average are not independent.

f_o	f_t
20	16.7
45	44.4
35	38.9
40	33.3
90	88.9
70	77.8
15	25.0
65	66.7
70	58.3
450	450.0

$\chi^2 = 9.584$
$\nu = 4$
$\chi^2_{0.01} = 13.277$
$\chi^2_{0.05} = 9.488$

Therefore, H_0 would be rejected at the 0.05 significance level, but would be accepted at the 0.01 significance level

6. a. H_0: Preferences for ice creams are independent of buyer's sex
H_1: Preferences for ice creams are not independent of buyer's sex

$\chi^2 = 3.72$
$v = (2 - 1)(3 - 1) = 2$
$\chi^2_{0.05} = 5.991$
Hence, accept H_0 at $\alpha = 0.01$.

b. $\chi^2 = 20.582$
$v = (2 - 1)(4 - 1) = 3$
$\chi^2_{0.01} = 11.345$
Hence, reject H_0 at $\alpha = 0.01$.

Exercises 8.4

2. H_0: There is no real difference among subcontractors in average number of defectives per batch.
H_1: There is a real difference among subcontractors in average number of defectives per batch.

Analysis of variance table:

Source of Variation	Sum of Squares	Degrees of Freedom	Mean Square
Between contractors	430	2	215
Within contractors	510	12	42.5
Total	940	14	

$F(2, 12) = \frac{215}{42.5} = 5.06$
$F_{0.01}(2, 12) = 6.93$

Since $5.06 < 6.93$, accept H_0. Conclude that there is no real difference among contractors in average number of defectives per batch.

4. H_0: There are no differences among the various types of deals
H_1: There are difference among the various types of deals

Analysis of variance table:

Source of Variation	Sum of Squares	Degrees of Freedom	Mean Square
Between types of deal	1,400	4	350.0
Within types of deal	1,691	95	17.8
Total	3,091	99	

$F(4, 95) = \frac{350}{17.8} = 19.663$
$F_{0.05}(4, 95) \simeq 2.47$

Since $19.663 > 2.47$, reject H_0, and conclude that there was a real difference among these types of acquisition deals.

Review Exercises for Chapters 6 through 8

2. a. $\sigma_{\bar{x}} = \dfrac{0.4}{\sqrt{100}} = 0.04$

$\bar{x} \pm 2.58\sigma_{\bar{x}} = 3.2 \pm 2.58(0.04) = 3.097$ to 3.303 mg

b. $n = \dfrac{(2.58)^2(0.4)^2}{(0.05)^2} = 426$

4. a. $H_0\colon \mu_1 - \mu_2 = 0 \qquad \alpha = 0.05$
$H_1\colon \mu_1 - \mu_2 \neq 0$

$$s_{\bar{x}_1 - \bar{x}_2} = \sqrt{\frac{9^2}{400} + \frac{12^2}{400}} = 0.75$$

Critical $\bar{x}_1 - \bar{x}_2 = 0 \pm 1.96(0.75) = \pm 1.47$
Since $20.0 - 24.0 = -4 < -1.47$, reject H_0 and conclude that there is a difference in average coffee consumption in the two regions.

b. True, because of symmetry in the power curve.

c. False. The size of the critical region decreases.

d. False. $\alpha + \beta \neq 1$. However, for a fixed sample size, increasing α does bring about decreases in probabilities of Type II errors.

6. a. $H_0\colon p \geqslant 0.90 \qquad \alpha = 0.05$
$H_1\colon p < 0.90$

$$\sigma_{\bar{p}} = \sqrt{\frac{(0.90)(0.10)}{100}} = 0.03$$

Critical $\bar{p} = 0.90 - 1.65(0.03) = 0.85$

Decision Rule:

1. If $\bar{p} \geqslant 0.85$, accept H_0
2. If $\bar{p} < 0.85$, reject H_0

Conclusion: Since $\bar{p} = 0.88 > 0.85$, accept H_0.
That is, we cannot reject Youth Magazine's statement.
Note that although the burden of proof is usually placed on a claim to demonstrate itself by having the claim in the alternative hypothesis, because of the wording of the exercise, the claim (Youth Magazine's statement) is in the null hypothesis.

b. $z_0 = \dfrac{0.85 - 0.87}{\sqrt{\dfrac{(0.87)(0.13)}{100}}} = -0.59$

$P(\bar{p} > 0.85 \mid p = 0.87) = P(z > 0.59) = 0.2224 + 0.5000 \simeq 0.72$

c. $n = \dfrac{(1.65)^2(0.20)(0.80)}{(0.02)^2} = 1{,}089$

8. a. $s_{\bar{x}} = \dfrac{\$1{,}000}{\sqrt{4}} = \$500$

for $v = 3$, $t = 3.182$ (95% confidence)
Confidence Interval
$\$11{,}000 \pm 3.182(\$500) = \$9{,}409$ to $\$12{,}591$
The assumptions are

1. The four new employees are a simple random sample of all new employees.
2. The population of all new employees' salaries is normally distributed.

b. $1.96 \dfrac{\sigma}{\sqrt{n}} = \dfrac{1}{3}\sigma$

$\sqrt{n} = 3(1.96) = 5.88$
$n = (5.88)^2 = 34.6 \simeq 35$

c. Assume that by the Central Limit Theorem e is normally distributed with mean μ_e and standard deviation σ_e.

$$e = \frac{X_1 + 2X_2 + 2X_3 + X_4}{6}$$

$$\sigma_e^2 = \frac{\sigma^2(X_1) + 2^2\sigma^2(X_2) + 2^2\sigma^2(X_3) + \sigma^2(X_4)}{6^2}$$

But $\sigma^2(X_1) = \sigma^2(X_2) = \sigma^2(X_3) = \sigma^2(X_4) = \sigma^2$. That is, all observations come from the same probability distribution. Therefore,

$$\sigma_e^2 = \tfrac{10}{36}\sigma^2 = \tfrac{5}{18}\sigma^2$$

$$\sigma_e = \sqrt{\tfrac{5}{18}}\sigma = 0.5270(\$1{,}000) = \$527$$

95% Confidence Interval
$\$11{,}200 \pm 1.96(\$527) = \$11{,}200 \pm \$1{,}033 = 10{,}167$ to $\$12{,}233$

10. a. $\bar{p} = \frac{32}{100} = 0.32$

b. $\bar{p} = 0.32$ is an estimate of $\mu_{\bar{p}}$, $s_{\bar{p}}^2 = \dfrac{(0.32)(0.68)}{100}$

$= 0.0022$ is an estimate of $\sigma_{\bar{p}}^2$.

c. $\bar{p} \pm 2.33 s_{\bar{p}} = 0.32 \pm (2.33)(0.047) = 0.32 \pm 0.11$
$= 0.21$ to 0.43

d. $n = \dfrac{(1.645)^2(0.40)(0.60)}{(0.02)^2} = 1623.6 \simeq 1624$

12. a. $H_0: \mu_1 \leqslant \mu_2 \qquad \alpha = 0.05$
$H_1: \mu_1 > \mu_2$

$$\hat{s}^2 = \frac{(15)(78) + (15)(84)}{16 + 16 - 2} = 81$$

$$s_{\bar{x}_1 - \bar{x}_2} = 9\sqrt{\frac{1}{16} + \frac{1}{16}} = 3.182$$

$$t = \frac{74 - 68}{3.182} = 1.89$$

$t_{0.05,30}$ in a one-tailed test $= 1.697$
Since $1.89 > 1.697$, reject H_0 and conclude that the mini-course is worthwhile.

b. $H_0: \mu_1 - \mu_2 \leqslant 2 \qquad \alpha = 0.05$
$\mu_1 - \mu_2 > 2$
Critical $\bar{x}_1 - \bar{x}_2 = 2 + 1.697(3.182) = 7.40$
Since $74 - 68 = 6 < 7.40$, accept H_0. At $\alpha = 0.05$, we would not conduct the mini-course.

c. False d. True e. True

14. a. $$z_1 = \frac{400 - 500}{50} = -2$$

$$z_2 = \frac{450 - 500}{50} = -1$$

$P(400 \leqslant x \leqslant 450) = P(-2 \leqslant z \leqslant -1) = 0.4772 - 0.3413 = 0.1359$

b. $$\sigma_x = \frac{50}{\sqrt{100}} = 5$$

$$\bar{x} = 505,\ z = \frac{505 - 500}{500} = 1$$

$P(495 \leqslant \bar{x} \leqslant 505) = P(-1 \leqslant z \leqslant 1) = 2(0.3413) = 0.6826$

c. $$\sigma_p = \sqrt{\frac{(0.10)(0.90)}{100}} = 0.03$$

$$\bar{p}_1 = 0.16,\ z_1 = \frac{0.16 - 0.10}{0.03} = 2$$

$$\bar{p}_2 = 0.13,\ z_2 = \frac{0.13 - 0.10}{0.03} = 1$$

$P(0.13 \leqslant \bar{p} \leqslant 0.16) = P(1 \leqslant z \leqslant 2) = 0.4772 - 0.3414 = 0.1359$

16. a. $10 \pm 1.96 \dfrac{6}{\sqrt{36}} = 8.04\%$ to 11.96%

b. $n = \dfrac{(1.96)^2(0.06)^2}{(0.01)^2} = 138.3 \simeq 139$

c. $P\left(z_0 < \dfrac{x_0 - 0.10}{0.06}\right) = 0.10$

$-1.28 = \dfrac{x_0 - 0.10}{0.06}$

$x_0 = 0.023 = 2.3\%$

18. a. H_0: $\mu_1 \leqslant 50$ $\quad \alpha = 0.05$
$\mu_1 > 50$

Critical value of $\bar{x} = 50 + 1.711 \dfrac{\sqrt{466}}{\sqrt{25}} = 57.39$

Since $\bar{x}_1 = 62 > 57.39$, reject H_0. The claim that the mean rating for the first advertisement exceeds 50 is supported by the data.

b. H_0: $\mu_3 = \mu_4$ $\quad \alpha = 0.05$
$\mu_3 \neq \mu_4$

$\hat{s}^2 = \dfrac{24(422) + 24(508)}{25 + 25 - 2} = 465$

Critical values of $\bar{x}_3 - \bar{x}_4 = \pm 2.01\sqrt{465}\sqrt{\frac{1}{25} + \frac{1}{25}} = \pm 12.2594$. Since $\bar{x}_3 - \bar{x}_4 = -14$, reject H_0. The data support the claim that the mean ratings of the third and fourth advertisements are different.

c. H_0: $\mu_1 = \mu_2 = \mu_3 = \mu_4$
H_1: Not all μ_i are equal

Analysis of variance table

Source of Variation	Sum of Squares	Degrees of Freedom	Mean Square
Between advertisements (SSA)	3,000	3	1,000
Within advertisements (SSE)	43,584	96	454
Total	46,584	99	

Calculations:

$SSA = 25[(62 - 60)^2 + (56 - 60)^2 + (54 - 60)^2 + (68 - 60)^2]$
$= 3{,}000$

$SSE = 24[466 + 420 + 422 + 508] = 43{,}584$

$F(3, 99) \simeq 2.70$
$F = \frac{1{,}000}{454} = 2.21$
Since $2.21 < 2.70$, retain H_0. The data do not support the claim that the mean ratings of the four advertisements are not all the same.

20. a. H_0: $p \leqslant 0.20 \qquad \alpha = 0.50$
H_1: $p > 0.20$

$$\text{Critical } \bar{p} = 0.20 + 1.65\sqrt{\frac{(0.20)(0.80)}{100}} = 0.266$$

Decision Rule
1. If $\bar{p} > 0.266$, reject H_0
2. If $\bar{p} \leqslant 0.266$, retain H_0

Conclusion: Since $0.30 > 0.266$, reject H_0. Yes, there is a statistically significant result indicating that the company should proceed with the manufacturing of the prototype.

b. H_0: $\mu \leqslant 100 \qquad \alpha = 0.01$
H_1: $\mu > 100$

$$\text{Critical } \bar{x} = 100 + (2.33)\frac{20}{\sqrt{25}} = 109.32$$

Since $105 < 109.32$, accept H_0. Manufacturing should not begin.

c. $P(\text{Accept } H_0 | \mu = 110) = P(x \leqslant 109.32 | \mu = 110)$

$$= P\left(z \leqslant \frac{109.32 - 110}{20/\sqrt{25}}\right) = P(z \leqslant 0.17) = 0.4325$$

22. a. H_0: Reactions are the same as ten years ago $\quad \alpha = 0.01$
H_1: Reactions are not the same as ten years ago

$$\chi^2 = \frac{(80 - 40)^2}{40} + \frac{(80 - 100)^2}{100} + \frac{(40 - 60)^2}{60} = 50.67$$

$\chi^2_{0.01,2} = 9.21$
Since $50.67 > 9.21$, reject H_0 and conclude that student reactions have changed from those of ten years ago.

b. H_0: Students' perceptions of degree of difficulty of the program are independent of student backgrounds.
H_1: Not so

$$\chi^2 = \frac{(30 - 20)^2}{20} + \frac{(20 - 20)^2}{20} + \cdots + \frac{(15 - 20)^2}{20} = 33.125$$

$\chi^2_{0.05,4} = 9.488$
Since $33.125 > 9.488$, reject H_0. Conclude that student perceptions of degree of difficulty of the program are not independent of student backgrounds.

CHAPTER 9

Exercises 9.4

2. a. The method of least squares fits a regression line that imposes the following requirement: The sum of the squares of the deviations of the observed values of the dependent variable from the corresponding computed values on the regression line must be a minimum.

b. (1) A linear relation exists between $\mu_{y.x}$ and X.
(2) The Y values are independent of one another.
(3) The conditional probability distributions of Y given X are normal.
(4) The conditional standard deviations $\sigma_{y.x}$ are equal for all values of X.

4. a. $a = Y$ intercept $= 20$ (thousands of cans of beer)

$$b = \text{slope} = \frac{60 - 20}{30 - 0} = 1.33 \text{ cans of beer per hot dog}$$

$$\hat{Y} = 20 + 1.33X$$

6. a. $$b = \frac{3{,}800 - (45)(17.5)(5)}{15{,}150 - (45)(17.5)^2} = -0.100$$

$$a = 5 - (-0.1)(17.5) = 6.75$$

b. $\hat{Y} = 6.75 - 0.1X$
$\hat{Y} = 6.75 - 0.1(22.5) = 4.5\%$

Exercises 9.5

2. a. $Y =$ First-year sales (in millions of dollars)
$X =$ Advertising expenditures (in millions of dollars)

$$b = \frac{1271.2 - (14)(1.3)(55)}{29.22 - (14)(1.3)^2} = \frac{270.2}{5.56} = 48.597$$

$$a = 55 - (49.597)(1.3) = -8.176$$
$$\hat{Y} = -8.176 + 48.597X$$

b. $$s_{Y.X} = \sqrt{\frac{56476 - (-8.176)(770) - (48.597)(127.12)}{12}}$$

$$= \sqrt{82.918} = 9.106 \text{ (millions of dollars)}$$

c. $$s_{\hat{Y}} = s_{Y.X}\sqrt{\frac{1}{14} + \frac{(1 - 1.3)^2}{29.22 - (18.2)^2/14}} = (9.106)\sqrt{0.0876}$$

$$= 2.695 \text{ (million of dollars)}$$

The estimated interval is $40.421 \pm (2.179)(2.695) = 40.421 \pm 5.872$, or from 34.549 to 46.293.

d. $s_{IND} = s_{Y.X}\sqrt{1 + \frac{1}{14} + \frac{(1 - 1.3)^2}{29.22 - (18.2)^2/14}} = (9.106)\sqrt{1.0876}$

$= 9.496$ (millions of dollars)

The estimated interval is $40.421 \pm (2.179)(9.496) = 40.421 \pm 20.692$, or from \$19.729 million to \$61.113 million

e. Factors affecting the width of intervals are

(1) The larger the sample size n the smaller are the estimated standard error of the conditional mean and the standard error of forecast and the narrower are the widths of the intervals.

(2) The greater the deviation of X from $\bar{X}$, the greater the standard error and the wider are the intervals for the given X value.

(3) The larger the estimated standard error of estimate $s_{Y.X}$, the larger are the confidence and prediction intervals.

(4) The more variability there is in the sample of X values, the smaller will be the standard errors and the narrower will be the width of confidence and prediction intervals.

Exercises 9.8

2. a. The figure 3 is the value at which the regression line crosses the Y axis. It represents an extrapolated value of \$300 of sales when the number of calls on prospects is zero. The figure 25 represents the estimated difference of \$2,500 in sales for each additional sales call on prospects; the relationship is direct.

b. $r_c^2 = 1 - \frac{625}{7,225} = 0.91$

This means that 91% of the variation in sales is explained by the regression equation that relates sales and calls on prospects by salespersons.

c. s_Y is the standard deviation of the Y values in the sample measured around the mean $\bar{Y}$. $s_{Y.X}$ is the standard deviation of the Y values in the sample measured around the corresponding $\hat{Y}$ values on the regression line. s_r is the standard deviation of the sampling distribution of the r values.

d. $$t = \frac{0.95}{\sqrt{\frac{1 - 0.91}{94}}} = 30.7$$

The critical t value at the 0.05 level is approximately 1.99. Since $30.7 > 1.99$, we reject H_0 and conclude that a linear relationship exists between sales and the number of calls on prospects.

4. a. $r_c^2 = 1 - \frac{s_{Y.X}^2}{s_Y^2} = 1 - \frac{25}{144} = 0.8264$

$r_c = \sqrt{0.8264} = 0.909$

The use of r values tends to give an impression of a stronger relationship than is actually present. Moreover, the use of r^2 is convenient because it can be interpreted as a proportion or percentage figure.

b. $$s_b = \frac{s_{Y.X}}{\sqrt{\Sigma(x - \bar{x})^2}} = \frac{5}{\sqrt{10{,}000}} = 0.05$$

$$t = \frac{b - B}{s_b} = \frac{0.3 - 0}{0.05} = 6$$

60 degrees of freedom
Since $6 > 2.66$, we reject at the 10% level of significance the null hypothesis that the population regression coefficient is zero.

6. a. For two weeks whose production levels differed by 1,000 units, the estimated difference in unit cost is \$0.10.

b. $r_c^2 = 1 - (s_{Y.X}^2/s_Y^2) = 1 - (0.8)^2/(1.5)^2 = 0.7156$
For the sample of 102 weeks, 89.76% of the variation in unit cost was explained by the regression equation that related unit cost to weekly production level.

c. $s_b = 0.8/\sqrt{1{,}650} = 0.0197$
$t = (-0.1 - 0)/0.0197 \doteq -5.076$, with 100 degrees of freedom. Since $|-5.076| > 2.631$, we conclude that the estimated regression coefficient is significantly different from zero at the 1% level.

d. The Y intercept ($a = 12.5$) represents the fixed cost that must be borne at a production level of zero.

e. $\hat{Y} = 12.5 - (0.1)(10) = 11.5$

$$s_{\text{IND}} = (0.8)\sqrt{1 + \frac{1}{102} + \frac{(45)^2}{1{,}650}} = 0.8088$$

A 95% prediction interval for cost per unit for a production level of 10,000 units is $11.5 \pm (1.99)(0.8088) =$ \$9.89 to \$13.11. Therefore, a cost of \$13.30 per unit would be considered unusually high for a production level of 10,000 units.

CHAPTER 10

Exercises 10.13

2. a. Since $F = 2306.98$, and $F_{0.01,3,96} = \simeq 3.98$, we reject the null hypothesis of no relationship between the dependent and independent variables. Also, the R^2 value of 0.9866 is very high, especially for cross-sectional data. These facts are evidence that the model is a good fit to the data. Since the t-value for the slope coefficients of all variables greatly exceed $t_{0.01,96} \simeq 2.617$, all independent variables are statistically significant.

b. $\hat{Y} = 12 + 5.575(1) + 9.8125(1) - 22.25(-2) = 71.89$ sales per capita
90% confidence interval
$t_{0.10,96} \simeq 1.66$, $S_{1.234} = 0.12$
$\hat{Y} \pm tS_{1.234} = 71.89 \pm 1.66(0.12)$
$= 71.69$ to 72.09 sales per capita

c. Add a dummy variable that takes on the value zero when there are no rebates and the value one when there are rebates. The coefficient of the dummy variable measures the increase in sales per capita because of rebates.

d. For the purpose of analyzing separate effects, it would be desirable to choose levels for which the advertising expenditures variables are as close to independent as possible. In this case, the correlation coefficient between newspaper advertising (variable 3) and promotion (variable 4) is 0.9, and the correlation coefficient between TV and promotion is 0.8. Because of this high degree of collinearity, we should be critical of the choices of levels of the different types of advertising.

4. a. $$F = \frac{2.45/(5-1)}{0.735/(27-5)} = 18.33$$

$F_{0.05;4,22} = 2.82$
Since $18.33 > 2.82$, we reject the null hypothesis of no relationship between the manufacturer's sales and all of the independent variables considered collectively.

b. Variable X_1: Yes, the minus sign is consistent with the usual price-demand inverse relationship. That is, as the price of the manufacturer's calculator increases, sales tend to decrease.
Variable X_2: Yes, the plus sign makes sense. With an increase in the competitor's price, the manufacturer's calculator sales tend to increase.
Variable X_3: Yes, the plus sign is reasonable. The manufacturer's calculator sales would tend to vary directly with advertising expenditures.
Variable X_4: Yes, the plus sign is reasonable. The maufacturer's calculator sales would tend to vary directly with the size of the college student population.

c. $$R^2 = 1 - \frac{\Sigma(Y - \hat{Y})^2}{\Sigma(Y - \bar{Y})^2} = 1 - \frac{0.735}{0.735 + 2.45} = 0.769$$

Adjusted for degrees of freedom

$$R^2 = 1 - \frac{\Sigma(Y - \hat{Y})^2/(n-k)}{\Sigma(Y - \bar{Y})^2/(n-1)} = 1 - \frac{0.735/(27-5)}{3.185/(27-1)} = 0.727$$

d. $b_3 \pm t_{0.10,22}s_{b_3} = 1.25 \pm 1.717(0.29) = 0.752$ to 1.748

e. Holding constant the effect of the other independent variables, a \$1.00 increase in the competitor's price is estimated to raise the manufacturer's sales by 0.58 thousands or by 580 calculators.

f. $\hat{Y} = -8.79 - 0.47(20) + 0.58(17.5) + 1.25(15) + 1.34(7.5) = 13.26$
90% prediction interval

$$\hat{Y} \pm t_{0.10,22}S_{Y.1234} \left(\text{where } S_{Y.1234} = \sqrt{\frac{0.735}{27-5}} = 0.1828\right)$$

$= 13.26 \pm 1.717(0.1828) = 12.946$ to 13.574 (thousands of calculators)

g. A dummy variable could be used for the absence or presence of the gift item. For example:
Let $X_5 = 0$ if the gift item was not given
$X_5 = 1$ if the gift item was given

6. a. This type of result could occur because X_2 and X_3 are correlated with one another (collinearity) and are also correlated with Y.

b. The F-test indicates only that we are better off using the predictor than not using it. A significant F-test does not imply that precise predictions will be made in terms of narrow confidence intervals.

c. The Durbin-Watson test is not valid for regression analyses of cross-sectional data. The test is valid for time series regression analyses.

8. a. The method of selection of the sample could cause considerable bias. The five outliers might have resulted in a second-degree parabola type of relationship. Note that the resulting linear fit is weak.

b. $$R^2 = \frac{17913301.69}{17913301.69 + 3462790.873} = 0.838$$

c. The slope coefficient of -0.2682 means that a country with \$100 higher per capita income (in constant U.S. dollars) than another country is estimated to have 26 fewer movie theatres, holding constant the effect of variables 3 and 6.
90% confidence interval for slope coefficient for variable 6
$b = -0.2681 \quad \nu = 25 - 4 = 21 \quad t = 1.721 \quad s_b = 0.3$
$-0.2681 \pm 1.721(0.3) = -0.7844$ to 0.2482

d. $$F = \frac{17913301.69/(4-1)}{3462790.873/(25-4)} = 36.2$$

$F_{0.01;3,21} = 4.87$
Since $36.2 > 4.87$, we reject the null hypothesis of no relationship between the number of movie theatres and the independent variables considered collectively.

e. The calculation yields the value of R_c^2

$$R_c^2 = 1 - \frac{S_{1.236}^2}{s_1^2} = 1 - \frac{(406.07241)^2}{890670.5233} = 0.8149$$

Note that this result is somewhat lower than $R^2 = 0.838$ calculated in part (b).

Computer Review Exercises for Chapters 9 and 10

2. a. $\hat{Y} = 55.967 + 0.0015935X_1 + 15.09X_2 - 0.2683X_3 + 0.0004192X_4$
Selected computer output:

VARIATE CORRELATIONS (r VALUES)

	Y	X1	X2	X3
X1	0.978			
X2	0.993	0.981		
X3	0.969	0.996	0.976	
X4	0.980	0.937	0.979	0.929

SLOPE COEFFICIENTS AND STANDARD ERRORS:

	COEFF.	STD. ERROR	t
Y-INTERCEPT	55.967	2.158	25.94
X1	0.0015935	0.0005777	2.76
X2	15.09	11.67	1.29
X3	-0.2683	0.1601	-1.68
X4	0.0004192	0.0001611	2.60

b. The estimated employment in manufacturing index increases by 15.09 points with each one point increase in the implicit gross national product price deflator, after netting out the effect of the other independent variables in the regression equation.

c. $v = n - k = 20 - 5 = 15$ degrees of freedom
$t_{0.05,15} = |2.131|$
Therefore, the slope coefficients for X_1 and X_4 are significant, and those for X_2 and X_3 are not significant.

d. The positive signs for value added in manufacturing (X_1), implicit gross national product price deflator (X_2), and fixed investment (X_4) seem reasonable because increases in these variables tend to indicate higher levels of industrial activity. Hence, we can expect higher levels of employment in manufacturing as firms hire more workers. The negative sign for the wage index for industrial establishments (X_3) is somewhat surprising because both wage levels and employment tend to increase during periods of expanding industrial activity. The high positive correlation coefficient ($r = 0.969$) confirms the expected direct correlation between these two variables. Since the slope coefficient for the wage index does not differ significantly from zero, the negative sign is not taken as indicative of an inverse relationship.

e. (See following printout)

ANALYSIS OF VARIANCE TABLE

EFFECTS	SUM OF SQUARES	DEGREES OF FREEDOM	MEAN SUM OF SQUARES	F
REGRESSION	5536.761	4	1384.190	484.660
ERROR	42.841	15	2.856	
TOTAL	5579.602	19		

Since $F = 484.660 > F_{0.05;4,15} = 3.06$, reject the null hypothesis of no relationship between the dependent variable and the independent variables.

f. $R^2 = 99.2\%$, uncorrected for degrees of freedom. About 99.2% of the variation in employment in manufacturing has been accounted for by the regression equation relating the dependent variable to the four independent variables.

g. There is a great deal of multicollinearity among the independent variables. All of the two variable correlation coefficients between pairs of independent variables exceed 0.92. Furthermore, the correlation coefficients of the dependent variable with each of the independent variables are 0.969 or greater. All of these series have increased concommitantly throughout the 20-year period.

4. a. The computer output shown here is from a MINITAB regression program.

```
--REGRESS Y IN C1 ON 4 PREDICTORS IN C2, C3, C4, C5

THE REGRESSION EQUATION IS
Y =     1.42 + 0.137 X1 + 0.0102 X2
   + 0.0073 X3 - 0.775 X4
```

	COLUMN	COEFF.	ST. DEV. OF COEFF.	T-RATIO = COEFF./S.D.
	--	1.4188	0.4433	3.20
X1	C2	0.137057	0.005285	25.93
X2	C3	0.01023	0.06200	0.16
X3	C4	0.00726	0.01028	0.71
X4	C5	-0.7751	0.2141	-3.62

```
THE ST. DEV. OF Y ABOUT REGRESSION LINE IS
S = 0.6921
WITH ( 75 - 5) = 70 DEGREES OF FREEDOM
```

```
R-SQUARED = 95.0 PERCENT
R-SQUARED = 94.7 PERCENT, ADJUSTED FOR D.F.

ANALYSIS OF VARIANCE

   DUE TO                DF             SS          MS=SS/DF
REGRESSION                4       640.5542          160.1386
RESIDUAL                 70        33.5332            0.4790
TOTAL                    74       674.0875

FURTHER ANALYSIS OF VARIANCE
SS EXPLAINED BY EACH VARIABLE WHEN ENTERED IN THE ORDER GIVEN

   DUE TO                DF             SS
REGRESSION                4       640.5542
C2                        1       630.4330
C3                        1         1.6275
C4                        1         2.2137
C5                        1         6.2799

--DESCRIBE C1-C5

C1   N = 75   MEAN =   6.5173    ST. DEV. =   3.02
C2   N = 75   MEAN = 37.947      ST. DEV. = 19.4
C3   N = 75   MEAN =   2.7733    ST. DEV. =   1.55
C4   N = 75   MEAN = 37.520      ST. DEV. = 10.9
C4   N = 75   MEAN =   0.52000   ST. DEV. =   0.503

--CORRELATION COEFF. BETWEEN C1-C5

          C1          C2          C3          C4
C2     0.967
C3     0.431       0.399
C4     0.611       0.572       0.491
C5    -0.612      -0.527      -0.472      -0.604
```

b. Variables X_2 (family size) and age of the highest income earner (X_3) were deleted. The t values for the slope coefficients of these variables were 0.16 and 0.71. For 70 degrees of freedom, the slope coefficients were not significantly different from zero ($\alpha = 0.05$). Selected computer output is shown for the following three variables:

Y = Annual food expenditures (in thousands of dollars)

X_1 = Annual income (in thousands of dollars)

X_2 = Home owner (0) or renter (1) status

```
--REGRESS Y IN C1 ON 2 PREDICTORS IN C2, C5

THE REGRESSION EQUATION IS
Y =     1.70 + 0.139 X1 - 0.852 X2

                                              ST. DEV.          T-RATIO =
          COLUMN             COEFF.           OF COEFF.         COEFF./S.D.
          --                 1.6989           0.2607             6.52
X1        C2                 0.138659         0.004828          28.72
X2        C5                -0.8525           0.1864            -4.57

THE ST. DEV. OF Y ABOUT REGRESSION LINE IS
S = 0.6855
WITH ( 75 - 3) = 72 DEGREES OF FREEDOM

R-SQUARED = 95.0 PERCENT
R-SQUARED = 94.8 PERCENT, ADJUSTED FOR D.F.

ANALYSIS OF VARIANCE

 DUE TO               DF              SS                MS=SS/DF
REGRESSION             2          640.2584              320.1292
RESIDUAL              72           33.8291                0.4698
TOTAL                 74          674.0875

FURTHER ANALYSIS OF VARIANCE
SS EXPLAINED BY EACH VARIABLE WHEN ENTERED IN THE ORDER GIVEN

 DUE TO               DF              SS
REGRESSION             2          640.2584
C2                     1          630.4330
C5                     1            9.8254
```

c. (1) Approximate 95% prediction interval for food expenditures for family number 20.

$$\begin{aligned}\hat{Y} &= 1.70 + 0.139X_1 - 0.852X_2\\ &= 1.70 + 0.139(13) - 0.852(1)\\ &= 2.655 \text{ (in thousands of dollars)}\end{aligned}$$

The approximate 95% prediction interval is

$$\begin{aligned}\hat{Y} \pm 1.96\, S_{Y.12} &= 2.655 \pm 1.96(0.6855)\\ &= 2.655 \pm 1.344\\ &= 1.31 \text{ to } 4.00 \text{ (thousands of dollars)}\end{aligned}$$

(2) Approximate 95% prediction interval for the slope coefficient for the income variable

$$b \pm 1.96\, s_b = 0.139 \pm 1.96\,(0.004828)$$
$$= 0.1295 \text{ to } 0.1485$$

Interpretations: Annual food expenditures for a family with \$13,000 income and renting a home is estimated (with 95% confidence) to be included in the range of \$1,310 and \$4,000. For a family with \$1,000 greater income than another, annual food expenditures can be expected (with 95% confidence) to be from \$129.50 to \$148.50 higher than for the lower income family. It is assumed that the two families are either both home owners or both renters.

d. The value of R^2 in the three variable model, unadjusted for degrees of freedom is 95.0%. The two variable correlation coefficient between annual food expenditures and annual income is 0.967. The coefficient of determination for these two variables is $(0.967)^2 = 93.5\%$. Therefore, income alone explains almost all of the variation in the dependent variable, food expenditures. The addition of the home ownership or renter status variable only increases the unadjusted R^2 value from 93.5% to 95.0%.

CHAPTER 11

Exercises 11.4

2. a. $a = \dfrac{\Sigma Y}{n} = \dfrac{224.1}{5} = 44.82$

$$b = \frac{\Sigma xY}{\Sigma x^2} = \frac{49.7}{10} = 4.97$$

$Y_t = 44.82 + 4.97x \quad (x = 0 \text{ in } 1984)$

Trend values (millions of dollars): 34.88, 39.85, 44.82, 49.79, 54.76

b. No, because the time period of five years is too short to get a good description of trend.

c. $\left(\dfrac{Y - Y_t}{Y_t}\right)(100) = \left(\dfrac{44 - 44.82}{44.82}\right)(100) = -1.83\%$

The relative cyclical residual for 1984 is -1.83%, indicating that the actual sales figure is 1.83% below the trend figure because of cyclical and irregular factors.

4. $a = \frac{\Sigma Y}{n} = \frac{1,891,483}{14} = 135,105.92$

$b = \frac{\Sigma xY}{\Sigma x^2} = \frac{1,031,543}{910} = 1133.56$

$Y_t = 135,105.92 + 1133.56x$
$x = 0$ in mid 1968
x is in half-year intervals;
Y is in thousands

Year	Projected Figure (in thousands)
1986	177,048
1987	179,315
1988	181,582
1989	183,849
1990	186,116

6. a. $a = \frac{\Sigma Y}{n} = \frac{347}{5} = 69.40$

$b = \frac{\Sigma xY}{\Sigma x^2} = \frac{406}{10} = 40.6$

$Y_t = 69.40 + 40.6x$
$x = 0$ in 1984
Y is in million dollars

b. The trend level for assets in 1984 is $69.40 million. The estimated increase in the trend of assets is $40.6 million per year.

c. A straight-line trend is probably a poor fit, because the series roughly doubles every year. Hence, this is a series characterized by a roughly constant *percentage rate of change.* A straight-line trend is appropriate for series characterized by constant *amounts of change per* unit time.

Exercises 11.5

2. a. Roughly, the seasonal index of 101 means that sales in October are typically 101% of average monthly sales for Expo Corporation. More precisely, the effect of seasonality in October is to raise the sales of that month by 1% above what they would have been in the absence of seasonal influence.

b. In the multiplicative model of time series anslysis, the original monthly figure is viewed as consisting of the product of the trend level, the cyclical relative (combined effect of cyclical and irregular factors), and the seasonal index. In symbols,

$$Y = (Y_t)\left(\frac{Y/SI}{Y_t}\right)(SI)$$

For November 1987

$Y_t = 214 + (0.32)(52) = 230.64$

$\frac{Y/SI}{Y_t} = \frac{243.64}{230.64} = 1.0564$

$\frac{Y}{SI} = \frac{268}{1.10} = 243.64$

$SI = 1.10$

Hence, $(230.64)(1.0564)(1.10) = 268$

4. a. Production of Amareck, Inc., is increasing at a decreasing rate.

b.

Seasonally adjusted figures

January	February	March
$\frac{40}{1.10} = 36.36$	$\frac{38}{1.00} = 38$	$\frac{36}{0.90} = 40$

January: $\text{Log } Y_t = 1.5500 + 0.0135(1) - 0.0007(1)^2$

$= 1.5628$

$Y_t = 36.54$

January: $\frac{O/SI}{Y_t} = \frac{36.36}{36.54} = 100.00\%$

February: $\text{Log } Y_t = 1.5500 + 0.0135(2) - 0.0007(2)^2$

$= 1.5742$

$Y_t = 37.52$

February: $\frac{O/SI}{Y_t} = \frac{38}{37.52} = 101.3\%$

March: $\text{Log } Y_t = 1.5500 + 0.0135(3) - 0.0007(3)^2$

$= 1.5842$

$Y_t = 38.39$

March: $\frac{O/SI}{Y_t} = \frac{40}{38.39} = 104.2\%$

No. Production of Amareck, Inc., increased cyclically from January to February and from February to March 1988.

c. No, because the time span used to compute the trend equation is far too short.

6. a.

Year	Percentage of Moving Averages by Quarter I	II	III	IV
1980	93.2	102.3	108.7	95.9
1981	~~94.8~~	101.0	109.5	94.8
1982	93.6	~~100.8~~	107.7	97.4
1983	~~90.1~~	~~103.4~~	107.8	96.5
1984	92.7	102.7	108.4	96.8
1985	94.0	101.1	~~105.2~~	~~93.1~~
1986	91.3	102.6	~~109.9~~	95.8
1987	92.2	103.1	107.9	~~97.7~~
Modified Means	92.8	102.1	108.3	96.2

Total Modified Means = 399.4

Adjustment Factor = $\frac{400}{399.4} = 1.0015$

	I	II	III	IV
Seasonal Indices:	92.9	102.3	108.5	96.3

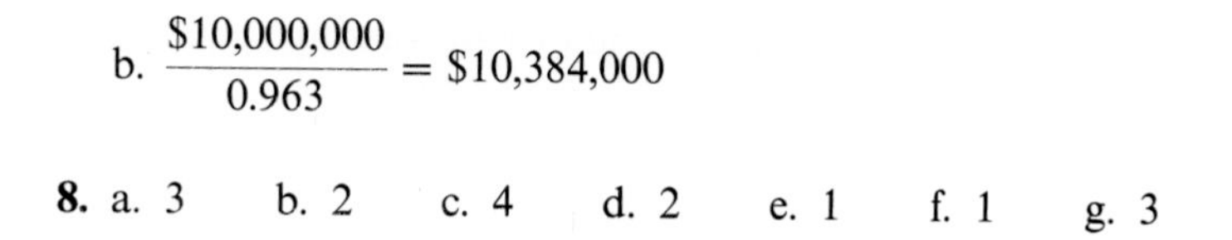

b. $\dfrac{\$10{,}000{,}000}{0.963} = \$10{,}384{,}000$

8. a. 3 b. 2 c. 4 d. 2 e. 1 f. 1 g. 3

Exercises 11.6

2.

Day	Next Forecast $w = 0.1$	Next Forecast $w = 0.7$
1	$10 = F_2$	$10 = F_2$
2	$10 = F_3$	$10 = F_3$
3	$10 = F_4$	$10 = F_4$
4	$11 = F_5$	$17 = F_5$
5	$11.9 = F_6$	$19.1 = F_6$
6	$12.71 = F_7$	$19.73 = F_7$
7	$13.439 = F_8$	$19.919 = F_8$
8	$14.0951 = F_9$	$19.9757 = F_9$
9	$14.68559 = F_{10}$	$19.99271 = F_{10}$
10	$15.217031 = F_{11}$	$19.997813 = F_{11}$

4.

Year	Next Forecast $w = 0.4$
1975	$F_{76} = \$4.23$
1976	$F_{77} = \$4.406$
1977	$F_{78} = \$4.4996$
1978	$F_{79} = \$4.62376$
1979	$F_{80} = \$4.662256$
1980	$F_{81} = \$4.7773536$
1981	$F_{82} = \$4.87441216$
1982	$F_{83} = \$5.084647296$
1983	$F_{84} = \$5.270788378$
1984	$F_{85} = \$5.642473027$
1985	$F_{86} = 5.885483816$
1986	$F_{87} = 6.051290290$

The forecast of \$6.05 for 1987 is 96.3% of the actual figure of \$6.28.

CHAPTER 12

Exercises 12.2

2. a. Weighted aggregates price index with base period weights for 1975 on 1982 base:

$$\frac{\Sigma P_{75}Q_{82}}{\Sigma P_{82}Q_{82}} \cdot 100 = \frac{(\$35)(3) + (\$40)(2) + (\$8.50)(4)}{(\$50)(3) + (\$50)(2) + (\$12.50)(4)} \cdot 100 = \frac{\$219}{\$300} \cdot 100$$

$$= 73.0$$

For 1985 on 1982 base:

$$\frac{\Sigma P_{85}Q_{82}}{\Sigma P_{82}Q_{82}} \cdot 100 = \frac{(\$65)(3) + (\$65)(2) + (\$16)(4)}{(\$50)(3) + (\$50)(2) + (\$12.50)(4)} \cdot 100 = \frac{\$389}{\$300} \cdot 100$$

$$= 129.7$$

b. The individual store: $\frac{129.7}{73.0} = 1.78$
National Sporting Goods Index: $\frac{185}{100} = 1.85$
Therefore, prices for the individual store have increased 78% while prices on a national level for sporting goods stores have increased by 85% from 1975 to 1985.

Exercises 12.3

2. a. For 1985 on a 1984 base,

$$\frac{\Sigma\left(\frac{P_{85}}{P_{84}}\right)P_{84}Q_{84}}{\Sigma P_{84}Q_{84}} \cdot 100 = \frac{\Sigma P_{85}Q_{84}}{\Sigma P_{84}Q_{84}} \cdot 100 = \frac{\$17.0325}{\$15.77} \cdot 100 = 108.0$$

For 1986 on a 1984 base

$$\frac{\Sigma\left(\frac{P_{86}}{P_{84}}\right)P_{84}Q_{84}}{\Sigma P_{84}Q_{84}} \cdot 100 = \frac{\Sigma P_{86}Q_{84}}{\Sigma P_{84}Q_{84}} \cdot 100 = \frac{\$17.7975}{\$15.77} \cdot 100 = 112.9$$

b. Same numerical results as in part (a).

4. a. The apparently paradoxical results occur because an unweighted arithmetic mean of relatives offsets equal percentage increases and decreases, regardless of the bases from which the percentages were computed. For example, if there are only two commodities, one which increased 5% and one which decreased 5%, the unweighted arithmetic mean of 95 and 105 would be 100 implying no change in the average level of prices. However, it is possible that the prices of the two commodities might differ greatly in order of magnitude.

b. Weighted aggregates price index with base period weights

for 1986 on 1984 base: $\frac{\Sigma P_{86}Q_{84}}{\Sigma P_{84}Q_{84}} \cdot 100 = \frac{11.1}{11.3} \cdot 100 = 98.2$

c. The 98.2 figure indicates that in 1985 it would have cost 98.2% of what it cost in 1984 to purchase the same quantities of these two commodities as were consumed in 1984.

Exercises 12.6

2. a. For 1985 on a 1984 base

$$\frac{\Sigma\left(\frac{Q_{85}}{Q_{84}}\right)Q_{84}P_{84}}{\Sigma Q_{84}P_{84}} \cdot 100 = \frac{\Sigma Q_{85}P_{84}}{\Sigma Q_{84}P_{84}} \cdot 100 = \frac{\$18.7425}{\$15.77} \cdot 100 = 118.8$$

For 1986 on a 1984 base

$$\frac{\Sigma\left(\frac{Q_{86}}{Q_{84}}\right)Q_{84}P_{84}}{\Sigma Q_{84}P_{84}} \cdot 100 = \frac{\Sigma Q_{86}P_{84}}{\Sigma Q_{84}P_{84}} \cdot 100 = \frac{\$19.0250}{\$15.77} \cdot 100 = 120.6$$

b. Same numerical results as in part(a)

4.

	Disposable Income ($ billions)	Disposable Income in 1980 Constant-dollars ($ billions)
1970	$100	$\$100 \div \frac{90}{120} = \$133.3$
1975	160	$\$160 \div \frac{100}{120} = \$192.00$
1980	180	$\$180 \div \frac{120}{120} = \$180.00$
1985	220	$\$220 \div \frac{125}{120} = \$211.20$

CHAPTER 13

Exercises 13.2

2. H_0: $p = 0.50$ $\quad \sigma_{\bar{p}} = \sqrt{\frac{(0.50-0.50)}{223}} = 0.033$

H_1: $p \neq 0.50$ $\quad z = \frac{0.547-0.500}{0.033} = 1.42$

$\alpha = 0.05$

Since $1.42 < 1.96$, we retain the null hypothesis that the proportion above 13.7 pounds is 0.50.

4. H_0: $p \leqslant 0.50$ $\sigma_{\bar{p}} = \sqrt{\frac{(0.50)(0.50)}{446}} = 0.024$
H_1: $p > 0.50$ $\bar{p} = \frac{296}{446} = 0.66$
$\alpha = 0.02$ $z = \frac{0.66 - 0.50}{0.024} = -6.67$

Since $|6.67| > |2.05|$, we reject the null hypothesis that $p \leqslant 0.50$. Hence, we conclude that more than 50% of U.S. citizens felt that there had been a decrease in U.S. influence in world affairs.

6. a.

City	Rank (+)	Rank (−)
Atlanta		5
Boston		2
Chicago		9
Detroit	7	
Los Angeles		3
Miami		1
New Orleans		4
New York		10
Philadelphia		8
San Francisco		6
	$T = 7$	48

Wilcoxon Test: $n = 10$, $T = 7$, $T_{0.05} = 10$. Since $T = 7 < T_{0.05} = 10$, we reject the null hypothesis of identical population distributions. Hence, we conclude that the product improvement led to a decrease in the number of product returns.

b.

$d = x_2 - x_1$	$(d - \bar{d})^2$
−68	246.49
−52	1,004.89
−166	6,773.29
+119	41,087.29
−57	712.89
−38	2,088.49
−59	610.09
−246	26,341.29
−162	6,130.89
−108	590.49
	85,586.10

$\bar{d} = -837/10 = -83.7$

$$s_d = \sqrt{\frac{\Sigma(d - \bar{d})^2}{n - 1}} = \sqrt{\frac{85,586.10}{10 - 1}} = 97.52$$

$$s_{\bar{d}} = \frac{s_{\bar{d}}}{\sqrt{n}} = \frac{97.52}{\sqrt{10}} = 30.84$$

$$t = \frac{\bar{d} - 0}{s_{\bar{d}}} = \frac{-83.7}{30.84} = -2.71$$

$v = n - 1 = 10 - 1 = 9$
$t_{0.05} = -1.883$ (Table A-6 under heading 0.10)
Since $t = -2.71 < t_{0.05} = -1.833$, we reject the null hypothesis of no difference between the average numbers of returns before and after the product improvement. We conclude that the average number of returns was lower after the product improvement.

Exercises 13.3

2.

City	X_1	X_2	$d = X_2 - X_1$	Rank (+)	Rank (−)
Los Angeles	22	30	8	9	
San Francisco	16	19	3	5.5	
Philadelphia	15	13	−2		3.5
New York	32	28	−4		7
Miami	18	17	−1		1.5
St. Louis	10	10	0		
Chicago	15	17	2	3.5	
Dallas	25	28	3	5.5	
Baltimore	17	16	−1		1.5
Boston	9	14	5	8	
				31.5	13.5

$n = 9$; $T = 13.5$; and $T_{0.05} = 8$. Since $T_{0.05} = 8 < T = 13.5$, we cannot reject the null hypothesis of identical population distributions. Therefore, we conclude that the promotional campaign was not accompanied by an increase in number of sales.

4. $\mu_T = 60(60 + 1)/4 = 915$
$\sigma_T = \sqrt{60(60 + 1)(120 + 1)/24} = 135.84$
$z = (532 - 915)/135.84 = -2.82$
Since $-2.82 < -1.96$, we reject the null hypothesis of identical population distributions.

Exercises 13.4

2. $R_1 = 214$ $\quad n_1 = 15$
$R_2 = 251$ $\quad n_2 = 15$
$U = 131$
$\mu_U = 112.5$ $\quad \alpha = 0.01$

$$\sigma_U = \sqrt{\frac{(15)(15)(31)}{12}} = 24.1$$

$$z = \frac{131 - 112.5}{24.1} = 0.77$$

Since $0.77 < 2.58$, we conclude that there is not a significant difference in the average level of sales in the two samples.

4. $R_1 = 183 \qquad n_1 = 15$
$R_2 = 282 \qquad n_2 = 15$
$U = 162$
$\mu_U = 112.5$
$\sigma_U = \sqrt{(15)(15)(31)/12} = 24.11$

$$z = \frac{162 - 112.5}{24.11} = 2.05$$

Since $2.05 > 1.96$, we reject the hypothesis of no difference between the average capital expenditure levels of the two industries.

Exercises 13.5

4. $n_1 = 24, n_2 = 26, r = 20$

$$\mu_r = \frac{2(24)(26)}{24 + 26} + 1 = 25.96$$

$$\sigma_r = \sqrt{\frac{2(24)(26)[2(24)(26) - 24 - 26]}{(24 + 26)^2(24 + 26 - 1)}} = 3.49$$

$$z = \frac{20 - 25.96}{3.49} = -1.71$$

Since $-1.71 > -1.96$ and $-1.71 > -2.58$, we cannot reject the hypothesis of randomness of runs above and below the mean 4.5 at both the 0.05 and 0.01 levels of significance.

Exercises 13.6

2.

Number of defectives:	0	2	3	4	6	7	8	10	12	13	14	15	16	17	18	19	20
Rank:	1	2	3	4	5	6	7	8	9	10	11	12	13	14	15	16	17

Number of defectives:	21	22	23	24	25	27	28	29	30	31
Rank:	18	19	20	21	22	23	24	25	26	27

$\nu = 4 - 1 = 3 \qquad \chi^2_{0.01} = 11.345$

$$K = \frac{12}{27(27+1)}\left(\frac{29^2}{6} + \frac{162^2}{8} + \frac{87^2}{7} + \frac{100^2}{6}\right) - 3(27+1)$$

$= 13.915$

Since $K = 13.915 > 11.345$, we reject the hypothesis of identically distributed populations.

Exercises 13.7

2. $r_r = 1 - \dfrac{6(208)}{10(10^2 - 1)} = -0.26$

4. $r_r = 1 - \dfrac{6(15)}{10(10^2 - 1)} = 0.91$

CHAPTER 14

Exercises 14.4

2. P (Major appliance) $= 0.11$
P (Home appliance) $= 0.22$
P (Small appliance) $= 0.67$

4. Expected profit (Preparing bid) = (0.1)($175,000) + (0.9)(−$10,000) = $8,500
Therefore, R.B.A., Inc., should prepare the bid.

6. Expected profit (order 0) = 0
Expected profit (1) = (0.15)(−$1,000) + (0.85)($5,000) = $4,100
Expected profit (2) = (0.15)(−$2,000) + (0.15)($4,000) + (0.7)($10,000)
= $7,300
Expected profit (3) = (0.15)(−$3,000) + (0.15)($3,000) + (0.2)($9,000)
+ (0.5)($15,000)
= $9,300
Expected profit (4) = (0.15)(−$4,000) + (0.15)($2,000) + (0.2)($8,000)
+ (0.2)($14,000) + (0.3)($20,000)
= $10,000
Expected profit (5) = (0.15)(−$5,000) + (0.15)($1,000) + (0.2)($7,000)
+ (0.2)($13,000) + (0.15)($19,000) + (0.15)($25,000)
= $10,000
The optimal stocking level is 4 coats.
The expected profit is $10,100.

8. Expected profit from \$17 million cost $= (0.3)(15) + (0.2)(20) + (0.4)(25) + (0.1)(40) - 17$
$= \$5.5$ million

Expected profit from \$27 million cost $= (0.3)(30) + (0.2)(40) + (0.4)(50) + (0.1)(80) - 27$
$= \$18$ million

Therefore, the oil company should spend the extra \$10 million on the refined equipment.

Exercises 14.6

2. The expected value of perfect information is the expected opportunity loss of the optimal act under uncertainty.

4. (In units of \$1000)
EOL $A_1 = (0.2)(12) + (0.4)(3) + (0.3)(2) + (0.1)(0) = 4.2$
EOL $A_2 = (0.2)(0) + (0.4)(7) + (0.3)(10) + (0.1)(10) = 6.8$
EOL $A_3 = (0.2)(4) + (0.4)(0) + (0.3)(8) + (0.1)(9) = 4.1$
EOL $A_4 = (0.2)(17) + (0.4)(5) + (0.3)(0) + (0.1)(3) = 4.7$
Therefore, EVPI = minimum EOL = 4.1
Under certainty, the expected payoff is

$$(0.2)(32) + (0.4)(35) + (0.3)(30) + (0.1)(35) = 28.9$$

Under uncertainty, the expected payoff of the best act, A_3, is

$$(0.2)(28) + (0.4)(25) + (0.3)(22) + (0.1)(26) = 24.8$$

EVPI $= 28.9 - 24.8 = 4.1$

6. *Opportunity Loss Table*

Outcome	A_1 Publish	A_2 Do Not Publish
Success	0	8
Failure	4	0

EOL$(A_1) = (\frac{1}{3})(0) + (\frac{2}{3})(4) = \$2.67$ million
EOL$(A_2) = (\frac{1}{3})(8) + (\frac{2}{3})(0) = \$2.67$ million
EVPI = \$2.67 million
The publisher should be indifferent as to publishing or not publishing the book.

8. *Under Certainty:* Expected return $= (0.7)(\$500) + (0.3)(\$800) = \$590$
Under Uncertainty: Expected return (Guess A) $= (0.7)(\$500) + (0.3)(-\$300)$
$= \$260$

Expected return (Guess B) = (0.7)(−\$40) + (0.3)(\$800) = \$212
EVPI = \$590 − \$260 = \$330
Alternatively,
EOL (Guess A) = (0.7)(\$0) + (0.3)(\$1,100) = \$330
EOL (Guess B) = (0.7)(\$540) + (0.3)(\$0) = \$378
Hence, EVPI = \$330

10. a. Expected gain (Open new region) = (0.3)(\$100,000) + (0.4)(\$10,000) + (0.3)(−\$80,000) = \$10,000

Yes, open the new region.

b. Under certainty, the expected gain is (0.3)(\$100,000) + (0.4)(\$10,000) + (0.3)(0) = \$34,000
EVPI = \$34,000 − \$10,000 = \$24,000

12. a. (In \$10,000 units)
EOL (A_1) = (0.5)(1) + (0.2)(0) + (0.1)(3) + (0.2)(3) = 1.4
EOL (A_2) = (0.5)(2) + (0.2)(1) + (0.1)(0) + (0.2)(2) = 1.6
EOL (A_3) = (0.5)(3) + (0.2)(2) + (0.1)(1) + (0.2)(0) = 2.0
EOL (A_4) = (0.5)(0) + (0.2)(4) + (0.1)(3) + (0.2)(3) = 1.7

b. EVPI = 1.4

c. Since A_1 has the lowest EOL, the optimal decision would be A_1.

Exercises 14.8

2. a. (0.4)(10) + (0.6)(2) > 5. Thus prefer AE combination.
b. (0.5)(8) + (0.5)(2) = 5. Thus, indifferent.
c. (0.3)(10) + (0.7)(3) > 5. Thus, prefer AD combination.
d. (0.4)(8) + (0.6)(2) < 5. Thus, prefer C.

4. a. Expected gain (100% interest) = (0.8)(−\$50,000) + (0.1)(\$130,000) + (0.1)(\$430,000) = \$16,000
Expected gain (50% interest) = (0.8)(−\$25,000) + (0.1)(\$65,000) + (0.1)(\$215,000) = \$8,000
Expected gain (don't drill) = 0
The best act is to drill with 100% interest.

b. Expected utility of 100% interest is

$$U(100\%) = (0.8)(-30) + (0.1)(60) + (0.1)(200) = 2$$

Expected utility of 50% interest is

$$U(50\%) = (0.8)(-10) + (0.1)(25) + (0.1)(120) = 6.5$$

The best act is to drill with 50% interest.

6. Expected utility (market drug) = (0.99)(9,000) + (0.1)(−900,000) = −90
Therefore, the FDA should not accept the drug.

CHAPTER 15

Exercises 15.2

2.

State of Nature	$P(S)$	$P(X\|S)$	$P(S)P(X\|S)$	Revised Probabilities $P(S\|X)$
S_1	0.15	0.90	0.135	0.260
S_2	0.25	0.80	0.200	0.385
S_3	0.30	0.45	0.135	0.260
S_4	0.20	0.20	0.040	0.077
S_5	0.10	0.10	0.010	0.019
			0.520	

4. No, since after sampling the best act remains the same, the actual value of the sample information was zero.

6. No. Although both prior and sample information indicate that S_1 rather than S_2 is true, whether act A_1 is better than A_2 depends on the relative size of their payoffs. It is conceivable that the expected payoff of act A_2 exceeds that of act A_1.

8.

θ	$P_0(\theta)$	$P(X\|\theta)$	$P_0(\theta)P(X\|\theta)$	$P_1(\theta\|X)$
Higher rates	0.45	0.10	0.0450	0.144
No change	0.35	0.25	0.0875	0.280
Lower rates	0.20	0.90	0.1800	0.576
			0.3125	

$$\begin{aligned}\text{Prior expected profit} &= (-\$15{,}000)(0.45) \\ &\quad + (\$5{,}000)(0.35) \\ &\quad + (\$18{,}000)(0.20) \\ &= -\$1{,}400\end{aligned}$$

Before considering the additional information, her best action would be to sell immediately.

$$\begin{aligned}\text{Posterior expected profit} &= (-\$15{,}000)(0.144) \\ &\quad + (\$5{,}000)(0.280) \\ &\quad + (\$18{,}000)(0.576) \\ &= \$9{,}608\end{aligned}$$

After considering the additional information, her best action would be to sell immediately.

Exercises 15.3

2.

μ	$P_0(\mu)$	$P(X = 6\|\mu)$	$P_0(\mu)P(X = 6\|\mu)$	$P_1(\mu)$
5	0.6	0.146	0.088	0.58
6	0.4	0.160	0.064	0.42
			0.152	

4. a. EOL (accept) $= (0.4)(0) + (0.3)(100) + (0.2)(300) + (0.1)(600) = 150$

b.

$P(X = 4\|p)$	$P_1(p)$
0.0898	0.24
0.2182	0.43
0.1896	0.25
0.1304	0.08

$$\text{EOL (accept)} = (0.24)(0) + (0.43)(100) + (0.25)(300) + (0.08)(600) = 166$$

6. a. Benefit (Dodd) $= (785{,}000)(\$10) - (15{,}000)(\$50) + (1 - x)(15{,}000)(\$10) = \$7{,}250{,}000 - \$150{,}000x$

Benefit (Todd) $= (1{,}108{,}000)(\$10) - (92{,}000)(\$50) + (1 - x)(92{,}000)(\$10) = \$7{,}400{,}000 - \$920{,}000x$

b. For Dodd County,

$$E_0(\theta) = (0.3)(0.05) + (0.4)(0.10) + (0.3)(0.15) + (0.0)(0.20) = 0.10$$

For Todd County,

$$E_0(\theta) = (0.0)(0.05) + (0.3)(0.10) + (0.4)(0.15) + (0.3)(0.20) = 0.15$$

$$E_0\text{ (benefit to Dodd)} = \$7{,}250{,}000 - \$150{,}000(0.10) = \$7{,}235{,}000$$

$$E_0\text{ (benefit to Todd)} = \$7{,}400{,}000 - \$920{,}000(0.15) = \$7{,}262{,}000$$

Therefore, according to this comparison, it would be slightly preferable to build the reservoir in Todd County.

c. Dodd County:

θ	$P_0(\theta)$	$P(X = 3\|p = \theta, n = 20)$	Joint Probability	$P_1(\theta)$
0.05	0.3	0.0596	0.01788	0.11
0.10	0.4	0.1901	0.07604	0.45
0.15	0.3	0.2428	0.07284	0.44
			0.16676	

$$E_1(\theta) = (0.11)(0.05) + (0.45)(0.10) + (0.44)(0.15)$$
$$= 0.117$$

$$E_1 \text{ (benefit to Dodd)} = \$7{,}250{,}000 - \$150{,}000(0.117)$$
$$= \$7{,}232{,}450$$

Todd County:

θ	$P_0(\theta)$	$P(X = 5 \mid p = \theta, n = 20)$	Joint Probability	$P_1(\theta)$
0.10	0.3	0.0319	0.00957	0.09
0.15	0.4	0.1029	0.04116	0.40
0.20	0.3	0.1746	0.05238	0.51
			0.10311	

$$E_1(\theta) = (0.09)(0.10) + (0.40)(0.15) + (0.51)(0.20)$$
$$= 0.171$$

$$E_1 \text{ (benefit to Todd)} = \$7{,}400{,}000 - \$920{,}000(0.171)$$
$$= \$7{,}242{,}680$$

Therefore, the comparison of expected net benefits indicates that it is still slightly preferable to build the reservoir in Todd County.

8. Prior mean = (0.10)(0.25) + (0.20)(0.30) + (0.30)(0.45) = 0.220
For part (a),
Posterior mean = (0.10)(0.15) + (0.20)(0.45) + (0.30)(0.40) = 0.225
For part (b),
Posterior mean = (0.10)(0.01) + (0.20)(0.89) + (0.30)(0.10) = 0.209

CHAPTER 16

Exercises 16.2

2.

θ	$P(\theta)$	$P(X \mid \theta)$	$P(Y \mid \theta)$	$P(\theta)P(X \mid \theta)$	$P(\theta)P(Y \mid \theta)$	$P(\theta \mid X)$	$P(\theta \mid Y)$
θ_1	0.2	0.7	0.3	0.14	0.06	0.467	0.086
θ_2	0.8	0.2	0.8	0.16	0.64	0.533	0.914
				0.30	0.70	1.000	1.000

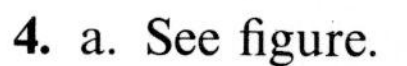

4. a. See figure.

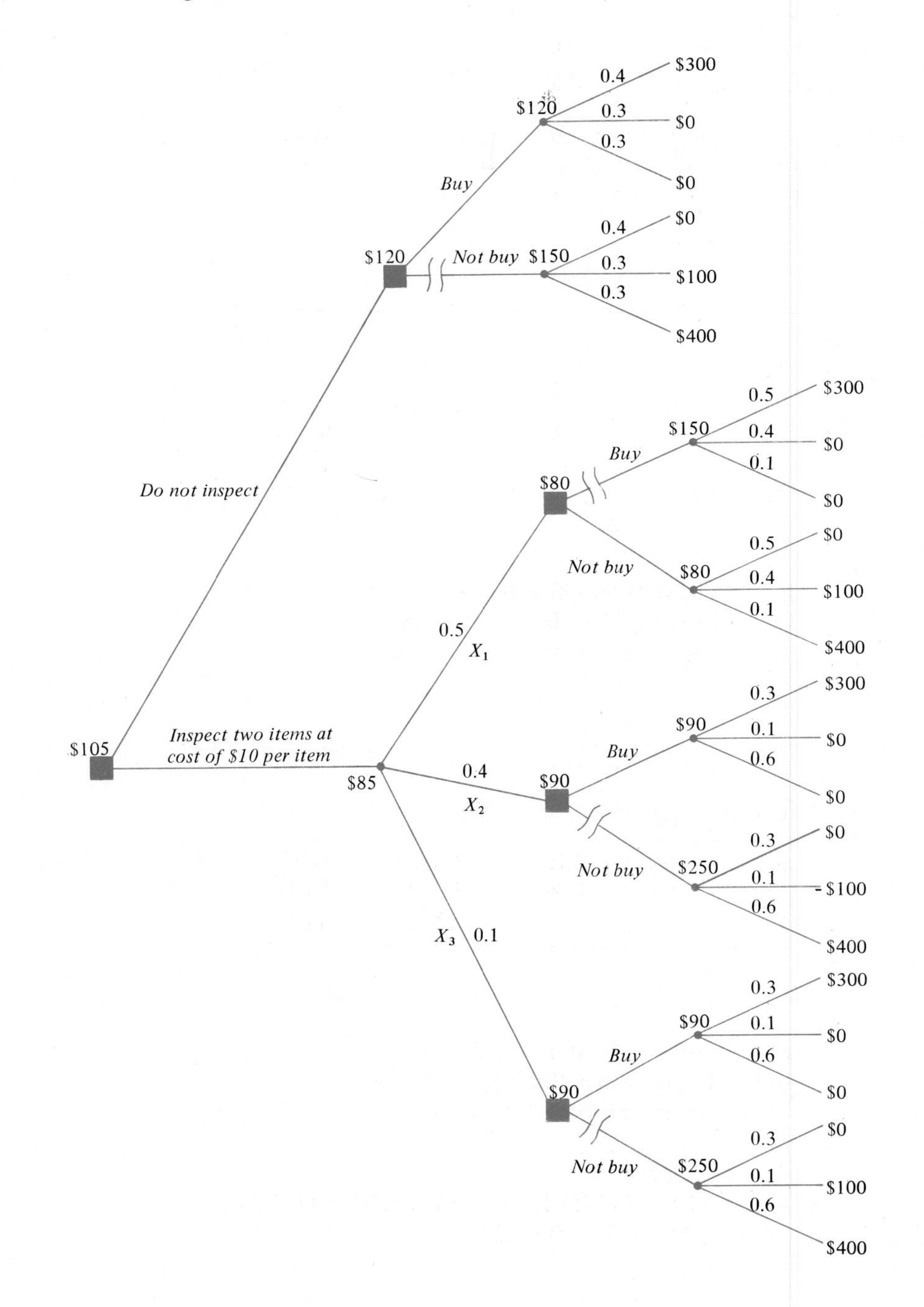

c. EVPI (before) = \$120
   EVPI (after) = \$85
d. EVSI = \$120 − \$85 = \$35
e. ENGS = \$35 − \$20 = \$15

6. In this exercise money figures are in tens of thousands of dollars.

a. (1) EOL (sell) = (0.7)(8) = 5.6
   EOL (do not sell) = (0.3)(8) = 2.4

(2)

State of Nature	$P(\theta)$	$P(X_1\|\theta)$	$P(\theta)P(X_1\|\theta)$	$P(\theta\|X_1)$
θ_1	0.7	0.85	0.595	0.768
θ_2	0.3	0.60	0.180	0.232
			0.775	1.000

EOL (sell) = (0.768)(8) = 6.144
EOL (do not sell) = (0.232)(8) = 1.856
Therefore, the better act is not to sell.

(3)

State of Nature	$P(\theta)$	$P(X_2\|\theta)$	$P(\theta)P(X_2\|\theta)$	$P(\theta\|X_2)$
θ_1	0.7	0.15	0.105	0.467
θ_2	0.3	0.40	0.120	0.533
			0.225	1.000

EOL (sell) = (0.467)(8) = 3.736
EOL (do not sell) = (0.533)(8) = 4.264
Therefore, the better act is to sell.

(4) $P(X_1) = 0.775$
   $P(X_2) = 0.225$

(5) EVSI = 2.4 − (0.775)(1.856) + (0.225)(3.736) = 0.121

b. ENGS = \$1,210 − \$1,000 = \$210

8. EVSI = 12 − 4.82 = 7.18 (in tens of thousands of dollars)
(EVSI = 71,800)

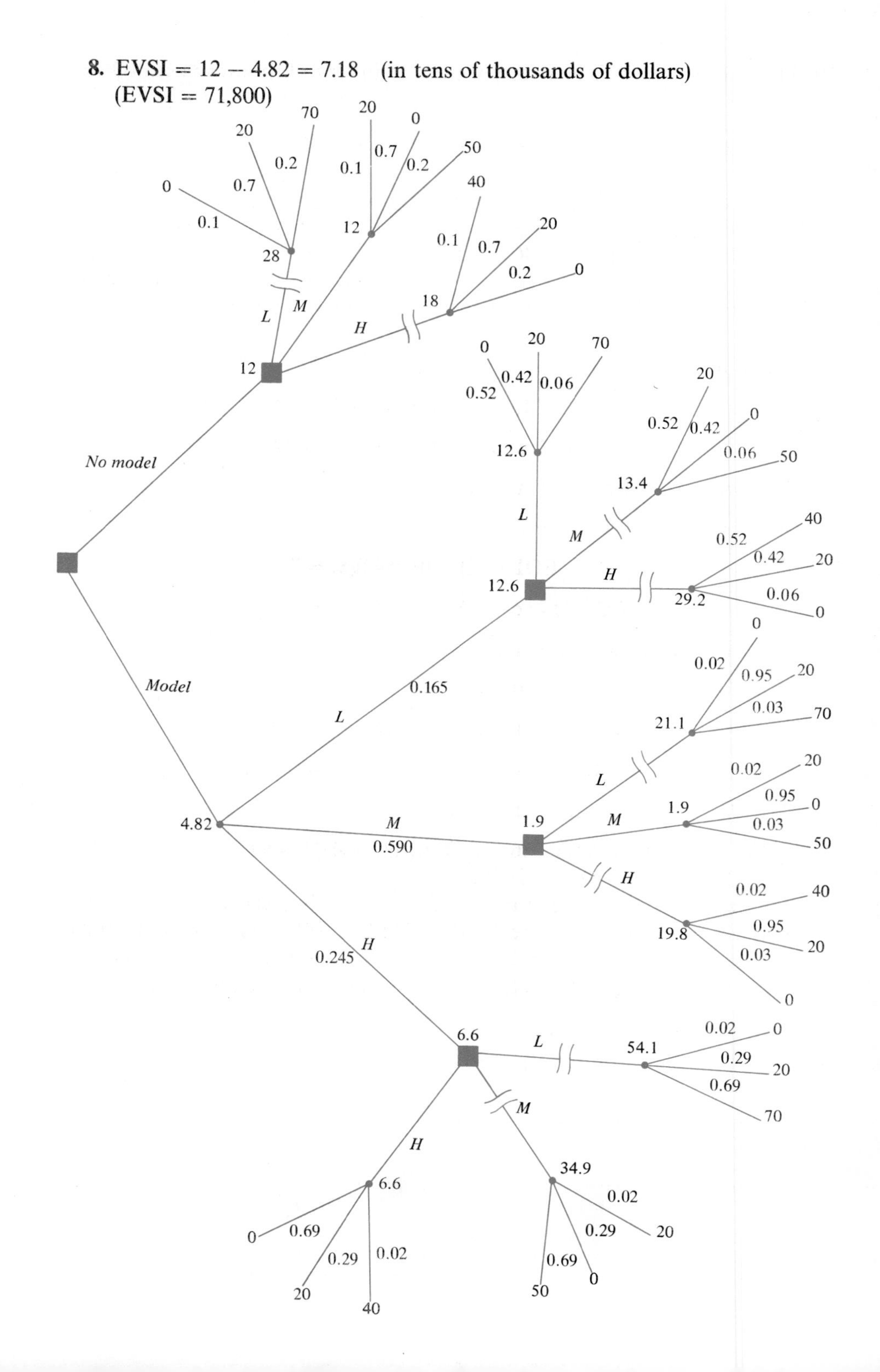

10. In this exercise, EOL monetary values are in tens of thousands of dollars.

State of Nature	$P(\theta)$	$P(X_1\|\theta)$	$P(X_2\|\theta)$	$P(\theta)P(X_1\|\theta)$	$P(\theta)P(X_2\|\theta)$	$P(\theta\|X_1)$	$P(\theta\|X_2)$
θ_1	0.95	0.85	0.15	0.8075	0.1425	0.9642	0.8769
θ_2	0.05	0.60	0.40	0.0300	0.0200	0.0358	0.1231
				$P(X_1) = 0.8375$	$P(X_2) = 0.1625$	1.0000	1.0000

Without a study:

$$\text{EOL (sell)} = (0.95)(8) = 7.6$$

$$\text{EOL (do not sell)} = (0.05)(8) = 0.4$$

Therefore, the better act is not to sell.

With information X_1 from the study:

$$\text{EOL (sell)} = (0.9642)(8) = 7.7136$$

$$\text{EOL (do not sell)} = (0.0358)(8) = 0.2864$$

Therefore, the better act is not to sell.

With information X_2 from the study:

$$\text{EOL (sell)} = (0.8769)(8) = 7.0152$$

$$\text{EOL (do not sell)} = (0.1231)(8) = 0.9848$$

Therefore, the better act is not to sell.

$\text{EVSI} = 0.4 - [(0.8375)(0.2864) + (0.1626)(0.9848)] = \$0$
$\text{ENGS} = \$0 - \$1{,}000 = -\$1{,}000$
In exercise 6, on the other hand, ENGS = \$210. In this exercise, EVSI = 0 because the study leads to the decision "Do Not Sell" regardless of the study outcome. Since "Do Not Sell" is the better action without a study and the study results cannot change this decision, sample information has no value.

Exercises 16.4

2.

Sample Outcome	s_1	s_2	s_3	s_4	s_5	s_6	s_7	s_8
X_1	a_1	a_1	a_1	a_1	a_2	a_2	a_2	a_2
X_2	a_1	a_1	a_2	a_2	a_1	a_1	a_2	a_2
X_3	a_1	a_2	a_1	a_2	a_1	a_2	a_1	a_2

a_1 denotes "buy lot"
a_2 denotes "do not buy lot"

s_3 is the optimal strategy.

4. L = Low M = Medium H = High

Strategy s_5(L, M, M)

States of Nature	Opportunity Losses (in $00,000) Marketing Plan Used			Probability of Action s_5(L, M, M)			Risk (conditional expected loss)
	L	M	H	L	M	M	
L	0	2	4	0.85	0.10	0.05	0.30
M	2	0	2	0.10	0.80	0.10	0.20
H	7	5	0	0.05	0.10	0.85	5.10

$$R(s_5;\ \text{L}) = (0.85)(0) + (0.10)(2) + (0.05)(2) = 0.3$$
$$R(s_5;\ \text{M}) = (0.10)(2) + (0.8)(0) + (0.10)(0) = 0.2$$
$$R(s_5;\ \text{H}) = (0.05)(7) + (0.10)(5) + (0.85)(5) = 5.1$$

Strategy s_{23}(L, M, H)

States of Nature	Opportunity Losses (in $00,000) Marketing Plan Used			Probability of Action s_{23}(L, M, H)			Risk (conditional expected loss)
	L	M	H	L	M	H	
L	0	2	4	0.85	0.10	0.05	0.40
M	2	0	2	0.10	0.80	0.10	0.40
H	7	5	0	0.05	0.10	0.85	0.85

$$R(s_{23};\ \text{L}) = (0.85)(0) + (0.10)(2) + (0.05)(4) = 0.40$$
$$R(s_{23};\ \text{M}) = (0.10)(2) + (0.80)(0) + (0.10)(2) = 0.40$$
$$R(s_{23};\ \text{H}) = (0.05)(7) + (0.10)(5) + (0.85)(0) = 0.85$$

Strategy s_{15}(M, M, H)

States of Nature	Opportunity Losses (in $00,000) Marketing Plan Used			Probability of Action s_{15}(M, M, H)			Risk (conditional expected loss)
	L	M	H	M	M	H	
L	0	2	4	0.85	0.10	0.05	2.10
M	2	0	2	0.10	0.80	0.10	0.20
H	7	5	0	0.05	0.10	0.85	0.75

$R(s_{15}; \text{L}) = (0.85)(2) + (0.10)(2) + (0.05)(4) = 2.10$
$R(s_{15}; \text{M}) = (0.10)(0) + (0.80)(0) + (0.10)(2) = 0.20$
$R(s_{15}; \text{H}) = (0.05)(5) + (0.10)(5) + (0.85)(0) = 0.75$

Expected Opportunity Losses of the Three Strategies
(in hundreds of thousands of dollars)

Strategy	Expected Opportunity Loss
s_5	$(0.1)(0.3) + (0.7)(0.2) + (0.2)(5.1) = 1.19$
s_{15}	$(0.1)(2.1) + (0.7)(0.2) + (0.2)(0.75) = 0.50$
s_{23}	$(0.1)(0.4) + (0.7)(0.40) + (0.2)(0.85) = 0.49$

thus, s_{23} is the optimal strategy.

6. Calculation of risks or conditional expected opportunity losses:

Losses for Strategies s_3 and s_5
Strategy $s_3(a_1, a_2, a_2)$

Events	Opportunity Loss of Wrong Act	Probability of Wrong Act	Risk (conditonal expected loss)
θ_1	1,300	0.25	325
θ_2	300	0.09	27
θ_3	800	0.07	56

Strategy $s_5(a_1, a_1, a_2)$

Events	Opportunity Loss of Wrong Act	Probability of Wrong Act	Risk (conditonal expected loss)
θ_1	1,300	0.10	130
θ_2	300	0.59	177
θ_3	800	0.16	128

Expected Opportunity Losses of the Two Strategies

Strategy	Expected Opportunity Loss
s_3	$(0.35)(325) + (0.30)(27) + (0.35)(56) = 141.45$
s_5	$(0.35)(130) + (0.30)(177) + (0.35)(128) = 143.40$

Thus, s_3 is the optimal strategy.

Exercises 16.6

2. ENGS = $\$200 - \$167.14 - \$9.00 = \$23.86$

Decision Rule

If 0 defective, then ship
If 1 defective, then ship
If 2 defective, then scrap
If 3 defective, then scrap

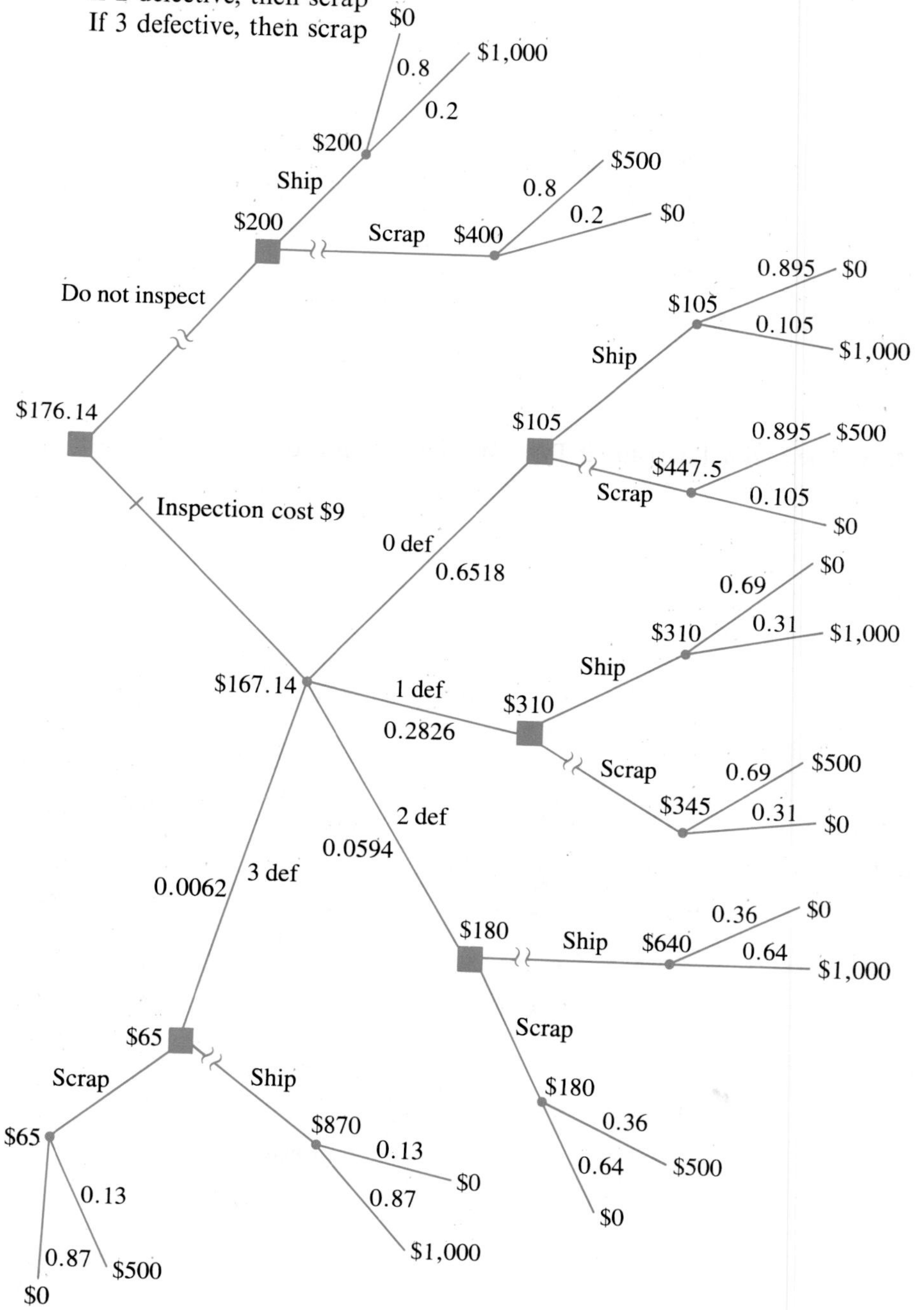

CHAPTER 17

Exercises 17.2

2. Strategy s_7 satisfies the requirement that α be less than or equal to 0.05.

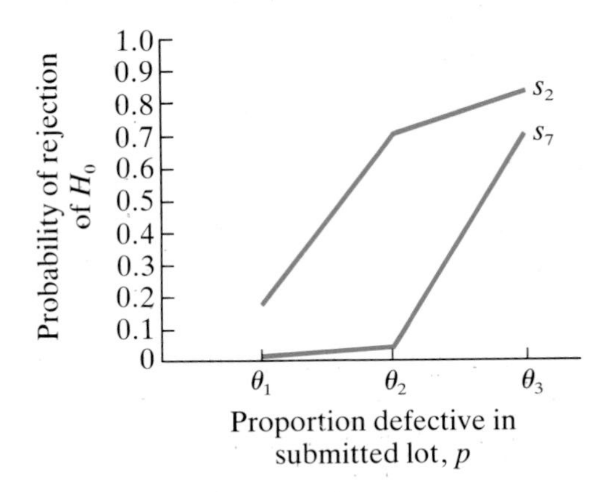

4. a. The optimal decision if Type M information is observed is to reject the shipment.

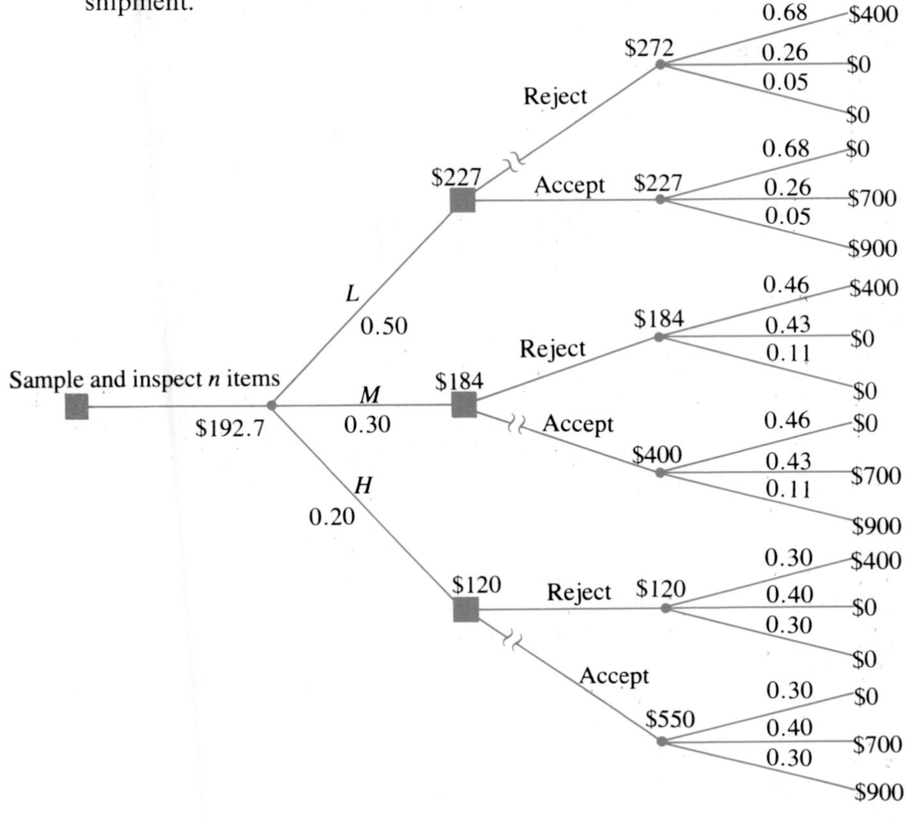

b. Strategy s_3 meets the criterion, but s_4 does not. In part (a), s_3 has been shown to be nonoptimal. Failure to consider opportunity losses results in the selection of a nonoptimal strategy.

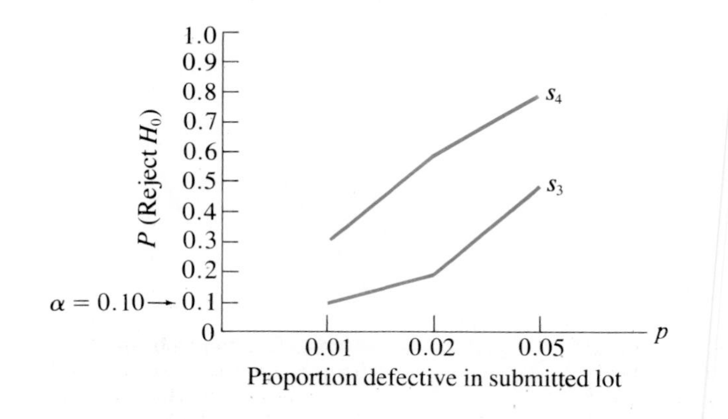

Index

B 7 C 8 D 9 E 0 F 1 G 2 H 3 I 4 J 5

TABLE 4-1

Random digits

98389	95130	36323	33381	98930	60278	33338	45778	86643	78214
17245	58145	89635	19473	61690	33549	70476	35153	41736	96170
01289	68740	70432	43824	98577	50959	36855	79112	01047	33005
98182	43535	79938	72575	13602	44115	11316	55879	78224	96740
59266	39490	21582	09389	93679	26320	51754	42930	93809	06815
42162	43375	78976	89654	71446	77779	95460	41250	01551	42552
50357	15046	27813	34984	32297	57063	65418	79579	23870	00982
11326	67204	56708	28022	80243	51848	06119	59285	86325	02877
55636	06783	60962	12436	75218	38374	43797	65961	52366	83357
31149	06588	27838	17511	02935	69747	88322	70380	77368	04222
25055	23402	60275	81173	21950	63463	09389	83095	90744	44178
35150	34706	08126	35809	57489	51799	01665	13834	97714	55167
61486	33467	28352	58951	70174	21360	99318	69504	65556	02724
44444	86623	28371	23287	36548	30503	76550	24593	27517	63304
14825	81523	62729	36417	67047	16506	76410	42372	55040	27431
59079	46755	72348	69595	53408	92708	67110	68260	79820	91123
48391	76486	60421	69414	37271	89276	07577	43880	08133	09898
67072	33693	81976	68018	89363	39340	93294	82290	95922	96329
86050	07331	89994	36265	62934	47361	25352	61467	51683	43833
84426	40439	57595	37715	16639	06343	00144	98294	64512	19201
41048	26126	02664	23909	50517	65201	07369	79308	79981	40286
30335	84930	99485	68202	79272	91220	76515	23902	29430	42049
33524	27659	20526	52412	86213	60767	70235	36975	28660	90993
26764	20591	20308	75604	49285	46100	13120	18694	63017	85112
85741	22843	16202	48470	97412	65416	36996	52391	81122	95157

Source: Rand Corporation, *A Million Digits with 100,000 Normal Deviates* (New York: Free Press, 1955), excerpt from page 387.